<u>PLASTICITY</u>

<u>PLASTICITY</u>:

<u>FUNDAMENTALS AND GENERAL RESULTS</u>

J. B. Martin

The MIT Press
Cambridge, Massachusetts, and London, England

Publisher's Note

The aim of this format is to close the time gap between the
preparation of certain works and their publication in book
form. A large number of significant though specialized manu--
scripts make the transition to formal publication either after
a considerable delay or not at all. The time and expense of
detailed text editing and composition in print may act to prevent publication or so to delay it that currency of content is
affected.

The text of this book has been photographed directly from the
author's typescript. It is edited to a satisfactory level of
completeness and comprehensibility though not necessarily to
the standard of consistency of minor editorial detail present
in typeset books issued under our imprint.

The MIT Press

Library of Congress Cataloging in Publication Data

Martin, John Brand, 1937-
   Plasticity : fundamentals and general results.

   Includes indexes.
   1. Plasticity. I. Title.
TA418.14.M34   531'.3825   75-14160
ISBN 0-262-13114-5

To the memory of my father

I have attempted in this volume to give a coherent account of
basic concepts and general results in the theory of plasticity
which on the one hand are sufficiently broad that they are applicable to all aspects of solid and structural mechanics and
on the other hand are now considered to be well established
and well understood. To this end concepts and theorems are
presented in a notation which is appropriate for a wide range
of problems, including trusses, beams and frames, plates and
shells, plane stress and plane strain, and general continuum
problems. Emphasis is placed on demonstrating the generality
of the principles and their applicability in a variety of problem types, rather than on developing a variety of solutions
in a specific class of problems. In order to remain within
the confines of well established ideas, attention has been limited to isothermal, time-independent static problems in which
the small displacement assumptions are applicable. In consequence, important current developments in certain aspects of
the theory of plasticity are not covered, such as the study of
dynamic and other time-dependent problems, finite deformations
and plastic instability. It is believed, however, that a tho-
rough understanding of established fundamentals will provide a
basis for the ready appreciation of new developments.

It is intended that the volume can be used for study at
graduate or senior undergraduate level. An understanding of
elastic behavior of solids or structures is assumed. It will
be sufficient if this knowledge is confined to one aspect of
elastic behavior, such as structural analysis or stress analysis, since the volume can be read selectively as the extension
to plasticity of elastic studies in any one of these aspects.

The work is divided into six parts. In Part I the governing equations for a variety of problem types are given, and a
suitable notation for the discussion of all of these problems
is introduced. The derivation of elastic-plastic constitutive
relations is then considered from the viewpoint of the generalization of macroscopic behavior observed in a tension test.
This is followed by a study of the broad characteristics of
elastic-plastic structural response. In Part II a number of
specific solutions for elastic-plastic problems are treated
from first principles, and a number of commonly used idealizations are introduced. The study of general theorems occupies
Parts III - V. Part III deals with limit analysis and its
consequences, Part IV with periodic loading and shakedown in
elastic, perfectly plastic problems, and Part V with deformation theory and incremental analysis. In Part VI the thermodynamic approach to the formulation of constitutive relations

is treated.

Equations and figures are numbered independently in each Chapter. Bibliographical notes are deferred until the ends of Chapters, except in Part I where all references appear at the end of Chapter III. This course is followed because it is generally difficult, and sometimes impossible, to ascribe proper credit for the development of ideas in parallel with a presentation which attempts to integrate these ideas into a coherent whole.

My own understanding of the principles which are set out in this volume is largely derived from my association, as a student, colleague and co-worker, with a number of individuals who have done much to develop and clarify the theory of plasticity. Neither success nor failure in achieving my objective in this volume reduces my indebtedness to them. In a broad sense, I wish to acknowledge the influence of W.Prager (who suggested that this volume should be written), D.C.Drucker, J.Heyman and P.S.Symonds. I am also indebted to R.J.Clifton, L.B.Freund, J. Kestin, F.A.Leckie, A.R.S.Ponter and J.R.Rice for many hours of detailed discussion on topics which form part of this work.

I have also received a great deal of assistance in the detailed work of preparing this volume. Z.Bilek has provided me with numerous references from the Soviet literature. B.D.Reddy, P.Carter and T.B.Griffin have assisted in the checking of the final manuscript. Mrs.J.Bardsley and Mrs.E.Fonseca were responsible for the first typed draft of the manuscript. I am very greatly indebted to Mrs.D.F.Murcott, who with great skill and inexhaustible patience prepared the final manuscript for publication. The diagrams were drawn by K.Martin.

Serious work on this volume was begun during a sabbatical leave spent at the University of Leicester in 1969-1970, and I am grateful to F.A.Leckie for the arrangements which among other things gave me this opportunity. Work continued at Brown University during the years 1970-1972, and was completed at the University of Cape Town during 1973-1974. The assistance of G.v.R.Marais in making arrangements to complete the final manuscript is gratefully acknowledged.

November 1974

PART I

BASIC CONCEPTS IN PLASTICITY

Chapter 1

INTRODUCTION

1.1 Preliminary Remarks

In this monograph we shall be concerned exclusively with the
structural problem in plasticity. The structural problem can
be defined as follows: from controlled laboratory experiments
on simple bodies composed of a given material, we wish to be
able to predict the mechanical behavior of any given body also
composed of that material and subjected to specified loads and
supported in a specified manner. The concepts of stress (internal force) and strain (measures of deformation which vanish
everywhere in the body during a rigid body displacement)play a
central role in the formulation of this problem.

From considerations of dynamics we can determine relations
between stresses and stress gradients in the body, and relations between the stresses at the boundary of the body and the
external forces. Stress fields which satisfy these dynamical
restrictions will be said to be *dynamically admissible.*Kinematic considerations relate strains and displacement gradients.
It may also be necessary to impose compatibility conditions
which ensure that these strain-displacement relations are integrable. A strain field which satisfies these conditions
will be said to be *kinematically admissible.* Both the dynamic
and kinematic requirements are independent of the characteristics of the material of which the body is composed. However,
they do depend in form on the configuration of the body under
consideration because of the mathematical and computational
convenience in adopting particular coordinate systems for particular configurations. It is also common to simplify the
problem when the body is composed of thin sheets of material
(plates and shells) or an assemblage of curved or straight
bars with cross section dimensions small compared to their
length. In the first case the thickness dimension is

neglected in formulating the problem, and only two spatial coordinates appear in the theory. In the second case both cross
section dimensions are neglected, and only one spatial coordinate is considered. Generalized stresses, generalized strains
and generalized displacements must be introduced, and the form
of the dynamic and kinematic equations differs from that in the
three-dimensional theory.

The relations between stress and strain characterize the
material under discussion, and will be referred to as *constitutive relations*. These relations must be determined experimentally, usually by testing specimens in which the stresses
and strains are homogeneous to at least a very close approximation. Relations between generalized stresses and generalized strains in shell and bar structures are often derived
from the constitutive relations for three-dimensional bodies,
but it remains true that the fundamental information must be
experimentally determined. The constitutive relations may
involve measurable physical quantities other than stress and
strain, such as temperature and time, or internal parameters
which cannot be measured directly. The effects of these internal parameters can sometimes be more conveniently expressed in terms of history of stress and strain, or memory of
past mechanical events inherent in the material.

It is normally found, however, that in certain limited
ranges of behavior (i.e., within certain limits of stress,
strain rate, temperature and so on) materials exhibit certain
dominant characteristics. Since we must always be concerned
with the ease with which solutions of the structural problem
can be found, it is our usual practice to categorize these
dominant characteristics into classes of behavior. Elastic
behavior is the simplest class from the point of view of
generating solutions to the structural problem. Plastic

behavior is another class, as are viscoelasticity, thermoelasticity, creep and viscoplasticity. This type of classification is entirely mathematical, and in consequence we must state explicitly what we mean by any particular class of behavior. Nothing can compel real materials to behave according to our definitions. Indeed, a common engineering material such as steel can exhibit elastic behavior under limited stresses and at moderate temperatures, viscoelastic behavior under high frequency, low amplitude vibrations, nonlinear creep at high temperatures, and viscoplastic behavior under high strain rates. It is essential, therefore, that we determine for each material the conditions under which it can sensibly be assumed to exhibit the dominant characteristics of any particular category of behavior. It is also essential that we appreciate the limits of accuracy which are required in any particular structural problem whose solution is desired. Having done so, we treat only those characteristics of the material which contribute significantly to the structural behavior under the pertaining conditions.

This approach to the description of material behavior is phenomenological. No attempt is made to explain why materials behave as they do, or more precisely in this context, why they exhibit certain dominant characteristics in certain ranges of behavior. This does not imply that a knowledge of the underlying mechanisms involved in deformation does not play an important part in suggesting particular forms for constitutive relations, or in suggesting the limits within which certain idealizations of behavior remain valid.

It is the purpose of this volume to present an introduction to fundamental concepts and general results in that range of mechanical behavior which is known as plasticity. We shall be particularly concerned with those concepts and those results which are common to all forms of the mathematical problem (i.e., all configurations of the body under discussion) rather than with particular problems. For this reason we shall adopt a notation and description of the dynamic and kinematic equations which can be applied to all coordinate systems, and to three-dimensional bodies, plates and shells, and bar structures. There is not sufficient space available to develop these equations from first principles. It will be assumed that the reader is familiar with elastic analysis in at least one of the problem types in structural or continuum mechanics. In the following four sections we shall present the relevant equations for four problem types; straight bars loaded in one plane, flat plates loaded transversely, plane stress, and three-dimensional continuum equations treated with cartesian tensors. These presentations follow an identical pattern; in some instances a word for word repetition will be found. A more general notation will then be introduced, and its relation to the four systems described specifically. In addition, all examples presented in the volume will be drawn from the four particular problems outlined in detail.

In order to limit the scope of the monograph, first to that which can be reasonably covered in limited space, and second to that part of the field which is most developed, we shall deal only with *linear* dynamic and kinematic equations and further consider only *static loading under isothermal conditions.* Thus we introduce the classical small displacement assumptions, which require on one hand that strain magnitudes and thus displacement gradients are very small, and on the other that the dynamic equations can be written without significant error in the original or undeformed configuration rather than in the deformed configuration of the body

resulting from the imposition of loads. Static loading implies that all mechanical processes are carried out very
slowly, and that acceleration terms are not present in the
dynamic equations. The term dynamically admissible can thus be
replaced by *statically admissible*.

It will then be our task, starting in Chapter 2, to construct
a general framework for plastic constitutive relations, and to
discuss the implications and general results available for
structural behavior. It must be re-emphasized that this process involves a mathematical definition of a class of behavior
which we will call *plasticity*. While everyone will agree on
the fundamental characteristics of plastic behavior, all
schools of thought may not agree on details of the mathematical
definition of the class. Some writers may prefer more general
or less general frameworks, or definitions which lead to some
minor differences. This is an inevitable consequence of our
approach, and it cannot be said that any framework, provided it
is self-consistent, is right while another in wrong. Some may
be preferable for certain types of problems, but unsuitable for
others. In this monograph we cannot cover more than one framework because of space limitations, and it is hoped that the
reader will be encouraged to read other texts and papers.

It is also important that our framework should be such that
we can incorporate within it varying degrees of idealization
of mechanical behavior. We should be able to describe the
material under quite general conditions. However, we also
wish to be able to use simpler descriptions if we are interested only in predicting certain aspects of structural behavior. Any theory or framework which produces more information than is needed for the purposes at hand is wasteful in
terms of effort. Calculations in plasticity, in particular,
benefit from judicious idealization in that many important

characteristics of structural behavior can be predicted without making use of the full constitutive relations.

1.2 Bar Structures

We shall first consider one of the simplest structural elements. A straight bar, as shown in Fig.1, is subjected to end forces and couples, and to transverse distributed forces acting along its length. All forces lie in the x-y plane, and all couples act about axes perpendicular to the x-y plane. The end forces are for convenience resolved into components perpendicular and parallel to the bar. The external forces acting on the bar must themselves be in equilibrium.

The internal forces, or generalized stresses at a cross section distance x from A are obtained by cutting the bar at that cross section (Fig.2). In general a moment M, a shear force V and an axial force N must act on each side of the cut in order that the right and left hand portions of the bar can

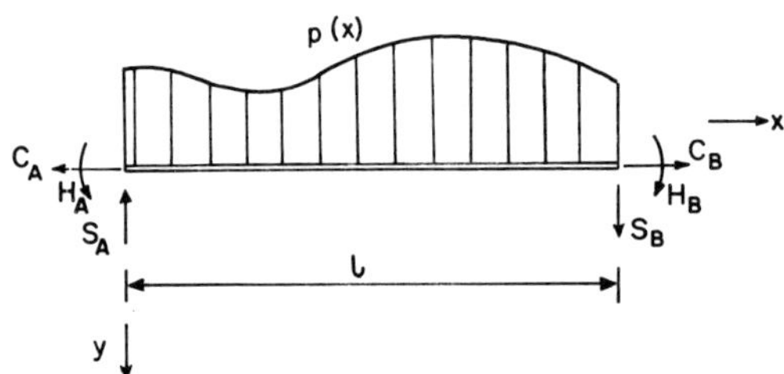

Figure 1. Bar element subjected to external forces

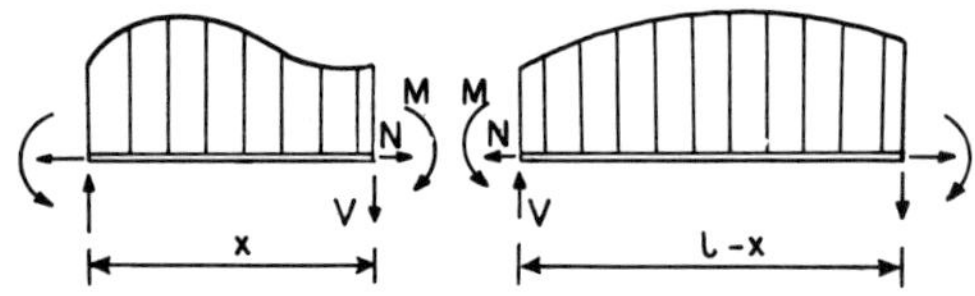

Figure 2. Internal forces in a bar element

remain in equilibrium. These forces vanish when the bar is
rejoined. It should be noted that when x = 0 the forces on
the right-hand side of the cut coincide with the forces at the
end of the bar A in Fig.1. Similarly, the forces on the left-
hand side of the cut coincide with the forces at B in Fig. 1
when x = ℓ.

Equilibrium relations involving the generalized stress components M, V and N can be derived by considering the equilibrium of either portion of the bar in Fig. 2, or they can be
put in differential form. The differential equations of equilibrium can be obtained from equilibrium relations for an
element of bar formed by making cuts at x and x + dx, as shown
in Fig. 3. With the sign convention as given, these equations
are

$$\frac{dN}{dx} = 0, \tag{1}$$

$$V = -\frac{dM}{dx}, \tag{2}$$

$$-\frac{dV}{dx} = \frac{d^2M}{dx^2} = p(x). \tag{3}$$

N is thus constant. In the special case p(x) = 0 the shear
force V is constant and M is a linear function of x.   The

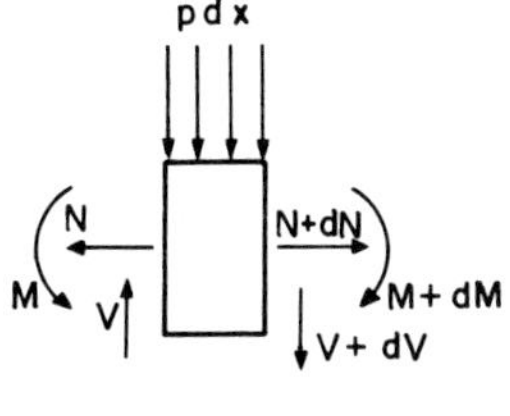

Figure 3. Infinitesimal slice of a bar element

boundary condition in the integration of these equations are the forces at the end of the bar, as shown in Fig. 1.

It is necessary to introduce three generalized displacement components in order to describe the deformation of the bar. The components are the x and y components of the actual displacement, which we will denote by u and v respectively, and the rotation of the cross section θ. The three generalized strain components ε, γ and κ associated with N, V and M respectively are then given by

$$\varepsilon = \frac{du}{dx} , \tag{4}$$

$$\gamma = \frac{dv}{dx} - \theta , \tag{5}$$

$$\kappa = \frac{d\theta}{dx} . \tag{6}$$

We shall make the common assumption that the shear strain γ is zero. This is an idealization of the constitutive relations, and is valid when shear forces are small as is expected in long bars. Putting $\gamma = 0$ in Eq.(5) we see that

$$\theta = \frac{dv}{dx} , \tag{7}$$

and hence that in Eq.(6)

$$\kappa = \frac{d^2v}{dx^2} . \tag{8}$$

We can assemble more complex structures from the straight bar elements discussed above. The only additional conditions which must be imposed are those of equilibrium of forces at the joints (i.e., the end forces and moments in the bars must add vectorially to the external forces acting at that joint) and continuity of displacement (again in a vectorial sense) and of rotation θ.

A particularly simple case is the truss, which consists of an assemblage of bars with pinned joints which are free to rotate. This implies that the end couples in each element are both zero. If external loads are applied only at the joints, $p(x)$ is zero in each bar and it follows from Eqs. (2) and (3) that $V = M = 0$ in all bars. The axial force N is the only component of generalized stress which is not zero.

In a structure composed of an assemblage of bars, a generalized stress distribution will be said to be statically admissible if the equilibrium equations (1), (2) and (3) are satisfied at all points, and if the external and joint equilibrium requirements are also satisfied. Similarly, a generalized strain and displacement field is kinematically admissible if the strain-displacement equations (4) and (8) are satisfied.

Statically and kinematically admissible fields will satisfy the principle of virtual work. Since a variety of single bar coordinate systems will be present in an assemblage of bars, we shall introduce a single spatial parameter s which locates all cross sections in the structure. Similarly, it is convenient to introduce the displacement vector $\underline{u}$ (which we previously resolved into two components u and v) and the distributed force vector $\underline{p}(s)$, which is always normal to the bar on which it acts.

Suppose that we are given loads on the structure which consist of distributed loads $\underline{p}(s)$ and point loads $\underline{P}(s_k)$ and point couples $C(s_k)$ acting at isolated points s_k ($k=1,\ldots,n$) which we treat as joints. Suppose that we determine a distribution of generalized stresses $N(s)$, $V(s)$ and $M(s)$ such that all the equilibrium requirements are satisfied. This is a statically admissible system.

Further, suppose that we determine for this structure a set of generalized displacements and generalised strains $\underline{u}^*(s)$, $\theta^*(s)$, $\varepsilon^*(s)$ and $\kappa^*(s)$ which are kinematically admissible, i.e., they satisfy the strain-displacement relations (4) and (8) and all continuity requirements at joints. Asterisks have been introduced to emphasize that the two systems are totally independent; the kinematic quantities need not be the deformations produced by the loads.

The principle of virtual work then states that;

$$\int_S \underline{p}(s)\underline{u}^*(s)ds \; + \; \sum_{k=1}^{n} \underline{P}(s_k)\underline{u}^*(s_k) \; + \; \sum_{k=1}^{n} C(s_k)\theta^*(s_k)$$

$$= \int_S N\varepsilon^*ds \; + \; \int_S M\kappa^*ds \; . \tag{9}$$

The symbol S indicates that integration is carried out over the whole structure.

In the particular bar problem outlined above a standard structural problem is formulated in the following way. The distributed load $\underline{p}(s)$ must be given for all s. We can, of course, specify $\underline{p}(s) = 0$ over all or part of the structure. At the joints, denoted by the points s_k (k=1, ..., n), either the component of $\underline{P}$ or the component of $\underline{u}$ in each of the two mutually perpendicular directions must be specified. Similarly, either the external couple C or the rotation θ must be specified at each joint.

If the material of which the structure is composed is linear elastic, the most general linear relation between the components of generalized stress and generalized strain will be of the form

$$M = D_{11}\kappa + D_{12}\varepsilon ,$$
$$N = D_{21}\kappa + D_{22}\varepsilon . \tag{10}$$

Shear is treated as a reaction which does not affect the curvature or the axial extension. The work done in changing the generalized strains κ,ε to $\kappa+d\kappa$, $\varepsilon+d\varepsilon$ is given by

$$dW = Md\kappa + Nd\varepsilon \tag{11}$$

The *strain energy* associated with strains $\bar{\kappa}$, $\bar{\varepsilon}$ is given by

$$W(\bar{\kappa},\bar{\varepsilon}) = \int dW = \int_0^{\bar{\kappa}} Md\kappa + \int_0^{\bar{\varepsilon}} Nd\varepsilon . \tag{12}$$

This relation can be integrated, and is a function of κ,ε only. It must have a homogeneous quadratic form,

$$W = A_1\kappa^2 + A_2\kappa\varepsilon + A_3\varepsilon^2 . \tag{13}$$

The stress–strain relation can be derived directly from the strain energy function. From equations (12) and (13)

$$M = \frac{\partial W}{\partial \kappa} = 2A_1\kappa + A_2\varepsilon ,$$

$$N = \frac{\partial W}{\partial \varepsilon} = A_2\kappa + 2A_3\varepsilon . \tag{14}$$

Comparing equations (10) and (14), we see that the coefficients D_{12} and D_{21} must be equal; thus the stiffness matrix must be symmetric.

We can in the same way introduce a *complementary energy* function associated with generalized stresses $\bar{M}$, $\bar{N}$, given by

$$\Omega(\bar{M},\bar{N}) = \int_0^{\bar{M}} \kappa dM + \int_0^{\bar{N}} \varepsilon dN . \tag{15}$$

Again, for a linear relation between the generalized stresses and generalized strains Ω must be a homogeneous quadratic form

$$\Omega = B_1 M^2 + B_2 MN + B_3 N^2 \; . \tag{16}$$

By differentiating equation (15) and using (16), we find the
form

$$\kappa = \frac{\partial \Omega}{\partial M} = C_{11}M + C_{12}N \; ,$$

$$\varepsilon = \frac{\partial \Omega}{\partial N} = C_{21}M + C_{22}N \; . \tag{17}$$

The compliance matrix, which is the inverse of the stiffness
matrix, is thus also symmetric.

We demand that both $W(\kappa,\varepsilon)$ and $\Omega(M,N)$ should be positive
definite functions of their respective arguments. This is a
reasonable physical restriction, and it also ensures that the
solution of any structural problem is unique. We shall further restrict ourselves in most examples to beams which are
symmetric about a plane through the center line of the cross
section and normal to the plane of loading.

In these cases the application of a bending moment does not
cause an extension of the center line, and the application of
an axial force does not cause a change in curvature. The coefficients D_{12} and C_{12} in equations (10) and (17) are thus
zero, and the elastic relations are normally put in the form

$$\kappa = \frac{M}{EI} \; , \quad \varepsilon = \frac{N}{AE} \; . \tag{18}$$

where E, I and A are respectively Young's modulus for the material, the area moment of inertia of the cross section and
the area of the cross section.

1.3 Flat Circular Plates Subjected to Axisymmetric Transverse Loading

In the plate problem we consider thin sheets of material
whose thickness is small compared to the other dimensions of
the plate. For our purpose we shall assume that the plates

are circular and that the loading and support conditions are
axisymmetric. We shall adopt a polar coordinate system shown
diagrammatically in Fig.4.

A small element of the plate lying between the lines θ and
$\theta+d\theta$ and between the circular arcs r and r+dr will be subjected to internal forces or generalized stresses as shown in Fig.
5. M_r is the radial bending moment per unit length,and M_θ the
circumferential bending moment per unit length. As a result of
the symmetry, only one shear force Q per unit length is present. The transverse load is p per unit area. These quantities
are functions of r only. Equilibrium considerations for the
element provide the differential equations of equilibrium;

$$\frac{d}{dr}\left(rM_r\right) = M_\theta - rQ_r \ . \tag{19}$$

$$\frac{d}{dr}\left(rQ_r\right) = pr \ . \tag{20}$$

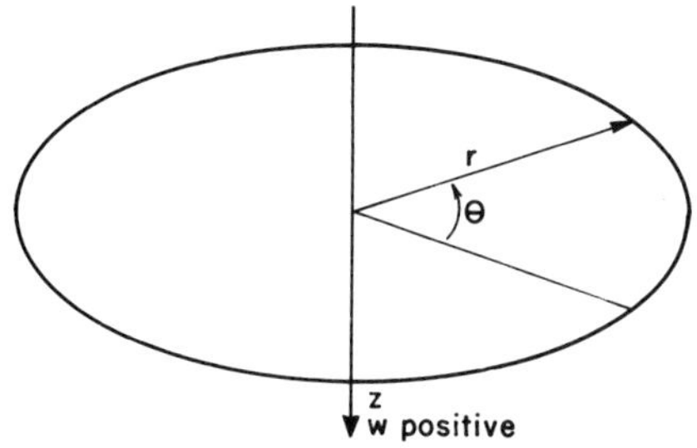

Figure 4. Coordinate system for axisymmetric plates

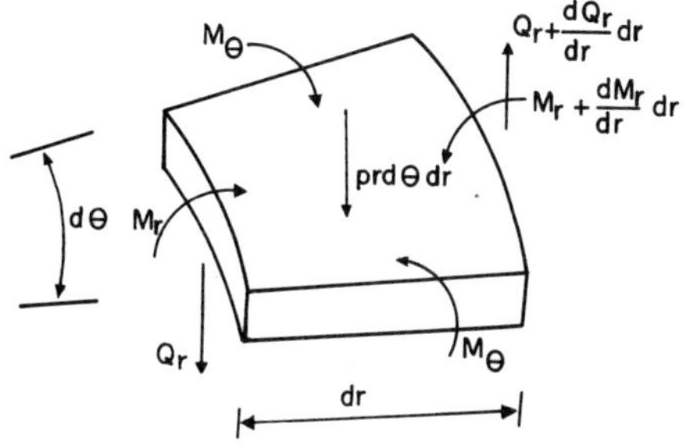

Figure 5. Infinitesimal element of an axisymmetric plate

We shall neglect shear deformations of the plate from the outset, and the displacement of the plate is then characterized by the transverse displacement $w(r)$. The generalized strains are the radial and circumferential curvatures κ_r and κ_θ, given by

$$\kappa_r = - \frac{d^2 w}{dr^2} \, , \tag{21}$$

$$\kappa_\theta = - \frac{1}{r} \frac{dw}{dr} \, . \tag{22}$$

On any circular boundary of the plate the external forces (a transverse force and a radial couple) are given directly by the generalized stresses. A set of generalized stresses, transverse forces and boundary forces will be said to be statically admissible if the equilibrium equations (19) and (20) are satisfied for all r. A set of curvatures and displacements is kinematically admissible if the strain-displacement relations (21) and (22) are satisfied.

The principle of virtual work relates any statically admissible set and any kinematically admissible set. For a plate of radius R we may write

$$2\pi \int_0^R p(r) w^*(r) r dr \; + \; 2\pi R M_r(R) \left. \frac{dw^*}{dr} \right|_{r=R} \; - \; 2\pi R Q(R) w^*(R)$$

$$= \; 2\pi \int_0^R (M_r \kappa_r^* \; + \; M_\theta \kappa_\theta^*) r dr \; . \tag{23}$$

The statically admissible quantities and the kinematically admissible quantities (marked by asterisks) in this equation are independent.

A standard problem is formulated in this context by specifying the transverse load $p(r)$ and either $M_r(R)$ or $(dw/dr)_{r=R}$ and either the shear force $Q(R)$ or $w(R)$ at the boundaries

of the plate. Two common boundary conditions are encountered;
the simply supported plate where $M_r(R) = 0$, $w(R) = 0$ and the
clamped plate where $w(R) = (dw/dr)_{r=R} = 0$.

If the plate is composed of an isotropic linear elastic material and it is uniform, the relation between the curvatures
and the moments will have the form

$$\kappa_r = \frac{1}{D} (M_r - \nu M_\theta) \, ,$$

$$\kappa_\theta = \frac{1}{D} (M_\theta - \nu M_r) \, , \tag{24}$$

where ν is Poisson's ratio, $D = Et^3/12$, E is Young's modulus
and t is the plate thickness. These equations may be inverted
to give

$$M_r = \frac{D}{(1-\nu^2)} (\kappa_r + \nu\kappa_\theta) \, ,$$

$$M_\theta = \frac{D}{(1-\nu^2)} (\kappa_\theta + \nu\kappa_r) \, . \tag{25}$$

The strain energy per unit area of the plate is given by

$$W(\kappa_r, \kappa_\theta) = \int (M_r d\kappa_r + M_\theta d\kappa_\theta)$$

$$= \frac{D}{2(1-\nu^2)} (\kappa_r^2 + 2\nu\kappa_r\kappa_\theta + \kappa_\theta^2) . \tag{26}$$

It can readily be confirmed that

$$M_r = \frac{\partial W}{\partial \kappa_r} \, , \quad M_\theta = \frac{\partial W}{\partial \kappa_\theta} \, . \tag{27}$$

Similarly, the complementary energy per unit area of the
plate is given by

$$\Omega(M_r, M_\theta) = \int (\kappa_r dM_r + \kappa_\theta dM_\theta)$$

$$= \frac{1}{2D} (M_r^2 - 2\nu M_r M_\theta + M_\theta^2) \ . \tag{28}$$

Again, we note that

$$\kappa_r = \frac{\partial \Omega}{\partial M_r} \ , \qquad \kappa_\theta = \frac{\partial \Omega}{\partial M_\theta} \ . \tag{29}$$

From more general considerations (cf. Section 1.5) it is required that $-1 \leq \nu \leq 1/2$. Thus positive definiteness of D is sufficient to ensure that W and Ω are positive definite functions of their respective arguments, and this, in turn, is sufficient to guarantee the uniqueness of the solution of elastic problems.

1.4 Plane Stress Problems

In the plane stress problem we consider thin sheets of material which are subjected to loads and imposed displacements at the boundaries of the sheet and in the plane of the sheet. Such a sheet is shown diagrammatically in Fig. 6, with a cartesian coordinate system. The sheet is assumed to occupy an area D with a boundary curve L. An element of material enclosed between lines x, x+dx, y, y+dy and of unit thickness is assumed to be acted upon by stress components $\sigma_x(x,y)$, $\sigma_y(x,y)$, $\tau_{xy}(x,y)$ and $\tau_{yx}(x,y)$ which do not vary across the thickness (i.e. they are not functions of z). These stress components are shown diagrammatically in Fig. 7. A body force $\underline{F}$ per unit volume may also act on the element. The body force has components F_x in the x direction and F_y in the y direction, and we require that the component in the z direction, F_z is zero. The equilibrium equations are

$$\frac{\partial \sigma_x}{\partial x} + \frac{\partial \tau_{yx}}{\partial y} + F_x = 0 \quad ,$$

$$\frac{\partial \tau_{xy}}{\partial x} + \frac{\partial \sigma_y}{\partial y} + F_y = 0 \quad .$$

(30)

In addition, from the requirement that there should be zero torque on an element, we have

$$\tau_{yx} = \tau_{xy} \quad .$$

(31)

If the unit outward normal vector at a point on the boundary line is $\underline{\nu}(x,y)$ with components ν_x, ν_y, and the external force is $\underline{p}(x,y)$ with components p_x, p_y, the force and stress components on the boundary are related by means of the equations

$$p_x = \sigma_x \nu_x + \tau_{xy} \nu_y \quad ,$$

$$p_y = \tau_{yx} \nu_x + \sigma_y \nu_y \quad .$$

(32)

A particle with coordinates (x,y) before deformation of the sheet will suffer a displacement $\underline{u}$ when the loads are applied. We concern ourselves only with the components of displacement which lie in the x,y plane, and assume that these components

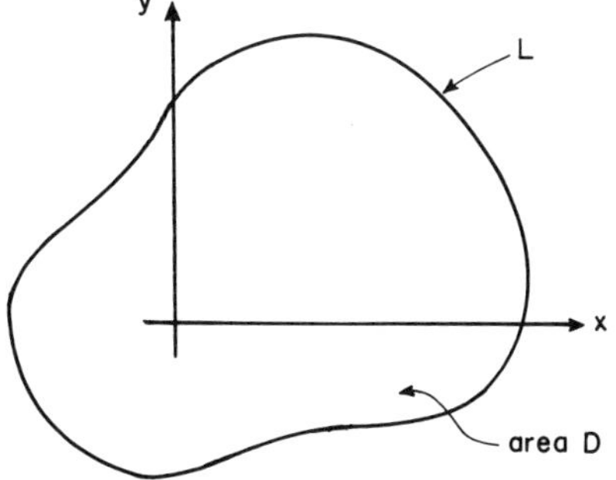

Figure 6. Coordinate system for plane stress problems

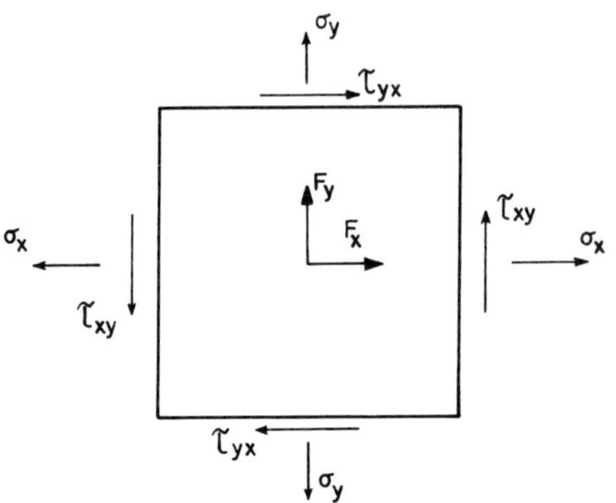

Figure 7. Infinitesimal element in plane stress

are functions of x and y only.   Hence the displacement field $\underline{u}$ is described by components $u(x,y)$ and $v(x,y)$. The strain-displacement relations give the manner in which the strain components $\varepsilon_x(x,y)$, $\varepsilon_y(x,y)$ and $\gamma_{xy}(x,y) = \gamma_{yx}(x,y)$ are derived from the displacement field;

$$\varepsilon_x = \frac{\partial u}{\partial x} \; , \quad \varepsilon_y = \frac{\partial v}{\partial y} \; , \quad \gamma_{xy} = \frac{\partial u}{\partial y} + \frac{\partial v}{\partial x} \; . \tag{33}$$

If we are given the strain components in the sheet and wish to find the displacements to within a rigid body motion, we must ensure that the strain components are integrable.   For this purpose we introduce a compatibility condition which is necessary and sufficient for the integrability of the strains in a simply connected domain;

$$\frac{\partial^2 \varepsilon_x}{\partial y^2} + \frac{\partial^2 \varepsilon_y}{\partial x^2} = \frac{\partial^2 \gamma_{xy}}{\partial x \partial y} \; . \tag{34}$$

In the case of a sheet with holes, three further discrete conditions must be introduced for each hole.

In the context of this problem, a statically admissible field is a set of boundary forces $\underline{p}(x,y)$, body forces $\underline{F}(x,y)$ and stress components $\sigma_x(x,y)$, $\sigma_y(x,y)$, $\tau_{xy}(x,y)$ which satisfy

the equilibrium relations (30) and the boundary relations (32).
A kinematically admissible field is a set of displacements
$\underline{u}(x,y)$ and strain components $\varepsilon_x(x,y)$, $\varepsilon_y(x,y)$, $\gamma_{xy}(x,y)$ which
satisfies the strain-displacement relations (33) and the com-
patibility conditions. The principle of virtual work gives a
relation between any statically admissible field and any kinematically admissible field;

$$\int_L \underline{p} \cdot \underline{u}^* \, dxdy + \int_D \underline{F} \cdot \underline{u}^* \, dxdy = \int_D (\sigma_x \varepsilon_x^* + \sigma_y \varepsilon_y^* + \tau_{xy} \gamma_{xy}^*) \, dxdy \; . \tag{35}$$

Asterisks have been attached to the kinematic quantities to
emphasize that the two fields may be completely independent;
$\underline{u}^*$ need not be the displacements caused by the loads $\underline{p}$ and
body forces $\underline{F}$.

Standard boundary value problems in plane stress are formulated in the following way. At each point on the boundary
either the force component or the displacement component must
be prescribed in both of two perpendicular directions in the
plane of the sheet. These directions need not coincide with
the coordinate directions. In addition, the body forces must
be prescribed at all points in the plane. In order to solve
the problem we must determine the stresses, displacements and
strains throughout the body. The solution must satisfy the
imposed boundary conditions, and must be both statically and
kinematically admissible. Additional information required to
solve the problem is obtained from the mechanical behavior of
the material of which the sheet is composed in the form of relations between stress and strain components.

As an illustration, the stress-strain relations for an
isotropic elastic sheet are

$$\varepsilon_x = \frac{\sigma_x}{E} - \frac{\nu}{E}\sigma_y \;,$$

$$\varepsilon_y = \frac{\sigma_y}{E} - \frac{\nu}{E}\sigma_x \;, \tag{36}$$

$$\gamma_{xy} = \frac{2(1+\nu)}{E}\tau_{xy} \;,$$

where E is Young's modulus and ν is Poisson's ratio. These equations may be inverted to give the stresses as functions of the strains;

$$\sigma_x = \frac{E}{(1-\nu^2)}(\varepsilon_x + \nu\varepsilon_y) \;,$$

$$\sigma_y = \frac{E}{(1-\nu^2)}(\varepsilon_y + \nu\varepsilon_x) \;, \tag{37}$$

$$\tau_{xy} = \frac{E}{2(1+\nu)}\gamma_{xy} \;.$$

The strain energy density (i.e. the strain energy per unit volume or the strain energy per unit area for a sheet of unit thickness) is given by

$$W(\varepsilon_x,\varepsilon_y,\gamma_{xy}) = \frac{E}{2(1-\nu^2)}\{\varepsilon_x^2 + 2\nu\varepsilon_x\varepsilon_y + \varepsilon_y^2\} + \frac{E}{4(1+\nu)}\gamma_{xy}^2 \tag{38}$$

The complementary energy density is

$$\Omega(\sigma_x,\sigma_y,\tau_{xy}) = \frac{1}{2E}\{\sigma_x^2 - 2\nu\sigma_x\sigma_y + \sigma_y^2\} + \frac{(1+\nu)}{E}\tau_{xy}^2 \tag{39}$$

From more general considerations, we require that $E > 0$, $-1 \leq \nu \leq 1/2$ (cf. Section 1.5). This ensures that the strain and complementary energy densities are positive definite, and that the solutions to elastic boundary value problems of the form set out above are unique.

1.5 Three-Dimensional Continua

In the three-dimensional problem we consider bodies of arbitrary shape. The fundamental equations will be written for a cartesian coordinate system, and indicial notation with the conventional summation convention will be used. The volume of the body will be denoted by V, and the area of the bounding surfaces by A.

An element of material enclosed between the planes x_1, $x_1 + dx_1$, x_2, $x_2 + dx_2$, x_3 and $x_3 + dx_3$ is acted upon by the components of the stress tensor $\sigma_{ij}(x_1, x_2, x_3)$ with $i = 1,2,3$. These stress components are shown diagrammatically in Fig.8. A body force $F_i(x_1, x_2, x_3)$ per unit volume may also act on the element. The equilibrium equations, obtained by considering the forces acting on this element of material, are

$$\frac{\partial \sigma_{ij}}{\partial x_j} + F_i = 0 \ , \tag{40}$$

$$\sigma_{ij} = \sigma_{ji} \ . \tag{41}$$

The stress σ_{ij} is thus symmetric, and its tensorial nature may be confirmed by investigating rotations of the coordinate system. If the unit outward normal vector at a point on the surface is given by ν_i, the external forces on the boundary p_i are related to the stress tensor at that point by the equation

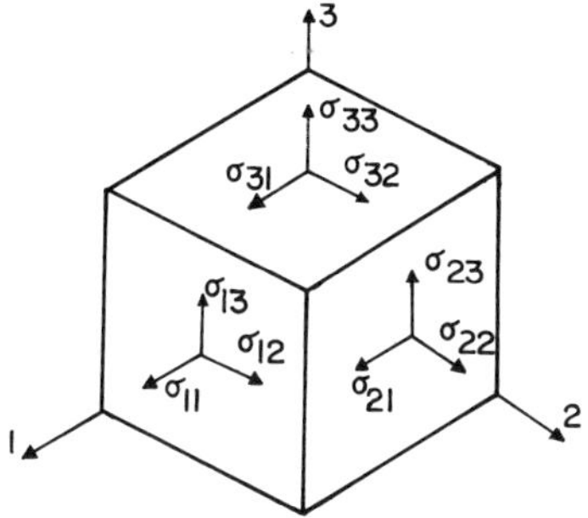

Figure 8.   Infinitesimal continuum element

$$p_i = \sigma_{ij}\nu_j \quad . \tag{42}$$

A set of boundary forces p_i, stresses σ_{ij} and body forces F_i will be termed statically admissible if they satisfy equations (40) and (41).

A particle with coordinates x_i in the undeformed configuration of the body undergoes a displacement u_i as deformation occurs. The strain tensor ε_{ij} is derived from the displacement field u_i by means of the strain-displacement relations,

$$\varepsilon_{ij} = \frac{1}{2}\left(\frac{\partial u_i}{\partial x_j} + \frac{\partial u_j}{\partial x_i}\right) \quad . \tag{43}$$

A strain field is integrable (i.e. equation (43) can be integrated to determine the displacements to within a rigid body motion) if the compatibility conditions are satisfied;

$$\frac{\partial^2 \varepsilon_{ij}}{\partial x_k \partial x_l} + \frac{\partial^2 \varepsilon_{kl}}{\partial x_i \partial x_j} = \frac{\partial^2 \varepsilon_{ik}}{\partial x_j \partial x_l} + \frac{\partial^2 \varepsilon_{jl}}{\partial x_i \partial x_k} \quad . \tag{44}$$

Additional discrete relations must be added when cavities occur in the body. A set of displacements u_i and strains ε_{ij} will be termed kinematically admissible if equations (43) and (44) are satisfied.

The principle of virtual work may be written for any statically admissible set and any kinematically admissible set.

$$\int_A p_i u_i^* \; dA + \int_V F_i u_i^* \; dV = \int_V \sigma_{ij}\varepsilon_{ij}^* \; dV \quad . \tag{45}$$

Asterisks have been introduced to emphasize that the statically admissible set may be completely independent.

A standard boundary value problem in this context is formulated by prescribing the body forces F_i at every point in the body, and prescribing either a boundary force component or a boundary displacement component in each of the three mutually perpendicular directions at every point on the surface. In order to solve the boundary value problem we must determine $\sigma_{ij}(x_i)$, $\varepsilon_{ij}(x_i)$ and $u_i(x_i)$ such that the boundary conditions are satisfied and the solution is both statically and kinematically admissible. Further information is provided by the mechanical behavior of the body in the form of constitutive equations involving ε_{ij} and σ_{ij}.

As an example, the constitutive equations for an *isotropic elastic material* may be written as

$$\sigma_{ij} = \lambda \varepsilon_{kk} \delta_{ij} + 2\mu \varepsilon_{ij} \ , \tag{46}$$

where λ, μ are Lamé's constants and δ_{ij} is the Kronecker delta. It will often be convenient to introduce the *stress* and *strain deviators*, s_{ij} and e_{ij} respectively, defined by

$$s_{ij} = \sigma_{ij} - \frac{1}{3} \sigma_{kk} \delta_{ij} \ , \tag{47}$$

$$e_{ij} = \varepsilon_{ij} - \frac{1}{3} \varepsilon_{kk} \delta_{ij} \ . \tag{48}$$

Contracting equation (46), it is found that the volume strain ε_{kk} and the mean hydrostatic tension $\sigma_{kk}/3$ are related by

$$\frac{\sigma_{kk}}{3} = (\lambda + \frac{2\mu}{3}) \varepsilon_{kk} \tag{49}$$

Equations (46 - 49) together show that the deviators are also directly related;

$$s_{ij} = 2\mu e_{ij} \tag{50}$$

The strain energy density $W(\varepsilon_{ij})$ is given by

$$W(\varepsilon_{ij}) \;=\; \frac{\lambda}{2}\,\varepsilon_{kk}\varepsilon_{jj} \;+\; \mu\varepsilon_{ij}\varepsilon_{ij} \tag{51}$$

Equation (46) can be inverted to give strain as a function of stress. This relation is usually given in the form

$$\varepsilon_{ij} \;=\; \frac{(1+\nu)}{E}\,\sigma_{ij} \;-\; \frac{\nu}{E}\,\sigma_{kk}\,\delta_{ij} \tag{52}$$

where E is Young's modulus and ν is Poisson's ratio. The complementary energy density $\Omega(\sigma_{ij})$ may then be written as

$$\Omega(\sigma_{ij}) \;=\; \frac{(1+\nu)}{2E}\,\sigma_{ij}\sigma_{ij} \;-\; \frac{\nu}{2E}\,\sigma_{kk}\sigma_{jj} \tag{53}$$

On physical grounds we require that the strain energy density should be positive definite. It may be shown that this follows if the coefficients in equations (49) and (50) are positive definite, i.e. if

$$\lambda + \frac{2\mu}{3} \;>\; 0 \;,\quad \mu \;>\; 0\;. \tag{54}$$

This also places restrictions on the values of E and ν; by considering the inversion process we can show that

$$\frac{2\mu(1+\nu)}{3(1-2\nu)} \;=\; \lambda + \frac{2\mu}{3} \;,\quad E \;=\; 2\mu(1+\nu)\;. \tag{55}$$

Comparing equation (55) with the restrictions of equation (54) we must have

$$-1 \;\leq\; \nu \;\leq\; \frac{1}{2}\;,\quad E > 0. \tag{56}$$

As a result of these restrictions the complementary density is also positive definite, and elastic boundary value problems of the type set out in this section have unique solutions.

It should be noted that the plane stress problem of Section 1.4 is a special case of this more general problem obtained by setting $\sigma_{33}(=\sigma_z)$, $\sigma_{13}(=\tau_{xz})$, $\sigma_{23}(=\tau_{yz})$ and $F_3(=F_z)$ equal to

zero, and assuming that nonzero stress components do not vary
in the 3 (or z) direction. However, because the displacement
component $u_3(= u_z)$ and the strain component $\varepsilon_{33}(= \varepsilon_{zz})$ are ignored, the compatibility equations are not properly satisfied.
Nevertheless, the plane stress assumptions can give very good
approximations for appropriately loaded thin plates.

1.6 Notation for the General Discussion of Structural Behavior

The particular formulations of the structural problem given in
the preceding four sections differ quite considerably in some
aspects, but they have several important features in common.
These features are in fact common to all formulations which are
based on the same idealizations, namely the classical small
displacement assumption and static loading. It is our intention in this monograph to carry out a discussion of mechanical
behavior based on just these common features in order that we
can emphasize fundamental ideas and derive general results.

Much of the basic part of the formulation of constitutive
equations can be based on work concepts. General results for
structures and bodies will be derived by means of the principle of virtual work, which makes use of the concepts of statically admissible fields and kinematically admissible fields
rather than the particular equilibrium and strain-displacement
relations in any one situation. If we then adopt a notation
for our discussion which enables us to express work relations
and the principle of virtual work in general terms,we can proceed with the formulation of plastic constitutive relations and
the derivation of general results for structures without being
limited to particular cases of the structural problem. This notation will be developed by considering in turn those features

which the four structural problems discussed above have in
common, and generalizing each in turn.

In each problem the state of stress, or generalized stress,
at a point in the body is described by a number of stress components. Once the coordinate system for the problem has been
chosen it is sufficient for us to consider these stresses as
the components of a vector in an n-dimensional space. The
number n is not fixed, but is permitted to take a value appropriate to any particular problem to which we wish to apply our
results. We shall denote this *stress vector* by Q_j, where j=1,
2,...,n. In the bar problem outlined in Section 1.2, there
are three generalized stress components, and hence n = 3. The
order in which we choose them is not important; we can put
Q_1 = M, Q_2 = N, Q_3 = V. In the plate problem of Section 1.3,
n = 3 again and we could take Q_1 = M_r, Q_2 = M_θ and Q_3 = Q. In
the plane stress problem we might put Q_1 = σ_x, Q_2 = σ_y and
Q_3 = τ_{xy}. In the continuum formulation of Section 1.5 there
are nine stress components, and hence n = 9. We may put
Q_1 = σ_{11}, Q_2 = σ_{22} and so on. We can also make use of the
symmetry of the stress tensor, which gives σ_{12} = σ_{21}, σ_{23} =
σ_{32}, σ_{31} = σ_{13} and use a stress vector with the six independent components of σ_{ij}. Either choice is acceptable, although
some modification of the strain vector will be necessary.

The introduction of a *stress space* is a very useful device.
We can consider that the stress Q_j is represented by a stress
point in an n-dimensional space with coordinates $Q_1,Q_2,...,Q_n$.
A two-dimensional stress space is shown in Fig. 9. The trajectory traced out by the stress point as the stress changes
will be termed the *stress path*.

In the four problems above, as in any formulation of the
structural problem, the state of strain, or generalized strain

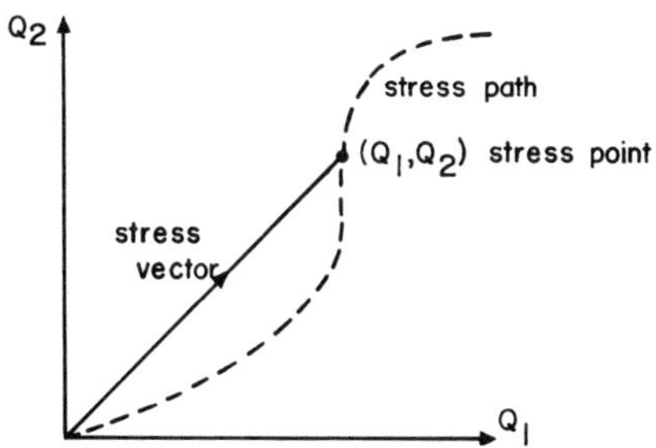

Figure 9. Stress space for n=2

at a point in the body is described by a number of strain components equal to the number of stress components (remembering
that we have assumed the shear strains in the beam and plate
problems to be zero). We shall denote these strains as the
components of an n-dimensional *strain vector* q_j with j=1,2,..
..,n. We order the strain components in such a way that Q_1,
q_1, Q_2, q_2 and so on are associated in a work expression. The
increment of work done per unit generalized volume when the
strains are increased from q_j to $q_j + dq_j$ is given by

$$dW = \sum_{j=1}^{n} Q_j dq_j = Q_1 dq_1 + Q_2 dq_2 + \ldots + Q_n dq_n \ . \tag{57}$$

The work increment dW is thus the scalar product of the vectors Q_j and dq_j. The summation convention for subscripts
will be adopted: when like subscripts appear in an expression
this indicates that they must be summed over the range 1 to n.
The summation signs can then be dropped, and equation (57) is
written concisely in the form

$$dW = Q_j dq_j \ . \tag{58}$$

In the bar problem, therefore, with the choice $Q_1 = M_1$,
$Q_2 = N$, $Q_3 = V$ we put $q_1 = \kappa$, $q_2 = \varepsilon$ and by assumption $q_3 = 0$.
In this problem the work increment given by equation (58) is

$$dW = Q_j dq_j = Md\kappa + Nd\varepsilon \quad . \tag{59}$$

The appropriate choices for $q_1, \ldots, q_n$ in the plate and plane stress problems can similarly be demonstrated. It should be noted that in the general continuum problem in cartesian tensors the work done when the strain is changed by $d\varepsilon_{ij}$ (with $i,j = 1,2,3$) is

$$\sigma_{ij}d\varepsilon_{ij} = \sigma_{11}d\varepsilon_{11} + \sigma_{12}d\varepsilon_{12} + \sigma_{13}d\varepsilon_{13} + \sigma_{21}d\varepsilon_{21} + \sigma_{22}d\varepsilon_{22}$$
$$+ \sigma_{23}d\varepsilon_{23} + \sigma_{31}d\varepsilon_{31} + \sigma_{32}d\varepsilon_{32} + \sigma_{33}d\varepsilon_{33} \quad . \tag{60}$$

If we choose to set $n = 9$, with $Q_1 = \sigma_{11}, \ldots, Q_9 = \sigma_{33}$ we let $q_1 = \varepsilon_{11}, \ldots, q_9 = \varepsilon_{33}$, and equations (58) and (60) become identical. If we wish to make use of the symmetry of σ_{ij} ($\sigma_{12} = \sigma_{21}, \sigma_{23} = \sigma_{32}, \sigma_{31} = \sigma_{13}$) and set $n = 6$, we must let the strain component associated with σ_{12} be given by $\varepsilon_{12} + \varepsilon_{21}$ = $2\varepsilon_{12}$, and so on. This difference of a factor of two is the distinction between tensorial and engineering shear strains.

If the material is linear elastic, the most general relation between the strain and stress components can be written in the form

$$q_j = C_{jk}Q_k \quad . \tag{61}$$

The stiffness matrix C_{jk} must be symmetric, i.e.

$$C_{jk} = C_{kj} \quad , \tag{62}$$

and we restrict the matrix to be positive definite. As outlined in the discussion of the four problems above we can write the strain energy associated with strains $\bar{q}_j$,

$$W(\bar{q}_j) = \int_0^{\bar{q}_j} Q_j dq_j \quad . \tag{63}$$

The strain energy W will be a homogeneous quadratic function
of the strain components q_1, ..., q_n. Differentiation of
equation (63) gives

$$Q_j = \frac{\partial W}{\partial q_j} \cdot \tag{64}$$

Similarly, the complementary energy associated with
stresses $\bar{Q}_j$ is given by

$$\Omega(\bar{Q}_j) = \int_0^{\bar{Q}_j} q_j \, dQ_j \cdot \tag{65}$$

The complementary energy $\Omega(Q_j)$ will likewise be a homogeneous
quadratic function of the components Q_1, Q_2, ...,Q_n. We may
note that the equation

$$\Omega(Q_j) = C \; , \tag{66}$$

where C is a positive constant, is the equation of a hypersurface, or level surface of Ω in the stress space. This is
shown diagrammatically in the two-dimensional case in Fig.10.
The strain components, obtained by differentiation of equation
(65),

$$q_j = \frac{\partial \Omega}{\partial Q_j} \tag{67}$$

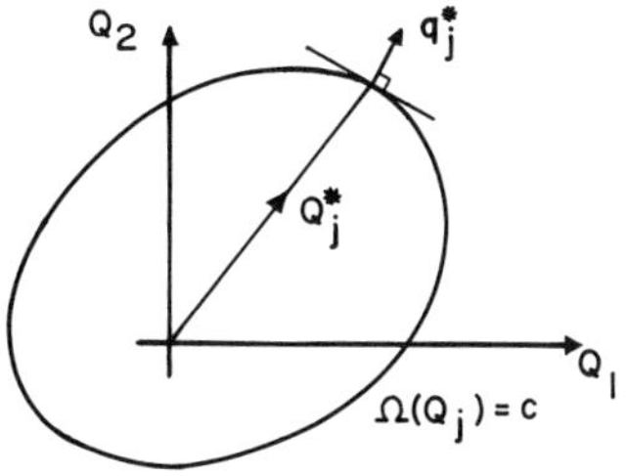

Figure 10. Level surface of constant complementary energy for
an elastic material

are the components of the *gradient* of the *scalar field* Ω. Consider a stress Q_j^* which is such that $\Omega(Q_j^*) = C$. The strain q_j^* given by equation (67) with $Q_j = Q_j^*$ is a vector which may be plotted in the stress space (with q_1^* as the component in Q_1 direction, etc.) with its origin at the stress point. This vector will be normal to the level surface $\Omega(Q_j) = C$ at that point, as shown in Fig.10 for the two-dimensional case. Further, because of their quadratic form, the level surfaces of Ω will be convex (i.e. a line joining any two states of stress which lie on or within a given level surface will not cut the surface).

The components of the stress vector Q_j and the strain vector have different dimensions in various formulations of the problem, and may have different dimensions in any one formulation, as in the bar structure problem. However, in each problem the scalar product $Q_j q_j$ is composed of the sum of terms of like dimensions. The dimension of this product is that of *specific work*, i.e. work per unit length in the bar problem, per unit area in the plate problem, per unit area in the plane stress problem (where the thin sheet is assumed to be of unit thickness) and per unit volume in the three-dimensional case. If a quantity with dimensions of specific work is integrated over the whole structure, in each case the integral has the dimensions of work. Let us represent the space variables with respect to which the integration is carried out by the single symbol V. In our general notation, then, the strain energy integrated over the whole structure is written as

$$\int_V W(q_j)\,dV \quad .$$

Thus in the three-dimensional cartesian formulation,

$dV = dx_1 dx_2 dx_3$ and integration is carried out over the whole volume of the body. In the plane stress problem, where all quantities are assumed to be constant over the unit thickness, $dV = dxdy$ and the integration is carried out over the area of the sheet. In the plate problem in polar coordinates, $dV = rd\theta dr$ and the integration is carried out over the whole area of the plate. In the bar problem, $dV = dx$, and we integrate over the whole length of the structure.

Applied forces in structural problems are components of a vector in three-dimensional space, although in some problems we limit the directions in which the forces can act. Similarly, applied couples are also the components of a vector. In the four problems outlined the coordinate systems in which the force and couple vectors are expressed is defined for that particular problem, but we can in all cases adequately represent the forces by the vector $\underline{p}$ and the couples by the vector $\underline{c}$. In any structural problem the forces and couples $\underline{p}$ and $\underline{c}$ will be functions of the space variables. In the bar problem $\underline{p}$ and $\underline{c}$ are functions of the distance along the bar x. In the plate problem they are functions of r and θ, although by the assumption of axial symmetry the θ dependence vanishes. In the plane stress problem, the forces are functions of x and y, but we assume that the component of $\underline{p}$ in the z direction is zero, and applied forces act only on the boundary which is defined by a relation between x and y. In the three-dimensional problem $\underline{p}$ is a function of x,y and z, and the boundary surface is defined by a relation between x,y and z. In any problem, therefore, there are at most two independent space variables defining the surface. We shall denote the appropriate space variables on the surface by s, and write $\underline{p} = \underline{p}(s)$, $\underline{c} = \underline{c}(s)$ to denote dependence on these space variables.

In the bar and plate problems we encounter point loads and

line loads. It is not necessary for us to introduce a special
symbol to represent these loads, or point couples or line
couples. Such loads and couples are simply delta functions in
$\underline{p}(s)$ and $\underline{c}(s)$, i.e. $\underline{p}(s)$ and $\underline{c}(s)$ take infinitely large values
at a point or along a line on the surface. The integral across
this discontinuity in the function $\underline{p}(s)$ or $\underline{c}(s)$ is finite, and
is the value of the point load or couple. In particular prob-
lems, however, we will continue to treat point or line loads
or couples directly.

The displacements in any structural problem, as exemplified
in the four cases discussed in detail, can also be treated as
the components of a vector $\underline{u}$ in three-dimensional space. In
those cases where applied couples are present, we must also
treat rotation components. Under the classical assumptions of
small displacements and displacement gradients, rotations are
infinitesimal and can therefore be treated as the components
of a rotation vector $\underline{\theta}$.

In a manner similar to stress-strain products, the scalar
products $\underline{p}(s).\ \underline{u}(s)$ and $\underline{c}(s).\ \underline{\theta}(s)$ on the surface of the body
always have the dimensions of specific work, i.e. work per
unit length when s represents one independent space variable
and work per unit area when s represents two independent space
variables. If these scalar products are integrated over the
surface, the integral will have the dimensions of work. Such
integrals will be written as

$$\int_S \underline{p}(s).\ \underline{u}(s)dS \quad \text{and} \quad \int_S \underline{c}(s).\ \underline{\theta}(s)dS.$$

We can further simplify our general notation by dropping the
terms $\underline{c}(s)$ and $\underline{\theta}(s)$, and regarding $\underline{p}(s)$ and $\underline{u}(s)$ as *generalized forces* and *generalized displacements*. The generalized
force $\underline{p}(s)$ thus may represent as many as three components of

force and three components of a couple, and the generalized
displacement $\underline{u}(s)$ may represent three components of displacement and three components of rotation.

Body force components, in the cases where they are present,
can also be treated as the components of a vector $\underline{F}$. This
vector is a function of the space coordinates, as are the
generalized stresses, generalized strains and displacements.
The scalar product of the body force $\underline{F}$ and the actual displacement $\underline{u}$ will again have dimensions of specific work. The
integral of this product over the body will have the dimensions
of work, and will be denoted by

$$\int_V \underline{F} \cdot \underline{u} dV \quad .$$

In deriving general results for structural behavior it will
not be necessary for us to refer to the particular forms of the
equilibrium equations or of the strain-displacement relations
which must be used in particular problems. However, we demand
that the equilibrium equations and the relations between
stresses and applied forces at boundaries should be *linear* in
the components of $\underline{p}$, $\underline{F}$ and Q_j and their spatial derivatives.
Similarly we demand that the strain-displacement relations
should be *linear* in the components of $\underline{u}$ and q_j. These restrictions will always be met if particular problems are formulated
under the classical small displacement assumptions.

Linearity of the equilibrium and strain-displacement relations will permit the principle of virtual work to be written.
If the components of $\underline{p}$, $\underline{F}$ and Q_j satisfy the appropriate equilibrium equations they will be said to be *statically admis-
sible*. If the components of $\underline{u}^*$ and q_j^* satisfy the strain-
displacement relations they will be said to be *kinematically
admissible*. For *any* statically admissible set $\underline{p}$, $\underline{F}$, Q_j and

any kinematically admissible set $\underline{u}^*$, q_j^*, we can write the principle of virtual work,

$$\int_S \underline{p}(s) \cdot \underline{u}^*(s)dS + \int_V \underline{F} \cdot \underline{u}^* dV = \int_V Q_j q_j^* dV \ . \qquad (68)$$

This equation is the general form of the particular equations given in Section 1.2 (equation 9), Section 1.3 (equation 23), Section 1.4 (equation 35) and Section 1.5 (equation 45).

Structural problems are set, again as exemplified in the four cases discussed in detail, by specifying at each point on the surface either the component of $\underline{p}(s)$ or the associated component of $\underline{u}(s)$. Thus, when the scalar product $\underline{p}(s) \cdot \underline{u}(s)$ is expanded, it consists of the sum of products of two terms, one of which is given as data and one of which must be determined by solving the problem. In discussing the integral

$$\int_S \underline{p}(s) \cdot \underline{u}(s)dS$$

it is often convenient to divide the total surface S into two parts. We denote by S_p all points where force components are given, and include in the integral

$$\int_{S_p} \underline{p}(s) \cdot \underline{u}(s)dS$$

all terms in the expanded scalar product which contain specified force components. Similarly, we denote by S_u all points where displacement components are given, and include in the integral

$$\int_{S_u} \underline{p}(s) \cdot \underline{u}(s)dS$$

all terms in the expanded scalar product which contain specified displacement components. Thus

$$\int_S \underline{p}(s) \cdot \underline{u}(s) dS = \int_{S_p} \underline{p}(s) \cdot \underline{u}(s) dS + \int_{S_u} \underline{p}(s) \cdot \underline{u}(s) dS . \qquad (69)$$

All points in the structure must appear in S_p or S_u or in both. The structural problem itself may then be phrased in the following way; we are given $\underline{p}(s) = \hat{\underline{p}}(s)$ on S_p, $\underline{u}(s) = \hat{\underline{u}}(s)$ on S_u and $\underline{F} = \hat{\underline{F}}$ on V. We are then required to determine $p(s)$ on S_u, $\underline{u}$ on V and S_p, and Q_j, q_j on V. The given forces on S_p are usually referred to as the *loads*, and the forces which must be determined on S_u as the *reactions*.

In much of the development of general results for **structures** we shall make use of special classes of statically and kinematically admissible sets. If $\underline{p}(s)$, $\hat{\underline{F}}$ and Q_j are admissible and, in addition, $\underline{p}(s) = \hat{\underline{p}}(s)$ on S_p we shall term $p(s)$, $\hat{\underline{F}}$ and Q_j a *pertinent statically admissible set*. Since the forces on S_u and the stresses which are the actual solution to the problem must satisfy the equilibrium conditions, it is clear that the solution is a member of the class of *pertinent statically admissible sets*. In a similar way, if $\underline{u}$, q_j, are kinematically admissible and $\underline{u}(s) = \hat{\underline{u}}(s)$ on S_u, $\underline{u}$, q_j will be termed a *pertinent kinematically admissible set*. The displacements and strains which make up the actual solution to the problem must be a member of the class of *pertinent kinematically admissible sets*.

The concepts of pertinent statically and kinematically admissible fields together with the convexity of the functions $W(q_j)$ and $\Omega(Q_j)$ lead us directly to the classical potential and complementary energy theorems in *elasticity*. Since at a later point in the monograph we shall introduce results which can be considered as generalizations of these theorems it is convenient to outline the derivation of the classical theorems at this point.

We first express the convexity of the strain energy function $W(q_j)$ and the complementary energy function $\Omega(Q_j)$ in mathematical terms. Since for a linear elastic material $W(q_j)$ is homogeneous and quadratic in the components of q_j, the mathematical expression follows directly provided that $W(q_j)$ is positive definite. We introduce two arbitrary states of strain q_j^a, q_j^b and let

$$q_j = q_j^a - q_j^b \; . \tag{70}$$

It follows then that

$$W(q_j) = W(q_j^a - q_j^b) \;=\; W(q_j^a) - C_{jk}q_j^a q_k^b + W(q_j^b) \; . \tag{71}$$

Making use of the linearity of the relations, we may put

$$C_{jk}q_j^a q_k^b \;=\; Q_j^b q_j^a \; , \tag{72a}$$

and

$$2W(q_j^b) \;=\; Q_j^b q_j^b \; , \tag{72b}$$

or alternatively

$$W(q_j^b) \;=\; Q_j^b q_j^b - W(q_j^b) \; . \tag{72c}$$

Substituting equations (72a) and (72c) into (71), and noting that

$$W(q_j) \geq 0 \; , \tag{73}$$

we find that

$$W(q_j^a) - W(q_j^b) \geq (q_j^a - q_j^b)Q_j^b \; . \tag{74}$$

Since

$$Q_j^b \;=\; \left.\frac{\partial W}{\partial q_j}\right|_{q_j=q_j^b} \; , \tag{75}$$

equation (74) can also be written in the form

$$W(q_j^a) - W(q_j^b) \geq (q_j^a - q_j^b) \left.\frac{\partial W}{\partial q_j}\right|_{q_j = q_j^b} . \qquad (76)$$

Inequality (76) is in fact the mathematical expression of the convexity of $W(q_j)$. It should be noted that (76) follows from (73) if $W(q_j)$ is quadratic, i.e. if the stress-strain relation is linear. The classical potential energy theorem follows directly from (76); consequently it holds for a linear elastic material for which $W(q_j) \geq 0$, as for a nonlinear elastic material which admits a strain energy potential (of the form of equation 75) and for which (76) holds.

In a similar manner we can introduce two arbitrary stress states Q_j^a, Q_j^b and show that if $\Omega(Q_j)$ is homogeneous and quadratic in the components of Q_j and $\Omega(Q_j) \geq 0$,

$$\Omega(Q_j^a) - \Omega(Q_j^b) \geq (Q_j^a - Q_j^b)q_j^b . \qquad (77)$$

Alternatively, this inequality may be written as

$$\Omega(Q_j^a) - \Omega(Q_j^b) \geq (Q_j^a - Q_j^b) \left.\frac{\partial \Omega}{\partial Q_j}\right|_{Q_j = Q_j^b} . \qquad (78)$$

The classical complementary energy theorem is derived from (78); it consequently also holds for a nonlinear elastic material for which

$$q_j = \frac{\partial \Omega}{\partial Q_j} \qquad (79)$$

and for which inequality (78) holds for all choices of Q_j^a and Q_j^b.

Let us now consider the structural problem characterized by $\hat{\underline{p}}(s)$ on S_p, $\hat{\underline{u}}(s)$ on S_u, $\underline{\hat{F}}$ on V. Let the solution to this problem for an elastic material be given by $\underline{p}(s)$ on S_u, $\underline{u}$ on S_p and V, Q_j and q_j. These quantities constitute, by definition, both a pertinent statically admissible field and a pertinent kinematically admissible field. In addition, Q_j and q_j satisfy the constitutive relations.

Let $\underline{u}^c$ on S_p and V, q_j^c be any pertinent kinematically admissible set for this problem. Both q_j^c and q_j are defined at each point in the body; hence (associating q_j^c with q_j^a and q_j with q_j^b) from (74) or (76)

$$W(q_j^c) - Q_j q_j^c \geq W(q_j) - Q_j q_j \quad . \tag{80}$$

This inequality applies at each point in the body and may thus be integrated over the body. From the principle of virtual work

$$\int_V Q_j q_j^c \, dV = \int_{S_p} \hat{\underline{p}} \cdot \underline{u}^c \, dS \;+\; \int_{S_u} \underline{p} \cdot \hat{\underline{u}} \, dS \;+\; \int_V \hat{\underline{F}} \cdot \underline{u}^c \, dV \quad , \tag{81a}$$

and

$$\int_V Q_j q_j \, dV = \int_{S_p} \hat{\underline{p}} \cdot \underline{u} \, dS \;+\; \int_{S_u} \underline{p} \cdot \hat{\underline{u}} \, dS \;+\; \int_V \hat{\underline{F}} \cdot \underline{u} \, dV \quad . \tag{81b}$$

Thus, integrating inequality (80) over the volume, substituting from equations (81) and eliminating the integral over S_u which appears on each side of the expression,

$$\int_V W(q_j^c) \, dV - \int_V \hat{\underline{F}} \cdot \underline{u}^c \, dV - \int_{S_p} \hat{\underline{p}} \cdot \underline{u}^c \, dV \geq \int_V W(q_j) \, dV - \int_V \hat{\underline{F}} \cdot \underline{u} \, dV$$

$$- \int_{S_p} \hat{\underline{p}} \cdot \underline{u} \, dS \quad . \tag{82}$$

This is the classical elastic minimum potential energy theorem. It may also be expressed by defining the potential energy for any pertinent kinematically admissible field,

$$U_p(\underline{u}^c,\ q_j^c)\ =\ \int_V W(q_j^c)dV\ -\ \int_V \hat{\underline{F}}\cdot\underline{u}^c dV\ -\ \int_{S_p} \hat{\underline{p}}\cdot\underline{u}^c dS\ ,\tag{83}$$

and noting that U_p takes its least value when q_j^c is the actual strain field or the solution of the problem.

Now let $\underline{p}^s(s)$ on S_u, Q_j^s be any pertinent statically admissible field. In a formally identical manner, making use of inequalities (77) or (78), we may derive the dual result, the classical minimum complementary energy theorem. We find that

$$\int_V \Omega(Q_j^s)dV\ -\ \int_{S_u} \underline{p}^s\cdot\hat{\underline{u}}dS\ \geq\ \int_V \Omega(Q_j)dV\ -\ \int_{S_u} \underline{p}\cdot\hat{\underline{u}}dS\ .\tag{84}$$

Alternatively we may define the complementary energy of the structure for any pertinent statically admissible field

$$U_c(\underline{p}^s,Q_j^s)\ =\ \int_V \Omega(Q_j^s)dV\ -\ \int_{S_u} \underline{p}^s\cdot\hat{\underline{u}}dS\ ,\tag{85}$$

and note that U_c takes its least value when Q_j^s is the actual stress field or the solution of the problem.

The dual nature of the theorems can be emphasized by noting that

$$Q_j q_j\ =\ \int Q_j dq_j\ +\ \int q_j dQ_j\ =\ W(q_j)\ +\ \Omega(Q_j)\ .\tag{86}$$

Eq.(81b) may thus be written as

$$\int_V W(q_j)dV\ -\ \int_{S_p} \hat{\underline{p}}\cdot\underline{u}dS\ -\ \int_V \hat{\underline{F}}\cdot\underline{u}dV\ =\ -\int_V \Omega(Q_j)dV\ +\ \int_{S_u} \underline{p}\cdot\hat{\underline{u}}dS,\tag{87}$$

or, equivalently

$$U_p(\underline{u},q_j)\ =\ -U_c(\underline{p},Q_j).\tag{88}$$

The energy theorems may thus be written as the continued inequality

$$U_p(\underline{u}^c,q_j^c)\ \geq\ U_p(\underline{u},q_j)\ =\ -U_c(\underline{p},Q_j)\ \geq\ -U_c(\underline{p}^s,Q_j^s)\ .\tag{89}$$

Chapter 2

THE FRAMEWORK OF PLASTIC CONSTITUTIVE RELATIONS

2.1 Plastic Behavior in Simple Tension

In this Chapter we shall introduce our study of plasticity by
discussing the relation between any one component of generalized stress, say Q_1, and the associated component of generalized
strain q_1. The characteristics of the mechanical behavior that
will be presented are those which are observed experimentally,
and correspond to a test on a specimen in which a *homogeneous
stress state* (i.e. a stress state which is not a function of
the spatial coordinates) has been established. As a particular
example we may refer to a bar subjected to an axial force. If
sufficient care is taken in applying the axial force, this test
can be considered as a satisfactory experiment for three of the
particular problems given above. Treating the specimen as a
bar, the generalized stress N is homogeneous. If the bar is in
fact a thin sheet, and the x axis is taken as coincident with
the center line of the bar, plane stress conditions can be set
up in which σ_x is constant and σ_y, τ_{xy} are zero. Alternatively,
if the bar is not a thin sheet it can be treated as a three-
dimensional body where σ_{11} is constant but all other components
of σ_{ij} are zero, again with the x axis coinciding with the center line of the bar. If the bar is composed of a *homogeneous
material* (i.e. one whose properties do not vary with the spatial coordinates) and the test is carried out under isothermal
conditions, a homogeneous state of stress will produce a homogeneous state of strain. In this experiment, as in any others
used to determine constitutive relations, the stress state can
be determined from the external forces using equilibrium considerations alone, and the external loads are measured. The
displacements, or the displacement gradients, are measured independently and the strains are derived from these measurements.

It will be assumed that a testing machine is available in
which either the external forces acting on the specimen or the
crosshead displacement can be controlled independently. It will
be indicated whether it is important to distinguish between a
load controlled or strain controlled experiment.

We shall be interested only in the general characteristics of
the results of the experiments on the bar, rather than in precise numerical values of the data which would be measured when
particular materials are tested. It is the general characteristics which must be identified first in order to construct a
framework for plastic constitutive relations. Once the framework has been constructed we can specialize to particular materials and give quantitative relations between stress and strain.
These relations will in fact identify those parameters which
should be precisely measured in an experiment.

In order to illustrate what is meant by general characteristics we can consider briefly linear elastic behavior. Most materials which are of technological interest which we normally
classify as plastic exhibit linear elastic behavior under certain conditions, so that the general characteristics of linear
elastic behavior must be incorporated in plasticity. Leaving
aside for the present the limits of validity of elastic stress-
strain relations, the bar subjected to an axial force is said to
be composed of a linear elastic material if it is found that the
strain is a linear function of stress under all conditions. This
linear relation can be written in the form

$$q = CQ \tag{1}$$

where the subscript 1 is dropped for the present. Thus Q, q
always lie on a straight line on a stress-strain diagram. On
carrying out experiments on various materials which exhibit

elastic behavior, we would observe that C takes different values in different experiments, but that C is invariably positive. The general characteristics of linear elastic behavior under isothermal conditions are thus described by equation (1), with $C > 0$.

There are in fact some implicit characteristics which are often not emphasized in discussing linear elastic behavior, but which are significant in comparing elastic and plastic behavior. Let the stress be changes from Q^a to Q^b, resulting in a strain change from q^a to q^b. Then, from equation (1),

$$q^b - q^a = C(Q^b - Q^a) \ . \tag{2}$$

The change in strain is thus a linear function of the change in stress. This equation also implies that a change in strain cannot occur if there is no change in stress, and that the change in strain is not a function of the *rate* at which the stress changes from Q^a to Q^b. These characteristics are referred to as *time independence*. Further, the change in strain is not a function of the *stress path* between Q^a and Q^b, e.g. it is the same if Q is changed monotonically from Q^a to Q^b or if it passes through some intermediate value Q^c which would not be achieved in a monotonic change. This characteristic is termed *path independence*.

Consider now an experiment in which a bar of, say, an aluminum alloy or cold-rolled steel is subjected to a *monotonically increasing* tension force. The resulting relation between stress and strain would have the form shown in Fig.1. The relation would initially be *linear*. At A the slope of the curve begins to decrease monotonically, and eventually failure of the specimen occurs at B.

Failure of the specimen is usually (but not always) associated with the development of an instability in the specimen

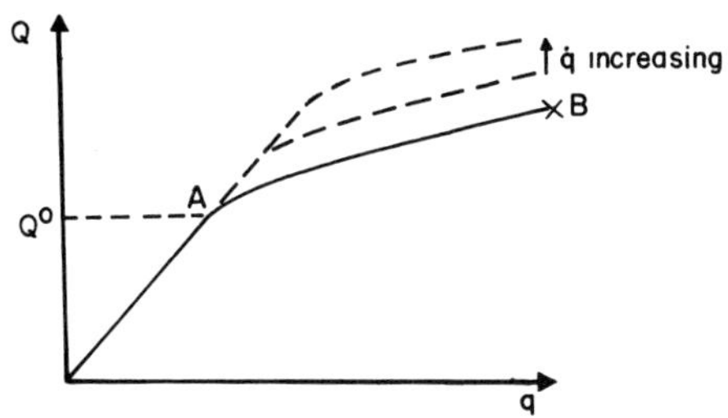

Figure 1. Monotonic loading of an aluminum bar

which leads to a non-homogeneous strain distribution. Depending on whether the testing machine imposes load or displacement, there may be a region in which the slope of the curve is
negative. The ability of the material to incur comparatively
large strains without failure (typically 10-200 times the
strain at the end of the linear region) is an important factor
in the applicability of plasticity to practical problems. This
characteristic is termed the *ductility* of the material.

If this test is carried out very slowly it is called a *static test*. If the stress were suddenly held constant after
considerable strain beyond the end of the linear range had
taken place, we would usually observe that the strain would
continue to increase slowly with time, although at low temperatures at least, the strain would eventually reach a steady
value not much greater than the value when the stress was
first held constant. This behavior is known as *creep*, and is
exhibited by most metals to some degree. In a testing machine which imposes strain at a constant rate, stress-strain
curves such as that in Fig. 1 can be produced by testing identical specimens at different rates. It is normally found that
while the linear range is not affected, the remaining parts of
the curve tend to move upwards as the strain rate increases.
In some metals this effect is scarcely discernible; in others

an increase in the strain rate of 100, say, could produce increases in stress at a given strain between 20% (for many aluminum alloys) and 100% (mild steel). Most metals, therefore,
are not time independent.

Returning to static tests, consider a test on a specimen in
which the stress is increased monotonically to some value Q^*
and then decreased monotonically to zero. An important distinction must be drawn between two cases. If Q^* is less than
some critical value Q^o, the behavior is elastic. An increase
and decrease in the load is shown diagrammatically in Fig.2(a);
the double arrows indicate that a point with the coordinates of
stress and strain moves up and down the same line during the
loading program. If, however, Q^* is greater than Q^o, the behavior is as indicated in Fig. 2(b). The stress curve during
the period when the load is increased is the same as that for
the monotonically increasing load (Fig.1), but when the load is

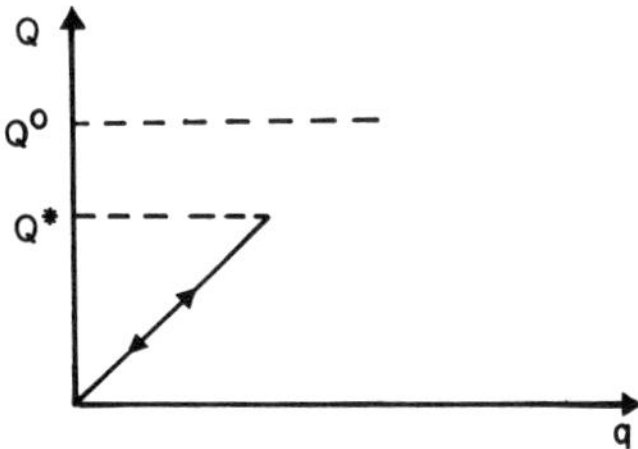

Figure 2(a).Addition and removal of stress Q^*; $Q^* < Q^o$

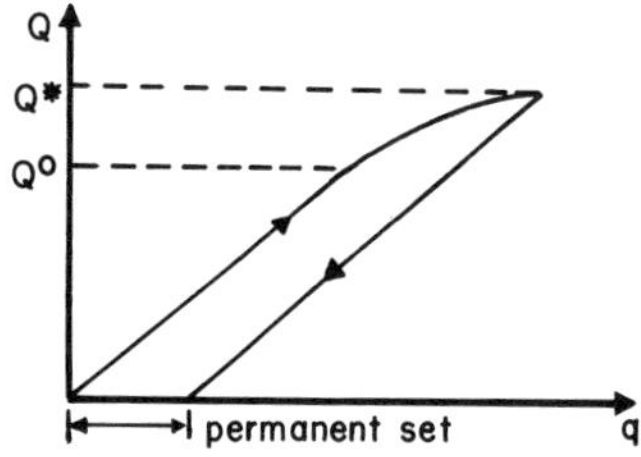

Figure 2(b).Addition and removal of stress Q^*; $Q^* > Q^o$

decreased the point whose coordinates are stress and strain
moves down a line which is parallel to the linear region of the
part of the curve traced out during the monotonic increase in
load. When the stress is again zero at the end of the program
the strain is not zero; there is a residual strain which for
the present we shall term the *permanent set*.

The value of the stress Q^O which distinguishes the behavior
under this loading program is called the *yield stress*. The second class of behavior exhibits one of the major characteristics of plasticity as opposed to elasticity; the addition and
removal of stress produces a residual strain in the specimen.
It may be noted at this point that the experimental determination of the yield stress is difficult. In the loading program
just described it is possible to determine whether the yield
stress has been exceeded (i.e. $Q^* > Q^O$) only if a permanent
set can be measured. On the other hand, permanent sets when Q^*
is only very slightly larger than Q^O may be very small, and too
small to be measured accurately by strain gages or extensometers. The best measurement of Q^O is generally the value of Q^*
which produces the smallest measurable permanent set.

If, having completed the loading program described above
(monotonic increase of Q from zero to Q^*, followed by a monotonic decrease to zero), we again increase the stress monotonically in the same specimen, the result will generally be of
the form shown in Fig.3. The stress strain curve follows the
linear relation up to a value of stress just less than Q^*,
traces out a slight curved region, or knee, and then follows a
line essentially the same as that given in Fig. 1 for the monotonic loading case. Thus the interruption of the monotonic
loading curve caused by decreasing the stress from Q^* to zero
only affects subsequent strains very slightly, and mostly at

stress values around Q^*. In some specimens a slight loop may be associated with the reduction in stress from Q^* and the subsequent increase. In many cases, however, neither the loop nor the knee is discernible.

We have so far discussed only behavior in tension. If we were to carry out the same loading programs in compression (suitably eliminating any geometric effects) we will obtain qualitatively the same behavior. Figure 4 shows the results of monotonic loading tests carried out in compression and tension plotted in the same diagram; let the yield stress in tension be Q^{o+} and the yield stress in compression be Q^{o-}. This indicates that a specimen in its initial state will behave elastically provided that $Q^{o-} \leq Q \leq Q^{o+}$. If the stress is increased monotonically from zero to Q^{*+}, and then decreased monotonically into compression, linear behavior will persist until some value of stress Q^{*-} is reached, when further nonlinear behavior may

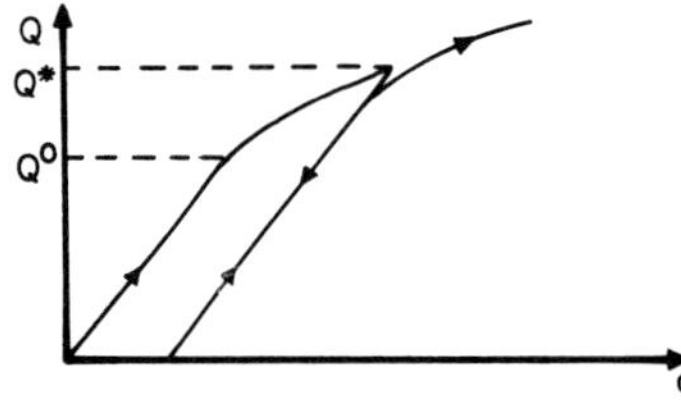

Figure 3. Addition and removal of stress Q^* ($Q^* > Q^o$) followed by a monotonic increase in stress

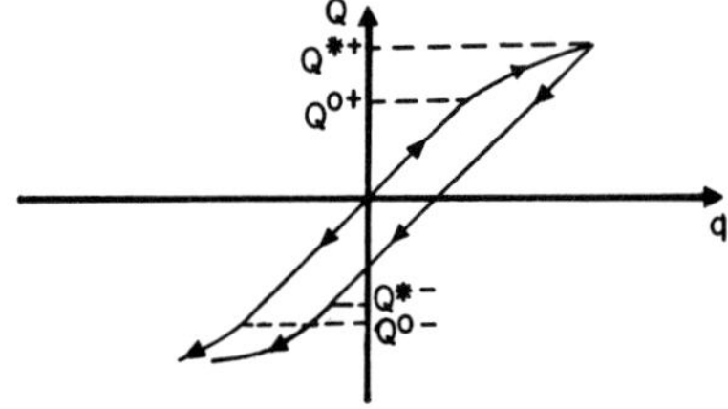

Figure 4. Yielding in tension and compression

occur. Q^{*-} is a new yield stress in compression, and will, in
general, be different from Q^{o-}. If $Q^{o-} < Q^{*-}$ (i.e. Q^{o-} is a
greater compressive stress than Q^{*-}) this is called the *Bauschinger effect*.

We can continue to subject the specimen to ever more complex programs of loading, and the strain response will be cor-
respondingly complex. From the point of view of our subsequent development, however, we can note that two important
characteristics are present. If, after some program of loading,
we are either monotonically increasing or monotonically decreasing the stress, and we suddenly decrease or increase the
stress respectively, we enter into a range of linear behavior.
If we continue to decrease or increase the stress, linear behavior will continue until some new yield stress is reached,
and nonlinear behavior will resume. If on the other hand we
revert back to a monotonic increase or decrease we return to
nonlinear behavior at the value of stress at which the monotonic increase or decrease was interrupted to within a close
approximation.

It is evident from our discussion that the material under
discussion is *path dependent.* The strain is not a function of
stress alone, but depends on the previous loading program or
stress history as well. This is exemplified by the simple case
of zero stress, when permanent sets of differing magnitudes can
be established by varying histories in which the stress starts
and finishes at zero.

Some aluminum alloys exhibit a markedly different behavior
from that shown in Fig. 1 when a specimen is subjected to a
monotonically increasing strain. As shown in Fig. 5, the
stress-strain curve has a linear region, a short region of
fairly high curvature and then a flat region where strain

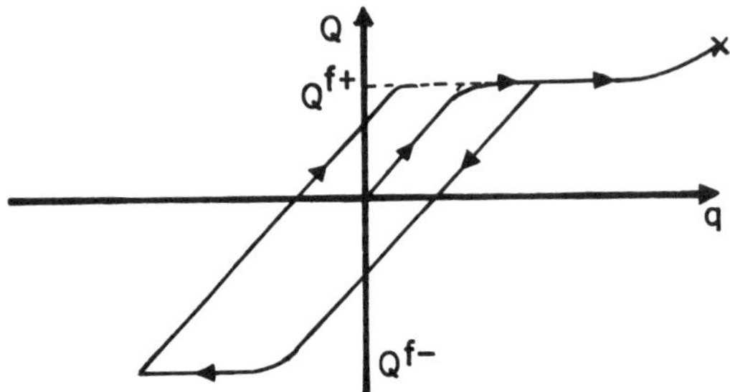

Figure 5. Aluminum bar showing plastic flow

increases at sensibly constant stress. This flat region is
terminated by failure, possibly with a slight change in slope
before failure occurs. The region of high curvature may be very
very short in some instances, approaching the discontinuity in
slope indicated by the dotted lines. If the stress is decreased monotonically from a point in the flat region, the
stress-strain curve is again linear until the yield stress in
compression is reached, when the same type of behavior is again
observed. The behavior in the flat region is generally referred to as *plastic flow*, and is in some senses analogous to flow
in a fluid; however, the strain rate during flow is not in many
metals strongly dependent on the stress. This can again be observed by carrying out monotonic loading tests on specimens
using a machine in which the strain rate can be held constant.
For small strain rates (say less than 10 sec^{-1}) variations in
the strain rate at which the test is carried out produce very
little change in the stress at which plastic flow occurs. Increases in strain rate of one or two orders of magnitude lead
to increases in the flow stress between a few percent (or even
zero in some cases) and a hundred percent. This is very different from a Newtonian fluid, where stress and strain rate are
linearly related.

It is also observed in these materials that the flow stress

is to a fairly close approximation independent of the loading
program. If the stress were to be increased again after flow
in compression (Fig. 5), flow would again commence at the same
tension flow stress which characterized flow during the first
monotonic increase in the loading program. The more common materials whose stress-strain curve is shown in Fig. 1 are generally referred to as *hardening materials* to distinguish them
from materials in which flow occurs (Fig.5).

Mild steel is another material which exhibits plastic flow.
Monotonic increase in the tension strain produces a stress-
strain curve of the form shown in Fig. 6. The material behaves
linearly until an *upper yield stress* Q^{oU} is reached. The stress
then drops suddenly, and plastic flow occurs at a lower yield
stress Q^{oL}. The range of flow changes to one of distinct hardening, and finally failure occurs. The upper yield stress can
be associated with an internal,or microstructural, instability.
It tends to become less significant as the tension strain rate
is increased (in mild steel the flow stress is comparatively
sensitive to strain rate), and is not observed in compression.
In a machine which imposes stress rather than strain the phenomenon is again less marked.

The qualitative descriptions of the experiments outlined
above are by no means comprehensive, but they will suffice to

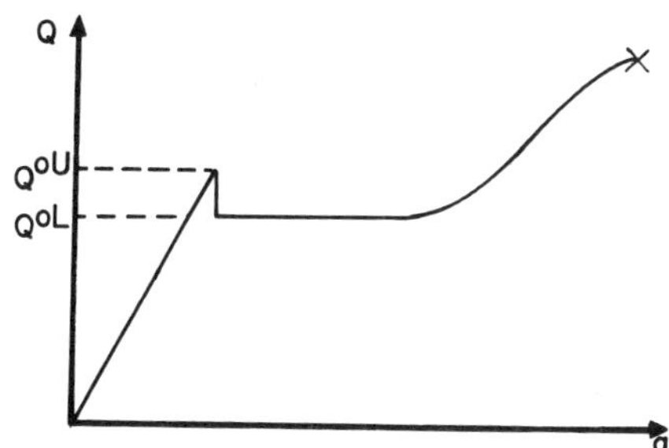

Figure 6. Monotonic straining of a mild steel bar

permit us to set out the assumptions and restrictions which characterize the type of behavior which constitutes the *theory of plasticity*. This, together with generalization of the concepts to deal with complex stress states in which more than one component of the stress Q_j is nonzero, is our task in this chapter. Before proceeding, however, it is instructive to consider some further points on the experiments which are carried out and the type of material with which we expect to deal.

It is of some importance to realize that in the bar experiment described in this section, as in any experiment used to investigate constitutive relations, strain is not measured in an absolute sense. We can measure only changes in displacement from the position in which the test specimen was first set up. Thus we can measure changes in strain from the original state of strain. It is for convenience only that we call the original strain zero; the original state of strain cannot be measured. As an example, suppose we increase and decrease the stress in the bar so as to produce a permanent set. If this specimen is given to another experimenter who was ignorant of the loading program to which the specimen had been subjected, he would carry out experiments assuming that the strain was zero at the beginning of the test. His first experiment would produce results like those of the second monotonic increase in stress in Fig. 3, with the stress axis displaced to the right. However, he would not observe qualitatively different behavior in any program of tests he carried out. This is of general significance, because almost all technologically important metals are produced by processes which involve mechanical deformation, such as hot or cold rolling, drawing, forging or extrusion. These produce permanent sets in the material, and in analyzing the response of a body or structure to loads we must then recognize that the material has been subject to some previous complex loading program, the details of which we do not know.

Most metals of technological importance have a microstructure which is *polycrystalline*, i.e. they are composed of an assemblage of small crystals with random orientation. The stress-strain curves so far discussed are typical of polycrystalline metals. The behavior of a bar which is composed of a *single crystal* may not resemble the experimental results described above, but we shall defer a discussion of single crystals until a later stage.

Polycrystalline materials are clearly not homogeneous on the level of size of the single crystal grains. However, if the grains are very much smaller than the shortest distance on the specimen over which displacement changes can be measured in an experiment and the orientation of the crystal axes of the grains is random, we can never detect the inhomogeneity on the macroscopic level at which the experiments are conducted. We are quite justified in assuming that the material is macroscopically homogeneous.

Mechanical deformation involving permanent sets will produce anisotropy in a polycrystal. It is also convenient to think of a material which has never been subjected to mechanical deformation in the past. Such a material will be *isotropic* on the macroscopic scale if it is composed of randomly oriented grains on the microstructural level. Thus, bar specimens cut with different orientations from a large block of material will show quantitatively identical behavior when subjected to a program of loading. Materials produced for commercial uses are seldom isotropic because of the processing to which they are

subjected. Isotropic polycrystalline materials can be produced for laboratory purposes by various heat treatments which
are designed to erase the mechanical effects of previous loading programs. Polycrystalline materials in which no effects of
previous loading history can be discerned will be said to be in
the *virgin state*.

2.2 Generalization of Results in Simple Tension

The objective of this chapter is the formulation of a framework for the description of plastic constitutive relations. The
need to predict the behavior of bodies composed of polycrystalline metals under static loading at room temperatures will
provide the original motivation for the construction of this
framework in the first instance, although it is, of course,
possible that other materials can be fitted into the resulting
framework under appropriate conditions. The axial force tests
on metals discussed in a qualitative manner in Section 2.1 will
consequently be used as a guide. It must be emphasized again,
however, that it is not our intention to describe exactly, in
every respect, the mechanical behavior of polycrystalline metals. We wish to include only significant features of the behavior, and to exclude the insignificant. There is one major
reason for idealizing the relations in this way; we wish to
make the solution of the structural problem as simple as possible, obtaining solutions which are only as good as they need
to be for the purposes at hand and not complicated by effects
which increase the labor required in the determination of the
solution but do not significantly affect either its accuracy
or its usefulness. Of course, what is insignificant in one
problem may well be significant in another. Much depends on
what in particular we are looking for in the solution. This
difficulty in deciding in advance what features are not needed

introduces a certain ambiguity into the problem of constructing
a general framework for plastic constitutive relations which
can only be eliminated if we succeed in including every possible effect or phenomenon, and this in turn will lead to structural problems which are intractable. Idealization of the real
mechanical behavior is necessary, and is perfectly legitimate,
provided we appreciate one important factor. The adoption of
idealizations and assumptions about the behavior of the material leads us to define a hypothetical material. This hypothetical material approximates reality under certain conditions.
First, we must appreciate the conditions under which the hypothetical material can be used, since the hypothetical material
itself cannot predict when it is invalid as an approximation to
reality; this appreciation must derive from experience, common
sense and experimental confirmation, all depending on the particular structural problem we are trying to solve and the information we wish to select from the solution.

The idealized, or hypothetical, material with which we are
concerned in this monograph is a *time-independent plastic
material*. Idealization of the behavior of the polycrystalline
metals discussed in Section 2.1 become *axioms* or *postulates* in
the mathematical characterization of plastic behavior. These
axioms can be conveniently grouped into three categories which
progressively limit the mechanical behavior of the hypothetical
material. The first category consists primarily of the restriction to *time-independent, path-dependent* behavior under isothermal conditions. The second category deals with the relation
between elastic and plastic behavior (elasticity will always be
treated as a special class of plastic behavior). The third
group consists of further postulates regarding plastic behavior
which are important in formulating constitutive equations,

ensuring unique solutions to plastic structural problems and establishing general theorems regarding plastic structural behavior.

Throughout this Chapter we shall be discussing a *homogeneously stressed body* or an element of such a body. This homogeneously stressed body or element is conceived as a generalization of the specimen of Section 2.1, i.e., we presume that we can measure or impose stress or strain changes independently in order to obtain information about the constitutive relations.

The first broad idealization of material behavior which we shall adopt is concerned with *failure* or *fracture* of the material. We shall assume that in plastic materials the *ductility* is *unlimited*. This implies on one hand that no loading program will cause failure of the material, and on the other hand that strains of unlimited magnitude can be incurred. This is a severe limitation in a general sense, since failure in almost all materials occurs by some mechanism such as fracture, cleavage, embrittlement, etc. which depends strongly on stress history and strain magnitudes. These effects are of major importance when sharp cracks are present in the body. Geometric effects, such as necking in a tensile specimen, can also contribute to failure under some circumstances. However, since we are confining our attention to linear strain-displacement relations and equilibrium equations, the geometric effects are already excluded, and indeed strain magnitudes are limited except in small regions where the configuration is such that large displacement gradients may exist over a very small region (again, an example is the immediate neighborhood of a crack tip).

In consequence, therefore, we limit ourselves *in practice* to the consideration of structural problems in which failure or fracture is not a consideration. *In theory,* our idealized material will not fracture under any conditions.

The second broad idealization is that of *time independence*. In the context of the bar test described in Section 2.1, we assume that the strain q resulting from any program of loading which results in a final stress Q does not depend on the *rate* at which the loading program is carried out. We can generalize this idea by considering a stress path in a stress space (shown diagrammatically in Fig. 7). Let the strain associated with stress Q_j^a be q_j^a. Let the stress be changed to Q_j^b so that the stress point traces out a specified path, and let the final strain be q_j^b. Time-independence requires that q_j^b should not be a function of the rate of change of stress $\dot{Q}_j$ as it changes from Q_j^a to Q_j^b . Thus q_j^b does not depend on the rate at which the stress point traverses the stress path. It may, however, depend on the stress path; in Fig. 7 the strain obtained by changing the stress along the path indicated by a full line may be different from the strain obtained by following the dotted path. The material is thus *path-dependent*.

The consequence of this assumption in practice is that phenomena such as rate sensitivity (i.e., the dependence of the stress-strain curve on the strain rate) and creep do not appear in time independent plasticity. These effects are important in many structural problems where dynamic effects are present (such as wave propagation) and where the strain rates may be

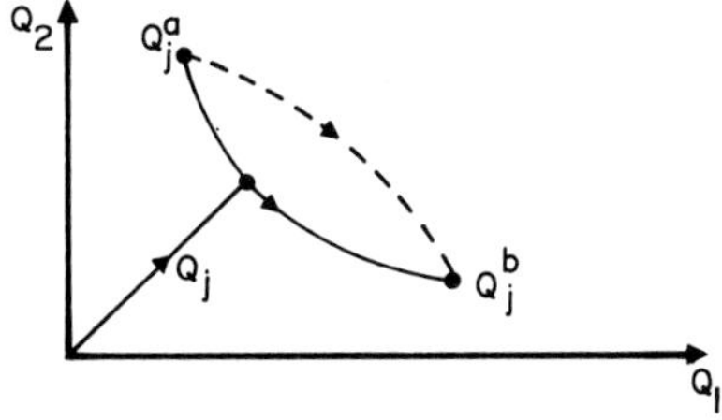

Figure 7. Stress path in stress space

large and vary considerably in space and time, or in structures
subjected to loads for considerable times at high temperatures
(such as engine parts or pressure vessels containing hot li-
quids). We effectively confine ourselves to structures in
which load rates are always very small or to materials which
are truly time-independent under the given loading.

We should note that it is still possible under this assumption for the strain to change under constant stress, as indeed
it does during flow. However, we demand that such flow should
be *inviscid*, i.e., that the strain is not a function of time. If
the strain rate $\dot{q}$ is a function of the stress Q at constant
stress, integration of the relation will give strain as a function of time. As an example, let

$$\dot{q} = \mu Q \tag{3}$$

where μ is a constant. If $q(0) = q^*$, and Q is constant, we see
that the strain at time t is given by

$$q(t) = q^* + \mu Q t \tag{4}$$

Strain is thus an explicit function of time, and the material
is not inviscid. Some consequences of the assumption of inviscid flow will be considered in the following section; as our
development proceeds it will be seen that the strain rate $\dot{q}$ in
inviscid flow is *not* determined by the stress Q but by other
factors usually involving the whole structural problem. From
the point of view of constitutive equations, it will be seen
that the stress level determines whether or not inviscid flow
can occur. Some other restrictions will be placed on the direction of the strain rate vector $\dot{q}_j$ during flow, but its magnitude cannot be predicted by inviscid constitutive relations.

Time cannot appear explicitly as a variable in the constitutive equations of any problem in time independent plasticity. It
may, in many instances, be convenient to introduce time and to

speak of rates of change, but this is done simply as a means of
indicating the order of events. Time could be measured in se-
conds, hours or months without affecting the result. Alternatively, the sign of the rates of change is all that is of interest.

In addition to the assumptions of unlimited ductility and
time independence, we also limit ourselves to the study of
bodies under *isothermal conditions*, as given earlier. We shall
not be concerned with structures in which temperature gradients
occur, nor shall we be concerned for the present with the effect of temperature on the constitutive relations.

2.3 Yield Surfaces

In this section we shall proceed by considering first the
case of *hardening materials*, i.e. materials in which flow or
change in strain at constant stress does not occur. We shall
then modify our conclusions to deal with *flow*, the case where
changes in strain can take place at constant stress.   In all
cases we shall be considering homogeneously stressed elements.

It was noted in Section 2.1 that in a bar test on a *hardening material* it is always possible to enter into a range of
linear behavior by abruptly reversing the loading. This is
shown diagrammatically in Fig. 8. Suppose that the load is increased monotonically to Q^{a+}, and then reduced.   A linear region on the stress–strain diagram is entered, and if the stress
is decreased monotonically the lower limit of the linear region
is found to be Q^{a-}.   In order to simplify our description we
shall make some further assumptions about the behavior of the
material. First, we shall assume that the linear region is in
fact a *linear elastic* region. Second, we shall assume that the
boundaries of the linear elastic region, the stresses Q^{a+} and
Q^{a-}, are not affected by any stress changes in which $Q^{a-} \leq Q \leq$

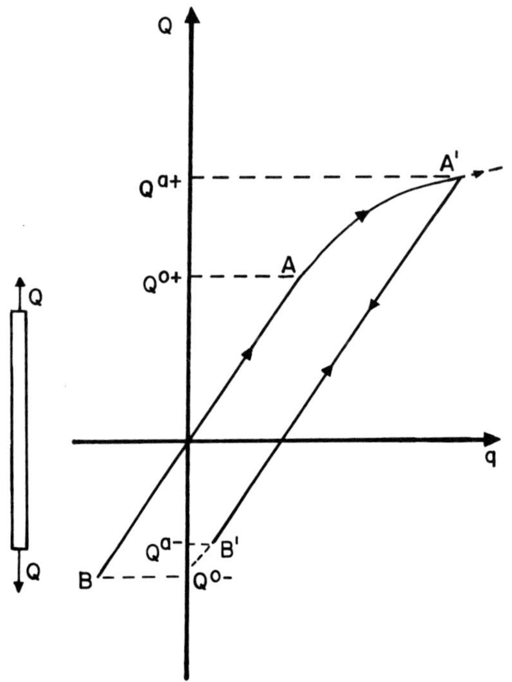

Figure 8. Stress-strain relation for a bar in tension
Q^{a+}, i.e., the boundaries are not altered when the stress does
not exceed Q^{a+} in tension or Q^{a-} in compression.

The consequences of these assumptions may be stated briefly.
By linear elastic behavior we imply that *changes in strain are
linear functions of changes in stress.* Thus, if the stress is
changed monotonically from Q^s to Q^r, leading to a change in
strain from q^s to q^r,

$$q^r - q^s = C(Q^r - Q^s) \; . \tag{5}$$

Furthermore, elastic behavior is *path independent,* so that
equation (5) holds even if the stress change is not monotonic,
provided that the stress remains within the elastic range. The
second assumption implies that an *elastic region* $Q^{a-} \leq Q \leq Q^{a+}$
is set up and that the boundaries of this region Q^{a-}, Q^{a+} cannot change unless we impose stress states $Q > Q^{a+}$ or $Q < Q^{a-}$.

This involves some idealization of real behavior. If, for
example, we were to increase the stress monotonically, from a
value within the elastic range, linear behavior would prevail
until Q^{a+} was reached, whereafter the stress-strain curve
would follow the dotted line in Fig.8. In many real materials
the end of the linear range may occur slightly below Q^{a+} (see
Section 2.1, Fig.3). In terms of overall structural behavior,
however, the consequences of this assumption are generally
slight. In addition, the slight hysteresis loops associated
with an increase, decrease and subsequent increase in stress
are also excluded. Here again the consequences of the assumption are negligible in all but the most exceptional circumstances.

It is convenient to distinguish between the stress changes
which might be imposed when the stress is on the boundary of
the elastic region, i.e. $Q = Q^{a+}$ or $Q = Q^{a-}$. If the stress is
changed so that the stress tends to move out of the elastic region defined by $Q^{a-} \leq Q \leq Q^{a+}$, the process is called *loading*.
If the stress is changed so that the stress value moves into
the elastic region, the process is called *unloading*. Thus
loading involves an increase of stress when $Q = Q^{a+}$ and a decrease when $Q = Q^{a-}$, and unloading involves a decrease in Q
when $Q = Q^{a+}$ and an increase when $Q = Q^{a-}$. The boundaries of
the elastic region, Q^{a+} and Q^{a-} are called the *yield stresses*.

We note then that if the stress does not lie *within* the
linear elastic region, we can always *enter into the elastic region by unloading*. Thus we assume that the stress point on the
stress-strain curve always lies *either inside or on the boundary of the elastic region*. We thus imply that the *elastic region follows the stress when loading occurs*. The yield stresses
must therefore be functions of *stress history;* since we have

already assumed that the stress history does not affect the
yield stresses when the stresses are inside the elastic region,
the yield stresses must then be functions only of *those parts
of the stress history during which loading occurs.*

Finally, we assume that the slope of all linear elastic regions of the stress-strain curve are identical. Thus the coefficient C in equation (5) must be a constant which is unaffected by stress history. This is again sensibly true for
polycrystalline materials; very slight changes in the elastic
coefficients have been observed but they are not of sufficient
magnitude to affect the general characteristics of the structural problem. Once this assumption has been adopted it is
seen that the *initial elastic region* and the *initial yield
stresses* found at the beginning of a test on a specimen have no
special significance, apart from the fact that they are the
initial values.

The fixed coefficient C also suggests a means of characterizing strain changes in the elastic regions and during loading
more conveniently. We consider the strain to be the sum of two
terms,

$$q = q^e + q^p , \tag{6}$$

where q^e is called the *elastic component of strain* (or the
elastic strain) and q^p is called the *plastic component* of
strain (or the plastic strain). q^e is *defined* by the equation

$$q^e = CQ , \tag{7}$$

where C is the same elastic coefficient that appears in equation (5). q^p is then *defined* by equation (6); it is the *difference* between the total strain q and the elastic strain q^e.
We can then make the following statement. *The plastic strain
q^p changes only during loading,* i.e., only when the yield
stress changes as a result of the stress change.   During

unloading, and when Q lies in the elastic range, q^p must be
constant. The relation between q^e and Q is an elastic one, and
hence q^e is both time and path independent. If the stress is
changed from Q^s to Q^r in such a way that the stress is always
within the elastic region and loading never occurs, the strain
change is given by

$$q^r - q^s = (q^{er} + q^{pr}) - (q^{es} + q^{ps})$$
$$= q^{er} - q^{es} \quad \text{since} \quad q^{ps} = q^{pr} . \tag{8}$$

However, from equation (7)

$$q^{er} - q^{es} = CQ^r - CQ^s$$
$$= C(Q^r - Q^s) , \tag{9}$$

which is consistent with the result given in equation (5).

If the stress is zero, equation (6) gives

$$q = q^p ,$$

since the elastic component of strain is zero. If by a stress
program which starts and ends at $Q = 0$ we cause a change in
plastic strain Δq^p, Δq^p is the permanent set observed in the
material. In any specimen at the beginning of a test (with $Q =
0$) we do not have any means of measuring q^p. We can in fact
measure only changes in q^p from its initial value. For convenience we generally assume that q^p is zero in the initial
state.

It is already apparent that we are forced to discuss changes
in stress and changes in strain. It will, in fact, be seen
that we shall further be restricted to consider infinitesimally
small changes in stress and strain; these are the first indica-
tions of the *incremental* nature of plasticity.

We shall proceed now to reconsider these idealizations and
their description in terms of a more general situation where

more than one stress component is nonzero. The stresses will be represented by Q_j and we shall be concerned with stress histories which trace out paths in the stress space.

We assume first that for any element of material in a given state of stress (following some stress history) there exists a *domain* or *region* in the *stress space* such that the behavior is elastic and path independent if the stress point lies within this region. The *elastic region* is bounded by a *hypersurface* in the stress space, as shown diagrammatically in two dimensions in Fig. 9. The points where this hypersurface cuts the Q_1 axis, say, give the yield stresses for the case when Q_1 is the only nonzero component of stress. The hypersurface is called the *yield surface*.

As before we express the strain q_j as the sum of an elastic strain q_j^e and a plastic strain q_j^p,

$$q_j = q_j^e + q_j^p .\tag{10}$$

The elastic strain q_j^e is *defined* as a linear function of the stresses,

$$q_j^e = C_{jk} Q_k ,\tag{11}$$

where C_{jk} is a symmetric matrix of *constant* coefficients. The plastic strain q_j^p is defined by equation (10) as the difference

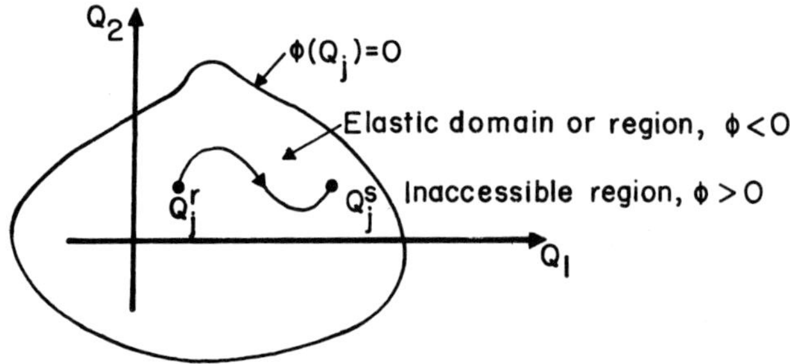

Figure 9. Yield surface in stress space

between the total strain q_j and the elastic strain q_j^e. When
the stress path lies within the yield surface, we require that
the plastic strain q_j^p does not change. Hence the strain
changes associated with the stress change from Q_j^s to Q_j^r, as
shown in Fig. 9, are for such a path given by

$$q_j^r - q_j^s = q_j^{er} - q_j^{es} = C_{jk}Q_k^r - C_{jk}Q_k^s$$

$$= C_{jk}(Q_j^r - Q_j^s). \tag{12}$$

The strain change is thus a function of the stress change alone
when the path lies inside the elastic region.

Provided that the stress point lies within or on the boundary of the elastic region it is assumed that the yield surface does not change. We will further assume that the elastic
region is simply connected, i.e. that any two stress states
which lie in the elastic region can be joined by a path which
lies in the elastic region. These two assumptions are generalizations of assumptions made in idealizing the results of the
bar test.

Both the plastic strain and the yield surface may be functions of the stress path or the stress history. Since we have
assumed that stress paths which lie inside the yield surface do
not change the yield surface or the plastic strain, it is clear
that only certain parts of the stress history contribute to
these quantities. We shall refer to that part of the previous
stress path which contributes to the position of the yield surface and to the plastic strain as the *recorded history* of mechanical behavior. The material is not affected by, and does
not record, stress changes which occur in the elastic region.

It is not necessary to attempt to quantify the recorded history at this point. It must be emphasized, however, that while
the term "recorded history" may be an appealing and graphical

description of the dependence on past events in the mechanical
history of the specimen, we cannot imply that the material is
in any sense endowed with a memory. The recorded history must
consist of the instantaneous values of a number of physical parameters, and the term "recorded history" itself is applicable
only if we recognize that it expresses the observation that the
instantaneous values of the parameters are known in principle
if the entire program of stressing to which the material has
been subjected in the past is known. The physical parameters
which we have referred to may consist of measurable macroscopic
mechanical quantities, such as the components of the plastic
strain or work terms, or they could be made up of a number of
internal variables which are themselves related to the mechanical behavior. From our present phenomenological viewpoint we
do not limit the character of these parameters or the recorded
history.

Let us denote the parameters which constitute the recorded
history by H_α $(\alpha = 1, \ldots, \eta)$. Then, whenever we observe our
homogeneously stressed element, we associate with the element a
particular imposed stress Q_j and a particular recorded history
H_α which in principle may be determined from the previous history of stress. The stress Q_j and the recorded history H_α ,
together with the temperature T, define the *state* of the material element.

The stress and the concept of recorded history are useful in
characterizing the yield surface in mathematical terms. Let us
assume that we may assign to each state Q_j, H_α a scalar quantity. This quantity is then a function of the stress and recorded history, and will be termed the *yield function* ϕ. We may
write

$$\phi = \phi(Q_j, H_\alpha) \ . \tag{13}$$

We then differentiate between *accessible states* and *inaccessible states*. Inaccessible states are those sets of Q_j, H_α which cannot be achieved in an actual material element by *any* program of stressing. Accessible states can be achieved by *some* program of stressing. The yield function is arranged so that for all accessible states $\phi < 0$ and for all inaccessible states $\phi > 0$. A multi-dimensional space with sufficient axes for the components of Q_j and H_α may be imagined; the accessible states are bounded by the *hypersurface* $\phi = 0$. We assume that the accessible region is simply connected, i.e., two accessible states may be joined by a line which does not pass into the inaccessible region. This is shown diagrammatically in Fig.10 although it is clear that a two-dimensional representation is not adequate for a proper description.

Partly because the recorded history H_α has not been quantified, and partly because we are concerned with the response of the material element to a given stress program, we normally work only with the *stress space*. The stress space is a subspace of the more general Q_j, H_α space in which H_α is *fixed and is the current recorded history in the element*. We may now refer back to Fig.9 for a diagram of the stress subspace. The projection of the surface $\phi = 0$ for H_α fixed becomes the *yield*

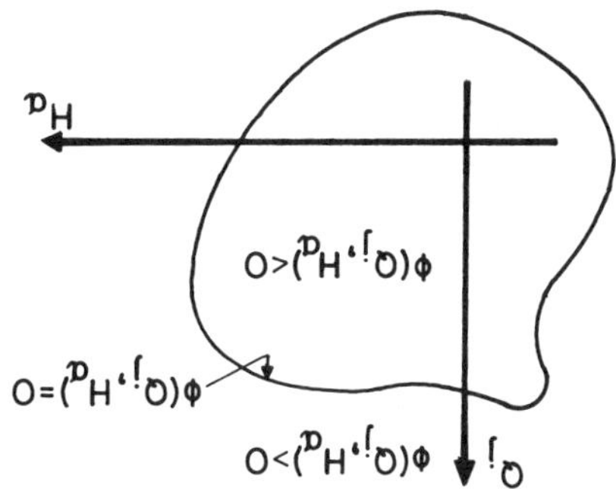

Figure 10. Yield function in Q_j, H_α space

surface; it is characterized by

$$\phi(Q_j, H_\alpha)_{H_\alpha \text{ constant}} = 0 \ . \tag{14}$$

The projections of the accessible and inaccessible regions for the fixed values of H_α now lie respectively within and without the yield surface. States within or on the yield surface are accessible *without* change in H_α, and are characterized by

$$\phi(Q_j, H_\alpha)_{H_\alpha \text{ constant}} \leq 0 \ . \tag{15}$$

It is clear then that the projection of the accessible region is the *elastic region in stress space;* changes in stress imposed in such a way that the stress point moves within this region do not lead to changes in the recorded history. Since any actual state Q_j, H_α must lie within the accessible region, it is also clear that the actual stress point always lies within or on the boundary of the elastic region in stress space.

On the other hand, the projection of the inaccessible region indicates that *stress states such that*

$$(Q_j, H_\alpha)_{H_\alpha \text{ constant}} \geq 0 \ . \tag{16}$$

cannot be achieved without changes in the recorded history. It will not be possible to achieve certain stress states at all, even with changes in the recorded history; nevertheless, the qualification in the previous sentence still stands. If it is possible to achieve a stress state in the projection of the inaccessible region, a change in H_α must occur.

In this section we shall continue to emphasize situations in which the recorded history is regarded as constant by writing ϕ in the form given in equations (14), (15) and (16); in future discussion, however, we shall write simply $\phi(Q_j)$ when we wish to denote that the recorded history is temporarily fixed.

The consequences of the restriction given in equation (16)

are important when the state of stress is such that the stress
point lies on the yield surface. Let us consider a state Q_j,
H_α for which $\phi = 0$, and change the stress by a small amount to
a new value $(Q_j + dQ_j)$. There will always exist possible stress
increments dQ_j for which the response will be elastic. These
are the stress increments for which the new stress point $Q_j + dQ_j$
lies *within* the yield surface. Consistent with the same phenomenon in the bar tests, we refer to the application of such a
stress increment as *unloading*. Figure 11 illustrates diagrammatically the unloading increments.

Since the recorded history does not change along an elastic
path, unloading can easily be formulated in terms of the yield
function. If, as we have assumed, Q_j lies on the yield surface

$$\phi(Q_j, H_\alpha) = 0 \quad . \tag{17}$$

The new stress point lies within the yield surface, so that

$$\phi(Q_j + dQ_j, \ H_\alpha) < 0 \quad . \tag{18}$$

Comparison of (17) and (18) implies that

$$d\phi = d\phi\Big|_{H_\alpha \ \text{constant}} < 0 \tag{19}$$

for a stress increment dQ_j which causes unloading.

On the other hand, a stress increment dQ_j imposed on a

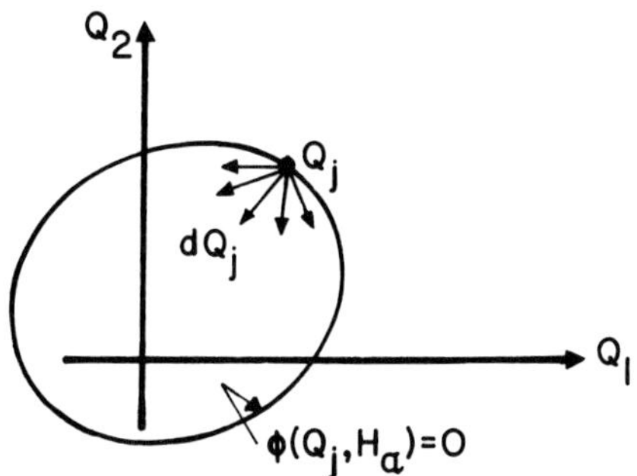

Figure 11. Unloading increments in stress space

stress point on the yield surface for which

$$\left. d\phi \right|_{H_\alpha \text{ constant}} > 0 \ , \tag{20}$$

must lead to a change in the recorded history and *in the projection of the yield function in the stress space*. The application of such a stress increment we refer to as *loading*. The new stress point $(Q_j + dQ_j)$ will still lie on the yield surface, as shown diagrammatically in Fig.12, so that for loading

$$\phi(Q_j + dQ_j; \ H_\alpha + dH_\alpha) \ = \ 0 \ . \tag{21}$$

Comparing equations (17) and (21), we see that in actuality a stress increment which causes loading leads to a change in the yield function $d\phi$ given by

$$d\phi \ = \ 0 \ . \tag{22}$$

We can write

$$d\phi \ = \ \left. d\phi \right|_{H_\alpha \text{ constant}} + \left. d\phi \right|_{Q_j \text{ constant}} \ , \tag{23}$$

breaking a change in ϕ into the sum of a part due to the change in stress with recorded history constant and to the change in recorded history with stress constant. In view of equation (20), equations (22) and (23) imply that

$$\left. d\phi \right|_{Q_j \text{ constant}} = \left. -d\phi \right|_{H_\alpha \text{ constant}} < 0 \ . \tag{24}$$

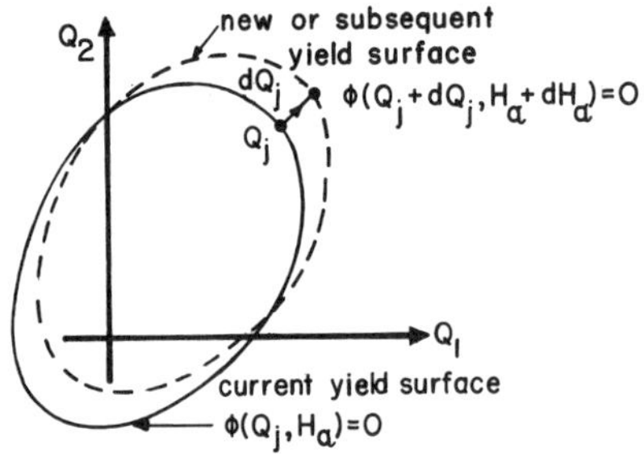

Figure 12. Loading increments in stress space

This last inequality is consistent with the addition and subsequent removal of a stress increment which causes loading. The yield surface will change during the addition of the increment, but the removal of the same increment will constitute unloading. Inequality (24) assures us that the yield function will be negative after the addition of a loading increment and its removal, so that the stress point which lay on the yield surface initially now lies within it.

If we now reconsider this postulated behavior in the more complete space in which the components of Q_j and H_α are plotted we can obtain a more general perspective. Referring to Fig. 10 (which is necessarily a rather limited picture of the behavior) the loci along which the state point is permitted to move is limited severely. For state points such as $\phi(Q_j, H_\alpha) < 0$ the recorded history H_α remains constant and the state point moves only in the stress subspace. The only circumstances under which H_α can change is when loading occurs. The state point must lie on the surface $\phi(Q_j, H_\alpha) = 0$, and *during loading the state point moves along the surface* $\phi(Q_j, H_\alpha) = 0$.

At this point we could characterize the recorded history H_α in specific terms, and continue on this basis with the derivation of a set of constitutive equations. However, one of the major concerns in the mathematical theory of constitutive equations for plasticity has been the continued development of a general framework with recourse only to further assumptions on a macroscopic level. We shall continue along these lines, leaving the recorded history largely uncharacterized, and thus retain our ability to construct a variety of models of plastic behavior in which the physical interpretation of the recorded history parameters may differ.

Let us assume now that the yield function ϕ is continuously differentiable with respect to the components of both stress Q_j and recorded history H_α. This implies that the hypersurface $\phi = 0$ in the complete Q_j, H_α space and the yield surface in stress space are smooth surfaces with uniquely determined tangent hyperplanes at each point; sharp corners or lines of discontinuity in the derivatives $\partial\phi/\partial Q_j$ and $\partial\phi/\partial H_\alpha$ are excluded. We shall see later that such discontinuities have both mathematical and physical significance, and must be specially treated.

If the yield function is continuously differentiable we can further characterize the loading and unloading conditions. It is clear that for an infinitesimal stress increment dQ_j imposed on a state for which $\phi(Q_j, H_\alpha) = 0$,

$$d\phi\Big|_{H_\alpha \text{ constant}} = \frac{\partial\phi}{\partial Q_j} dQ_j \quad , \tag{25}$$

where the summation rule for like subscripts is implied.

Comparing equations (19) and (25), we see that for *unloading*

$$\phi(Q_j, H_\alpha) = 0 \quad \text{and} \quad \frac{\partial\phi}{\partial Q_j} dQ_j < 0 \quad , \tag{26a}$$

while for *loading*

$$\phi(Q_j, H_\alpha) = 0 \quad \text{and} \quad \frac{\partial\phi}{\partial Q_j} dQ_j > 0 \quad . \tag{26b}$$

A further possibility, not apparent in the uniaxial bar test, occurs when

$$\phi(Q_j, H_\alpha) = 0 \quad \text{and} \quad \frac{\partial\phi}{\partial Q_j} dQ_j = 0 \quad . \tag{26c}$$

This corresponds to a stress increment which lies in the yield surface and will be termed *neutral loading*. No change in the recorded history is required to accommodate neutral loading. Note that this is consistent with equation (24); dH_α is identically zero for neutral loading.

When *loading* occurs, we may make use of the continuity of

the derivatives of ϕ to write a relation between dQ_j and the
associated change in recorded history dH_α; equation (23) and
(24) become

$$d\phi = \frac{\partial \phi}{\partial Q_j}\, dQ_j + \frac{\partial \phi}{\partial H_\alpha}\, dH_\alpha = 0 \ ,$$

i.e.

$$\frac{\partial \phi}{\partial H_\alpha}\, dH_\alpha = -\frac{\partial \phi}{\partial Q_j}\, dQ_j < 0 \ , \tag{27}$$

when $\phi(Q_j, H_\alpha) = 0$ and $\dfrac{\partial \phi}{\partial Q_j}\, dQ_j > 0$.

The summation rule is adopted for the subscript α in these expressions, with summation over $\alpha = 1, 2, \ldots, \eta$.

Before we pass on to further topics, it might be appropriate
to add a note concerning usage of the term *yield surface* in
this volume. Throughout this monograph we will continue to
speak of the *yield surface as the hypersurface which bounds the
elastic region in stress space for a particular recorded history* H_α. At the beginning of a program of stressing the yield
surface is generally referred to as the *initial yield surface*.
In most cases we are unable to characterize the recorded history at the beginning of any stress program, and for convenience we often assume that there is no initial recorded history.
In such cases the initial yield surface is referred to as the
virgin yield surface. Yield surfaces generated by changes in
recorded history during a stress program are referred to as
subsequent yield surfaces.

Remembering that we are still dealing exclusively with hardening materials, we may now proceed to consider the framework
in which we can express the plastic component of strain q_j^p .
Changes in plastic strain occur, as we have seen, *only when the*

recorded history changes. This loose restriction can be met adequately for our purposes if we put

$$dq_j^p = h_{j\alpha}(Q_j, H_\alpha)dH_\alpha \qquad (28)$$

In adopting a functional dependence of this type we are assuming that $dq_j^p = 0$ if $dH_\alpha = 0$, and that the plastic strain q_j^p can be obtained by integrating along the stress path with the plastic strain remaining unchanged when the behavior is elastic. It is not implied that the expression given on the right-hand side of equation (28) is a perfect differential; indeed, the plastic strain q_j^p cannot depend on the current stress if it is to remain constant during elastic behavior. If, however, the plastic strain q_j^p is a function of H_α,

$$q_j^p = q_j^p(H_\alpha) \ , \qquad (29a)$$

it follows that

$$dq_j^p = \frac{\partial q_j^p}{\partial H_\alpha} dH_\alpha \ . \qquad (29b)$$

This is a special case of equation (28) in which $h_{j\alpha}$ depends on H_α and $h_{j\alpha}dH_\alpha$ is a perfect differential. Further, we do not exclude the possibility that the H_α are the plastic strains q_j^p themselves; in this case $h_{j\alpha}$ simply becomes an identity matrix.

Returning to our general framework, it is clear from equation (28) and our earlier discussion that

$$dq_j^p = 0 \quad \text{when} \quad \phi(Q_j, H_\alpha) < 0 \ ,$$

$$(30)$$

$$\text{or} \quad \phi(Q_j, H_\alpha) = 0 \quad \text{and} \quad d\phi\Big|_{H_\alpha \text{ constant}} \leq 0 \ ,$$

since under these conditions dH_α is zero.

Equations (30) enable us to determine conditions under which

$dq_j^p = 0$ when Q_j, H_α and dQ_j are known. To obtain a further
expression for the case when $dq_j^p \neq 0$, we will first take the
view that when loading occurs dH_α is a consequence of the
stress increment dQ_j. The changes dH_α may clearly also depend
on the point Q_j, H_α on the yield surface at which the loading
increment is imposed. Combining this with equation (28), we
can write with sufficient generality that

$$dq_j^p = dq_j^p(Q_j, H_\alpha, dQ_j) \quad ,$$

when $\phi(Q_j, H_\alpha) = 0$ and $d\phi\big|_{H_\alpha \text{ constant}} > 0$.   (31)

Remembering that we have restricted ourselves to hardening
behavior, in which by definition changes in plastic strain cannot occur when the stress remains constant, equation (31) must
be such that dq_j^p is zero when all the components of the stress
increment dQ_j are zero. Hence the expression on the right-hand
side of equation (31) cannot contain terms which are independent of the components of dQ_j. We can expand this restriction
by requiring that dq_j^p should not depend on terms which are of
degree zero in the components of dQ_j; if this were so, dq_j^p
would be in part independent of the magnitude of the stress increment dQ_j and consequently as we let dQ_j tend to zero, while
preserving the ratios between the components, dq_j^p would not
necessarily vanish. Furthermore, since dQ_j is conceived as an
infinitesimal increment in stress, terms which are of second
and higher order in the components of dQ_j on the right-hand
side of equation (31) can be neglected in comparison with the
first-order terms.

Thus we arrive at the conclusion that in the case of hardening behavior dq_j^p is *homogeneous and of degree one in the
components of* dQ_j. Note that this does not imply that dq_j^p is

linear in the components of dQ_j; dq_j^p could depend on terms which consist of combinations of the components of dQ_j provided that the combinations are of degree one. The elimination of terms of degree zero and of degree two and higher in dQ_j from equation (31) permits us to define without difficulty the plastic strain rate $\dot{q}_j^p$. It follows directly from the result that dq_j^p is homogeneous and of degree one in dQ_j that

$$\dot{q}_j^p = \frac{dq_j^p}{dt} = \lim_{dt \to 0} \frac{1}{dt} \{ dq_j^p (Q_j, H_\alpha, dQ_j) \}$$

$$= \dot{q}_j^p (Q_j, H_\alpha, \dot{Q}_j) \ . \tag{32}$$

Equation (32) shows that the *requirements of time independence are satisfied*, since it implies that $\dot{q}_j^p$ is homogeneous and of degree one in $\dot{Q}_j$. This means that if we modify the stress rate $\dot{Q}_j$ by multiplying by a factor a, the plastic strain rate is modified by the same factor,

$$\dot{q}_j^p (Q_j, H_\alpha, a\dot{Q}_j) = a\dot{q}_j^p (Q_j, H_\alpha, \dot{Q}_j) \ . \tag{33}$$

We are thus assured that if we integrate equation (31) or (32) along a path in stress space the stress rate at any point on the path does not affect the change in plastic strain. That dq_j^p is homogeneous and of degree one in dQ_j is an important general result and, since we have not made use of the smoothness of the yield surface, it applies even when there are discontinuities in the derivatives $\partial\phi/\partial Q_j$ and $\partial\phi/\partial H_\alpha$.

Essentially the same result can be obtained by considering the relation between dH_α and dQ_j. Using identical reasoning, it can be argued that dH_α is homogeneous and of degree one in the components of dQ_j. If this is then combined with equation (28), it immediately follows that dq_j^p is homogeneous and of degree one in the components of dQ_j.

This alternative argument can provide further information in
the case where $\phi(Q_j, H_\alpha)$ is continuously differentiable, since
equation (27) provides a relation between dH_α and dQ_j during
loading. However, this is a *scalar* equation, and cannot be
solved for dH_α except in the case where there is only one recorded history parameter, say H. Under these circumstances

$$dH = -\frac{1}{\partial\phi/\partial H}\frac{\partial\phi}{\partial Q_j}dQ_j \; . \tag{34}$$

More generally, we shall assume that during loading dQ_j *con-*
trols the magnitude of the vector dH_α, through equation (27),
and *that the direction of* dH_α *depends only on the state point*
Q_j, H_α and not on dQ_j. Thus, we put

$$dH_\alpha = bt_\alpha(Q_j, H_\alpha) \; , \tag{35}$$

where b is a scalar. b is determined by substituting equation
(35) into equation (28). This gives

$$b = -\frac{1}{t_\alpha\dfrac{\partial\phi}{\partial H_\alpha}}\frac{\partial\phi}{\partial Q_j}dQ_j \; . \tag{36}$$

Substituting back from equation (36) into equation (35)

$$dH_\alpha = -\frac{t_\alpha}{t_\beta\dfrac{\partial\phi}{\partial H_\beta}}\frac{\partial\phi}{\partial Q_j}dQ_j \; . \tag{37}$$

When the recorded history is more completely characterized
this result can be utilized. For the present we note that on
substituting equation (37) into equation (28),

$$dq_j^p = -\frac{h_{j\alpha}t_\alpha}{t_\beta\dfrac{\partial\phi}{\partial H_\beta}}\frac{\partial\phi}{\partial Q_k}dQ_k \; . \tag{38}$$

Formerly we noted that dq_j^p was homogeneous and of degree one in the components of dQ_j. We see now, through equation (38), that when $\phi(Q_j, H_\alpha)$ *is continuously differentiable* dq_j^p *is homogeneous and linear in the scalar quantity* $(\partial\phi/\partial Q_j)dQ_j$. Furthermore since this scalar quantity is itself homogeneous and linear in the components of dQ_j, dq_j^p *is homogeneous and linear in the components of* dQ_j.

Since $h_{j\alpha}$, t_α and $\partial\phi/\phi H_\alpha$ each depend only on the state Q_j, H_α, equation (38) can be simplified and written in the form

$$dq_j^p = f_j(Q_j, H_\alpha)\frac{\partial\phi}{\partial Q_k}dQ_k \; , \tag{39}$$

where f_j is an n-dimensional vector valued function of its arguments. In this representation the *direction* of the vector dq_j^p depends only on the current state Q_j, H_α, while the magnitude of the vector dq_j^p depends also on the components of dQ_j. We further assume that we may put

$$f_j = G(Q_j, H_\alpha)\frac{\partial g}{\partial Q_j}(Q_j, H_\alpha) \; , \tag{40}$$

where $G = G(Q_j, H_\alpha)$ is a scalar hardening coefficient and $g = g(Q_j, H_\alpha)$ is referred to as the *plastic potential*. The direction of the plastic strain rate increment is thus governed by the partial derivatives of the plastic potential.

It should be noted that dq_j^p vanishes when $(\partial\phi/\partial Q_j)dQ_j = 0$, i.e. when neutral loading occurs. This result, which is essential from the point of view of equations (29), played an important role in the development of theories of plasticity, and is sometimes referred to as the *consistency condition*.

Let us now summarize our progress. We have considered only hardening behavior, and we have established in part a framework for the description of the constitutive relations for *hardening materials.* The strain q_j is divided into an elastic and a plastic part,

$$q_j = q_j^e + q_j^p \ . \tag{41a}$$

The elastic component of strain dq_j^e is a function of the stress Q_j,

$$q_j^e = C_{jk} Q_k \ , \tag{41b}$$

where C_{jk} is a matrix whose coefficients are constant. We have introduced a yield function ϕ such that

$$\phi(Q_j, H_\alpha) \leq 0 \tag{41c}$$

for all accessible states. Then, assuming that the partial derivatives $\partial\phi/\partial Q_j$ and $\partial\phi/\partial H_\alpha$ are continuous, we have

$$dq_j^p = 0 \quad \text{for } \phi(Q_j, H_\alpha) < 0$$

$$\text{or } \phi(Q_j, H_\alpha) = 0 \quad \text{and} \quad \frac{\partial\phi}{\partial Q_j} dQ_j \leq 0 \ , \tag{42a}$$

and

$$dq_j^p = G(Q_j, H_\alpha) \frac{\partial g}{\partial Q_j} (Q_j, H_\alpha) \frac{\partial\phi}{\partial Q_k} dQ_k$$

$$\text{for } \phi(Q_j, H_\alpha) = 0 \quad \text{and} \quad \frac{\partial\phi}{\partial Q_j} dQ_j \geq 0 \ . \tag{42b}$$

It is seen that we can combine the neutral loading condition

$$\phi = 0 \ , \quad \frac{\partial\phi}{\partial Q_j} dQ_j = 0 \tag{43}$$

with either the unloading condition (equations 42a) or the loading condition (equations 42b). Again, this is the consistency condition, and it provides a smooth transition between the behavior in loading and unloading.

In addition, we need the expression for the change in H_α,

$$dH_\alpha = - \frac{t_\alpha}{t_\beta \dfrac{\partial \phi}{\partial t_\beta}} \frac{\partial \phi}{\partial Q_j} \, dQ_j \tag{44}$$

where $t_\alpha = t_\alpha(Q_j, H_\alpha)$. This equation enables us to obtain the change in recorded history, and hence to determine the recorded history for the next increment in stress and to compute the next increment in plastic strain.

Let us now include within our framework the possibility that *flow* can take place in the material; specifically we can admit that the *strain can change when the stress remains constant.* For the present we shall not concern ourselves with the conditions under which flow can occur; these conditions are of course of major importance since flow takes place only under limited circumstances (see Section 2.1). Our immediate interest is confined to the question of what can be said regarding the constitutive equations for flow when it does occur.

Consistent with the framework which has so far been constructed, when flow takes place at constant stress the elastic strain q_j^e remains constant, and hence only the plastic strain q_j^p changes. Plastic strain changes can occur only when the stress point lies on the yield surface, and hence the state must be such that $\phi(Q_j, H_\alpha) = 0$. It is clear that the stress point must continue to lie on the yield surface as flow occurs; however, we have also assumed that the *yield surface* does not change when the stress point lies within or on the yield surface. If the dependence of ϕ on H_α ($\alpha = 1, \ldots, \eta$) is such that ϕ depends on η *linearly independent* recorded history parameters, we are led to the conclusion that the yield surface can remain fixed only if there is no change in the recorded history H_α. However, if $dq_j^p = h_{j\alpha} dH_\alpha$, as given in equation (28)

no change in plastic strain can occur while the stress remains
constant; we have in fact used this statement in discussing
hardening behavior.

We can resolve this dilemma by recognizing that the recorded
history parameters H_α, which were first introduced in the yield
function ϕ, represent only a subset of the actual internal parameters which contribute to the *total* plastic strain. While
this subset of internal parameters serves with the stress to
define the state of the material with reference to the yield
function, the plastic strain q_j^p may depend on all the internal
parameters, and not simply on the H_α subset as given in equation (28). Let us then introduce additional internal para-
meters J_ζ ($\zeta = 1, \ldots, \mu$) such that H_α and J_ζ are linearly independent and together make up the complete set of internal
parameters. We consequently widen our definition of an infinitesimal increment of plastic strain, putting

$$dq_j^p = h_{j\alpha}\, dH_\alpha + \bar{h}_{j\zeta}\, dJ_\zeta \quad . \tag{45}$$

Until further restrictions are imposed, we may consider that

$$\bar{h}_{j\zeta} = \bar{h}_{j\zeta}(Q_j, H_\alpha, J_\zeta) \quad . \tag{46}$$

However, $h_{j\alpha}$ remains a function of Q_j and H_α, and not of J_ζ.
The first term on the right-hand side of equation (45) may then
be identified as a plastic strain increment resulting from hardening behavior; it must be accompanied by a change in the
yield surface. The second term may be identified as an increment of plastic strain resulting from flow; when flow takes
place under constant stress the yield surface does not change,
and the state Q_j, H_α does not change.

From our present considerations there is no way in which we

can predict either the *magnitude* of dJ_ζ or its direction (i.e.
the ratios of its components).   There is consequently very
little we can do at this point to further characterize the
change in plastic strain dq_j^p which occurs as a result of flow;
certainly neither the magnitude nor the direction of dq_j^p can
be predicted. It is usual, however, for us to recognize that
the plastic strain increment may to some extent depend on Q_j
and H_α (i.e. on the current state) and possibly on J_ζ. We
may put

$$dq_j^p = \sum_{r=1}^{m} \beta_r \bar{f}_{jr}(Q_j, H_\alpha, J_\zeta) \ , \tag{47}$$

where the functions $\bar{f}_{jr}$ $(r = 1, \ldots, m)$ are some unspecified
number of vector valued functions and β_r $(r = 1, \ldots, m)$ are
scalars of unspecified magnitude. Notice that if there exists
only one function $\bar{f}_j$ (i.e. $m = 1$) it provides the direction of
dq_j^p, but the magnitude of dq_j^p would still be unspecified. More
generally, with $m \geq 2$ and the β_r unspecified, neither the direction nor the magnitude of dq_j^p is fixed. We could also put

$$\bar{f}_{jr} = \frac{\partial \bar{g}_r}{\partial Q_j} \tag{48a}$$

where $\bar{g}_r = \bar{g}_r(Q_j, H_\alpha, J_\zeta)$, $r = 1, \ldots, m,$ $\tag{48b}$

are *plastic potential* functions.

   While equation (47) provides us with very little information
at present, it is a form which we shall be able to further restrict as a result of additional assumptions. However, we
shall not be able to predict completely the magnitude of the
plastic strain increment during flow, and in some cases its
precise direction, *on the basis of the constitutive relations*

alone.

Equation (47) does satisfy one important restriction; it is *time independent.* If the change in plastic strain dq_j^p occurs in time dt, it is possible to conceive of the unspecified scalars β_k as

$$\beta_k = \dot{\beta}_k \, dt \tag{49}$$

where $\dot{\beta}_k$ are themselves unspecified scalars. It follows then that the plastic strain rate $\dot{q}_j^p$ is

$$\dot{q}_j^p = \lim_{dt\to 0} \frac{dq_j^p}{dt} = \dot{\beta}_k \bar{f}_{jk}(Q_j,H_\alpha,J_\zeta) \; . \tag{50}$$

An attempt to integrate this relationship over some interval will not lead to a plastic strain change that is an explicit function of time because the $\dot{\beta}_k$ are not specified. (Note that if the $\dot{\beta}_k$ were provided explicitly in the constitutive relation the material would be time dependent.)

So far we have discussed the possibility of flow at constant stress; this requires that the stress point should be on the yield surface or $\phi(Q_j, H_\alpha) = 0$. Changes in plastic strain, whether due to hardening behavior or flow, cannot occur when the stress point lies within the yield surface, $\phi(Q_j, H_\alpha) < 0$, or during an incremental change in stress which constitutes unloading. However, until we further define the conditions under which flow can occur we must leave open the possibility that changes in plastic strain due to flow (i.e. due to changes in J_ζ) can occur during *neutral loading* or *loading*, since during these processes the stress point remains on the yield surface. A discussion of these conditions will be deferred until Section 2.6. In the following section we shall introduce a further set of restrictions on the constitutive relations. These assumptions will permit us to identify the plastic potentials g and $\bar{g}$ which were introduced in this Section.

2.4 Uniqueness and Stability Postulates

It is a convenient feature of any theory of mechanical behavior that the structural problems which we set ourselves should have *unique solutions* and exhibit *stable equilibrium configurations* when these factors are to be expected in the corresponding physical problem. It must be recognized that if the real body we are attempting to model may deform in a non-unique manner, or may assume unstable equilibrium configurations, no amount of idealization of the mathematical problem will compel it to do otherwise. The consequences of limiting the mathematical theory to exhibit only certain types of mechanical behavior are thus always to limit the class of real problems to which the theory can be applied.

We have already confined ourselves to simple problems, in a relative sense, by assuming that the equilibrium and kinematic relations are linear. We will confine ourselves still further by making additional assumptions which guarantee that our idealized material exhibits properties which are *sufficient* to ensure *unique* and *stable* solutions to structural problems in which the equilibrium and kinematical relations are linear. A further benefit of these assumptions is that they allow us to establish a number of general theorems which are useful in determining complete or approximate solutions to many structural problems.

Even though the motivation for the additional restrictions we shall make for our class of 'simple' materials and 'simple' problems is clear, there still remain a number of ways in which the restrictions can be formulated. Some sets of assumptions

which have been proposed are equivalent (i.e. they imply each
other) and some are not. The assumptions themselves can be
phrased in purely mathematical terms, in physical terms on a
phenomenological level, or can be based on the physics of
mechanical behavior on the microstructural level. We will
choose at this point to make assumptions about the relation between stress and strain which are physically motivated but on
the phenomenological level. The consequences of the assumptions which will be made in this section may not be immediately
apparent, but will be developed in the remainder of this chapter and in succeeding chapters. The scheme which we shall follow is equivalent to, and borrows greatly from, concepts introduced by D.C.Drucker. Although we shall deviate somewhat from
Drucker's original exposition, we shall follow his ideas in
presenting the additional assumptions as generalizations of
some dominant aspects of the behavior of the bar described in
Section 2.1. We shall make two assumptions regarding the behavior of time-independent materials under isothermal conditions, treating each in turn.

We may observe in the bar tests described for aluminum (ex-
cluding any geometric effects associated with failure) that a
monotonic change in stress produces a *monotonic change in
strain of the same sign*. Alternatively, the point with the
coordinates of stress and strain traces out a curve of *positive
slope* during a *monotonic* change in stress. Thus, suppose that
a bar is subjected to stress Q^a with associated strain q^a (Fig.
13). If the stress is increased monotonically to Q^b, the
strain also increases monotonically to q^b (remembering that the
signs of stress and strain are such that the product of positive stress and positive strain has the dimensions of specific
work and is positive).

Similarly, if the stress is decreased monotonically to Q^c, the strain decreases monotonically to q^c. This statement can be characterized in the following way. If the stress is changed monotonically (increasing or decreasing) from Q^a to Q^b, accompanied by changes in strain from q^a to q^b, then

$$(Q^b - Q^a)(q^b - q^a) \geq 0 . \tag{51}$$

We wish to preserve this feature in our theory of plasticity and, with suitable generalization, inequality (51) will be accepted as a condition which the material must satisfy for any choice of Q^a and Q^b. By making this assumption we exclude the type of behavior shown in the stress-strain curves of Fig. 14. If we choose two stress states Q^a and Q^b on those parts of the curve where the slope is negative, (51) will not hold. Some important real materials are affected; for example if we wish

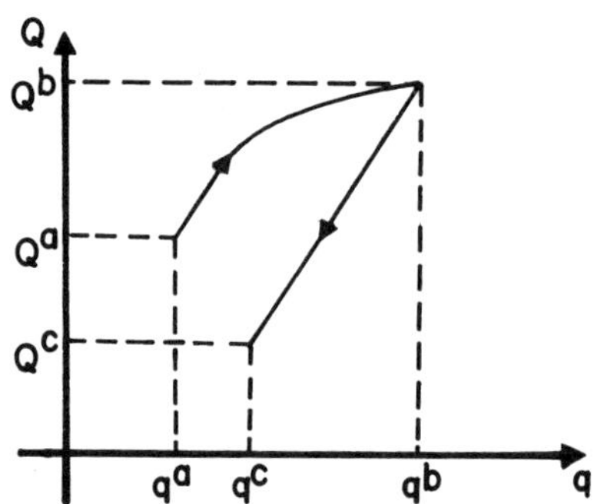

Figure 13.   Stress-strain curve for loading and unloading

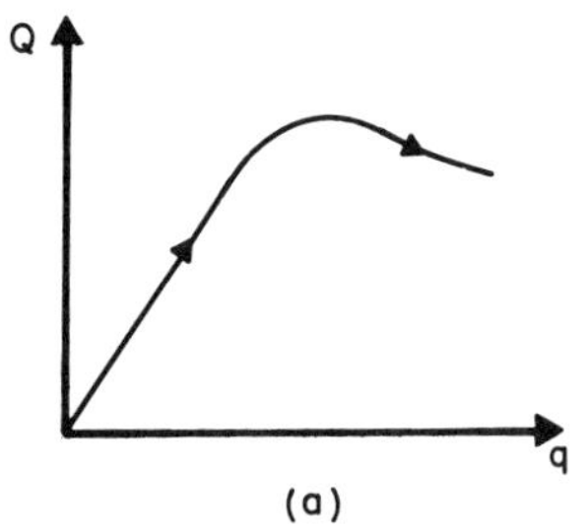

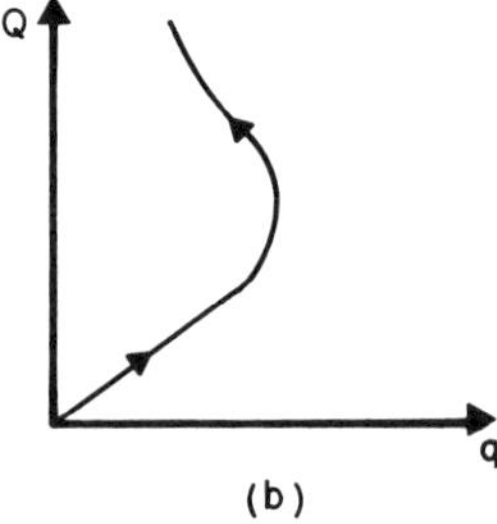

Figure 14.   Unstable stress-strain curves

to model the mild steel shown in Section 2.1, Fig. 6, the upper yield point phenomenon will have to be neglected.

In generalizing (51) to a form suitable for elements of material subjected to a number of stress components, we may observe that a single stress component changing monotonically
traces out a *straight line* in the stress space coincident with
one of the axes. We accept this as the important feature of
the monotonic stress change, and then state the *first postulate*
in the following way. If the stress is changed from Q_j^a to Q_j^b
in such a way that the stress point moves monotonically along a
straight line path in the stress space (Fig. 15), we require
that

$$(Q_j^b - Q_j^a)(q_j^b - q_j^a) \geq 0 \quad , \tag{52}$$

where q_j^a, q_j^b are the strains associated with Q_j^a and Q_j^b respectively.

This statement applies equally well if a stress increment
dQ_j is imposed on the stress Q_j^a, leading to a strain increment
dq_j. An infinitesimal increment is by definition applied monotonically along a straight line path; hence (52) requires that

$$dQ_j dq_j \geq 0 \quad . \tag{53}$$

As expected, (53) implies (52), but two other useful pieces
of information can be derived if this is demonstrated.

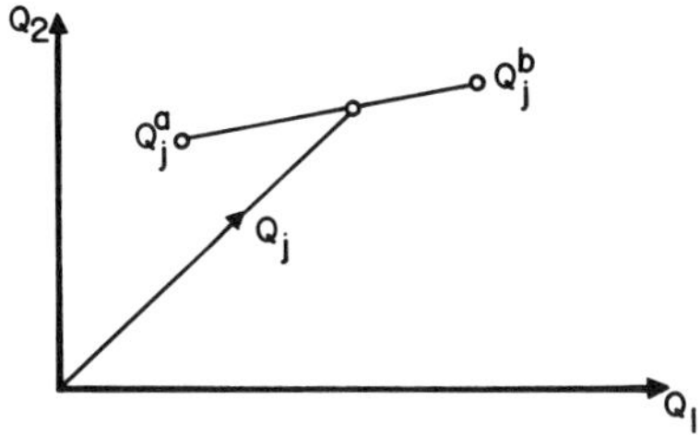

Figure 15. Straight line path in stress space

Returning to Fig. 15, let any stress on the straight line path be denoted by Q_j. The monotonic progress of the stress point Q_j along the straight line path can be characterized by

$$Q_j = Q_j^a + \lambda(Q_j^b - Q_j^a) \ ,\tag{54}$$

where λ increases monotonically from zero to one. Thus

$$0 \le \lambda \le 1 \ ,\tag{55}$$

$$d\lambda \ge 0 \ .$$

An increment of stress at some point on the stress path is then

$$dQ_j = d\lambda(Q_j^b - Q_j^a) \ .\tag{56}$$

Applying (53), it is seen that at any point on the path

$$d\lambda(Q_j^b - Q_j^a)dq_j \ge 0 \ .\tag{57}$$

Since both λ and $d\lambda$ are non-negative, it follows from (57), (55) and (54) that

$$(Q_j^b - Q_j^a)dq_j \ge \lambda(Q_j^b - Q_j^a)dq_j = (Q_j - Q_j^a)dq_j \ge 0 \ .\tag{58}$$

These inequalities may be integrated along the straight line path from Q_j^a, q_j^a to Q_j^b, q_j^b. This leads to

$$(Q_j^b - Q_j^a)(q_j^b - q_j^a) \ge \int_{q_j^a}^{q_j^b} (Q_j - Q_j^a)dq_j \ge 0 \ .\tag{59}$$

The first term shows simply that (53) implies (52). The second term will be referred to as the *net work* along the straight path, i.e. it is the specific work done by the *net*

stresses, which are the stresses imposed *in addition* to Q_j^a.
Further, we may make use of the identify

$$\int_{q_j^a}^{q_j^b} (Q_j - Q_j^a)dq_j + \int_{Q_j^a}^{Q_j^b} (q_j - q_j^a)dQ_j = (Q_j^b - Q_j^a)(q_j^b - q_j^a) \quad . \quad (60)$$

Together with (59), equation (60) leads to the conclusion that

$$\int_{Q_j^a}^{Q_j^b} (q_j - q_j^a)dQ_j \geq 0 \quad . \tag{61}$$

This term will be referred to as the *net complementary work*
along the straight line path.

The consequences of the first postulate can thus be summarized as follows. For a monotonic change in stress along a
straight line path in stress space, the produce of the change
in stress and the change in strain, the net work and the net
complementary work along the path are each non-negative.

The second postulate is motivated by another feature which
is commonly observed in the bar test. Consider a test in which
the load is increased, after some previous stress program, from
Q^a to Q^b, and then reduced again to Q^a. It is expected that
the slope of the curve in the elastic unloading region is
greater than that in the plastic region, so that unloading occurs in the manner shown in Fig.16(a) rather than that shown in
Fig.16(b). It can be noted that both responses satisfy the
first postulate, since the slope is everywhere positive. If the
increase of stress occurs in an elastic region, the slopes
during the increase and decrease of stress are of course the
same. The second postulate will be framed in such a way that
the behavior shown in Fig.16(b) is excluded from the theory

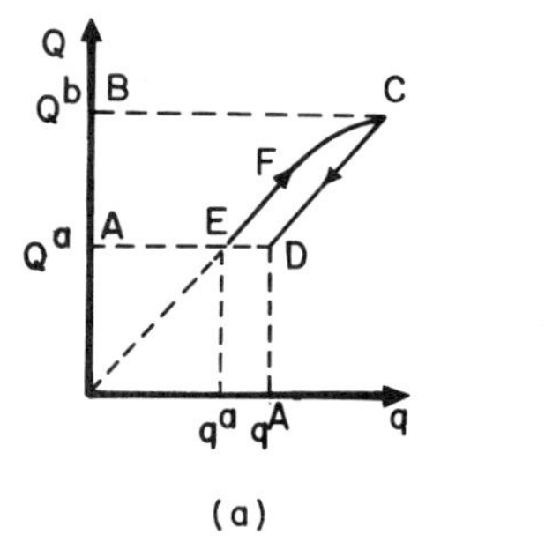
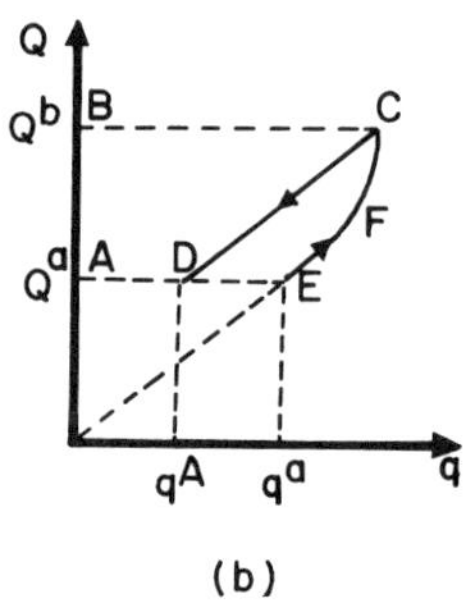

Figure 16. Behavior on unloading

of plasticity. These are again a variety of ways of achieving this, and since the result will be generalized to the multi-stress case, the various methods have somewhat different implications.

We shall formulate the postulate in terms of a *cycle in stress*, i.e. a stress program in which the initial and final stresses are the same. This should not be confused with a true cycle (in the thermodynamic sense) in which the material is returned to its original state. It should also be observed that in a cycle of stress, starting and ending with stress Q^a, the initial strain q^a and the final strain q^A are not necessarily the same, as shown in Fig.16(a).

Since stress is treated as the independent variable, it is natural to first formulate the second postulate in terms of the *complementary work* around the cycle. Consider then

$$\oint q\,dQ$$

applied to the cycles $Q_j^a \rightarrow Q_j^b \rightarrow Q_j^a$ in Figs.16(a) and 16(b). In Fig.16(a), as the stress is increased from Q_j^a to Q_j^b, the complementary work is given by the area AEFCB. As the stress is reduced to Q^a, the area BCDA is subtracted, since in this range of stress the stress increment dQ and the strain q have opposite signs. For this cycle, therefore,

$$\oint q\,dQ \;\leq\; 0 \;. \tag{62}$$

The equality sign is used to include cycles in which the stress path lies completely in the elastic region. For elastic cycles the initial and final strains are the same and the complementary work integral around the cycle is zero. In Fig.16(b), the complementary work along the part of the cycle Q^a to Q^b is again given by the positive area AEFCB. As the stress is decreased to Q^a, the area BCDA is subtracted. Clearly the complementary work around the cycle is positive.

The sign of the complementary work integral allows us therefore to distinguish between the behavior shown in Figs.16(a) and 16(b), and it is in a form which can be readily generalized. However, before we adopt inequality (62) as a restriction which our materials must satisfy, we must make a careful statement about the stress cycle to which it is to be applied. The intent of the restriction, i.e. to distinguish between the types of behavior shown in Figs.16(a) and 16(b), can be served adequately by assuming that inequality (62) is satisfied for *all stress cycles in which only an infinitesimal increment in the plastic strain occurs* (including cycles in which the plastic strain remains unchanged). This means that we apply inequality (62) only to stress cycles in which the initial and final strain differ at most by an infinitesimal quantity. On the other hand, it is attractive to assume that inequality (62) applies to *all stress cycles*, placing no limitation on the magnitude of the plastic strain change which may occur during the cycle. This is a much stronger assumption, but it is satisfied by many simple materials; its mathematical consequences are sufficiently important that it is reasonable to include it in a class of restrictive assumptions in the theory.

We shall consequently state the *second postulate* as follows.

It is assumed that for a *cycle in stress* the complementary work
is non-positive i.e.

$$\oint q_j dQ_j \;\leq\; 0 \;. \tag{63}$$

A cycle in stress is a stress program in which the initial and
final values of the stress are the same, as shown diagrammatically in Fig. 17. If we limit the cycles to those in which the
change in plastic strain Δq_j^p is infinitesimal or zero, we will
refer to inequality (63) as the *second postulate in its weak
form*. If on the other hand we place no limitations on Δq_j^p, we
will refer to inequality (63) as the *second postulate in its
strong form*.

The implications of this postulate are wider than might
first appear, and some commonly observed phenomena are excluded
from the theory. For example, in a bar test in which the
stress is increased to some stress Q^a above the yield stress,
reduced to Q^b, and subsequently increased again monotonically.
Yield may occur before the stress Q^a is reached on reloading,
as shown and discussed in Section 2.1, Fig.3. The complemen-
tary work around the cycle $Q^a \rightarrow Q^b \rightarrow Q^a$ is positive, and thus
violates the postulate. The material is thus constrained by
the postulate to behave elastically until the stress reaches
Q^a. This is not inconvenient, of course, because we have already adopted this assumption in Section 2.3.

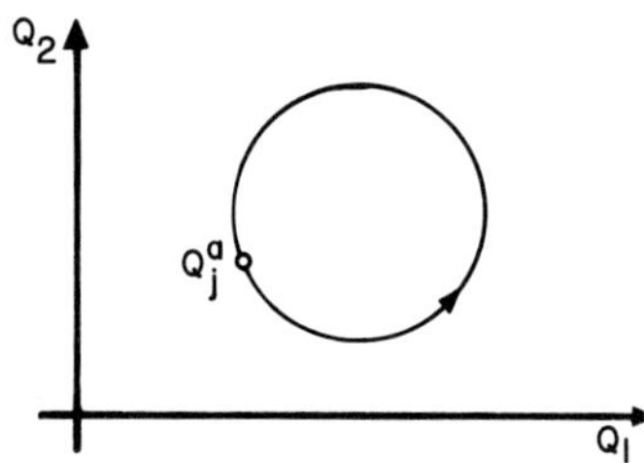

Figure 17. Cycle in stress space.

Some other important general implications of the *second postulate in its strong form* can be drawn immediately. Along any path in the stress space (straight or curved) with initial stress and strain Q_j^a, q_j^a and final stress and strain Q_j^b, q_j^b,

$$\int_{Q_j^a}^{Q_j^b} q_j \, dQ_j + \int_{q_j^a}^{q_j^b} Q_j \, dq_j = Q_j^b q_j^b - Q_j^a q_j^a \ . \tag{64}$$

If this identity is applied to a cycle in stress space, with initial stress and strain Q_j^a, q_j^a and final stress and strain Q_j^a, q_j^A, we obtain

$$\oint q_j \, dQ_j + \int_{q_j^a}^{q_j^A} Q_j \, dq_j = Q_j^a q_j^A - Q_j^a q_j^a \ . \tag{65}$$

The second postulate does *not* imply that the work $\int Q_j \, dq_j$ for a cycle in stress is necessarily positive, since $(Q_j^a \, q_j^A - Q_j^a q_j^a)$ may be either positive or negative. This is not a disturbing conclusion, since it is easy to construct a stress cycle in which the work is negative in a bar test on a material which satisfies our postulates and is thus well behaved. However, (65) may be rewritten as

$$\oint q_j \, dQ_j + \int_{q_j^a}^{q_j^A} (Q_j - Q_j^a) \, dq_j = 0 \ . \tag{66}$$

The second postulate (63) in its strong form thus implies that the *net work around any cycle in stress is non-negative.*

A stronger result can be obtained by combining the first postulate and the second postulate in its strong form. Consider

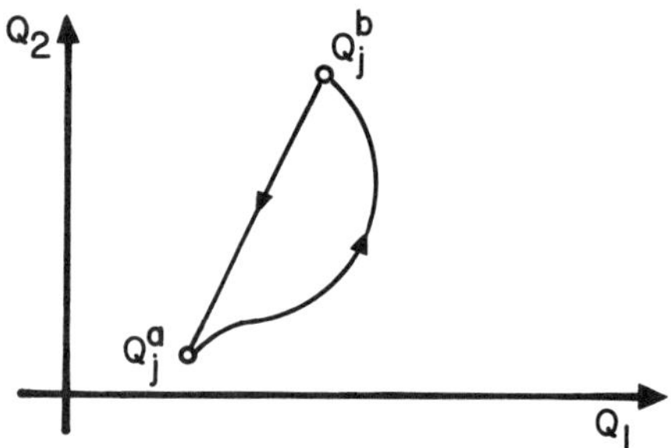

Figure 18. Completion of cycle by straight line path

the stress program shown in Fig. 18. After some previous stress history, the stress is changed from Q_j^a to Q_j^b by *any* path. The cycle is then completed by following a straight line path from Q_j^b to Q_j^a. It can always be *imagined* that such a closure can be effected after any stress change, since the mechnical behavior during the change from Q_j^a to Q_j^b is not affected by *subsequent* stress programs. The complementary work integral around the cycle may be broken into two parts

$$\oint q_j dQ_j \;=\; \int_{Q_j^a}^{Q_j^b} q_j dQ_j \;+\; \int_{Q_j^b}^{Q_j^a} q_j dQ_j \;\le\; 0 \;. \tag{67}$$

Let the initial strain at Q_j^a be q_j^a, and the strain at Q_j^b be q_j^b. We make use of the two identities

$$\int_{Q_j^a}^{Q_j^b} q_j dQ_j \;+\; \int_{q_j^a}^{q_j^b} Q_j dq_j \;=\; Q_j^b q_j^b \;-\; Q_j^a q_j^a \;, \tag{68}$$

and

$$\int_{Q_j^b}^{Q_j^a} q_j dQ_j \;=\; \int_{Q_j^b}^{Q_j^a} (q_j - q_j^b) dQ_j \;+\; (Q_j^a - Q_j^b) q_j^b \;. \tag{69}$$

Substituting these identities in equation (67), we obtain

$$\int_{q_j^a}^{q_j^b} (Q_j - Q_j^a)dq_j \;\geq\; \int_{Q_j^b}^{Q_j^a} (q_j - q_j^b)dQ_j \;. \tag{70}$$

The net work integral on the left-hand side of inequality (70) is carried out along any path from Q_j^a to Q_j^b. The net complementary work integral on the right-hand side is carried out along a *straight line path* from Q_j^b to Q_j^a; hence, from the first postulate, equation (61), it is itself non-negative. Thus, for a change in stress from Q_j^a to Q_j^b *along any path* the net work is non-negative,

$$\int_{q_j^a}^{q_j^b} (Q_j - Q_j^a)dq_j \;\geq\; 0 \;. \tag{71}$$

We can see immediately that, if $Q_j^b = Q_j^a$ and the stress path is in fact a cycle in stress, we recover the second postulate in its strong form in that the net work around a cycle of stress is non-negative, and through the identity given in equation (66), the complementary work around the cycle in stress is non-positive.

If we choose the path to be a straight line path from Q_j^a to Q_j^b, inequality (71) implies part of inequality (59), but does not directly give the first postulate. However, if we choose, in addition, that $Q_j^b = Q_j^a + dQ_j$, where dQ_j is an infinitesimal increment, we can write, to a close approximation,

$$\int_{q_j^a}^{q_j^a + dq_j} (Q_j - Q_j^a)dq_j \;=\; \tfrac{1}{2}\, dQ_j dq_j \;. \tag{72}$$

We can see that (71) and (72) imply inequality (53); it has already been shown that inequality (53) implies the first postulate. *Consequently the assumption that the net work along any path in stress space is non-negative implies both the first postulate and the second postulate in its strong form*, and can thus be used in their place.

Before proceeding to consider the limitations imposed on plastic constitutive relations by the postulates given in this section, it is of interest to note briefly their implications for *elastic materials*, where stress and strain are governed directly by a *path-independent* relation. In the context of a bar test (i.e. where only one component of stress is nonzero) the first postulate implies that the stress-strain curve has a positive slope. If we adopt a strict inequality in (52), strain must be a *single-valued* function of stress. The greater than or equal to sign permits the limiting case of an infinite or zero slope; while this is not to be expected in conventional elastic materials it is not inconsistent with the concept of elastic behavior,

In the context of a bar test the second postulate (inequality 62) does not provide any direct information, since the complementary work integral around *any* cycle of stress must be zero. However, if we regard the second postulate (inequality 63) as taking the form

$$\oint q_j \, dQ_j \; = \; 0 \tag{73}$$

for an *elastic* material, it becomes a necessary and sufficient condition that a potential function $\Omega(Q_j)$,

$$\Omega(Q_j) \; = \; \int_0^{Q_j} q'_j \, dQ'_j \tag{74}$$

should exist, and that

$$q_j = \frac{\partial \Omega}{\partial Q_j} \, . \tag{75}$$

Without providing a detailed argument, it also provides a necessary and sufficient condition for the existence of a potential function of strain, $W(q_j)$, where

$$W(q_j) = \int_o^{q_j} Q_j' dq_j' \tag{76}$$

and

$$Q_j = \frac{\partial W}{\partial q_j} \, . \tag{77}$$

Returning again to the first postulate, we see that if

$$\int_{q_j^a}^{q_j^b} (Q_j - Q_j^a) dq_j \geq 0 \tag{78a}$$

and

$$\int_{Q_j^a}^{Q_j^b} (q_j - q_j^a) dQ_j \geq 0 \tag{78b}$$

for a straight line path, it also follows by virtue of the path independence characteristic of *elastic* materials that these inequalities apply to *any* path between the initial and terminal states Q_j^a, q_j^a and Q_j^b, q_j^b respectively. Inequalities (78a) and (78b) may be written as

$$\int_{q_j^a}^{q_j^b} Q_j\, dq_j \;\geq\; (q_j^b - q_j^a)Q_j^a \;\;,$$ (79a)

$$\int_{Q_j^a}^{Q_j^b} q_j\, dQ_j \;\geq\; (Q_j^b - Q_j^a)q_j^a \;\;.$$ (79b)

Further, making use of equations (74) – (77), inequalities (79a) and (79b) become

$$W(q_j^b) - W(q_j^a) \;\geq\; (q_j^b - q_j^a)\left.\frac{\partial W}{\partial q_j}\right|_{q_j = q_j^a} \;\;,$$ (80a)

$$\Omega(Q_j^b) - \Omega(Q_j^a) \;\geq\; (Q_j^b - Q_j^a)\left.\frac{\partial \Omega}{\partial Q_j}\right|_{Q_j = Q_j^a} \;\;.$$ (80b)

These inequalities were introduced in Section 1.6; they show that the strain energy function $W(q_j)$ and the complementary energy function $\Omega(Q_j)$ are *convex*. It is evident that our present conclusion that the postulates adopted in this section imply the convexity of $W(q_j)$ and $\Omega(Q_j)$ is thus consistent with the elastic relations discussed in Chapter 1.

The convexity requirements are also consistent with elastic deformations which may take place *within* the theory of plasticity, since the postulates are applicable also to paths which lie entirely within the yield surface. We have required that

$$q_j^e \;=\; C_{jk}Q_k \;\;,$$ (81)

where C_{jk} is a positive definite symmetric matrix. It is clear then that we can introduce an *elastic complementary energy* $\Omega^e(Q_j)$ defined by

$$\Omega^e(Q_j) = \int q_j^e \, dQ_j \; . \tag{82}$$

Then $\Omega^e(Q_j)$ will be a *quadratic* function of the components of
the generalized stress, and will certainly be convex. Further,

$$q_j^e = \frac{\partial \Omega^e}{\partial Q_j} \; . \tag{83}$$

Similarly there will exist a quadratic, convex and positive
definite *elastic strain energy function,* $W^e(q_j^e)$, where

$$W^e(q_j^e) = \int Q_j \, dq_j^e \; , \tag{84}$$

and

$$Q_j = \frac{\partial W^e}{\partial q_j^e} \; . \tag{85}$$

The existence of these functions simply confirms that when
we consider mechanical behavior in which yielding never occurs
the theory of plasticity reduces to the theory of linear elasticity. When $q_j^p \equiv 0$, q_j^e is identical to the total generalized
strain q_j, and the superscripts in equations (82) - (85) can be
omitted.

In closing, it is appropriate to mention an alternative assumption suggested by A.A.Ilyushin. This assumption is essentially the dual of the second postulate: it is assumed that

$$\oint Q_j \, dq_j \geq 0 \tag{86}$$

for a *cycle in strain.* From the standpoint of the classical
theory, Ilyushin's postulate is equivalent to the second postulate in its weak form when it is applied to strain cycles in
which the change in plastic strain is infinitesimal. Applied
to any strain cycle it is weaker than the second postulate in

its strong form. Since it is implied by inequality (71), but
does not in itself lead to any particularly strong results, we
shall not use it in preference to the second postulate given in
this section.

Finally, it must be emphasized that while the assumptions
outlined in this section are physically motivated, in the sense
that we have selected two dominant characteristics of a uniaxial experiment on a bar, they must be regarded as *mathematical restrictions* on the relation between stress Q and strain
q in the framework for plastic constitutive relations which we
are constructing. Thus, while restrictions are placed on work
and complementary work

$$\int Q_j dq_j \quad \text{and} \quad \int q_j dQ_j$$

it is understood that we are concerned with the relation between stress and strain given by the constitutive equations
rather than the concept of work in the thermodynamic sense.

The postulates given in this section have been developed
separately because they have important direct consequences in
structural behavior, and they permit us to obtain a number of
general results without reference to the specific form of the
constitutive relations. In our development, however, we shall
proceed first to consider the restrictions placed on the form
of the constitutive relations as a consequence of the adoption
of these additional assumptions.

2.5 Convexity of the Yield Surface and the Normality Rule

The assumptions introduced in Sections 2.3 and 2.4 are to a
large extent independent of each other, and each sets limits on
the properties of the time-independent, path-dependent material
adopted in Section 2.2. We are now in a position to further
refine the general framework for constitutive relations derived

in Section 2.3 by incorporating the restrictions imposed by the postulates adopted in Section 2.4.

Consider an element of material after the execution of some stress program. Let the yield surface be as indicated in Fig. 19, and let the stress be Q_j^a. We now suppose that the material element is subject to a particular stress cycle. The stress point follows some elastic path from Q_j^a to a stress state $\bar{Q}_j$ on the yield surface (i.e. $\phi(\bar{Q}_j) = 0$). At this point an infinitesimal increment of the plastic component of strain $d\bar{q}_j^p$ occurs, either as the result of a loading increment $d\bar{Q}_j$ (hardening behavior) or as a result of flow at constant stress. Thereafter unloading occurs and the stress changes back to Q_j^a by an elastic path. We now evaluate the complementary work around this cycle. The strain changes are elastic along the paths Q_j^a to $\bar{Q}_j$ and $(\bar{Q}_j + dQ_j)$ to Q_j^a. Let the plastic strain at the beginning of the cycle be q_j^{pa}. Thus, for the elastic path from Q_j^a to $\bar{Q}_j$,

$$q_j = q_j^e + q_j^{pa} , \tag{87}$$

where $q_j^e = C_{jk} Q_{jk}$ and q_j^{pa} is constant. For the elastic path from $(\bar{Q}_j + d\bar{Q}_j)$ to Q_j^a, the strain is given by

$$q_j = q_j^e + (q_j^{pa} + d\bar{q}_j^p) , \tag{88}$$

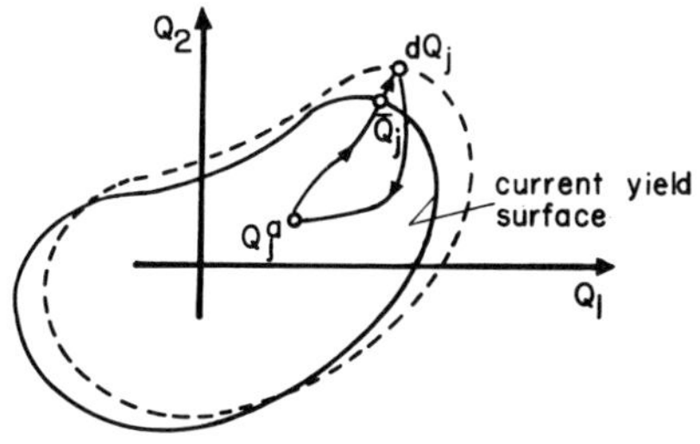

Figure 19. Stress cycle leading to a small change in q_j^p

where q_j^{pa} and $d\bar{q}_j^{p}$ are constant.

The complementary work integral around the cycle may be divided into three parts.

$$\oint q_j dQ_j = \int_{Q_j^a}^{\bar{Q}_j} (q_j^e + q_j^{pa}) dQ_j \;+\; \int_{\bar{Q}_j}^{\bar{Q}_j + d\bar{Q}_j} (q_j^e + q_j^p) dQ_j$$

$$+ \int_{\bar{Q}_j + d\bar{Q}_j}^{Q_j^a} (q_j^e + q_j^{pa} + d\bar{q}_j^p) dQ_j \tag{89}$$

Collecting like strain terms, we may rewrite equation (89) as

$$\oint q_j dQ_j = \oint q_j^e dQ_j + \oint q_j^{pa} dQ_j + \int_{\bar{Q}_j}^{\bar{Q}_j + d\bar{Q}_j} (q_j^p - q_j^{pa}) dQ_j$$

$$+ \int_{\bar{Q}_j + d\bar{Q}_j}^{Q_j^a} d\bar{q}_j^p dQ_j \tag{90}$$

The first integral in the expression vanishes, since the elastic complementary energy $\Omega(Q_j) = \int q_j^e dQ_j$ has the same value at the beginning and end of the cycle. Similarly, the second term vanishes since q_j^{pa} is constant and $Q_j = \oint dQ_j$ has the same value at the beginning and end of the cycle. The third term involves the integration of the change in plastic strain *increment* along the incremental stress path; it is of order $dQ_j d\bar{q}_j^p$ (i.e. of second order in the increments) and may be neglected by comparison with first order terms if we assume that the increments are infinitesimal. The fourth term may be integrated, since $d\bar{q}_j^p$ in the integrand is constant. Thus,

$$\oint q_j dQ_j = (Q_j^a - \bar{Q}_j - d\bar{Q}_j)d\bar{q}_j^p \tag{91}$$

The product $dQ_j d\bar{q}_j^p$ in this expression is also neglected as
being of second order. Finally, we apply the second postulate,
which requires that the complementary work integral around a
cycle should be non-positive. For our material, therefore, for
a cycle commencing at any stress Q_j^a for which $\phi(Q_j^a) \leq 0$, following an elastic path to any stress $\bar{Q}_j$ on the yield surface,
causing a plastic strain increment $d\bar{q}_j^p$, and then unloading and
following an elastic path back to Q_j^a, the second stability postulate requires

$$\oint q_j dQ_j = (Q_j^a - \bar{Q}_j)d\bar{q}_j^p \leq 0 \quad ,$$

or

$$(\bar{Q}_j - Q_j^a)d\bar{q}_j^p \geq 0 \quad . \tag{92}$$

In discussing the implications of inequality (92) it is convenient to plot the increment of plastic strain $d\bar{q}_j^p$ in the
stress space with its origin at the stress point $\bar{Q}_j$. $d\bar{q}_j^p$ is
then a vector in the stress space whose component in the Q_1
direction is $d\bar{q}_1^p$, and so on. Similarly, $(\bar{Q}_j - Q_j^a)$ is a vector
in the stress space joining the stress points $\bar{Q}_j$ and Q_j^a. This
is shown diagrammatically in Fig.20. Inequality (92) may then

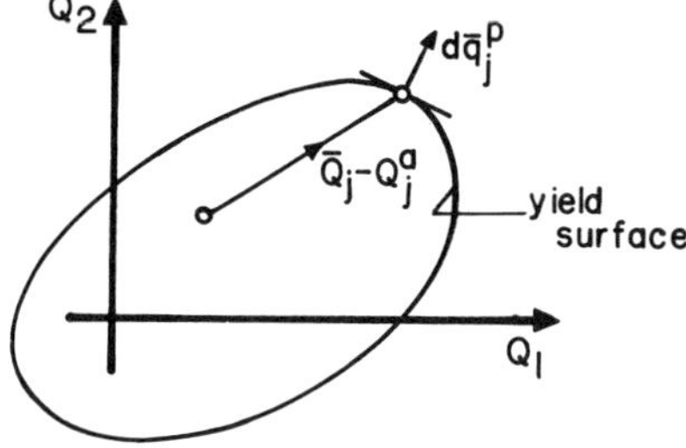

Figure 20.Increment of plastic strain plotted in stress space

be interpreted in turn as requiring that the scalar product of the vectors $(\bar{Q}_j - Q_j^a)$ and $d\bar{q}_j^p$ should be non-negative. This can only occur if the angle between the vectors in their positive senses is less than or equal to $\pi/2$.

This result has two extremely important consequences. First (92) requires that the *yield surface is convex*, i.e. the straight line joining any two stress points lying inside or on the yield surface must itself lie completely within or on the yield surface. Secondly, restrictions are placed on the direction of the increment of the plastic strain . The gradient of the yield function $\partial\phi/\partial Q_j$ may be interpreted as a vector which has the direction of the outward normal to the level surface of ϕ at the point in the stress space at which it is evaluated. Inequality (92) requires that $d\bar{q}_j^p$ has the direction of the outward normal to the yield surface (the level surface $\phi = 0$) at the stress point $\bar{Q}_j$ if $\partial\phi/\partial Q_j$ is a continuous function (i.e. if the direction of the outward normal changes continuously as the yield surface is traversed; such yield surfaces are sometimes called *regular*). Thus we may write

$$d\bar{q}_j^p = \lambda \left. \frac{\partial\phi}{\partial Q_j}\right|_{Q_j = \bar{Q}_j} , \qquad (93)$$

where $\lambda \geq 0$ to ensure the direction is that of the outward normal. If the stress increment occurs at a point on the yield surface where the outward normal is discontinuous, the plastic strain increment must be bounded in direction by the outward normal vectors at all adjacent points on the yield surface. Points on the yield surface where such discontinuities occur are called *singular points* or *corners*.

In order to demonstrate these results consider a typical yield surface. We assume that the results are not true, and

make use of geometrical arguments to show that this assumption
is untenable. In doing so, we can choose any stress points $\bar{Q}_j$
and Q_j^a, since the argument by which inequality (92) was established does not place any restrictions on the stresses except
that $\phi(\bar{Q}_j) = 0$ and $\phi(Q_j^a) \leq 0$. Assume first that the yield surface is convex, but that the plastic strain increment does not
have the direction of the outward normal. Thus, in Fig. 21
$d\bar{q}_j^p$ occurs as a result of a stress increment dQ_j imposed on
stress $\bar{Q}_j$. It is clear that the choice of stress Q_j^a shown in
Fig. 21 will lead to a violation of inequality (92). By a process of elimination,in which the two-dimensional representation
of Fig. 21 can be easily generalized to n dimensions, it is
seen that only when the increment $d\bar{q}_j^p$ has the direction of the
outward normal to the yield surface will it never be possible
to choose a stress Q_j^a such that inequality (92) is not satisfied.

If $\bar{Q}_j$ does lie at a singular point, the result is not quite
as restrictive. The elimination process leads us to the conclusion that $d\bar{q}_j^p$ may have any direction within the *fan* or *hypercone* defined by the outward normals at all the points *adjacent* to $\bar{Q}_j$ on the yield surface. This is shown diagrammatically
in Fig. 22.

Assume next that the plastic strain increment satisfies the
restrictions set down, but the yield surface is not convex. As

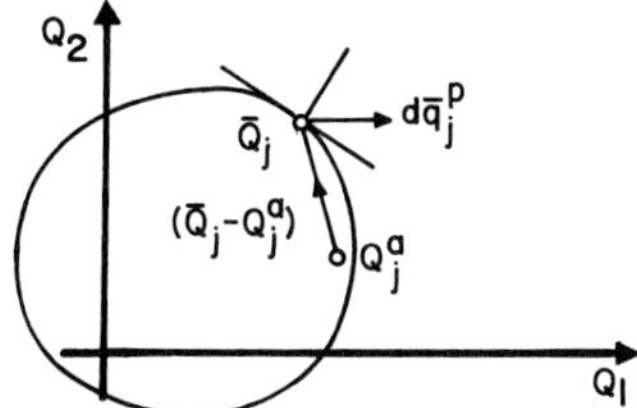

Figure 21. Convex yield surface without normality

shown in Fig. 23, choose $\bar{Q}_j$ to lie in a non-convex part of the surface. Q_j^a can then be chosen, as shown in Fig. 23, such that (92) does not hold. Again, a process of elimination leads to the conclusion that the surface must be convex if it is to be possible to choose any $\bar{Q}_j$ and Q_j^a such that inequality (92) is always true. This also holds if $\bar{Q}_j$ lies at a singular point on the yield surface.

If now both the requirements of convexity and the restrictions on the direction of dq_j^p are relaxed, it is clear that it will always be possible to choose $\bar{Q}_j$, $d\bar{q}_j^p$ and Q_j^a on any given yield surface so that inequality (92) is not satisfied.

Let us now compare the consequences of inequality (92) with the results derived in Section 2.3. Assume first that the yield surface is smooth and contains no singular points (i.e. ϕ is continuous and continuously differentiable). Equation (93) expresses the requirements of the normality rule for any

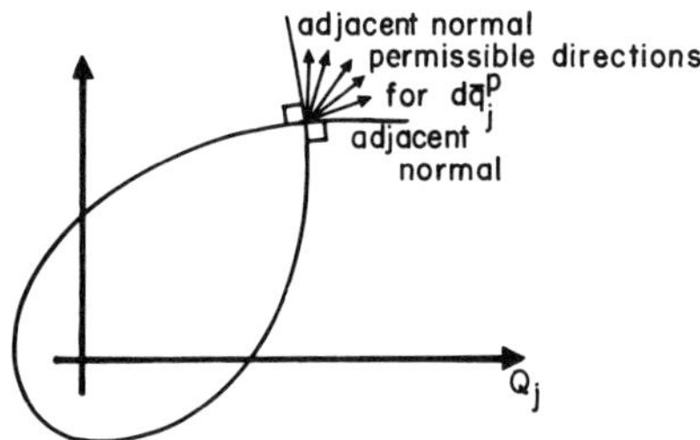

Figure 22. Plastic strain increment at a singular point

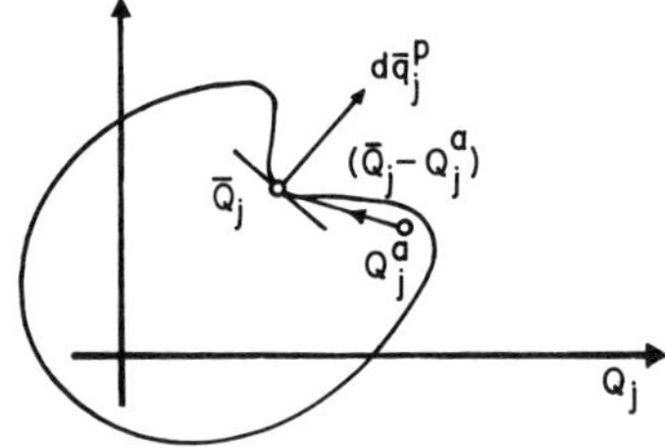

Figure 23. Normality without convexity

incremental change in the plastic strain;

$$dq_j^p = \lambda \frac{\partial \phi}{\partial Q_j} \quad , \quad \lambda \geq 0 \quad . \tag{93}$$

For *hardening behavior*, by investigating the quantities on which dq_j^p may depend and the implications of time independence, we determined that (Section 2.3, equations 42)

$$dq_j^p = G(Q_j, H_\alpha) \frac{\partial g}{\partial Q_j} (Q_j, H_\alpha) \frac{\partial \phi}{\partial Q_k} dQ_k \tag{94}$$

when $\phi(Q_j) = 0$ and $\dfrac{\partial \phi}{\partial Q_j} dQ_j \geq 0$.

Equations (93) and (94) describe the same phenomenon. The important differences between them is that λ in equation (93) is a scalar and the *direction* of dq_j^p is given in terms of $\partial \phi / \partial Q_j$. In equation (94), dq_j^p is linear and homogeneous in $(\partial \phi / \partial Q_k) dQ_k$ and depends on an unknown vector function $\partial g / \partial Q_j$. Thus, if both statements must hold, we can put

$$\lambda = G(Q_j, H_\alpha) \frac{\partial \phi}{\partial Q_k} dQ_k \quad \text{and} \quad \frac{\partial g}{\partial Q_j} = \frac{\partial \phi}{\partial Q_j} \quad , \tag{95}$$

identifying the plastic potential with the yield function. Further, $\lambda \geq 0$ and since $(\partial \phi / \partial Q_k) dQ_k$ is positive definite for hardening, $G > 0$. Substituting equation (95) into equation (93), we obtain a more restrictive form for the increment of the plastic strain component for hardening behavior and for a loading increment;

$$dq_j^p = G \frac{\partial \phi}{\partial Q_j} \frac{\partial \phi}{\partial Q_k} dQ_k \quad , \quad G = G(Q_j, H_\alpha) > 0 \quad . \tag{96}$$

The geometrical significance of the scalar $(\partial \phi / \partial Q_k) dQ_k$ is also apparent. $\partial \phi / \partial Q_j$ is a vector normal to the yield surface in

the outward direction (although it is not in general a unit
vector). Thus $(\partial\phi/\partial Q_k)dQ_k$ is a measure of the component of dQ_j
normal to the yield surface, and it follows that dq_j^p depends
linearly on this normal component of dQ_j, and is zero when it
is zero or negative. Some further significant consequences can
be drawn from equation (94). First, for a *loading increment*,
we could write equation (96) as

$$dq_j^p \;=\; A_{jk}dQ_k \;\;,\tag{97a}$$

where

$$A_{jk} \;=\; G\,\frac{\partial\phi}{\partial Q_j}\,\frac{\partial\phi}{\partial Q_k}\;\;.\tag{97b}$$

A_{jk} can be treated as a matrix of coefficients which for any
one point on the yield surface is independent of the stress or
strain increment. Thus there is a *linear* relation between dq_j^p
and dQ_j at any one point on the yield surface, and it is clear
from the form of equation (97b) that A_{jk} is symmetric, i.e.

$$A_{jk} \;=\; A_{kj}\tag{98}$$

The scalar product of dQ_j and $d\bar{q}_j^p$, can be written, from equation (96), as

$$dQ_j dq_j^p \;=\; G\,\frac{\partial\phi}{\partial Q_j}\,\frac{\partial\phi}{\partial Q_k}\,dQ_k dQ_j$$

$$\;=\; G\left(\frac{\partial\phi}{\partial Q_j}dQ_j\right)^2 \;\geq\; 0 \;\;,\tag{99}$$

since G is non-negative. The first postulate of Section 2.4 is
consequently satisfied in its incremental form (Section 2.4,
equation 53);

$$dQ_j dq_j \;=\; dQ_j(dq_j^e + dq_j^p)$$

$$= C_{jk} dQ_j dQ_k + G \left(\frac{\partial \phi}{\partial Q_j} dQ_j\right)^2 \geq 0 \quad , \tag{100}$$

since we have already required that C_{jk} should be positive definite. A further result, of importance in establishing uniqueness for hardening behavior, follows from the linearity of the dependence of dq_j^p on stress increments dQ_j imposed on a given stress Q_j. Consider two such *alternative* increments, $dQ_j^{(1)}$ and $dQ_j^{(2)}$, leading *respectively* to total strain increments $dq_j^{(1)}$ and $dq_j^{(2)}$. By dividing each increment into its elastic and plastic parts we may write

$$(dQ_j^{(1)} - dQ_j^{(2)})(dq_j^{(1)} - dq_j^{(2)}) = (dQ_j^{(1)} - dQ_j^{(2)})(dq_j^{e(1)} - dq_j^{e(2)})$$

$$+ (dQ_j^{(1)} - dQ_j^{(2)})(dq_j^{p(1)} - dq_j^{p(2)}) \quad . \tag{101}$$

Because of the linearity of the elastic relation,

$$(dQ_j^{(1)} - dQ_j^{(2)})(dq_j^{e(1)} - dq_j^{e(2)}) = (dQ_j^{(1)} - dQ_j^{(2)})(C_{jk} dQ_k^{(1)} - C_{jk} dQ_k^{(2)})$$

$$= C_{jk}(dQ_j^{(1)} - dQ_j^{(2)})(dQ_k^{(1)} - dQ_k^{(2)})$$

$$= \geq 0 \quad . \tag{102}$$

If $\phi(Q_j) < 0$ the stress point lies inside the elastic region and stress increments cannot produce plastic strain increments; the last term in equation (101) is zero, and it follows that the expression on the left-hand side of equation (101) is non-negative. The same conclusion holds if $\phi(Q_j) = 0$ and both stress increments $dQ_j^{(1)}$ and $dQ_j^{(2)}$ constitute unloading or neutral loading. If on the other hand, $\phi(Q_j) = 0$ and both stress increments represent loading, we may use the linearity of equation (96) to write

$$(dQ_j^{(1)} - dQ_j^{(2)})(dq_j^{p(1)} - dq_j^{p(2)}) = G \left\{ \frac{\partial \phi}{\partial Q_j} (dQ_j^{(1)} - dQ_j^{(2)}) \right\}^2$$

$$\geq 0 \quad . \tag{103}$$

Again, the term on the left-hand side of equation (101) consists of the sum of two non-negative terms, and is in consequence itself non-negative. The only remaining case we must consider is that when one increment, say $dQ_j^{(1)}$, constitutes loading, and the other, $dQ_j^{(2)}$, constitutes unloading or neutral loading. $dq_j^{p(2)}$ is then zero, and

$$(dQ_j^{(1)} - dQ_j^{(2)})dq_j^{p(1)} = G \frac{\partial \phi}{\partial Q_j} \frac{\partial \phi}{\partial Q_k} dQ_k^{(1)}(dQ_j^{(1)} - dQ_j^{(2)}) \quad . \tag{104}$$

Because $dQ_j^{(1)}$ is a loading increment,

$$\frac{\partial \phi}{\partial Q_k} dQ_k^{(1)} > 0 \quad . \tag{105}$$

Now consider

$$\frac{\partial \phi}{\partial Q_j} (dQ_j^{(1)} - dQ_j^{(2)}) = \frac{\partial \phi}{\partial Q_j} dQ_j^{(1)} - \frac{\partial \phi}{\partial Q_j} dQ_j^{(2)} \quad . \tag{106}$$

The first term is positive, as in equation (105). $dQ_j^{(2)}$ is an unloading increment or a neutral loading increment, and hence

$$\frac{\partial \phi}{\partial Q_j} dQ_j^{(2)} \leq 0 \quad . \tag{107}$$

The expression on the right-hand side of equation (106) is hence non-negative, and the expression in equation (104) consists of the product of three non-negative terms and is itself non-negative. We see then from equation (101) that for *any* two stress increments $dQ_j^{(1)}$, $dQ_j^{(2)}$ which might each be imposed on *any* stress state Q_j

$$(dQ_j^{(1)} - dQ_j^{(2)})(dq_j^{(1)} - dq_j^{(2)}) \geq 0 \quad . \tag{108}$$

It will be seen in Chapter 3 that inequality (108) is a *suffi-cient condition* for uniqueness in structural problems in the incremental sense.

In the case of *flow*, when a plastic strain increment may occur under constant stress for $\phi(Q_j) = o$, equation (93) still holds when the yield function is continuously differentiable;

$$dq_j^p = \lambda \frac{\partial \phi}{\partial Q_j} \quad ,$$

where λ is a non-negative scalar. However, in Section 2.3 (equations 47 and 48) we also came to the conclusion that the plastic strain increment during flow could be written in the form

$$dq_j^p = \sum_{r=1}^{m} \beta_r \frac{\partial \bar{g}_r}{\partial Q_j} (Q_j, H_\alpha, J_\zeta) \tag{109}$$

where β_r are unspecified scalars. Comparison of equations (93) and (109) lead us to the conclusion that both representations can hold if we have the *flow rule*

$$dq_j^p = \lambda \frac{\partial \phi}{\partial Q_j} \tag{110}$$

where λ is a *non-negative but otherwise unspecified scalar.* When flow occurs under constant stress, $dQ_j = 0$ and hence

$$dQ_j dq_j^p = 0 \quad . \tag{111}$$

The first postulate of Section 2.4 will again be satisfied.

In this instance we shall not identify the plastic potential functions $\bar{g}_r$ with the yield surface, since we have yet to

discuss the conditions under which flow can occur. Nevertheless, it is apparent that the derivatives of the plastic potential functions coincide with the derivatives of the yield function with respect to stress when flow *does* occur.

The cases we have dealt with up to this point are based on
the assumption that $\partial\phi/\partial Q_j$ and hence the direction of the normal vector to the yield surface is uniquely determined at each
point on the yield surface. Let us now consider cases where
corners or singular points may be present. In the case of flow
no further assumptions are really necessary to characterize the
plastic strain increments when the normal to the yield surface
is not uniquely defined; it can be accepted that the direction
of dq_j^p is not specified uniquely but is only required to lie
between the normals at adjacent points on the yield surface, as
in Fig. 22. In the case of hardening, however, indeterminacy
of the direction of the plastic strain increment causes problems in establishing sufficient conditions for uniqueness, and
cannot be accepted. To overcome this we make a further assumption regarding the origin of corners; this assumption resolves
the difficulty in the case of hardening, but adds no restrictions in the case of flow although it is consistent with our
remarks in this paragraph.

We assume that *corners or singular points can only arise as
a result of the existence of two or more independent continuously differentiable yield functions.* It will be seen that
this assumption is an attractive mathematical description of
the behavior in the vicinity of corners; there are also strong
physical reasons for adopting it, as we shall see later.

As a simple example, assume that two independent, smooth
yield functions ϕ_1 and ϕ_2 exist. In order that the stress
point should lie in the elastic region we require that *both*

$\phi_1(Q_j) < 0$ and $\phi_2(Q_j) < 0$. The yield surface or boundary of
the elastic region $\phi = 0$ is made up of the inner envelope of
the two surfaces $\phi_1 = 0$ and $\phi_2 = 0$. As sketched in Fig. 24,
corners appear at the intersections of the two yield surfaces.
The plastic strain increment dq_j^p is assumed to be the sum of
two parts, dq_j^{p1}, dq_j^{p2} determined independently from the yield
functions ϕ_1, ϕ_2 assuming that each has a continuous derivative. It is clear from the diagram that the sum of the increments dq_j^{p1} and dq_j^{p2} which may result from a stress increment
which constitutes loading on both yield surfaces at a corner
will always lie in the fan defined by adjacent normals, which
are now simply the directions of dq_j^{p1} and dq_j^{p2}. The yield surface itself (i.e. the inner envelope of the surfaces $\phi_1 = 0$ and
$\phi_2 = 0$) must be convex. A sufficient condition for convexity
of this envelope is that both the surfaces $\phi_1 = 0$ and $\phi_2 = 0$
should be convex. There are again good physical reasons for
assuming that this should be true, and we will proceed on this
assumption.

More generally we admit the existence of m independent con-
vex yield functions ϕ_1, ϕ_2, ..., ϕ_m. For the rth yield function we require that

$$\phi_r(Q_j, H_\alpha) \leq 0 \qquad r = 1, \ldots, m \tag{112}$$

The plastic strain increment is given by

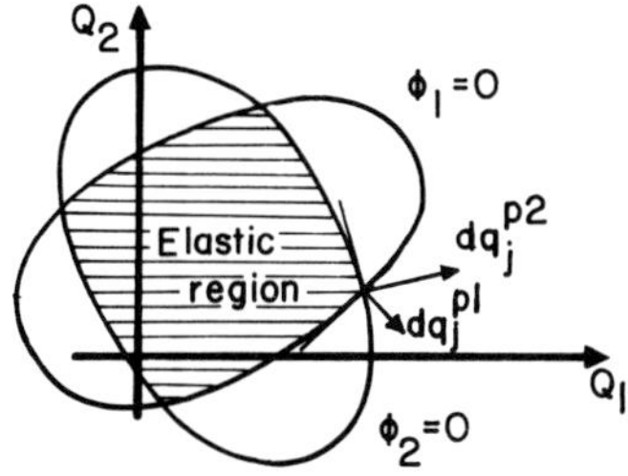

Figure 24. Two independent yield functions

$$dq_j^p = dq_j^{p1} + dq_j^{p2} + \ldots + dq_j^{pr} + \ldots + dq_j^{pm} \; . \tag{113}$$

In the case of hardening behavior,

$$dq^{pr} = 0 \quad \text{if} \quad \phi_r(Q_j) < 0$$

$$\text{or} \quad \phi_r(Q_j) = 0 \quad \text{and} \quad \frac{\partial \phi_r}{\partial Q_j} dQ_j \leq 0 \; , \tag{114a}$$

$$dq_j^{pr} = G_r \frac{\partial \phi_r}{\partial Q_j} \frac{\partial \phi_r}{\partial Q_k} dQ_k$$

$$\text{if} \quad \phi_r(Q_j) = 0 \quad \text{and} \quad \frac{\partial \phi_r}{\partial Q_j} dQ_j \geq 0. \tag{114b}$$

The summation convention does not apply to subscripts r in these equations. The scalar coefficients

$$G_r = G_r(Q_j, H_\alpha) \tag{115}$$

are each positive definite.

The plastic strain increment for any increment in stress is uniquely determined, and the restrictions imposed in Sections 2.3 and 2.4 are all satisfied. Note specifically that dq_j^p is homogeneous and of order one in dQ_j, and that the direction of dq_j^p lies between adjacent normals at a singular point. Further, we may show by identical arguments which led to inequalities (100) and (108) that

$$dQ_j dq_j \geq 0 \tag{116}$$

for any stress increment, and that

$$(dQ_j^{(1)} - dQ_j^{(2)})(dq_j^{(1)} - dq_j^{(2)}) \geq 0 \tag{117}$$

for two *alternative* stress increments $dQ_j^{(1)}$, $dQ_j^{(2)}$ imposed on the same state of stress and leading to strain increments

$dq_j^{(1)}$, $dq_j^{(2)}$ respectively.

This representation may be used without loss in generality for *flow*. Equations (112) and (113) remain applicable. The restrictions on the direction of the plastic strain increment and the restrictions imposed in Section 2.3 (cf. equation 109) are both satisfied if we assume that when flow may occur at constant stress

$$dq_j^{pr} = 0 \quad \text{if} \quad \phi_r(Q_j) \; < \; 0 \quad , \tag{118a}$$

$$dq_j^{pr} = \lambda_r \frac{\partial \phi_r}{\partial Q_j} \quad \text{if} \quad \phi_r(Q_j) \; = \; 0 \quad . \tag{118b}$$

Again, the summation convention does not apply to the subscript r. The scalars λ_r (r = 1, ..., m) are each non-negative but otherwise unspecified. It still follows that for flow at constant stress,

$$dQ_j dq_j^p \; = \; 0 \quad . \tag{119}$$

In many situations in this monograph it will be burdensome to write out the full set of relations given in equations (112) – (118). In discussing general results we shall frequently use the simpler result for a single continuously differentiable yield function and indicate that the more complex constitutive relations embodying a number of independent yield functions can be substituted by an extension of the appropriate argument. In particular, we shall continue to represent the *yield surface* (i.e. the boundary of the elastic domain) by the simple expression $\phi = 0$. It is clear that this means that $\phi_r = 0$ for one or more values of r, while $\phi_r < 0$ for all other values of r.

Apart from some further restrictions which will be placed on the conditions under which flow can occur, the *general framework* for plastic constitutive relations has now been developed.

To the expressions for plastic strain increments (equations 114
and 118) we add the elastic strain increment to obtain the total strain increment;

$$dq_j = dq_j^e + dq_j^p$$

$$= C_{jk}dQ_j + dq_j^p \qquad (120)$$

In Chapter 3 it will be shown that this framework provides sufficient conditions for uniqueness and stability in structural
problems where the equilibrium and kinematic relations are
linear.

In developing this general framework it may be noticed that
the first postulate of Section 2.4 has not been used explicitly;
indeed, it is seen that the plastic constitutive relations
based on the assumptions of Sections 2.2, 2.3 and the second
postulate of Section 2.4 lead to the statement

$$dQ_j dq_j^p \geq 0 \qquad (121)$$

for all stress increments, including of course those for which
$dq_j^p = 0$. The importance of the first postulate, however, appears in establishing that

$$dQ_j dq_j^e \geq 0 \quad , \qquad (122)$$

as indicated in the final paragraphs of Section 2.4. It then
follows from equations (121) and (122) that

$$dQ_j dq_j \geq 0 \quad , \qquad (123)$$

which is consistent with the first postulate, and thus means
that the first postulate does not impose any further restrictions on dq_j^p .

Another important point to be noted is that the *second*

postulate in its weak form is sufficient to establish inequality (92), since we discussed only cycles in which an infinitesimal change in the plastic strain occurs. It does not necessarily follow that if inequality (92) is satisfied, the second postulate in its strong form is also satisfied without *further restrictions* on $\phi(Q_j, H_\alpha)$ and $G(Q_j, H_\alpha)$.

This is an appropriate point to discuss briefly an alternative physically motivated basis for the derivation of the constitutive equations given by R.Hill. In this basis we assume that the elastic strain energy function is convex or, since it is quadratic, that it is positive definite. In addition to this, Hill adopts inequality (92) directly, terming it the *principle of maximum plastic work*. Inequality (92) may be written as

$$\bar{Q}_j d\bar{q}_j^p \;\geq\; Q_j^a d\bar{q}_j^p \;,\tag{124}$$

where $\phi(\bar{Q}_j) = 0$ and $\bar{Q}_j$ is the stress at which the plastic strain increment $d\bar{q}_j^p$ occurs, and Q_j^a is any stress for which $\phi(Q_j^a) \leq 0$. Inequality (124) may then be interpreted as requiring that for a *given plastic strain increment* $d\bar{q}_j^p$ the associated stress, $\bar{Q}_j$, is distinguished from all other accessible stress states compatible with the current recorded history H_α by the restriction that the *plastic work* $Q_j d\bar{q}_j^p$ during the increment takes its greatest value. A physical argument can be given to support this maximum principle for polycrystalline metals.

Since it is conventional to give the constitutive equations for hardening materials in the form of equation (96) or equations (114) with only the explicit requirement that ϕ is convex and G is non-negative, the physical assumptions of Drucker and Hill are equivalent in the sense that we have so far used them.

More formally, the first postulate and the second postulate of Section 2.4 in its weak form imply the convexity of the strain energy function and the principle of maximum plastic work, and the converse is true.

As we have already mentioned, all the commonly used relations for plastic constitutive equations do also satisfy the second postulate of Section 2.4 in its strong form. We shall not attempt to write out the further restrictions on ϕ and G which are necessary to ensure that the complementary work is negative for all stress cycles, but in Part VI we will discuss the reasons why our conventional models automatically incorporate these restrictions. Some of the consequences of the second postulate in its strong form will be discussed in Chapter 3 but we shall only use the postulate as an *essential* requirement in Part V.

2.6 Limit Surfaces

In discussing plastic strain changes which occur as a result of flow we have seen that while flow in a uniaxial bar test can occur only under special conditions, it can be both preceded and succeeded by hardening behavior in a test in which the stress or strain is increased monotonically. We shall now make the further assumption that *the stress at which flow occurs, in tension say, represents a limiting value of the tension stress which can be applied to the specimen.* Similarly the stress at which flow occurs in compression represents a limiting value of the compression stress which can be applied to the specimen. Such behavior will be termed *perfectly plastic behavior* or *limit behavior* (*ideal plasticity* is also used, but some writers reserve this expression for perfect plasticity with time effects).

This assumption is a very convenient one when we deal with

large strains which occur as a result of flow. Some aspects of
the behavior of a body or structure can as a result be obtained
at the cost of very little effort.   In a bar test the typical
stress-strain curves obtained for mild steel and some aluminum
alloys (Fig.25) are approximated by the curves shown in Figs.
26(a) and 26(b) respectively. The full implications of this
idealization will become apparent as the theory is developed,
but again it must be emphasized that any idealization of this
kind can be made only when the properties of the structural behavior which are sought in the solution render it appropriate.

We shall further assume that the limit stresses in the bar
test are *path-independent*, and thus that they cannot be altered
by mechanical deformation of the specimen. Flow can take place
only when the stress is equal to the limit stress in tension or
compression. Since we have placed no limit on the magnitude of

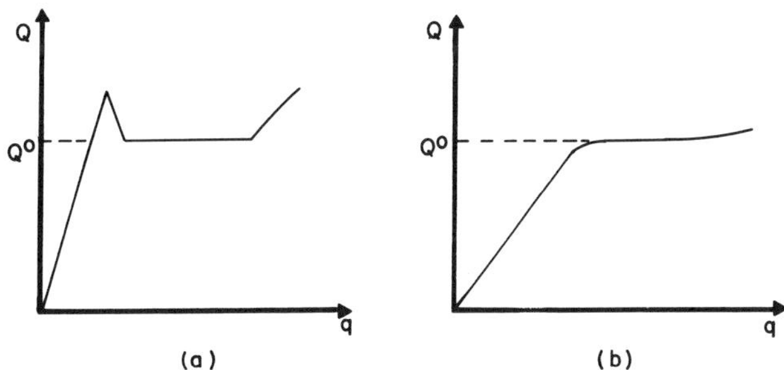

Figure 25. Observed stress-strain curves

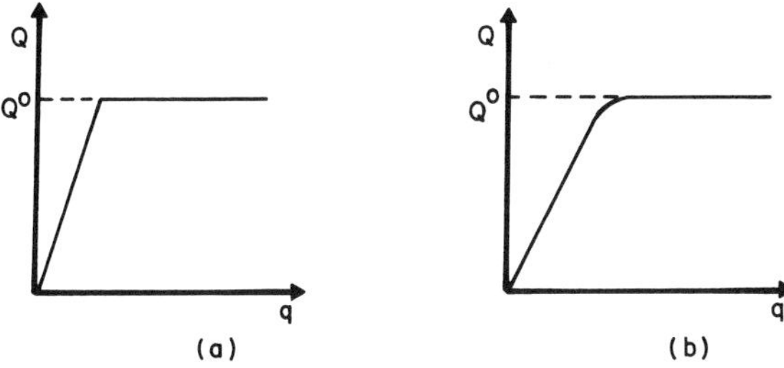

Figure 26. Idealized stress-strain curves

the plastic strain changes which can occur, in principle we permit the plastic strain to become infinitely large as a result of flow. This can provide an alternative explanation of the significance of the limit stress; a stress greater than the limit stress in tension, say, can be achieved only when the strain is infinite. It is not, of course, necessary that limit stresses should exist; if there are no limit stresses (or the limit stresses are themselves infinite) the material will never exhibit flow.

We have already assumed that when flow occurs that set of internal parameters or that part of the recorded history H_α which governs hardening behavior is not affected. This implies that the *magnitude of the plastic strain change which occurs as a result of flow at the limit stress has no effect on the subsequent hardening behavior.*

We may now proceed to characterize this idealized behavior for situations in which more than one component of Q_j is non-zero. It is convenient to introduce a *limit function* $\psi(Q_j)$ such that for all *accessible stress states* $\psi(Q_j) \leq 0$, and for all *inaccessible stress states* $\psi(Q_j) > 0$. The boundary of the *accessible region in stress space* is given by the equation

$$\psi(Q_j) = 0 \ . \tag{125}$$

Equation (125) may be interpreted as the equation of a hypersurface in the stress space, as shown diagrammatically in Fig. 27, and will be referred to as the *limit surface.* In accordance with our assumption above ψ is path independent and depends on Q_j alone; specifically it does not depend on the H_α.

Since we impose the requirements of the limit function *in addition* to the requirements of the yield function, we see that the domain of the accessible *states* Q_j, H_α is now further limited by the requirement that $\psi(Q_j) \leq 0$. Thus whereas in a material which exhibits only hardening behavior any state Q_j,H_α for which $\phi(Q_j,H_\alpha) \leq 0$ can be achieved by some stress path, in a material which exhibits flow only those *states* Q_j,H_α for which $\phi(Q_j,H_\alpha) \leq 0$ *and* $\psi(Q_j) \leq 0$ can ever be achieved. Clearly the stress state $Q_j = 0$ is always accessible. Also, taking into account the requirements of $\phi(Q_j,H_\alpha) \leq 0$ and $\psi(Q_j) \leq 0$, stress states in the elastic region are always accessible and consequently the elastic region is constrained to lie within the limit surface.

Flow, or plastic strain changes at constant stress, make take place only when $\psi = 0$ *and the stress point lies on the limit surface.* Whether or not flow may occur consequently depends only on the stress state Q_j, and not on the internal parameters or recorded history H_α which govern hardening behavior.

We may now illustrate the possible behavior diagrammatically (Fig.27). The limit surface, being a function only of Q_j, is a

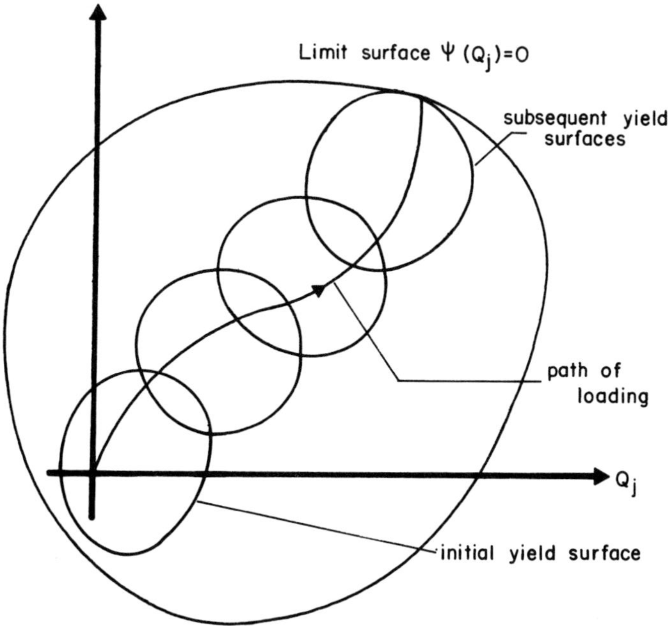

Figure 27. Limit surface and yield surfaces

fixed surface in the space. The yield surface $\phi = 0$ may move around within the limit surface, but may never lie outside of it, i.e. we cannot have $\phi(Q_j, H_\alpha) \leq 0$ when $\psi(Q_j) > 0$ for any given stress Q_j. For stress states and stress increments such that $\phi = 0$, $(\partial\phi/\partial Q_j)dQ_j > 0$ and $\psi < 0$, hardening behavior occurs, but the plastic strain increment is always zero for neutral loading. When the stress point is on the limit surface, $\psi = 0$, *that part of the limit surface must also bound the yield surface*, and $\phi = 0$. Under these circumstances flow may take place under constant stress. When flow does take place, the plastic strain increment dq_j^p must still be normal to the *yield surface* at a non-singular point, or between adjacent normals at a singular point.

We have noted a previous assumption which requires that ϕ and the incremental stress-plastic strain relations associated with hardening are not affected by any increment in the plastic strain which occurs as a result of flow. This leads to the conclusion that $\psi(Q_j)$ is a *convex* function and that the *flow rule* applies to the limit surface, i.e. at a non-singular point on the limit surface,

$$dq_j^p = \lambda \frac{\partial\psi}{\partial Q_j} , \tag{126}$$

where λ is non-negative unspecified scalar. At a singular point the direction of dq_j^p is bounded by adjacent outward normals to the limit surface. To demonstrate this result, consider a stress cycle (Fig.28) which commences at Q_j^a, the material having been subjected to any previous stress history. From Q_j^a the stress path leads to a point on the limit surface, $\bar{Q}_j$ where $\psi(\bar{Q}_j) = 0$, and returns to Q_j^a, completing the cycle. The increment of plastic strain which takes place as a result of flow at $\bar{Q}_j$, $d\bar{q}_j^p$, is of unspecified magnitude. Assume first that it is

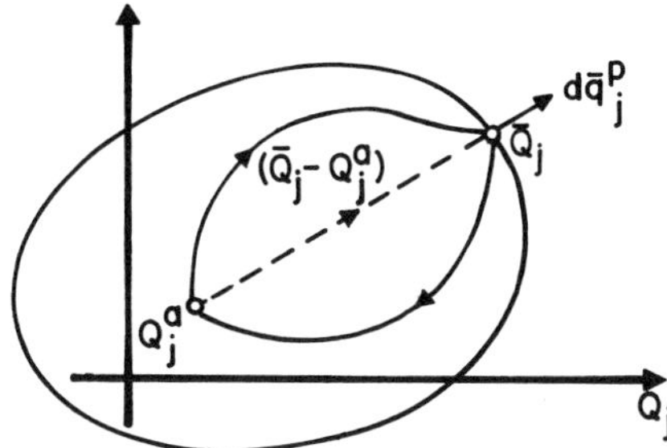

Figure 28.   Stress cycle involving flow

zero.   Changes in both the elastic and plastic strains may take
place during the cycle, but if the second postulate of Section
2.4 in its strong form applies,

$$F = \oint q_j dQ_j \leq 0 \; . \tag{127}$$

Now assume that $d\bar{q}^p_j$ is non-zero. The total strain along the
part of the path Q^a_j to $\bar{Q}_j$ is of course unchanged; between $\bar{Q}_j$
and Q^a_j it is the *former value plus the increment* $d\bar{q}^p_j$, since we
have assumed that the response for $\psi < 0$ is not otherwise affected by $d\bar{q}^p_j$. Hence the complementary work integral around
the cycle may be written as

$$\oint q_j dQ_j = F + (Q^a_j - \bar{Q}_j)d\bar{q}^p_j \leq 0 \; . \tag{128}$$

The integral F is non-positive, as given in (127). However,
$d\bar{q}^p_j$ is of unspecified magnitude and of fixed (or nearly fixed)
direction because it must be normal to the yield surface. The
scalar $(Q^a_j - \bar{Q}_j)d\bar{q}^p_j$ is thus of fixed sign, but unspecified magnitude, whereas F is fixed for the cycle. Under these circumstances, (128) can only hold true if $(Q^a_j - \bar{Q}_j)d\bar{q}^p_j$ is non-
positive itself, i.e.

$$(\bar{Q}_j - Q^a_j)d\bar{q}^p_j \geq 0 \; . \tag{129}$$

This inequality holds for any stress $\bar{Q}_j$ such that $\psi(\bar{Q}_j) = 0$ and any stress Q_j^a such that $\psi(Q_j^a) < 0$. By the same reasoning that was used in Section 2.5 in the case of the yield surface, inequality (129) can be shown by a process of elimination to imply that $\psi(Q_j) = 0$ is convex, and that, for a non-singular point on the limit surface, $d\bar{q}_j^p$ has the direction of the outward normal to the limit surface (equation 126).

At a singular point on the limit surface, the increment $d\bar{q}_j^p$ is unspecified except that it lies between the outward normals at adjacent points on the limit surface. Alternatively, it could be assumed that the accessible region is defined by a set of independent limit functions $\psi_1(Q_j)$, $\psi_2(Q_j)$, ..., $\psi_m(Q_j)$. The plastic strain increment $d\bar{q}_j^p$ for constant stress is given by

$$d\bar{q}_j^p = d\bar{q}_j^{p1} + d\bar{q}_j^{p2} + \ldots + d\bar{q}_j^{pm} \tag{130}$$

where

$$d\bar{q}_j^p = \lambda_r \frac{\partial \psi_r}{\partial Q_j} \quad \text{for} \quad \psi_r = 0 \ ,$$

$$d\bar{q}_j^p = 0 \quad \text{for} \quad \psi_r < 0 \ , \tag{131}$$

and λ_1, λ_2, ..., λ_m are unspecified non-negative scalars. Stress states such that $\psi(Q_j) > 0$ are not admitted.

A comparison of equations (126), (130) and (131) with the framework derived in Section 2.3 (equations 47 and 48) show that the plastic potential $\bar{g}_j(Q_j,H_\alpha,J_\zeta)$ for flow can be identified with the limit function $\psi(Q_j)$. The plastic potential for flow consequently depends only on stress and not on the internal parameters H_α and J_ζ.

Up to this point we have referred to plastic strain increments which occur as a result of flow at constant stress and as a result of hardening during stress increments which constitute loading on the yield surface. In this discussion the contributions to the plastic strain resulting from the two mechanisms of flow and hardening are easily identifiable and no opportunity for confusion arises. However, we must now take into account the possibility of contributions to the plastic strain increment which result from flow during a stress increment. In this discussion it will be easier for us temporarily to use a different notation for plastic strain increments due to flow, say dq_j^{pf}, and due to hardening, say dq_j^{ph}. Referring back to Section 2.3, equations (45) and (46), we put

$$dq_j^{pf} = \bar{h}_{j\zeta}dJ_\zeta \quad , \quad dq_j^{ph} = h_{j\alpha}dH_\alpha \tag{132}$$

Flow can occur only when $\psi(Q_j) = 0$; thus if we consider a stress increment dQ_j imposed on a stress state Q_j for which $\psi(Q_j) = 0$, we can readily establish that flow cannot occur for a stress increment which constitutes *unloading from the limit surface*, since the stress point leaves the limit surface. Hence

$$dq_j^{pf} = 0 \quad \text{for} \quad \psi = 0 \quad \text{and} \quad \frac{\partial\psi}{\partial Q_j} dQ_j < 0 \quad , \tag{133}$$

$$\text{and} \quad \psi < 0 \quad .$$

However, when the stress point moves *along* the limit surface, constituting neutral loading with respect to the limit surface, flow may occur during the increment. Thus

$$dq_j^{pf} = \lambda \frac{\partial\psi}{\partial Q_j} \quad \text{for} \quad \psi = 0 \quad \text{and} \quad \frac{\partial\psi}{\partial Q_j} dQ_j = 0 \quad , \tag{134}$$

where $\lambda \geq 0$. This condition includes the case $dQ_j = 0$. When there are a number of independent yield surfaces $\psi_r (r = 1,...,m)$ the generalization is straightforward.

We must admit the possibility that under certain circumstances neutral loading on the limit surface may constitute hardening with respect to the yield surface; this is the only case in which plastic strain increments due to hardening and flow can occur *simultaneously*. In such circumstances

$$dq_j^p = dq_j^{pf} + dq_j^{ph} \; , \qquad (135)$$

and we make use of the appropriate constitutive equations for each part.

We do wish to establish that under all conditions

$$dQ_j dq_j \; \geq \; 0 \qquad (136)$$

for a stress increment dQ_j, and

$$(dQ_j^{(1)} - dQ_j^{(2)})(dq_j^{(1)} - dq_j^{(2)}) \; \geq \; 0 \qquad (137)$$

for two alternative stress increments $dQ_j^{(1)}$, $dQ_j^{(2)}$ imposed on the same stress state. Making use of equation (135), we can write

$$dQ_j dq_j \; = \; dQ_j dq_j^e + dQ_j dq_j^{ph} + dQ_j dq_j^{pf} \; . \qquad (138)$$

We have already established that the first and second terms on the right-hand side of equation (138) are each non-negative; it follows then that if the last term is non-negative inequality (136) holds for *all* stress increments. When flow does not occur $dq_j^{pf} = 0$, and when it occurs under constant stress, $dQ_j = 0$. During neutral loading on the limit surface, whether or not the stress point lies at a singular point on the limit surface, dQ_j is normal to dq_j^{pf}. Hence for all stress increments

$$dQ_j dq_j^{pf} \; = \; 0 \; , \qquad (139)$$

and inequality (136) holds for all stress increments.

Similarly, for the case of two alternative stress increments imposed on the same stress state, we put

$$(dQ_j^{(1)} - dQ_j^{(2)})(dq_j^{(1)} - dq_j^{(2)}) = (dQ_j^{(1)} - dQ_j^{(2)})(dq_j^{e(1)} - dq_j^{e(2)})$$

$$+ \; (dQ_j^{(1)} - dQ_j^{(2)})(dq_j^{ph(1)} - dq_j^{ph(2)})$$

$$+ \; (dQ_j^{(1)} - dQ_j^{(2)})(dq_j^{pf(1)} - dq_j^{pf(2)}) \; . \qquad (140)$$

Again, we have already shown that the first and second terms on the right-hand side of equation (140) are each non-negative; consequently if we can show that the last term is non-negative inequality (137) will hold for all alternative stress increments.

First, suppose that both $dQ_j^{(1)}$ and $dQ_j^{(2)}$ constitute neutral loading increments. Each lies in the limit surface, and hence $(dQ_j^{(1)} - dQ_j^{(2)})$ also lies in the limit surface. Since both $dq_j^{pf(1)}$ and $dq_j^{pf(2)}$ are normal to the limit surface, $(dq_j^{pf(1)} - dq_j^{pf(2)})$ is also normal to the limit surface. $(dQ_j^{(1)} - dQ_j^{(2)})$ and $(dq_j^{pf(1)} - dq_j^{pf(2)})$ are orthogonal vectors, and their scalar product is zero. If one stress increment, $dQ_j^{(1)}$ say, constitutes neutral loading, and the other unloading, we have

$$dq_j^{pf(1)} \; = \; \lambda^{(1)} \frac{\partial \psi}{\partial Q_j} \; , \quad dq_j^{pf(2)} \; = \; 0 \; . \qquad (141)$$

From equation (139)

$$dQ_j^{(1)} dq_j^{pf(1)} \; = \; 0 \; , \qquad (142)$$

and from the unloading condition

$$dQ_j^{(2)} dq_j^{pf(1)} \; = \; \lambda^{(1)} \frac{\partial \psi}{\partial Q_j} dQ_j^{(2)} \; \leq \; 0 \; , \qquad (143)$$

since $\lambda^{(1)} \geq 0$. Consequently

$$(dQ_j^{(1)} - dQ_j^{(2)}) dq_j^{pf(1)} \; \geq \; 0 \; , \qquad (144)$$

which implies inequality (137) for this case. Finally, if both $dQ_j^{(1)}$ and $dQ_j^{(2)}$ constitute unloading, $dq_j^{p(1)} = dq_j^{p(2)} = 0$ and (137) is again valid.

By an extension of this argument it can be shown that inequality (144) is also true when the stress point lies on a corner in the limit surface. It follows then that inequality (137) holds for *all* alternative stress increments imposed on the same stress state.

In the remainder of this volume we shall not find it necessary to distinguish between dq_j^{pf} and dq_j^{ph}, and will be able to refer to both increments as dq_j^{p} without ambiguity. The reader should bear in mind, however, that two distinct mechanisms do contribute to the plastic strain.

In the special case where the *initial yield surface* coincides completely with the limit surface it follows that the yield function is independent of the history of loading. In this case the *only* plastic strain changes which take place result from flow, and flow cannot affect the yield surface. Hardening behavior does not occur at all, and the behavior within the yield or limit surface is elastic. This type of behavior is a useful idealization in many problems, and will be referred to as *elastic, perfectly plastic behavior.*

Finally, we may note that in this section we have made use of the second postulate of Section 2.4 in its strong form. This can be circumvented by adopting inequality (129) directly; it might be referred to as a *principle of maximum plastic work for flow,* and would state that for a given *flow increment* $d\bar{q}_j^p$ the associated stress $\bar{Q}_j$, which can be chosen from all *accessible stress states*, is such that the plastic work $\bar{Q}_j d\bar{q}_j^p$ is a maximum. This modified principle of plastic work for flow is then a sufficient further requirement to ensure that inequalities (136) and (137) are satisfied under all conditions.

Alternatively, we may note that the second postulate of Section 2.4 in its strong form requires that the integral F in equation (127) should be non-positive, while a sufficient condition to show that the limit surface is convex and that the flow rule (equations 126 and 137) holds is that the contribution to the complementary work integral around a cycle made by an infinitesimal increment of the plastic strain resulting from flow is non-positive. Thus, if we do not wish to invoke the second postulate of Section 2.4 in its strong form, we can add to the second postulate in its weak form a further postulate that

$$\oint q_j^{pf} dQ_j \;\leq\; 0 \tag{145}$$

for cycles in which the change in the plastic strain due to flow q_j^{pf}, is infinitesimal. This alternative postulate does not require that F should be non-positive, but does lead to inequality (129). This additional postulate (or the principle of maximum plastic work for flow) can be regarded as playing the same role for limit surfaces as the second postulate of Section 2.4 in its weak form does for yield surfaces. This distinction need not be made for elastic, perfectly plastic materials, where the yield surface and the limit surface coincide and plastic strain increments due to hardening do not occur.

Chapter 3

GENERAL ASPECTS OF THE BEHAVIOR OF ELASTIC-PLASTIC STRUCTURES

3.1 Uniqueness and Stability in Elastic-Plastic Structures

The general framework for constitutive relations which we have
devised as a result of our assumptions in Chapter 2 is *sufficient* to guarantee uniqueness and stability in an incremental
sense for the limited class of structural problems under consideration. Using the notation of Section 1.6, let us assume
that a body has been subjected to a certain loading program, as
a result of which there exists in the body a stress field and a
strain field whose instantaneous values and previous history
are assumed known. The yield function is defined at each point
in the body; in general it will vary from one point to another
because the stress histories will differ. At some points we
may have $\phi(Q_j) < 0$, and at others $\phi(Q_j) = 0$. At some points the
stress may lie on a limit surface, so that $\psi(Q_j) = 0$.

We suppose now that we impose infinitesimally small changes
of load, body force and imposed displacement on the body. These
will be characterized by load changes $d\hat{\underline{p}}(s)$ on part of the surface S_p, body force changes $d\hat{\underline{F}}(s)$ on V, and displacement changes
$d\hat{\underline{u}}(s)$ on the remainder of the surface S_u. In order to determine the response of the structure to these increments we must
determine the stress changes $dQ_j(s)$, the strain changes $dq_j(s)$,
the displacement changes $d\underline{u}(s)$ everywhere except S_u, and the
changes in the reactions $d\underline{p}(s)$ on S_u. We observe that $d\hat{\underline{p}}$ on S_p,
$d\underline{p}$ on S_u, $d\underline{F}$, dQ_j must constitute a pertinent statically admissible field, and $d\underline{u}$ on S_p and V, $d\hat{\underline{u}}$ on S_u, dq_j must constitute
a pertinent kinematically admissible field.   The relation be-
tween dq_j and dQ_j must fall within the framework we have set out,
and we require that the stress state Q_j and the entire previous
history of stress be known at each point in order that the

constitutive relations are fully determined.

Let us assume that the solution of this incremental problem is not unique. There must then exist at least two distinct solutions which we will characterize by $dp^{(1)}$ on S_u, $du^{(1)}$ everywhere except on S_u, $dQ_j^{(1)}$, $dq_j^{(1)}$ and $dp^{(2)}$, $du^{(2)}$, $dQ_j^{(2)}$, $dq_j^{(2)}$. Each solution satisfies all the requirements demanded in the previous paragraph. Because of the linearity of the equilibrium relations the difference between the statically admissible sets contained in the two solutions is itself a statically admissible field. This difference may be characterized by forces $(dp^{(1)} - dp^{(2)})$, which are zero on S_p where $dp^{(1)} = dp^{(2)} = d\hat{p}$, zero body forces $(d\hat{F} - d\hat{F} = 0)$, and stresses $(dQ_j^{(1)} - dQ_j^{(2)})$. Similarly, because of the linearity of the strain-displacement relations, the difference between the kinematically admissible fields contained in the two solutions is also kinematically admissible. This field is characterized by displacements $(du^{(1)} - du^{(2)})$, which are zero on S_u where $du^{(1)} = du^{(2)} = d\hat{u}$, and strains $(dq_j^{(1)} - dq_j^{(2)})$. We may then apply the principle of virtual work to the two sets of differences;

$$\int_S (dp^{(1)} - dp^{(2)}) \cdot (du^{(1)} - du^{(2)}) dS$$

$$= \int_V (dQ_j^{(1)} - dQ_j^{(2)})(dq_j^{(1)} - dq_j^{(2)}) dV \quad . \tag{1}$$

The left-hand side of this equation is zero, for, when we introduce the boundary conditions on S_p and S_u,

$$\int_S (dp^{(1)} - dp^{(2)}) \cdot (du^{(1)} - du^{(2)}) dS$$

$$= \int_{S_p} (d\hat{p} - d\hat{p}) \cdot (du^{(1)} - du^{(2)}) dS$$

$$+ \int_{S_u} (dp^{(1)} - dp^{(2)}) \cdot (d\hat{u} - d\hat{u}) dS = 0 \quad . \tag{2}$$

Thus, from equation (1),

$$\int_V (dQ_j^{(1)} - dQ_j^{(2)})(dq_j^{(1)} - dq_j^{(2)})dV = 0 \ . \tag{3}$$

The stress increments $dQ_j^{(1)}$, $dQ_j^{(2)}$ are at each point in the body *alternative* stress increments imposed on the *same* stress Q_j and recorded history H_α. In such a situation, whether $\phi < 0$ or $\phi = 0$, whether $(\partial\phi/\partial Q_j)dQ_j$ is greater than, equal to or less than zero, or whether $\psi(Q_j) = 0$ and $(\partial\psi/\partial Q_j)dQ_j \leq 0$, we have (as set out in Section 2.5, equation 108 and equation 117 and Section 2.6, equation 137),

$$(dQ_j^{(1)} - dQ_j^{(2)})(dq_j^{(1)} - dq_j^{(2)}) \geq 0 \tag{4}$$

at *each* point in the body. Since the integrand is non-negative, the integral of equation (3) can be zero if and only if the integrand is identically zero at *each* point in the body. Thus

$$(dQ_j^{(1)} - dQ_j^{(2)})(dq_j^{(1)} - dq_j^{(2)}) = 0 \tag{5}$$

at each point. This term can be broken up into two parts by substituting the sum of elastic and plastic strains for the total strains. The plastic part

$$(dQ_j^{(1)} - dQ_j^{(2)})(dq_j^{p(1)} - dq_j^{p(2)}) \tag{6}$$

is either positive or zero for given $dQ_j^{(1)}$ and $dQ_j^{(2)}$. However, the elastic part

$$(dQ_j^{(1)} - dQ_j^{(2)})(dq_j^{e(1)} - dq_j^{e(2)})$$

$$= C_{jk}(dQ_j^{(1)} - dQ_j^{(2)})(dQ_k^{(1)} - dQ_k^{(2)}) \tag{7}$$

is a quadratic form and is zero if and only if $dQ_j^{(1)} = dQ_j^{(2)}$. Equation (5), therefore, can only be satisfied when

$dQ_j^{(1)} = dQ_j^{(2)}$. As a consequence, there cannot exist two distinct stress increment fields which are solutions of the incremental structural problem set out above. Because of the linearity of the relation between stress and elastic strain, $dq_j^{e(1)} = dq_j^{e(2)}$ if $dQ_j^{(1)} = dQ_j^{(2)}$, and the elastic part of the strain field is unique. Similarly, if hardening occurs at any point $dq_j^{p(1)} = dq_j^{p(2)}$ if $dQ_j^{(1)} = dQ_j^{(2)}$ because of the one to one relation between the plastic strain increment and the stress increment (Section 2.5, equation 96). However, if at any point $\psi(Q_j) = 0$ and $(\partial\psi/\partial Q_j)dQ_j = 0$, $dq_j^{p(1)}$ and $dq_j^{p(2)}$ are not necessarily equal if $dQ_j^{(1)} = dQ_j^{(2)}$. Alternatively, the plastic strain increment is not fully determined during neutral loading on the limit surface. Hence we must admit the possibility of non-uniqueness in plastic strain increments taking place as a result of flow. The displacement increments will also be non-unique. In this case, however, the difference between the two incremental strain fields $(dq_j^{(1)} - dq_j^{(2)})$ must be kinematically admissible, and hence non-uniqueness in the strain increments can arise only when the differences between the plastic strain increments due to flow in two distinct solutions are themselves kinematically admissible with the displacement increment differences $(du^{(1)} - du^{(2)})$. This type of behavior we refer to as flow in the body or structure, and we shall consider it in greater detail later in this section.

The incremental uniqueness apparent here is consistent with the concept of path dependent constitutive relations. If we wish to take an unloaded body and impose on it loads $\hat{p}$ on S_p, body forces $\hat{F}$ and displacements $\hat{u}$ on S_u, we must expect that with a path dependent material the response of the structure will depend on the manner in which the loads and displacements are applied, i.e. *on the path of external loading*. We can determine the solution only if we follow incrementally along

the loading path, starting with zero load. At the beginning of
each increment in load we must know the current loads, displacements, stresses, strains and the stress history at each
point. After the application of the increment we can update
the yield function and the constitutive relations, and are then
ready to solve another incremental problem. In this sense the
lack of uniqueness which may occur when the body undergoes flow
is not disturbing, since the recorded history H_α is not affected at any point. The yield condition and constitutive relations are not affected, and the *next* incremental problem *can be
solved*.

By solving a succession of incremental loading problems,
therefore, we can determine the response of the structure to
finite load changes. It can be noted at this point that in order to carry out this process the *initial* condition of the body
must be known. During preparation or manufacture, as has been
pointed out previously, most bodies of technological interest
are subjected to complex mechanical processes, and the body has
therefore been subjected to some previous, generally non-
homogeneous, stress history. Since we usually have no means of
characterizing this history of previous mechanical deformation,
we are invariably forced to make assumptions about the initial
state of the body before we can proceed to solve a structural
problem. It is frequently assumed that the material of which
the body is composed is in its *virgin state*.

We have so far discussed the response of the body when incremental changes in the loads take place. Let us now study
more carefully the *special* case in which *changes in the dis-
placement field* $d\underline{u}$ *take place when the loads* $\hat{\underline{p}}$ *on* S_p, *body
forces* $\hat{\underline{F}}$ *and displacements* $\hat{\underline{u}}$ *on* S_u *remain constant.* We define
such behavior as *flow of the body or structure.*

Let us assume that during a short period the stresses change from Q_j to $Q_j + dQ_j$. In this same period let the displacement changes and strain changes be $d\underline{u}$ and dq_j. Since $d\underline{u}$ ($d\underline{u} = 0$ on S_u) and dq_j are kinematically admissible, and since Q_j, $Q_j + dQ_j$ are both in equilibrium with loads $\hat{\underline{p}}$ on S_p and body forces $\hat{\underline{F}}$, by the principle of virtual work

$$\int_{S_p} \hat{\underline{p}} \cdot d\underline{u}\, dS + \int_V \hat{\underline{F}} \cdot d\underline{u}\, dV = \int_V Q_j dq_j dV \quad , \tag{8a}$$

$$\int_{S_p} \hat{\underline{p}} \cdot d\underline{u}\, dS + \int_V \hat{\underline{F}} \cdot d\underline{u}\, dV = \int_V (Q_j + dQ_j) dq_j dV \quad . \tag{8b}$$

Subtracting (8a) from (8b), we obtain

$$\int_V dQ_j dq_j dV = 0 \quad . \tag{9}$$

In the first postulate of Section 2.4 we require that

$$dQ_j dq_j \geq 0 \quad , \tag{10}$$

and hence the integral in equation (9) can be zero if and only if the integrand is zero at each point in the body. Hence

$$dQ_j dq_j = dQ_j dq_j^e + dQ_j dq_j^p = 0 \tag{11}$$

during flow. We note first that if $dQ_j \neq 0$ then $dQ_j dq_j$ is positive definite and equation (11) us not satisfied. This follows because

$$dQ_j dq_j^p \geq 0 \tag{12}$$

for all dQ_j, and

$$dQ_j dq_j^e = C_{jk} dQ_j dQ_k \tag{13}$$

is a quadratic form which is zero if and only if $dQ_j = 0$. *Hence the stresses remain constant when the loads remain constant*; this holds whether flow occurs or not, but it is certainly true when flow takes place under constant load.

Secondly, if flow is to occur we must have nonzero strain changes taking place in the structure. If the material is elastic or in a hardening regime $dq_j^p = 0$ if $dQ_j = 0$; nonzero plastic strain increments can only occur when the stress point lies on the limit surface in stress space $\psi(Q_j) = 0$ and the plastic strain increment is the result of flow. This must be true at each point in the body; hence *when flow occurs in the body the total strain increments* dq_j *are made up entirely of plastic strain increments which occur as a result of flow in the material,* i.e. *at points in the body where* $\psi(Q_j) = 0$.

We may now investigate the stress field and the plastic strain increment field when flow occurs. Under given loads, let us assume that flow can occur with two distinct solutions $Q_j^{(1)}$, $dq_j^{p(1)}$ and $Q_j^{(2)}$, $dq_j^{p(2)}$. As before, the stress field $(Q_j^{(1)} - Q_j^{(2)})$ must be in equilibrium with zero loads on S_p and zero body forces, and $(dq_j^{p(1)} - dq_j^{p(2)})$ must be kinematically admissible and derivable from a displacement increment field which is zero on S_u. Hence by the principle of virtual work,

$$0 = \int_V (Q_j^{(1)} - Q_j^{(2)})(dq_j^{p(1)} - dq_j^{p(2)})dV \quad . \tag{14}$$

Since $\psi(Q_j) \leq 0$ at every point in the body, from our earlier result which implies convexity and normality for the limit surface (Section 2.6, equation 129), we may write

$$(Q_j^{(1)} - Q_j^{(2)})dq_j^{p(1)} \geq 0 \quad , \tag{15a}$$

$$(Q_j^{(2)} - Q_j^{(1)})dq_j^{p(2)} \geq 0 \quad . \tag{15b}$$

The sum of inequalities (15a) and (15b) gives the integrand of

equation (14); hence, since the integrand is non-negative ,
equation (14) is satisfied if and only if

$$(Q_j^{(1)} - Q_j^{(2)})(dq_j^{p(1)} - dq_j^{p(2)}) \; = \; 0 \tag{16}$$

at each point in the body.

There are three conditions under which equation (16) can be
satisfied. First, if either $dq_j^{p(1)}$ or $dq_j^{p(2)}$ is nonzero and
the limit surface $\psi = 0$ is *strictly* convex (i.e. with no flat
regions in the surface), (16) can only hold if $Q_j^{(1)} = Q_j^{(2)}$. The
stresses are thus uniquely determined at these points. If the
yield surface does contain flat regions, $(Q_j^{(1)} - Q_j^{(2)})$ may be
normal to $(dq_j^{p(1)} - dq_j^{p(2)})$, thus fulfilling equation (16), if
$Q_j^{(1)}$ and $Q_j^{(2)}$ both lie on the same flat. This situation is
shown diagrammatically in Fig. 1. The third alternative is
that $dq_j^{p(1)} = dq_j^{p(2)} = 0$. In this case $Q_j^{(1)}$, $Q_j^{(2)}$ may be any
stress states which are such that $\psi < 0$.

In these last two cases, therefore, $Q_j^{(1)}$ is not necessarily
equal to $Q_j^{(2)}$, and hence the stress field is not unique in
these regions of the body. The strain increments and the displacement increments are similarly not uniquely determined.This
may appear at first sight to be a little disappointing; however,
it must be emphasized that *the loading history has not been introduced* into the discussion. It is not surprising that the

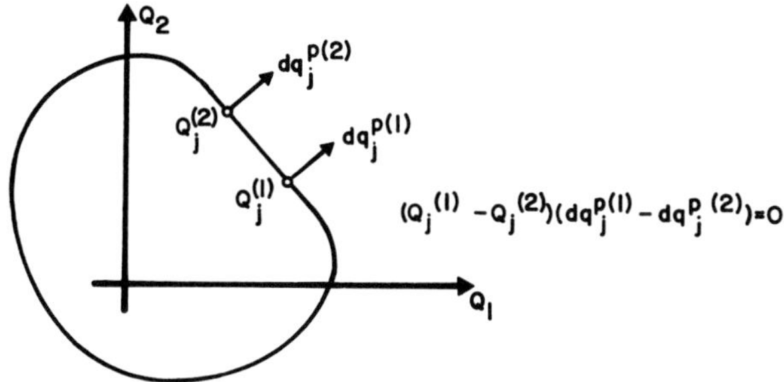

Figure 1. Limit surface with a flat region

stress field is not uniquely determined in parts of the body during flow; this simply means that the stresses in these regions are dependent on the history of loading. What is significant is that the stress field in parts of the body is unique and *independent* of the history of loading. The fact that we have limited uniqueness of the stress field implies that some other properties of the behavior of the body or structure during flow may also be independent of path. This is of great importance and we shall return to discuss it in greater length in Section 3.2 and Part III.

In the case where an increment of load takes the body or structure from one loading state where flow of the structure may occur to another loading state where flow may also occur, it is apparent that flow may occur *during* the load increment. This is just the situation encountered earlier, in discussion of the uniqueness of the incremental solution, where the plastic strain increments due to flow in the material may be non-unique. It is in fact the only incremental problem in which the strain increment field will be non-unique.

It may be noted at this point that all the results we have so far discussed depend on inequalities (4),(12) and (15). As pointed out in Sections 2.5 and 2.6, the first postulate and the second postulate in its weak form (including the amendment for flow) are sufficient conditions to establish these inequalities. Alternatively, the convexity of the elastic strain energy function and the principle of maximum plastic work(again including the amendment for flow) are sufficient conditions. The second postulate of Section 2.4 in its strong form is not essential to establish uniqueness for incremental problems.

The general framework for constitutive relations which has been developed as a result of our assumptions in Sections 2.2

through 2.6 is also *sufficient* to guarantee the *stability* of
the equilibrium of a body or structure at each instant during a
program of loading. As the loading evolves the structure passes
through a series of states in which it is in static equilibrium
with the instantaneous loads; we can show that each of these
states is one of *stable equilibrium* (except in some special
cases where neutral equilibrium may be found). Our assumptions
regarding structural and material behavior are such that states
of unstable equilibrium are not given by the analysis.

Before proceeding to establish this result for *elastic, plastic* structures we should examine briefly the concept of states
of stable, neutral and unstable equilibrium in *elastic* structures with conservative loads. The loading is conservative if
its intensity or direction at any point is not a function of
the deformation of the body. Thus if a point on a body subjec-
ted to a constant load $\underline{p}$ undergoes a displacement change $\underline{u}$ as a
result of load changes in other parts of the body, the work
done by the load $\underline{p}$ is $\underline{p}.\underline{u}$.

Let an elastic body or structure be subjected to loads $\hat{\underline{p}}(s)$
on S_p, body forces $\hat{\underline{F}}(s)$ on V and displacement $\hat{\underline{u}}(s)$ on S_u. The
given loads are conservative. Let the response of the body be
characterized by stresses $Q_j(s)$, strains $q_j(s)$, displacements
$\underline{u}(s)$ on S_p and V, and reactions $\underline{p}(s)$ on S_u. The set of forces
and stresses $\hat{\underline{p}}(s)$, $\underline{p}(s)$, $\hat{\underline{F}}(s)$, $Q_j(s)$ is certainly statically
admissible and in equilibrium; we are now concerned with the
nature of the equilibrium state.

The classical *static* approach to stability is framed in terms
of the *potential energy* U_p of the body. The potential energy of
the body is the sum of the energy stored mechanically as strain
energy and the potential energy of the conservative loads and
body forces, with the undeformed state taken as datum.The total

potential energy of the system described in the preceding paragraph is given by

$$U_p(\underline{u},\ q_j) = \int_V W(q_j)dV - \int_{S_p} \underline{\hat{p}}.\underline{u}dS - \int_V \underline{\hat{F}}.\underline{u}dV \ . \tag{17}$$

We now suppose that the system is disturbed, so that the displacements are changed to $(\underline{u}+\delta\underline{u})$ on S_p and V, and the strains become $(q_j+\delta q_j)$. The potential energy of the disturbed system is

$$U_p(\underline{u}+\delta\underline{u},q_j+\delta q_j) = \int_V W(q_j+\delta q_j)dV - \int_{S_p}\underline{\hat{p}}.(\underline{u}+\delta\underline{u})dS - \int_V\underline{\hat{F}}.(\underline{u}+\delta\underline{u})dV \ . \tag{18}$$

The equilibrium of the system is described as *stable* if for *all* $\delta\underline{u}$, δq_j,

$$U_p(\underline{u}+\delta\underline{u},\ q_j+\delta q_j) \ > \ U_p(\underline{u},\ q_j) \ . \tag{19}$$

If, for some $\delta\underline{u}$, δq_j,

$$U_p(\underline{u}+\delta\underline{u},\ q_j+\delta q_j) \ = \ U_p(\underline{u},\ q_j) \ , \tag{20}$$

the system is described as being in *neutral equilibrium* with respect to these disturbances. Finally, if for *some* $\delta\underline{u}$, δq_j,

$$U_p(\underline{u}+\delta\underline{u},\ q_j+\delta q_j) \ < \ U_p(\underline{u},\ q_j) \tag{21}$$

the system is said to be *unstable*.

It has been shown in Section 1.6 that for an elastic body in which the equilibrium and kinematic equations are linear (i.e. one in which the principle of virtual work is applicable) and which satisfies the restriction set down in Section 2.4 (equations 52 or 53), (19) holds whenever $\delta\underline{u}$, δq_j are nonzero, (20) holds if and only if $\delta\underline{u} = 0$, $\delta q_j = 0$ and (21) does not hold under any circumstances. This means that an *analysis* carried out under the assumptions set out *always* provides an equilibrium

state which is stable. This result can be obtained in a
variety of ways, and we shall give an alternative proof when
the case of elastic, plastic materials has been considered.

The simple classical result given in equations (19),(20) and
(21) must be generalized to include plastic behavior. At this
point we shall provide a mechanically motivated generalization.
In Part VI we shall see, from strict thermodynamic arguments,
that the arguments given are indeed sufficient for stability in
a rigorous sense. In order to achieve this generalization it
is necessary to consider in greater detail the concepts which
lead to the criteria of stability given in equation (19).

It is assumed that a body is stationary, and hence in equilibrium, under given loads, as illustrated for a simple model
in Figs. (2a), (3a) and (4a). This equilibrated system is then
disturbed; we imagine it is done by *disturbing forces*, which we
will conceive of as applied by an *external agency*, while the
original conservative loads remain constant. We further assume
that *as the external agency applies the disturbing forces the
structure passes through a series of equilibrium configurations*;
thus no inertia forces are necessary to ensure that dynamic requirements are satisfied. These intermediate equilibrium states
are shown diagrammatically in Figs. (2b), (3b) and (4b). When
the disturbance has been fully applied, the external agency is
abruptly removed, so that the structure is left in a disturbed
configuration, and more importantly cannot *remain* in that configuration because it is *not in equilibrium* unless the final
value of the disturbing forces applied by the external agency
happened to be zero. This state is shown diagrammatically in
Figs. (2c), (3c) and (4c).

The stability of the original equilibrium configuration is
usually characterized by the behavior of the structure after

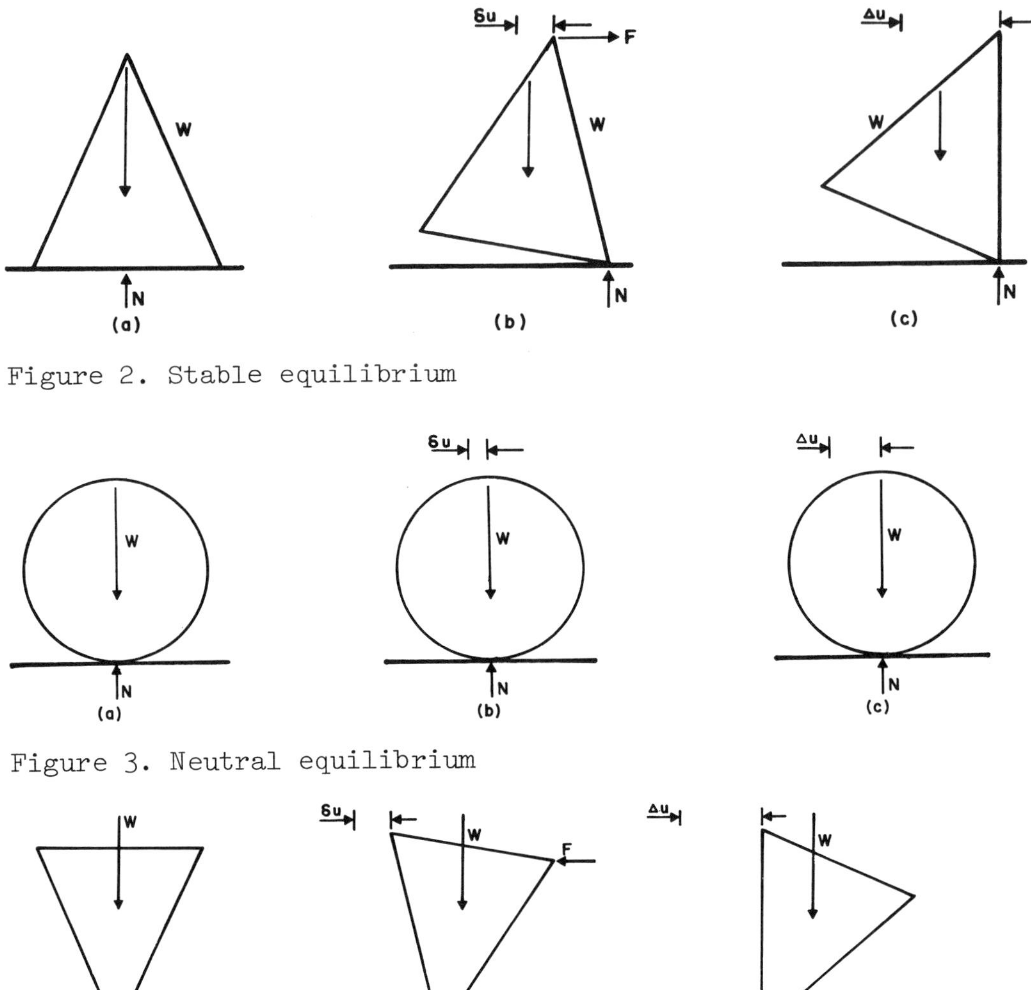

Figure 2. Stable equilibrium

Figure 3. Neutral equilibrium

Figure 4. Unstable equilibrium

the external agency is abruptly removed. In the special case,
shown by Fig.(3c), where the structure is in equilibrium after
the external agency is removed, the structure will remain in
its disturbed configuration. The original state is said to be
one of *neutral equilibrium*.

In both other cases the structure will accelerate when the
external agency is removed. In one case, shown in Fig.(2c), the

initial motion tends to return the structure to the configuration it held before the external agency was applied. In the other case (Fig.4c), the tendency is to move away from the original configuration. If the system is elastic the *stable* structure shown in Fig.(2c) will oscillate about its original equilibrium configuration. During this oscillation the kinetic energy in the system will always be bounded from above by the increase in potential energy between the initial equilibrium state and the state just after the external agency is removed (i.e. the difference in potential energy between Figs.2a and 2c). It is just this change in potential energy which appears in equation (19).

Alternatively, if the change in potential energy is negative, as between Figs.(4a) and (4c) and as given by equation (20), the kinetic energy cannot be bounded. It may increase indefinitely, or it may be limited when the structure finds an alternative stable equilibrium position about which it can oscillate. However, in the absence of an adjacent stable equilibrium position, the maximum value of the kinetic energy can exceed the difference in the potential energy before and after the disturbance caused by the external agency.

Qualitatively, at least, these concepts of stability can equally well be applied to inelastic systems. The behavior after the removal of the external agency will be different; in particular, in the stable case the structure will no longer oscillate about its original configuration if plastic strain takes place during the application of the external agency or after its removal. Nevertheless the kinetic energy will still be bounded, and the new equilibrium position will not be far removed from the original one. In the unstable case the kinetic energy is not bounded, and the structure will either continue to accelerate or move to an adjacent stable equilibrium position.

The other important difference between the elastic and inelastic cases is that we can no longer categorize these two types of behavior in terms of the potential energy before and after the action of the external agency in the inelastic case. However, we can do so in terms of the *intermediate behavior*, i.e. the behavior during the period when the external agency is applied. Remembering that the external agency is such that *equilibrium is maintained during the application of the disturbing force*, it is clear that in the *stable* case (Fig. 2b) the external agency does *positive work* as it is applied, in the *neutral equilibrium* case it does *zero* work (Fig.3b), and in the *unstable* case (Fig.4b) it does *negative* work.

We are thus now in a position to formulate a stability criterion for elastic-plastic structures which is an acceptable generalization of the criterion for elastic structures. Consider a structure subjected to instantaneous loads $\hat{\underline{p}}(s)$ on S_p, body forces $\hat{\underline{F}}(s)$ on V and displacements $\hat{\underline{u}}(s)$ on S_u. The history of loading and the initial conditions were given, so that the solution to the problem and the stress history at each point is known. Let the solution at this instant be characterized by stresses $Q'_j(s)$, strains $q'_j(s)$, reactions $\underline{p}'(s)$ on S_u and displacements $\underline{u}'(s)$ on S_p and V. In contrast to the elastic case outlined above, the primes have been added to the quantities characterizing the solution to avoid confusion during integration. Let $\delta\hat{\underline{p}}(s)$ on S_p, $\delta\hat{\underline{F}}(s)$ on V, be the disturbing forces applied by the external agency. Their initial value is zero, and their final values are given by $\Delta\hat{\underline{p}}(s)$, $\Delta\hat{\underline{F}}(s)$. The displacements on S_u remain constant during this period (i.e. $\delta\hat{\underline{u}} = 0$ on S_u). We assume that the structural response is characterized by stresses $Q_j' + \delta Q_j$, strains $q_j' + \delta q_j$ and displacements $\underline{u}' + \delta\underline{u}$; the final values of δQ_j, δq_j, $\delta\underline{u}$ are given by ΔQ_j, Δq_j, $\Delta\underline{u}$.

The stability criterion is now written as follows; *if for all disturbing forces* $\Delta\underline{\hat{p}}$ *on* S_p, $\underline{\hat{F}}$ *on* V,

$$\int_{S_p} dS \int_{\underline{u}'}^{\underline{u}'+\Delta\underline{u}} \delta\underline{\hat{p}}.d\underline{u} + \int_{V} dV \int_{\underline{u}'}^{\underline{u}'+\Delta\underline{u}} \delta\underline{\hat{F}}.d\underline{u} \geq 0 \tag{22}$$

the structure is in stable or neutral equilibrium. If a strict inequality appears for all $\Delta\underline{\hat{p}}$, $\Delta\underline{\hat{F}}$ in (22) the structure is always stable.

First, we can readily show that the expression on the left-hand side of (22) is equivalent to the *potential energy change in elastic structures*. We put

$$\delta\underline{\hat{p}} = (\underline{\hat{p}} + \delta\underline{\hat{p}}) - \underline{\hat{p}} \; . \tag{23}$$

Equating work done by the external loads to energy stored in the elastic structure,

$$\int_{S_p} dS \int_{\underline{u}'}^{\underline{u}'+\Delta\underline{u}} (\underline{\hat{p}}+\delta\underline{\hat{p}}).d\underline{u} + \int_{V} dV \int_{\underline{u}'}^{\underline{u}'+\Delta\underline{u}} (\underline{\hat{F}}+\delta\underline{\hat{F}}).d\underline{u}$$

$$= \int_{V} dV \int_{q_j'}^{q_j'+\Delta q_j} (Q'+\delta Q_j)dq_j$$

$$= \int_{V} W(q_j'+\Delta q_j)dV - \int_{V} W(q_j')dV \; . \tag{24}$$

In addition,

$$\int_{S_p} dS \int_{\underline{u}'}^{\underline{u}'+\Delta\underline{u}} \underline{\hat{p}}.d\underline{u} + \int_{V} dV \int_{\underline{u}'}^{\underline{u}'+\Delta\underline{u}} \underline{\hat{F}}.d\underline{u}$$

$$= \int_{S_p} \hat{\underline{p}} \cdot (\underline{u}' + \Delta\underline{u}) \, dS + \int_{V} \hat{\underline{F}} \cdot (\underline{u}' + \Delta\underline{u}) \, dV - \int_{S_p} \hat{\underline{p}} \cdot \underline{u}' \, dS - \int_{V} \hat{\underline{F}} \cdot \underline{u}' \, dS \quad . \tag{25}$$

Substitution of equations (24) and (25) into (22) leads to the same criterion of stability that was given in equations (17) and (18) for stable and neutral equilibrium in elastic bodies.

We now consider bodies or structures in which the equilibrium and kinematic equations are linear (implying that the principle of virtual work is applicable) and materials which satisfy the assumptions set out in this chapter.

Using the principle of virtual work for the statically admissible set $\delta\hat{\underline{p}}$, $\delta\hat{\underline{F}}$, δQ_j and the kinematically admissible set $d\underline{u}$, δq_j, we have

$$\int_{S_p} \delta\hat{\underline{p}} \cdot d\underline{u} \, dS + \int_{V} \delta\hat{\underline{F}} \cdot d\underline{u} \, dV = \int_{V} \delta Q_j \, dq_j \, dV \quad . \tag{26}$$

Integrating this equation over the period when the external agency is applied, equation (26) becomes

$$\int_{S_p} dS \int_{\underline{u}'}^{\underline{u}'+\Delta\underline{u}} \delta\hat{\underline{p}} \cdot d\underline{u} + \int_{V} dV \int_{\underline{u}'}^{\underline{u}'+\Delta\underline{u}} \delta\hat{\underline{F}} \cdot d\underline{u} = \int_{V} dV \int_{q_j'}^{q_j'+\Delta q_j} \delta Q_j \, dq_j \quad . \tag{27}$$

The left-hand side of this equation is the expression appearing in the stability criterion (22). We can now show that the integrand of the expression on the right-hand side of (27) is nonnegative. This can be done in two ways. The first, which we refer to as *stability in the small,* is to treat $\Delta\hat{\underline{p}}$, $\Delta\hat{\underline{F}}$ as infinitesimally small. ΔQ_j, Δq_j are then assumed to be infinitesimally small in consequence. We can then carry out the integration on the right-hand side of equation (27) using the mean value of δQ_j, which is $\frac{1}{2} \Delta Q_j$. Thus

$$\int_{q_j'}^{q_j'+\Delta q_j} \delta Q_j \, dq_j = \frac{1}{2} \Delta Q_j \Delta q_j \quad . \tag{28}$$

The strain increment Δq_j occurs as a consequence of the stress increment ΔQ_j; the first stability postulate (Section 2.4, equation 53) requires that

$$\Delta Q_j \Delta q_j \geq 0 \quad . \tag{29}$$

In fact $\Delta Q_j \Delta q_j = 0$ only when $\Delta Q_j = 0$; this can occur in the trivial case when the load increment is zero, but it can also occur when flow is taking place in the structure under *constant load*. Apart from this case, a comparison of equations (27), (28), (29) and the stability criterion (22) shows that the *structures we are considering are always in stable equilibrium* for infinitesimally small disturbances as a consequence of the first postulate. The exception is the case of flow in the structure, when we have *neutral equilibrium*.

Alternatively, we need not limit the magnitude of $\Delta \hat{\underline{p}}$, $\Delta \hat{\underline{F}}$, and we can still show *stability in the large*. Putting

$$Q_j = Q_j' + \delta Q_j$$

and

$$Q_j^* = Q_j' + \Delta Q_j \quad , \tag{30}$$

the right-hand side of equation (27) becomes

$$\int_V dV \int_{q_j'}^{q_j^*} (Q_j - Q_j') \, dq_j \quad . \tag{31}$$

As a consequence of the first postulate and the second postulate in its strong form (Section 2.4, inequality 71)

$$\int_{q_j'}^{q_j^*} (Q_j - Q_j')dq_j \;\geq\; 0 \tag{32}$$

along *any* stress path. Substituting (31) and (32) into (27) and
(22), we see again that the structure is stable for disturbances of any magnitude (again with the exception of a structure
undergoing flow, which is in a state of neutral equilibrium).
In this respect the second postulate in its strong form en-
hances the stability of the system beyond that provided by the
first.

It was pointed out in Sections 2.5 and 2.6 that the convexity of the elastic strain energy function and the principle of
maximum plastic work are together sufficient to establish the
first postulate of Section 2.4; they are consequently sufficient for stability in the small. The second postulate of
Section 2.4 in its weak form does not contribute to stability
(except insofar as it is sufficient together with the convexity
of the strain energy function to justify the first postulate).

The uniqueness and stability exhibited by structural problems as a consequence of our assumptions is a *desirable* feature
of the theory in the sense that it removes many difficulties
which might otherwise be encountered in the computations. However, it must again be emphasized that it is *justifiable* only
if the physical problem warrants it; the theory cannot be used
in situations where non-uniqueness or instability (or both) are
expected on physical grounds because it is incapable of describing these phenomena.

It should be noted that the uniqueness and stability properties in the structural problems we are discussing arise partly
because of the linearity of the equilibrium and strain-

displacement relations (implicit in the use of the principle of
virtual work) and partly because of the assumptions regarding
the material. It is convenient to divide these concepts into
geometrical stability and *intrinsic stability*. Uniqueness and
stability may be lost in a real structure because the equilibrium (in particular) and kinematic equations are not linear.
The most common phenomena in this class are the buckling of
bars, plates and shells as a result of geometry changes leading
to nonlinear equilibrium equations. Alternatively, in some
structures the small displacement assumptions may be applicable,
but the material may not be intrinsically stable, and as a result the whole structure may be unstable under certain conditions. Mild steel (in the vicinity of the upper yield stress),
and concrete and soils under some circumstances are materials
which fall into this category. One consequence of the assumptions which have been used to construct the framework for plastic constitutive relations is that problems solved using these
relations will never exhibit intrinsic instability; if we relax
the assumptions regarding linearity of the equilibrium and
kinematic relations we may, of course, encounter geometric instability.

3.2 Generalized Loads and the Behavior in Load Space

In discussing the stability of structures it is observed that
the *structure* exhibits the same general mechanical characteristics as the *material* of which it is composed. This is not in
fact surprising, since the homogeneously stressed and strained
body on which experiments are carried out to determine constitutive relations is in fact just a simple structure. It is instructive to investigate this relation between the structure
and the material in greater detail. However, it is convenient
to reduce the generality of the structural problem to some

extent to simplify the discussion. We have already seen the importance of the path of loading in structural problems. In the most general case the loads $\hat{\underline{p}}$ on the surface and the body forces $\hat{\underline{F}}$ throughout the volume can be varied in an infinite number of ways. In practical situations, on the other hand, the loading on a body is controlled by a limited number of independent parameters. We shall assume, therefore, that the structural problem is set in the following way. Let the *displacements* $\hat{\underline{u}}$ *on* S_u *be given to be zero*. Let the forces $\hat{\underline{p}}(s)$ on S_p be written as

$$\hat{\underline{p}}(s) = \Gamma_1 \, \underline{\pi}_1(s) + \Gamma_2 \, \underline{\pi}_2(s) + \ldots + \Gamma_\mu \, \underline{\pi}_\mu(s) \tag{33}$$

where Γ_1, Γ_2, $\ldots$, Γ_μ are independent parameters with the dimensions of force or couple, and $\underline{\pi}_1(s)$, $\underline{\pi}_2(s)$, $\ldots$, $\underline{\pi}_\mu(s)$ are vector functions indicating the distribution of the load. These functions may include delta functions to characterize point loads. Further, let the body forces be written as

$$\hat{\underline{F}} = \Gamma_{\mu+1} \, \underline{f} \tag{34}$$

where $\Gamma_{\mu+1}$ is again a scalar with the dimensions of force and $\underline{f}$ is a vector function defined, where appropriate, at every point in the body. This representation covers a wide variety of problems, examples of which are given in Fig. 5.

Having the loads defined by a finite number of load parameters Γ_α $(\alpha = 1, \ldots, \mu+1)$, we may represent any state of loading by a point in an $(\mu+1)$-dimensional *load space*, as shown in Fig. 6. The trajectory of the *load point* as changes occur is the *load path*. A straight line load path radially outwards from the origin will be termed a *proportional loading path*, since the Γ_α always bear the same ratio to each other.

If the body is subject to displacements u*(s), not

necessarily caused by the loads but with $\underline{u}^* = 0$ on S_u, the
work done by the constant loads during the deformation is

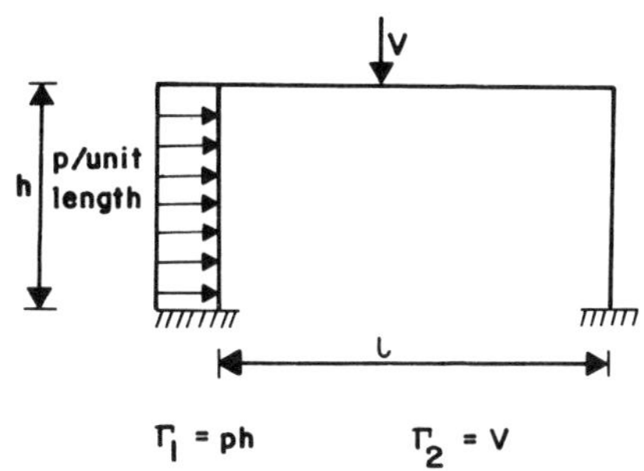

Figure 5(a) Generalized loads for portal frame

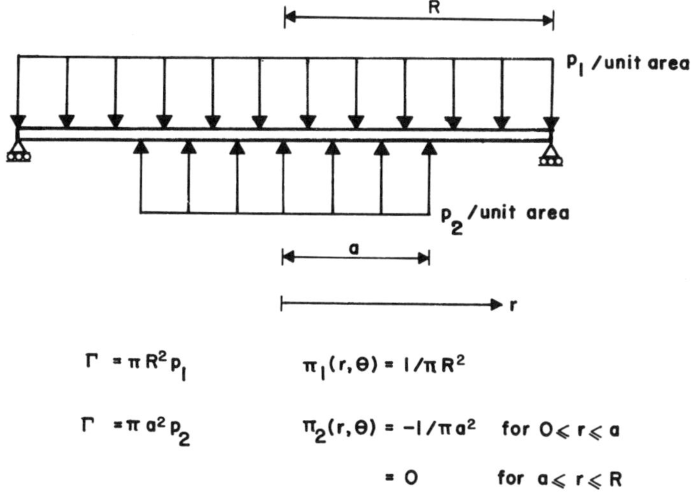

$\Gamma = \pi R^2 P_1$ $\pi_1(r,\theta) = 1/\pi R^2$

$\Gamma = \pi a^2 P_2$ $\pi_2(r,\theta) = -1/\pi a^2 \quad \text{for } 0 \leqslant r \leqslant a$

$\qquad\qquad\qquad\qquad = 0 \qquad\quad \text{for } a \leqslant r \leqslant R$

Figure 5(b).Symmetrically loaded simply supported circular plate

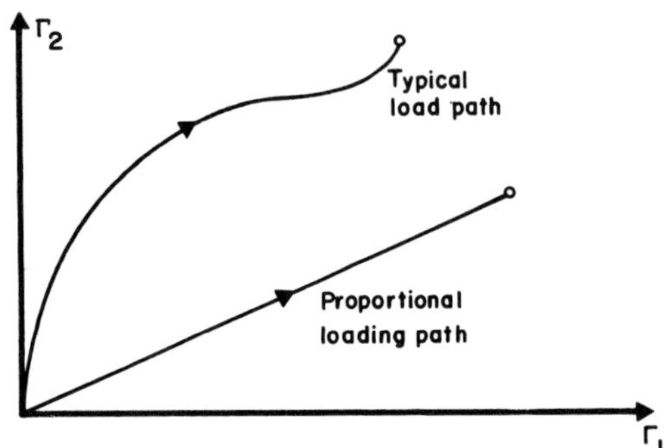

Figure 6. Load space

$$\int_S \underline{p} \cdot \underline{u}^* dS + \int_V \underline{F} \cdot \underline{u}^* dV = \Gamma_1 \int_S \underline{\pi}_1 \cdot \underline{u}^* dS + \Gamma_2 \int_S \underline{\pi}_2 \cdot \underline{u}^* dS +$$

$$\dots + \Gamma_\mu \int_S \underline{\pi}_\mu \cdot \underline{u}^* dS + \Gamma_{\mu+1} \int_V \underline{f} \cdot \underline{u}^* dV \quad . \tag{35}$$

We introduce *generalized displacements* ξ_α^* ($\alpha = 1, \dots, \mu+1$) such that

$$\xi_\alpha^* = \int_S \underline{\pi}_\alpha \cdot \underline{u}^* dS \qquad \alpha = 1, \dots, \mu \tag{36a}$$

$$\xi_{\mu+1}^* = \int_V \underline{f} \cdot \underline{u}^* dV \tag{36b}$$

These equations indicate that the generalized displacements are weighted averages of the actual displacements, with the weighting functions being the load distribution functions. The principle of virtual work now becomes

$$\int_S \hat{\underline{p}} \cdot \underline{u}^* dS + \int_V \hat{\underline{F}} \cdot \underline{u}^* dV = \Gamma_\alpha \xi_\alpha^* = \int_V Q_j q_j^* dV \quad , \tag{37}$$

where the summation rule is adopted for the Greek subscripts. Equation (37) can be written provided that the forces defined by Γ_α and the stresses Q_j are statically admissible, and the displacements represented by ξ_α^* and the strains q_j^* are kinematically admissible.

At the initiation of a structural problem, when the loads

are zero, we usually assume that the material is in its virgin state, since we have no means of determining this previous stress history at each point in the body. The generalized strains ξ_α are then also initially zero.

We may observe first that the stability postulates hold in the load space. Consider first a body subjected to loads Γ_α^a in which, as a consequence of a succession of incremental solutions along the load path, the generalized displacements, stresses and strains are ξ_α^a, Q_j^a, q_j^a. Let an infinitesimal increment of load $d\Gamma_\alpha$ be added to these loads, producing in turn infinitesimal increments $d\xi_\alpha$, dQ_j, dq_j. By the principle of virtual work

$$d\Gamma_\alpha d\xi_\alpha \;=\; \int_V dQ_j dq_j dV \;\;. \tag{38}$$

However, the first postulate of stability requires that $dQ_j dq_j \geq 0$; it follows therefore that

$$d\Gamma_\alpha d\xi_\alpha \;\geq\; 0 \;\;. \tag{39}$$

By an argument identical to that used in Section 2.4, we may then show that along a *straight line path* in the load space between Γ_α^a and Γ_α^b,

$$(\Gamma_\alpha^b - \Gamma_\alpha^a)(\xi_\alpha^b - \xi_\alpha^a) \;\geq\; 0 \;\;, \tag{40a}$$

$$\int_{\xi_\alpha^a}^{\xi_\alpha^b} (\Gamma_\alpha - \Gamma_\alpha^a) d\xi_\alpha \;\geq\; 0 \;\;, \tag{40b}$$

$$\int_{\Gamma_\alpha^a}^{\Gamma_\alpha^b} (\xi_\alpha - \xi_\alpha^a) d\Gamma_\alpha \;\geq\; 0 \;\;. \tag{40c}$$

If we impose a load change from Γ_α^a to Γ_α^b along *any path*, causing changes in generalized displacement from ξ_α^a to ξ_α^b, changes in stress from Q_j^a to Q_j^b and changes in strain from q_j^a to q_j^b, we have, by the principle of virtual work,

$$\int_{\xi_\alpha^a}^{\xi_\alpha^b} (\Gamma_\alpha - \Gamma_\alpha^a)d\xi = \int_V dV \int_{q_j^a}^{q_j^b} (Q_j - Q_j^a)dq_j \tag{41}$$

From the extended result using both the first and second postulates (Section 2.4, equation 71)

$$\int_{q_j^a}^{q_j^b} (Q_j - Q_j^a)dq_j \geq 0 \ . \tag{42}$$

It follows that

$$\int_{\xi_\alpha^a}^{\xi_\alpha^b} (\Gamma_\alpha - \Gamma_\alpha^a)d\xi_\alpha \geq 0 \ . \tag{43}$$

Along any such path in the load space we may write the identity

$$\int_{\xi_\alpha^a}^{\xi_\alpha^b} (\Gamma_\alpha - \Gamma_\alpha^a)d\xi_\alpha + \int_{\Gamma_\alpha^a}^{\Gamma_\alpha^b} (\xi_\alpha - \xi_\alpha^a)d\Gamma_\alpha = (\Gamma_\alpha^b - \Gamma_\alpha^a)(\xi_\alpha^b - \xi_\alpha^a) \ . \tag{44}$$

Suppose now that the load change is in fact a *cycle* in the load space, i.e. $\Gamma_\alpha^b = \Gamma_\alpha^a$. The right-hand side of equation (44) is zero, so that as a consequence of inequality (43)

$$\oint (\xi_\alpha - \xi_\alpha^a)d\Gamma_\alpha = \oint \xi_\alpha d\Gamma_\alpha \leq 0 \ . \tag{45}$$

It would normally be expected that the first loading increment applied to the structure would lead to an elastic response, with strain changes throughout the body being elastic. Since we are concerned with an incremental problem involving small changes in load, generalized displacements, stress and strain, this problem does not involve the initial stresses and strains. Suppose that the elastic incremental problem gives

$$d\xi_\alpha = K_{\alpha\beta} d\Gamma_\beta \quad . \tag{46}$$

Using this solution, we may define *elastic generalized displacements* ξ_α^e given by

$$\xi_\alpha^e = K_{\alpha\beta} \Gamma_\beta \quad . \tag{47}$$

The difference between the actual generalized displacements ξ_α and ξ_α^e can then be termed the *plastic generalized displacements* ξ_α^p. Provided we remain within a yield surface $\Phi(\Gamma_\alpha) = 0$ in the load space, the generalized displacement changes are elastic (meaning, again, that no plastic strain increments occur anywhere in the body). When *loading* in the load space occurs, $(\partial\Phi/\partial\Gamma_\alpha)d\Gamma_\alpha > 0$, plastic strain increments occur and $d\xi_\alpha^p$ is not zero. Note, however, that $d\xi_\alpha^p$ will in general be made up of both elastic and plastic *strain* increments in the body. Making use of arguments identical to those used for the yield surface in stress space, we can deduce from the second stability postulate in load space (inequality 45) that $\Phi(\Gamma_\alpha)$ is convex and that $d\xi_\alpha^p$ is normal to the yield surface at a nonsingular point (Fig. 7). The yield surface will change as a result of a plastic displacement increment, as in the stress space, and it is clear that the behavior is qualitatively identical. Results of this type, concerning the yield surface in load space and the plastic displacement increment vector $d\xi_\alpha^p$ are of great

importance in studying the form of plastic constitutive relations. We shall return to this problem in Part VI. They are
complex in detail, however, and do not contribute very greatly
to the solution of problems.

Of much greater immediate importance is the concept of a
limit surface in *load space*. Suppose that the material is such
that there is a limit surface in stress space, so that flow in
the structure can be expected under certain conditions. Imagine
as shown diagrammatically in Fig. 8, that a limit surface $\Psi(\Gamma_\alpha)$
= 0 is established in the load space by carrying out an infinite number of proportional loading tests on identical structures. The load point moves along a radial straight line, and
the point where flow occurs identifies the limit surface.

We can first show that this limit surface is *convex* and that
a *flow rule* holds for the increments in $d\xi_\alpha^p$ during flow. We
have established earlier, in Section 3.1, that during flow with

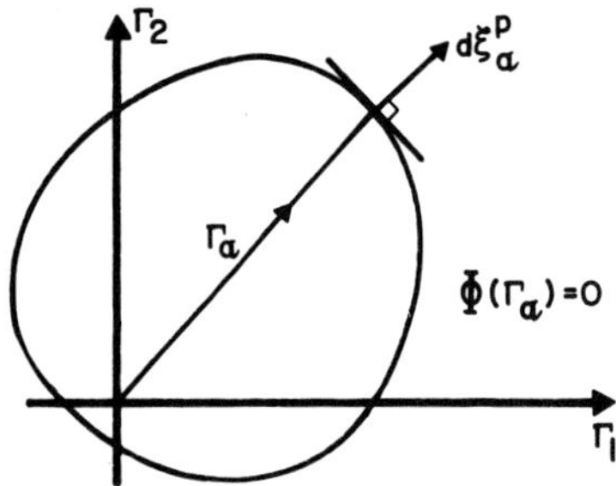

Figure 7. Yield surface in load space

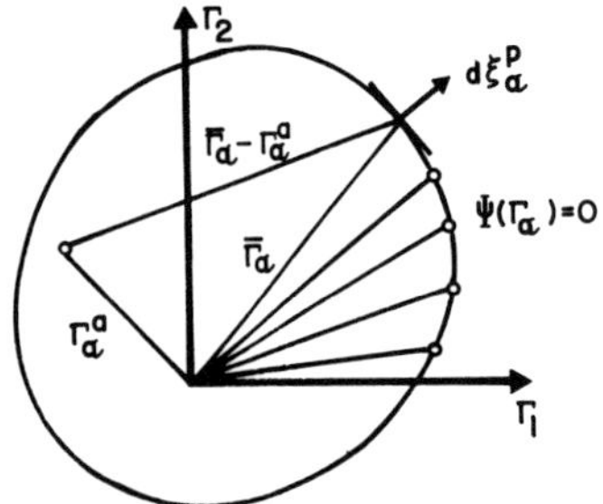

Figure 8. Limit surface generated by proportional loading

$d\xi_\alpha^p \neq 0$, $d\Gamma_\alpha = 0$, $d\xi_\alpha^p$ corresponds to a strain increment field
in which the elastic strain changes are zero. Let flow take
place under loads $\bar{\Gamma}_\alpha$ such that $\Psi(\bar{\Gamma}_\alpha) = 0$. Let the plastic generalized displacement increments be $d\bar{\xi}_\alpha^p$, the stresses be $\bar{Q}_j$
(note that $\psi(\bar{Q}_j) \leq 0$) and the plastic strain increments be
$d\bar{q}_j^p$. Consider another load state Γ_α^a, also reached by a proportional loading path, such that $\Psi(\Gamma_\alpha^a) \leq 0$, so that it lies on or
within the limit surface. Let the stresses in this structure
be Q_j^a, and note also that $\psi(Q_j^a) \leq 0$. As a result of the linearity of the equilibrium equations, the loads $(\bar{\Gamma}_\alpha - \Gamma_\alpha^a)$ and the
stresses $(\bar{Q}_j - Q_j^a)$ are statically admissible. Hence by the principle of virtual work,

$$(\bar{\Gamma}_\alpha - \Gamma_\alpha^a)d\bar{\xi}_\alpha^p = \int_V (Q_j - Q_j^a)d\bar{q}_j^p dV \quad . \tag{48}$$

However, as shown in the discussion on limit surfaces in stress
space (Section 2.6, equation 129),

$$(\bar{Q}_j - Q_j^a)d\bar{q}_j^p \geq 0 \tag{49}$$

for $d\bar{q}_j^p \neq 0$ and for $d\bar{q}_j^p = 0$. From (48) and (49) it follows,
therefore, that

$$(\bar{\Gamma}_\alpha - \Gamma_\alpha^a)d\bar{\xi}_\alpha^p \geq 0 \quad . \tag{50}$$

We may again use the elimination argument used in the discussion of yield and limit surfaces to show that (50) can hold for
any $\bar{\Gamma}_\alpha$, Γ_α^a only if $\Psi(\Gamma_\alpha)$ is a *convex surface*, and if $d\bar{\xi}_\alpha^p$ is
normal to the limit surface at a nonsingular point, and between
adjacent normals at a singular point.

When flow takes place in a structure under loads Γ_α it is
always true that $\Gamma_\alpha d\xi_\alpha^p$ is positive. This follows from an internal and external energy dissipation balance,

$$\Gamma_\alpha d\xi^p_\alpha = \int_V Q_j dq^p_j dV \quad .$$

(51)

Since the limit surface in the stress space is convex and contains the origin, and since the flow rule applies,

$$Q_j dq^p_j > 0 \quad \text{for} \quad dq^p_j \neq 0$$

(52)

during flow. This result is confirmed by the convexity of the limit surface in load space and the flow rule, which itself implies that $\Gamma_\alpha d\xi^p_\alpha > 0$ since the limit surface contains the origin.

We next ask the following question. Is it possible, either by altering the path of loading or by changing the initial stress distribution, for the structure to flow at a load Γ_α which is such that $\Psi(\Gamma_\alpha) \neq 0$? Alternatively, is it possible to generate a limit surface different to that shown in Fig. 8 by following nonproportional loading paths? To answer this question, we assume that flow may occur at loads such that $\Psi(\Gamma_\alpha) \neq 0$. Consider first flow at a load point Γ^b_α which lies outside the limit surface generated by proportional loading. Let the associated stresses be Q^b_j during flow, with $\psi(Q^b_j) \leq 0$. We may rewrite equation (48) with Γ^b_α, Q^b_j substituted for Γ^a_α, Q^a_j, and $\bar{\Gamma}_\alpha$ again representing any load on the limit surface at which flow occurs. Following the same argument that led to inequality (50),

$$(\bar{\Gamma}_\alpha - \Gamma^b_\alpha) d\bar{\xi}^p_\alpha \geq 0 \quad .$$

(53)

Now, as shown diagrammatically in Fig. 9, it is clearly possible to choose $\bar{\Gamma}_\alpha$ such that (53) is violated when Γ^b_α is outside the limit surface. In fact, it was not necessary for us to require that flow occurs under loads Γ^b_α; inequality (53) indicates that loads such that $\Psi(\Gamma_\alpha) \geq 0$ cannot be achieved without

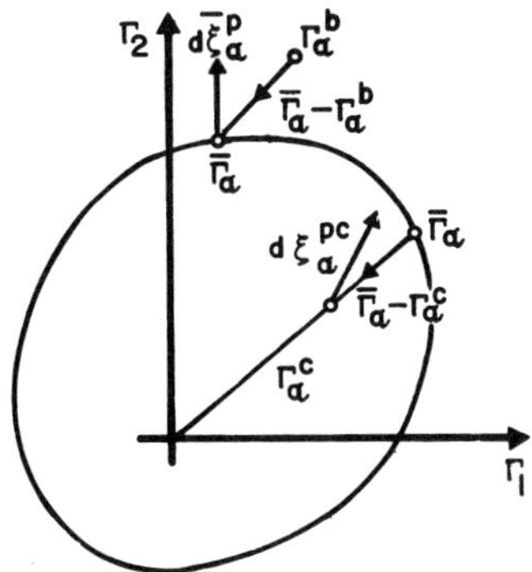

Figure 9. Uniqueness of the limit surface in load space

contradicting our assumptions.

Assume next that flow occurs at loads Γ_α^c such that Γ_α^c is within the limit surface of Fig. 8. Let the stresses be Q_j^c, and the plastic generalized displacement and plastic strain increments be $d\xi_\alpha^{pc}$ and dq_j^{pc}. As before, let $\bar{\Gamma}_\alpha$ be a load point on the limit surface, with stresses $\bar{Q}_j$ obtained as a result of proportional loading. Using the principle of virtual work in the same manner as before, we can write

$$(\Gamma_\alpha^c - \bar{\Gamma}_\alpha) d\xi_\alpha^{pc} = \int_V (Q_j^c - \bar{Q}_j) dq_j^{pc} dV \geq 0 \ . \tag{54}$$

We may choose $\bar{\Gamma}_\alpha$ as the load which lies on the yield surface where the radial line through the origin and Γ_α^c cuts the limit surface, as shown diagrammatically in Fig. 9. Remembering that $\Gamma_\alpha^c d\xi_\alpha^{pc} > 0$, it is clear then that

$$(\Gamma_\alpha^c - \bar{\Gamma}_\alpha) d\xi_\alpha^{pc} \leq 0 \tag{55}$$

if Γ_α^c lies inside the limit surface. This is in direct contradiction to inequality (54).

Thus we arrive at the conclusion that *flow in the structure can occur only on a convex limit surface* $\Psi(\Gamma_\alpha) = 0$, *and that this limit surface does not depend either on the path of loading or on the initial stress and strain distribution in the body.*

This is an extremely important practical result for a number of reasons. First, the limit surface in many structures indicates the maximum load under which the structure is serviceable and will perform its function. Secondly, the independence of the limit surface on the path of loading removes difficulties in practical structures where the exact path of loading may not be known or where initial stresses cannot be determined. Third, we shall see later that the determination of the limit surface is a comparatively simple problem in plasticity; one which is certainly a great deal simpler than incremental solutions for load increments within the limit surface in load space.

We have explored the relations between generalized loads and generalized displacements to a degree sufficient to demonstrate that (provided that the equilibrium and kinematic relations are linear) the general framework of the relations between Γ_α and ξ_α are *identical* with the general framework of the relations between generalized stress Q_j and generalized strain q_j. This provides an internal consistency within the theory which is extremely important in practical applications.

Consider, for example, a beam problem. The generalized stresses appropriate to this problem will be bending moment M, shear force S and axial force N. Assuming that shear deformations are negligible, the appropriate generalized strains are curvature κ and axial strain ε. If the development of this chapter is applied directly to this problem, we make phenomenological assumptions regarding the relations between κ, ε and M, N and set up a framework for the constitutive relations. Measurements on beam elements must then provide the specific form for this framework.

On the other hand we could also consider the determination of the relations between κ, ε and M, N as another *structural*

problem on a different level; a level in which the generalized
stresses and strains are the three-dimensional continuum stress
and strain tensors σ_{ij} and ε_{ij}, and M, N and κ, ε are the generalized loads and generalized displacements. Phenomenological
assumptions and measurements are then made for the relations
between σ_{ij} and ε_{ij}, and the constitutive relations for the beam
problem are arrived at by computation.

The internal consistency of the theory assures us that we
arrive at the same framework for the constitutive equations for
the beam problem by either approach. Depending on specific assumptions, the details may be different, but the essential features are not. In more general terms, the theory of plasticity
embraces a hierarchy of structural problems in which the behavior of the elements of which it is supposed that the structure
is composed is reflected in the behavior of the entire structure
or the entire collection of elements.

This viewpoint can aid our thinking in the construction of
the framework of constitutive relations in many ways. For example, consider the nature of the yield function $\Phi(\Gamma_\alpha)$ in load
space. Let us suppose that under a given set of loads Γ^*_α the
stress field in the structure is $Q^*_j(s)$. It is evident that
$\Phi(\Gamma^*_\alpha) \leq 0$, $\phi(Q^*_j) \leq 0$. It is further evident that if the loads
are changed by $\Delta\Gamma_\alpha$ to a new value $(\Gamma^*_\alpha + \Delta\Gamma_\alpha)$ the load point will
remain within the yield surface in load space provided that the
stress remains within the yield surface in stress space at *each*
point within the structure. If the stress changes resulting
from $\Delta\Gamma_\alpha$ are *elastic*, we may introduce an *influence function*
$I_{j\alpha}(s)$ which gives the stress change ΔQ_j at the point s in the
structure or body as a linear function of the load change,

$$\Delta Q_j(s) = I_{j\alpha}(s)\Delta\Gamma_\alpha \; . \tag{56}$$

Thus the behavior of the entire structure will continue to be
elastic provided that

$$\phi(Q_j + \Delta Q_j) \;=\; \phi(Q_j + I_{j\alpha}\Delta\Gamma_\alpha) \leq 0 \tag{57}$$

at *each* point in the body. It is now clear that we may plot
the surface (with the load point Γ_α as origin)

$$\phi(Q_j + I_{j\alpha}\Delta\Gamma_\alpha) \;=\; 0 \tag{58}$$

for each point in the body in *load space*. The yield surface in
load space, $\Phi(\Gamma_\alpha) = 0$, must be the *inner envelope* of all the
individual yield surfaces given in equation (58). Just how
many points in the body will contribute to the inner envelope
will depend on the body and the loading.   It may be a very
large number, or it may be small if there are a limited number
of critical points in the body at which the stresses have peak
values. In the limit of the former case we can expect a yield
surface with a continuously turning tangent; in the latter the
yield surface is made up of the envelope of a limited number of
instantaneously independent yield surfaces, and corners can be
expected. If the surface $\Phi(\Gamma_\alpha) = 0$ may be considered as a yield
surface in stress space for a structural problem on a *different
level* of approximation, it is then evident that the existence
of corners is perfectly reasonable, and further that our
characterization of corners as the intersection of independent
yield surfaces is justifiable.

3.3 The Rigid-Plastic Constitutive Relations

In many problems of technological interest it is convenient to
further idealize the constitutive relations by neglecting entirely the *elastic components of strain* in the constitutive relations. In certain problems this idealization can lead to a
significant simplification of the structural problem, and is

justifiable when the nature of the solution, or those aspects
of the solution which are of interest, are dictated primarily
by the plastic deformation which occurs in the body. We have
already encountered one such situation in dealing with the limit
surface in load space, as we shall see presently.

Two cases are of major interest; the idealizations we shall
refer to as *rigid-perfectly plastic* and *rigid-hardening* behavior. In each case the elastic components of strain are zero.
This can be achieved by imagining that all the coefficients of
the stiffness matrix C_{jk} in the relation

$$q_j^e = C_{jk} Q_k \tag{59}$$

are identically zero, so that $q_j^e = 0$ for all Q_j. The total
strain q_j is then identical to the plastic component of strain
q_j^p, and we can dispense with the superscript p.

In the *rigid-perfectly plastic material* we adopt a convex
limit function $\psi(Q_j)$, or a number of independent limit functions
$\psi_r(Q_j)$ (r = 1, ..., m), which is independent of recorded history. The limit surface $\psi = 0$ or the inner envelope of the independent limit surfaces, now encloses a domain in the stress
space which we refer to as the *rigid region*. For a smooth yield
surface the constitutive equations are

$$dq_j = 0 \quad \text{for} \quad \psi < 0$$
$$\text{or} \quad \psi = 0 \quad \text{and} \quad \frac{\partial \psi}{\partial Q_j} dQ_j < 0 \ ,$$

$$\tag{60}$$

$$dq_j = \lambda \frac{\partial \psi}{\partial Q_j} \quad \text{for} \quad \psi = 0 \quad \text{and} \quad \frac{\partial \psi}{\partial Q_j} dQ_j = 0 \ ,$$

where λ is a non-negative but otherwise unspecified scalar.
These equations are identical to those given in Section 2.6 for
flow with $C_{jk} = 0$ and $q_j^p = q_j$. The generalization to a number

of independent yield surfaces follows immediately; let

$$q_j = q_j^1 + q_j^2 + \dots + q_j^m \ .\tag{61}$$

Then

$$dq_j^r = 0 \quad \text{for} \quad \psi_r < 0$$

$$\text{or} \quad \psi_r = 0 \quad \text{and} \quad \frac{\partial \psi_r}{\partial Q_j} dQ_j < 0 \ ,$$

$$dq_j^r = \lambda_r \frac{\partial \psi_r}{\partial Q_j} \quad \text{for} \quad \psi_r = 0 \quad \text{and} \quad \frac{\partial \psi_r}{\partial Q_j} dQ_j = 0 \ ,\tag{62}$$

where λ_r are again non-negative but otherwise unspecified scalars.

Thus in a rigid, perfectly plastic material changes in the total strain occur only as a result of flow when the stress point lies on the yield or limit surface. For stress states or stress increments which lie within the rigid region the *total strain* remains constant. This kind of behavior can most conveniently be considered as the limit of an elastic, perfectly plastic material when the elastic constants C_{jk} become zero.

Consider now a body or structure composed of a rigid, perfectly plastic material, and subjected to generalized loads Γ_α. As we have been led to suspect in Section 3.2, the description of the behavior in the load space will reflect the constitutive rquations for the material. We thus expect to find a *limit surface* $\Psi = 0$ in *load space* for this body. In Section 3.2 we discussed in some detail the limit surface for an elastic-plastic material in which flow can occur (i.e. an elastic-plastic material which has associated with it a limit surface $\psi(Q_j) = 0$). This discussion can be used directly to establish an extremely important result: within the framework of our present discussion (i.e. for linear equilibrium and kinematic relations and the conventional small displacement theory) two identical bodies subjected to identical generalized loads Γ_α and differing only in that one is composed of an *elastic-plastic material* with a limit surface $\psi(Q_j) = 0$ on which flow can occur, and the other is composed of a *rigid, perfectly plastic material* with the *same* limit surface $\psi(Q_j) = 0$, will have *identical limit surfaces in load* space $\Psi(\Gamma_\alpha) = 0$. This applies equally well if either the limit surface in stress space or in load space is made up of a number of independent yield surfaces. Within the limit surface in load space the behavior will be markedly different; in the elastic-plastic body changes in the generalized displacements ξ_α can occur as a result of changes in the generalized loads Γ_α as consequences of elastic and plastic strains in the body. In the rigid, perfectly plastic body the behavior within the limit surface in load space will be *rigid*, i.e. changes in the generalized displacements ξ_α can occur *only* as a result of flow in the body when $\Psi(\Gamma_\alpha) = 0$ and neutral loading in load space occurs.

To establish this result we make use of arguments which are very close to those given in Section 3.2 for elastic-plastic materials in which flow may occur. Only a general outline will be given; the details are left as an exercise for the reader. First, we note that flow in an elastic-plastic body can occur when the *stress increments are zero* and the plastic strain increments, whenever they occur, *are a result of flow in the material*. Thus the only part of the elastic-plastic constitutive equations which are used are those associated with flow at the limit surface in stress space. Consequently, if we can find a statically admissible stress field and a kinematically admissible *plastic* strain increment field which satisfies the flow rule for the elastic-plastic body for certain loads Γ_α, this also represents a solution for the rigid-perfectly plastic body with an identical limit surface in stress space. Hence flow can occur in the rigid, perfectly plastic body for the same generalized loads Γ_α. Carrying out this argument for each point on the surface $\Psi = 0$ for the elastic-plastic body, we establish an identical surface in load space for which flow can occur in the rigid-perfectly plastic body. We must then ask whether flow can occur in the rigid-perfectly plastic body for generalized loads Γ_α which do not lie on the limit surface.This argument is again formally identical to the argument used to establish that the limit surface for an elastic-plastic body is unique. The basis of the argument (Section 3.2, inequalities 49 and 50) is unchanged, and we show that if flow occurs for generalized loads which lie either within or without the surface $\Psi = 0$ the basic inequalities must be violated.

Finally we note that in the rigid, perfectly plastic body if *any* changes in the generalized displacements occur they *must* be associated with plastic strain changes in the body associated with flow of the material; they thus constitute flow of the body. However, flow of the body can only occur on the limit surface, and hence we are led to the conclusion that no changes in the generalized displacements ξ_α can occur in the rigid, perfectly plastic body when the generalized loads lie inside the limit surface in load space; the *body* is *rigid* within the limit surface.

In Section 3.1 we also discussed in detail to what extent the stress and strain increment fields during flow could be determined by consideration of flow alone in the elastic-plastic body. This argument led us to the conclusion that we could not expect to determine these fields completely or

uniquely from a consideration of the conditions during flow. In
an elastic-plastic body, of course, an incremental analysis beginning in a known initial state will lead to a unique solution
for all ranges of mechanical behavior including flow. In a rigid, perfectly plastic body, however, no deformation takes
place while the load point remains within the limit surface.
There are no kinematic relations which contribute to the incremental analysis prior to flow, and as a consequence the stress
field in the body corresponding to loads in the rigid region in
load space cannot be uniquely determined (except in the very
special case where the structure is statically determinate).
Thus, to summarize, the stress field in non-unique and the
strain changes are zero in a rigid-perfectly plastic body or
structure when the load point lies within the limit surface in
load space. When the load point lies on the limit surface,flow
may occur and some information about the stress and strain increment field may be obtained. In certain cases we may be able
to determine these fields completely, but as a general rule
some degree of non-uniqueness may be encountered.

The comparison of the limit surface in load space for rigid,
perfectly plastic bodies and elastic, plastic bodies with an
identical limit surface in stress space has been the subject of
some controversy in the literature. There is no doubt that in
our present context, where we have *explicitly* required that the
classical small displacement assumptions are applicable, the
two limit surfaces will be identical. We should note, however,
that had we first formulated our structural problem without the
classical small displacement assumptions these arguments do not
go through. Consider a body subjected to monotonically increasing proportional loads. If the body is rigid-perfectly plastic
no deformation will occur until the initiation of flow. This

means that the limit problem is formulated unambiguously in terms of the *original geometry* of the problem. In an elastic-plastic body, however, elastic and plastic deformation will occur before flow commences, and the limit problem is formulated in a *deformed configuration*; only if the initial and the deformed configuration are sufficiently close that the differences can be ignored (which is precisely the small displacement assumption) will the two problems be identical. Further, if the small displacement assumptions are not applicable, the surface on which flow is initiated may be path dependent. Questions regarding the conditions under which flow continues, and of stability, may also be present.

This issue is of considerable technological importance because there is an important class of problems which, if they are considered as elastic-plastic problems, do not fall within our present limitations, while when considered as rigid, perfectly plastic problems can be treated within the scope of our basic framework. Among the best known examples in this class are the *processes of steady state drawing and extrusion*. In steady state flow, we define the stress in the current configuration and make use of the rate of deformation rather than the strain increments. The equilibrium relations and the relations between rate of deformation and velocity are linear and the boundary conditions are independent of time. Thus the basic equations are formally identical to the equations of the small displacement theory we have so far employed. It should also be noted that many manufacturing processes are not steady state problems; forging is an example. We may still idealize the problem as rigid, perfectly plastic, but the geometry of the problem may change during the application of the load.

The constitutive relations for a *rigid-hardening* or *rigid-plastic* material have not been used as frequently as those of a

rigid, perfectly plastic material in technical applications,
although they have been studied intensively in general terms.
For a smooth yield surface the constitutive equations are

$$dq_j = 0 \quad \text{for} \quad \phi = 0$$

$$\text{or} \quad \phi < 0 \quad \text{and} \quad \frac{\partial \phi}{\partial Q_j} dQ_j \leq 0 \quad ,$$

$$\tag{63a}$$

$$dq_j = G \frac{\partial \phi}{\partial Q_j} \frac{\partial \phi}{\partial Q_k} dQ_k$$

$$\text{for} \quad \phi = 0 \quad \text{and} \quad \frac{\partial \phi}{\partial Q_j} dQ_j \geq 0 \quad ,$$

where

$$G = G(Q_j, H_\alpha) \tag{63b}$$

is non-negative, and

$$\phi = \phi(Q_j, H_\alpha) \tag{63c}$$

is a yield function leading to a convex yield surface $\phi = 0$.
The generalization of these relations for a number of yield
surfaces follows the usual lines we have established. Since we
shall not discuss the rigid-plastic problem in great detail
this generalization is left to the reader as an exercise.

The boundary value problem associated with a rigid-plastic
body is incremental. By following arguments similar to those
used in the rigid, perfectly plastic problem we may establish
that there will exist a *yield surface in load space* $\phi = 0$
(which may be either smooth or contain corners). For load
states within this load space the behavior of the body or structure will be *rigid*, i.e. there will be no changes in the generalized displacements ξ_α. Changes in the generalized displacements $d\xi_\alpha$ will follow changes in the generalized loads $d\Gamma_\alpha$ only

when the load point lies on the yield surface in load space and
the load increment leads to a new load point outside the current yield surface in load space.

Difficulties arise in this incremental problem as a result of
lack of uniqueness in the stress fields in many rigid-plastic
problems. The uniqueness arguments of Section 3.1 are applicable, and it will be remembered that uniqueness follows from establishing that

$$(dQ_j^{(1)} - dQ_j^{(2)})(dq_j^{(1)} - dq_j^{(2)}) \;=\; 0 \tag{64}$$

at each point in the structure or body for two assumed solutions
$dQ_j^{(1)}$, $dq_j^{(1)}$ and $dQ_j^{(2)}$, $dq_j^{(2)}$. In the elastic plastic case consideration of the *elastic components of strain* (Section 3.1,
equation 7) leads us to the conclusion that (64) cannot hold
unless $dQ_j^{(1)} = dQ_j^{(2)}$, thus establishing that two distinct solutions cannot exist. However, if $dq_j^{(1)}$, $dq_j^{(2)}$ are *plastic
strains*, and the elastic components are zero, there exist possibilities that $dQ_j^{(1)}$ and $dQ_j^{(2)}$ may be distinct while equation
(64) is satisfied. Thus the stress increment field may be non-
unique; on the other hand it is clear that whenever parts of a
body remain rigid such non-uniqueness is inevitable. It is
not possible to satisfy equation (64) with $dq_j^{(1)} \neq dq_j^{(2)}$, and
the strain increment field is unique. This ensures that the
change in recorded history can be computed, and the next incremental problem can therefore be solved. The uniqueness results, while not as strong as those for elastic-plastic bodies,
do permit the response of the structure to a path of loading to
be computed.

3.4 Historical and Bibliographical Remarks

In Part I we have attempted to present a systematic development
of a set of constitutive relations governing time-independent

inelastic behavior, together with some motivation and justification in terms of the associated structural problems. In the remainder of this volume we shall be primarily concerned with a brief study of particular applications of this theory (Part II), with the derivation of general theorems of structural behavior which follow as a consequence of the framework for the constitutive relations (Parts III-V) and a more detailed study of the underlying physical basis of the theory in the context of polycrystalline metals (Part VI).

It has been emphasized that this development is to some extent a personal view of the subject matter; other writers might develop a general framework by an approach which is different in either detailed aspects or in overall conception. More importantly in the present context, the systematic development presented in this chapter is not a chronologically ordered outline of major contributions to the theory of plasticity. Like almost all general theories, plasticity originated in the study of particular situations and passed through successive stages in which its generality was extended. The development of constitutive equations, in particular, has a long and sometimes erratic history, including independent descriptions of the same important ideas, controversies which the passage of time and an increase in understanding have shown to be unimportant, and controversies which have not yet been resolved. This makes it very difficult to trace and to ascribe correctly the origin of many important concepts which have become part of the fabric of the theory.

With this in mind we have chosen deliberately to omit references to the origins of the concepts presented in Part I before this point; an attempt will now be made to give a brief, but necessarily incomplete, account of major developments. It is hoped that the inadequacies of this approach, particularly in regard to the omission of references to writers who presented their contributions in particular contexts and whose work was subsequently superseded by more general descriptions, will be counterbalanced by references to important reviews and monographs which at various stages in the development of the theory have been significant in crystallizing ideas and which contain detailed bibliographies.

The origins of the mathematical theory of plasticity can be traced to Tresca [1864] who first postulated a yield condition for the continuum problem. St.Vénant [1870] introduced constitutive relations for rigid, perfectly plastic materials in plane stress, and these relations were generalized to three dimensional continua by Levy [1870]. While these relations represent the earliest attempts to provide constitutive relations for plastic solids they are not acceptable in the framework for constitutive relations derived in this chapter, since St. Vénant and Levy used the yield condition given by Tresca together with a flow rule (or a plastic potential) which gives strain rates which are not necessarily normal to the yield surface. Von Mises [1913], still dealing with rigid, perfectly plastic continua, introduced an alternative yield condition which together with the flow rule of the St. Vénant-Levy theory provided an acceptable set of constitutive relations. Von Mises [1928] generalized the theory of rigid, perfectly plastic solids, discussing the relation between the direction of the plastic strain rate and a yield surface whose derivatives with respect to stress are continuous. Elastic strains were incorporated by Prandtl [1924] (in two dimensions) and Reuss [1930] (in three dimensions), providing constitutive relations for an elastic, perfectly plastic material based on Von Mises' yield

condition. The appropriate flow rule for Tresca's yield condition, which contains singular points (i.e. corners or discontinuities in the derivatives with respect to stress) was
discussed by Reuss [1932], [1933].

The development of constitutive relations for hardening materials in incremental form proceeded more slowly, since greater
emphasis was placed on problems involving flow in the years before 1940. Prandtl [1928], for example, attempted to formulate general laws for time independent hardening relations, and
Melan [1938a] gave incremental relations for hardening solids
with a smooth yield surface.

While the few references given here note some of the major
contributors to the modern theory of plasticity in the years
preceding 1940, there was a great deal of activity in the field
particularly in the years 1920-1940. A great deal of important
experimental work was carried out (see, for example, the review
paper by Drucker [1956a]) and many important boundary value
problems were considered. The first sytematic treatment of
plasticity was given by Nadai [1931]. Some progress was also
made in the study of beam problems in these early years. The
relation between bending moment and curvature of a beam bent
into the inelastic range was discussed by Ewing [1899], Kazincky
[1914] and Maier-Liebnitz [1936].

The fifteen years after 1940 saw the most intensive period
of development of basic concepts in what is now regarded as the
classical theory of plasticity. A number of systematic treatments of plasticity appeared which describe these developments
in great detail. The comprehensive monographs by Hill [1950]
(concerned with both hardening and flow) and by Prager and Hodge
[1951] (dealing exclusively with perfectly plastic solids) are
particularly valuable in following the development of continuum

plasticity. Systematic treatments of continuum plasticity were
also given by Nadai [1950], Freudenthal [1950] and Westergaard
[1952] at about the same time. The behavior of elastic, perfectly plastic beams and framed structures also received considerable attention during this period. The early monographs by
van den Broek [1948], Neal [1956] and Baker, Horne and Heyman
[1956] describe this work in detail. The volume by Baker,
Horne and Heyman, in particular, provides a comprehensive account of the development of the theory in relation to civil engineering structures.

More recent texts in plasticity also give brief accounts of
the development of plastic constitutive equations together with
applications in various classes of problems. Applications to
beams and framed structures are given by Beedle, Thürlimann and
Ketter [1955], Heyman [1957], Beedle [1958], Hodge [1959], Massonet and Save [1965] and Baker and Heyman [1969]. Hodge [1963]
and Massonet and Save [1963] deal with plate and shell problems.
Prager [1959] and Calladine [1969] have given elementary unified treatments of one, two and three dimensional continuum
problems. Treatments which emphasize continuum problems include Hoffman and Sachs [1953], Phillips [1956], Johnson and
Mellor [1962], Lin [1968], Mendelson [1968], Kachanov [1969],
Sokolovskij [1969] and Knets [1971].

A number of important review articles have also appeared.The
reader is referred, in particular, to the articles by Prager
[1955], [1957], Hill [1956a], Hodge [1958], [1960], Kliushnikov
[1958], [1959], Freudenthal and Geiringer [1958], Koiter [1960],
Naghdi [1960] and Drucker [1960].

Returning to the development of the specific ideas in Part I
it can be seen that Melan [1938a], [1938b] arrived at a general
framework for plastic constitutive relations which is

essentially identical to that presented here for smooth yield
surfaces. Prager [1949] independently gave the same framework.
The yield function and the loading and unloading conditions
were precisely formulated (the case of neutral loading was first
discussed by Handelmann, Lin and Prager [1947]). The relationship between the convexity of the yield surface, the normality
of the plastic strain rate increment and the uniqueness of the
associated boundary value problem was clearly recognised. The
subsequent work on constitutive equations essentially provided
a generalization to less restrictive yield functions and dependence on recorded history, a simplification of the assumptions
which define the class of plastic materials, and the isolation
of important concepts which consolidate the internal consistency of the theory. The uniqueness and stability postulates
used in Chapter 2 were introduced by Drucker [1951] and ampli-
fied in further papers (e.g. Drucker [1964], Gemmerling [1964].
Postulates which provide assumptions which play an equivalent
role in the development of the framework of constitutive relations have been given by Hill [1948a], and extended in a study
of polycrystalline aggregates by Bishop and Hill [1951], and by
Illyushin [1961], [1963]. A detailed discussion and comparison
of postulates of this type for both small and finite deformations has been given by Hill [1968].

Other important discussions on the development of constitutive relations have been given by Hill, Lee and Tupper [1947],
Hill [1948b], Taylor [1947] and Hill [1950].

The nature of the incremental stress-strain relations in the
presence of singularities in the yield surface or discontinuities in the direction of the normal vector to the yield surface
was not satisfactorily resolved until the work of Koiter [1953a]
and Sanders [1954] for hardening materials and Koiter [1953b]

and Prager [1953] for perfectly plastic materials. The introduction of independent convex smooth yield surfaces and the concept of a plastic strain increment made up of contributions from each active yield surface ended a period of considerable controversy in which supposedly alternative frameworks for constitutive relations which admitted corners in the yield surface were advanced. These alternative theories are now recognized to fall within the conventional framework.

Uniqueness theorems for the elastic, plastic incremental problem appear to have been first established by Melan [1938a], [1938b] for both hardening and perfectly plastic materials with some limiting assumptions, and extended by Prager [1949], Greenberg [1949] and Bauer [1948]. Generalization to the case of more than one yield function was achieved by Koiter [1953a]. Further discussions of uniqueness and stability have been given by Drucker [1956b] and Hill [1958].

Consideration of the limit surface in load space (or points on this surface) for perfectly plastic materials originated with Kazincky [1914] and Kist [1917]. Further contributions were made by Gvozdev [1938], Feinberg [1948] and Baker [1949] and his collaborators for the case of framed structures (see also Baker, Horne and Heyman [1956] for further references). Precise definitions of the limit surface for more general problems were given by Greenberg and Prager [1951] and Drucker, Greenberg and Prager [1951], [1952] from the point of view of elastic-plastic materials and by Hill [1951], [1952] from the point of view of rigid-plastic materials. Rigid, perfectly plastic and rigid-plastic problems have been discussed in great detail by Hill (see for example, Hill [1948a], [1950], [1956], [1957]) and by Ivlev [1966].

It is hoped that this brief and of necessity incomplete

review of major contributions to the classical theory of plasticity will provide the reader with sufficient material to gain an appreciation of the historical development of the subject and the various points of view which has become part of the fabric of the modern theory. The most notable omissions from this review are those of workers in the Soviet Union; the reader is referred to the translations of volumes by Sokolovskij [1955], [1969], Ilyushin [1956], Kachanov [1969] and Knets [1971] to gain some appreciation of Soviet work during the period in which the developments described in this section took place.

Finally, it should be remembered that in Chapters 2 and 3 we have attempted to present a brief picture of all aspects of the theory of plasticity, including the development of constitutive relations, uniqueness, stability and structural behavior in order to emphasize the interdependence of any one aspect of the theory with each of the others. In the remainder of this volume, we shall enlarge on various aspects of this discussion. Many contributions which have been referred to in this review also covered topics not dealt with in this section; many important contributions have not yet been discussed. A full appreciation of the significant contributions to plasticity really demands that we consider the theory of plasticity in a much broader context than we have so far done.

References

J.F.Baker   1949   "A review of recent investigations
                                 into the behaviour of steel frames in
                                 the plastic range", J.Inst.Civ.Eng.,
                                 31, 188.

J.F.Baker and   1969   *Plastic Analysis of Frames*,Cambridge.
J.Heyman

J.F.Baker,   1956   *The Steel Skeleton*, Vol.2, Cambridge.
M.R.Horne and
J.Heyman

F.B.Bauer — 1948 — Ph.D.Thesis, Brown University.

L.S.Beedle — 1958 — *Plastic Design of Steel Frames*, Wiley (N.Y.).

L.S.Beedle, B.Thürlimann and R.L.Ketter — 1955 — *Plastic Design in Structural Steel*, American Institute of Steel Construction (N.Y.).

J.F.W.Bishop and R.Hill — 1951 — "A Theory of Plastic Distortion of a Polycrystalline Aggregate under Combined Stresses", Phil.Mag.(7) $\underline{42}$,414.

J.A.van den Broek — 1948 — *Theory of Limit Design*, Wiley (N.Y.).

C.R.Calladine — 1969 — *Engineering Plasticity*, Pergamon Press.

D.C.Drucker — 1951 — "A More Fundamental Approach to Plastic Stress-Strain Relations", Proc. 1st U.S.Nat.Congr.Appl.Mech., 487.

D.C.Drucker — 1956a — "Stress-Strain Relations in the Plastic Range of Metals — Experiments and Basic Concepts", Chapter 4 in *Rheology, Theory and Applications*, Edited by F.R.Eirich, Academic Press (N.Y.).

D.C.Drucker — 1956b — "On Uniqueness in the Theory of Plasticity", Quart.Appl.Math., $\underline{14}$, 35.

D.C.Drucker — 1960 — "Plasticity", *Structural Mechanics*, Proc. 1st Symp. on Naval Struc.Mech., Edited by N.Goodier and N.J.Hoff, Pergamon Press.

D.C.Drucker — 1964 — "On the Postulate of Stability of Material in the Mechanics of Continua", J.de Mécanique, $\underline{3}$, 235.

D.C.Drucker, H.J.Greenberg and W.Prager — 1951 — "The Safety Factor of an Elastic-Plastic Body in Plane Strain", J.Appl.Mech., $\underline{18}$, 371.

D.C.Drucker, H.J.Greenberg and W.Prager — 1952 — "Extended Limit Design Theorems for Continuous Media", Quart.Appl.Math., $\underline{9}$, 351.

J.A.Ewing — 1899 — *The Strength of Materials*, Cambridge.

S.M.Feinberg — 1948 — "The Principle of Limiting Stress", Prik.Mat.Mekh., $\underline{7}$, 85.

A.M.Freudenthal — 1950 — *The Inelastic Behavior of Engineering Structures and Materials*, Wiley (N.Y.).

A.M.Freudenthal   1958   "The Mathematical Theories of the
and H.Geiringer   Inelastic Continuum", *Handbuch der
                               Physik*, <u>6</u>, Springer (Berlin).

G.A.Gemmerling   1964   "On the postulate of plasticity",
                               Zhurn.Prikl.Mekh.Techn.Fiz., <u>1</u>, 80.

H.J.Greenberg   1949   "Complementary Minimum Principles for
                               an Elastic-Plastic Material", Quart.
                               Appl.Math., <u>7</u>, 85.

H.J.Greenberg and   1951   "Limit Design of Beams and Frames",
W.Prager   Proc.ASCE, <u>77</u> (Separate 59).

A.A.Gvozdev   1938   "The Determination of the Value of
                               the Collapse Load for Statically In-
                               determinate Systems Undergoing Plas-
                               tic Deformations", Akad.Nauk.Moscow-
                               Leningrad, 19-33. Translated by R.M.
                               Haythornthwaite, Int.J.Mec.Sci., <u>1</u>,
                               322, 1960.

G.H.Handelman,   1947   "On the Mechanical Behavior of Metals
C.C.Lin and   in the Strain-Hardening Range", Quart.
W.Prager   Appl.Math., <u>4</u>, 397.

J.Heyman   1957   *Plastic Design of Portal Frames*,
                               Cambridge.

R.Hill   1948a   "A Variational Principle of Maximum
                               Plastic Work in Classical Plasticity",
                               Quart.J.Mech.Appl.Math., <u>1</u>, 18.

R.Hill   1948b   "A Theory of the Yielding and Plastic
                               Flow of Anisotropic Metals", Proc.
                               Roy.Soc.London, <u>A 193</u>, 281.

R.Hill   1950   *Mathematical Theory of Plasticity*,
                               Oxford University Press.

R.Hill   1951   "On the State of Strain in a Plastic-
                               Rigid Body at the Yield Point", Phil.
                               Mag., <u>42</u>(7), 868.

R.Hill   1952   "A Note on Estimating the Yield Points
                               in a Plastic-Rigid Body", Phil.Mag.,
                               <u>43</u>(7), 353.

R.Hill   1956a   "The Mechanics of Quasi-Static Defor-
                               mation in Metals", *Surveys in Mechan-
                               ics* (Edited by E.K.Batchelor and R.M.
                               Davies) 7-31, Cambridge.

R.Hill   1956b "On the Problem of Uniqueness of
                             the Theory of a Rigid-Plastic Solid
                             I", J.Mech.Phys.Sol., $\underline{4}$, 247.

R.Hill   1957   "Stability of Rigid-Plastic Solids",
                             J.Mech.Phys.Sol., $\underline{5}$, 1.

R.Hill   1958   "A General Theory of Uniqueness and
                             Stability in Elastic-Plastic Solids",
                             J.Mech.Phys., $\underline{6}$, 238.

R.Hill   1968   "On Constitutive Relations for Simple
                             Materials", J.Mech.Phys.Sol., $\underline{16}$, 229.

R.Hill, E.H.Lee   1947   "The Theory of Combined Plastic and
and S.J.Tupper   Elastic Deformation with Particular
                             Reference to a Thick Tube Under In-
                             ternal Pressure", Proc.Roy.Soc. Lon-
                             don, $\underline{A\ 191}$, 278.

P.G.Hodge,Jr.   1958   "The Mathematical Theory of Plasti-
                             city", *Elasticity and Plasticity* by
                             J.N.Goodier and P.G.Hodge,Jr. (N.Y.).

P.G.Hodge,Jr.   1959   *Plastic Analysis of Structures*,
                             McGraw-Hill (N.Y.).

P.G.Hodge,Jr.   1960   "Boundary Value Problems in Plasti-
                             city", *Plasticity*, Proc. 2nd Symp. on
                             Naval Struc.Mech., Edited by E.H.Lee
                             and P.S.Symonds, 297, Pergamon Press
                             (N.Y.).

P.G.Hodge,Jr.   1963   *Rotationally Symmetric Plates and
                             Shells*, Prentice Hall (Englewood
                             Cliffs, N.J.).

O.Hoffman and   1953   *Introduction to the Theory of Plasti-
G.Sachs   city for Engineers*, McGraw-Hill (N.Y.).

A.A.Ilyushin   1956   *Plasticity*,Eyrolles (Paris).

A.A.Ilyushin   1961   "On the Postulate of Plasticity",
                             Prik.Mat.Mekh., $\underline{25}$, 503.

A.A.Ilyushin   1963   *Plasticity*, Akademija Nauk. USSR.,
                             (Moscow).

D.D.Ivlev   1966   *The Theory of Ideal Plasticity*, Nauka
                             (Moscow).

W.Johnson and   1962   *Plasticity for Mechanical Engineers*,
P.B.Mellor   Van Nostrand (Princeton).

L.M.Kachanov   1969   *Introduction to the Theory of Plasti-
                               city*, Nauka (Moscow). Republished as
                               *Foundations of the Theory of Plasti-
                               city*, North-Holland, 1971

G.von Kazincky   1914   "Tests with Fixed end Beams",
                               Betonszemle, Vol.2.

N.C.Kist   1917   Inaugural Dissertation, Technical
                               University, Delft.

V.D.Kliushnikov   1958   "On Plasticity Laws for Work Harden-
                               ing Materials", TPMM (Translation of
                               Prikl.Mat.Mekh) $\underline{22}$, 129.

V.D.Kliushnikov   1959   "On the Possibility of the Construc-
                               tion of the Plasticity Relations",
                               Prikl.Mat.Mekh, $\underline{23}$, 282.

I.V.Knets   1971   *Basic Present Trends in the Mathema-
                               tical Theory of Plasticity*, Zinatne
                               (Riga).

W.T.Koiter   1953a   "Stress-Strain Relations, Uniqueness
                               and Variational Theorems for Elastic-
                               Plastic Materials with a Singular
                               Yield Surface", Quart.Appl.Math., $\underline{11}$,
                               350.

W.T.Koiter   1953b   "On Partially Plastic Thick-Walled
                               Tubes", C.B.Biezeno Anniversary Vol-
                               ume on Applied Mechanics, 233
                               (Haarlem).

W.T.Koiter   1960   "General Theorems for Elastic-Plastic
                               Solids", *Progress in Solid Mechanics*,
                               Vol.1, Edited by I.N.Sneddon and R.
                               Hill, Chapter 4, North Holland Press.

M.Levy   1870   "Mémoire sur les Equations Générales
                               des Mouvements Intérieurs des Corps
                               Solides Ductiles au delà des Limites
                               où l'Elasticité Pourrait les Ramener
                               à leur Premier Etat", C.Rend., Paris
                               $\underline{70}$, 1323.

T.H.Lin   1968   *Theory of Inelastic Structures*,
                               Wiley (N.Y.).

H.Maier-Liebnitz   1936   "Test Results, Their Interpretation
                               and Application", Prelim.Publ. Int.
                               Assoc.for Bridge and Struc.Eng., 2nd
                               Congress, Berlin.

C.E.Massonet and   1965   *Plastic Analysis and Design*, Vol.1:
M.A.Save   Beams and Frames, Blaisdell (Boston).

C.E.Massonet and   1963   *Calcul Plastique des Constructions:*
M.A.Save   Vol. II, Structures Spéciales,
                              C.B.L.I.A. (Bruxelles). Republished
                              as *Plastic Analysis and Design of
                              Plates, Shells and Disks* (Vol. 15 in
                              Series in Applied Mathematics and
                              Mechanics) North-Holland.

E.Melan   1938a   "Zur Plastizitat des Raumlichen Kon-
                              tinuums", Ing.Arch., $\underline{9}$, 116.

E.Melan   1938b   "Der Spannungszustand eines Hencky-
                              Mises'schen Kontinuums bei Veränder-
                              licher Belastung", Sitz.Ber.Ak.Wiss.
                              Wien, IIa, $\underline{147}$, 73.

A.Mendelson   1968   *Plasticity, Theory and Application*,
                              Macmillan(N.Y.).

R.von Mises   1913   "Mechanik der festen Körper im
                              plastisch deformablen Zustant",
                              Göttingen Nachr.Math.Phys. Kl, 582.

R.von Mises   1928   "Mechanik der plastischen Forman-
                              derung von Kristallen", ZAMM, $\underline{8}$, 161.

A.Nadai   1931   *Plasticity*, McGraw-Hill (N.Y.).

A.Nadai   1950   *The Theory of Flow and Fracture of
                              Solids*, McGraw-Hill (N.Y.).

P.M.Naghdi   1960   "Stress-Strain Relations in Plasti-
                              city and Thermoplasticity", *Plasti-
                              city*, Proc. 2nd Symp.on Naval Struc.
                              Mech., Edited by E.H.Lee and P.S.
                              Symonds, 121.

B.G.Neal   1956   *The Plastic Methods of Structural
                              Analysis*, Chapman and Hall (London).

A.Phillips   1956   *Introduction to Plasticity*, Ronald
                              Press (N.Y.).

W.Prager and   1951   *Theory of Perfectly Plastic Solids*,
P.G.Hodge,Jr.   Wiley (N.Y.) 1951. Republished by
                              Dover (N.Y.) 1968.

W.Prager   1953   "On the Use of Singular Yield Condi-
                              tions and Associated Flow Rules",
                              J.Appl.Mech., $\underline{20}$, 317.

W.Prager	1955	"The Theory of Plasticity - A Survey of Recent Achievements", James Clayton Lecture, Proc.Inst.Mech.Engr., $\underline{169}$, 41.
W.Prager	1949	"Recent Developments in the Mathematical Theory of Plasticity", J.Appl. Phys., $\underline{20}$, 235.
W.Prager	1957	"An Introduction to the Concepts and Principles of Plasticity", Special Lectures delivered at the Seminar in Appl.Math., Boulder, Colorado, July 1957.
W.Prager	1959	*An Introduction to Plasticity*, Addison-Wesley (Reading, Mass.). Translation of *Probleme der Plastizitätstheorie*, Birkhauser Verlag (Basel) 1955.
L.Prandtl	1924	"Spannungsverteilung in Plastischen Körpern", Proc.1st Int.Congr.Appl. Mech., 43, Delft.
L.Prandtl	1928	"Ein Gedankenmodell zur Kinetischen Theorie fester Körpern", ZAMM, $\underline{8}$, 55.
A.Reuss	1930	"Berucksichtigung der Elastischen Formanderungen in der Plastizitätstheorie", ZAMM, $\underline{10}$, 266.
A.Reuss	1932	"Fliesspotential oder Gleitebenen?", ZAMM, $\underline{12}$, 15.
A.Reuss	1933	"Vereinfachte Berechnung der Plastischen Formänderungsgeschwindigkeiten bei Voraussetzung der Schubspannungsfliessbedingung", ZAMM, $\underline{13}$, 356.
J.L.Sanders	1954	"Plastic Stress-Strain Relations Based on Linear Loading Functions", Proc. 2nd U.S.Nat.Congr.Appl.Mech., 455.
V.V.Sokolovskij	1955	*Theorie der Plastizität*, VEB Verlag (Berlin). (Originally published in Russian in 1945.)
V.V.Sokolovskij	1969	*The Theory of Plasticity*, Vyshaja Shkola (Moscow).

G.I.Taylor   1947   "A Connection between the Criterion
                               of Yield and the Strain Ratio Rela-
                               tionship in Plastic Solids", Proc.
                               Roy.Soc., A 191, 441.

H.Tresca   1864   "Mémoire sur l'écoulements des Corps
                               Solides Soumis à des Fortes Pressions",
                               C.Rend., Paris, 59, 754.

B.de Saint Venant   1870   "Mémoire sur l'Etablissement des
                               Equations Différentielles des Mouve-
                               ments Intérieurs Opérés dans les
                               Corps Solides Ductiles au delà des
                               Limites où l'Elasticité Pourrait les
                               Ramener à leur Premier Etat", C.Rend.
                               Paris, 70, 473.

H.M.Westergaard   1950   *Theory of Elasticity and Plasticity*,
                               Harvard University Press (Cambridge,
                               Mass.).

PARTICULAR ASPECTS OF PLASTIC BEHAVIOR

INITIAL YIELD SURFACES FOR POLYCRYSTALLINE METALS

4.1 Summary of the General Form of the Plastic Constitutive Equations

In Chapter 2 we limited the form of the constitutive relations by making a number of simplifying assumptions. These relations will be summarized briefly, with a slight change in notation to effect some economy in writing out the equations.

The total strain q_j is divided into an elastic component q_j^e and a plastic component q_j^p. The elastic component of strain is given by

$$q_j^e = C_{jk} Q_k \; , \tag{1}$$

where C_{jk} is symmetric and positive definite.

Plastic strain increments due to *hardening* behavior are governed by the yield function $\phi(Q_j, H_\alpha)$. When $\phi(Q_j, H_\alpha)$ is continuously differentiable,

$$dq_j^p = c \, G \, \frac{\partial \phi}{\partial Q_j} \frac{\partial \phi}{\partial Q_k} \, dQ_k \; , \tag{2}$$

where

$$G = G(Q_j, H_\alpha) \tag{3}$$

is a scalar hardening coefficient, and

$$c = 1 \quad \text{if} \quad \phi = 0 \quad \text{and} \quad \frac{\partial \phi}{\partial Q_j} dQ_j \geq 0 \; ,$$

$$c = 0 \quad \text{if} \quad \phi < 0$$

$$\text{or} \quad \phi = 0 \quad \text{and} \quad \frac{\partial \phi}{\partial Q_j} dQ_j \leq 0 \; . \tag{4}$$

If the yield function is not continuously differentiable, we assume the existence of a number of independent continuously differentiable yield functions ϕ_r (Q_j, H_α) $(r = 1, \ldots, m)$. It follows then that $\phi < 0$ if $\phi_r < 0$ for all r, $\phi = 0$ if $\phi_r = 0$ for one or more values of r and $\phi_r < 0$ for the remaining values of r, and $\phi > 0$ if $\phi_r > 0$ for some value of r. The plastic strain increment dq_j^p is assumed to be given by

$$dq_j^p = dq_j^{p1} + dq_j^{p2} + \ldots + dq_j^{pm} , \tag{5}$$

where

$$dq_j^{pr} = c_r \, G_r \, \frac{\partial \phi_r}{\partial Q_j} \, \frac{\partial \phi_r}{\partial Q_k} \, dQ_k . \tag{6}$$

The summation convention does not apply to the subscript r. In equation (6)

$$G_r = G_r \, (Q_j, H_\alpha) \tag{7}$$

are hardening coefficients and

$$c_r = +1 \quad \text{if} \quad \phi_r = 0 \text{ and } \frac{\partial \phi_r}{\partial Q_j} \, dQ_j \geq 0 ,$$

$$c_r = 0 \quad \text{if} \quad \phi_r < 0$$

$$\text{or} \quad \phi_r = 0 \text{ and } \frac{\partial \phi_r}{\partial Q_j} \, dQ_j \leq 0 . \tag{8}$$

The yield surface $\phi(Q_j, H_\alpha) = 0$ is required to be convex; where a number of independent yield functions $\phi_r(Q_j, H_\alpha)$ are present we assume that each hypersurface $\phi_r(Q_j, H_\alpha) = 0$ is convex. The hardening coefficients $G(Q_j, H_\alpha)$ (equation 3) and $G_r(Q_j, H_\alpha)$ (equation 7) are assumed to be positive definite.

Whether or not plastic strain increments can occur as a result of flow is governed by the limit function $\psi(Q_j)$. If the limit function is continuously differentiable

$$dq_j^p = \lambda \frac{\partial \psi}{\partial Q_j} \quad , \tag{9}$$

where

$$\lambda > 0 \quad \text{if} \quad \psi = 0 \quad \text{and} \quad \frac{\partial \psi}{\partial Q_j} dQ_j = 0 \quad ,$$

$$\lambda = 0 \quad \text{if} \quad \psi < 0$$

$$\text{or} \quad \psi = 0 \quad \text{and} \quad \frac{\partial \psi}{\partial Q_j} dQ_j < 0 \quad . \tag{10}$$

If the limit function is not continuously differentiable, we assume the existence of a number of independent continuously differentiable limit functions $\psi_r(Q_j)(r = 1, \ldots, m)$. It follows then that $\psi < 0$ if $\psi_r < 0$ for all r, $\psi = 0$ if $\psi_r = 0$ for one or more values of r and $\psi_r < 0$ for all remaining r, and $\psi > 0$ if $\psi_r > 0$ for any r. The plastic strain increment dq_j^p is assumed to be given by

$$dq_j^p = dq_j^{p1} + dq_j^{p2} + \ldots + dq_j^{pm} \quad , \tag{11}$$

where

$$dq_j^{pr} = \lambda_r \frac{\partial \psi_r}{\partial Q_j} \quad . \tag{12}$$

The summation convention, it will be remembered, does not apply to the subscript r. Further,

$$\lambda_r \gtreqless 0 \quad \text{if} \quad \psi_r = 0 \quad \text{and} \quad \frac{\partial \psi_r}{\partial Q_j} dQ_j = 0 \quad ,$$

$$\lambda_r = 0 \quad \text{if} \quad \psi_r < 0$$

$$\text{or} \quad \psi_r = 0 \quad \text{and} \quad \frac{\partial \psi_r}{\partial Q_j} dQ_j < 0 \quad . \tag{13}$$

The limit surface $\psi = 0$ is required to be convex; where a number of independent limit functions $\psi_r(Q_j)$ are present we assume that each hypersurface $\psi_r(Q_j) = 0$ is convex. Stress states such that $\psi(Q_j) > 0$ (and consequently $\psi_r(Q_j) > 0$) are not admitted.

It is evident that while we already have general results regarding incremental uniqueness for the structural problem for any set of relations which fall within this framework, we must develop explicit expressions for ϕ, G, ψ and C_{jk} before we can proceed with the solution of any particular problem. Although we have been successful in limiting the general form of the constitutive relations, equations (1) - (13) still represent extremely complex mechanical behavior. Extensive testing programs would be required to determine the dependence of ϕ and G on recorded history; indeed it is not clear that we could completely characterize this dependence for all conceivable histories in any particular material. If we bear in mind that our major purpose in the structural problem is to determine constitutive relations which can be used in determining analytical or numerical solutions, we are led to the conclusion that further progress in regard to specific problems must depend on further approximations and the construction of more specific models.

In Part II we shall develop some of the commonly used specific idealizations of the constitutive relations and explore their application in some simple structural problems from the four classes set out in Chapter 1. We shall be concerned primarily with idealizations which are based on experimental results for *polycrystalline metals*, under isothermal conditions, and will proceed in a rather piecemeal fashion, considering in turn significant facets of the mechanical behavior, methods of characterizing them in a mathematical form, and applications in simple problems.

4.2 Hydrostatic Stress States and Plastic Volume Change in Metals

One of the most important features of the study of plastic deformation in metals is the observation that *changes in hydrostatic pressure* on an element of material *do not cause changes in plastic strain* over a very wide range of magnitudes of hydrostatic pressure or tension. This is confirmed both by experiment and by our understanding of the microstructural mechanism of plastic deformation on the scale of the crystal lattice. In very few technological problems does the hydrostatic pressure lie outside of this range; it might occur, for example, in hypervelocity impact problems where one mass of metal strikes another at very high velocity, causing the material to behave more like a fluid than a solid for very short times. In this problem, as in others where very large hydrostatic pressures occur, the theory of plasticity presented in this monograph would not be applicable because of rate effects, thermal effects, large deformations and other factors. It is eminently justifiable therefore, that we should assume that changes in hydrostatic pressure or tension never cause changes in plastic strains.

We can best investigate the effects of this assumption by considering an element of material (or a homogeneously stressed body) in a *cartesian coordinate system which coincides with the principal axes.* An element of material subjected to principal stresses σ_I, σ_{II} and σ_{III} in the x, y and z directions respectively is shown in Figure 1(a). The appropriate stress space for this state of stress (shown in Figure 1(b)) is three dimensional, and the stress point has coordinates σ_I, σ_{II}, σ_{III}. A state of pure hydrostatic tension or compression is

Figure 1. Principal stress state

characterized by $\sigma_I = \sigma_{II} = \sigma_{III}$; the stress point corresponding to such states must therefore lie on a line in the stress
space which passes through the origin and which makes equal
angles with the positive σ_I, σ_{II} and σ_{III} directions. This
line is parallel to the unit vector $\underline{v}$ which has components
which are equal

$$v_I = v_{II} = v_{III} = \frac{1}{\sqrt{3}} \ . \tag{14}$$

If we consider a stress state σ_I, σ_{II}, σ_{III} and impose a
stress increment which consists of a pure hydrostatic tension,
the stress increment is given by a vector of length $d\gamma$, say,
with the direction of the unit vector given in equation (14).
This stress increment is thus given by

$$d\sigma_I = \frac{1}{\sqrt{3}}\,d\gamma, \quad d\sigma_{II} = \frac{1}{\sqrt{3}}\,d\gamma, \quad d\sigma_{III} = \frac{1}{\sqrt{3}}\,d\gamma \tag{15}$$

As a result of this stress increment, the stress point will
move along a line *parallel to the pure hydrostatic tension
line.*

If we now introduce a yield surface into the stress space,
we see that as a result of the assumption that changes in hydrostatic tension or pressure do not cause plastic strain
changes, the yield surface must be such that for any stress

state σ_I, σ_{II}, σ_{III} which lies on the yield surface a stress increment of the type described by equation (15) can never cause *loading*. Identifying σ_I with Q_1, σ_{II} with Q_2 and σ_{III} with Q_3, we may not permit the condition

$$\frac{\partial\phi}{\partial Q_j}\,dQ_j \;\equiv\; \frac{\partial\phi}{\partial\sigma_I}\frac{d\gamma}{\sqrt{3}} \;+\; \frac{\partial\phi}{\partial\sigma_{II}}\frac{d\gamma}{\sqrt{3}} \;+\; \frac{\partial\phi}{\partial\sigma_{III}}\frac{d\gamma}{\sqrt{3}}$$

$$= \left(\frac{\partial\phi}{\partial\sigma_I} + \frac{\partial\phi}{\partial\sigma_{II}} + \frac{\partial\phi}{\partial\sigma_{III}}\right)\frac{d\gamma}{\sqrt{3}} \;>\; 0$$

(16)

In addition we cannot permit *unloading*, since if we had the expression in equation (16) less than zero, we could change the sign of $d\gamma$ and have loading. Thus, *if the stress point σ_I, σ_{II}, σ_{III} lies on the yield surface, the stress increment of equation (15) must constitute neutral loading.*

We can study the implications of this statement in two ways. First, looking at it geometrically, for any state of stress on the yield surface the stress increment of equation (15) must lie *in the yield surface*; it is evident, therefore, that the yield surface must be a *prism generated by a line which is parallel to the unit vector* $\nu_I = \nu_{II} = \nu_{III} = 1/\sqrt{3}$, i.e., parallel to the pure hydrostatic tension line. This prism is of course open ended, so that hydrostatic stress increments (equation 15) of any magnitude always lead to a stress point for which $\phi \leq 0$ i.e., either inside or on the yield surface. It is convenient to look at this prism from a point at infinity on the pure hydrostatic tension line. We would then see the prism as a closed curve, and the axes of the stress space as lines each separated by an angle of 120°, as shown diagrammatically in Figure 2. Alternatively, this figure can be thought of as the *projection* of the prism and the axes on a plane which passes through the origin and is perpendicular to the pure hydrostatic tension line. This plane is often referred to as the π-*plane*.

In order to describe the dependence of the yield function on stress, it is convenient to consider the *stress vector* (the position vector of the stress point in the stress space) as the sum of two components; one component lying in the π-plane and one component perpendicular to the π-plane and parallel to the pure hydrostatic tension line (Figure 3). The component perpendicular to the π-plane has the direction of the unit vector $\nu_I = \nu_{II} = \nu_{III} = 1/\sqrt{3}$. Its magnitude is given by the scalar product of the stress vector, which has components σ_I, σ_{II}, σ_{III} and the unit vector $\nu_I = \nu_{II} = \nu_{III} = 1/\sqrt{3}$; this is $(\sigma_I + \sigma_{II} + \sigma_{III})/\sqrt{3}$. Next we resolve this vector into three components in the σ_I, σ_{II}, σ_{III} directions. These three components are equal, and each is given by

$$\bar{\sigma} = \frac{1}{3}(\sigma_I + \sigma_{II} + \sigma_{III}) \ . \tag{17}$$

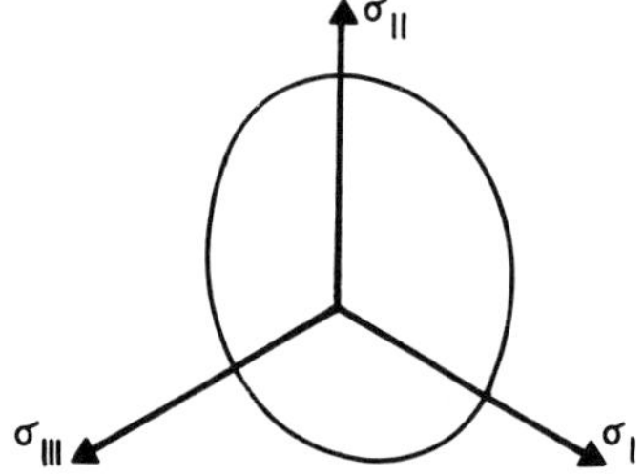

Figure 2. Yield surface projected onto the π-plane

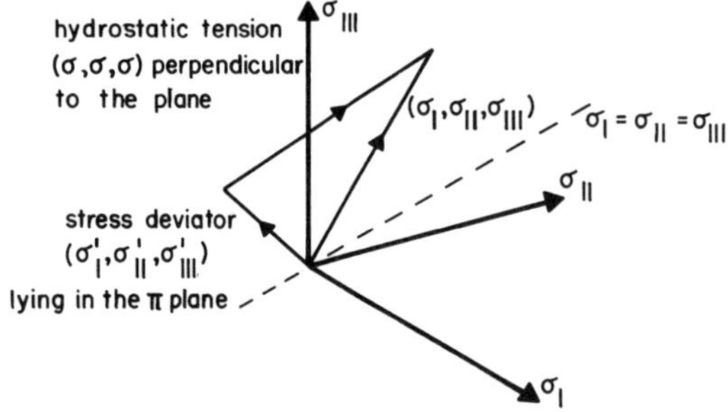

Figure 3. Stress deviator and hydrostatic tension

The component $\bar{\sigma}$ is the mean value of the principal stresses
σ_I, σ_{II}, σ_{III}; it is called the *mean hydrostatic tension*. The
component of the stress vector which lies in the π-plane is
found by subtracting the mean hydrostatic tension vector $\bar{\sigma}$, $\bar{\sigma}$,
$\bar{\sigma}$ from the stress vector σ_I, σ_{II}, σ_{III}. We refer to this as
the *stress deviator*, and its components in the three coordinate
directions in the stress space are

$$\sigma_I' = \sigma_I - \bar{\sigma} = \frac{2}{3}\sigma_I - \frac{1}{3}\sigma_{II} - \frac{1}{3}\sigma_{III} \quad ,$$

$$\sigma_{II}' = \sigma_{II} - \bar{\sigma} = \frac{2}{3}\sigma_{II} - \frac{1}{3}\sigma_I - \frac{1}{3}\sigma_{III} \quad , \qquad (18)$$

$$\sigma_{III}' = \sigma_{III} - \bar{\sigma} = \frac{2}{3}\sigma_{III} - \frac{1}{3}\sigma_I - \frac{1}{3}\sigma_{II} \quad .$$

It is clear from this representation that a stress point which
lies on the π-plane is such that $\bar{\sigma} = 0$.

If we again view the prismatic yield surface from a point of
infinity on the pure hydrostatic tension line, we see that
whether or not the stress point lies on the yield surface *depends only on the stress deviator* and not on the mean hydrostatic tension, since changes in hydrostatic tension are not apparent from this viewpoint. We can thus further limit the
functional dependence of ϕ for an element of material;

$$\phi = \phi\left(\sigma_I', \sigma_{II}', \sigma_{III}', H_\alpha\right) \quad . \qquad (19)$$

It should be noted that we are still bound by the requirement
that ϕ should be a convex function. Regarding convexity as the
requirement that any two points which lie within or on the
yield surface can be joined by a line which does not lie outside the yield surface, this requirement reduces to that of
convexity of the projection of the yield surface on the π-plane

(Figure 2).

Whether plastic strain increments occur as a result of hardening behavior or flow, the plastic strain increment at a nonsingular point on the yield surface is given by $dq_j^p = \lambda \partial\phi/\partial Q_j$ (cf. Section 2.5, equation 93). In the present context this gives

$$d\varepsilon_I^p = \lambda \frac{\partial\phi}{\partial\sigma_I} \quad , \quad d\varepsilon_{II}^p = \lambda \frac{\partial\phi}{\partial\sigma_{II}} \quad , \quad d\varepsilon_{III}^p = \lambda \frac{\partial\phi}{\partial\sigma_{III}} \quad , \tag{20}$$

where ε_I^p, ε_{II}^p, ε_{III}^P are the plastic extensions per unit length in the three principal coordinate directions. We have already noted that a stress increment parallel to the unit vector ν_I $\nu_{II} = \nu_{III} = 1/\sqrt{3}$ represents neutral loading. Hence (see equation 16)

$$\frac{1}{\sqrt{3}} \left(\frac{\partial\phi}{\partial\sigma_I} + \frac{\partial\phi}{\partial\sigma_{II}} + \frac{\partial\phi}{\partial\sigma_{III}} \right) = 0 \quad . \tag{21}$$

Comparing equations (20) and (21), it is clear that

$$d\bar{\varepsilon}^p = d\varepsilon_I^p + d\varepsilon_{II}^p + d\varepsilon_{III}^p = 0 \quad . \tag{22}$$

The sum $\bar{\varepsilon}^p$ is the plastic volume strain; consequently there is *no volume change associated with plastic strain.* This result has been confirmed experimentally in metals, and is also consistent with the microstructural mechanism by which plastic strain occurs. Plastic deformation appears in an element of material as a *distortion* of the element i.e., a change in *shape* without a change in volume.

4.3 Shear Stress on a Plane

We have seen that for a given recorded history the yield function depends on the stress deviator only. Since plastic changes can be caused only when the stress point lies on the

yield surface, this suggests that the *shear stresses on various
planes through the element* are the chief factor in determining
whether the stress point lies on the yield surface and whether
plastic strain increments can be induced by further loading or
through flow. This comes about because the shear stress acting
on any plane through the material is unaffected by the imposition of hydrostatic tension or compression. To determine the
stresses acting on any plane through the material, as opposed
to the stresses on the planes normal to the three principal
axes as shown in Figure 1(a), we consider a tetrahedron made up
of the plane we wish to consider and the three planes perpendicular to the principal axes; this tetrahedron is shown diagrammatically in Figure 4. Let the unit vector normal to the
inclined plane be $\underline{\mu}$, with components μ_x, μ_y and μ_z. If the
area of the face of the tetrahedron formed by the inclined
plane is ds, it follows that the other three faces of the tetrahedron have areas $\mu_x$ds, $\mu_y$ds and $\mu_z$ds respectively. As shown
in the figure, the *forces* acting on the three planes of the
tetrahedron perpendicular to the coordinate axes are $\sigma_I\mu_x$ds,
$\sigma_{II}\mu_y$ds and $\sigma_{III}\mu_z$ds in the negative coordinate directions
only. The force acting on the inclined plane is a vector $\underline{T}$
which we resolve into two components; $\sigma_n$ds perpendicular to the
plane and $\sigma_s$ds lying in the plane. σ_n is then the normal stress
on the plane and σ_s is the shear stress.

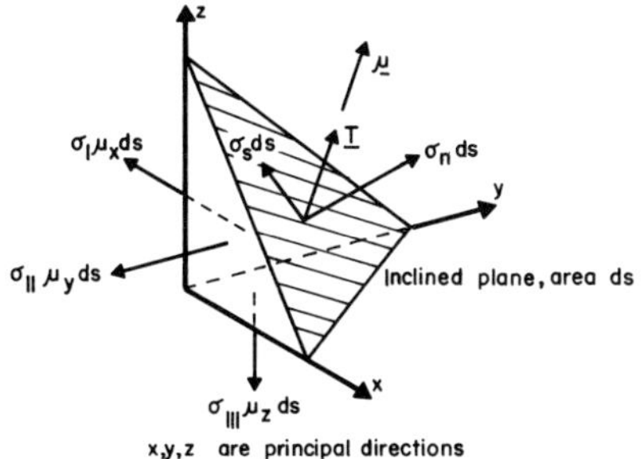

Figure 4. Forces acting on an element of material

Equilibrium requirements applied to the tetrahedron show that the force vector $\underline{T}$ acting on the inclined plane has components $(\sigma_I \mu_x ds, \sigma_{II} \mu_y ds, \sigma_{III} \mu_z ds)$. The component of this force normal to the plane is found by taking the scalar product of the force and the unit vector normal to the plane;

$$\sigma_n ds = \underline{T} \cdot \underline{\mu}$$

$$= (\sigma_I \mu_x^2 + \sigma_{II} \mu_y^2 + \sigma_{III} \mu_z^2) ds \quad . \tag{23}$$

The component of $\underline{T}$ lying in the plane is then given by

$$\sigma_s^2 \, ds^2 = \underline{T} \cdot \underline{T} - \sigma_n^2 \, ds^2 \quad . \tag{24}$$

Cancelling the common factor ds^2, the shear stress on the inclined plane is given by

$$\sigma_s^2 = (\sigma_I^2 \mu_x^2 + \sigma_{II}^2 \mu_y^2 + \sigma_{III}^2 \mu_z^2)$$

$$- (\sigma_I \mu_x^2 + \sigma_{II} \mu_y^2 + \sigma_{III} \mu_z^2)^2 \quad . \tag{25}$$

This expression holds for any set of principal stresses σ_I, σ_{II}, σ_{III}. Suppose now that we add to these stresses a hydrostatic tension (σ, σ, σ) leading to a net set of principal stresses $(\sigma_I + \sigma, \sigma_{II} + \sigma, \sigma_{III} + \sigma)$. For this new state of stress, the shear stress on the inclined plane is given by

$$\sigma_s^2 = \{(\sigma_I + \sigma)^2 \, \mu_x^2 + (\sigma_{II} + \sigma)^2 \, \mu_y^2 + (\sigma_{III} + \sigma)^2 \, \mu_z^2\}$$

$$- \{(\sigma_I + \sigma)\mu_x^2 + (\sigma_{II} + \sigma)\mu_y^2 + (\sigma_{III} + \sigma)\mu_z^2\}^2 \quad . \tag{26}$$

Consider in turn the two expressions on the right-hand side of this equation. First

$$\{(\sigma_I + \sigma)^2 \mu_x^2 + (\sigma_{II} + \sigma)^2 \mu_y^2 + (\sigma_{III} + \sigma)^2 \mu_z^2\}$$

$$= (\sigma_I^2 \mu_x^2 + \sigma_{II}^2 \mu_y^2 + \sigma_{III}^2 \mu_z^2) \qquad (27)$$

$$+ 2\sigma(\sigma_I \mu_x^2 + \sigma_{II} \mu_y^2 + \sigma_{III} \mu_z^2) + \sigma^2 \quad,$$

after making use of the property of the unit vector $\underline{\mu}$,

$$\mu_x^2 + \mu_y^2 + \mu_z^2 = 1 \quad. \qquad (28)$$

The second term is written as

$$\{(\sigma_I \mu_x^2 + \sigma_{II} \mu_y^2 + \sigma_{III} \mu_z^2) + \sigma\}^2$$

$$= (\sigma_I \mu_x^2 + \sigma_{II} \mu_y^2 + \sigma_{III} \mu_z^2)^2 \qquad (29)$$

$$+ 2\sigma(\sigma_I \mu_x^2 + \sigma_{II} \mu_y^2 + \sigma_{III} \mu_z^2) + \sigma^2 \quad.$$

Subtracting equation (29) from equation (27), and comparing
the result with equation (25), it is evident that σ_s^2 is unchanged by an addition of hydrostatic tension to the principal
stress state. The shear stress on the inclined plane is consequently a function of the stress deviator only.

The earliest investigations into the nature of the yield
surface were strongly concerned with plastic deformation as the
limit of the range of behavior of the theory of elasticity. We
shall follow this chronological development and consider in the
remainder of this Chapter two important hypotheses which were
advanced concerning the *initial yield surface.*

These hypotheses assumed that the material was in its *virgin
state,* partly because of the difficulty in characterizing the
initial state in real materials. As a consequence the initial

yield function depends only on stress; there are no recorded history parameters.

4.4 The von Mises Initial Yield Condition

The first hypothesis is termed the *von Mises yield condition*. It assumes that plastic deformation becomes possible when the shear stress on a particular inclined plane, the *octahedral plane* which is equally inclined to the three coordinate directions with $\mu_x = \mu_y = \mu_z = 1/\sqrt{3}$, reaches a magnitude k. Normally we write this in a quadratic form, so that the *initial yield function* is

$$\phi(\sigma_I, \sigma_{II}, \sigma_{III}) = \sigma_{so}^2 - k^2 \tag{30}$$

where σ_{so} in the *octahedral shear stress*. The equation of the *initial yield surface* is then

$$\sigma_{so}^2 - k^2 = 0 \quad . \tag{31}$$

We may obtain an expression for σ_{so}^2 from equation (25) by putting $\mu_x = \mu_y = \mu_z = 1/\sqrt{3}$. This gives

$$\sigma_{so}^2 = \frac{1}{3} (\sigma_I^2 + \sigma_{II}^2 + \sigma_{III}^2) - \frac{1}{9} (\sigma_I + \sigma_{II} + \sigma_{III})^2 \quad . \tag{32}$$

As a special case of σ_s^2, σ_{so}^2 is a function of the stress deviator. If we return to Figure 3, we see that the stress deviator is a vector in the stress space which lies in the π-plane with components σ_I', σ_{II}', σ_{III}'. The square of the *magnitude* of this vector is, using equation (18),

$$(\sigma_I')^2 + (\sigma_{II}')^2 + (\sigma_{III}')^2$$

$$= \frac{1}{9} \{(2\sigma_I - \sigma_{II} - \sigma_{III})^2 + (2\sigma_{II} - \sigma_I - \sigma_{III})^2 \tag{33}$$

$$+ (2\sigma_{III} - \sigma_{II} - \sigma_I)^2\} \quad .$$

Through straight-forward algebraic manipulations which we
leave to the reader, it may be shown that

$$\sigma_{so}^{2} = \frac{1}{3}\{(\sigma_{I}')^{2} + (\sigma_{II}')^{2} + (\sigma_{III}')^{2}\} \ . \tag{34}$$

Noting again that the stress deviatior is a vector lying in the
π-plane, comparison of equations (31) and (34) show that the
projection of the von Mises yield surface on the π-plane is a
circle whose origin coincides with the origin of the stress
space and whose radius is $\sqrt{3}$ k, as shown in Figure 5. The complete yield surface, described by equations (31) and (32) is a
cylinder whose axis coincides with the pure hydrostatic tension line in the stress space.

4.5 The Tresca Initial Yield Condition

The second hypothesis is termed the *Tresca yield condition*. It
assumes that plastic deformation first becomes possible when
the shear stress on *any* inclined plane reaches a magnitude $\bar{k}$.
In alternative terms, plastic deformation is not possible (i.e.
the yield function is negative) when the greatest value of σ_s
for any value of μ_x, μ_y, μ_z is less than $\bar{k}$. It is necessary,
therefore, to determine the greatest value of σ_s, and we do
this by returning to equation (25) and looking for *stationary
values* of σ_s^2 for fixed σ_I, σ_{II}, σ_{III} and varying μ_x, μ_y, μ_z.
In view of equation (28), μ_x, μ_y and μ_z are not independent

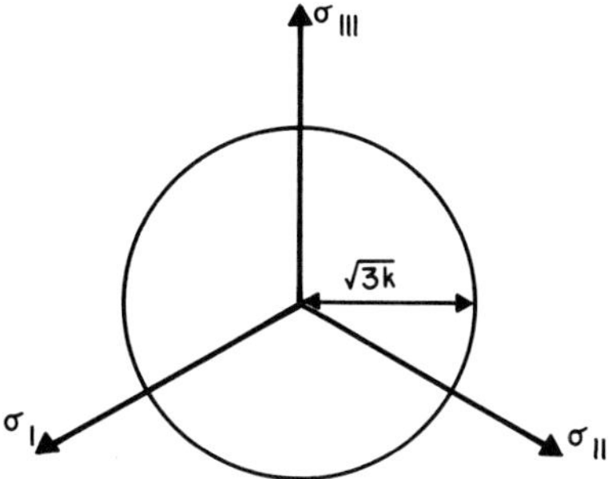

Figure 5.   The von Mises Yield Surface in the π-plane

parameters; let us put

$$\mu_z^2 = 1 - \mu_x^2 - \mu_y^2 \quad , \tag{35}$$

and substitute this value into equation (25). We then obtain

$$\sigma_s^2 = \mu_x^2(\sigma_I^2 - \sigma_{III}^2) + \mu_y^2(\sigma_{II}^2 - \sigma_{III}^2) + \sigma_{III}^2$$
$$\tag{36}$$
$$- \left[\mu_x^2(\sigma_I - \sigma_{III}) + \mu_y^2(\sigma_{II} - \sigma_{III}) + \sigma_{III}\right]^2 \quad .$$

The quantity σ_s^2 is stationary when

$$\frac{\partial \sigma_s^2}{\partial \mu_x} = \frac{\partial \sigma_s^2}{\partial \mu_y} = 0 \quad . \tag{37}$$

Carrying out these partial differentiations, we obtain two simultaneous equations for μ_x and μ_y;

$$\mu_x(\sigma_I - \sigma_{III}) \left[(\sigma_I + \sigma_{III}) - 2 \{\mu_x^2(\sigma_I - \sigma_{III})\right.$$
$$\tag{38a}$$
$$\left. + \mu_y^2(\sigma_{II} - \sigma_{III}) + \sigma_{III}\}\right] = 0 \quad ,$$

$$\mu_y(\sigma_I - \sigma_{III}) \left[(\sigma_{II} + \sigma_{III}) - 2 \{\mu_x^2(\sigma_I - \sigma_{III})\right.$$
$$\tag{38b}$$
$$\left. + \mu_y^2(\sigma_{II} - \sigma_{III}) + \sigma_{III}\}\right] = 0 \quad .$$

One solution to these equations is given by $\mu_x = \mu_y = 0$, and hence, from equation (35), $\mu_z = 1$. The plane defined by this normal vector is perpendicular to the z axis, and on this plane the normal stress is σ_{III}, the shear stress zero. Hence this stationary value corresponds to a minimum of σ_s^2. An alternative solution can be obtained by putting $\mu_x = 0$, so that (38a) is satisfied, and the contents of the bracket in equation (38b) equal to zero,

$$(\sigma_{II} + \sigma_{III}) = 2[\mu_y^2(\sigma_{II} - \sigma_{III}) + \sigma_{III}] \ . \tag{39}$$

Solving for μ_y, and determining μ_z from equation (35), we see that

$$\mu_x = 0, \ \mu_y = \frac{1}{\sqrt{2}}, \ \mu_z = \frac{1}{\sqrt{2}} \tag{40}$$

gives a stationary value of σ_s^2. The appropriate value of σ_s^2 is found by substituting equation (40) into equation (36) or equation (25);

$$\sigma_s^2 = \frac{1}{4}(\sigma_{II} - \sigma_{III})^2 \ . \tag{41}$$

We can now repeat this process by substituting for μ_x^2 and μ_y^2 into equation (25) in place of μ_z^2 (cf. equation 35). We find two other minimum values corresponding to $\sigma_s = 0$ when $\underline{\mu}$ has components (1, 0, 0) and (0, 1, 0). Two other stationary values also found,

$$\sigma_s^2 = \frac{1}{4}(\sigma_I - \sigma_{II})^2 \ \text{when} \ \underline{\mu} = (\frac{1}{\sqrt{2}}, \frac{1}{\sqrt{2}}, 0) \ , \tag{42}$$

$$\sigma_s^2 = \frac{1}{4}(\sigma_I - \sigma_{III})^2 \text{when} \ \underline{\mu} = (\frac{1}{\sqrt{2}}, 0, \frac{1}{\sqrt{2}}) \ . \tag{43}$$

The three planes on which the non-zero stationary values of σ_s^2 occur are respectively at $45°$ to two axes and parallel to the third, as shown in Figure 6. The maximum value of σ_s^2 must be the largest of these stationary values.

We can now construct the *Tresca yield function* by observing

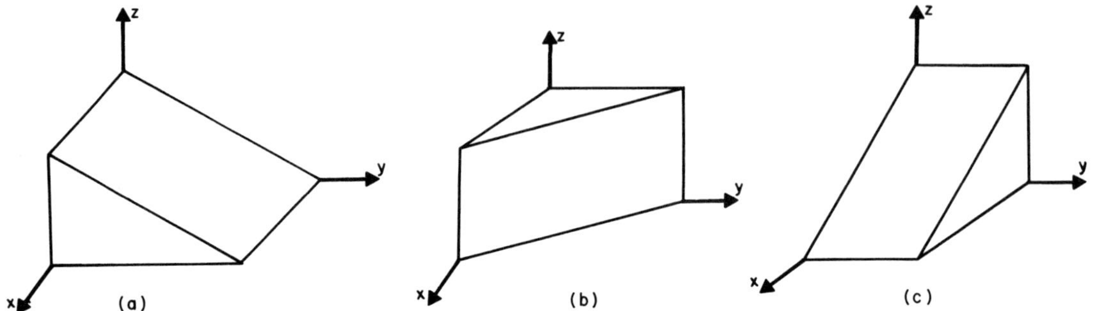

Figure 6. Planes on which σ_s^2 is stationary

that if the yield function is non-positive provided that the largest of the three stationary values of σ_s^2 (equations 41, 42 and 43) is less than or equal to $\bar{k}^2$, it must be non-positive if *each* of the stationary values itself is less than or equal to $\bar{k}^2$. Taking into account the plus or minus sign introduced when we write an expression for σ_s rather than σ_s^2, the Tresca yield function actually consists of *six independent yield functions* ϕ_r $(r = 1, \ldots, 6)$,

$$\phi_{1,2} = \pm (\sigma_I - \sigma_{II}) - 2\bar{k} \ ,$$

$$\phi_{3,4} = \pm (\sigma_{II} - \sigma_{III}) - 2\bar{k} \ , \tag{44}$$

$$\phi_{5,6} = \pm (\sigma_{III} - \sigma_I) - 2\bar{k} \ .$$

The *Tresca yield surface* is thus a piecewise continuously differentiable surface in the principal stress space; it is the surface of the domain enclosed by the *six planes*

$$\pm (\sigma_I - \sigma_{II}) = 2\bar{k} \ ,$$

$$\pm (\sigma_{II} - \sigma_{III}) = 2\bar{k} \ , \tag{45}$$

$$\pm (\sigma_{III} - \sigma_I) = 2\bar{k} \ .$$

In order to determine the projection of this yield surface

on the π-plane we can employ the following procedure. Consider first the two planes

$$+ (\sigma_I - \sigma_{II}) = 2\bar{k}, \quad - (\sigma_I - \sigma_{II}) = 2\bar{k}.$$

These equations are independent of σ_{III}, hence the planes are parallel to the σ_{III} axis in the principal stress space. The projection of these planes on the $(\sigma_I - \sigma_{II})$ plane are two straight lines of unit slope with intercepts $2\bar{k}$ and $-2\bar{k}$ on the σ_I axis. It is clear that the plane $(\sigma_I - \sigma_{II}) = 0$ is parallel to the yield planes; it also passes through the origin and the σ_{III} axis. Next we note that the line $\sigma_I = \sigma_{II} = \sigma_{III}$ lies in the plane $(\sigma_I - \sigma_{II}) = 0$. Thus, if we view the stress space from a point at infinity on the pure hydrostatic tension line $\sigma_I = \sigma_{II} = \sigma_{III}$, the plane $(\sigma_I - \sigma_{II}) = 0$ appears as a *line* which coincides with the σ_{III} axis. Hence the projections of the planes $\pm (\sigma_I - \sigma_{II}) = 2\bar{k}$ are *parallel* to the σ_{III} axis on the π-plane, and symmetrically disposed on either side of it (Figure 7). We can show in a similar way that in the π-plane the projections of the planes $\pm (\sigma_{II} - \sigma_{III}) = 2\bar{k}$ are parallel to the σ_I axis, and the projections of the planes $\pm (\sigma_I - \sigma_{III}) = 2\bar{k}$ are parallel to the σ_{II} axis. The complete projection of the *yield surface* on the π-plane is thus the hexagonal envelope of the projections of the six planes of equation (45). The origin of the stress space lies at the center of this hexagon. The

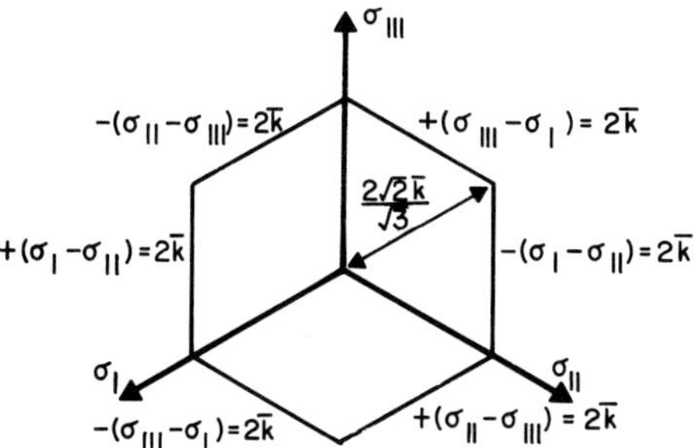

Figure 7. Tresca yield surface in the π-plane

yield surface is therefore a *hexagonal prism* whose center line coincides with the pure hydrostatic tension line.

In order to determine the distance from the origin to a vertex of the hexagon in the π-plane, we note that the principal stress state ($\sigma_I = 2\bar{k}$, $\sigma_{II} = 0$, $\sigma_{III} = 0$) lies on the yield surface at a vertex, since it lies on two of the six planes of equation (45). The σ_I axis is inclined at an angle to the π-plane whose cosine is $\sqrt{2}/\sqrt{3}$; this is obtained by considering first the cosine of the angle between the σ_I axis and the normal to the π-plane, which is the scalar product of the unit vectors $(1, 0, 0)$ and $(1/\sqrt{3}, 1/\sqrt{3}, 1/\sqrt{3})$. Consequently the projection of the vector joining the origin to the stress state $(2\bar{k}, 0, 0)$ on the π-plane is $2\sqrt{2}\bar{k}/\sqrt{3}$.

In order to compare the von Mises and Tresca initial yield conditions some relation between k and $\bar{k}$ must be assumed. It is usual, although by no means necessary, to assume that $\sqrt{3}k = 2\sqrt{2}\bar{k}/\sqrt{3}$, or $k = 2\sqrt{2}\bar{k}/3$, so that the radius of the von Mises cylinder is equal to the projection of a line joining the origin to a point on the vertex of the Tresca prism. It is evident then, with this normalization, that *the von Mises cylinder circumscribes the hexagonal prism of the Tresca yield surface.*

4.6 Consequences of Isotropy

The construction of the von Mises and Tresca yield surfaces contains two important implicit assumptions which we shall investigate briefly. First, we note that the octahedral plane and the plane on which the maximum shear stress occurs have a *fixed orientation* with respect to the principal axes x, y, z. On the other hand, the principal axes are not required to have any fixed orientation with respect to a set of axes X, Y, Z fixed in the larger body from which the element of material shown in Figure 1 is taken. According to our formulation, if a principal stress state σ_I, σ_{II}, σ_{III} with principal directions x, y, z lies on the yield surface, then a principal stress state σ_I, σ_{II}, σ_{III} with different principal axes x', y', z' (i.e., differently oriented with X, Y, Z) also lies on the yield surface. Alternatively, we have *assumed* that k and $\bar{k}$ are not functions of the orientation of the principal axes x, y, z with respect to X, Y, Z. In turn, this implies that the body has the same relevant physical properties in all direc- tions in the X, Y, Z coordinate system. The material under discussion is consequently *isotropic.*

In order to understand the implications of the assumptions of isotropy on the projection of the yield surface on the π- plane we note that a necessary condition for isotropy is that we can relabel the coordinate axes x, y, z in any order. Thus, if a principal stress state (σ_I = a, σ_{II} = b, σ_{III} = c) lies on the yield surface, the *five other principal stress states* (a, c, b), (c, a, b), (c, b, a), (b, a, c), (b, c, a) *must also lie on the yield surface.* Consider one pair of these stress states, (a, b, c) and (b, a, c). Plotting these points in the plane σ_{III} = c in the principal stress space, it is readily ascertained that the stress points (a, b, c) and (b, a, c) are symmetrically disposed about the plane σ_I = σ_{II}. We have al- ready seen that the projection of the σ_I = σ_{II} plane on the π-plane is a line coinciding with the σ_{III} axis. Hence the projections of the points (a, b, c) and (b, a, c) and the pro- jection of the yield surface must be symmetric about the pro- jection of the σ_{III} axis in the π-plane. Similarly, by con- sidering other pairs of the six principal stress combinations which lie on the yield surface, it may be seen that the projec- tion of the yield surface in the π-plane must also be symmetric about the projections of the σ_I and σ_{II} axes. Figure 8 shows

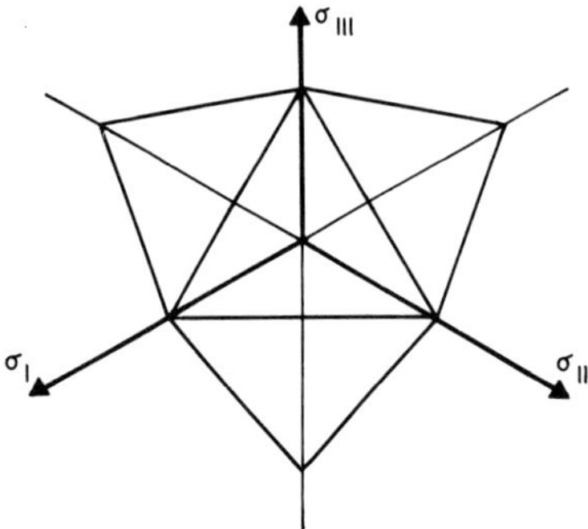

Figure 8. Isotropic yield surfaces

two convex yield surfaces in the π-plane which imply isotropic
materials.

The second important implicit assumption concerns the direction in which σ_s acts on the inclined plane. We have computed
σ_s simply on the assumption that it is the component of $\underline{T}$
(Figure 4) which lies in the inclined plane without regard to
its direction. Alternatively, we have assumed that k and $\bar{k}$
are not functions of the direction in which σ_s acts. One result of this assumption is the following. If the principal
stress state (a, b, c) leads to a shear stress σ_s on any inclined plane, the stress state (-a, -b, -c) leads to a shear
stress on the same inclined plane of the same magnitude σ_s but
opposite direction. Thus if the stress point (a, b, c) lies
on the yield surface, (-a, -b, -c) also lies on the yield surface. In a *uniaxial stress state* where b = c = 0, this is
equivalent to the statement that the *yield stresses in tension
and compression are equal.*

The implications of this assumption for the projection of
the yield surface in the π-plane can be determined as follows.
We first note that we can add hydrostatic tension to a stress
point on the yield surface and still leave it on the yield
surface. To the principal stress states (a, b, c) and (-a, -c,
-b), both of which lie on the yield surface if we make use of

isotropy to interchange σ_{II} and σ_{III} in the second stress
state, we add respectively the hydrostatic tensions

$$\left(- \frac{b+c}{2}, - \frac{b+c}{2}, - \frac{b+c}{2}\right) \text{ and } \left(\frac{b+c}{2}, \frac{b+c}{2}, \frac{b+c}{2}\right) \quad .$$

We then get two further stress states

$$\left(a - \frac{b+c}{2}, \frac{b-c}{2}, \frac{c-b}{2}\right) \text{ and } \left(\frac{b+c}{2} -a, \frac{b-c}{2}, \frac{c-b}{2}\right).$$

The projection of each of these stress states on the π-plane
lies on the projection of the yield surface. However, we have
simply changed the sign of σ_I alone in order to get one stress
state from the other; this means that the projections of these
two stress points and the projection of the yield surface are
symmetric about a line in the π-plane perpendicular to the
projection of the σ_I axis. In a similar way it can be shown
that the projection of the yield surface on the π-plane is also
symmetric about lines perpendicular to the projections of the
σ_{II} and σ_{III} axes. Each of these lines are projections of
states of stress in which one principal stress is zero and the
other two are equal in magnitude and opposite in sign.

Thus the yield surface of a material which is isotropic and
whose yielding behavior is invariant under a reversal of sign
of the stresses exhibits a high degree of symmetry in the
principal stress space. It is symmetric about six planes which
divide the yield surface into twelve segments; if one of these
segments is known, the others can be reproduced. Both the von
Mises and Tresca yield surfaces observe this symmetry. Neither
of these two assumptions is inconsistent for a polycrystalline
metal in its virgin state. It is composed of a very large number of randomly oriented crystals, and can thus be expected to
be isotropic. Initial yielding in a single crystal under

uniaxial stress is found to depend only on the magnitude of
the shear stress on those crystal planes where sliding can occur when the material is in its virgin state. Although the
mechanism of plastic deformation is more complicated, it is
then to be expected that the yield stresses in tension and
compression in a virgin polycrystal are equal, and this is confirmed experimentally.

We can also use the symmetry of the projection of the yield
surface in the π-plane to make some remarks about the probable
accuracy of the von Mises and Tresca yield conditions for an
isotropic virgin polycrystalline material. Figure 9 shows the
π-plane with the six axes of symmetry of the yield surface
drawn. If we add to this the requirement that the yield
stresses in uniaxial tension and compression in each of the
three coordinate directions in the stress space must all be
equal, we obtain six points (shown as heavy dots) through which
the yield surface must pass. Finally, the yield surface must
be convex. The yield surface can be convex and satisfy the
other requirements only if it lies *between* the two figures
drawn in Figure 9. The inner figure is the Tresca yield condition; it can be thought of as the inner limit of convexity.
The von Mises circle circumscribes the inner figure and inscribes the outer figure. The greatest discrepancy between the
outer hexagon and the Tresca hexagon is 25%, and between the

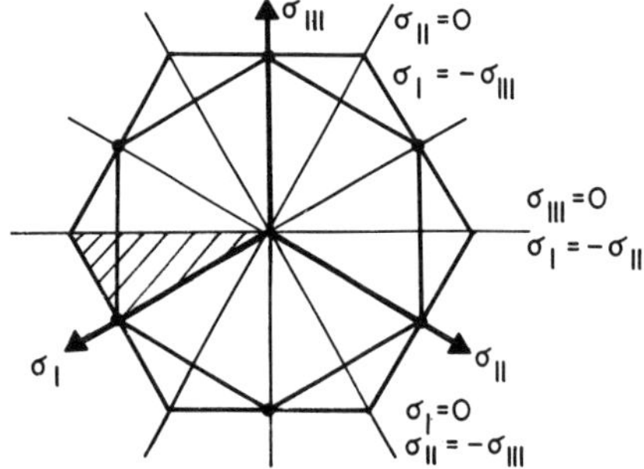

Figure 9. Limiting polygons for the initial yield surface

von Mises circle and the Tresca hexagon, about 13.5%. It may
be expected, therefore, that the Tresca and von Mises yield
surfaces cannot be greatly different from the measured initial
yield surface in an isotropic material. If a different normalization procedure is used the discrepancy between the von
Mises or Tresca yield condition and any actual initial yield
surface can be still further reduced.

After some plastic deformation has taken place it can no
longer be expected that the material will be isotropic or that
the tension and compressive yield stresses will be the same.
In Chapter 5, we shall consider some experimental results which
show initial and subsequent yield surfaces for some stress histories in particular materials, and some simple idealizations
regarding the dependence of the yield function on recorded history will be constructed.

4.7 Historical and Bibliographical Remarks

The effects of hydrostatic pressure on yielding and plastic deformation are best known from the classical work of Bridgman
[1923], [1947], [1952]. A critical review of the experiments
by Bridgman and other investigators has been given by Crossland
[1954].

The maximum shear stress condition for yielding was proposed
by Tresca [1864]. von Mises [1913] introduced the yield condition with which we have associated his name. It should be
noted that von Mises considered this criterion as a mathematically simple idealization of Tresca's yield condition; in a
geometrical sense this can be seen in a comparison between the
cylinder and the hexagonal prism in the principal stress space,
and the simplicity will be better appreciated when we consider
situations in which the principal axis directions are not
known. Physical interpretations can be ascribed; the von Mises

yield condition was first written in terms of the octahedral
shear stress by Nadai [1937]. We have preferred this approach,
in the principal stress space at least, because the magnitude
of the octahedral shear stress is directly proportional to the
magnitude of the stress deviator.

It is also possible to write the von Mises condition as a
limit on the strain energy which can be stored in distortion
[Hencky 1924]. The idea of using the strain energy as a yield
condition can be traced to Maxwell [1856] and Beltrami [1885].
However, they suggested that the total strain energy be taken
into account. Huber [1904] anticipated Hencky's observation
by suggesting a yield condition which employed the total or
distortional strain energy depending on whether the hydrostatic
tension was positive or negative. Mohr [1914] suggested a more
general yield condition of which the von Mises condition is a
special case.   The von Mises condition is consequently sometimes referred to as the Huber-Mises condition or as the Mohr
condition.

References

E.Beltrami   1885   Rendiconti 1st Lomb. (2), $\underline{18}$, 704.

P.N.Bridgman   1923   "The compressibility of thirty me-
                         tals as a function of pressure and
                         temperature", Proc.Am.Acad.Art.Sci.,
                         $\underline{58}$, 163.

P.N.Bridgman   1947   "The effect of hydrostatic pressure
                         on the fracture of brittle substan-
                         ces", J.Appl.Phys., $\underline{18}$, 246.

P.N.Bridgman   1952   *Studies in Large Plastic Flows and
                         Fracture with Special Emphasis on
                         the Effects of Hydrostatic Pressure,*
                         McGraw-Hill (N.Y.).

B.Crossland   1954   "The effects of fluid pressure on
                         the shear properties of metals",
                         Proc.Inst.Mech.Engr., $\underline{169}$, 935.

H.Hencky   1924   "Zur theorie plastischer Deforma-
                            tionen und der hierdurch in Material
                            hervorgerufenen Nach spannungen",
                            Proc.1st Int.Congr.Appl.Mech.(Delft)
                            312.

M.T.Huber   1904   Czasopismo Techniczne (Lwow), 22,
                            81.

J.C.Maxwell   1856   Letter to W.Thompson, Dec.18, 1956.
                            (see Sir Joseph Larmor, *Origins of
                            Clerk Maxwell's Electric Ideas,* Cam-
                            bridge, 31, 1937).

R. von Mises   1913   "Mechanik der festen Körper in plas-
                            tisch deformablen Zustand",
                            Göttinger Nachr.Math.Phys.Kl., 582.

O.Mohr   1914   *Abhandhingen aus dem Gebiete der
                            technischen Mechanik,* 2nd Ed., N.
                            Ernst and Son (Berlin)

A.Nadai   1937   "Plastic behavior of metals in the
                            strain hardening range", J.Appl.
                            Phys., 8, 205.

H.Tresca   1864   "Mémoire sur l'écoulement des corps
                            solides soumis à des fortes pres-
                            sions", C.Rend.Paris, 59, 754.

PLASTIC BEHAVIOR UNDER PLANE STRESS CONDITIONS

5.1 Initial and Subsequent Yield Surfaces in Tension-Torsion

The simplest experimental arrangement which can be used to investigate plastic constitutive relations under complex stress states is that of a thin tube (i.e. a tube in which the thickness is small compared to the diameter). The tube, as shown in Fig.1(a), can be subjected to tension, torsion and internal pressure with comparative ease. Provided that the tube is sufficiently thin, variations in stress across the cross section can be neglected, and provided that the external forces are carefully applied, the stresses will not vary in the circumferential or axial directions. In consequence a state of homogeneous plane stress is set up in the tube, with the stresses shown in Fig. 1(b). These stresses can be determined from the axial force N, the torque T and the internal pressure p by means of equilibrium calculations alone. If R is the radius of the tube, and t its thickness (t << R), we have

$$\sigma_x = \frac{N}{2\pi R t} \quad , \quad \sigma_y = \frac{pR}{2t} \quad , \quad \tau_{xy} = \frac{T}{2\pi R^2 t} \quad . \tag{1}$$

The strains (and strain increments) ε_x, ε_y and γ_{xy} can then be measured independently, usually by means of resistance strain gauges or extensometers.

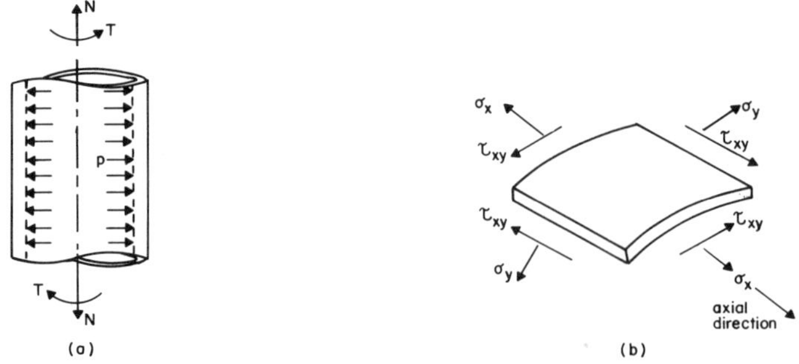

Figure 1. Tubular specimen

Any combination of σ_x, σ_y and τ_{xy} and any history of loading can in principle be obtained in this experiment. In practice, however, difficulties with buckling arise when the direct stresses σ_x and σ_y are compressive, and experiments are normally carried out with a tension force N and internal pressure p rather than external pressure. Although buckling provides a severe limitation, a good deal of important information can nevertheless be obtained.

For isotropic materials the tension-torsion test (internal pressure zero) has considerable significance. States of stress which lie in the first quadrant of the σ_x, τ_{xy} plane are certainly obtainable (see Fig. 3). The principal stresses associated with these states of stress may readily be computed. We are not concerned at present with the planes on which the principal stresses act. Letting σ_I be the largest principal stress,

$$\sigma_I = \frac{\sigma_x}{2} + \frac{1}{2}\sqrt{\sigma_x^2 + 4\tau_{xy}^2} \ , \tag{2a}$$

$$\sigma_{II} = \frac{\sigma_x}{2} - \frac{1}{2}\sqrt{\sigma_x^2 + 4\tau_{xy}^2} \ , \tag{2b}$$

$$\sigma_{III} = 0 \ . \tag{2c}$$

When σ_x is positive, and $\tau_{xy} = 0$, we have $\sigma_I = \sigma_x$, $\sigma_{II} = \sigma_{III} = 0$. When σ_x is zero, we have $\sigma_I = -\sigma_{II} = \tau_{xy}$, $\sigma_{III} = 0$. These two lines may be plotted in the principal stress space (Fig.2). It may readily be checked that the principal stresses corresponding to all other combinations of σ_x, τ_{xy} in the first quadrant lie in the plane $\sigma_{III} = 0$ between the two lines whose equations are given above. This area is shaded in Fig.2. The projection of this shaded area on the π-plane

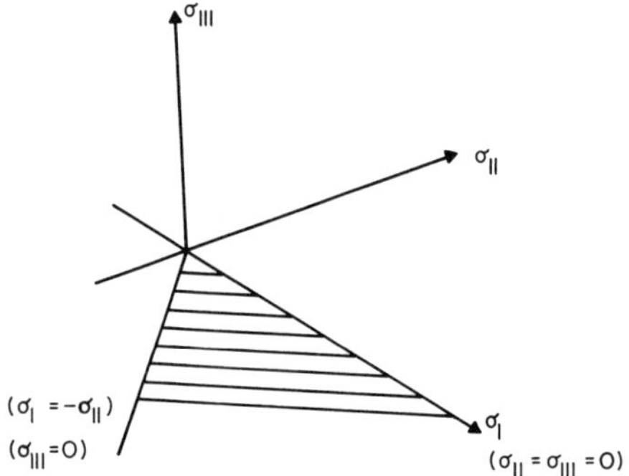

Figure 2. Stress states provided by the tension-torsion test

will also lie between the projections of the lines $\sigma_{II} = \sigma_{III}$ = 0 and $\sigma_{I} = -\sigma_{II}$, $\sigma_{III} = 0$ on the π-plane. This projected area is shown shaded in Fig. 9, Chapter 4. It covers entirely one of the twelve sectors into which the π-plane is divided by the constraints of isotropy and equal yield stresses in tension and compression. Thus, for an isotropic material the tension-torsion test will give sufficient information to enable the complete initial yield surface in the principal stress space to be constructed. If it is to be expected, for some special reason, that the subsequent yield surfaces will also be isotropic, the tension-torsion test will provide sufficient information for a complete description of the constitutive relations for a plastic material.

Using the expressions for the principal stresses given in equations (2) the hypothetical von Mises and Tresca yield surfaces can be constructed from the expressions given in Chapter 4. The von Mises initial yield condition (Chapter 4, equations 31 and 32) is

$$\sigma_{so}^{2} - k^{2} = 0 \, , \tag{3}$$

where

$$\sigma_{so}^{2} = \tfrac{1}{3} (\sigma_{I}^{2} + \sigma_{II}^{2} + \sigma_{III}^{2}) - \tfrac{1}{9} (\sigma_{I} + \sigma_{II} + \sigma_{III})^{2} \, . \tag{4}$$

Substituting from equations (2), this expression reduces to

$$\sigma_x^2 + 3\tau_{xy}^2 = \frac{9}{2} k^2 \ . \tag{5}$$

The Tresca initial yield condition is given by the six equations (Chapter 4, equations 45)

$$\pm(\sigma_I - \sigma_{II}) = 2\bar{k} \ ,$$

$$\pm(\sigma_{II} - \sigma_{III}) = 2\bar{k} \ , \tag{6}$$

$$\pm(\sigma_{III} - \sigma_I) = 2\bar{k} \ .$$

The largest and smallest principal stresses decide which of these six yield surfaces will be operative in any one situation; in this case initial yielding occurs when

$$+(\sigma_I - \sigma_{II}) = 2\bar{k} \ .$$

Substituting from equations (2a) and (2b),the Tresca initial yield condition becomes

$$\sigma_x^2 + 4\tau_{xy}^2 = 4\bar{k}^2 \ . \tag{7}$$

As in Chapter 4, it is usual to normalize the von Mises and Tresca yield conditions by introducing the initial yield stress in simple tension σ_o so that

$$\sigma_o^2 = \frac{9}{2} k^2 = 4\bar{k}^2 \ . \tag{8}$$

Finally, then, the two initial yield surfaces become

$$\sigma_x^2 + 3\tau_{xy}^2 = \sigma_o^2 \quad \text{(von Mises)} \ , \tag{9a}$$

$$\sigma_x^2 + 4\tau_{xy}^2 = \sigma_o^2 \quad \text{(Tresca)} \ . \tag{9b}$$

These are the equations of ellipses in the stress space σ_x, τ_{xy}, as shown in Fig. 3. As a result of our normalization, the greatest difference between the two curves occurs on the τ_{xy} axis. The von Mises yield equation predicts that the initial yield stress in pure shear $(\sigma_x = 0)$ is given by $\sigma_o/\sqrt{3}$, while the Tresca equation predicts first yield under pure shear when the shear stress is $\sigma_o/2$. It should be emphasized that this normalization is arbitrary, and it is conventional to normalize with respect to the pure tension yield stress only for studies of the initial yield surface. In any particular problem the normalization could be carried out with respect to the stress state (or stress states) which are expected to predominate in that particular situation, so that the effect of using an approximation to the initial yield surface, or the effect of the choice of yield surfaces, would be minimised.

Let us consider very briefly the results of some representative experiments carried out on thin tubes of polycrystalline metals. In this context these experimental results are presented only to show broad trends which we shall use to construct idealized models of hardening behavior. A detailed study of experimental results is, of course, extremely important in the development of the theory, but is beyond the scope of this monograph. Detailed consideration of experimental results is further complicated by difficulties in determining the exact

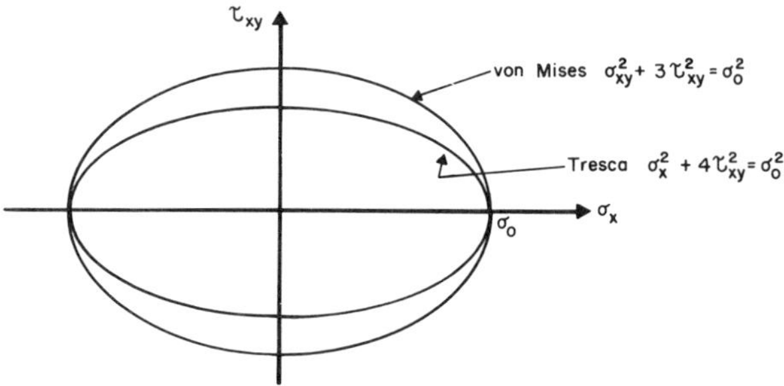

Figure 3. Initial yield surfaces

point on any loading path at which yielding occurs. Different
investigators have used slightly different definitions of the
yield stress. Another experimental difficulty is that of determining the complete yield surface. The yield surface can be
probed at different points from within the elastic region, but
each probe must be carried beyond the yield surface under investigation. Slight changes in the yield surface can thus be
expected with each probe, and consequently a yield surface determined in this manner can only be approximate. Alternatively,
different specimens can be used, each subjected to different
stress histories. Inevitable small differences between specimens again lead to an approximation of any particular yield
surface. In our discussion of the experiments many of the
particular methods used by the various investigators will not
be given. The reader is invited to study the original papers
to gain a better understanding of the experimental methods used
in plasticity.

The experiments of Taylor and Quinney are probably the best
known of the early studies carried out in experimental plasticity. With regard to yield surfaces, Taylor and Quinney carried out tests on a number of identical specimens of each of a
number of metals using the loading program shown in Fig. 4.
The specimen was loaded first in tension only, up to a stress

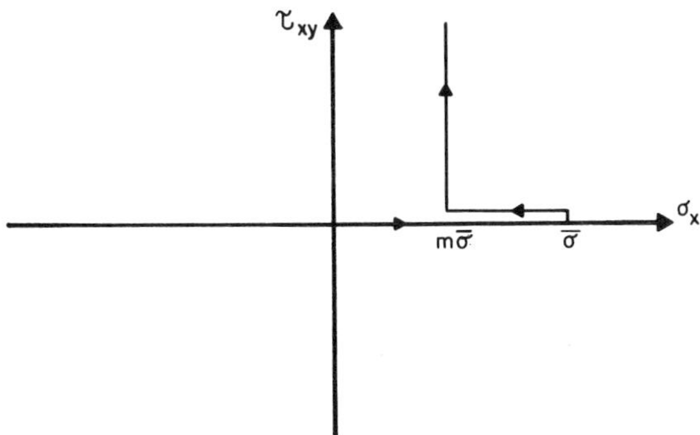

Figure 4. Stress path used by Taylor and Quinney

$\bar{\sigma}$ which was greater than the initial yield stress (i.e., plastic deformation took place during this part of the loading program). The tension stress was then decreased (i.e., unloading took place) until the tension stress was m $\bar{\sigma}$ (m < 1). At this point the torsion stress was increased monotonically with the tension stress held constant. The torsion yield stress was then determined, giving a point on the yield surface. It is evident that the yield surface defined by this procedure is the subsequent yield surface associated with tension loading from the origin to the stress $\bar{\sigma}$. It is not the initial yield surface.

Yield surfaces determined by Taylor and Quinney are shown in Figs. 5(a), 5(b) and 5(c) for copper, aluminum and mild steel. In each diagram the ellipses of equations (9a) and (9b) are plotted with $\bar{\sigma}$ substituted for σ_o. It is seen that the agreement between the experimental points and the von Mises condition is fairly good. This agreement is confirmed in other studies of initial yield surfaces, although many results lie between the two ellipses and consequently do not show that the Tresca condition is as far from the experiments as the results of Taylor and Quinney.

The fact that the yield surfaces of Fig. 5 are not initial yield surfaces but subsequent yield surfaces is significant, for the results imply that the yield surface grows uniformly as the tension stress is increased monotonically, and therefore that the subsequent yield surface is given by

$$\sigma_x^2 + a\tau_{xy}^2 \;=\; \bar{\sigma}^2 \text{ with } \bar{\sigma} \;\geq\; \sigma_o \;. \tag{10}$$

In this equation a = 3 for the von Mises yield equation, and a = 4 in the Tresca case. $\bar{\sigma}$ is the maximum tension stress reached in the monotonic loading program. This is shown

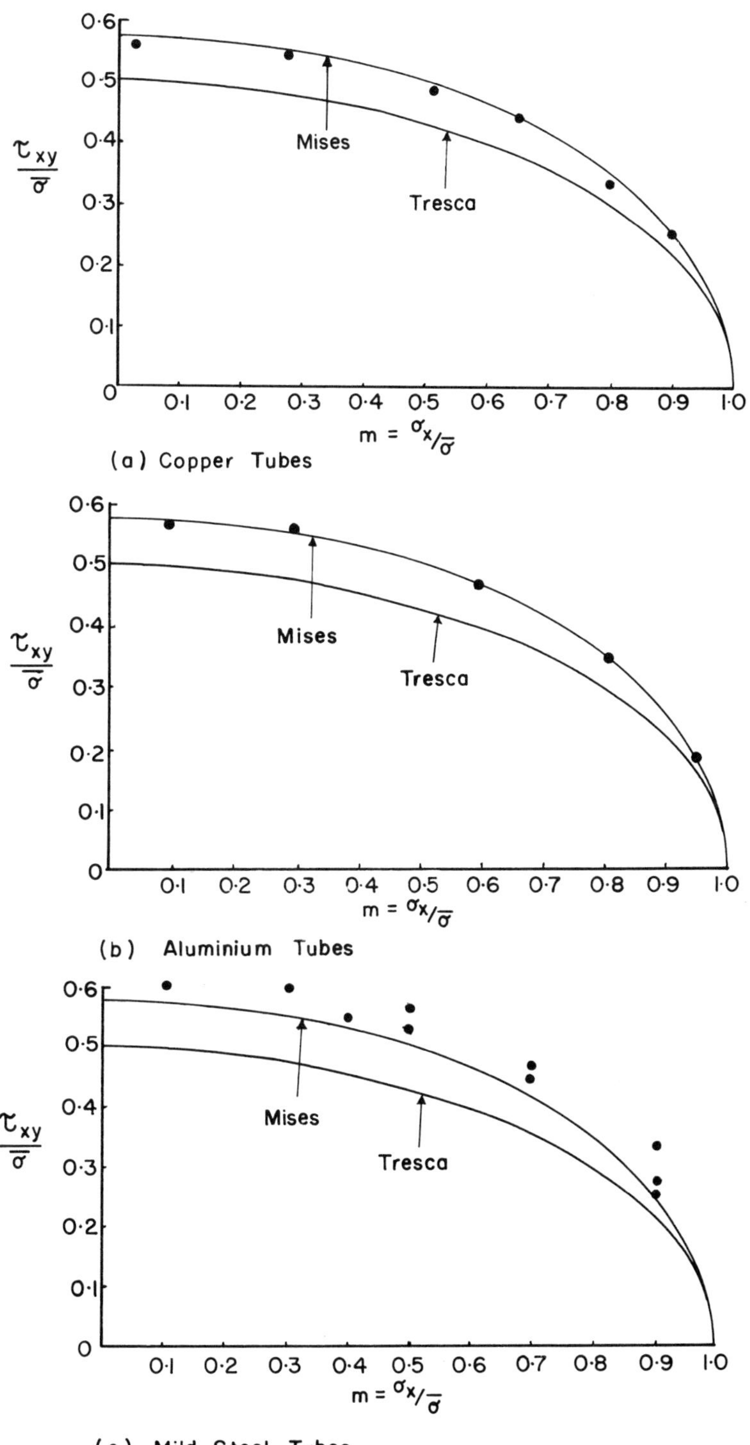

Figure 5.   Results of Taylor and Quinney [1931]

diagrammatically in Fig. 6.

A generalization of this result for subsequent yield surfaces associated with monotonic tension loading is known as *isotropic hardening*, and has played a very significant part in the development of the theory of plasticity. It is assumed that the subsequent yield surface retains the same *shape* as the initial yield surface but merely increases in size following *any* path of loading. It can, of course, only expand and not contract, since unloading from the yield surface results in elastic behavior and the yield function ϕ cannot change. Thus, following equation (10), the yield function ϕ is given by

$$\phi \;=\; \sigma_x^2 \,+\, a\tau_{xy}^2 \,-\, \bar{\sigma}^2 \tag{11}$$

for all loading paths. $\bar{\sigma}$ must be redefined, and we do this by noting that along a path in which *continuous loading* takes place $\phi = 0$ at the stress point, except when the stress point lies inside the initial yield surface. Thus along this loading path

$$\bar{\sigma}^2 \;=\; \sigma_x^2 \,+\, a\tau_{xy}^2 \;. \tag{12}$$

If unloading now takes place, $\bar{\sigma}$ must remain constant. Clearly then for a general stress path $\bar{\sigma}^2$ is the largest previous value

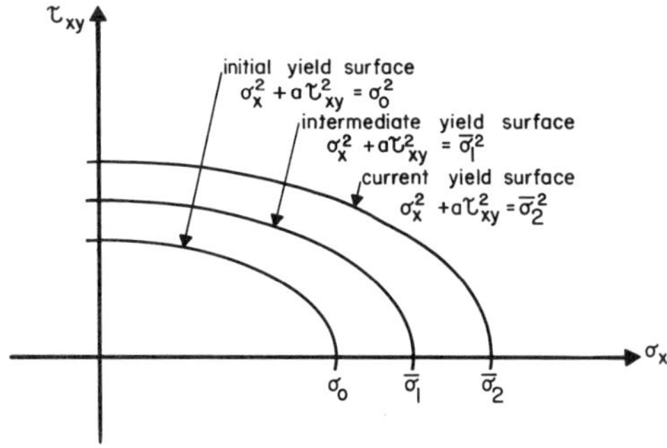

Figure 6. Isotropic Hardening

of $(\sigma_x^2 + a\tau_{xy}^2)$ reached in the previous stress history, except that $\bar{\sigma}^2 \geq \sigma_o^2$. This material can be regarded as a *stress hardening material*, since the recorded history H_α $(\alpha = 1)$ consists of the maximum value of the function of stress $(\sigma_x^2 + a\tau_{xy}^2)$ provided that it is greater than σ_o^2. In terms of the yield equations written in the forms of equations (3) and (7) it is seen that k becomes a function of the maximum previous value of $(\sigma_x^2 + a\tau_{xy}^2)$. Yield is still governed by the shear stress on the octohedral plane or the plane of maximum shear stress without regard to the orientation of the planes or the direction of the shear stress on the plane. The yield function and the yield surface consequently remain isotropic.

The experiments of Ivey on an aluminum alloy show a different type of behavior. In this case the initial yield surface and subsequent yield surfaces following monotonic loading in pure shear were determined. The yield surfaces are shown in Fig.7. The initial yield surface is fairly close to the von Mises ellipse. Several significant features of the behavior can be noted.

First, there is a pronounced Bauschinger effect in shear, so that the subsequent yield surfaces do not cut the shear axis at points which are equal and opposite in sign. Indeed, the second and third subsequent yield surfaces do not contain the origin. Second, the maximum value of the tension stress which can be supported elastically remains almost constant when the yield surfaces are compared. In Fig. 7 the width of the yield surface is unchanged. Thirdly, the shape of the yield surface changes greatly.

If we concentrate on the first and second of these features of this behavior, and neglect for the present the third, we can construct a simple idealization of the change in the yield function. Speaking very broadly, the first and second features

indicate that under monotonic loading in shear in this material
the yield surface is displaced along the shear axis without rotation. If we also assume that the shape and size of the surface remain constant, the subsequent yield surface is the initial ellipse with its origin displaced along the τ_{xy} axis by a
distance $(\tau_{xy} - \sigma_o/\sqrt{a})$ where τ_{xy} is the current stress in the
monotonic loading program. If we put

$$\hat{\tau} = \tau_{xy} - \sigma_o/\sqrt{a} \quad , \tag{13}$$

the equation of the subsequent yield surface is

$$\sigma_x^2 + a(\tau_{xy} - \hat{\tau})^2 = \sigma_o^2 \quad . \tag{14}$$

A generalization of this behavior is commonly used in plasticity and is referred to as *kinematic hardening*. The subsequent yield function is taken to be

$$\phi = (\sigma_x - \hat{\sigma})^2 + a(\tau_{xy} - \hat{\tau})^2 - \sigma_o^2 \quad . \tag{15}$$

The coefficients a and σ_o remain constant, so that the recorded
history $H_\alpha (\alpha = 1, 2)$ is given by the stresses $\hat{\sigma}$ and $\hat{\tau}$ which are
the coordinates of the center of the ellipse. It is necessary

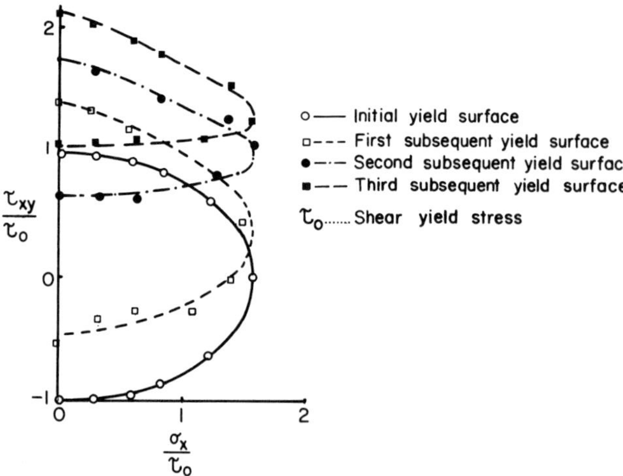

Figure 7. Results of Ivey [1961]

to categorize the way in which $\hat{\sigma}$ and $\hat{\tau}$ depend on the stress path. Recorded history is not affected by unloading or neutral loading. The loading condition may be written in terms of stresses σ_x and τ_{xy}, i.e.

$$\frac{\partial\phi}{\partial Q_j}\, dQ_j \;=\; \frac{\partial\phi}{\partial\sigma_x}\, d\sigma_x + \frac{\partial\phi}{\partial\tau_{xy}}\, d\tau_{xy} \;. \tag{16}$$

It is evident then that

$$\left.\begin{array}{l} d\hat{\sigma} = 0 \\[2ex] d\hat{\tau} = 0 \end{array}\right\} \text{when } \phi = 0 \text{ and } \frac{\partial\phi}{\partial\sigma_x}\, d\sigma_x + \frac{\partial\phi}{\partial\tau_{xy}}\, d\tau_{xy} \;\leq\; 0 \;,$$
$$\text{or } \phi < 0 \;. \tag{17}$$

Now consider a loading increment $d\sigma_x$, $d\tau_{xy}$ imposed on a stress state σ_x, τ_{xy} which lies on the yield surface $\phi = 0$ (equation 15). Loading requires that

$$\frac{\partial\phi}{\partial Q_j}\, dQ_j = \frac{\partial\phi}{\partial\sigma_x}\, d\sigma_x + \frac{\partial\phi}{\partial\tau_{xy}}\, d\tau_{xy} \;>\; 0 \;. \tag{18}$$

Since the stress point remains on the yield surface during loading, the change in ϕ, $d\phi$, must be zero. We have then that

$$d\phi = \left.\frac{\partial\phi}{\partial Q_j}\, dQ_j\right|_{H_\alpha \text{ constant}} + \left.\frac{\partial\phi}{\partial H_\alpha}\, dH_\alpha\right|_{Q_j \text{ constant}} \;=\; 0 \;. \tag{19}$$

It follows from equation (15) that

$$\frac{\partial\phi}{\partial Q_j}\, dQ_j = \frac{\partial\phi}{\partial\sigma_x}\, d\sigma_x + \frac{\partial\phi}{\partial\tau_{xy}}\, d\tau_{xy} = 2(\sigma_x - \hat{\sigma})d\sigma_x + 2a(\tau_{xy} - \hat{\tau})d\tau_{xy} \;, \tag{20a}$$

and

$$\frac{\partial\phi}{\partial H_\alpha}\, dH_\alpha = \frac{\partial\phi}{\partial\hat{\sigma}}\, d\hat{\sigma} + \frac{\partial\phi}{\partial\hat{\tau}}\, d\hat{\tau} = -2(\sigma_x - \hat{\sigma})\, d\hat{\sigma}_x - 2a(\tau_{xy} - \hat{\tau})d\hat{\tau} \;. \tag{20b}$$

Substituting equations (20a) and (20b) into equation (19), we obtain one linear equation relating $d\hat{\sigma}$, $d\hat{\tau}$ and $d\sigma_x$, $d\tau_{xy}$;

$$(\sigma_x-\hat{\sigma})d\sigma_x + a(\tau_{xy}-\hat{\tau})d\tau_{xy} = (\sigma_x-\hat{\sigma})d\hat{\sigma} + a(\tau_{xy}-\hat{\tau})d\hat{\tau} \quad . \qquad (21)$$

One further equation is needed if we are to solve for $d\hat{\sigma}$ and $d\hat{\tau}$. Following our procedure in Chapter 2, Section 2.3, we assume that the direction of dH_α is a function of Q_j and H_α alone. Let us choose a simple plausible relation; we assume that the direction of dH_α *coincides* with the direction of $\partial\phi/\partial H_\alpha$. It is most convenient to write this relation as

$$\frac{dH_1}{dH_2} = \frac{\partial\phi/\partial H_1}{\partial\phi/\partial H_2} \quad , \qquad (22a)$$

i.e. that the ratio of the components of dH_α coincides with the ratio of the components of $\partial\phi/\partial H_\alpha$. In our present context equation (22a) becomes

$$\frac{d\hat{\sigma}}{d\hat{\tau}} = \frac{\partial\phi/\partial\hat{\sigma}}{\partial\phi/\partial\hat{\tau}} = \frac{(\sigma_x-\hat{\sigma})}{a(\tau_{xy}-\hat{\tau})} \quad . \qquad (22b)$$

We may now solve equations (21) and 22b) simultaneously to give

$$d\hat{\sigma} = \frac{(\sigma_x-\hat{\sigma})\{(\sigma_x-\hat{\sigma})d\sigma_x + a(\tau_{xy}-\hat{\tau})d\tau_{xy}\}}{\{(\sigma_x-\hat{\sigma})^2 + a^2(\tau_{xy}-\hat{\tau})^2\}} \quad , \qquad (23a)$$

$$d\hat{\tau} = \frac{a(\tau_{xy}-\hat{\tau})\{(\sigma_x-\hat{\sigma})d\sigma_x + a(\tau_{xy}-\hat{\tau})d\tau_{xy}\}}{\{(\sigma_x-\hat{\sigma})^2 + a^2(\tau_{xy}-\hat{\tau})^2\}} \quad , \qquad (23b)$$

for $\phi(\sigma_x,\tau_{xy},\hat{\sigma},\hat{\tau}) = 0$ and $\{(\sigma_x-\hat{\sigma})d\sigma_x + a(\tau_{xy}-\hat{\tau})d\tau_{xy}\} \geq 0$. Equations (15), (17) and (23) now characterize the yield function and the yield surface completely.

A simple physical model for kinematic hardening can be constructed if equations (23) are reinterpreted. On the σ_x, τ_{xy}

plane (Fig. 8) suppose that the ellipse of equation (15) is physically represented by a rigid metal frame which is constrained to move on the plane without rotation with respect to the σ and τ axes. Let the stress point be interpreted as a pin perpendicular to the plane which may engage the metal frame and which moves in response to the applied loading. Further, assume that the contact between the pin and the frame is frictionless, so that the pin applies only normal forces to the frame. The pin must always lie inside the frame (corresponding to $\phi \leq 0$). The frame will move only when the pin is in contact with the frame and the pin displacement has a component in the outward normal direction ($\phi = 0$ and $(\partial\phi/\partial Q_j)dQ_j > 0$). Because of the rotation constraint, the center of the ellipse will move in the outward normal direction through a distance equal to the component of motion of the pin in that direction. The vector with components $\partial\phi/\partial\sigma_x$, $\partial\phi/\partial\tau_{xy}$ is normal to the yield surface and the rigid frame. Thus the *unit* outward normal vector has components

$$\frac{(\sigma_x-\hat{\sigma})}{\sqrt{(\sigma_x-\hat{\sigma})^2 + a^2(\tau_{xy}-\hat{\tau})^2}} \quad , \quad \frac{a(\tau_{xy}-\hat{\tau})}{\sqrt{(\sigma_x-\hat{\sigma})^2 + a^2(\tau_{xy}-\hat{\tau})^2}} \quad .$$

The magnitude of the motion of the center of the ellipse is

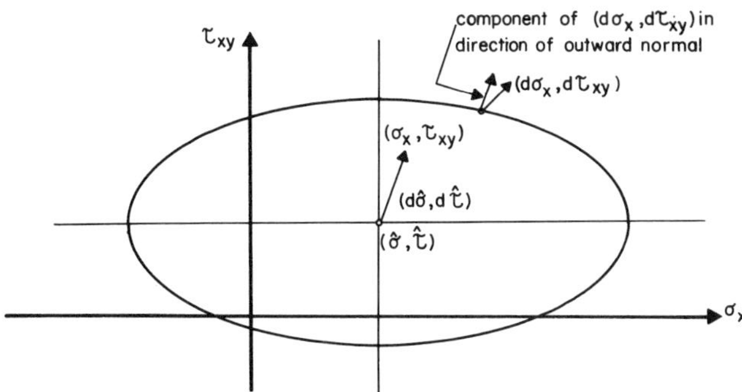

Figure 8. Kinematic hardening model

given by the scalar product of the vector with components $(d\sigma_x, d\tau_{xy})$ and the unit outward normal vector. $d\hat{\sigma}$, $d\hat{\tau}$ are then found by multiplying this magnitude by the two components of the outward normal vector, giving equations (23).

It is also evident that a relation between $d\hat{\sigma}$, $d\hat{\tau}$ and the plastic strain increments $d\varepsilon_x^p$, $d\gamma_{xy}^p$ can be obtained. We shall return to this relation later in this section when the strains are discussed more fully.

The kinematic hardening model describes to some extent the motion of the yield surface as a result of a history of stress, but it does not describe the changes of shape of the yield surface found in Fig. 7. One interesting feature of these shape changes is the development of a pointed region with its apex at the stress point where loading took place. This agrees to some extent with the predictions of a more sophisticated theory of plasticity, generally known as *slip theory*, where it is assumed slip leading to plastic deformation may take place along any plane as a consequence of the shear stress on that plane (rather than on the plane of maximum shear stress or the octahedral plane as in the Tresca and von Mises conditions). The initial critical values of the shear stress on any plane are assumed to be identical, so that the initial yielding is isotropic and satisfies the Tresca condition. Subsequently, however, the dependence of the critical shear stress on stress history on any one plane is assumed to be independent of the history on any other plane. Thus, even under monotonic loading in shear, say, slip occurs on an increasing number of planes as the stress increases, rather than on just the plane of maximum shear stress. The predicted yield surface corresponding to the stress path of Ivey's experiments is of the form of Fig. 9, with a sharp corner developing at the stress point during loading. This theory is discussed in greater detail in Chapter 7.

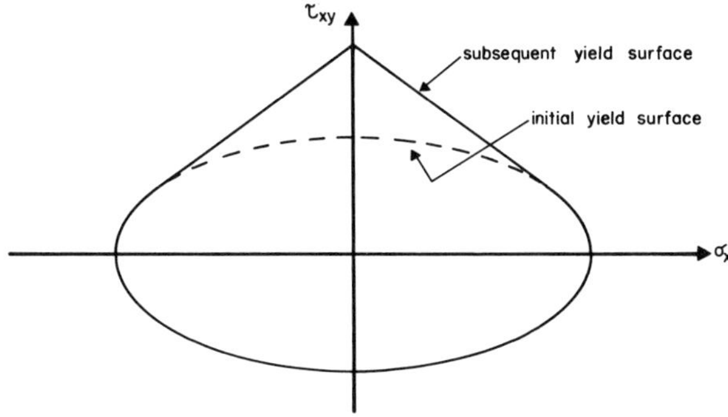

Figure 9. Subsequent yield surface given by slip

A number of experimental programs have been directed at establishing whether or not sharp corners are in fact developed
in the yield surface for polycrystalline metals. The present
evidence is not entirely conclusive, but it is apparent that
such corners do not develop for all materials. Figures 10(a)
and 10(b) show subsequent yield surfaces determined by Bertsch
and Findley for a polycrystalline aluminum alloy. These experimental results were all determined from one specimen, with
the yield surface being probed to established the surface. The
combined stress paths (i.e. paths inclined to the σ_x axis) actually consisted of small increments of tension and shear
stresses alternatively. A region of higher curvature is de-
finitely established in the region of the stress point at the
end of the loading path in the second, third and fourth subsequent yield surfaces. However, these regions are much less
sharp than predicted by slip theory.

Partly as a result of a lack of space in a short monograph,
we shall not develop in sufficient detail for applications more
sophisticated models of hardening than the two given, isotropic
and kinematic hardening. In the applications that are given
they will serve adequately to illustrate the methods by which
plastic analyses are carried out. It is also true, however,
that more sophisticated models are more complex. While the

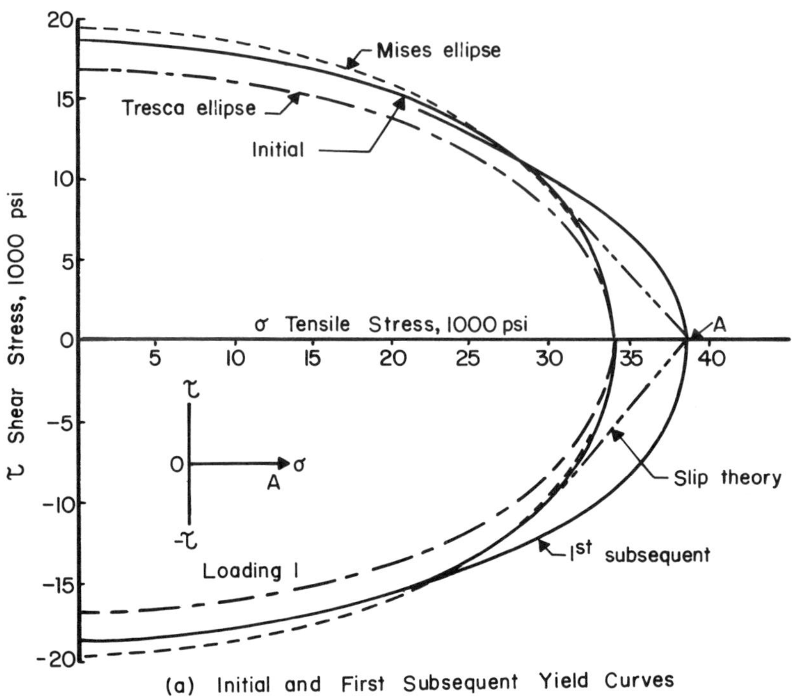

(a) Initial and First Subsequent Yield Curves

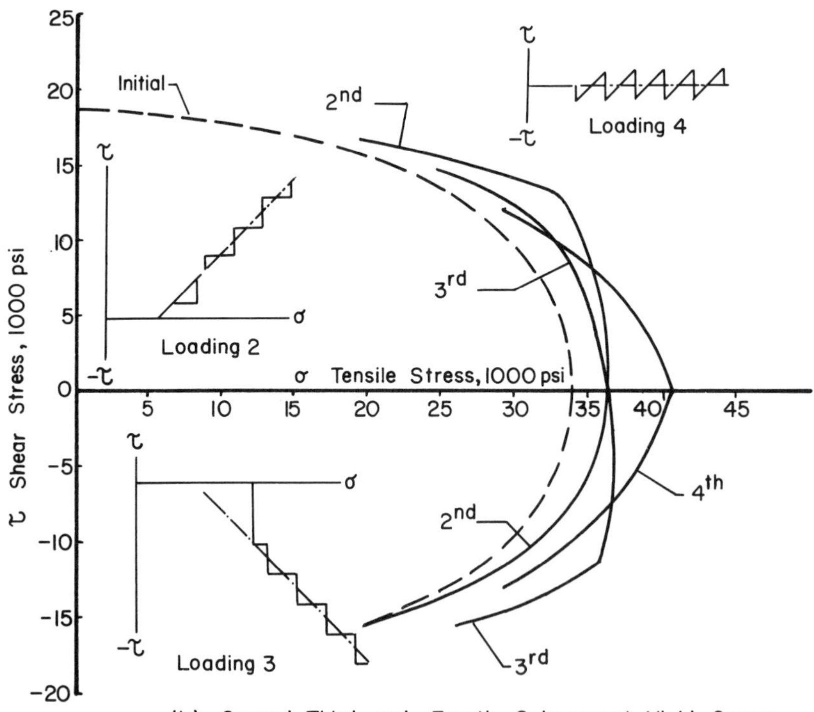

(b) Second, Third and Fourth Subsequent Yield Curves

Figure 10. Results of Bertsch and Findley [1962]

isotropic and kinematic hardening models are in many respects
inadequate to provide a satisfactorily accurate description of
plastic behavior when the plastic constitutive equations are
studied in their own right, they are in many cases sufficiently
accurate and sufficiently tractable to permit adequate structural analyses to be carried out

5.2 The Isotropic Hardening Model

Having discussed at some length the characterization of the
yield surface, we may now turn to the incremental stress-strain
relations for combined tension and shearing under plane stress
conditions.

The appropriate form of the general relations (see Chapter
4, equations 1-4) will be as follows. As before, superscripts
e and p indicate the elastic and plastic components of strain.
The elastic strain increments are given by

$$d\varepsilon_x^e = \frac{1}{E}\, d\sigma_x \quad , \qquad d\gamma_{xy}^e = \frac{2(1+\nu)}{E}\, d\tau_{xy} \quad , \tag{24}$$

and the plastic strain increments by

$$d\varepsilon_x^p = cG\, \frac{\partial\phi}{\partial\sigma_x} \left\{ \frac{\partial\phi}{\partial\sigma_x}\, d\sigma_x + \frac{\partial\phi}{\partial\tau_{xy}}\, d\tau_{xy} \right\} \quad ,$$

$$\tag{25}$$

$$d\gamma_{xy}^p = cG\, \frac{\partial\phi}{\partial\tau_{xy}} \left\{ \frac{\partial\phi}{\partial\sigma_x}\, d\sigma_x + \frac{\partial\phi}{\partial\tau_{xy}}\, d\tau_{xy} \right\} \quad ,$$

where $\phi = \phi(\sigma_x, \tau_{xy}, H_\alpha)$,

$$G = G(\sigma_x, \tau_{xy}, H_\alpha) \quad ,$$

$c = +1$ when $\phi = 0$ and $\left(\dfrac{\partial\phi}{\partial\phi_x}\, d\sigma_x + \dfrac{\partial\phi}{\partial\tau_{xy}}\, d\tau_{xy} \right) \geq 0$,

$c = 0$ when $\phi < 0$

or $\phi = 0$, and $\left(\dfrac{\partial\phi}{\partial\phi_x}\, d\sigma_x + \dfrac{\partial\phi}{\partial\tau_{xy}}\, d\tau_{xy} \right) \leq 0$.

The basic incremental form of these relations, with the implication of plastic strain increments normal to the yield surface and proportional to the component of the stress increment in the direction of the normal, has been studied experimentally by a number of investigators, including Ivey and Bertsch and Findley. Very satisfactory agreement has been found for hardening metals, as would be expected since the assumptions on which the results were based are in good (even though not exact) agreement with the observed behavior. We shall not discuss the experiments concerning the incremental form of the relations, but pass on to the characterization of the scalar function G. An experimental result for some particular loading paths (normally including monotonic loading in tension or shear) must be used to provide fundamental data for the determination of G. The important question to be considered involves the number of such tests which must be carried out before the plastic strains associated with other paths can be predicted with reasonable accuracy.

An assumption which leads to a particularly simple form of the increment plastic constitutive relations is that G has the same value at all points on any one subsequent yield surface. This assumption is identified particularly with isotropic hardening, since if G depends on stress and stress history in the same way as the yield function ϕ the incremental constitutive relations will also be isotropic. Repeating equation (11), the yield function for an isotropic hardening material in our present context is

$$\phi = \sigma_x^2 + a\tau_{xy}^2 - \bar{\sigma}^2 \ . \qquad (26)$$

The term $\bar{\sigma}^2$ is the largest previously attained value of $(\sigma_x^2 + a\tau_{xy}^2)$, provided that this value is larger than the square of the initial yield stress in tension, σ_o^2. Suppose then that we put

$$G = G(\bar{\sigma}^2) \ . \tag{27}$$

The coefficient G is then a function only of the single recorded history parameter $\bar{\sigma}^2$. The important feature of equation (27) is that G can be determined from a *single experiment involving a monotonic loading path.*

Let us suppose that an experiment is carried out in which the tension stress σ_x is increased monotonically while $\tau_{xy} = 0$. The stress-strain curve for this given stress history must be measured, and σ_o must be determined. Provided that $\sigma \leq \sigma_o$, the total strain ε_x will be given by

$$\varepsilon_x = \frac{1}{E} \sigma_x \ . \tag{28}$$

When $\sigma_x > \sigma_o$ (again, for this particular stress history) the strain increments are given by (see equations 25, 26 and 27);

$$d\varepsilon_x = d\varepsilon_x^e + d\varepsilon_x^p$$

$$= \frac{d\sigma_x}{E} + G(\bar{\sigma}^2) \frac{\partial \phi}{\partial \sigma_x} \left(\frac{\partial \phi}{\partial \sigma_x} d\sigma_x \right)$$

$$= \left(\frac{1}{E} + 4\sigma^2 G \right) d\sigma_x \ . \tag{29}$$

With monotonic loading $\phi = 0$ when $\sigma \geq \sigma_o$, and hence from equation (26) $\sigma^2 = \bar{\sigma}^2$. It is probably most convenient to reduce the monotonic stress-strain curve to an incremental form, writing

$$d\varepsilon_x = \frac{1}{E_T} d\sigma_x \ . \tag{30}$$

The tangent modulus E_T is the slope of the monotonic stress-strain curve, and can clearly be written as $E_T = E_T(\sigma^2) = E_T(\bar{\sigma}^2)$. Supposing that this function is measured from the curve, we can equate equations (29) and (30) to give

$$\frac{1}{E} + 4\bar{\sigma}^2 G(\bar{\sigma}^2) = \frac{1}{E_T(\bar{\sigma}^2)} \quad ,$$

or

$$G(\bar{\sigma}^2) = \frac{1}{4\bar{\sigma}^2} \left\{ \frac{1}{E_T(\bar{\sigma}^2)} - \frac{1}{E} \right\} \quad . \tag{31}$$

The hardening coefficient $G(\bar{\sigma}^2)$ is thus obtained as a function of the measured quantities $E_T(\bar{\sigma}^2)$ and E.

The shear strain-shear stress curve may now be predicted using these values of $G(\bar{\sigma}^2)$. For a monotonic increase in τ_{xy} with $\sigma_x = 0$, from equations (25), (26) and (31)

$$d\gamma_{xy} = \frac{2(1+\nu)}{E} d\tau_{xy} \quad \text{for} \quad \tau_{xy}^2 \leq \frac{\sigma_o^2}{a} \quad ,$$

$$\text{and } d\gamma_{xy} = d\gamma_{xy}^e + d\gamma_{xy}^p$$

$$= 2\left(\frac{1+\nu}{E}\right) d\tau_{xy} + G(\bar{\sigma}^2) \frac{\partial\phi}{\partial\tau_{xy}} \left(\frac{\partial\phi}{\partial\tau_{xy}} d\tau_{xy}\right)$$

$$= \left[\frac{2(1+\nu)}{E} + a\left\{\frac{1}{E_T(\bar{\sigma}^2)} - \frac{1}{E}\right\}\right] d\tau_{xy} \tag{32}$$

$$\text{for} \quad \tau_{xy}^2 \geq \frac{\sigma_o^2}{a} \quad \text{and} \quad \tau_{xy}^2 = \frac{\bar{\sigma}^2}{a} \quad .$$

These equations can now be integrated to obtain the relation between γ_{xy} and τ_{xy} for monotonic loading in shear.

Ivey performed experiments on a silicon-aluminum alloy using the isotropic hardening relations given in equations (25), (26) and (31). Figure 11 shows a measured shear stress-shear strain curve for monotonic loading and a curve derived from a tension test on the same material. The agreement can be seen to be moderately good. Figures 12(a) and 12(b) compare measured and predicted total strains for more complex stress histories on the same material. Again, the agreement is reasonably good considering the simplicity of the model which is being used.

A physical interpretation can be given to the relation $G = G(\bar{\sigma}^2)$ because $\bar{\sigma}^2$ can be related to the plastic work W^P. An increment in the plastic work dW^P is defined as

$$dW^P = \sigma_x d\varepsilon_x^p + \tau_{xy} d\gamma_{xy}^p \quad . \tag{33}$$

The plastic work W^P can change only when $d\varepsilon_x^p$, $d\gamma_{xy}^p$ are nonzero, and hence only during loading. Substituting from equations (25) and (26), the increment in the plastic work becomes

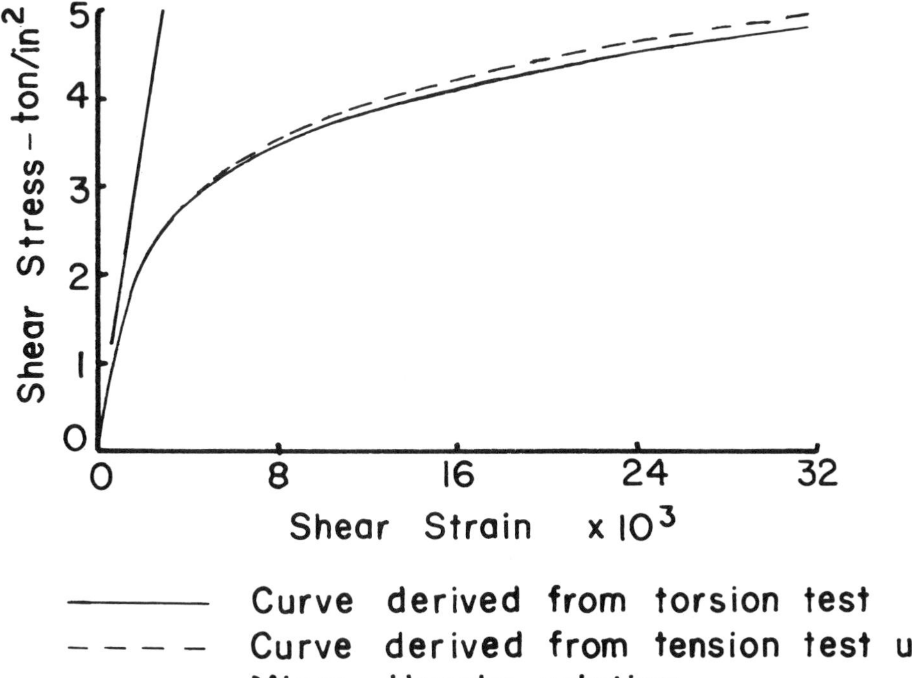

Figure 11.Plastic strain measurements due to Ivey [1961]

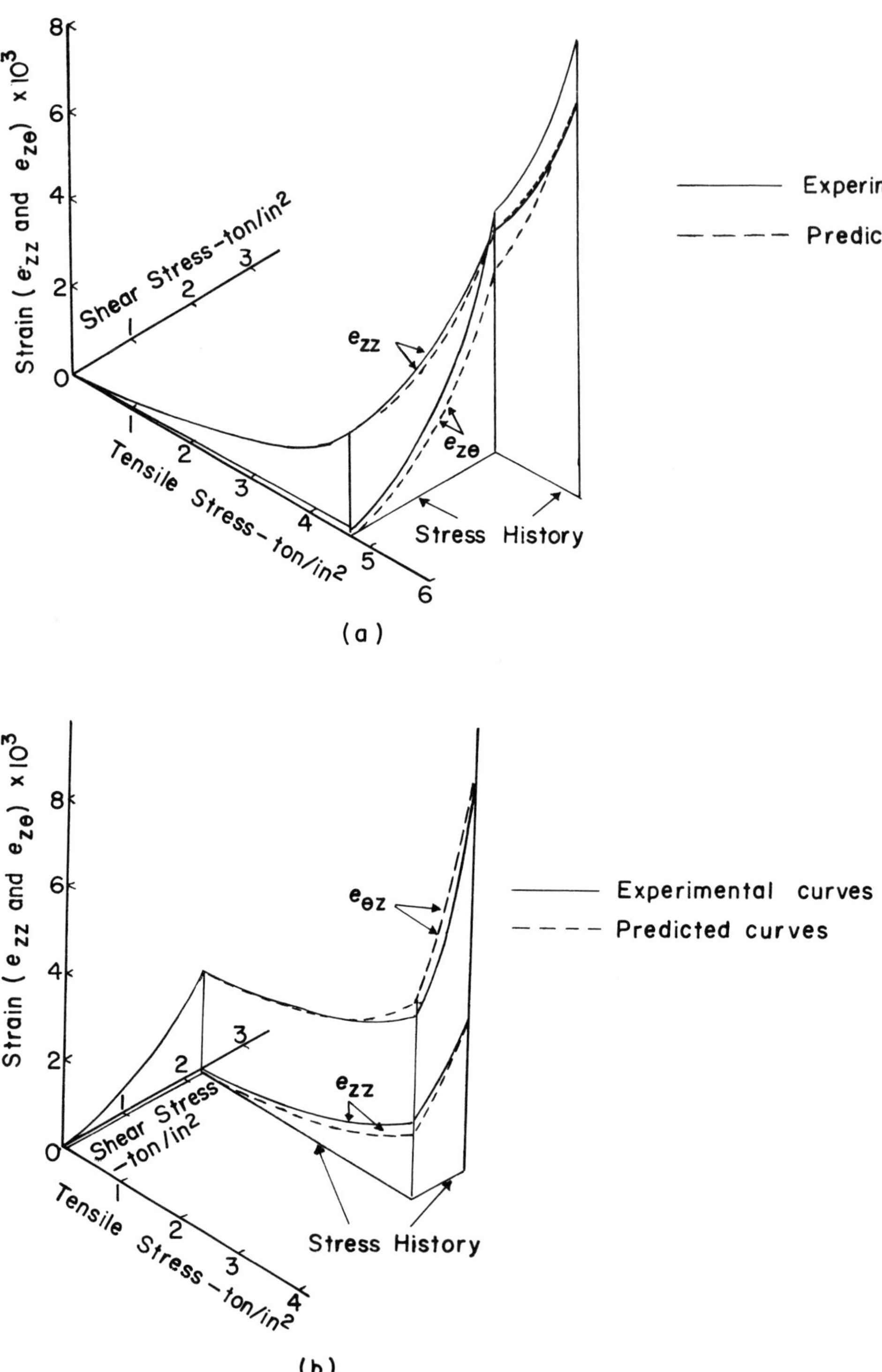

Figure 12. Plastic strain measurements due to Ivey [1961]

$$dW^P = 4G(\bar{\sigma}^2)(\sigma_x^2 + a\tau_{xy}^2)(\sigma_x d\sigma_x + a\tau_{xy}d\tau_{xy})$$

$$= 2G(\bar{\sigma}^2)\bar{\sigma}^2 d(\bar{\sigma}^2) \quad , \tag{34}$$

where we make use of the requirement that $\phi = 0$ and $\sigma_x^2 + a\tau_{xy}^2 = \bar{\sigma}^2$ during loading. Equation (34) can be integrated along any path from the virgin state ($W^P = 0$, $\bar{\sigma}^2 = \sigma_o^2$) to the current state;

$$W^P = \int_{\sigma_o^2}^{\bar{\sigma}^2} 2G(\bar{\sigma}^2)\bar{\sigma}^2 d(\bar{\sigma}^2) \quad . \tag{35}$$

This integral equation can, in principle, be solved for $\bar{\sigma}^2$, so that it is seen that there exists a one to one relation between W^P and $\bar{\sigma}^2$, and that the yield function of equation (26) could be written as

$$\phi = \phi(\sigma_x, \tau_{xy}, W^P) \quad . \tag{36}$$

In this expression W^P becomes the recorded history, and the recorded history clearly changes only when ϵ_x^p, σ_{xy}^p change. This material could be referred to as a *work hardening material*.

A particularly simple case which is of interest is that when G is a constant (i.e. independent of $\bar{\sigma}^2$). The tangent modulus for a monotonic tension test is then given by (equation 31)

$$\frac{1}{E_T(\bar{\sigma}^2)} = \frac{1}{E} + 4\bar{\sigma}^2 G \quad \text{for} \quad \bar{\sigma}^2 \geq \sigma_o^2 \tag{37}$$

The monotonic tension stress-strain curve can then be obtained by integrating equation (30), giving

$$\epsilon_x = \frac{\sigma_x}{E} \quad \text{for} \quad \sigma_x \leq \sigma_o \quad , \tag{38a}$$

$$\epsilon_x = \frac{\sigma_x}{E} + \frac{4}{3}G(\sigma_x^3 - \sigma_o^3) \quad \text{for} \quad \sigma_x \geq \sigma_o \quad . \tag{38b}$$

This curve has a distinct discontinuity in slope at $\sigma = \sigma_o$.

As an example of the computations which lead to plastic strains for given paths, let us consider an isotropically hardening material in which G is constant, and compare the plastic strains for the two paths shown in Fig. 13. In both cases the material is initially in the virgin state, and both paths terminate at the stress point ($\sigma_x = 2\sigma_o$, $\tau_{xy} = 2\sigma_o$). One path is a radial path, while in the other the tension stress is increased monotonically from 0 to $2\sigma_o$ with the shear stress zero, and then the shear stress is increased monotonically with the tension stress σ_x held constant. Both paths involve continuous loading outside of the initial yield surface, and the final subsequent yield surfaces will be the same.

For the radial path, $\sigma_x = \tau_{xy}$, and hence first yield occurs when

$$\bar{\sigma}^2 = \sigma_x^2 + a\tau_{xy}^2 = \sigma_x^2(1+a) = \sigma_o^2 . \tag{39}$$

Let us choose a = 3 (von Mises initial yield surface), so that first yield occurs for

$$\sigma_x = \frac{\sigma_o}{2} . \tag{40}$$

Substituting G = constant and $\sigma_x = \tau_{xy}$, $d\sigma_x = d\tau_{xy}$ (for the radial path) into equations (25), (26) and (27), we obtain

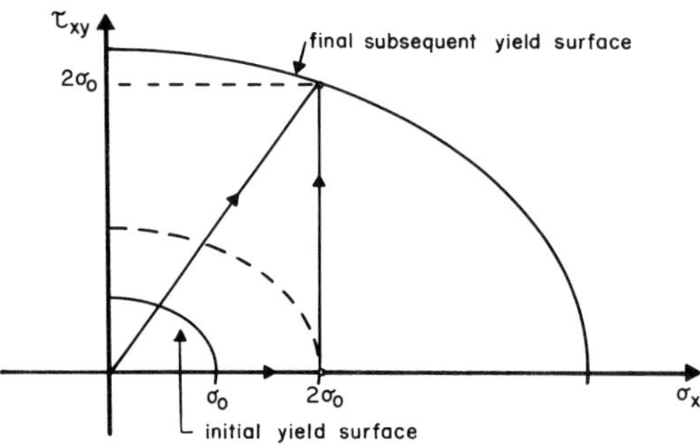

Figure 13. Comparison of plastic strains for two stress paths

$$d\varepsilon_x^p = 4G\sigma_x(\sigma_x d\sigma_x + 3\sigma_x d\sigma_x) = 16G\sigma_x^2 d\sigma_x$$

$$\tag{41}$$

$$d\gamma_{xy}^p = 12G\sigma_x(\sigma_x d\sigma_x + 3\sigma_x d\sigma_x) = 48G\sigma_x^2 d\sigma_x$$

The final plastic strain is obtained by integrating these equations, which now include the precise stress path, between the limits $\sigma_o/2$ and $2\sigma_o$. Thus, at the end of the path,

$$\varepsilon_x^p = \frac{16}{3}G\sigma_x^3\Big|_{\sigma_o/2}^{2\sigma_o} = 42G\sigma_o^3 \quad ,$$

$$\tag{42}$$

$$\gamma_{xy}^p = \frac{48}{3}G\sigma_x^3\Big|_{\sigma_o/2}^{2\sigma_o} = 126G\sigma_o^3 \quad .$$

For the second path we integrate in two parts. The monotonic increase in σ_x gives $\tau_{xy} = d\tau_{xy} = 0$, and hence

$$d\varepsilon_x^p = 4G\sigma_x(\sigma_x d\sigma_x) = 4G\sigma_x^2 d\sigma_x \quad ,$$

$$\tag{43}$$

$$d\gamma_{xy}^p = 0 \quad .$$

First yield occurs for $\sigma_x = \sigma_o$, so the final plastic strain at the end of the first part of the path is

$$\varepsilon_x^p = \frac{4}{3}G\sigma_x^3\Big|_{\sigma_o}^{2\sigma_o} = \frac{28}{3}G\sigma_o^3 \quad ,$$

$$\tag{44}$$

$$\gamma_{xy}^p = 0 \quad .$$

The subsequent yield surface at the end of this part of the path is the dotted curve shown in Fig. 13. If we now embark on the second part of the path, with $\sigma_x = 2\sigma_o$, $d\sigma_x = 0$, $d\tau_{xy} > 0$

the first increment will constitute neutral loading. Subsequent increments, however, lead to hardening behavior and because of the consistency condition we can treat the plastic deformation as continuous in this case. We now have

$$d\varepsilon_x^p \;=\; 8G\sigma_o(3\tau_{xy}d\tau_{xy}) \;=\; 24G\sigma_o\tau_{xy}d\tau_{xy} \;,$$

$$d\gamma_{xy}^p = 12G\tau_{xy}(3\tau_{xy}d\tau_{xy}) = 36G\tau_{xy}^2 d\tau_{xy} \;. \tag{45}$$

The total plastic strain at the end of the path is found by integrating equations (45) between the limits $\tau_{xy} = 0$ (when first yield occurs) and $\tau_{xy} = 2\sigma_o$, and adding to this the plastic strain accumulated during the first part of the path (equations 44). Hence

$$\varepsilon_x^p \;=\; 12G\sigma_o\tau_{xy}^2\Big|_0^{2\sigma_o} + \frac{28}{3}\,G\sigma_o^3 \;=\; \frac{172}{3}\,G\sigma_o^3 \;,$$

$$\gamma_{xy}^p = \frac{36}{3}\,G\,\tau_{xy}^3\Big|_0^{2\sigma_o} + 0 \qquad = 96G\sigma_o^3 \;. \tag{46}$$

Comparing equations (42) and (46), it is seen that a quite considerable difference in the plastic strains is found.

Another idealization of interest is that which leads to a monotonic tension stress-strain curve which consists of two straight lines. This approximation, often referred to as a bilinear stress-strain curve, is a useful one, for with a modification in the initial yield stress it can be used to fit an actual stress-strain curve which has a rounded knee but which again becomes almost linear as the strain is increased, as shown in Fig. 14. In the idealization the tangent modulus E_T is constant for $\sigma > \sigma_o$, and hence, if the material hardens isotropically, we have from equation (31),

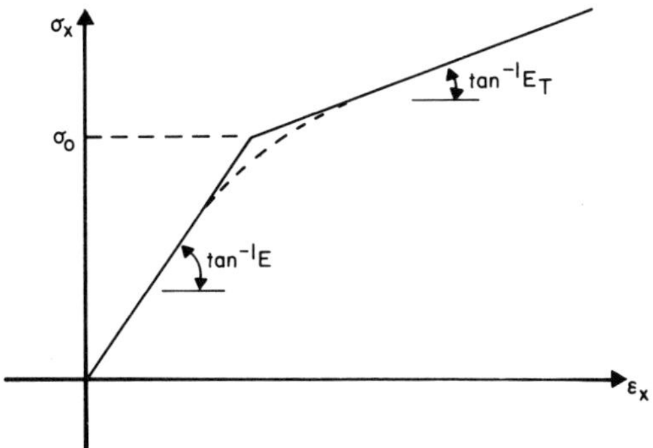

Figure 14. Bilinear stress-strain curve

$$G(\bar{\sigma}^2) \;=\; \frac{1}{4\bar{\sigma}^2}\,\{\frac{1}{E_T} - \frac{1}{E}\} \; . \tag{47}$$

It is often convenient to define a plastic modulus E_p by the equation

$$\frac{1}{E_T} \;=\; \frac{1}{E} + \frac{1}{E_p} \tag{48}$$

so that if E_T and hence E_p is constant,

$$G(\bar{\sigma}^2) \;=\; \frac{1}{4\bar{\sigma}^2 E_p} \; . \tag{49}$$

5.3 The Kinematic Hardening Model

We have up to this point discussed only the plastic strains associated with isotropic hardening. Let us now treat simple idealizations for kinematic hardening. Equations (25) give the general form of the relations, where we now take the subsequent yield surface of equation (15),

$$\phi \;=\; (\sigma_x - \hat{\sigma})^2 + a(\tau_{xy} - \hat{\tau})^2 - \sigma_o^2 \;=\; \phi(\sigma_x, \tau_{xy}, \hat{\sigma}, \hat{\tau}) \tag{50a}$$

where σ_o is constant. The changes in $\hat{\sigma}$, $\hat{\tau}$ for an increment are given by equations (23), which are repeated for convenience

$$d\hat{\sigma} \;=\; \frac{(\sigma_x-\hat{\sigma})\{(\sigma_x-\hat{\sigma})d\sigma_x \;+\; a(\tau_{xy}-\hat{\tau})d\tau_{xy}\}}{\{(\sigma_x-\hat{\sigma})^2 \;+\; a^2(\tau_{xy}-\hat{\tau})^2\}} \;,$$

$$(50b)$$

$$d\hat{\tau} \;=\; \frac{a(\tau_{xy}-\hat{\tau})\{(\sigma_x-\hat{\sigma})d\sigma_x \;+\; a(\tau_{xy}-\hat{\tau})d\tau_{xy}\}}{\{(\sigma_x-\hat{\sigma})^2 \;+\; a^2(\tau_{xy}-\hat{\tau})^2\}} \;,$$

for $\phi = 0$ and $\{(\sigma_x-\hat{\sigma})d\sigma_x \;+\; a(\tau_{xy}-\hat{\tau})d\tau_{xy}\} \geq 0$.

$d\hat{\sigma} \;=\; d\hat{\tau} \;=\; 0$ otherwise.

In general, G depends on the stress and the recorded history, so that

$$G \;=\; G(\sigma_x, \tau_{xy}, \hat{\sigma}, \hat{\tau}) \;. \qquad (51)$$

From equations (25) and (50a),

$$d\varepsilon_x^p \;=\; 4cG(\sigma_x-\hat{\sigma})\{(\sigma_x-\hat{\sigma})d\sigma_x \;+\; a(\tau_{xy}-\hat{\tau})d\tau_{xy}\} \;,$$

$$(52)$$

$$d\gamma_{xy}^p \;=\; 4cGa(\tau_{xy}-\hat{\tau})\{(\sigma_x-\hat{\sigma})d\sigma_x \;+\; a(\tau_{xy}-\hat{\tau})d\tau_{xy}\} \;,$$

where

$$c \;=\; +1 \quad \text{for } \phi = 0 \text{ and } \{(\sigma_x-\hat{\sigma})d\sigma_x \;+\; a(\tau_{xy}-\hat{\tau})d\tau_{xy}\} \geq 0 \;,$$

$$c \;=\; 0 \quad \text{for } \phi < 0$$

$$\text{or } \phi = 0 \text{ and } \{(\sigma_x-\hat{\sigma})d\sigma_x \;+\; a(\tau_{xy}-\hat{\tau})d\tau_{xy} \leq 0 \;.$$

Comparison of equations (50b) and (52) show similarities in the expressions for the increments of plastic strain and the origin of the yield ellipse. In fact, we may write,

$$\frac{d\varepsilon_x^p}{4G\{(\sigma_x-\hat{\sigma})^2+a^2(\tau_{xy}-\hat{\tau})^2\}} \;=\; d\hat{\sigma} \;,\qquad \frac{d\gamma_{xy}^p}{4G\{(\sigma_x-\hat{\sigma})^2+a^2(\tau_{xy}-\hat{\tau})^2\}} \;=\; d\hat{\tau} \;. \quad (53)$$

It is evident that if G has the functional dependence of equation (51) the total plastic strains ε_x^p, γ_{xy}^p may be obtained by integrating along any path traced out by $\hat{\sigma}$, $\hat{\tau}$ and the stress path, beginning with the virgin state when $\hat{\sigma} = \hat{\tau} = 0$. This integration will give

$$\varepsilon_x^p = \varepsilon_x^p(\sigma_x, \tau_{xy}, \hat{\sigma}, \hat{\tau}) \ ,$$

$$\gamma_{xy}^p = \gamma_{xy}^p(\sigma_x, \tau_{xy}, \hat{\sigma}, \hat{\tau}) \ . \tag{54}$$

In principle, these equations may be solved to give

$$\hat{\sigma} = \hat{\sigma}(\sigma_x, \tau_{xy}, \varepsilon_x^p, \gamma_{xy}^p) \ ,$$

$$\hat{\tau} = \hat{\tau}(\sigma_x, \tau_{xy}, \varepsilon_x^p, \gamma_{xy}^p) \ . \tag{55}$$

Thus we are dealing with a material in which the recorded history can be interpreted as the plastic strain, so that the yield function of equation (50) can be expressed as

$$\phi = \phi(\sigma_x, \tau_{xy}, \varepsilon_x^p, \gamma_{xy}^p) \ . \tag{56}$$

Such a material could be referred to as a *strain-hardening material*.

The simplest material idealization which falls into the class defined by equation (52) is that where

$$G(\sigma_x, \tau_{xy}, \hat{\sigma}, \hat{\tau}) = \frac{G_o}{\{(\sigma_x - \hat{\sigma})^2 + a^2(\tau_{xy} - \hat{\tau})^2\}} \tag{57}$$

where G_o is a constant. Substituting equation (51) into equation (53), it is seen that equation (55) becomes

$$\hat{\sigma} = \frac{\varepsilon_x^p}{4G_o} \quad , \quad \hat{\tau} = \frac{\gamma_{xy}^p}{4G_o} \quad . \tag{58}$$

It is also evident that in a *monotonic loading test in simple tension* the assumption of equation (51) leads to a bilinear stress-strain curve. Noting that the ellipse translates along the σ_x axis in this case, so that $(\sigma_x - \hat{\sigma}) = 0$ during loading, equations (52) and (57) give

$$d\varepsilon_x^p = 4G_o \, d\sigma_x \quad , \quad d\gamma_{xy}^p = 0 \quad . \tag{59a}$$

It follows then that

$$\frac{1}{E_p} = 4G_o \quad . \tag{59b}$$

5.4 Yield Surfaces Made up of Two or More Yield Functions

The behavior of an element in plane stress under tension and shear discussed in detail above has been characterized by smooth yield surfaces. It is of interest to consider briefly the structure of the equations when two or more independent yield functions exist and corners are present in the yield surface. In order to illustrate this case, let us consider an element subjected to two direct stresses, σ_x, σ_y with $\tau_{xy} = 0$. This is the case of biaxial tension or compression. σ_x, σ_y are in fact the principal stresses, with the third principal stress zero as a result of the plane stress assumptions. The Tresca yield surface(equation 6)becomes

$$\pm(\sigma_x - \sigma_y) = \sigma_o, \quad \pm\sigma_x = \sigma_o, \quad \pm\sigma_y = \sigma_o \quad . \tag{60}$$

The inner envelope of these six independent equations is the six-sided figure shown in Fig. 15.

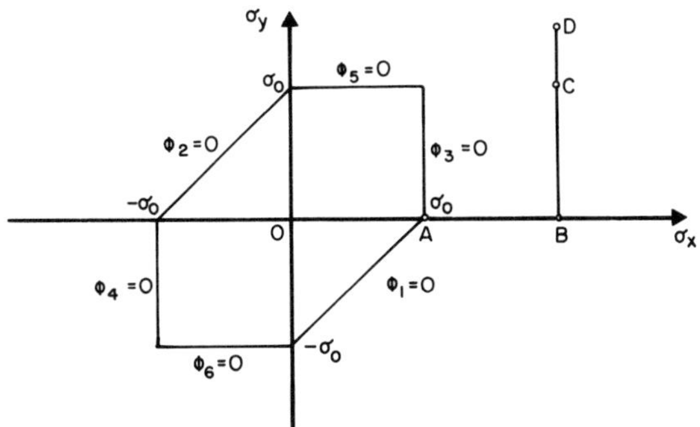

Figure 15. Tresca initial yield surface for biaxial tension

In constructing the yield functions we have the option of treating each of the six lines in Fig. 15 as an independent function, or of treating each *pair* of parallel lines as an independent function. The second option is simple and attractive, and implies that we associate with each plane on which the shear stress is stationary a set of recorded history parameters. A further simplifying assumption is that each pair of lines hardens either isotropically or kinematically; in each case we assign only *one* recorded history parameter to each *pair* of lines.

Thus for a model in which each pair of lines hardens isotropically the yield functions are

$$\phi_{1,2} = \overset{+}{\underset{-}{}}(\sigma_x - \sigma_y) - \bar{\sigma}_{12} \; , \tag{61a}$$

$$\phi_{3,4} = \overset{+}{\underset{-}{}}\sigma_x - \bar{\sigma}_{34} \; , \tag{61b}$$

$$\phi_{5,6} = \overset{+}{\underset{-}{}}\sigma_y - \bar{\sigma}_{56} \; . \tag{61c}$$

The lines $\phi_1 = 0$ and $\phi_2 = 0$ are parallel and in fact part of the same yield function: note that loading can never take place on $\phi_1 = 0$ and $\phi_2 = 0$ simultaneously. Similar remarks apply to the pairs $\phi_3 = 0$, $\phi_4 = 0$ and $\phi_5 = 0$, $\phi_6 = 0$. The recorded history parameters are $\bar{\sigma}_{12}$, $\bar{\sigma}_{34}$, $\bar{\sigma}_{56}$; $\bar{\sigma}_{12}$, $\bar{\sigma}_{34}$, $\bar{\sigma}_{56}$ are the larger of σ_o or the maximum previously attained value of $|\sigma_x - \sigma_y|$, $|\sigma_x|$

and $|\sigma_y|$ respectively.

A model in which the pairs of lines harden kinematically is defined by the yield functions

$$\phi_{1,2} = \pm(\sigma_x-\sigma_y) - (\sigma_o+H_{12}) \quad , \tag{62a}$$

$$\phi_{3,4} = \pm\sigma_x - (\sigma_o+H_{34}) \quad , \tag{62b}$$

$$\phi_{5,6} = \pm\sigma_y - (\sigma_o+H_{56}) \quad . \tag{62c}$$

The three recorded history parameters are H_{12}, H_{34} and H_{56}.

The subsequent development of these models follows the isotropic and kinematic hardening models we have already discussed. For illustrative purposes let us develop the kinematic hardening model. Some simplification can be achieved by rearranging the yield functions ϕ_1 and ϕ_2 so that the vector $(\partial\phi/\partial\sigma_x, \partial\phi/\partial\sigma_y)$ is a unit vector for all six yield functions. This can be done by redefining the yield functions of equations (62) as

$$\phi_{1,2} = \pm\frac{1}{\sqrt{2}}(\sigma_x-\sigma_y) - \frac{1}{\sqrt{2}}(\sigma_o+H_{12}) \quad , \tag{63a}$$

$$\phi_{3,4} = \pm\sigma_x - (\sigma_o+H_{34}) \quad , \tag{63b}$$

$$\phi_{5,6} = \pm\sigma_y - (\sigma_o+H_{56}) \quad . \tag{63c}$$

The total plastic strain is then given by

$$\varepsilon_x^p = \varepsilon_x^{p12} + \varepsilon_x^{p34} + \varepsilon_x^{p56} \quad ,$$

$$\varepsilon_y^p = \varepsilon_y^{p12} + \varepsilon_y^{p34} + \varepsilon_y^{p56} \quad . \tag{64}$$

Following the general formulation for the plastic strain increments (Chapter 4, equations 6-8) we may then write

$$d\varepsilon_x^{p12} \;=\; c_{12}\frac{G_{12}}{2}\,(d\sigma_x - d\sigma_y)\quad,$$

$$d\varepsilon_y^{p12} \;=\; -c_{12}\frac{G_{12}}{2}\,(d\sigma_x - d\sigma_y)\quad,$$

$$d\varepsilon_x^{p34} \;=\; c_{34}G_{34}d\sigma_x\quad,$$

$$d\varepsilon_y^{p34} \;=\; 0\quad,$$

$$d\varepsilon_x^{p56} \;=\; 0\quad,$$

$$d\varepsilon_y^{p56} \;=\; c_{56}G_{56}d\sigma_y\quad,$$

(65)

where

$$
c_{12} \;=\;
\begin{cases}
+1 & \text{if}\quad \phi_1 = 0\quad \text{and}\quad d\sigma_x - d\sigma_y \geq 0\quad,\\[2ex]
-1 & \text{if}\quad \phi_2 = 0\quad \text{and}\quad d\sigma_x - d\sigma_y \leq 0\quad,\\[2ex]
0 & \text{otherwise,}
\end{cases}
$$

$$
c_{34} \;=\;
\begin{cases}
+1 & \text{if}\quad \phi_3 = 0\quad \text{and}\quad d\sigma_x \geq 0\quad,\\[2ex]
-1 & \text{if}\quad \phi_4 = 0\quad \text{and}\quad d\sigma_x \leq 0\quad,\\[2ex]
0 & \text{otherwise}\quad,
\end{cases}
$$

(66)

$$
c_{56} \;=\;
\begin{cases}
+1 & \text{if}\quad \phi_5 = 0\quad \text{and}\quad d\sigma_y \geq 0\quad,\\[2ex]
-1 & \text{if}\quad \phi_6 = 0\quad \text{and}\quad d\sigma_y \leq 0\quad,\\[2ex]
0 & \text{otherwise.}
\end{cases}
$$

The non-negative hardening coefficients G_{12}, G_{34} and G_{56} must now be chosen. The simplest choice is $G_{12} = G_{34} = G_{56} = G_o$ where G_o

is a constant. This implies bilinear hardening on each pair of
lines, and because $(\partial\phi/\partial\sigma_x, \partial\phi/\partial\sigma_y)$ is a vector of unit magnitude for each line bounding the yield surface, it further im-
plies that the plastic modulus has the same magnitude for each
pair of lines.

   If we now study the changes in the yield function which occur during loading we can identify the recorded history parameters. For example, loading on the yield function ϕ_1 requires
that

$$d\phi_1 = (d\sigma_x - d\sigma_y) - dH_{12} = 0 \ . \tag{67}$$

Comparing equation (67) with the first of equations (65), it is
evident that

$$dH_{12} = \frac{2}{G_o} d\varepsilon_x^{pl2} \ . \tag{68}$$

Similarly loading on the yield function ϕ_2 requires that

$$d\phi_2 = -(d\sigma_x - d\sigma_y) - dH_{12} = 0 \ , \tag{69}$$

and comparison of equation (69) with the second of equations
(65) gives

$$dH_{12} = \frac{2}{G_o} d\varepsilon_y^{pl2} \ . \tag{70}$$

If the material is initially in its virgin state (i.e. all components of the plastic strain appearing in equations 64 are
zero initially) equations (68) and (70) can be integrated to
give

$$H_{12} = \frac{1}{G_o} (\varepsilon_x^{pl2} - \varepsilon_y^{pl2}) \ . \tag{71a}$$

Similar arguments give

$$H_{34} = \frac{1}{G_O} \varepsilon_x^{p34} \quad , \quad H_{56} = \frac{1}{G_O} \varepsilon_y^{p56} \quad . \tag{71b}$$

This kinematic hardening model is now complete in that the yield surface in stress space is completely defined, the incremental plastic strain relations are given in equations (65) and the change in the recorded history parameters after each increment of load can be computed.

While these relations are apparently complex, they are, in fact, extremely easy to use because the independent yield functions can be treated quite separately. As an example of the computation of plastic strains consider the response of an element to the stress path OABCD shown in Fig. 16. We treat each pair of yield functions in turn, obtaining the final plastic strain by adding the contributions at the end of the process.

Figure 16(a) shows the pair of lines $\phi_1 = 0$ and $\phi_2 = 0$. As

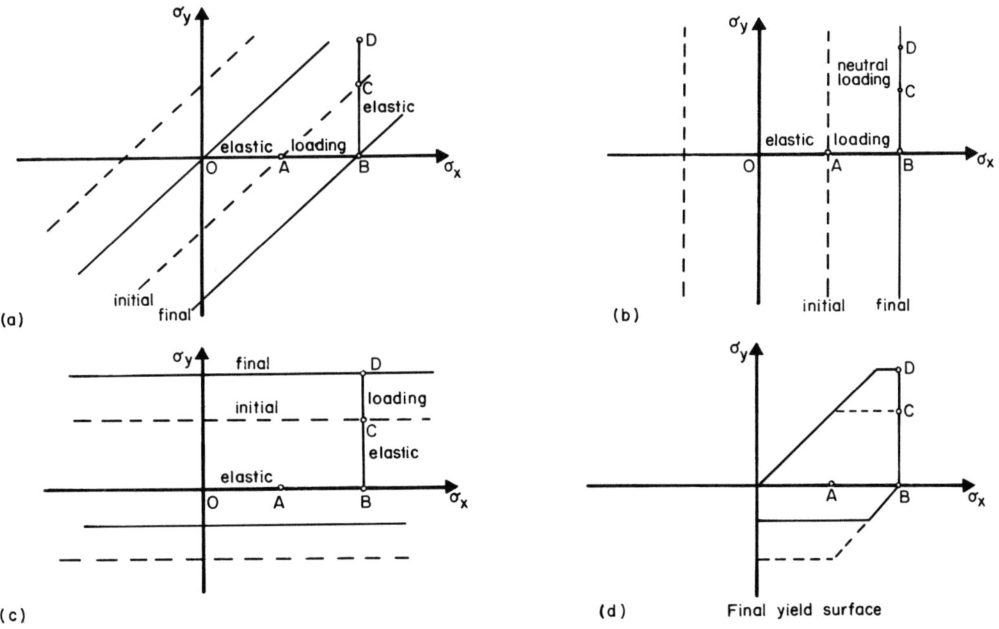

Figure 16. Subsequent yield surfaces for independent yield functions

plastic deformation occurs the lines translate without rotation, and the distance between the lines remains constant. Inspection shows that loading occurs only during the part AB of the stress path. Along AB $d\sigma_y = 0$, and hence from equations (65)

$$d\varepsilon_x^{p12} = \frac{G_o}{2}\, d\sigma_x \quad , \qquad d\varepsilon_y^{p12} = -\frac{G_o}{2}\, d\sigma_x \quad . \tag{72}$$

These equations may be integrated along the straight line path AB i.e. between $\sigma_x = \sigma_o$ and $\sigma_x = 2\sigma_o$ to give

$$\varepsilon_x^{p12} = \frac{G_o}{2}\, \sigma_o \quad , \qquad \varepsilon_y^{p12} = -\frac{G_o}{2}\, \sigma_o \quad , \tag{73}$$

as the total contribution to the final plastic strain at D from this mechanism.

Figure 16(b) shows the pair of lines $\phi_3 = 0$ and $\phi_4 = 0$; loading takes place again along AB. Hence from equations (65)

$$d\varepsilon_x^{p34} = G_o\, d\sigma_x \quad , \qquad d\varepsilon_y^{p34} = 0 \quad . \tag{74}$$

Integrating along AB the total contribution from this mechanism to the final plastic strain at D is

$$\varepsilon_x^{p34} = G_o\sigma_o \quad , \qquad \varepsilon_y^{p34} = 0 \quad . \tag{75}$$

Figure 16(c) shows the pair of lines $\phi_5 = 0$ and $\phi_6 = 0$. Loading occurs along CD, and it is readily seen that at D

$$\varepsilon_x^{p56} = 0 \quad , \qquad \varepsilon_y^{p56} = \frac{1}{2} G_o\sigma_o \quad . \tag{76}$$

The final plastic strain is then

$$\varepsilon_x^{p} = \varepsilon_x^{p12} + \varepsilon_x^{p34} + \varepsilon_x^{p56} = \frac{3}{2} G_o\sigma_o \quad ,$$

$$\varepsilon_y^{p} = \varepsilon_y^{p12} + \varepsilon_y^{p34} + \varepsilon_y^{p56} = 0 \quad . \tag{77}$$

The yield surface when the stress point reaches D is obtained simply by superposing the final independent yield surfaces in Fig. 16(a), (b) and (c) and drawing the inner envelope. This surface is shown in Fig. 16(d). The dotted lines show the yield surface after the stress path OAB. It should be noted that while the individual yield functions harden kinematically their combined effect does not mean that the yield surface in stress space translates without rotation or change in shape or size in stress space. This remark would also apply if the individual yield functions hardened isotropically; the complete yield surface in stress space would not harden isotropically because the individual yield functions are activitated in different ways.

5.5 Piecewise Linear Yield Surfaces

Difficulties in dealing with nonlinear yield functions in complex problems have motivated another type of approximation in which the yield function is deliberately linearized; such yield functions are termed *piecewise linear yield functions*. Linearization of this kind introduces corners into the yield surface, but they are generally handled differently from those which appear because of the existence of independent yield functions. In fact these linearized yield functions are an *a posteriori* approximation, in that we write out the behavior of the model of the constitutive relations under consideration first with a continuously differentiable yield function and only then linearize; this is fundamentally different from the introduction of independent yield surfaces and leads to quite different kinds of behavior.

As an example, let us consider a kinematically hardening model for biaxial plane stress ($\tau_{xy} = 0$) with a von Mises initial condition. The initial yield surface is obtained by substituting the principal stresses $\sigma_I = \sigma_x$, $\sigma_{II} = \sigma_y$, $\sigma_{III} = 0$ into equations (3) and (4); this leads to

$$\sigma_x^2 - \sigma_x\sigma_y + \sigma_y^2 = \sigma_o^2 \quad . \tag{78}$$

The initial yield surface is an ellipse, shown in Fig. 17(a). Subsequent yield surfaces are obtained by translating this ellipse without rotation or change in size or shape; to effect this translation we introduce recorded history parameters $\hat{\sigma}_x$, $\hat{\sigma}_y$ which are the coordinates of the center of the ellipse in the stress space, and put

$$\phi = (\sigma_x-\hat{\sigma}_x)^2 - (\sigma_x-\hat{\sigma}_x)(\sigma_y-\hat{\sigma}_y) + (\sigma_y-\hat{\sigma}_y)^2 - \sigma_o^2 \quad . \tag{79}$$

Noting that

$$\frac{\partial\phi}{\partial\sigma_x} = 2(\sigma_x-\hat{\sigma}_x) - (\sigma_y-\hat{\sigma}_y) \quad , \tag{80a}$$

$$\frac{\partial\phi}{\partial\sigma_y} = 2(\sigma_y-\hat{\sigma}_y) - (\sigma_x-\hat{\sigma}_x) \quad , \tag{80b}$$

the plastic strain increments are given by

$$d\varepsilon_x^p = cG \frac{\partial\phi}{\partial\sigma_x} \left\{\frac{\partial\phi}{\partial\phi_x} d\sigma_x + \frac{\partial\phi}{\partial\sigma_y} d\sigma_y\right\} \quad , \tag{81a}$$

$$d\varepsilon_y^p = cG \frac{\partial\phi}{\partial\sigma_y} \left\{\frac{\partial\phi}{\partial\sigma_x} d\sigma_x + \frac{\partial\phi}{\partial\sigma_y} d\sigma_y\right\} \quad , \tag{81b}$$

where $c = +1$ for $\phi = 0$ and $\dfrac{\partial\phi}{\partial\sigma_x} d\sigma_x + \dfrac{\partial\phi}{\partial\sigma_y} d\sigma_y \geq 0$,

$c = 0$ otherwise ,

and $G = G(\sigma_x, \sigma_y, \hat{\sigma}_x, \hat{\sigma}_y) \geq 0$,

The changes in the recorded history parameter must now be computed. During loading $d\phi = 0$, giving

$$\frac{\partial \phi}{\partial \sigma_x} d\sigma_x + \frac{\partial \phi}{\partial \sigma_y} d\sigma_y = -\left\{ \frac{\partial \phi}{\partial \hat{\sigma}_x} d\hat{\sigma}_x + \frac{\partial \phi}{\partial \hat{\sigma}_y} d\hat{\sigma}_y \right\} . \tag{82a}$$

Following the precedent set in the case of kinematic hardening in the tension-torsion case (equation 22a), let us assume that

$$\frac{d\hat{\sigma}_x}{d\hat{\sigma}_y} = \frac{\partial \phi / \partial \hat{\sigma}_x}{\partial \phi / \partial \hat{\sigma}_y} . \tag{82b}$$

Solving equations (82a) and (82b) simultaneously,

$$d\hat{\sigma}_x = \frac{\frac{\partial \phi}{\partial \sigma_x} \left\{ \frac{\partial \phi}{\partial \sigma_x} d\sigma_x + \frac{\partial \phi}{\partial \sigma_y} d\sigma_y \right\}}{\left\{ \left(\frac{\partial \phi}{\partial \sigma_x}\right)^2 + \left(\frac{\partial \phi}{\partial \sigma_y}\right)^2 \right\}} , \tag{83a}$$

$$d\hat{\sigma}_y = \frac{\frac{\partial \phi}{\partial \sigma_y} \left\{ \frac{\partial \phi}{\partial \sigma_x} d\sigma_x + \frac{\partial \phi}{\partial \sigma_y} d\sigma_y \right\}}{\left\{ \left(\frac{\partial \phi}{\partial \sigma_x}\right)^2 + \left(\frac{\partial \phi}{\partial \sigma_y}\right)^2 \right\}} . \tag{83b}$$

Note that

$$\left\{ \left(\frac{\partial \phi}{\partial \sigma_x}\right)^2 + \left(\frac{\partial \phi}{\partial \sigma_y}\right)^2 \right\}^{1/2}$$

is the magnitude of the vector $\partial \phi / \partial \sigma_x, \partial \phi / \partial \sigma_y$ whose direction is the outward normal to the yield surface. Thus the vector $d\hat{\sigma}_x$, $d\hat{\sigma}_y$ has the direction of the outward normal and a magnitude equal to the component of the vector $d\sigma_x$, $d\sigma_y$ in the direction of the normal.

Further, comparison of equations (81) and (83) shows that

$$\frac{1}{G} d\varepsilon_x^p = d\hat{\sigma}_x \left\{ \left(\frac{\partial \phi}{\partial \sigma_x}\right)^2 + \left(\frac{\partial \phi}{\partial \sigma_y}\right)^2 \right\}$$

$$\frac{1}{G} d\varepsilon_x^p = d\hat{\sigma}_y \left\{ \left(\frac{\partial \phi}{\partial \sigma_x}\right)^2 + \left(\frac{\partial \phi}{\partial \sigma_y}\right)^2 \right\} \tag{84}$$

Let us put

$$G = G_0 / \left\{ \left(\frac{\partial \phi}{\partial \sigma_x} \right)^2 + \left(\frac{\partial \phi}{\partial \sigma_y} \right)^2 \right\} , \tag{85}$$

where G_0 is a positive constant. We find that, upon integrating from the virgin state

$$\varepsilon_x^p = G_0 \, \hat{\sigma}_x , \qquad \varepsilon_y^p = G_0 \, \hat{\sigma}_y . \tag{86}$$

In addition, for monotonic loading in simple tension ($d\sigma_x > 0$, $\sigma_y = 0$),

$$d\varepsilon_x^p = G_0 \, d\sigma_x , \tag{87}$$

thus leading to a bilinear stress-strain curve in simple tension.

The model is now complete, and it can be seen that it involves dealing with a number of nonlinear and cumbersome expressions. The strain increment equations would have to be integrated numerically along all but some exceptional stress paths.

In constructing a piecewise linear approximation to this model we retain the notion that $\hat{\sigma}_x$, $\hat{\sigma}_y$ are the coordinates of a reference point for a yield surface which simply displaces in the stress space, and that equation (86) relates the plastic strains and $\hat{\sigma}_x$, $\hat{\sigma}_y$. We then proceed to linearize the yield surface; a typical linearization is shown in Fig. 17(b). The linearization may be carried out to any desired degree of approximation, but we retain a convex yield surface. It is now necessary to construct a relation between $d\hat{\sigma}_x$, $d\hat{\sigma}_y$ and $d\sigma_x$, $d\sigma_y$ when loading occurs, observing the general requirement that the plastic strain increment should be normal to the yield surface at a regular point (where the normal vector is uniquely defined)

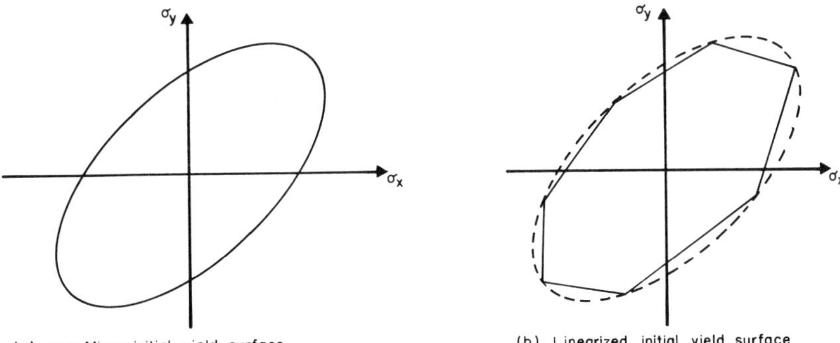

Figure 17.Construction of a piecewise linear yield surface

and between adjacent normals at a singular point (where the

normal vector is not uniquely defined).

Consider first the case when the stress point lies on a side

of the linearized surface. Let n_x, n_y be the components of the

unit normal vector. For loading characterized by $(n_x d\sigma_x +$

$n_y d\sigma_y) \geq 0$, the constraints are satisfied if we put,

$$d\hat{\sigma}_x = n_x(n_x d\sigma_x + n_y d\sigma_y) \quad,$$

$$d\hat{\sigma}_y = n_y(n_x d\sigma_x + n_y d\sigma_y) \quad.$$

(88)

Thus the vector $d\hat{\sigma}_x$, $d\hat{\sigma}_y$ has the direction of n_x, n_y and the

magnitude of the component of $d\sigma_x$, $d\sigma_y$ in the direction of n_x,

n_y. When loading occurs with the stress point at a corner we

must consider two separate cases; that where the stress incre-

ment vector $d\sigma_x$, $d\sigma_y$ lies *between* the adjacent normals to the

yield surface and that where the stress increment vector lies

outside of the fan between the adjacent normals. Let the adja-

cent *unit* normal vectors be n'_x, n'_y and n''_x, n''_y respectively. If

the stress increment vector lies outside the fan we use the

same rule for $d\hat{\sigma}_x$, $d\hat{\sigma}_y$ as at a regular point; $d\hat{\sigma}_x$, $d\hat{\sigma}_y$ has the

direction of the nearest adjacent normal and the magnitude of

the component of $d\sigma_x$, $d\sigma_y$ in the direction of that adjacent nor-

mal. However, if $d\sigma_x$, $d\sigma_y$ lies in the fan between adjacent nor-

mals we put

$$d\hat{\sigma}_x = d\sigma_x \quad , \quad d\hat{\sigma}_y = d\sigma_y \quad . \tag{89}$$

In this case $d\hat{\sigma}_x$, $d\hat{\sigma}_y$ has both the *direction* and the *magnitude* of $d\sigma_x$, $d\sigma_y$. Once these rules have been established we can easily compute the plastic strain increments from the incremental forms of equations (86),

$$d\varepsilon_x^p = G_o d\hat{\sigma}_x \quad , \quad d\varepsilon_y^p = G_o d\hat{\sigma}_y \quad . \tag{90}$$

Note that $d\varepsilon_x^p$, $d\varepsilon_y^p$ has the same direction as $d\hat{\sigma}_x$, $d\hat{\sigma}_y$, and consequently the requirements of the normality rule are met at both regular and singular points.

As an example of the calculation of the response of the model we may note that the Tresca initial yield condition for biaxial tension (Fig. 15) inscribes the von Mises initial yield condition (Fig. 17a). It may consequently *in our present context* be regarded as a suitable linearization of the von Mises initial condition. Subsequent yield surfaces may be formally regarded as the inner envelope of the pairs of lines

$$\phi_{1,2} = \pm \frac{1}{\sqrt{2}} \{(\sigma_x - \hat{\sigma}_x) - (\sigma_y - \hat{\sigma}_y)\} - \frac{1}{\sqrt{2}} \sigma_o = 0 \quad ,$$

$$\phi_{3,4} = \pm(\sigma_x - \hat{\sigma}_x) - \sigma_o = 0 \quad , \tag{91}$$

$$\phi_{5,6} = \pm(\sigma_y - \hat{\sigma}_y) - \sigma_o = 0 \quad .$$

Consider the response to the stress path OABC shown in Fig. 16, and given again in Fig. 18. Elastic behavior occurs along OA; when the stress point reaches A, $\phi = 0$. Loading occurs along AB, with the stress increment vector coinciding with the direction of one of the adjacent normals. Hence the reference point with coordinates $\hat{\sigma}_x$, $\hat{\sigma}_y$, which is initially at the origin and is always equidistant from each line, moves along the σ_x

axis. The subsequent yield surface when the stress point
reaches B is shown by dotted lines in Fig. 18. As the stress
point moves along BC we have neutral loading; further loading
occurs as the stress point moves along CD. It is apparent that
the reference point moves along a line parallel to the σ_y axis
during this loading phase. The subsequent yield surface when
the stress point reaches D is shown in Fig. 18.

The final coordinates of the reference point are seen to be

$$\hat{\sigma}_x = \sigma_o \quad , \quad \hat{\sigma}_y = \frac{1}{2}\sigma_o \quad . \tag{92}$$

It follows then from equation (86) that the final plastic
strains are

$$\varepsilon_x^p = G_o\sigma_o \quad , \quad \varepsilon_y^p = \frac{G_o\sigma_o}{2} \quad . \tag{93}$$

A comparison of the response of this model with that constructed from a Tresca initial yield condition and independent yield
functions shows a distinct difference in the final yield surface
(Figs. 16d and 18) and in the final plastic strains (equations
77 and 93).

It is clear from this simple illustrative calculation of the
response to a particular stress path that linearization can
lead to a significant simplification of the computational work;

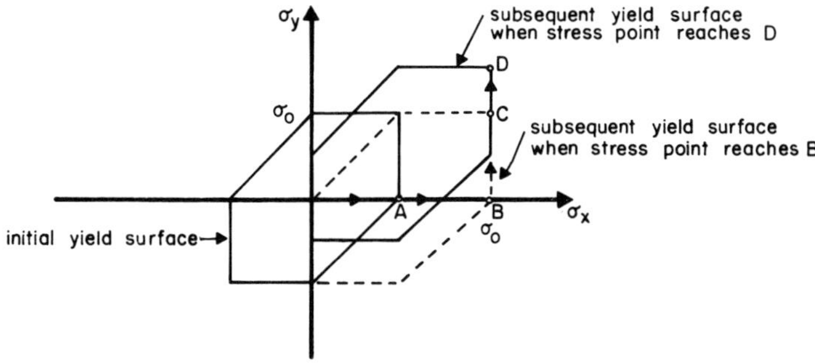

Figure 18. Kinematic hardening with a piecewise linear yield
surface

the reader is invited to calculate the response to this same
stress path of the model given in equations (79), (81) and (83).
Furthermore, all the essential requirements of the general
framework for plastic constitutive equations have been met, so
that the sufficient conditions for uniqueness and stability in
structural problems discussed in Chapter 3 have not been affected.

Although the point is trivial, it should be emphasized that
plastic constitutive relations for a single nonzero stress component can be constructed in the same way as the more complex
models we have discussed in this section. Suppose, for example,
that in the plane stress situation $\sigma_y = \tau_{xy} = 0$, and that we
are concerned only with the strain component ε_x^p. We are not
required to invoke the complete yield surface in the stress
space (i.e. to choose a particular yield function involving σ_y
and τ_{xy}), but we must choose a hardening relation. We can do
this by choosing two yield functions involving σ_x only; these
two yield functions characterize the two points at which the
more complete yield functions cut the σ_x axis in the stress
space.

For an isotropic hardening material, for instance, the yield
functions are

$$\phi_1 = +\sigma_x - \bar{\sigma}, \quad \phi_2 = -\sigma_x - \ddot{\sigma} , \tag{94}$$

where $\bar{\sigma}$ is positive and is the largest previously obtained va-
lue of $|\sigma_x|$, provided that this value is greater than the ini-
tial yield stress σ_o. The plastic strain increment $d\varepsilon_x^p$ is
given by

$$d\varepsilon_x^p = Gd\sigma_x \quad \text{for} \quad \phi_1 = 0 \quad \text{and} \quad d\sigma_x > 0$$

$$\text{and for} \quad \phi_2 = 0 \quad \text{and} \quad d\sigma_x < 0 , \tag{95}$$

$$d\varepsilon_x^p = 0 \quad \text{otherwise.}$$

The scalar G can be replaced by $1/E_p$, where E_p is the plastic modulus.

For a kinematic hardening material the two yield functions would be written as

$$\phi_1 = +(\sigma_x-\hat{\sigma}) - \sigma_o \, , \qquad \phi_2 = -(\sigma_x-\hat{\sigma}) - \sigma_o \, . \tag{96}$$

Using the same argument that led to equation (68), we put

$$d\hat{\sigma} = \frac{1}{G}\, d\varepsilon_x^p = E_p d\varepsilon_x^p \, . \tag{97}$$

If G and E_p are constants and the material is initially in the virgin state, the yield functions (70) could be rewritten as

$$\phi_1 = +(\sigma_x-E_p\varepsilon_x^p) - \sigma_o \, , \qquad \phi_2 = -(\sigma_x-E_p\varepsilon_x^p) - \sigma_o \, . \tag{98}$$

The plastic strain increment is then given by

$$d\varepsilon_x^p = \frac{1}{E_p}\, d\sigma_x \quad \text{for} \quad \phi_1 = 0 \quad \text{and} \quad d\sigma_x > 0$$

$$\qquad\qquad \text{and for} \quad \phi_2 = 0 \quad \text{and} \quad d\sigma_x < 0 \, , \tag{99}$$

$$d\varepsilon_x^p = 0 \quad \text{otherwise.}$$

Consideration of assumed hardening laws for uniaxial stress conditions renders it appropriate to introduce a cautionary statement about simple models of this kind, particularly kinematic hardening models. We have taken the view in this section that the possible stress states are known *a priori*, and can be limited to, say, tension and shear, biaxial tension, or in the most recent example, uniaxial tension. We have then proceeded to formulate simple hardening models for each of these cases. While each of these models is self-consistent and perfectly acceptable, the reader must not be misled into believing that the model for any simple set of stress states is a special case of

that *same* model for more complex stress states. This statement can best be illustrated by a comparison of uniaxial, biaxial and triaxial tension. Suppose that we are concerned with
a constitutive relation for simple tension ($\sigma_x \neq 0$, all other
stress components zero). We could choose to work in the σ_x
subspace, the σ_x, σ_y subspace or the σ_x, σ_y, σ_z subspace of
the principal stress space. The major difference between
these choices is that the first alternative permits us to pre-
dict changes in ε_x^p only, the second to predict changes in ε_x^p,
ε_y^p (and only by using the incompressibility condition changes
in ε_z^p), and the third to predict directly changes in all three
strain components. Suppose further that we adopt a kinematic
hardening model of the type given in equation (98) for uniaxial tension. This immediately tells us something about the
subsequent yield surfaces in the σ_x, σ_y space, for example. If
the initial yield surface is the von Mises surface (equation
78), and it moves in the stress space without change in size or
shape, the requirement that subsequent yield surfaces in the
σ_x, σ_y subspace should have a projection on the σ_x axis of the
form of equation (98) for loading along the σ_x axis means that
it should move in the manner shown in Fig. 19(a). The center
of the ellipse moves along the σ_x axis so that the difference
between the tension and compressive yield surfaces is always
given by $2\sigma_o$.

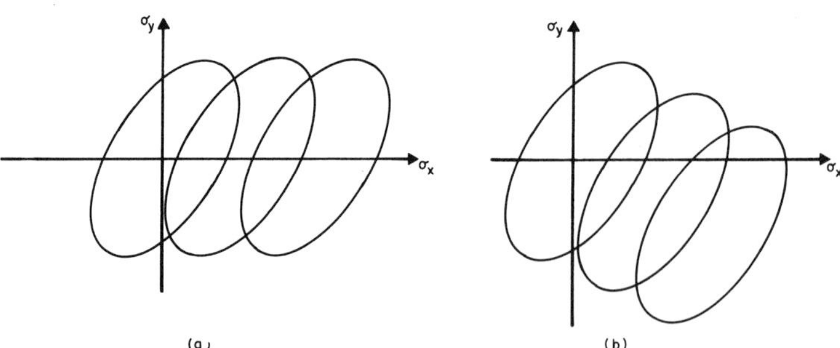

Figure 19. Comparison of kinematic hardening models

However, were we to assume that the behavior in the σ_x, σ_y space was governed by conventional kinematic hardening, we might assume that the motion of the center of the ellipse was in the same direction as the plastic strain increment. Loading along the σ_x axis would then lead to the subsequent yield surface shown in Fig. 19(b). The projection of these subsequent yield surfaces onto the σ_x axis is not described by equations (78) for uniaxial tension; the difference between the yield stresses in tension and compression is not fixed at $2\sigma_o$. By analogy we can argue that if we were to assume conventional kinematic hardening in the σ_x, σ_y, σ_z space we would not necessarily obtain the same response to stress paths in the σ_x, σ_y subspace as we would get if we adopted conventional kinematic hardening in the σ_x, σ_y space. These differences are not disturbing provided that we recognize their origin. It is advisable, however, to construct a model for the most general stress state which will be encountered in any application, and then to specialize to simpler stress states, rather than the converse. Models for more general stress states will be considered in Chapter 7.

5.6 Elastic, Perfectly Plastic Materials

In the case of an *elastic, perfectly plastic* material the initial yield surface becomes the limit surface, with plastic deformation taking place only when $\phi = 0$ and neutral loading occurs. Thus, for plane stress with σ_x and τ_{xy} acting on the element ($\sigma_y = 0$), the complete plastic constitutive relations are characterized by

$$\phi = \sigma_x^2 + a\tau_{xy}^2 - \sigma_o^2 \, ,$$

$$d\varepsilon_x^p = 2\lambda\sigma_x, \quad d\gamma_{xy}^p = 2\lambda a\tau_{xy} \tag{100}$$

$$\text{for} \quad \phi = 0 \quad \text{and} \quad \sigma_x d\sigma_x + a\tau_{xy} d\tau_{xy} = 0,$$

$$d\varepsilon_x^p = 0 , \qquad d\gamma_{xy}^p = 0$$

otherwise,

with the restriction that $\phi(\sigma_x, \tau_{xy}) > 0$ is not permitted. λ is an unspecified non-negative scalar which must be determined by other considerations. The constitutive equations in this form give the result

$$\frac{d\varepsilon_x^p}{d\gamma_{xy}^p} = \frac{\sigma_x}{a\tau_{xy}} , \tag{101}$$

without giving the magnitude of $d\varepsilon_x^p$ or $d\gamma_{xy}^p$.

When corners are present on the limit surface, further indeterminacy of the direction of the plastic strain increment vector is present. As an example, Fig. 20 shows the directions of the plastic strain increment vector for the case of biaxial tension with a Tresca yield surface, with the limitations in directions of the adjacent normals at the singular points. It may be remarked that linearization of a smooth limit surface as an approximation to simplify computations in any particular problem can be accomplished without difficulty; the linearized yield surface should be convex, and the conventional flow rule at regular and singular points is applied.

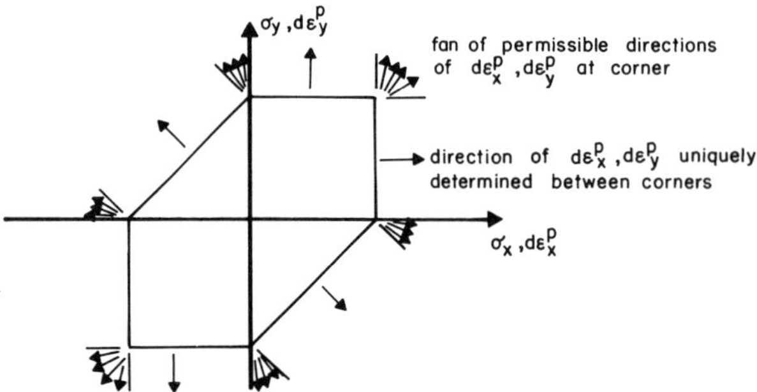

Figure 20. Elastic, perfectly plastic material with Tresca yield surface

The material presented in this Chapter is a brief survey of
some of the commonly used idealizations in plasticity. In the
following Chapter we shall consider some simple problems in bar
structures which make use of these idealizations. In Chapter 7
the plane stress idealizations will be generalized for continua
described in cartesian tensor notation, and further examples
will be given.

One final comment should be given on the choice of the idealization of the initial yield surface. The von Mises condition
has been seen to be closer to the actual initial yield surface
for many materials, although this is partly a result of the
manner in which the yield surfaces are normalized. In most
cases the choice depends on computational convenience rather
than on the marginal difference in accuracy. From this point
of view the von Mises condition has the advantages that the
principal stress directions need not be known (an important
point in bodies with non-homogeneous stress fields) and that it
does not contain any corners. However, it is quadratic and
therefore nonlinear, and this causes additional complexity
(which can be reduced by linearization). The Tresca condition
is piecewise linear if the principal stress directions are
known. This is not an advantage if the principal stress directions must be computed at each point in the body. If, on
the other hand, as a result of symmetry or other considerations,
the principal stress directions are known *a priori* in a particular problem it is generally much easier to use the Tresca
condition.

5.7 Historical and Bibliographical Remarks

The three experimental investigations referred to directly in
this brief outline of the development of simple models of plastic behavior in plane stress (Taylor and Quinney [1931], Ivey

[1961], Bertsch and Findley [1962] represent only a small sample of the extensive experimental studies carried out since the early years of this century. Some of this early work is reviewed by Nadai [1933] and Marin [1936]; more recent surveys have been given by Drucker [1956] and Michno and Findley [1972].

Experimental investigations have shown quite a wide variety in the behavior of metals, although some of these differences may be due to different definitions of yield and different experimental techniques. Among recent investigations, for example, Phillips and Gray [1961] and Bertsch and Findley [1962] report observations of the development of corners in subsequent yield surface in general (although not exact) agreement with the model involving independent yield functions and the more sophisticated slip theory. Ivey [1961], Parker and Kettlewell [1961], Paul, Chen and Lee [1962] and Jenkins [1965] note the absence of corners and report general agreement with kinematic hardening models. Szczepinski [1963] and Parker and Bassett [1964] report general agreement with the concept of a uniformly expanding yield surface and thus with the model of isotropic hardening. Miastkowski and Szczepinski [1965] observed a marked rotation of the yield surface.

The development of the isotropic hardening model arose directly from the study of isotropic initial yield surfaces; the material remains isotropic after plastic deformation takes place. The development of this model is described extensively by Prager [1948] and Hill [1950]. More recent developments in the construction of models have attempted to describe anistropy introduced by plastic deformation. The kinematic hardening model described in detail in this section derives from the work of Reuss [1935], Prager [1948] and Ishlinskii [1954]. Early attempts to include rotation and expansion or contraction of the yield surface together with a translation in stress space

were made by Drucker [1949] and Edelman and Drucker [1951].
Still more recently attempts have been made to incorporate more
firmly into the phenomenological theory some of the results
available from the study of the physics of plastic deformation.
These attempts began with the slip theory of Batdorf and
Budiansky [1949](which will be discussed in greater detail in
Chapter 7). Many of these relations can be used in the phenomenological framework through the device of independent yield
functions (Koiter [1953], Sanders [1954]).

The use of piecewise linear approximations to a yield surface in any particular model has been discussed extensively by
Prager [1953], [1956] and Hodge [1955], [1957]. The discrepancy between kinematic hardening models in stress spaces of
different dimensions was pointed out by Perrone and Hodge [1958]
and Shield and Ziegler [1958].

Further work on the generation of more sophisticated anisotropic models has continued: as examples the work of Baltov and
Sawczuk [1965], Mroz [1967], Eisenberg and Phillips [1969] and
Kadashevich and Novozhilov [1959], [1968] can be mentioned. Extensive reviews of the development of particular forms of the
constitutive relations have been given by Prager [1955], [1966]
and Naghdi [1960].

References

A.Baltov and A.Sawczuk	1965	"A rule of anisotropic hardening", Acta Mechanica, $\underline{1}$, 81.
S.F.Batdorf and B.Budiansky	1949	"A mathematical theory of plasti- city based on the concept of slip", NACA Technical Note 1871.
P.K.Bertsch and W.N.Findley	1962	"An experimental study of subsequent yield surfaces; corners, normality, Bauschinger and allied effects", Proc.4th U.S.Nat.Congr.Appl.Mech., ASME, 893.

D.C.Drucker   1949   "The significance of the criterion
                               for additional plastic deformation
                               of metals", Colloid Sci. (Rheology
                               Issue), $\underline{4}$, 299.

F.Edelman and   1951   "Some extensions of elementary
D.C.Drucker   plasticity theory", J.Franklin Inst.
                               $\underline{251}$, 581.

D.C.Drucker   1956   "Stress-strain relations in the
                               plastic range of metals - experi-
                               ments and basic concepts", Chapter
                               4 in *Rheology, Theory and Applica-
                               tions*. Edited by F.R.Eirich, Aca-
                               demic Press (N.Y.).

M.A.Eisenberg and   1969   "On nonlinear kinematic hardening",
A.Phillips   Acta Mech., $\underline{5}$, 1.

R.Hill   1950   *Mathematical theory of plasticity*,
                               Oxford University Press.

P.G.Hodge,Jr.   1955   "The theory of piecewise linear
                               isotropic plasticity", in *Deforma-
                               tion and Flow of Solids*(edited by
                               R.Grammel),Springer (Berlin),147.

P.G.Hodge, Jr.   1957   "A general theory of piecewise
                               linear plasticity based on shear",
                               J.Mech.Phys.Sol., $\underline{5}$, 242.

A.Iu.Ishlinskii   1954   "General theory of plasticity with
                               linear strain hardening", Ukr.Mat.
                               Zhurnal, $\underline{6}$, 314.

H.J.Ivey   1961   "Plastic stress-strain relations
                               and yield surfaces for aluminum
                               alloys", J.Mech.Eng.Sci., $\underline{3}$, 15.

D.R.Jenkins   1965   "Kinematic hardening in zinc-alloy
                               tubes", J.Appl.Mech., $\underline{32}$, 849.

Iu.I.Kadashevich   1959   "The theory of plasticity which
and V.V.Novozhilov   takes into account residual micro-
                               stresses", TPMM, $\underline{22}$, 104.

Iu.I.Kadashevich   1968   "On microstresses in the theory of
and V.V.Novozhilov   plasticity", Mekhanika Tverdovo
                               Tela, $\underline{3}$, 82.

W.Koiter   1953   "Stress-strain relations, uniqueness
                               and variational theorems for elastic-
                               plastic materials with a singular
                               yield surface", Q.Appl.Math.,$\underline{11}$,350.

J.Marin   1936   "Failure theories of materials sub-
                                   jected to combined stresses",Trans.
                                   ASCE, 101, 1162.

J.Miastkowski and   1965   "An experimental study of yield
W.Szcepinski   surfaces of prestrained brass",Int.
                                   J.Solids and Structures, 1, 189.

M.J.Michno,Jr. and   1972   "An historical perspective of yield
W.N.Findley   surface investigations for metals",
                                   Tech.Rept.N00014-0003/17, Division
                                   of Engineering,Brown University.

Z.Mroz   1967   "On the description of anistropic
                                   work hardening", J.Mech.Phys.Sol.,
                                   15, 163.

A.Nadai   1933   "Theories of Strength", Trans.ASME,
                                   55, A111.

P.M.Naghdi   1960   "Stress-strain relations in plas-
                                   ticity and thermoplasticity",
                                   Plasticity (edited by E.H.Lee and
                                   P.S.Symonds) Pergamon Press (New
                                   York), 121.

J.Parker and   1964   "Plastic stress-strain relation-
M.B.Bassett   ships - some experiments to derive
                                   a subsequent yield surface", J.Appl.
                                   Mech., 31, 676.

J.Parker and   1961   "Plastic stress strain relation-
J.Kettlewell   ships - further experiments on the
                                   effect of loading history", J.Appl.
                                   Mech., 28, 439.

B.Paul, W.Chen   1962   "An experimental study of plastic
and L.Lee   flow under stepwise increments of
                                   tension and torsion", Proc.4th U.S.
                                   Nat.Congr.Appl.Mech., ASME, 641.

N.Perrone and   1958   "Strain hardening solutions with
P.G.Hodge,Jr.   generalized kinematic models", Proc.
                                   3rd U.S.Nat.Congr.Appl.Mech., ASME,
                                   641.

A.Phillips and   1961   "Experimental investigation of cor-
G.A.Gray   ners in the yield surface", Trans.
                                   ASME (J.Basic Eng.), 83, 275.

W.Prager   1948   "The stress-strain laws of the mathe-
                                   matical theory of plasticity - a
                                   survey of recent progress", J.Appl.
                                   Mech., 15, 226.

W.Prager   1949   "Recent developments in the mathema-
                         tical theory of plasticity", J.Appl.
                         Phys., 20, 235.

W.Prager   1953   "On the use of singular yield con-
                         ditions and associated flow rules",
                         J.Appl.Mech., 20, 317.

W.Prager   1955   "The theory of plasticity - a survey
                         of recent achievements", Proc.Inst.
                         Mech.Engrs., 169, 41.

W.Prager   1956   "A new method of analyzing stresses
                         and strains in work-hardening plas-
                         tic solids", J.Appl.Mech., 23, 493.

W.Prager   1966   "Models of Plastic Behavior", Proc.
                         5th U.S.Nat.Congr.Appl.Mech., ASME,
                         435.

E.Reuss   1935   "Anistropy caused by strain", Proc.
                         4th Intl. Congr.Appl.Mech., Cam-
                         bridge, 241.

J.L.Sanders,Jr.   1954   "Plastic stress-strain relations
                         based on linear loading functions",
                         Proc.2nd U.S.Nat.Congr.Appl.Mech.,
                         ASME, 455.

R.T.Shield and   1958   "On Prager's hardening rule", ZAMP,
H.Ziegler   9a, 260.

W.Szczepinski   1963   "On the effect of plastic deforma-
                         tion on the yield condition", Arch.
                         Mech.Stos., 15, 275.

G.I.Taylor and   1931   "The plastic distortion of metals",
H.Quinney   Phil.Trans.Roy.Soc., 230A., 323.

Chapter 6

PLASTIC BEHAVIOR OF BAR STRUCTURES

6.1 Behavior of a Three Bar Truss

Bar structure problems (including trusses, beams and frames) are sufficiently easy to solve that they are a convenient vehicle to illustrate many aspects of plastic behavior. In this section we shall consider a variety of simple problems with this purpose in mind.

Consider first a plane truss composed of pin-jointed bars and subjected to loads which lie in the plane of the truss. In such trusses the only generalized stress which appears in any member will be the axial force N. The relation between the axial force N and the extension per unit length in the bars ε could be measured directly. On the other hand, it can be assumed that the axial force in the bar leads to a uniaxial stress σ which is constant over the cross-section of the bar. The axial force is then given by

$$N = \sigma A \; , \tag{1}$$

where A is the cross-sectional area of the bar. The relation between N and ε can thus be easily obtained from the relation between σ and ε.

In order to solve particular problems we must have a specific constitutive relation. Let us assume that we have an isotropically hardening material whose stress-strain curve for monotonically increasing stress is bilinear. The $N \sim \varepsilon$ relation for monotonically increasing N thus has the form of Figure 1.

The constitutive relations are given by

$$d\varepsilon^e = \frac{1}{AE} \, dN \tag{2a}$$

$$d\varepsilon^p \;=\; \frac{1}{AE_p}\, dN \quad \text{for } (N - \bar{N}) = 0 \text{ and } dN > 0$$

$$\text{and } (-N - \bar{N}) = 0 \text{ and } dN < 0 \quad , \tag{2b}$$

$$d\varepsilon^p \;=\; 0 \quad \text{otherwise} \quad ,$$

where E is Young's modulus, E_p is a plastic modulus, and $\bar{N}$ is N_o or the highest previously attained value of $|N|$, whichever is larger.

The simple truss shown in Figure 2 will be analyzed. The axial forces and strains in each bar will be uniform. Loads V and H are applied at joint D, and the vertical and horizontal components of the displacement of D are respectively u and v. The equilibrium relations for the axial forces N_A, N_B, N_C in bars AD, BD and CD are

$$N_A + 2N_B + N_C \;=\; 2V \quad , \tag{3a}$$

$$N_A - N_C \;=\; \frac{2}{\sqrt{3}}\, H \quad . \tag{3b}$$

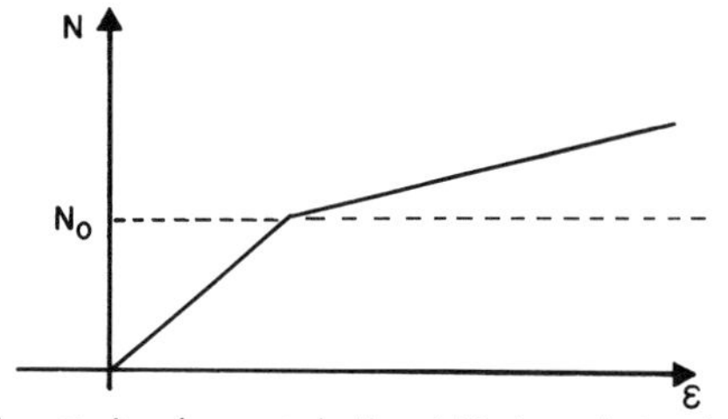

$N_o = \sigma_o A$ where σ_o is the yield stress in tension

Figure 1. Bilinear axial force-axial strain relation

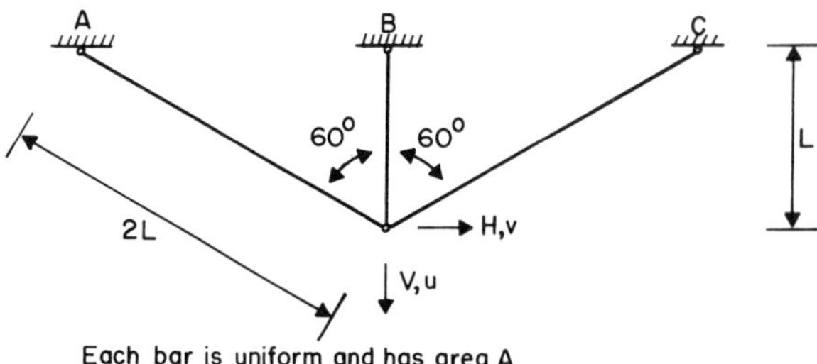

Each bar is uniform and has area A

Figure 2. Three bar truss

The relations between the strains ε_A, ε_B, ε_C in the three bars and the displacements u, v are

$$\varepsilon_A = \frac{1}{2L}\{\frac{u}{2} + \frac{\sqrt{3}}{2}v\} \quad , \tag{4a}$$

$$\varepsilon_B = \frac{u}{L} \quad , \tag{4b}$$

$$\varepsilon_C = \frac{1}{2L}\{\frac{u}{2} - \frac{\sqrt{3}}{2}v\} \quad . \tag{4c}$$

The displacement components u and v can be eliminated from these equations to provide a compatibility condition for the structure;

$$\varepsilon_A + \varepsilon_C = \frac{1}{2}\varepsilon_B \quad . \tag{5}$$

The elastic solution can be easily obtained by substituting into equation (5) the elastic relations,

$$\frac{N_A}{AE} = \varepsilon_A \quad , \quad \frac{N_B}{AE} = \varepsilon_B \quad , \quad \frac{N_C}{AE} = \varepsilon_C \quad , \tag{6}$$

and solving simultaneously equations (3a), (3b) and (5). This gives

$$N_A = \frac{V}{5} + \frac{1}{\sqrt{3}}H \quad , \quad N_B = \frac{4V}{5} \quad , \quad N_C = \frac{V}{5} - \frac{1}{\sqrt{3}}H \quad . \tag{7}$$

The displacements may then be found by using equations (4) and (5):

$$\frac{AEu}{L} = \frac{4}{5}V \quad , \quad \frac{AEv}{L} = \frac{4}{5}H \quad . \tag{8}$$

The initial yield surface in the *load space* may now be given. It is, in fact, a piecewise continuous surface made up of the inner envelope of the six equations

$$\pm N_A = N_0 \quad , \quad \pm N_B = N_0 \quad , \quad \pm N_C = N_0 \quad . \tag{9}$$

The initial yield surface is the six sided figure shown in the load space in Figure 3.

Now consider a load path which coincides with the V axis,so that H = 0. Equation (3b) gives $N_A = N_C$, and hence equation (3a) can be written in incremental form as

$$dN_A + dN_B = dV , \qquad dN_A = dN_C . \tag{10}$$

If $N_A = N_C$ the histories of loading in bars AD and CD must be identical, and consequently $\varepsilon_A = \varepsilon_C$. The compatibility condition in incremental form is thus

$$4d\varepsilon_A = d\varepsilon_B , \qquad d\varepsilon_A = d\varepsilon_C . \tag{11}$$

From the elastic solution we see that the first yielding occurs in bar BD when $N_B = +N_0$ and $V = 5N_0/4$. At this point we must introduce the plastic constitutive relations for bar N_B, since

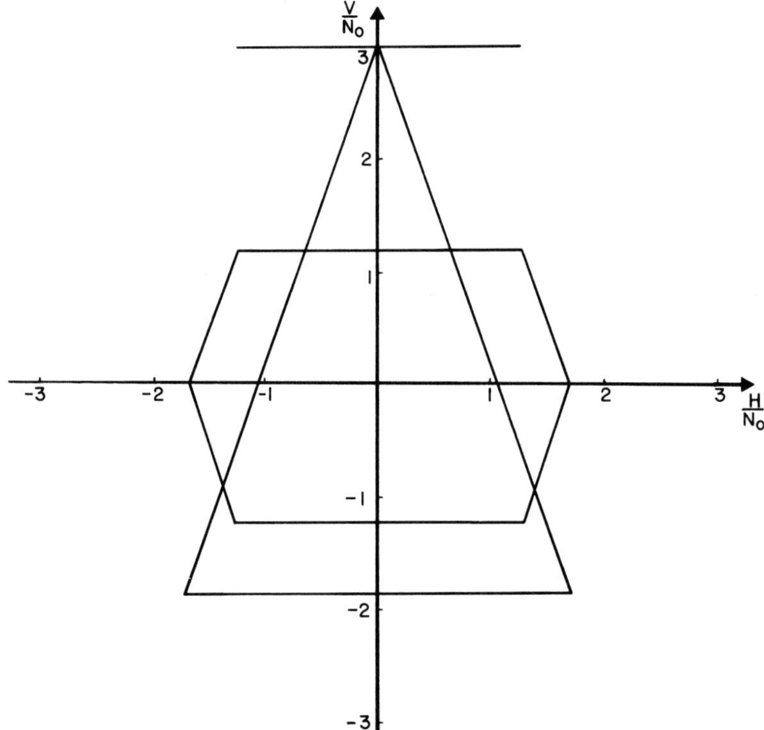

Figure 3. Yield surfaces in load space for three bar truss

we expect that loading will occur $(dN_B > 0)$ for a further increase in V. Thus the incremental constitutive relations become

$$d\varepsilon_A = \frac{dN_A}{AE} \text{ (elastic)} \quad,$$

$$d\varepsilon_B = \frac{dN_B}{AE} + \frac{dN_B}{AE_p} \text{ (for } dN_B > 0) \quad,$$

$$= \frac{dN_B}{AE}(1 + \gamma) \text{ with } \gamma = \frac{E}{E_p} \quad,$$

$$\tag{12}$$

$$d\varepsilon_C = \frac{dN_C}{AE} \quad.$$

Equations (10), (11) and (12) are solved simultaneously to give

$$dN_A = dN_C = \frac{1 + \gamma}{5 + \gamma} dV \quad,$$

$$dN_B = \frac{4}{5 + \gamma} dV \quad.$$

$$\tag{13}$$

This solution confirms that we have guessed correctly in assuming that $dN_B > 0$. It may be noted now that for continued
loading $(dV > 0)$, and provided that $-N_0 \le N_A = N_C \le N_0$, these
same equations apply for further increments since there will be
no changes introduced into equations (12). The increments of
equation (13) can thus be regarded as finite, rather than infinitesimal, increments in the generalized stress. At the beginning of this first plastic range, from the elastic solution

$$N_A = N_C = \frac{V}{5} = \frac{N_0}{4} \quad,$$

$$\tag{14}$$

so that $N_A = N_C = N_0$ when, from the first of equations (13)

$$dN_A = \frac{1 + \gamma}{5 + \gamma} dV = N_0 - \frac{N_0}{4} = \frac{3}{4} N_0 \quad, \quad \text{or}$$

$$V = \frac{3}{4}\frac{(5+\gamma)}{(1+\gamma)} N_0 + \frac{5}{4} N_0 \quad . \tag{15}$$

If we continue to increase V, we may expect all the bars to continue loading ($dN_A = dN_C > 0$, $dN_B > 0$), and hence the constitutive equations (12) are replaced by

$$d\varepsilon_A = \frac{dN_A}{AE}(1+\gamma) \quad ,$$

$$d\varepsilon_B = \frac{dN_B}{AE}(1+\gamma) \quad . \tag{16}$$

Solving simultaneously equations (10), (11) and (16), we find the second plastic range

$$dN_A = dN_C = \frac{dV}{5} \quad ,$$

$$dN_B = \frac{4}{5}dV \quad . \tag{17}$$

It again confirmed that $dN_A > 0$, $dN_B > 0$ for $dV > 0$. As before, this solution will hold for finite increments provided that $dV > 0$.

It is a comparatively easy matter to obtain the displacement u from equation (4b) and the solution to the problem. The load-displacement relation for monotonically increasing V is shown in Figure 4 for $\gamma = 1$, 4 and 9. Each curve is composed of three linear regions corresponding to elastic behavior (independent of γ), and the first and second plastic regions.

Suppose that we cease loading when $V = 3N_0$. If the load is held constant, and choosing for illustrative purposes $\gamma = 4$, the generalized stresses are

$$N_A = N_C = 1.08N_0 \quad , \quad N_B = 1.92N_0 \quad . \tag{18}$$

If we now decrease V, we expect that $dN_A < 0$, $dN_B < 0$ and

$dN_C < 0$, so that we return to incremental elastic behavior. It is instructive to determine the yield surface in load space, or the values of H, V for which we would find further yielding after unloading from $V = 3N_0$. For elastic behavior we use the incremental form of the elastic solution (equations 7)

$$dN_A = \frac{dV}{5} + \frac{dH}{\sqrt{3}} \;, \quad dN_B = \frac{4dV}{5} \;, \quad dN_C = \frac{dV}{5} - \frac{dH}{\sqrt{3}} \;. \qquad (19)$$

Provided that no further yielding occurs, these increments may be superimposed on the solution for $V = 3N_0$, $H = 0$, giving

$$N_A = 1.08\, N_0 + \frac{dV}{5} + \frac{dH}{\sqrt{3}} \;,$$

$$N_B = 1.92\, N_0 + \frac{4dV}{5} \;, \qquad (20)$$

$$N_C = 1.08\, N_0 + \frac{dV}{5} - \frac{dH}{\sqrt{3}} \;.$$

Returning to equations (2), the current yield stress in the

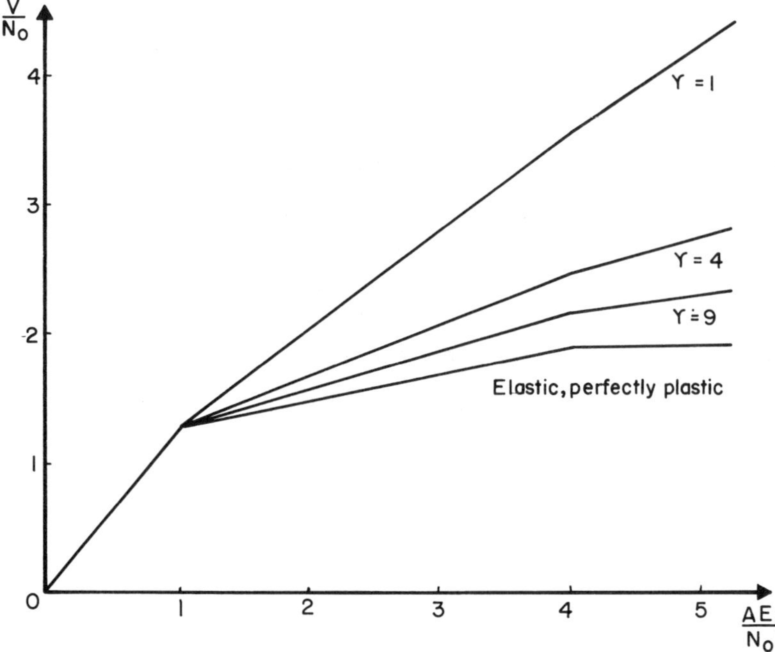

Figure 4. Load displacement curves for three bar truss

bars is given by the highest value of $|N|$ previously attained, since $|N| > N_0$ in all bars. Thus the current yield surface, from equations (18), is given by the inner envelope of the six lines

$$\pm N_A = 1.08 \, N_0, \quad \pm N_B = 1.92 \, N_0, \quad \pm N_C = 1.08 \, N_0 \quad . \quad (21)$$

Combining equations (20) and (21) the six lines may be plotted in the load space with the load point $V = 3N_0$, $H = 0$ as origin, or alternatively we can put

$$dV = V - 3N_0 \, , \quad dH = H \, , \tag{22}$$

to refer the equations to the origin. The inner envelope of these lines is the triangular figure plotted in Figure 3. The lines $N_A = -1.08 \, N_0$, $N_C = -1.08 \, N_0$ lie outside of this triangle and hence do not contribute to the envelope of the yield surface. The line $N_B = +1.92 \, N_0$ passes through the apex of the triangle and is parallel to the H axis. This example shows clearly that *although the material hardens isotropically, the yield surface in load space does not,* since the shape of the yield surface has been altered considerably.

Let us now reconsider this problem under the assumption that the bars are composed of an elastic, perfectly plastic material. The yield surface and limit surface coincide, so that flow can occur for $\pm N = N_0$. The monotonic stress strain curve has a plastic region given by the dash line in Figure 1. The hardening constitutive relation (equation 2) is replaced by

$$d\varepsilon^e = \frac{dN}{AE} \, ,$$

$$d\varepsilon^p = \lambda < N > \, , \tag{23}$$

where

$$< N > = \begin{cases} +1 & \text{when } N = N_0 \\ -1 & \text{when } N = -N_0 \\ 0 & \text{when } -N < N < N_0 \end{cases} ,$$

and $\lambda \geq 0$ but otherwise unspecified.

Bar forces such that $N > N_0$ or $N < -N_0$ are not admitted. The elastic solution (equations 7) and the initial yield surface remain the same as the hardening case for the same yield stress N_0. Suppose that, as before, we consider a monotonic increase in V with H = 0. First yield occurs in bar BD when $N_B = N_0$ and $V = 5N_0/4$. Thereafter, for dV > 0 we do not expect that $dN_B < 0$; the only other choice is $dN_B = 0$, and hence $N_B = N_0$. With $dN_B = 0$ the incremental equilibrium equations (10) become determinate, giving immediately

$$dN_A = dN_C = dV. \tag{24}$$

This enables us to determine the strain increments. Bars AD and CD are elastic ($N < N_0$) and hence

$$d\varepsilon_A = d\varepsilon_C = \frac{1}{AE} dN_A = \frac{1}{AE} dV \tag{25}$$

From the incremental compatibility condition (11) $d\varepsilon_B$ may then be found,

$$\lambda = d\varepsilon_B = 4d\varepsilon_A = \frac{4dV}{AE} . \tag{26}$$

This calculation illustrates that in situations where flow of structure does not take place, *the flow which occurs in parts of the structure is constrained by the elastic deformations in the remainder of the structure.* Equation (26) also confirms that λ is positive for dV > 0, justifying the assumption that $N_B = N_0$.

As in the hardening case, equation (24) will apply for all increments provided that $N_A = N_B \leq N_0$. At first yield of bar BD $N_A = N_0/4$ (equation 14), and hence

$$N_A = N_C = N_0 \quad \text{when} \quad V = 5N_0/4 + 3N_0/4 = 2N_0 \quad . \tag{27}$$

At this point we have $N_A = N_B = N_C = N_0$; if the load is held constant the equilibrium equation (3) is satisfied for $V = 2N_0$. If we study the incremental compatibility condition (11), we see that the plastic strain increments

$$d\varepsilon_A^p = d\varepsilon_C^p = \lambda \quad , \quad d\varepsilon_B^p = 4\lambda \quad , \tag{28}$$

satisfy these equations. The strain increments of equation (28) are possible for $\lambda > 0$ when $N_A = N_B = N_C = N_0$. Thus flow can take place in the structure for loads $V = 2N_0$, $H = 0$; this load state thus lies on the *limit surface* for the structure.

The load-displacement relation for monotonic loading can be obtained very easily from this analysis, and is shown in Figure 4, with the horizontal slope for $V = 2N_0$ indicating flow. Figure 4, showing as it does comparisons between load-displacement relations for monotonic loading for materials with different hardening rates, is significant in a technological sense. It is also fairly typical of structural behavior; load displacement rates for most structures subjected to monotonically increasing proportional loading will show similar variations as the hardening rate is altered. We can see in Figure 4 that the difference between the curves for the perfectly plastic material and the hardening rate $\gamma = 9$ is comparatively small; indeed the difference for $\gamma = 4$, which is a substantial hardening rate, is not very great. In many common structures where the simple theory we are discussing is applicable (i.e., when instability, fracture and other such effects are not important) the

serviceability of the structure is often governed by the deformations in the structure. In structures which are composed
of an elastic perfectly plastic material the limit surface can
be accepted as defining loads for which the deformations become large. This has a considerable practical advantage, since
the limit surface is independent of previous loading history,
and is comparatively easy to compute in general, as we shall
see in Part III. Further, the diagram in Figure 4 shows that
even if there is a small hardening rate (γ large) the deformations of the structure begin to get large at a load equal to
the limit load of that same structure when the material is assumed to be non-hardening (i.e., when it is assumed to be
elastic, perfectly plastic with the same yield stress). This
feature is extremely useful in a design situation when the additional effort required to carry out an incremental hardening
analysis cannot be justified.

In this particular problem the limit surface can be worked
out from first principles quite easily. To determine a load
on the limit surface we must find a set of stresses N_A, N_B, N_C
such that flow plastic strain increments can occur which satisfy the incremental compatibility condition. We have already
found one point on the limit surface, $V = 2N_0$, $H = 0$ with $N_A =
N_B = N_C = N_0$. Let us guess that flow can occur with $N_A = N_0$,
$N_C = -N_0$. From equilibrium equations (3a) and (3b), we see
that

$$H = \sqrt{3}\, N_0 \, , \qquad V = N_B \, . \tag{29}$$

It is required that $-N_0 \leq N_B \leq N_0$, so that equilibrium and the
yield condition are satisfied if $-N_0 \leq V \leq N_0$. The flow strain
increments in bars AD and CD will be

$$d\varepsilon_A = +\lambda \, , \quad d\varepsilon_B = 0 \, , \quad d\varepsilon_C = -\mu \, , \tag{30}$$

where λ, μ are non-negative numbers. Substituting these into
the incremental form of the compatibility equation (5), we
must be able to find λ, μ such that

$$\lambda - \mu = 0 \ . \tag{31}$$

This requires $\lambda = \mu$, and is clearly possible.

Another possibility is to choose $N_A = N_B = N_0$. The equilibrium equations (3) become

$$3N_0 + N_C = 2V \ , \tag{32}$$

$$N_0 - N_C = \frac{2}{\sqrt{3}} H \ .$$

Eliminating N_C, we get

$$V + \frac{H}{\sqrt{3}} = 2N_0 \ . \tag{33}$$

The axial force N_C must also lie between the upper and lower
yield stress; the solution for N_C provides the condition

$$-N_0 \leq N_C = V - \frac{H}{\sqrt{3}} - N_0 \leq N_0 \ . \tag{34}$$

Equation (33) is a line joining the points $V = 2N_0$, $H = 0$ and
$V = N_0$, $H = \sqrt{3}\, N_0$, and hence appears to complete the limit surface in the first quadrant. At the first of these points $N_C =
N_0$, and at the second $N_C = -N_0$. In between it will have inter-
mediate values, and consequently we have no yield violations on
this part of the line. The flow strain increments must be

$$d\varepsilon_A = + \lambda \ , \quad d\varepsilon_B = + \mu \ , \quad d\varepsilon_C = 0 \ , \tag{35}$$

where again λ, μ are non-negative. Substituted into the incremental form of equation (5), we require that

$$\lambda = \frac{1}{2} \mu \ , \tag{36}$$

which can be satisfied with $\lambda > 0$, $\mu > 0$. Note that the displacement increments du, dv, obtained by substituting equations
(35) and (36) into the incremental form of equations (4), are

$$du = 2 L \lambda , \quad dv = \frac{2}{\sqrt{3}} L \lambda . \tag{37}$$

It can readily be ascertained that this generalized displacement increment is normal to the line defining the operative part of the limit surface (equation 33). The load increment $dV = a$, $dH = - \sqrt{3}a$ lies in this line (i.e., is neutral
loading on the limit surface), and $(dVdu + dHdv) = 0$ with this
load increment and the generalized displacement increment of
equations (37).

The remainder of the limit surface in the load space can be
obtained in a similar way, or directly by symmetry arguments.
It is plotted as the outer figure in Figure 5. Also shown are

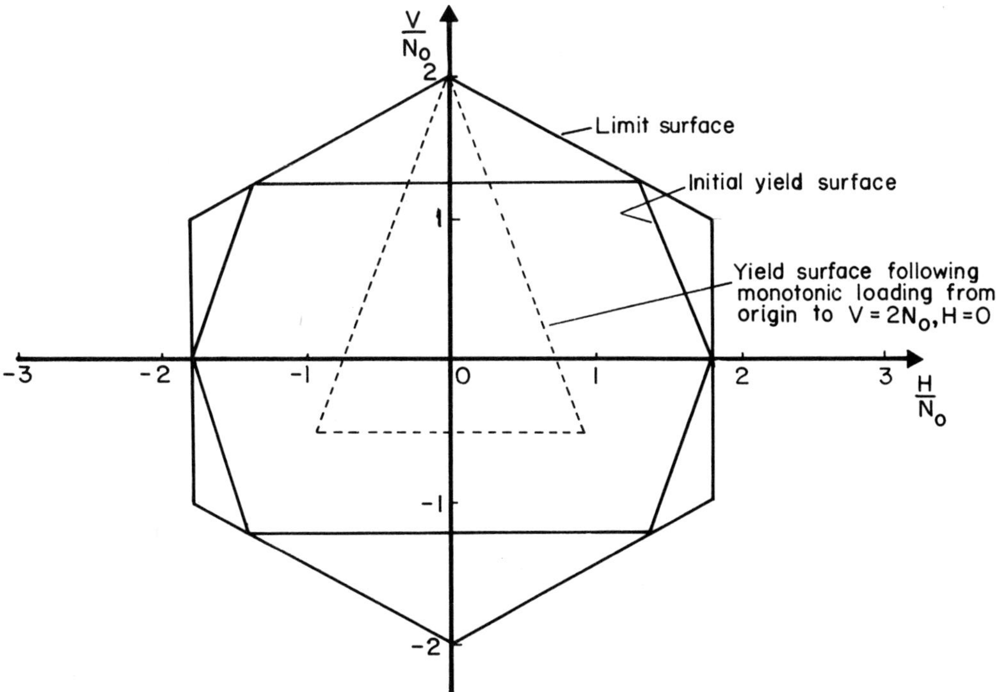

Figure 5. Limit surface for three bar truss

the initial yield surface, and, for interest, the yield surface
after loading along the V axis from the origin to the yield
surface.

6.2 Behavior of a Beam in Pure Bending

Let us now consider problems involving the bending of straight
bars or beams in which no axial loads act on the beam. We assume that shear strains are zero, and therefore we must determine only the relations between bending moment M and curvature
κ in order to proceed with the analysis of problems in this
class. This information can be obtained in two ways. Either
we can carry out experiments directly on beam elements under a
uniform bending moment and fit the results into the general
framework for plastic constitutive relations, or we can treat
the beam temporarily as three dimensional and calculate the
moment-curvature relations from information gained from the
type of experiment described in Chapter 5. In this second
course, which we shall follow here, we must actually solve a
structural problem (that shown in Figure 6) in order to determine the constitutive relations for the beam. In this structural problem the load is the moment M, and the generalized
displacement the rotation θ of the beam element between one end
and the other. However, the curvature κ will be uniform, and
since $\kappa = d\theta/dx$, κ is obtained simply by dividing the total

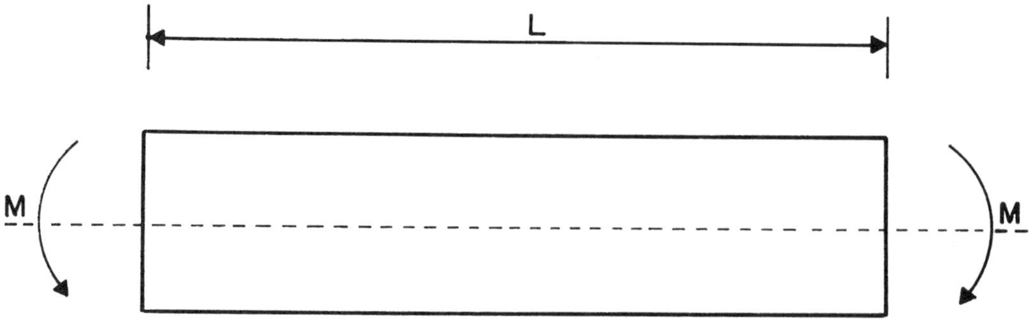

Figure 6. Beam in pure bending

rotation by the length of the beam L. Thus M, κ behave in the
same manner as M, θ, and we are assured (Chapter 3) that if
the constitutive relations for stress and strain fall within
the framework of plasticity, the load and generalized displacement relations also fall within this framework.

It is usual to use the strength of materials idealizations
for the beam calculations. We assume that plane sections remain plane, so that longitudinal *strains* vary linearly across
the cross section. For computational simplicity we shall deal
only with doubly symmetric cross sections i.e., beams which are
symmetric about the plane of loading and a plane perpendicular
to the plane of loading, as shown in Figure 7(a). In such
cases the neutral axis coincides with the centroidal axis of
the beam. When the curvatures are small the beam is bent into
a circular arc with radius $R = 1/\kappa$ under constant bending moment. Using the centroidal axis as an origin for coordinates
x, y on the cross section plane (Figure 7b), the longitudinal
strain component ε_z is given by

$$\varepsilon_z = y\kappa \ . \tag{38}$$

All stress components other than σ_z are assumed to be zero.
Because of the symmetry, the bending moment M can be written as

$$M = 2 \int_0^h b \, \sigma_z \, y \, dy \ . \tag{39}$$

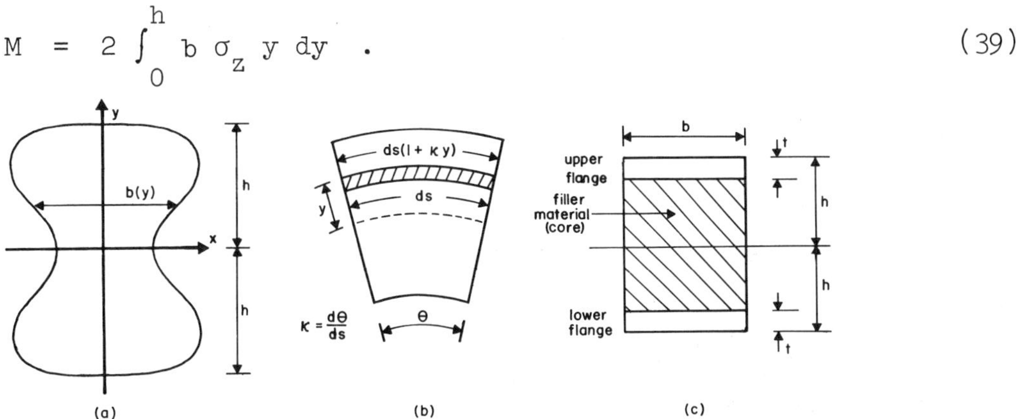

Figure 7. Bending of a beam element

The incremental forms of these relations can also be written;

$$d\varepsilon_z = y \, d\kappa \, , \tag{40}$$

$$dM = 2 \int_0^h b \, d\sigma_z \, y \, dy \, . \tag{41}$$

The relation between M and κ or between dM and dκ is obtained by eliminating σ_z, ε_z or dσ_z, dε_z from each pair of equations by means of the constitutive relations.

The simplest cross section which we can treat is the sandwich cross section (Figure 7c). The beam is composed of two thin plates of width b and thickness t, separated by a distance 2h by means of a filler material. The filler material does not contribute to the bending strength, but is assumed to carry shear forces when they are present. This section is, of course, highly idealized, but is approximated by an I or box section in which the webs are very thin.

If the thickness t is small compared to h, it can be assumed that the strain ε_z is constant in the face plates. The incremental strain relation (equation 40) hence gives

$$d\varepsilon_z = h \, d\kappa \tag{42}$$

in the tension flange. In the compression flange the strain has the same magnitude but opposite sign. The integral of equation (41) is replaced by

$$dM = 2 \, b \, t \, h \, d\sigma_z \, . \tag{43}$$

The significant feature of these equations is that dM, dκ are respectively directly proportional to dσ_z, dε_z. Consequently the relation between dM and dκ is of precisely the same form as that between dσ_z and dε^p if the initial yield stresses in tension and compression are equal in magnitude.

Suppose, for example, that the relation between σ_z and ε_z is such that a monotonic increase in σ_z gives a bilinear relation (of the same form as Figure 1), and the material hardens kinematically. The constitutive equations then have the form

$$d\varepsilon^e = \frac{d\sigma}{E} \,,$$

$$d\varepsilon^p = \frac{d\sigma}{E_p} \quad \text{for} \quad (\sigma - E_p \varepsilon^p) - \sigma_o = 0 \text{ and } d\sigma > 0 \,,$$

$$- (\sigma - E_p \varepsilon^p) - \sigma_o = 0 \text{ and } d\sigma < 0 \,, \tag{44}$$

$$d\varepsilon^p = 0 \quad \text{for} \quad (\sigma - E_p \varepsilon^p) - \sigma_o < 0 \,, \text{ or } (\sigma - E_p \varepsilon^p) - \sigma_o = 0 \text{ and } d\sigma < 0,$$

$$- (\sigma - E_p \varepsilon^p) - \sigma_o < 0, \text{ or } -(\sigma - E_p \varepsilon^p) - \sigma_o = 0 \text{ and } d\sigma > 0 \,.$$

The initial yield stresses in tension and compression both have magnitude σ_o if ε^p is also initially zero. Since strain increments in the upper and lower flange of the sandwich section also have the same magnitude, the histories in the upper and lower flanges are identical except in sign, and we need consider only the tension flange.

First, from equations (42), (43) and (44) we define an elastic curvature increment $d\kappa^e$ such that

$$d\kappa^e = \frac{d\varepsilon^e}{h} = \frac{d\sigma}{Eh} = \frac{dM}{2bth^2 E} = \frac{dM}{EI} \,, \tag{45}$$

where $I = 2bth^2$ is the second moment of area of the cross section about the neutral axis. Secondly we define

$$d\kappa^p = \frac{d\varepsilon^p}{h} \tag{46}$$

when $d\varepsilon^p \neq 0$. We are then assured that

$$d\kappa^e + d\kappa^p = \frac{d\varepsilon^e}{h} + \frac{d\varepsilon^p}{h} = \frac{d\varepsilon}{h} = d\kappa \,. \tag{47}$$

The conditions which specify whether or not $d\varepsilon^p = 0$ can be converted into relations between M and κ by using the integrated forms of equations (42) and (43). Defining

$$M_o = 2bth\,\sigma_o \tag{48}$$

as the moment at which first yielding of the cross section occurs, the complete incremental plastic curvature-moment relations become

$$d\kappa^p = \frac{dM}{E_p I} \text{ for } (M-E_p I\kappa_p) - M_o = 0 \text{ and } dM > 0 \; ,$$

$$\text{and } -(M-E_p I\kappa_p) - M_o = 0 \text{ and } dM < 0 \; , \tag{49}$$

$$d\kappa^p = 0 \text{ for } (M-E_p I\kappa_p) - M_o < 0, \text{ or } (M-E_p I\kappa_p) - M_o = 0 \text{ and } dM < 0,$$

$$\text{and } -(M-E_p I\kappa_p) - M_o < 0, \text{ or } -(M-E_p I\kappa_p) - M_o = 0 \text{ and } dM > 0 \; .$$

Comparison of equations (44) and (45) and (49) show that the moment-curvature relations for this cross-section replace σ, ε, E, E_p in equations (44) by M, κ, EI, $E_p I$ respectively.

This simple relation between moment-curvature and stress-strain relations does not hold true for more complex cross-sections, however. As an example of the added difficulty, consider a beam with a rectangular cross-section of constant width b and depth 2h. The strain distribution at any instant is shown in Figure 8(a). The stress distribution will be linear when the whole cross-section is elastic. The moment when first yield occurs will be

$$M_o = \frac{2}{3} bh^2 \sigma_o \tag{50}$$

from equation (39). The stress distribution at this instant is shown in Figure 8(b). Suppose now that the curvature κ is increased monotonically. Stress distributions at succeeding instants are shown in Figures 8(c) and 8(d). The moment must be found by integration (equation 39); it is evident that a simple relation no longer exists by which $d\kappa^e$ and $d\kappa^p$ may be defined, as in equations (45) and (46). Indeed, these integrations will show that the moment-curvature relation for monotonic increase in M or κ is no longer bilinear, but the slope will vary con tinuously, approaching the constant slope

$$\frac{1}{EI} + \frac{1}{E_p I}$$

asymptotically, where $I = 2bh^3/3$ is again the second moment of area of the cross-section about the neutral axis (Figure 9). If the curvature is now decreased monotonically, causing unloading, the stress *changes* will be linear across the cross-section.Further yielding will occur when the stress in the outer fiber reaches a value $(-\sigma_o + E_p \varepsilon^p)$. The stress distributions at the

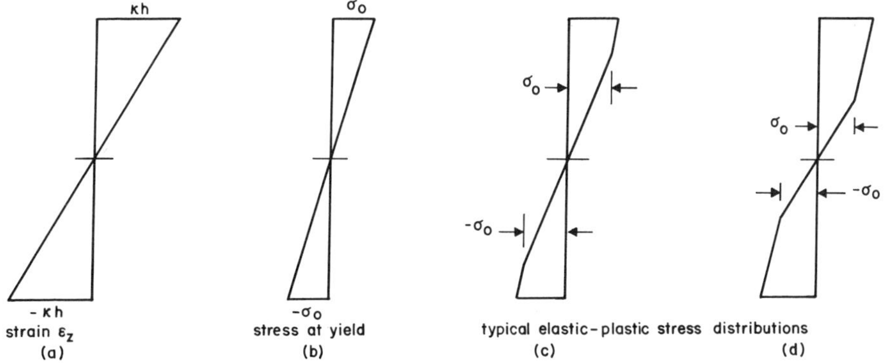

Figure 8. Stress distributions across beam depth

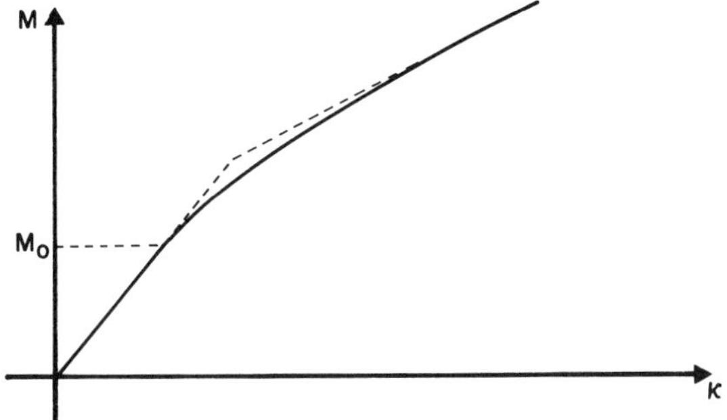

Figure 9. Moment-curvature relation

two ends of the new elastic range will differ in shape; only if
they had the same shape would we expect the elastic range always
to have the magnitude $2M_0$ as required for kinematic hardening.
This is sufficient to show that the moment-curvature relation
is neither bilinear for monotonic increase in M nor does it
harden kinematically.

These observations lead us to two possibilities for such
hardening relations. We can attempt to derive the incremental
moment-curvature relations in detail. However, they will
clearly be complex, and will add to the difficulty of carrying
out any structural analysis. Alternatively, we can compute the
moment-curvature relation for monotonically increasing M, idealize it so that it can be expressed in a tractable form, and
then assume a simple hardening rule on the basis of the stress-
strain behavior. For example, a bilinear kinematic hardening
relation could be substituted for the behavior of Figure 9.
This process will generally involve a redefinition of M_o, the
moment at which first yield occurs in the cross section, to the
intersection of the two dotted lines in Figure 9. The justification for this process is, in fact,identical to that for assuming simple hardening laws for stress-strain relations in the
first place (Chapter 5). It greatly simplifies the structural
problem, and provided the assumptions are such that the behavior of the structure would not be very sensitive to the degree
of idealization, the structural analysis will be accurate enough
for many purposes.

The same type of approximation is often also applied to
beams composed of an elastic, perfectly plastic material. This
material is characterized by the equations

$$d\varepsilon^e \;=\; \frac{d\sigma}{E} \;,$$

$$d\varepsilon^p = + \lambda \quad \text{when} \quad \sigma = \sigma_o \quad \text{and} \quad d\sigma = 0 \quad ,$$

$$d\varepsilon^p = - \lambda \quad \text{when} \quad -\sigma = \sigma_o \quad \text{and} \quad d\sigma = 0 \quad , \tag{51}$$

$$d\varepsilon^p = 0 \quad \text{when} \quad \sigma < \sigma_o \quad \text{or} \quad \sigma = \sigma_o \quad \text{and} \quad d\sigma < 0 \quad ,$$

$$\text{and} \quad -\sigma < \sigma_o \quad \text{or} \quad -\sigma = \sigma_o \quad \text{and} \quad d\sigma > 0 \quad .$$

In these equations λ is an unspecified non-negative quantity and σ_o is constant

For the same reasons as in the hardening case, the sandwich cross section of Figure 7(c) has an incremental moment-curvature relation of precisely the same form as equation (51), with M, κ and EI substituted for σ, ε and E, and $M_o = 2bth\,\sigma_o$. Thus, for the sandwich beam

$$d\kappa^e = \frac{dM}{EI} \quad ,$$

$$d\kappa^p = + \lambda \quad \text{when} \quad M = M_o \quad \text{and} \quad dM = 0 \quad ,$$

$$d\kappa^p = - \lambda \quad \text{when} \quad -M = M_o \quad \text{and} \quad dM = 0 \quad , \tag{52}$$

$$d\kappa^p = 0 \quad \text{when} \quad M < M_o \quad \text{or} \quad M = M_o \quad \text{and} \quad dM < 0 \quad ,$$

$$\text{and} \quad -M < M_o \quad \text{or} \quad -M = M_o \quad \text{and} \quad dM > 0 \quad .$$

Again, λ is an unspecified non-negative scalar. Values of M such that $M > M_o$ and $-M > M_o$ cannot be achieved. M_o and $-M_o$ are respectively the positive and negative limit moments.

In a rectangular section of width b and depth $2h$, however, the moment-curvature relation for monotonically increasing moment or curvature will exhibit hardening behavior before flow occurs. The moment when first yield occurs is given by equation (50). For a monotonically increasing curvature, the strain and stress distributions at subsequent instants are shown in Figure 10(a) and 10(b). The moment-curvature relation can be most easily expressed in terms of the parameter $\bar{y}$, which

decreases monotonically. The bending moment M is found from
the stress distribution, and is given by

$$M = b \sigma_o h^2 - \frac{b \sigma_o \bar{y}^2}{3} ,$$

$$\text{or} \quad \frac{M}{M_\ell} = 1 - \frac{1}{3} \left(\frac{\bar{y}}{h}\right)^2 ,$$

(53)

where $\quad M_\ell = \sigma_o bh^2$.   (54)

The strain at the elastic-plastic boundary $\bar{y}$ is the elastic
strain when $\sigma = \sigma_o$; hence

$$\kappa \bar{y} = \frac{\sigma_o}{E} ,$$

$$\text{or} \quad \frac{EI \kappa}{M_\ell} = \frac{2}{3(\bar{y}/h)} ,$$

(55)

where, as before, $I = 2bh^3/3$ is the second moment of area of
the cross section about the neutral axis. The moment-curvature
relation for monotonically increasing curvature is plotted in
Figure 11. It can be seen that the curve becomes asymptotic to
the line $M = M_\ell$, approaching this line fairly rapidly. M_ℓ is
the limit moment, and corresponds to the limiting stress distribution given in Figure 10(c). It is evident that this stress
distribution can be achieved only when the strains are infinite;
thus, in theory, flow can never occur, but it must be remembered that the simple theory we are using is limited to small
strains.

The asymptotic approach to the limit load observed in this
example is a common feature of structural behavior with elastic,
perfectly plastic materials, and is more usual than the case of
flow initiated when the deformations are finite, as in the case

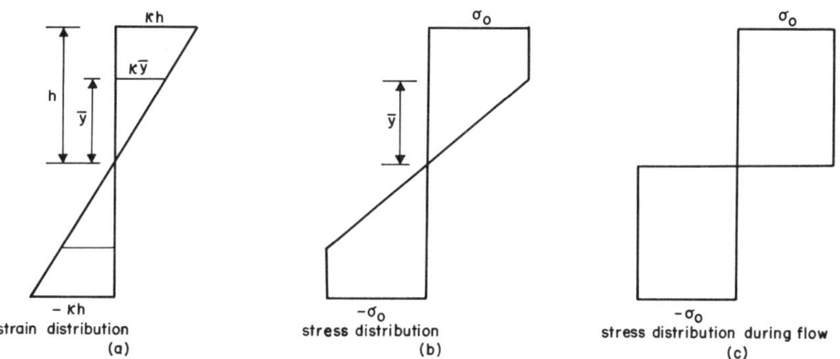

Figure 10. Stress distributions for a perfectly plastic beam

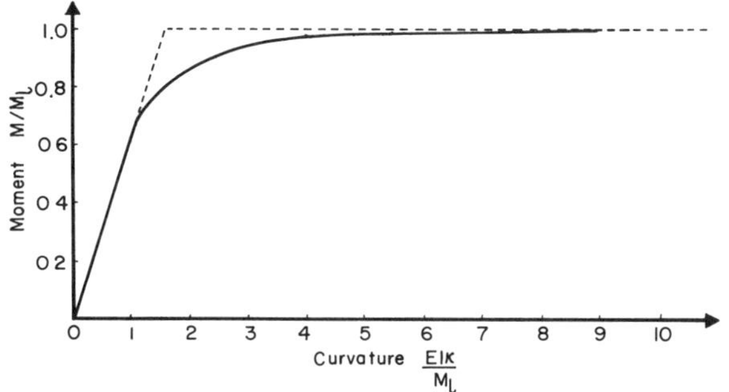

Figure 11. Moment-curvature relation for a perfectly plastic
beam

of the bar truss (Figure 4). From a practical or technological
point of view, however, the significance of the limit load is
not reduced by this phenomenon. It remains as the load (to a
close approximation) at which the deformations can be expected
to be, or begin to become, large, and therefore marks the limit
of serviceability of the structure. It is to some degree an
approximation, but its independence of the history of loading
means that it may be a better approximation than a carefully
carried out incremental analysis based on assumed initial conditions.

For the same reasons as given in the case of a hardening
moment-curvature relation, it is often convenient to idealize
the monotonic moment-curvature relation and adopt the dotted
line shown in Figure 11, so that the moment-curvature relation

(rather than the stress-strain relation) is elastic, perfectly plastic. The moment-curvature relations then take the form of equation (52), where M_o is now interpreted as the limit moment for the particular cross section under consideration (given by $M_o = M_\ell = \sigma_o bh^2$ as in equation (54) in the case of a rectangular beam). This approximation is, in fact, correct for an idealized sandwich beam, and the error for a rectangular beam is seen in Figure 11. For an I section, used commonly in building structures, it is a very good approximation since the web of an I beam contributes very little to the bending behavior. It should be noted that the idealization does not affect calculations of limit loads at which flow takes place, since the limiting moment is not affected by the idealization. If the idealization is used to compute displacements of beams rather than to attempt to calculate localized stresses or strains around the sections at which yielding and flow occur, very good results can be expected.

6.3 Simply Supported Beam Subjected to a Central Point Load

In order to study a simple example of a beam analysis, consider the simply supported beam of Figure 12(a), assuming that the moment-curvature relation has the kinematic hardening form of equations (45) and (49). The beam is statically determinate, with the moment $M(x)$ given by

$$M = \frac{Px}{2}, \quad 0 \leq x \leq \ell/2 \ . \tag{56}$$

The greatest moment in the beam occurs when $x = \ell/2$, and hence first yielding will occur when

$$\frac{P\ell}{4} = M_o \ . \tag{57}$$

While the beam behaves elastically, the curvatures can be

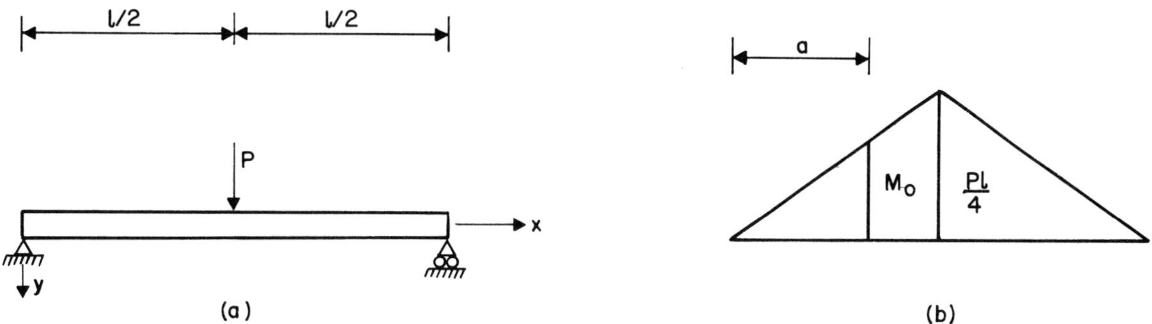

(a)   (b)

Figure 12. Simply supported beam

integrated to find the displacements $y(x)$. We have that

$$-\frac{d^2y}{dx^2} = \kappa = \frac{M}{EI} = \frac{Px}{2EI} \quad . \tag{58}$$

Integrating twice, and substituting the boundary conditions $y(0) = 0$ and $y'(\frac{\ell}{2}) = 0$ (the slope $y' = dy/dx$ must be zero at the center of the beam by symmetry), the displacements are given by

$$EIy = \frac{P\ell^3}{48}\left(\frac{3x}{\ell} - \frac{4x^3}{\ell^3}\right) \quad . \tag{59}$$

Letting $y_c = y(\frac{\ell}{2})$, and forming dimensionless variables,

$$\frac{12EIy_c}{M_o\ell^2} = \frac{P\ell}{4M_o} \quad , \tag{60}$$

so that the central diaplacement at first yield, when $P\ell/4M_o = 1$, can be easily calculated.

If P is increased monotonically, we see from equation (56) that $dM > 0$ everywhere for $dP > 0$. No unloading occurs anywhere in the beam, and we may consider the bending moment at any instant for which $P\ell/4M_o > 1$. The bending moment diagram is shown in Figure 12(b); let $M(a) = M_o$. We then have

$$\frac{Pa}{2} = M_o \,, \quad \text{or} \quad \frac{a}{\ell} = \frac{2M_o}{P\ell} \tag{61}$$

For increments of load P, the beam behaves elastically for $0 \leq x \leq a$ and plastically for $a \leq x \leq \ell/2$ (with dM > 0). The curvature increment, displacement increment relations must then be written in two parts;

$$-\frac{d^2}{dx^2}(dy) = d\kappa^e = \frac{Px}{2EI} \quad \text{for} \quad 0 \leq x \leq a \,, \tag{62a}$$

$$-\frac{d^2}{dx^2}(dy) = d\kappa^e + d\kappa^p = \frac{Px}{2EI}(1 + \gamma) \quad \text{for} \quad a \leq x \leq \ell/2 \,, \tag{62b}$$

where $\gamma = E_p/E$. Integrating twice, and using the boundary conditions $dy(0) = 0$, $(dy)'(\ell/2) = 0$ and equating the slope and displacement increments at $x = a$, we find that the central displacement increment is given by

$$d\left(\frac{12EIy_c}{M_o\ell^2}\right) = (1 + \gamma)\, d\left(\frac{P\ell}{4M_o}\right) - \gamma\left(\frac{4M_o}{P\ell}\right)^2 d\left(\frac{P\ell}{4M_o}\right) \,. \tag{63}$$

It can be seen from this equation that dy_c is the same as the elastic increment when $P\ell/4M_o = 1$; this is expected since $a = \ell/2$ and only the center point of the beam undergoes plastic deformation. A finite plastic curvature increment at a single point cannot affect the displacements which are obtained by integrating the curvatures. Also, when $P\ell/4M_o$ is large, the displacement increment is $(1 + \gamma)$ times the elastic increment, indicating that the load-displacement curve is asymptotic to a straight line. Equation (63) can be integrated, with initial conditions

$$\frac{12EIy_c}{M_o\ell^2} = 1 \quad \text{when} \quad \frac{P\ell}{4M_o} = 1 \,,$$

to give

$$\frac{12EIy_c}{M_o \ell^2} = 1 + (1 + \gamma)(\frac{P\ell}{4M_o} - 1) + \frac{\gamma}{2}\{(\frac{4M_o}{P\ell})^2 - 1\}$$

(64)

for $P\ell/4M_o \geq 1$.

The complete load, central displacement curve for monotonically increasing P is shown in Figure 13 for the case $\gamma = 9$.

If we consider the same beam with an elastic, perfectly plastic moment-curvature relation, as characterized by equation (52) with M_o giving the limit moment, the elastic behavior will be the same. First yield will occur when $P\ell/4M_o = 1$ and the moment at the center reaches M_o. From equation (56) a further increase in P must increase the central moment; however, this is not permitted by the constitutive relations since M_o is now a limit moment. Thus $P\ell/4M_o$ also gives the limiting value of

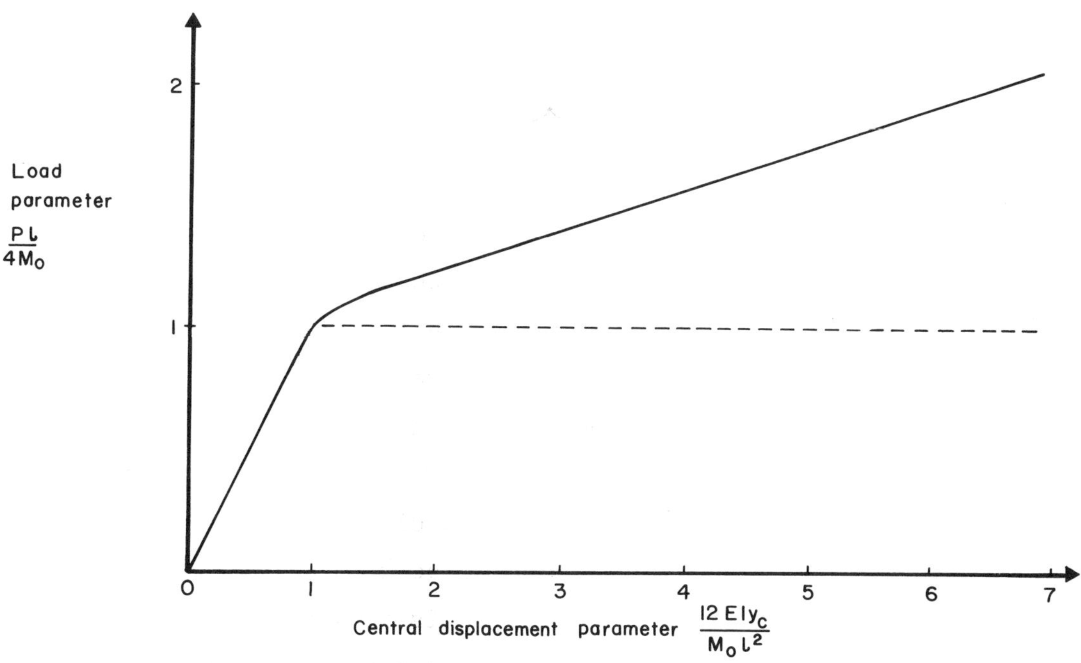

Figure 13. Load-displacement relation for a simply supported beam

the load. The mechanism by which flow occurs in this case is of great importance. Note that the moments must remain constant during flow (as shown in Chapter 3). Only the center section has $M = M_o$; at all other points $M < M_o$ and hence plastic deformation can take place only at the center section. However, if a *finite* curvature change occurs at a single point in the beam as a result of plastic deformation, the displacement field will not change. In order that the plastic curvature increment at the center should contribute to the displacement of the beam, *the plastic curvature increment at* $x = \ell/2$ *must be infinite*. The plastic curvature increment distribution during flow thus consists of a delta function at $x = \ell/2$. As a result of this delta function a jump in the rotation of adjacent cross sections of the beam, or in the slope of the beam develops at $x = \ell/2$.

 This jump is given by

$$[d\theta] \;=\; \int_{\ell/2^-}^{\ell/2^+} d\kappa^p \, dx \tag{65}$$

and will be finite. The increments in the displacement field during flow, i.e., the change in displacements which occur while P remains constant, are given in Figure 14(a). These displacements are associated only with the infinite plastic curvature increments at $x = \ell/2$. The total displacements of the beam have the form shown in Figure 14(b). The load-displacement curve will then follow the dotted line in Figure 13.

 This discontinuity in the slope of the beam is usually referred to as a *plastic hinge*. It is clearly a highly idealized picture of reality, and comes about because of two assumptions about the beam behavior. The first is that of treating the beam as a line element rather than as a three dimensional body.

This assumption is not valid where concentrated loads are applied, and it is at precisely such points that the bending moments reach peak values. Thus we cannot really expect that
the stresses in the beam reach maximum values across a single
cross section plane. The second assumption is that of replacing the actual moment-curvature relation by the elastic,
perfectly plastic relation. What actually happens is that in
a narrow region of the beam in the vicinity of the concentrated load the curvatures becomes very large as a consequence of
plastic strains in the beam. The actual plastic curvature increment distribution during the period when the displacements
increase markedly is of the form of Figure 15 rather than a
delta function.

The region over which the curvatures are large is in practice a very narrow region, of the order of the beam thickness
rather than the beam length. This can be confirmed by bending

Figure 14. Displacements of a simply supported beam

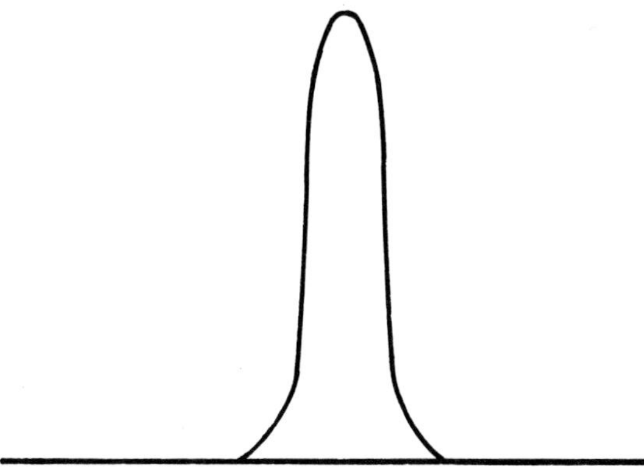

Figure 15. Actual curvature increments in the simply supported
beam

a piece of soft steel wire across a sharp corner such as a
table edge (line discontinuities or line hinges appear in plates
and shells, and these can be seen when a thin beverage can is
squeezed diametrically). Thus, while the hinge may not be a
good model of the details of stress and strain distributions,
*it is a very appropriate and very good means of modelling the
behavior at a region of high plastic deformation in a beam when
we are concerned with the displacements of the structure and
the computation of the limit load.* The presence of hinges
during flow also contributes to the simplicity of elastic, perfectly plastic analysis and limit analysis in bar structures,
since they appear whenever maxima or minima in the moments and
other generalized stresses occur and confine plastic deformation to a comparatively small number of points.

6.4 Fixed End Beam of an Elastic, Perfectly Plastic Material

As an example of the analysis of an elastic, perfectly plastic
statically indeterminate structure consider the uniform fixed
end beam shown in Figure 16(a). The moment-curvature relations
of equations (52) will be used. Because of the symmetry we
need consider only the range $0 \leq x \leq \ell/2$. The bending moment
M_A at the fixed end will be taken as the redundant. The bending moments may then be written as (Figure 16b)

$$M = -M_A + \frac{w\ell x}{2} - \frac{wx^2}{2} \; . \tag{66}$$

In the elastic range, $M = EI\kappa = -EIy''$, and the integration of
these equations gives

$$EIy' = M_A x - \frac{w\ell x^2}{4} + \frac{wx^3}{6} + A \; , \tag{67a}$$

$$EIy = \frac{M_A x^2}{2} - \frac{w\ell x^3}{12} + \frac{wx^4}{24} + Ax + B \; . \tag{67b}$$

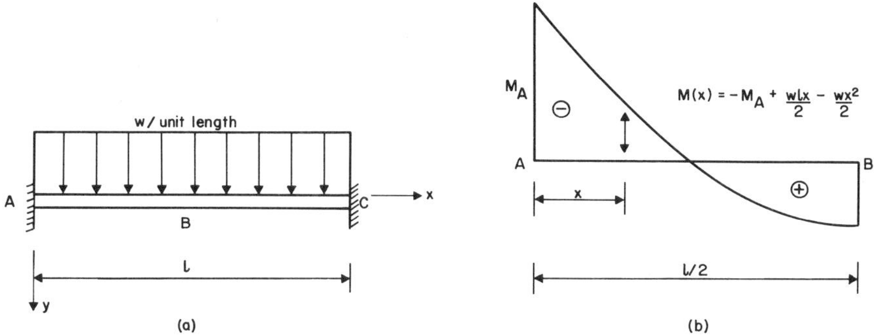

Figure 16. Fixed beam with a uniform load

The two constants of integration and the unknown moment M_A can
be evaluated from the three boundary conditions for the half
span $0 \leq x \leq \ell/2$;

$$y'(0) = 0 \ , \quad y'(\ell/2) = 0 \ , \quad y(0) = 0 \ , \tag{68a}$$

giving

$$M_A = \frac{w\ell^2}{12} \ , \quad A = 0 \ , \quad B = 0 \ . \tag{68b}$$

The largest moment occurs at $x = 0$, when $M = - M_A$, and hence
first yielding occurs when

$$- M_A = M_o \ , \quad \frac{w\ell^2}{12 M_o} = 1 \ . \tag{69}$$

In dimensionless form the central displacement $y_c = y(\ell/2)$ in
the elastic range can be written as

$$\frac{32 E I y_c}{M_o \ell^2} = \frac{w\ell^2}{12 M_o} \quad \text{for} \quad \frac{w\ell^2}{12 M_o} \leq 1 \ . \tag{70}$$

Thereafter, if the load is increased monotonically, plastic deformation may occur. It must be confined to a hinge at the
fixed end A; let the discontinuity of slope at A or the rotation
of the hinge be θ_A. We observe that in the remainder of the
beam the behavior is still elastic, so that equations (67a) and

(67b) remain valid. We now have $M_A = M_o$ and the slope at the end of the beam, as x tends to zero, will be θ_A. Substituting the second and third of the boundary conditions of equation (68a) into equations (67a) and (67b), we see that

$$\frac{2EI\theta_A}{M_o\ell} = \frac{2EIy'(0)}{M_o\ell} = (\frac{w\ell^2}{12M_o} - 1) \ , \tag{71a}$$

$$\frac{32EIy_c}{M_o\ell^2} = 5(\frac{w\ell^2}{12M_o}) - 4 \ . \tag{71b}$$

It is evident here that the unknown plastic strain (in this case expressed as a rotation) is determined from the deformation of the entire beam when flow occurs at one section. Equation (71a) shows that θ_A increases in the correct sense as w is increased monotonically, and hence the incremental moment-curvature relations are satisfied.

Equations (71a) and (71b) hold for monotonically increasing w provided that $w\ell^2/12M_o > 1$; they become invalid only when further yielding occurs. When $M_A = M_o$, the moment at the center of the beam M_c is given by

$$M_c = -M_o + \frac{w\ell^2}{8} \ ,$$

and further yielding occurs when $M_c = M_o$. This takes place when

$$\frac{w\ell^2}{12M_o} = \frac{4}{3} \ . \tag{72}$$

At this load discontinuities in slope may occur at both the fixed end and at the center of the beam. Flow may now take place, with the incremental displacement field taking the form shown in Figure 17. In such cases the sign of the discontinuity

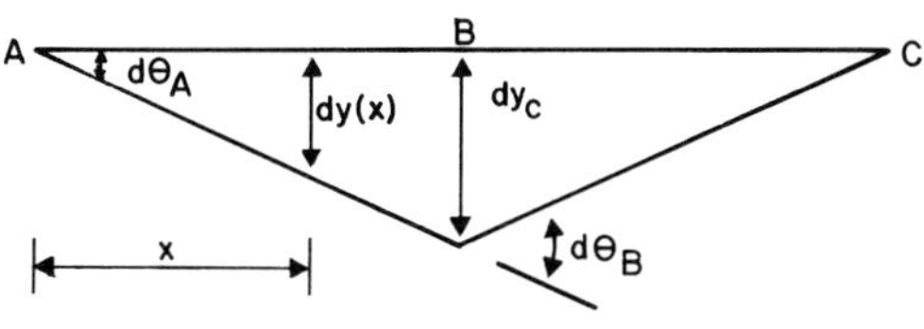

Figure 17. Incremental displacement field during flow

in slope must be checked carefully; we do indeed have a hinge
corresponding to negative curvature at the support and a hinge
corresponding to positive curvature at the center. Further,
increases in displacement cause increases of rotations of the
correct sign in the hinges. Thus the limit load is given by
equation (72), and the displacements may increase at constant
load.

As might be expected from the simple incremental displacement fields or *mechanisms of flow* in these beams it is comparatively easy to guess the mechanism of flow and determine the
limit load directly. For example, the uniform fixed end beam
of Figure 18(a) might be expected to flow with the mechanism of
Figure 18(b). This incremental displacement field is kinematically admissible and satisfies the displacement boundary conditions. From the incremental elastic, perfectly plastic moment-
curvature relations, the bending moment at A must be $-M_o$, at
B $+M_o$ and at C it must be $-M_o$. Figure 18(c) shows the beam in
two parts. Each part will be in equilibrium if the shear forces
have the magnitudes and signs shown, and the yield condition
$-M_o \leq M \leq M_o$ is everywhere satisfied. All the conditions for a
solution are thus met, and uniqueness assures us that the beam
must flow at this load. The external load is obtained by re-
joining the two parts of the beam, giving

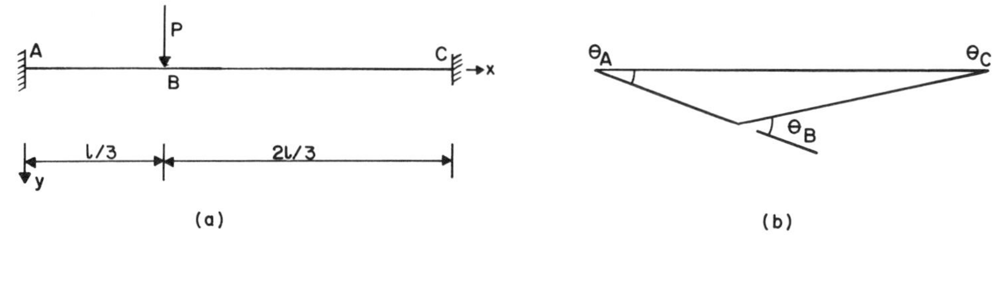

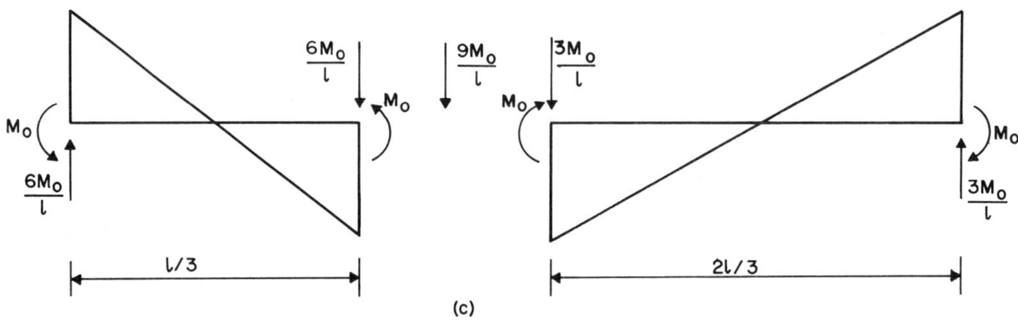

Figure 18.   Fixed beam with point load

$$P = \frac{9M_o}{\ell} \tag{73}$$

as the limit load.   Direct methods of determining the limit
load will be refined in Part III with the use of the theorems
of limit analysis.

6.5 Combined Bending and Axial Force

The previous examples in this section have considered bars subjected to either bending or axial force.   We shall consider now
yield and limit surfaces for a bar subjected to both bending
moment M and axial force N, as shown in Figure 19.

In the case of *hardening behavior*,the appropriate form of the
general constitutive equations (see Chapter 4, equations 1, 2
and 3) will be as follows.   The yield surface will have the form

Figure 19. Bar subjected to bending and axial force

$$\phi = \phi(M, N, H_\alpha) \ . \tag{74}$$

For a straight bar with a doubly symmetric cross section the elastic components of curvature κ and extension ε will be given by

$$\kappa^e = \frac{M}{EI} \ , \quad \varepsilon^e = \frac{N}{AE} \tag{75}$$

where, as before, E is Young's modulus, I is the second moment of area of the cross section about an axis through the centroid and perpendicular to the plane of bending, and A is the area of the cross section. The increments in the plastic components of curvature κ^p and the extension ε^p for a continuously differentiable yield function will be

$$d\kappa^p = c \ G \ \frac{\partial\phi}{\partial M} \left(\frac{\partial\phi}{\partial M} \, dM + \frac{\partial\phi}{\partial N} \, dN \right) \ ,$$

$$\tag{76}$$

$$d\varepsilon^p = c \ G \ \frac{\partial\phi}{\partial N} \left(\frac{\partial\phi}{\partial M} \, dM + \frac{\partial\phi}{\partial N} \, dN \right) \ ,$$

where

$$G = G(M, N, H_\alpha)$$

$$c = +1 \quad \text{for} \quad \phi = 0 \quad \text{and} \quad \frac{\partial\phi}{\partial M} \, dM + \frac{\partial\phi}{\partial N} \, dN > 0 \ ,$$

$$c = 0 \quad \text{for} \quad \phi < 0 \tag{77}$$

$$\text{or} \quad \phi = 0 \quad \text{and} \quad \frac{\partial\phi}{\partial M} \, dM + \frac{\partial\phi}{\partial N} \, dN \le 0 \ .$$

The initial yield surface for an isotropic virgin material can be readily computed, since we can superpose the results for bending and axial force obtained in earlier elastic analysis. Thus, for a *sandwich cross section* (Figure 7c), using the integrated form of equation 43) and the result that the stress due to axial force is simply force divided by area, the stresses in the upper sheet (σ_u) and the lower sheet (σ_ℓ) of the beam are

$$\sigma_u = \frac{M}{2bth} + \frac{N}{2bt} , \qquad \sigma_\ell = \frac{-M}{2bth} + \frac{N}{2bt} . \tag{78}$$

Yield may occur when either

$$\sigma_\ell = \pm\sigma_o \quad \text{or} \quad \sigma_u = \pm\sigma_o , \tag{79}$$

leading to four yield equations (and hence four independent yield functions)

$$\pm (M + h N) = M_o , \qquad \pm (-M + h N) = M_o , \tag{80}$$

where $M_o = 2bth\sigma_o$.

These four equations are plotted in the M-N space in Figure 20 to give the initial yield surface.

Similarly, for the rectangular cross section of width b and depth 2h discussed earlier, the maximum stresses σ_u and σ_ℓ which occur at the upper and lower outer fibers $(y = \pm h)$ are, using equation (50),

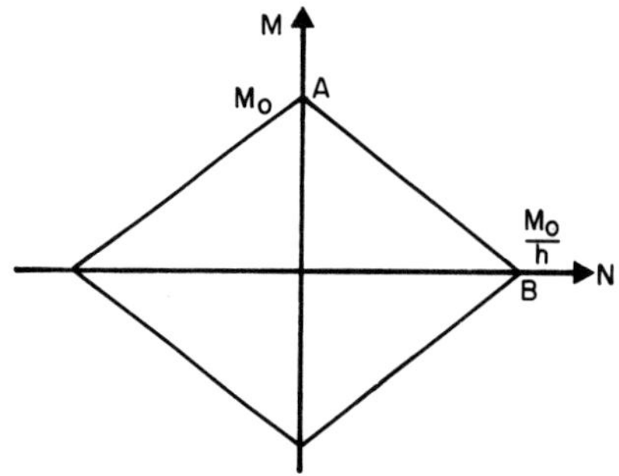

Figure 20. Sandwich cross section; yield and limit surface

$$\sigma_u = \frac{3\,M}{2bh^2} + \frac{N}{2bh} \,, \qquad \sigma_\ell = -\frac{3\,M}{2bh^2} + \frac{N}{2bh} \,. \qquad\qquad (81)$$

Equation (79) governs first yield, so that we again have four independent yield equations,

$$\pm \left(M + \frac{h}{3}\,N\right) = M_o \,, \qquad \pm \left(-M + \frac{h}{3}\,N\right) = M_o \,, \qquad (82)$$

where $M_o = \frac{2}{3}\,bh^2\,\sigma_o$.

These equations give the initial yield surface, which is the inner quadrilateral in Figure 21.

With respect to expressions for subsequent yield surfaces and for the scalar G we are in much the same situation as we were after expressing the initial yield surface under plane stress conditions. If we attempt to incorporate the exact behavior in bending which we investigated briefly above, we would have great difficulty and in any case develop expressions which

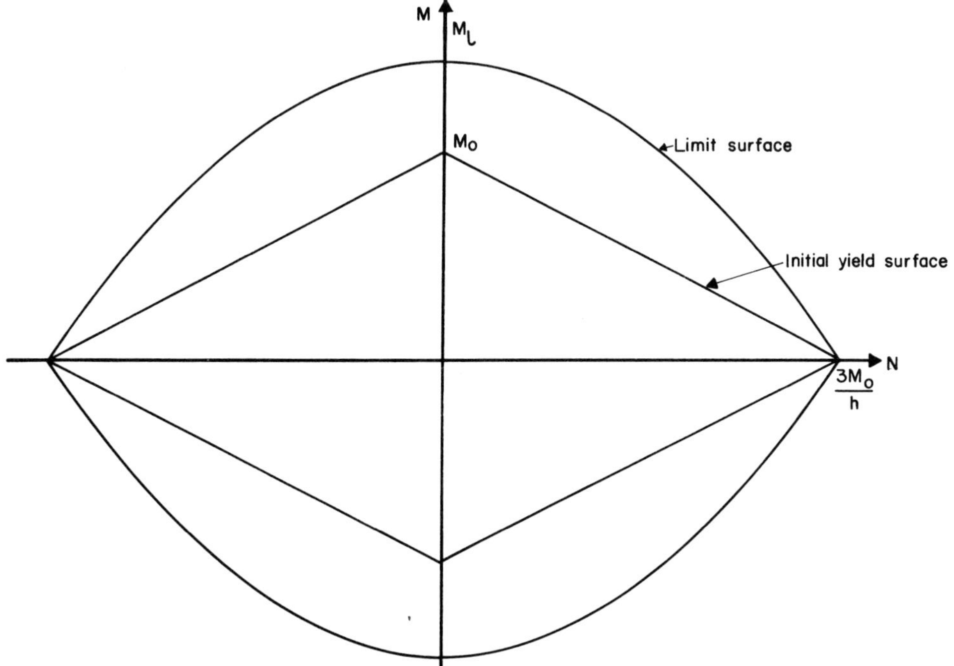

Figure 21. Rectangular cross section; yield and limit surface

would be difficult to use in actual structural problems. We
would consequently be led to adopt simple isotropic or kinematic hardening rules and appropriate expressions for G in precisely the same manner as was done for plane stress, and attempt to model the material as closely as possible. In this
process it may also be desirable to idealize the initial yield
surface.

For an *elastic, perfectly plastic material* the limit surface
$\psi(M, N)$ can also be readily computed by means of the approximations regarding the distribution of strain across the beam
which we have already introduced. In the *sandwich cross section*,
in fact, the limit surface $\psi(M, N)$ coincides with the yield
surface (as it does under pure tension and pure bending) and is
consequently given by the four equations (80). To confirm that
flow occurs under these conditions, consider the side AB of
Figure 20, which is given by the equation

$$M + hN = M_o \, . \tag{83}$$

The plastic generalized strain increments should then be

$$d\kappa^p = \lambda \frac{\partial \psi}{\partial M} = \lambda \, , \qquad d\varepsilon^p = \lambda \frac{\partial \psi}{\partial N} = \lambda h \, , \tag{84}$$

during flow when the stress point lies on the line AB between
the corners. In this case $\sigma_u = \sigma_o$ and $\sigma_\ell < \sigma_o$, so that the
strain increment varies linearly between some positive quantity
at the upper sheet and zero in the lower (Figure 22). The neutral axis thus coincides with the lower sheet. As shown in
Figure 22, this strain distribution may be considered as the
sum of two parts, one leading to pure bending and the other to
uniform extension. With the magnitudes given, and using equation (42), we see that

$$d\varepsilon^p = \lambda' \, , \qquad hd\kappa^p = \lambda' \, , \tag{85}$$

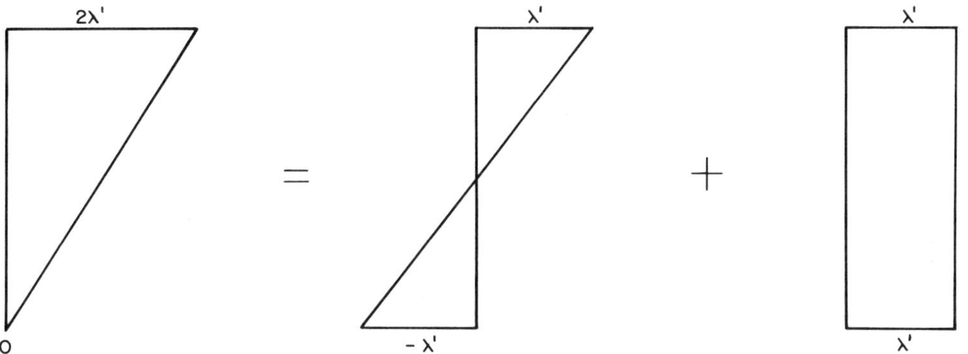

Figure 22. Strain distribution during flow in sandwich section
confirming that

$$\frac{d\kappa^p}{d\epsilon^p} = \frac{1}{h} \, , \tag{86}$$

or that equations (84) and (85) are identical with $\lambda h = \lambda'$.
This strain increment vector is normal to the line AB in Figure
20. The same process can be repeated for the other three sides
of the yield and limit surface. At the corners the plastic
strain contributions of two sides must be added, leading to a
strain increment vector which lies between adjacent normals.

In the sandwich cross section it can be noted that the neutral axis during flow coincides with either sheet except at the
corners. In the *rectangular cross section* more continuous behavior can be expected. We would postulate a plastic strain
increment distribution as shown in Figure 23(a); it is linear
so that plane sections remain plane, but the neutral axis is
displaced by a distance $\bar{y}$ from the centroid. For all strain
increments to result from *flow*, the associated stress distribution must be as shown in Figure 23(b). This stress distribution can in turn be divided into two parts, one associated
only with a net moment and one associated only with a net axial
force. Simple calculations show that

$$M = b\sigma_o(h^2 - \bar{y}^2) \, , \quad N = 2b\sigma_o\bar{y} \, . \tag{87}$$

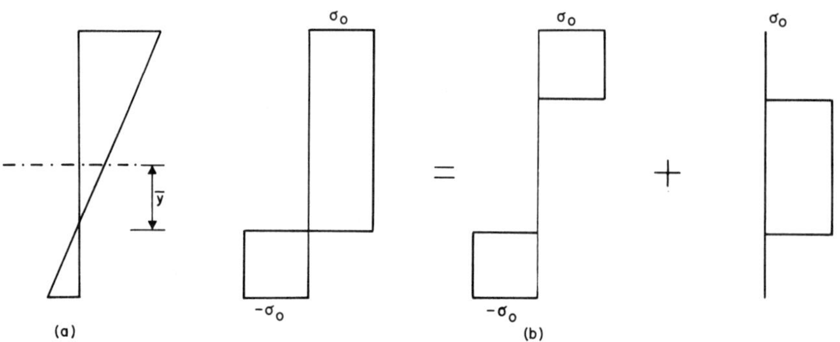

Figure 23. Strain and stress distribution during flow in a rectangular section

Eliminating $\bar{y}$, we obtain an expression for $\psi(M, N) = 0$.

$$M + \frac{N^2}{4b\sigma_o} = M_\ell \quad \text{for} \quad M > 0 \tag{88a}$$

where $M_\ell = bh^2\sigma_o$.

This expression holds only for $M > 0$; a second independent expression is obtained for $M < 0$;

$$-M + \frac{N^2}{4b\sigma_o} = M_\ell \quad \text{for} \quad M < 0 \quad . \tag{88b}$$

The resulting limit surface is shown in Figure 21.

The flow rule for stress states which satisfy, say, equation (88a) is now

$$d\kappa^p = \lambda \, , \quad d\epsilon^p = \frac{N}{2b\sigma_o} \lambda \, , \tag{89}$$

giving the ratio of the plastic strain increments

$$\frac{d\kappa_p}{d\epsilon_p} = \frac{2b\sigma_o}{N} \quad . \tag{90}$$

The reader is invited to confirm that these generalized plastic strain increment ratios are indeed predicted by the strain and

stress distributions of Figure 23. Corners occur at the points
on the N axis, where contributions from both limit surfaces may
occur. Here again the flow increments must be between adjacent
normals.

In many cases in elastic-plastic materials it is justifiable,
as in pure bending, to assume that the initial yield surface
coincides with the limit surface of equations (88), so that the
moment-curvature relation is also elastic, perfectly plastic.

6.6 Bibliographical Remarks

The applications described in this section represent some of
the simplest applications in the theory of plasticity and have
been described many times in the literature. The volumes by
Baker, Horne and Heyman [1956], Hodge [1959], Neal [1956],
Prager [1959], Baker and Heyman [1969] and Horne [1971] outline
many of these elementary applications in detail. It may be
pointed out that simple truss examples of the type discussed in
the first part of this section have been particularly valuable
in illustrating various facets of mechanical behavior, and have
been used repeatedly by W.Prager and many other writers. We
shall make further use of such examples in illustrating the
general theorems which will be discussed in later chapters. Of
the volumes listed above, that by Baker, Horne and Heyman, in
particular, contains an extensive review of experimental results for mild steel beams and frames.

References

J.F.Baker and   1969   *Plastic Analysis of Frames.*
J.Heyman   Cambridge.

J.F.Baker,   1956   *The Steel Skeleton,* Vol.2,
M.R.Horne and   Cambridge.
J.Heyman

P.G.Hodge,Jr.   1959   *Plastic Analysis of Structures,*
                                McGraw-Hill (N.Y.).

M.R.Horne   1971   *The Plastic Theory of Structures,*
                           Nelson.

B.G.Neal   1956   *The Plastic Methods of Structural
                           Analysis,* Chapman and Hall
                           (London).

W.Prager   1959   *An Introduction to Plasticity,*
                           Addison Wesley (Reading, Mass.).

Chapter 7

THREE-DIMENSIONAL CONTINUA

7.1 Initial Yield Surfaces

In discussing general three-dimensional bodies it is necessary
to represent stress and strain in tensor form to achieve some
economy in the characterization of the constitutive relations.
In this chapter we shall consider briefly the generalization of
the results given for plane stress (Chapter 5) using cartesian
tensor notation (see Section 1.5). The stress tensor is represented by the nine components of the symmetric stress tensor
σ_{ij}. The strain tensor ε_{ij} is likewise symmetric with nine
components, and is divided into elastic and plastic parts

$$\varepsilon_{ij} = \varepsilon_{ij}^{e} + \varepsilon_{ij}^{p} \ . \tag{1}$$

For an elastically isotropic material (Section 1.5, equation
13) the relation between the elastic component of strain and
the stress is given by

$$\varepsilon_{ij}^{e} = \frac{(1+\nu)}{E} \sigma_{ij} - \frac{\nu}{E} \sigma_{kk} \delta_{ij} \ , \tag{2}$$

where ν is Poisson's ratio and E is Young's modulus. The appropriate form of the plastic relations (Section 4.1, equations
2 and 3) is obtained by associating one component of Q_j, q_j^{p}
with each component of σ_{ij}, ε_{ij}^{p}, and is written in the form

$$d\varepsilon_{ij}^{p} = cG \frac{\partial\phi}{\partial\sigma_{ij}} \frac{\partial\phi}{\partial\sigma_{k\ell}} d\sigma_{k\ell} \ , \tag{3}$$

where

$$\phi = \phi(\sigma_{ij}, H_{\alpha}) \tag{4}$$

is the convex yield function,

$$G = G(\sigma_{ij}, H_{\alpha}) \geq 0 \ , \quad \text{and}$$

$$c \;=\; +1 \quad \text{for} \quad \phi = 0 \quad \text{and} \quad \frac{\partial \phi}{\partial \sigma_{ij}}\, d\sigma_{ij} \;\geq\; 0 \;,$$

$$c \;=\; 0 \quad \text{for} \quad \phi < 0$$

$$\text{or} \quad \phi = 0 \quad \text{and} \quad \frac{\partial \phi}{\partial \sigma_{ij}}\, d\sigma_{ij} \;\leq\; 0 \;. \tag{5}$$

We are again faced with the task of characterizing ϕ and G.

In the case of flow we introduce a convex limit function $\psi(\sigma_{ij})$, so that

$$d\varepsilon^{p}_{ij} \;=\; \lambda\, \frac{\partial \psi}{\partial \sigma_{ij}} \quad \text{when} \quad \psi = 0 \quad \text{and} \quad \frac{\partial \psi}{\partial \sigma_{ij}}\, d\sigma_{ij} \;=\; 0 \;,$$

$$d\varepsilon^{p}_{ij} \;=\; 0 \quad \text{when} \quad \psi < 0$$

$$\text{or} \quad \psi = 0 \quad \text{and} \quad \frac{\partial \psi}{\partial \sigma_{ij}}\, d\sigma_{ij} \;<\; 0 \;, \tag{6}$$

where λ is a non-negative but otherwise unspecified scalar.

Let us consider first the characterization of the initial yield surface for a virgin polycrystalline material. The initial yield surface (Chapter 4) was discussed for a three-dimensional body in terms of the principal stresses. Given an arbitrary stress tensor σ_{ij}, the principal stresses are obtained from the roots of the characteristic equation

$$\left| \sigma_{ij} - T\delta_{ij} \right| \;=\; 0 \tag{7}$$

which, when expanded, takes the form

$$T^{3} - I_{1}T^{2} + I_{2}T - I_{3} \;=\; 0 \;. \tag{8}$$

The three roots of this equation give the principal stresses σ_{I}, σ_{II}, σ_{III}, and the three principal directions μ_{i}^{I}, μ_{i}^{II}, μ_{i}^{III} are found by solving the three sets of three homogeneous equations

$$(\sigma_{ij} - T\delta_{ij})\mu_{i} \;=\; 0 \;, \tag{9}$$

where T takes the values of the roots of equation (8). The
three roots of equation (8) are real if σ_{ij} is symmetric, but
not necessarily distinct. We shall not discuss the details of
the cases where the roots are not distinct; in all cases we can
find three principal directions which are mutually orthogonal.

The coefficients I_1, I_2 and I_3 of the characteristic equation are *invariants* of the stress tensor, i.e., they are not
affected if the coordinate system is rotated with respect to
the body. The stress tensor σ_{ij} must of course be transformed
after such rotation. The coefficients are thus scalar properties of the stress tensor and are given by

$$I_1 = \sigma_{ii} \; ,$$

$$I_2 = \frac{1}{2} (\sigma_{ij}\sigma_{ij} - \sigma_{ii}\sigma_{jj}) \; , \tag{10}$$

$$I_3 = \frac{1}{6} (2\sigma_{ij}\sigma_{jk}\sigma_{ki} - 3\sigma_{ij}\sigma_{ij}\sigma_{kk} + \sigma_{ii}\sigma_{jj}\sigma_{kk}) \; .$$

We can further show that there are only three independent invariants of the stress tensor σ_{ij}. It is evident that the
principal stresses σ_I, σ_{II}, σ_{III} are functions *only* of the invariants I_1, I_2, I_3, since they are obtained from the characteristic equation (8).

In Chapter 4 we postulated that initial yielding in an isotropic virgin material depended only on the principal stresses
without regard for the relation between the principal directions in the material and any chosen fixed directions in the
body. In fact, in Chapter 4 we arrived at this in a different
manner by calculating the stresses on particular planes from
the principal stresses and directions and then assuming that
the orientation of these planes with respect to the body was
not important. However, this is equivalent to the more general statement that for the initial yield surface in an isotropic

body

$$\phi \;=\; \phi(\sigma_I,\; \sigma_{II},\; \sigma_{III}) \;.\tag{11}$$

We have just seen that σ_I, σ_{II}, σ_{III} are themselves functions of I_1, I_2, I_3. Consequently (11) can be rewritten; for a virgin isotropic material the yield function depends only on the invariants I_1, I_2, I_3

$$\phi \;=\; \phi(I_1,\; I_2,\; I_3) \;.\tag{12}$$

If we next specialize to a metal in which plastic deformation takes place as a result of slip we must introduce the requirement that volume changes in the body do not occur as a consequence of plastic deformation. In Chapter 4 we discussed this restriction only with reference to principal stress states. As a result of an increment in stress which causes loading, the increment in volume due to plastic deformation is

$$d\varepsilon^p_{kk} \;=\; d\varepsilon^p_{11} + d\varepsilon^p_{22} + d\varepsilon^p_{33} \;.\tag{13}$$

From equation (3), for the hardening case this becomes

$$d\varepsilon^p_{kk} \;=\; \left(G\,\frac{\partial\phi}{\partial\sigma_{ij}}\,d\sigma_{ij}\right)\left\{\frac{\partial\phi}{\partial\sigma_{11}} + \frac{\partial\phi}{\partial\sigma_{22}} + \frac{\partial\phi}{\partial\sigma_{33}}\right\}\tag{14}$$

$$=\; \left(G\,\frac{\partial\phi}{\partial\sigma_{ij}}\,d\sigma_{ij}\right)\frac{\partial\phi}{\partial\sigma_{kk}}$$

This increment is of course zero when $(\partial\phi/\partial\sigma_{ij})d\sigma_{ij} \le 0$. In the case of loading, where $(\partial\phi/\partial\sigma_{ij})d\sigma_{ij} > 0$, we can have $d\varepsilon^p_{kk} = 0$ if and only if

$$\frac{\partial\phi}{\partial\sigma_{kk}} \;=\; 0 \;.\tag{15}$$

Thus both initial and subsequent yield surfaces must be

independent of the mean hydrostatic tension $\sigma_{kk}/3$ if plastic
deformation is to lead only to distortion of an element of material. This suggests that the yield function ϕ is a function
only of the *stress deviator* s_{ij} defined by

$$s_{ij} = \sigma_{ij} - \frac{1}{3} \sigma_{kk} \delta_{ij} \; , \tag{16}$$

so that in general

$$\phi = \phi(s_{ij}, H_\alpha) \; . \tag{17}$$

The same arguments can be applied to the limit surface when
flow occurs; if flow is to lead to distortion without change in
volume we must have

$$\psi = \psi(s_{ij}) \tag{18}$$

in order that $d\varepsilon^p_{kk} = 0$.

As a result of these restrictions the plastic strain increments and the total integrated plastic strains will also be deviatoric. A strain deviator e_{ij} is defined by

$$e_{ij} = \varepsilon_{ij} - \frac{1}{3} \varepsilon_{kk} \delta_{ij} \; . \tag{19}$$

Since $\varepsilon^p_{kk} = 0$, $e^p_{ij} = \varepsilon^p_{ij}$ and we can use these quantities interchangeably.

Returning to the initial yield surface for a virgin polycrystalline metal (equation 12) we see that ϕ must be independent
of I_1 (equation 10) as a result of the incompressibility restriction, and that the expressions for I_2 and I_3 are greatly
simplified. In fact if we recognize that ϕ depends only on s_{ij}
rather than σ_{ij}, we can use the same arguments to show that for
an isotropic material ϕ must depend *only on the invariants of*
s_{ij}, denoted by J_1, J_2, J_3. Thus the initial yield surface for
an isotropic virgin polycrystalline metal has the form

$$\phi = \phi(J_2, J_3) , \tag{20}$$

where

$$J_1 = 0 ,$$

$$J_2 = \frac{1}{2} s_{ij} s_{ij} , \tag{21}$$

$$J_3 = \frac{1}{3} s_{ij} s_{jk} s_{ki} .$$

A further restriction which was added in Chapter 4 for the initial yield surface was the requirement that the value of the function ϕ should not change when the signs of each of the components of s_{ij} are reversed. J_2 is an even function of s_{ij}, but J_3 is an odd function, changing sign when the sign of s_{ij} is changed. Hence we require that ϕ *should be an even function of J_3 for a virgin polycrystalline metal.*

It is now evident that the von Mises initial yield function is a particular simple case of equation (20) in which

$$\phi = \phi(J_2) \tag{22}$$

and is independent of J_3. The shear stress on the octahedral plane may be shown to be

$$\sigma_{so}^2 = \frac{1}{3} s_{ij} s_{ij} = \frac{2}{3} J_2 \tag{23}$$

The von Mises initial yield function then becomes (Chapter 4, equation 30)

$$\phi = \frac{2}{3} J_2 - k^2 . \tag{24}$$

The Tresca initial yield function (Chapter 4. equations 44) can also be expressed in terms of J_2 and J_3. This yield function is

$$\phi = 4J_2^3 - 27J_3^2 - 36\bar{k}^2 J_2^2 + 96\bar{k}^4 J_2 - 64\bar{k}^6 . \tag{25}$$

This is quite obviously an extremely complex function, and
would be difficult to use in a structural problem. The remark made earlier holds true; in general, the Tresca yield
function is best expressed in terms of the principal stresses,
and is easiest to use when the principal directions are known
a priori as a result of symmetry or other considerations.

7.2 Idealized Hardening Rules

Generalization of the idealized hardening rules developed in
Chapter 5 can be readily carried out. For example, an isotropically hardening material with a von Mises initial yield surface would be characterized by the yield function

$$\phi = \frac{2}{3} J_2 - \frac{2}{3} \bar{J}_2 \tag{26}$$

where $\bar{J}_2$ is the largest previously attained value of J_2, provided that this value is greater than $\frac{3}{2} k^2$. To preserve isotropy the hardening coefficient G would be assumed to depend on
the invariants of s_{ij} and invariant properties of the history.
The simple case discussed in Chapter 5 (equation 27) would assume that

$$G = G(\frac{2}{3} \bar{J}_2) \quad . \tag{27}$$

The plastic constitutive relations for this isotropically hardening material then become

$$d\varepsilon_{ij}^p = de_{ij}^p = G \frac{\partial \phi}{\partial s_{ij}} \frac{\partial \phi}{\partial s_{k\ell}} ds_{k\ell}$$

$$\text{for} \quad \phi = 0 \quad \text{and} \quad \frac{\partial \phi}{\partial s_{k\ell}} ds_{k\ell} > 0 \quad , \tag{28}$$

$$d\varepsilon_{ij}^p = de_{ij}^p = 0 \text{ otherwise.}$$

Noting that

$$\frac{\partial}{\partial s_{ij}} \left(\frac{1}{2} s_{ij} s_{ij} \right) = s_{ij} \quad , \tag{29}$$

these expressions may be reduced to

$$d\varepsilon^{p}_{ij} = de^{p}_{ij} = \frac{4}{9} G(\frac{2}{3} J_2) s_{ij} dJ_2$$

$$\text{for} \quad \phi = 0 \quad \text{and} \quad dJ_2 > 0 \tag{30}$$

$$d\varepsilon^{p}_{ij} = de^{p}_{ij} = 0 \quad \text{otherwise.}$$

As before

$$G = G(\frac{2}{3} \bar{J}_2) = G(\frac{2}{3} J_2) \tag{31}$$

when $\phi = 0$ and loading takes place. G may again be determined
from a single test in which the stress is increased monotonically from zero and the virgin state. It should be noted that
in equation (3) the plastic strain increment vector de^{p}_{ij} has
the same direction as the deviatoric stress vector s_{ij}. In the
space of the stress deviator s_{ij} the initial and subsequent
yield surfaces are *hyperspheres* whose center coincides with the
origin. The stress vector s_{ij} is a radial vector whose direction consequently coincides with the direction of the normal to
the yield surface.

Other isotropic hardening rules may be constructed in a similar manner by assuming that both ϕ and G depend on the individual invariants J_2 and J_3. As before, it should be noted that
the yield surface in stress space will expand without change in
shape only if ϕ is homogeneous in the components of s_{ij}.

Rather than treat again the kinematic hardening idealization,
let us consider as an example a slightly more sophisticated
formulation which includes both the isotropic and kinematic
hardening models as special cases. We assume that

$$\phi = \phi(s_{ij}, e^p_{ij}, \alpha) \quad , \tag{32}$$

where α is the plastic work given by

$$\alpha = \int s_{ij} \, de^p_{ij} \tag{33}$$

over the entire stress path starting from the virgin state. In this formulation α and the components of e^p_{ij} become the recorded history parameters; both e^p_{ij} and α change only during loading, and remain constant during neutral loading and unloading. Equation (3) is applicable with the deviators s_{ij}, e^p_{ij} substituted for σ_{ij}, ε^p_{ij} for a polycrystalline metal; hence

$$de^p_{ij} = G \frac{\partial \phi}{\partial s_{ij}} \frac{\partial \phi}{\partial s_{k\ell}} ds_{k\ell}$$

$$\text{for} \quad \phi = 0 \quad \text{and} \quad \frac{\partial \phi}{\partial s_{ij}} ds_{ij} \geq 0 \quad , \tag{34}$$

$$de^p_{ij} = 0 \quad \text{otherwise.}$$

If we now consider a state of stress on the current yield surface, and impose an increment ds_{ij} which causes loading, we have

$$d\phi = \frac{\partial \phi}{\partial s_{k\ell}} ds_{k\ell} + \frac{\partial \phi}{\partial e^p_{ij}} de^p_{ij} + \frac{\partial \phi}{\partial \alpha} d\alpha = 0 \quad . \tag{35}$$

However, from equation (33)

$$d\alpha = s_{ij} \, de^p_{ij} \quad ,$$

and consequently equation (35) can be written in the form

$$\left(\frac{\partial \phi}{\partial e^p_{ij}} + \frac{\partial \phi}{\partial \alpha} s_{ij}\right) de^p_{ij} = - \frac{\partial \phi}{\partial s_{k\ell}} ds_{k\ell} \quad . \tag{36}$$

We have effectively written the *change* in recorded history in terms of the change in de_{ij}^p. Since we already have an expression for the direction of de_{ij}^p (equation 34), equation (36) places a restriction on the unknown function G. Substituting equation (34) into equation (36), we see that

$$G = - \frac{1}{\left(\dfrac{\partial\phi}{\partial e_{ij}^p} + \dfrac{\partial\phi}{\partial\alpha} s_{ij}\right) \dfrac{\partial\phi}{\partial s_{ij}}} \ . \tag{37}$$

In this case, therefore, we avoid the necessity of making an additional assumption concerning the direction of dH_α, and find an explicit relation between G and ϕ as a result. An additional restriction must be imposed on equation (37), because G must be non-negative. Hence when $\phi = 0$ we require that

$$\left(\frac{\partial\phi}{\partial e_{ij}^p} + \frac{\partial\phi}{\partial\alpha} s_{ij}\right) \frac{\partial\phi}{\partial s_{ij}} < 0 \ . \tag{38}$$

The special cases of kinematic and isotropic hardening can now be recovered. First, suppose that, with h a non-negative constant,

$$\phi = (s_{ij}-he_{ij}^p)(s_{ij}-he_{ij}^p) - 3k^2 \ . \tag{39}$$

This yield function is independent of α, and when $e_{ij}^p = 0$ it reduces to the von Mises initial yield surface. Inequality (38) is satisfied, since $\partial\phi/\partial\alpha = 0$, and during loading

$$\frac{\partial\phi}{\partial e_{ij}^p} \frac{\partial\phi}{\partial s_{ij}} = -4h(s_{ij}-he_{ij}^p)(s_{ij}-he_{ij}^p) = -12hk^2 \ . \tag{40}$$

Substituting into equations (37) and (34), the constitutive equation becomes

$$de_{ij}^p = \frac{1}{3hk^2} (s_{ij}-he_{ij}^p)(s_{k\ell}-he_{k\ell}^p)ds_{k\ell}$$

$$\text{for} \quad \phi = 0 \quad \text{and} \quad (s_{k\ell}-he^p_{k\ell})ds_{k\ell} > 0 \quad , \tag{41}$$

$$de^p_{ij} = 0 \quad \text{otherwise.}$$

This yield function will be recognized as the kinematic hardening model which leads to a bilinear monotonic uniaxial stress-strain curve when an elastic part is added.

As a second example, assume that

$$\phi = s_{ij}s_{ij} - 3k(k+g\alpha) \tag{42}$$

where g is a non-negative constant. The function ϕ is independent of e^p_{ij} (except insofar as it depends on α), and hence $\partial\phi/\partial e^p_{ij} = 0$. We note that the coefficients of the components of s_{ij} are constants in this function, so that $\phi = 0$ represents a convex yield surface which expands uniformly without change of shape and always contains the origin. It follows then that

$$s_{ij}\,\frac{\partial\phi}{\partial s_{ij}} = 2s_{ij}s_{ij} > 0 \quad . \tag{43}$$

Since

$$\frac{\partial\phi}{\partial\alpha} = -3kg < 0 \tag{44}$$

inequality (38) is satisfied. Determining G (equation 37), the constitutive relation becomes

$$de^p_{ij} = \frac{2s_{ij}}{9k^2 g(k+g\alpha)}(s_{k\ell}\,ds_{k\ell}) \tag{45}$$

$$\text{for} \quad \phi = 0 \quad \text{and} \quad s_{k\ell}\,ds_{k\ell} > 0 \quad ,$$

$$de^p_{ij} = 0 \quad \text{otherwise.}$$

This constitutive relation is in fact isotropically hardening. The relation between equations (26) and (42) and between equations (3) and (45) can be seen as follows. From equation (45), noting that $\phi = 0$ for $de^p_{ij} \neq 0$,

$$d\alpha = s_{ij}de^p_{ij} = \frac{2}{3kg}dJ_2 \; . \tag{46}$$

Integrating along those parts of any stress path for which $\phi = 0$ and loading occurs, it is evident that

$$\alpha = \frac{2}{3kg}\left(\bar{J}_2 - \frac{3}{2}k^2\right) \; , \tag{47}$$

where $\bar{J}_2$ is the largest value of J_2 achieved in the stress history provided that $\bar{J}_2 \geq 3k^2/2$. Equation (45) then becomes

$$de^p_{ij} = \frac{s_{ij}}{3kgJ_2}dJ_2 \tag{48}$$

$$\text{for } \phi = 0 \text{ and } dJ_2 > 0 \; ,$$

$$de^p_{ij} = 0 \text{ otherwise,}$$

remembering that $\bar{J}_2 = J_2$ when $\phi = 0$. This is a material in which, for example, the uniaxial stress-strain curve for monotonically increasing s_{12} with all other components of stress zero will be bilinear when the elastic strains are added.

The more general form of ϕ introduced in equation (32) will permit more complex idealizations of plastic constitutive relations; for example combinations of the kinematic hardening model and the uniformly expanding isotropic model could be constructed by putting

$$\phi = (s_{ij}-he^p_{ij})(s_{ij}-he^p_{ij}) - 3k(k+g'\alpha) \; . \tag{49}$$

With this assumption, the restriction (38) becomes

$$-4h\{(s_{ij}-he^p_{ij})(s_{ij}-he^p_{ij}) + \frac{3kg'}{2}s_{ij}(s_{ij}-he^p_{ij})\} < 0 \; . \tag{50}$$

The first term in the bracket of this equation is positive definite; it may be observed that the second term can be negative only when $s_{ij}\partial\phi/\partial s_{ij} < 0$. In turn, this is negative only when $s_{ij}\,de^p_{ij} < 0$, and consequently with a convex yield surface a sufficient (but by no means necessary) condition that (50) holds is that the subsequent yield surface should always contain the origin. Thus, we may require that

$$\phi(s_{ij} = 0) \leq 0$$

$$\text{or}\quad h^2 e^p_{ij} e^p_{ij} - 3k(k+g'\alpha) \leq 0 \quad. \tag{51}$$

This clearly holds when $e^p_{ij} = 0$ and $\alpha = 0$ (i.e., for the initial yield surface). We could, for example, continue the idealization by assuming that $\phi(s_{ij} = 0)$ decreases whenever plastic strain increments are nonzero. This requires that

$$d\phi\Big|_{s_{ij} = 0} < 0 \quad,$$

$$\text{or}\quad 2h^2 e^p_{ij} de^p_{ij} - 3kg'd\alpha < 0 \quad. \tag{52}$$

Substituting for $d\alpha$ and rearranging, this requirement becomes

$$2h\left(\frac{3kg'}{2h} s_{ij} - he^p_{ij}\right)de^p_{ij} \geq 0 \quad. \tag{53}$$

A further approximation can be added by assuming that

$$\frac{3kg'}{2h} = 1 \quad. \tag{54}$$

The expression in (53) then takes the form of $\dfrac{\partial\phi}{\partial s_{ij}} de^p_{ij}$, i.e., the scalar product of two vectors in the stress space which have the same direction, and the inequality is satisfied. The constitutive relations can now be recovered by substituting equation (54) into the expression for ϕ (equation 49). The resulting description has been obtained by employing three

assumptions all of which are sufficient conditions to meet the
imposed restrictions, and the generality of the original model
is decreased. It would be necessary before employing such a
model to investigate more completely what it implies, and to
consider whether it incorporates an adequate description of the
mechanical behavior for the purposes for which it will be used.
Consideration should also be given to whether it provides a
sufficient improvement of simpler models to justify additional
computational effort.

7.3 Slip Theory

As an example of a more sophisticated theory of plasticity
which leads to subsequent yield surfaces with singular points
or corners, we shall consider briefly the *slip theory* of plasticity. This theory, given originally by Batdorf and Budiansky,
was motivated by the observation that plastic deformation is
associated with slip along certain planes in the crystal lattice
which is governed by the shear stress acting on that plane. It
was assumed that in an isotropic polycrystalline aggregate a
close approximation to the real behavior could be obtained by
assuming that slip could occur along *all* planes in a homogeneously loaded element, and that the plastic strain increments
observed in a macroscopic experiment were the sums of these
slips.

Consider a homogeneously stressed element of material which
is subject to a current stress σ_{ij}, a given stress history originating in the virgin state, and an increment of stress $d\sigma_{ij}$.
The stress tensor is defined in a coordinate system x, y, z. We
consider a family of parallel planes through the element which
are normal to the unit vector n_i^a . One mechanism of plastic
deformation, therefore, is encountered when these planes slip
with respect to one another; the totality of mechanisms involves

all such families of parallel planes through the element.

Initially we consider only the slip of the planes normal to n_i^a in some particular direction with unit vector n_i^b. n_i^b lies in a plane normal to n_i^a, and for convenience we introduce a third unit vector n_i^c which is orthogonal to both n_i^a and n_i^b. This permits us to introduce a new coordinate system a,b,c whose direction cosines in the x,y,z coordinate system are given by n_i^a, n_i^b, n_i^c. The vector component of σ_{ij} which acts on any plane normal to n_i^a is given by $\sigma_{ij}n_j^a$; the scalar component of σ_{ij} which acts on the plane in the direction n_i^b is of course

$$\tau_{ab} = \sigma_{ij}n_j^a n_i^b \ . \tag{55}$$

We may now sketch an element of material in the a,b,c coordinate system (Figure 1). The shear stress component τ_{ab}, and the equal and opposite component acting beneath the element, can be conceived of as causing sliding along the planes marked with light lines, so that the element is deformed into the shape marked by dash lines. The aggregate effect is a plastic shear strain γ_{ab}^p.

Any reasonable uniaxial relationship between τ_{ab} and γ_{ab}^p can be adopted. We are specifically concerned with relationships which fall into the class defined in Chapter 2. As an example, we may assume that there exist initial yield stresses $\pm\tau_o$, that

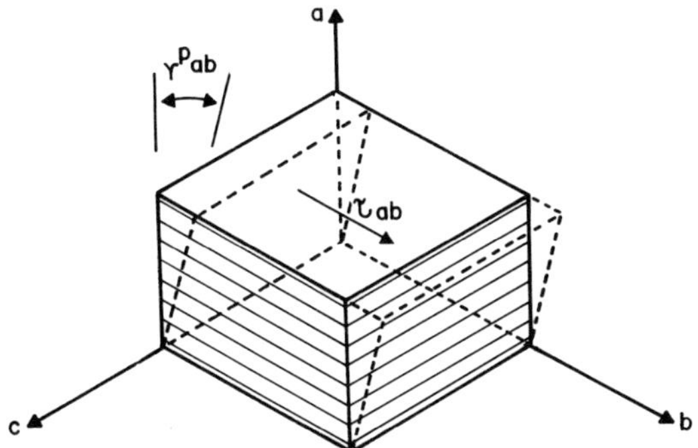

Figure 1. Contribution of one slip mechanism

the response for monotonically increasing τ_{ab} exhibits linear hardening, and that the material hardens kinematically so that the upper and lower yield surfaces are always separated by $2\tau_o$ (Figure 2). The constitutive relation for this element and for slip in this direction could thus be given by

$$d\gamma^p_{ab} = c_{ab} \frac{d\tau_{ab}}{G} ,$$
(56)

where G is a constant hardening coefficient and

$$c_{ab} = +1 \quad \text{for} \quad (\tau_{ab}-G\gamma^p_{ab}) - \tau_o = 0 \quad \text{and} \quad d\tau_{ab} \geq 0$$

$$\text{or} \quad -(\tau_{ab}-G\gamma^p_{ab}) - \tau_o = 0 \quad \text{and} \quad d\tau_{ab} \leq 0 ,$$

$$c_{ab} = 0 \quad \text{for} \quad (\tau_{ab}-G\gamma^p_{ab}) - \tau_o < 0$$

$$\text{or} \quad (\tau_{ab}-G\gamma^p_{ab}) - \tau_o = 0 \quad \text{and} \quad d\tau_{ab} \leq 0$$

$$\text{or} \quad -(\tau_{ab}-G\gamma^p_{ab}) - \tau_o < 0$$

$$\text{or} \quad -(\tau_{ab}-G\gamma^p_{ab}) - \tau_o = 0 \quad \text{and} \quad d\tau_{ab} \geq 0 .$$
(57b)

In order to generalize this result by changing to cartesian tensor notation we must recognize that a complementary shear stress component (Figure 3)

$$\tau_{ba} = \sigma_{ij} n^b_j n^a_i$$
(58)

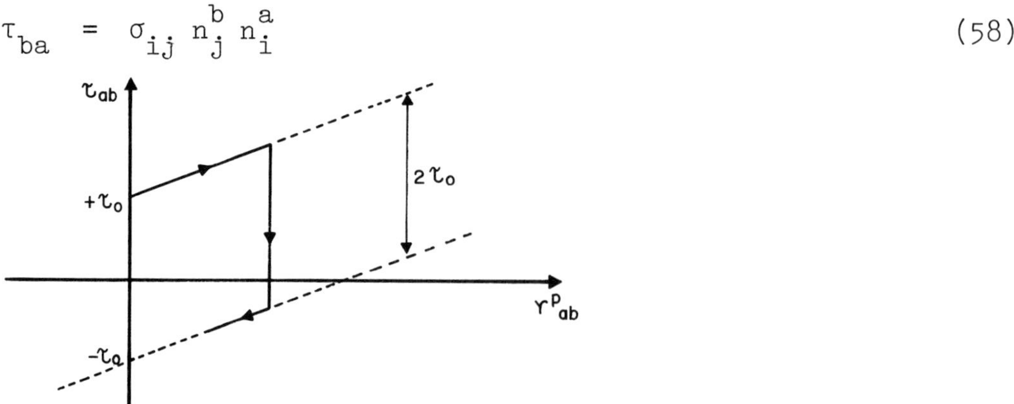

Figure 2. Relation between τ_{ab} and γ^p_{ab} for slip mechanism

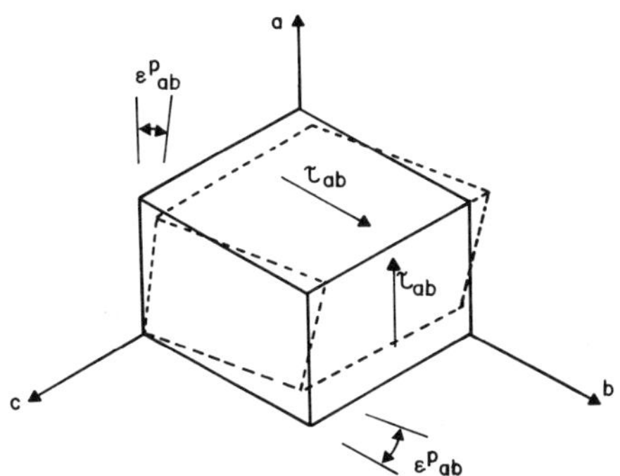

Figure 3. Tensorial shear stress and strain in the a,b,c,system

will be associated with τ_{ab}. Because σ_{ij} is symmetric

$$\tau_{ba} = \tau_{ab} \tag{59}$$

and it is convenient to put

$$\tau_{ab} = \frac{1}{2} \sigma_{ij}(n^a_i n^b_j + n^a_j n^b_i) \quad . \tag{60}$$

The shear strain in the a,b,c coordinate system must be re-
presented by the two equal components ε^p_{ab}, ε^p_{ba} where

$$\varepsilon^p_{ab} = \varepsilon^p_{ba} = \frac{1}{2} \gamma^p_{ab} \quad . \tag{61}$$

The strain tensor in the a,b,c coordinate system is then represented by the array

$$\bar{\varepsilon}^{pab}_{ij} = \begin{bmatrix} 0 & \varepsilon^p_{ab} & 0 \\ \varepsilon^p_{ab} & 0 & 0 \\ 0 & 0 & 0 \end{bmatrix} \quad . \tag{62}$$

It is now necessary to determine the corresponding strain tensor in the x,y,z coordinate system. If a vector has components
$\bar{x}_i$ in the a,b,c coordinate system, its components x_i in the
x,y,z system are given by

$$x_i = c_{ij}\bar{x}_j \quad , \tag{63a}$$

where

$$c_{ij} = \begin{bmatrix} n_1^a & n_1^b & n_1^c \\ n_2^a & n_2^b & n_2^c \\ n_3^a & n_3^b & n_3^c \end{bmatrix} . \tag{63b}$$

The corresponding tensor transformation is given by

$$\varepsilon_{ij}^{pab} = c_{ik}c_{j\ell}\bar{\varepsilon}_{k\ell}^{pab} . \tag{64}$$

The simple form of the tensor $\bar{\varepsilon}_{ij}^{pab}$ permits us to write this as

$$\varepsilon_{ij}^{pab} = \varepsilon_{ab}^{p}(n_i^a n_j^b + n_j^a n_i^b) . \tag{65}$$

The strain tensor ε_{ij}^{pab} is the contribution to the total plastic strain or slip on one plane (defined by n_i^a) in one direction n_i^b. We must now sum over all directions n_i^b associated with the one plane normal to n_i^a, and then over all planes. The direction n_i^b for fixed n_i^a is permitted to vary through the angle π, since any one direction on any plane can be denoted by a unit vector n_i^b or the equal and opposite unit vector $-n_i^b$. A similar restriction must be placed on the range of the unit vector n_i^a. It is convenient to consider a hemisphere of unit radius in the x,y,z coordinate system (Figure 4). At a point on the hemisphere the unit vector n_i^a may be taken to be normal to the hemisphere, so that the unit vectors n_i^b and n_i^c lie in the tangent plane. The contribution to the total plastic strain given by a range of values of n_i^a ranging over the solid angle $d\Omega$ represents the contribution of a small band of planes; thus the total plastic strain is given by

$$\varepsilon_{ij}^{p} = \int_H \int_{-\pi/2}^{+\pi/2} \varepsilon_{ab}^{p}(n_i^a n_j^b + n_j^a n_i^b)d\beta d\Omega , \tag{66}$$

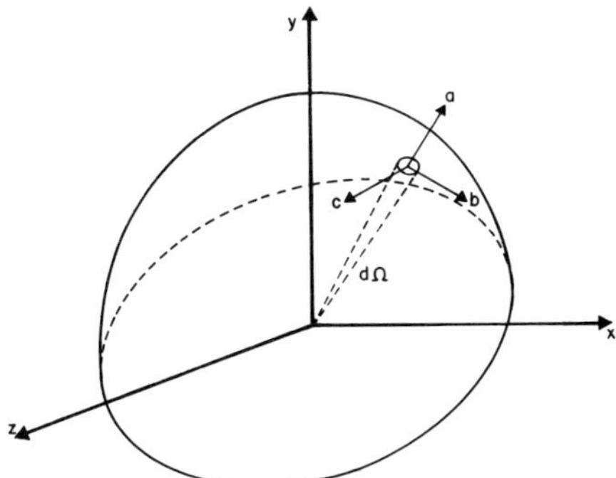

Figure 4. Integration over all slip directions

where dβ represents an infinitesimal band of slip directions on any one plane and H denotes the solid angle subtended at the center by the entire hemisphere.

For any given path of loading the plastic strain increments can be computed in terms of the stress increments by means of equations (56), (60), (61) and (66). The integrals which must be carried out are time consuming but tractable. We shall not attempt to carry out such integration in detail, but consider briefly some results. Before doing so, however, we may note that the slip theory results may be expressed in the framework derived in Chapter 2.

First, by substituting equation (6) into the conditions for c_{ab} in equations (57b) we see that there exists for each n_i^a and n_i^b a pair of yield functions

$$f_{ab}^+(\sigma_{ij}) = \{\tfrac{1}{2}\,\sigma_{ij}(n_i^a n_j^b + n_j^a n_i^b) - 2G\varepsilon_{ab}^p\} - \tau_o \ ,$$
$$f_{ab}^-(\sigma_{ij}) = -\{\tfrac{1}{2}\,\sigma_{ij}(n_i^a n_j^b + n_j^a n_i^b) - 2G\varepsilon_{ab}^p\} - \tau_o \ . \tag{67}$$

Thus, in stress space, there exist a pair of yield surfaces

$$f_{ab}^+ = 0 \ , \quad f_{ab}^- = 0 \tag{68}$$

associated with n_i^a, n_i^b. These surfaces are in fact a pair of parallel hyperplanes. When we consider all possible n_i^a, n_i^b we see that the elastic region is bounded by the inner envelope of all such hyperplanes. It is clear that the elastic region is convex, and this representation of the yield surface for the material is a particular case of the general assumption of independent yield surfaces given in Section 2.5 (equation 112).

The plastic strain contribution

$$\varepsilon_{ab}(n_i^a \, n_j^b + n_j^a \, n_i^b)$$

from slip associated with n_i^a, n_i^b has the direction of the outward normal to the hyperplane when the stress point lies on the hyperplane and loading occurs. The plastic strain increments associated with the yield function $f_{ab}^+(\sigma_{ij})$ (from Section 2.5, equation 113) take the form

$$d\varepsilon_{ij}^{pab} = \frac{c_{ab}}{G} \frac{\partial f_{ab}}{\partial \sigma_{ij}} \left(\frac{\partial f_{ab}}{\partial \sigma_{k\ell}} \, d\sigma_{k\ell} \right) , \qquad (69)$$

where

$$c_{ab} = +1 \quad \text{if} \quad f_{ab}^+ = 0 \quad \text{and} \quad \frac{\partial f_{ab}^+}{\partial \sigma_{k\ell}} \, d\sigma_{k\ell} > 0 ,$$

$$(70)$$

$$c_{ab} = 0 \quad \text{if} \quad f_{ab}^+ < 0 \quad \text{or} \quad f_{ab}^+ = 0 \quad \text{and} \quad \frac{\partial f_{ab}^+}{\partial \sigma_{k\ell}} \, d\sigma_{k\ell} < 0 .$$

First we see that

$$\frac{\partial f_{ab}^+}{\partial \sigma_{ij}} = \frac{1}{2} (n_i^a \, n_j^b + n_j^a \, n_i^b) , \qquad (71)$$

and that, from (60),

$$\frac{\partial f_{ab}^+}{\partial \sigma_{k\ell}} \, d\sigma_{k\ell} = \frac{1}{2} \, d\sigma_{k\ell}(n_k^a \, n_\ell^b + n_\ell^a \, n_k^b) = d\tau_{ab} . \qquad (72)$$

Substituting equations (71) and (72) into (69) we find that

$$d\varepsilon_{ij}^{pab} = d\varepsilon_{ab}^{p} (n_i^a n_j^b + n_j^a n_i^b) \ , \tag{73}$$

which is identical to equation (65). Similarly equations (70) reduce to equations (57b). The normality rule thus holds for each hyperplane; when all hyperplanes are considered the normality rule holds at a smooth point on the overall yield surface, and at a singular point formed by the intersection of a number of hyperplanes the total plastic strain increment must lie within the hypercone defined by the adjacent normals.

It is clear then that provided the relation between stress and plastic strain on any one plane in any one direction satisfies the assumptions of Chapter 2, slip theory falls within the general framework for plastic constitutive relations. The plastic strains γ_{ab}^{p} play the rôle of recorded history parameters in this particular example and express the effects of previous mechanical deformation. In all cases the initial yield surface will coincide with the Tresca initial yield condition, since plastic strain cannot take place on any one plane until the maximum shear stress on any plane reaches the value τ_o. Because the total plastic strain is made up of a sum of shears on the various planes on which slip has occurred, it is also evident that the plastic volume change is always identically zero. It is of course possible to introduce more sophisticated constitutive equations for the behavior on any one plane, including coupling between planes.

Slip theory is capable of predicting subsequent yield surfaces which are much closer to experimental observations than the simple isotropic and kinematic hardening rules. It is characteristic of slip theory that a path in stress space which involves continuous loading on all planes will always associate a singular point on the yield surface with the current stress point. As an example we show in Figures 5(a) and 5(b)

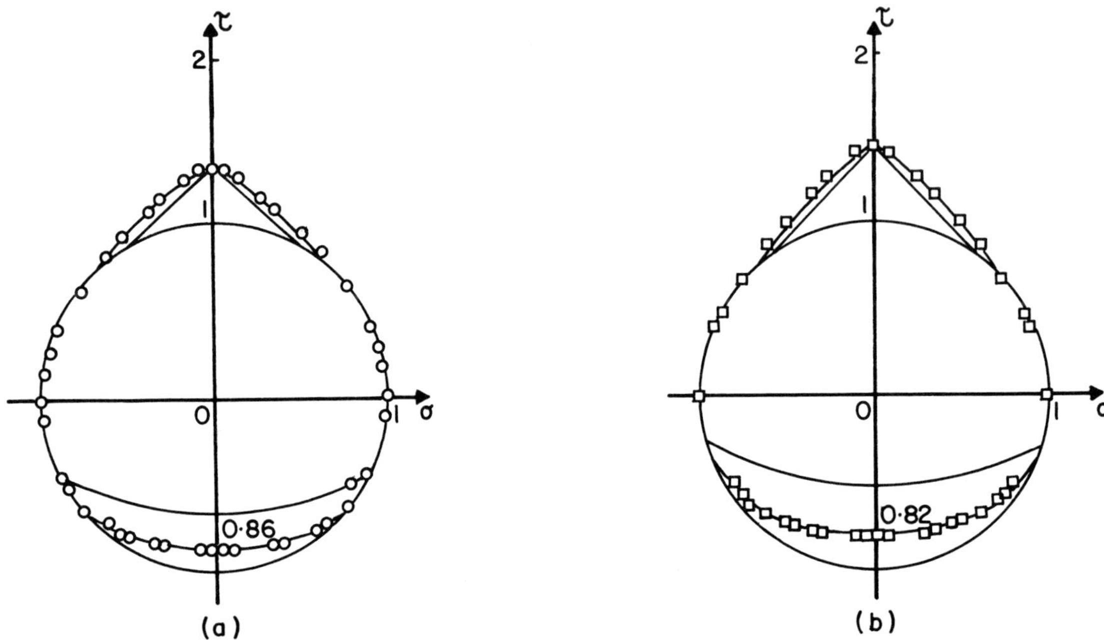

Figure 5. Comparison of experiments and slip theory predictions,
After Como and D'Agostino [1969]

comparisons between experimental and theoretical subsequent
yield surfaces in a tension-torsion test in plane stress. In
both cases the stress history consists of a monotonic increase
in the shear stress while the tension stress is zero. The experiment results are those of Naghdi, Essenburg and Koff, and
the theoretical results are given by Como and D'Agostino. The
computations were carried out numerically, with the integration
of equation (66) carried out with a number of divisions in ex-
cess of 5,000. The basic relation (56) was used in these calculations with the shear yield stress τ_o and the hardening coefficient chosen to agree with the experimental data.

From an overall point of view the agreement is very good,
although the increased accuracy is of course obtained at the
cost of a great deal of computational effort; greater effort,
in fact, than can normally be expended in the solution of

structural problems.

7.4 Torsion of a Circular Cylinder

As a simple example of the application of the plastic constitutive relations consider the problem of a circular cylinder of
radius R twisted by end couples T. As in the simple elastic
solution we solve this problem by an inverse process without
specifying exactly the distribution of the tractions on the
ends of the bar which constitute the torque T. The cylindrical
surface of the bar is stress free. Setting up a coordinate
system in which the z axis coincides with the center line of
the cylinder (Figure 6), we assume that the displacements u,v,
w of a generic point on a cross section plane in the x,y, and z
directions respectively are given by

$$u = yz\theta \ , \quad v = -xz\theta \ , \quad w = 0 \ . \tag{74}$$

The strain components are thus given by,

$$\varepsilon_{xx} = \varepsilon_{yy} = \varepsilon_{zz} = \varepsilon_{xy} = 0 \ ,$$

$$\varepsilon_{xz} = +y\theta, \quad \varepsilon_{yz} = -x\theta \ . \tag{75}$$

The strain increments are

$$d\varepsilon_{xz} = +yd\theta \ , \quad d\varepsilon_{yz} = -xd\theta \ . \tag{76}$$

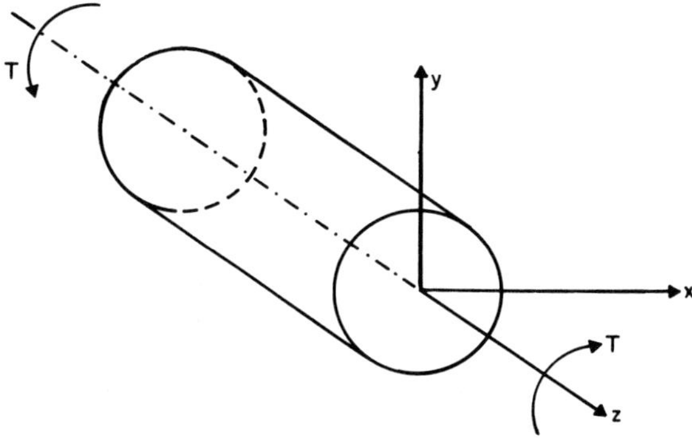

Figure 6. Torsion of a circular bar

It is now our object to determine the stress distribution and
the torque T as a function of θ. We shall confine ourselves to
the case where θ increases monotonically.

In the context of plasticity this problem is particularly
simple because of the severe kinematic constraints imposed by
equations (75) and (76). It can be seen that for any one point
on the cross section the ratio of ε_{xz} and ε_{yz} is constant; alternatively the strain path in strain space is a radial straight
line. For many simple idealizations of the plastic constitutive
relations straight line strain paths imply comparatively simple
stress paths; if these stress paths can be determined beforehand
we can derive the stress, total strain relation along this path
for use in the analysis and consequently avoid the necessity of
an incremental approach.

We shall assume that the cylindrical bar is composed of an
isotropically hardening material which satisfies the von Mises
initial yield condition. The elastic components of strain are
given by

$$\varepsilon_{xz}^{e} \;=\; (\frac{1+\nu}{E})\, \sigma_{xz} \;, \qquad \varepsilon_{yz}^{e} \;=\; (\frac{1+\nu}{E})\, \sigma_{yz} \;. \qquad (77)$$

Where the material behaves elastically all other components of
stress must be zero. Thus the elastic strain vector in the
stress space coincides in direction with the stress vector.

It will be remembered that the yield surface for an isotropically hardening material with a von Mises yield surface is
a hypersphere, so that the plastic strain increment coincides
in direction with the stress deviator. Consequently, combining
the elastic and plastic parts, a radial stress path in which
only σ_{xz} and σ_{yz} are nonzero must lead to total strain components ε_{xz} and ε_{yz} which always bear the same ratio to each
other.

To confirm this argument let us compute the strains associated with stresses $\overset{*}{\sigma}_{xz}$, $\overset{*}{\sigma}_{yz}$ which are achieved by a radial path. To this end we put

$$\sigma_{xz} = t\overset{*}{\sigma}_{xz} , \qquad \sigma_{yz} = t\overset{*}{\sigma}_{yz} . \tag{78}$$

with $0 \leq t \leq 1$, $dt > 0$.

The yield surface (cf. equation 26) can be written as

$$\phi = \sigma_{xz}^2 + \sigma_{yz}^2 - \bar{\sigma}^2 \tag{79}$$

provided that $\bar{\sigma} \geq \sigma_o$, so that σ_o governs initial yielding. We shall choose a hardening coefficient G which leads to a bi-linear stress strain relation for monotonically increasing σ_{xz} or σ_{yz} alone, so that

$$d\varepsilon_{xz}^p = \frac{4h\sigma_{xz}}{\bar{\sigma}^2} (\sigma_{xz}d\sigma_{xz} + \sigma_{yz}d\sigma_{yz}) ,$$

$$d\varepsilon_{yz}^p = \frac{4h\sigma_{yz}}{\bar{\sigma}^2} (\sigma_{xz}d\sigma_{xz} + \sigma_{yz}d\sigma_{yz}) , \tag{80}$$

for $\phi = 0$ and $(\sigma_{xz}d\sigma_{xz} + \sigma_{yz}d\sigma_{yz}) > 0$,

where h is a constant. Along a radial path, after first yield, continuous loading occurs and the unloading and neutral loading conditions are not needed.

Putting

$$\overset{*}{\sigma}_{xz}^{2} + \overset{*}{\sigma}_{yz}^{2} = \bar{\sigma}^{*2} , \tag{81}$$

equation (78) may be substituted into (80) to give

$$d\varepsilon_{xz}^p = 4h\overset{*}{\sigma}_{xz}dt , \qquad d\varepsilon_{yz}^p = 4h\overset{*}{\sigma}_{yz}dt . \tag{82}$$

First yield occurs when

$$\sigma_{xz}^2 + \sigma_{yz}^2 = \sigma_o^2 \; ,$$

i.e., when																				(83)

$$t^2 = \frac{\sigma_o^2}{\bar{\sigma}^{*2}} \; .$$

The total plastic strains associated with σ_{xz}^*, σ_{yz}^* are then found by integrating equations (82) between the limits $t=\sigma_o/\bar{\sigma}^*$ and 1;

$$\varepsilon_{xz}^{*p} = 4h\left(1 - \frac{\sigma_o}{\bar{\sigma}^*}\right)\sigma_{xz}^* \; , \qquad \varepsilon_{yz}^{*p} = 4h\left(1 - \frac{\sigma_o}{\bar{\sigma}^*}\right)\sigma_{yz}^* \; . \qquad (84)$$

The total strains are now given by the sum of the elastic and plastic parts. Using equation (77), and dropping the asterisks, the total strains associated with stresses σ_{xz}, σ_{yz} reached by a radial path are, for $\bar{\sigma} \geq \sigma_o$,

$$\varepsilon_{xz} = \left\{\left(\frac{1+\nu}{E}\right) + 4h\left(1 - \frac{\sigma_o}{\bar{\sigma}}\right)\right\}\sigma_{xz} \; ,$$

$$(85)$$

$$\varepsilon_{yz} = \left\{\left(\frac{1+\nu}{E}\right) + 4h\left(1 - \frac{\sigma_o}{\bar{\sigma}}\right)\right\}\sigma_{yz} \; .$$

As expected, these strain components bear a constant ratio to each other for σ_{xz}/σ_{yz} constant.

Having effectively removed the path dependence which characterizes plasticity, the solution may now proceed as if the material were nonlinear elastic. Expecting axial symmetry, we postulate a cylindrical elastic-plastic boundary at a radius S, so that for $r = \sqrt{x^2+y^2} < S$ we have $\phi < 0$ and the material is elastic with equations (77) governing, and for $r > S$ the material is elastic-plastic with equations (85) giving the total strains. We must of course have $S \leq R$; for $S = R$ the entire

cross section is elastic. Using equations (75), (77) and (85)
we have;

for $r \leq S$,

$$\sigma_{xz} = \frac{yE\theta}{(1+\nu)} \quad , \quad \sigma_{yz} = -\frac{xE\theta}{(1+\nu)} \quad , \tag{86a}$$

for $r \geq S$,

$$\sigma_{xz} = \frac{y\theta}{\{\frac{(1+\nu)}{E} + 4G(1-\frac{\sigma_o}{\bar{\sigma}})\}} \quad , \quad \sigma_{yz} = \frac{-x\theta}{\{\frac{(1+\nu)}{E} + 4G(1-\frac{\sigma_o}{\bar{\sigma}})\}} \quad . \tag{86b}$$

At this point the equilibrium requirements and the boundary
conditions on the cylindrical outer surface must be checked.
The equilibrium conditions reduce to

$$\frac{\partial \sigma_{xz}}{\partial z} = \frac{\partial \sigma_{yz}}{\partial z} = 0 \quad , \quad \frac{\partial \sigma_{xz}}{\partial x} + \frac{\partial \sigma_{yz}}{\partial y} = 0 \quad . \tag{87}$$

On the traction free boundary, $x^2+y^2 = R^2$, the unit normal vector has components $(x/R, y/R, 0)$. The stress boundary condition

$$\sigma_{ij}\nu_j = T_i$$

thus reduces to

$$\sigma_{xz}x + \sigma_{yz}y = 0 \quad . \tag{88}$$

It can readily be confirmed that the stresses of equation (86)
satisfy both these conditions for $S \leq R$.

   To complete the solution for the stress field as a function
of θ we must determine S and $\bar{\sigma}$. At the elastic-plastic boundary $x^2+y^2 = S^2$ we have

$$\sigma_{xz}^2 + \sigma_{yz}^2 = \sigma_o^2 \quad ,$$

which, using equations (86a), leads to

$$S = \frac{\sigma_o(1+\nu)}{E\theta} \cdot \qquad (89)$$

This implies that the bar behaves entirely elastically for

$$\theta \leq \frac{\sigma_o(1+\nu)}{ER} \cdot \qquad (90)$$

In the plastic region, i.e. for $S \leq r \leq R$ and $S < R$, we have $\phi = 0$ and hence

$$\sigma_{xz}^2 + \sigma_{yz}^2 = \bar{\sigma}^2 \; ,$$

which, using equations (86b), leads after some manipulation to

$$\bar{\sigma} = \frac{\theta r + 4h\sigma_o}{\frac{1+\nu}{E} + 4h} \cdot \qquad (91)$$

This expression permits us to eliminate $\bar{\sigma}$ and S from equations (86a) and (86b); we now have

$$\text{for } r \leq \frac{\sigma_o(1+\nu)}{E\theta} \; , \quad \sigma_{xz} = \frac{yE\theta}{(1+\nu)} \; , \quad \sigma_{yz} = -\frac{xE\theta}{(1+\nu)} \; , \qquad (92a)$$

$$\text{for } r \geq \frac{\sigma_o(1+\nu)}{E\theta} \; , \quad \sigma_{xz} = \frac{y(\theta r + 4h\sigma_o)}{r(\frac{1+\nu}{E} + 4h)} \; , \quad \sigma_{yz} = -\frac{x(\theta r + 4h\sigma_o)}{r(\frac{1+\nu}{E} + 4h)} \cdot \qquad (92b)$$

On the ends of the bar the tractions are given by the expression

$$T_i = \sigma_{ij}\nu_j \cdot$$

At the near end the unit normal vector has components $(0,0,1)$, so that

$$T_x = \sigma_{xz} \; , \quad T_y = \sigma_{yz} \; , \quad T_z = 0 \cdot \qquad (93)$$

The total couple acting on this face is then given by

$$T = \iint\limits_{A} (T_x y - T_y x)\, dA = \iint\limits_{A} (\sigma_{xz} y - \sigma_{yz} x)\, dA \quad . \tag{94}$$

After substituting from equations (92), this expression can be reduced to

$$T = 2\pi \int_{o}^{S} \frac{E\theta}{(1+\nu)}\, r^3 dr + 2\int_{S}^{R} \frac{(\theta r + 4h\sigma_o)}{(\frac{1+\nu}{E} + 4h)}\, r^2 dr \quad , \tag{95}$$

where the second integral vanishes for $S = R$.

Integration of equation (95) gives the required relation between T and θ for θ monotonically increasing

$$T = \frac{\pi}{2} \frac{ER^4 \theta}{(1+\nu)} \quad \text{for} \quad \theta \leq \frac{\sigma_o (1+\nu)}{ER} \quad , \tag{96a}$$

$$T = \frac{\pi}{2} \frac{\sigma_o^4 (1+\nu)^3}{E^3 \theta^3} + \pi \frac{R^4 \{\theta + \frac{16}{3} \frac{h\sigma_o}{R}\}}{2\,\{\frac{1+\nu}{E} + 4h\}} \tag{96b}$$

$$- \frac{\pi}{2} \frac{\sigma_o^4 (1+\nu)^3}{E^3 \theta^3} \frac{\{\frac{1+\nu}{E} + \frac{16}{3} h\}}{\{\frac{1+\nu}{E} + 4h\}} \quad \text{for} \quad \theta \geq \frac{\sigma_o (1+\nu)}{ER} \quad .$$

As θ becomes large, T asymptotically approaches the value of the second term on the right-hand side of equation (96b), which is linear in θ. This result can be most conveniently arranged in dimensionless form by introducing

$$\bar{T} = \frac{3T}{2\pi\sigma_o R^3} \,, \quad \bar{\theta} = \frac{ER\theta}{\sigma_o (1+\nu)} \,, \quad \beta = \frac{4hE}{1+\nu} \quad . \tag{97}$$

With these substitutions, equations (96) become

$$\bar{T} = \frac{3}{4} \bar{\theta} \qquad\qquad\qquad \text{for } \bar{\theta} < 1 \quad , \tag{98a}$$

$$\bar{T} = \frac{3\bar{\theta}}{4(1+\beta)} + \frac{\beta}{(1+\beta)} \{1 - \frac{1}{4\bar{\theta}^3}\} \text{ for } \bar{\theta} \geq 1 \quad . \tag{98b}$$

This relation is plotted in Figure 7 for $\beta = 5$; this case corresponds to a ratio of the elastic shear modulus to the plastic shear modulus of 5.

It is also of interest to consider the *rigid-plastic* analysis of this problem, with the elastic strains neglected. The
constitutive equation is assumed to contain only the contribution from the plastic strains. It was shown (equation 84) that
a straight line path in stress space leads to the proportional
plastic strains which would be required by equations (75).
Hence we now use equation (84) in place of equations (85), i.e.

$$\varepsilon_{xz} = \varepsilon_{xz}^{p} = 4h(1 - \frac{\sigma_o}{\bar{\sigma}})\, \sigma_{xz} \quad ,$$

$$(99)$$

$$\varepsilon_{yz} = \varepsilon_{yz}^{p} = 4h(1 - \frac{\sigma_o}{\bar{\sigma}})\, \sigma_{yz} \quad ,$$

for $\bar{\sigma} \geq \sigma_o$ with the strains being zero for $\bar{\sigma} < \sigma_o$. The stress

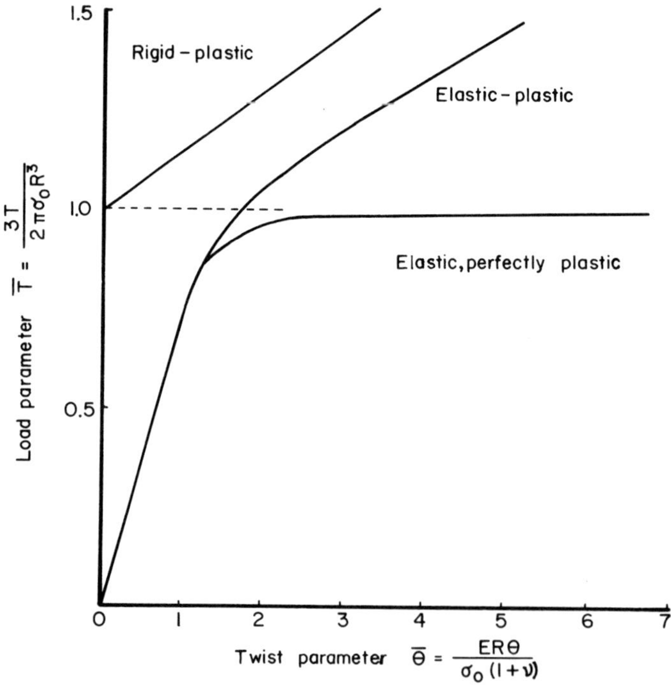

Figure 7. Torque-twist relations for circular bar

distribution is obtained from equations (75) and (99);

$$\sigma_{xz} = \frac{y\theta}{4h(1 - \frac{\sigma_o}{\bar{\sigma}})} \quad , \quad \sigma_{yz} = \frac{-x\theta}{4h(1 - \frac{\sigma_o}{\bar{\sigma}})} \quad . \qquad (100)$$

When θ is infinitesimally small σ_{xz} and σ_{yz} are still determined; this indicates that the severe kinematic constraints on
this problem render the stress field unique at the yield point
of the body (i.e., when first deformation occurs). There is no
rigid-plastic boundary to consider, since strain increments
take place at all points on the cross section, and hence it remains only to eliminate $\bar{\sigma}$. From the yield condition

$$\sigma_{xz}^2 + \sigma_{yz}^2 = \bar{\sigma}^2 \quad ,$$

and from equations (100) we find that

$$\bar{\sigma} = \sigma_o + \frac{r\theta}{4h} \quad . \qquad (101)$$

   Substituting (101) into equations (100) the stress distribution is obtained. These stresses may now be integrated to
give the torque T (equation 94):

$$T = 2\pi \int_o^R (4h \frac{\sigma_o}{4h} + r\theta)r^2 dr = \frac{2\pi\sigma_o}{3} R^3 + \frac{\pi}{8h} R^4 \theta \quad . \qquad (102)$$

When $\theta = 0$ this expression gives a finite value for T which is
the torque that when distributed on the ends of the bar in the
manner prescribed by the stress distribution (equation 100)
first causes yield in the rigid-plastic cylinder. Equation
(102) is linear in θ, and is plotted in Figure 7. Notice that
the slope of the rigid-plastic T $\sim$ θ line and the asymptotic
slope of the elastic-plastic curve do not coincide. This difference is due to the elastic contributions to the strain increments when

$$\sigma^2_{xz} + \sigma^2_{yz} > \sigma^2_o \; .$$

The analysis of the same torsion problem for an elastic-perfectly plastic material can be carried out in a very similar manner. Choosing the limit surface in stress space as

$$\psi \; = \; \sigma^2_{xz} + \sigma^2_{yz} - \sigma^2_o \; , \tag{103}$$

the constitutive equations are

$$\varepsilon^e_{xz} \; = \; \frac{1+\nu}{E} \, \sigma_{xz} \quad , \qquad \varepsilon^e_{yz} \; = \; \frac{1+\nu}{E} \, \sigma_{yz} \quad ,$$

$$\frac{d\varepsilon^p_{xz}}{d\varepsilon^p_{yz}} \; = \; \frac{\partial\psi/\partial\sigma_{xz}}{\partial\psi/\partial\sigma_{yz}} \; = \; \frac{\sigma_{xz}}{\sigma_{yz}} \; \text{for} \; \psi = 0 \; \text{and} \; (\sigma_{xz}\,d\sigma_{xz} + \sigma_{yz}\,d\sigma_{yz}) = 0, \tag{104}$$

$$d\varepsilon^p_{xz} \; = \; d\varepsilon^p_{yz} \; = \; 0 \quad \text{otherwise.}$$

It may again be observed that a straight line stress path from the origin which meets the limit surface with the stress point subsequently stationary produces the proportional strains required. Integrating along such a path the total strains are given by

$$\varepsilon_{xz} \; = \; (\frac{1+\nu}{E} + \lambda)\sigma_{xz} \quad , \qquad \varepsilon_{yz} \; = \; (\frac{1+\nu}{E} + \lambda)\sigma_{yz} \quad , \tag{105}$$

where λ is a non-negative monotonically increasing scalar which is zero for $\sigma^2_{xz} + \sigma^2_{yz} < \sigma^2_o$.

As in the elastic-plastic problem we introduce an elastic-plastic boundary given by $r = S$ $(S \leq R)$. The stresses are then given by the following expressions.

$$\text{For} \; r < S \quad \sigma_{xz} = \frac{yE\theta}{1+\nu} \quad , \qquad \sigma_{yz} = - \frac{xE\theta}{1+\nu} \quad , \tag{106a}$$

$$\text{for} \; r > S \quad \sigma_{xz} = \frac{y\theta}{(\frac{1+\nu}{E} + \lambda)} \quad , \qquad \sigma_{yz} = \frac{-x\theta}{(\frac{1+\nu}{E} + \lambda)} \quad , \tag{106b}$$

$$\text{and} \quad S = \frac{\sigma_o(1+\nu)}{E\theta} \quad . \tag{106c}$$

At all points in the plastic region $r > S$ the stresses must satisfy the yield or limit function (equation 103), and hence from equations (106b),

$$\frac{r\theta}{(\frac{1+\nu}{E} + \lambda)} = \sigma_o \quad , \tag{107}$$

$$\text{or} \quad \lambda = \frac{r\theta}{\sigma_o} - \frac{1+\nu}{E} \quad .$$

With $\lambda = 0$ for $r \leq S$, λ increases monotonically at any point as θ is increased. The expressions for σ_{xz} and σ_{yz} may also be simplified in the plastic region;

$$\sigma_{xz} = \frac{\sigma_o y}{r} \quad , \quad \sigma_{yz} = - \frac{\sigma_o x}{r} \quad \text{for} \quad r > S \quad . \tag{108}$$

The torque acting on the section is now found from equation (94). We have

$$T = 2\pi \int_o^S \frac{E\theta}{(1+\nu)} r^3 dr = \frac{\pi}{2} \frac{ER^4 \theta}{(1+\nu)} \quad \text{for} \quad \theta \leq \frac{\sigma_o(1+\nu)}{ER} \tag{109a}$$

$$T = 2\pi \int_o^S \frac{E\theta}{(1+\nu)} r^3 dr + 2\pi \int_S^R \sigma_o r^2 dr$$

$$= \frac{2\pi \sigma_o R^3}{3} - \frac{\pi \sigma_o^4 (1+\nu)^3}{6E^3 \theta^3} \quad \text{for} \quad \theta \geq \frac{\sigma_o(1+\nu)}{ER} \tag{109b}$$

The elastic response and first yield coincide with the results obtained for the hardening case. Thereafter, as θ increases monotonically T asymptotically approaches the value $2\pi \sigma_o R^3/3$. This is the limiting value of the torque T, and the $T \sim \theta$ curve is plotted in Figure 7. It may be noted that the

limit value of the torque for an elastic, perfectly plastic
analysis coincides with the yield value of the torque in the
rigid-plastic case.

In the problem of torsion of a prismatic bar the limit solution (i.e. the limit load for an elastic-plastic bar or the
load at which flow occurs in a rigid, perfectly plastic bar)
can be found directly with comparative ease for a simply connected but otherwise arbitrary cross section. It is again assumed that the only nonzero stress components are σ_{xz} and σ_{yz},
and that these components do not vary in the z direction along
the bar. The only remaining equilibrium equation (cf. equation
87) is

$$\frac{\partial \sigma_{xz}}{\partial x} + \frac{\partial \sigma_{yz}}{\partial y} = 0 \ .$$

This equation suggests the introduction of a stress function
$\Xi(x,y)$ such that

$$\sigma_{xz} = \frac{\partial \Xi}{\partial y} \ , \quad \sigma_{yz} = -\frac{\partial \Xi}{\partial x} \ . \tag{110}$$

The equilibrium equation is now satisfied. On the prismatic
traction free boundaries of the bar we must have (cf. equation
88)

$$\sigma_{xz}\nu_x + \sigma_{yz}\nu_y = 0 \ ,$$

or

$$\frac{\partial \Xi}{\partial y}\nu_x - \frac{\partial \Xi}{\partial x}\nu_y = 0 \ . \tag{111}$$

This condition expresses the requirement that $d\Xi/ds = 0$, where
s is measured along the boundary curve of the cross section.
Hence Ξ must be constant along the boundary, and we can for
convenience take $\Xi = 0$ on the boundary.

The yield condition

$$\sigma_{xz}^2 + \sigma_{yz}^2 = \sigma_o^2$$

now becomes, from equations (110),

$$\left(\frac{\partial \Xi}{\partial x}\right)^2 + \left(\frac{\partial \Xi}{\partial y}\right)^2 = \sigma_o^2 \quad . \tag{112}$$

This expression may be rewritten as

$$|\text{grad } \Xi| = \sigma_o \tag{113}$$

indicating that the stress function is a surface which has for all x,y a *maximum slope* equal to σ_o.

The differential equation (113) and the boundary condition $\Xi = 0$ define the stress function Ξ. The torque acting on the cross section may be determined by substituting equation (110) into equation (94);

$$T = \int_A \left(y \frac{\partial \Xi}{\partial y} + x \frac{\partial \Xi}{\partial x}\right) dA \quad . \tag{114}$$

Integrating by parts

$$T = \int_s \Xi(x\nu_x + y\nu_y)ds - 2 \int_A \Xi dA \tag{115}$$

where s is measured along the boundary curve and (ν_x, ν_y) are the components of the outward unit normal vector at a point on the boundary. However, since $\Xi = 0$ along the boundary the line integral vanishes and

$$T = -2 \int \Xi dA \tag{116}$$

The function Ξ will be a negative, but the torque can be interpreted as twice the magnitude of the volume contained between the surface Ξ and the x,y plane.

It remains to confirm that the plastic strain increments associated with the stress field derived from Ξ are integrable. In non-circular cross sections the displacement field of equation (74) will not be adequate to describe the deformation of the bar. For a prismatic bar with no displacement constraints at the free ends equations (74) may be altered to permit warping or distortion of the cross-section plane which is independent of z. Thus we permit, in incremental form for $d\theta > 0$,

$$du = yzd\theta \; , \quad dv = -xzd\theta \; , \quad dw = dw(x,y) \; . \tag{117}$$

The strain increments then become

$$d\varepsilon_{xz} = yd\theta + \frac{\partial}{\partial x}(dw) \; , \quad d\varepsilon_{yz} = -xd\theta + \frac{\partial}{\partial y}(dw) \; , \tag{118}$$

with all other components zero. It is then necessary to show that a suitable function dw can be found such that plastic strain increments described by equation (118) are compatible with the stress field of equation (110) and the constitutive equations.

It is possible to show that for simply connected cross sections the plastic strain increments derived from the stress function Ξ are always compatible. However, this is a fairly complex process which must take into account that solutions for Ξ are usually not continuously differentiable, leading to discontinuities in the stress field. Further, some care must be taken in constructing the stress function Ξ, especially in cross sections with reentrant corners. Full details of the argument will not be given; the reader is referred to the monograph by Prager and Hodge for an extensive treatment of this problem.

The stress functions for simple cross sections are, however, easily derived. For a circular cross section of radius R the stress function is cone with its base coinciding with the x,y, plane and pointing downwards. The slope of a generator must be σ_o, so that the height of the cone is $-\sigma_o R$. Twice the magnitude of the volume enclosed between the cone and the x,y plane is $2\pi\sigma_o R^2/3$, agreeing with the result obtained in equation (109b) for the limit load.

In the case of a square cross section, with a side dimension a, the stress function is a pyramid with a square base and height magnitude $a\sigma_o/2$. The limit torque is then given by $\sigma_o a^3/6$.

7.5 Plane Strain

Another idealization of the constitutive relations which is commonly used and is of importance involves the frequently encountered problem of *plane strain*. In problems where a body extends a theoretically infinite (or practically large) distance in, say, the z direction, and which has boundaries perpendicular to the x,y plane with the boundary conditions independent of z, the displacement field will have the form

$$u = u(x,y) , \quad v = v(x,y) , \quad w = 0 . \tag{119}$$

It can be seen immediately that

$$\varepsilon_{xz} = \varepsilon_{yz} = \varepsilon_{zz} = 0 . \tag{120}$$

This also holds in its incremental form, so that for an incremental change in the loading or in the displacement field during flow

$$d\varepsilon_{xz} = d\varepsilon_{yz} = d\varepsilon_{zz} = 0 . \tag{121}$$

If the material behaves elastically, substitution of the requirements of equation (101) into the elastic relations

$$\varepsilon_{ij} = (\frac{1+\nu}{E})\sigma_{ij} - \frac{\nu}{E}\sigma_{kk}\delta_{ij} \tag{122}$$

leads to the appropriate constitutive equations for plane strain;

$$\sigma_{xz} = \sigma_{yz} = 0 \;, \quad \sigma_{zz} = \nu(\sigma_{xx} + \sigma_{yy}) \;,$$

$$\varepsilon_{xx} = \frac{(1-\nu^2)}{E} \sigma_{xx} - \frac{\nu(1+\nu)}{E} \sigma_{yy} \;,$$

$$\varepsilon_{yy} = \frac{(1-\nu^2)}{E} \sigma_{yy} - \frac{\nu(1+\nu)}{E} \sigma_{xx} \;,$$

$$\varepsilon_{xy} = \frac{(1+\nu)}{E} \sigma_{xy} \;. \tag{123}$$

Alternatively, if we examine the constraints of equation (121) on plastic deformation, neglecting for the present the elastic strain or strain increments, the reduction of the constitutive equations is comparatively simple. Consider as an example the case of plastic flow. Since the material is incompressible, the strain increments are in deviator form, and we have

$$d\varepsilon^p_{ij} = de^p_{ij} = \lambda \frac{\partial \phi}{\partial s_{ij}} \tag{124}$$

If these strain increments are subject to the restrictions of equation (121), it is evident that we must have

$$\frac{\partial \phi}{\partial s_{zz}} = \frac{\partial \phi}{\partial s_{xz}} = \frac{\partial \phi}{\partial s_{yz}} = 0 \;. \tag{125}$$

For any particular function ϕ equations (125) provide three equations by means of which s_{xz}, s_{yz} and s_{zz} can be expressed in terms of the remaining three independent stress deviator components s_{xx}, s_{yy}, s_{xz}. If the yield function is isotropic, as might be expected in an elastic, perfectly plastic material, it can be shown that $s_{xz} = s_{yz} = s_{zz} = 0$. An isotropic yield function can be written in terms of the stress deviator invariants J_2, J_3 (equation 20). The three equations (125) then become

$$\frac{\partial \phi}{\partial s_{zz}} = \frac{\partial \phi}{\partial J_2} \frac{\partial J_2}{\partial s_{zz}} + \frac{\partial \phi}{\partial J_3} \frac{\partial J_3}{\partial s_{zz}} = 0 \ ,$$

$$\frac{\partial \phi}{\partial s_{xz}} = \frac{\partial \phi}{\partial J_2} \frac{\partial J_2}{\partial s_{xz}} + \frac{\partial \phi}{\partial J_3} \frac{\partial J_3}{\partial s_{xz}} = 0 \ , \qquad (126)$$

$$\frac{\partial \phi}{\partial s_{yz}} = \frac{\partial \phi}{\partial J_2} \frac{\partial J_2}{\partial s_{yz}} + \frac{\partial \phi}{\partial J_3} \frac{\partial J_3}{\partial s_{yz}} = 0 \ .$$

In general we do not expect that $\partial \phi / \partial J_2$ and $\partial \phi / \partial J_3$ will be zero. Further, for an arbitrary dependence of ϕ on J_2 and J_3, each of the six derivatives of J_2 and J_3 with respect to the stress components in equation (126) must be zero. Evaluating these expressions, we find first that

$$\frac{\partial J_2}{\partial s_{zz}} = \frac{\partial J_2}{\partial s_{xz}} = \frac{\partial J_2}{\partial s_{yz}} = 0 \qquad (127)$$

if and only if $s_{zz} = s_{xz} = s_{yz} = 0$.

The relevant partial derivatives of J_3 are

$$\frac{\partial J_3}{\partial s_{zz}} = s_{xz}^2 + s_{yz}^2 + s_{zz}^2 \ ,$$

$$\frac{\partial J_3}{\partial s_{xz}} = 2(s_{xy}s_{yz} + s_{xz}s_{zz} + s_{xz}s_{xx}) \ , \qquad (128)$$

$$\frac{\partial J_3}{\partial s_{yz}} = 2(s_{xy}s_{xz} + s_{yz}s_{zz} + s_{yz}s_{yy}) \ .$$

It is evident that these derivatives are each zero when $s_{zz} = s_{xz} = s_{yz} = 0$. In terms of the actual stresses these values of the deviator components give

$$s_{xz} = \sigma_{xz} = 0 \ ,$$

$$s_{yz} = \sigma_{yz} = 0 \ , \tag{129}$$

$$s_{zz} = \frac{2}{3}\sigma_{zz} - \frac{1}{3}(\sigma_{xx} + \sigma_{yy}) = 0 \ ,$$

i.e.

$$\sigma_{zz} = \frac{1}{2}(\sigma_{xx} + \sigma_{yy}) \ .$$

In the case of hardening materials equations (125) must also be satisfied if the plastic strain increments alone are required to satisfy the plane strain constraints (121). If the material hardens isotropically, equations (129) will also apply. However, if the material becomes anisotropic after initial yielding the particular yield function must be used in equations (125) to eliminate the stress deviator components s_{xz}, s_{yz} and s_{zz}. It should be noted, however, that in the case of a kinematic hardening material (see equation 39) where

$$\phi = (s_{ij} - he_{ij}^p)(s_{ij} - he_{ij}^p) - 3k^2 \ , \tag{130}$$

we have

$$\frac{\partial \phi}{\partial s_{zz}} = 2(s_{zz} - he_{zz}^p) \ ,$$

$$\frac{\partial \phi}{\partial s_{xz}} = 2(s_{xz} - he_{xz}^p) \ , \tag{131}$$

$$\frac{\partial \phi}{\partial s_{yz}} = 2(s_{yz} - he_{yz}^p) \ .$$

Since it is required that $e_{zz}^p = e_{yz}^p = e_{xz}^p = 0$, it follows that equation (125) is satisfied if and only if $s_{zz} = s_{xz} = s_{yz} = 0$, as in equation (129). This relation also holds for the more complex material described in equation (49).

Considered separately, therefore, the elastic and plastic
equations can be reduced for the plane strain case with comparative ease, and s_{zz}, s_{xz}, s_{yz} can be expressed in terms of the
remaining stress components unambiguously, i.e., in a manner
which does not depend in any way on the solution of the problem.
However, this is not possible when we consider elastic-plastic
behavior. Let us consider a body in plane strain composed of
an elastic-plastic hardening material as an example. Suppose
that an incremental change in the stress field takes place. In
those parts of the body where $\phi = 0$ or unloading takes place,
the incremental form of equations (123) must apply. In regions
where $\phi = 0$ and loading occurs, the *combined elastic and plastic strain increments* must satisfy equations (121). Thus we
have the restrictions

$$d\varepsilon_{zz} = \frac{(1+\nu)}{E}\, d\sigma_{zz} - \frac{\nu}{E}(d\sigma_{xx}+d\sigma_{yy}+d\sigma_{zz})+G\,\frac{\partial\phi}{\partial s_{zz}}\,\frac{\partial\phi}{\partial s_{ij}}\,ds_{ij} = 0 \quad ,$$

$$d\varepsilon_{xz} = \frac{(1+\nu)}{E}\, d\sigma_{xz} + G\,\frac{\partial\phi}{\partial s_{xz}}\,\frac{\partial\phi}{\partial s_{ij}}\,ds_{ij} = 0 \quad , \qquad\qquad (132)$$

$$d\varepsilon_{yz} = \frac{(1+\nu)}{E}\, d\sigma_{yz} + G\,\frac{\partial\phi}{\partial s_{yz}}\,\frac{\partial\phi}{\partial s_{ij}}\,ds_{ij} = 0 \quad .$$

In general the values of σ_{xz}, σ_{yz} and σ_{zz} and changes in these
values will depend on the remaining stresses and stress incre-
ments. This means that we cannot eliminate σ_{xz}, σ_{yz} and σ_{zz}
a priori, but must include them as functions of x and y in the
solution of the problem. Further, discontinuities in these
stress components can be expected on the elastic-plastic boundary. As a result, plane strain problems which may be solved
with comparative ease for an elastic or a rigid-plastic material
can become greatly complicated when elastic and plastic strains
are taken into account.

In suitable circumstances it is justifiable to reduce the complexity of the analytical elastic-plastic plane strain problem by assuming that the material is *incompressible* in the elastic range. Since plastic strains occur without change in volume, this assumption renders the material incompressible under all conditions. If we put $\varepsilon_{kk} = 0$ in equation (103), we must have

$$\frac{(1+\nu)}{E} \sigma_{kk} = 3\frac{\nu}{E}\sigma_{kk} \ ,$$

i.e.   (133)

$$\nu = \frac{1}{2} \ .$$

The elastic strains ε_{ij}^{e} are now identical to the elastic deviator strains e_{ij}^{e}, and resubstituting $\nu = 1/2$ into equation (122) we see that

$$e_{ij}^{e} = \frac{3}{2E}\sigma_{ij} - \frac{1}{2E}\sigma_{kk}\delta_{ij} = \frac{3}{2E}s_{ij} \ . \qquad (134)$$

If, in addition the yield function is isotropic, or if it hardens kinematically (see equations 131), it is evident that a sufficient condition that $d\varepsilon_{xz} = d\varepsilon_{yz} = d\varepsilon_{zz} = 0$ is given by $s_{xz} = s_{yz} = s_{zz} = 0$, since both the elastic and plastic components of these strain increments will each be zero. s_{xz}, s_{yz} and s_{zz} are thus again determined in terms of the other stress deviator components before the solution of the problem is attempted. This assumption thus greatly simplifies the formulation of the problem and its solution. It can be expected that in any problems the replacement of ν (which is 0.3 approximately for steel) by $\nu = 0.5$ will not greatly affect the solution, especially when the plastic strains are significantly larger than the elastic strains. However, each problem must be evaluated in order to determine whether the assumption is justified.

7.6 Historical and Bibliographical Remarks

Although the discussion of elastic-plastic constitutive relations applicable to continua have been divided for clarity into
three Chapters dealing respectively with initial yield conditions in terms of principal stresses (Chapter 4), plane stress
(Chapter 5) and the complete continuum description of the present Chapter, no such distinction can be found in the historical development of the important concepts. Consequently the
bibliographical remarks following Chapters 4 and 5 need not be
repeated. We shall limit ourselves to amplifying these previous remarks.

The composite model of plastic behavior from which both isotropic and kinematic hardening may be recovered is due to
Prager [1958]. This model has been extended by Birger and Dem-
janushko [1968].

In previous sections we have referred briefly to slip
theory. This theory was introduced by Batdorf and Budiansky
[1949], and was originally conceived as an alternative formulation to the classical incremental theory whose formulation
has been the subject of Chapter 2. That slip theory falls into
the classical formulation became apparent with a better understanding of singular yield surfaces, and was first clarified by
Koiter [1955] and Sanders [1955]. The physical basis of slip
theory has been continually developed; see, for example, Lin
[1954], [1958], [1960], Budiansky and Wu [1962] and Mandel
[1965]. Soviet work in this area is summarized by Knets [1971].
It is also very closely related to developments in plasticity
which make use of a thermodynamic basis; we shall discuss this
relation further in Part VI.

The experiments quoted in comparing the predictions of slip
theory to measured behavior are those of Naghdi, Essenburg and
Koff [1958], and the computations are due to Como and D'Agostino

[1969]. Further comparisons of this type have been given by Como and Grimaldi [1969].

The problem involving monotonic twisting of a circular bar has been treated many times as a simple illustrative example. The reader is referred to Prager and Hodge [1951] for a full discussion of early analytic work on the torsion of circular and prismatic bars. The direct analysis of rigid, perfectly plastic bars is due to Nadai [1931]; this method is commonly known as the *sand hill analogy*.

References

S.B.Batdorf and B.Budiansky	1949	"A mathematical theory of plasticity based on the concept of slip", NACA TN 1871.
I.A.Birger and I.V.Demjanushko	1968	"Teorii plastichnosti prineizoter-micheskom nagruzhenii", Mechanika Tverdovo Tela, $\underline{6}$, 70.
B.Budiansky and T.T.Wu	1962	"Theoretical predictions of plastic strains in polycrystals", Proc.4th U.S.Nat.Congr.Appl.Mech., ASME (N.Y.) $\underline{2}$, 1175.
M.Como and A.Grimaldi	1969	"Analytical formulation of the theoretical subsequent yield surfaces of metals with strain hardening and Bauschinger effects in an ideal tension-torsion test". Meccanica, $\underline{4}$,1.
M.Como and S.D'Agostino	1969	"Strain hardening plasticity with Bauschinger effect", Meccanica, $\underline{4}$, 146.
I.V.Knets	1971	*Basic Present Trends in the Mathematical Theory of Plasticity*, Zinatne (Riga).
W.T.Koiter	1955	"Stress-strain relations, uniqueness and variational principles for elastic-plastic materials with a singular yield surface", Quart.Appl. Math., $\underline{11}$, 350.

T.H.Lin   1954   "A proposed theory of plasticity
                             based on slips", Proc.2nd U.S.Nat.
                             Congr.Appl.Mech., ASME (N.Y.), 461.

T.H.Lin   1958   "On stress-strain relations based on
                             slips", Proc.3rd U.S.Nat.Congr.Appl.
                             Mech., ASME (N.Y.), 581.

T.H.Lin   1960   "On the associated flow rule of plas-
                             ticity based on crystal slip", J.
                             Franklin Inst., 270, 291.

J.Mandel   1965   "Généralisation de la Théorie de
                             plasticité de W.T.Koiter", Int.J.
                             Solids Structures, 1, 189.

A.Nadai   1931   *Plasticity*, McGraw-Hill (N.Y.).

P.M.Naghdi,   1958   "An experimental study of initial
F.Essenburg and   and subsequent yield surfaces in
W.Koff   plasticity", J.Appl.Mech., 25, 201.

W.Prager   1958   "Non-isothermal plastic deformation",
                             Proc.Konikl.Nederl.Akad. van Weten-
                             schappen, B61, 176

W.Prager and   1951   *Theory of perfectly plastic solids*,
P.G.Hodge,Jr.   Wiley (N.Y.). Reprinted by Dover
                             Publications, 1968.

J.L.Sanders,Jr.   1955   "Plastic stress-strain relations
                             based on linear loading functions", Proc.2n
                             Proc.2nd U.S.Nat.Congr.Appl.Mech.,
                             ASME (N.Y.), 455.

Chapter 8

TRANSVERSELY LOADED PLATE PROBLEMS

8.1 Constitutive Equations for a Sandwich Plate

As an illustration of elastic-plastic analysis in this class of
problems we shall consider one particular problem, that of a
simply supported circular sandwich plate subjected to a monotonically increasing uniformly distributed load. As in the examples given in earlier sections of this monograph, this problem is one which can be solved analytically with comparative
ease but it will nevertheless serve as an introduction to this
class of problems.

The idealized sandwich cross-section can be seen in the small
element of plate shown diagrammatically in Figure 1. It is supposed to consist of two thin sheets of metal, each of thickness
t, separated by a distance 2h by means of a filler material. The
filler material does not contribute to the bending strength of
the plate, but resists without deformation the shear forces in
the plate.

Because of the axial symmetry of the problem, we adopt polar
coordinates r, θ in the plane of the plate and note that the
transverse displacements w, measured positive downwards, are
given by (cf. Section 1.3)

$$w = w(r) \quad . \tag{1}$$

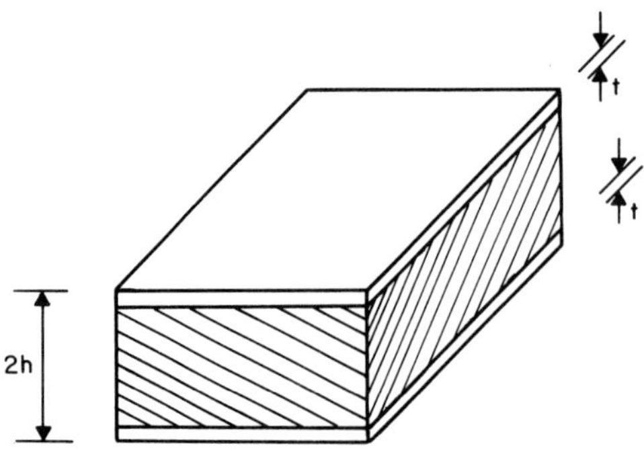

Figure 1. Section of the sandwich plate

The curvature rates in the r and θ directions are then given respectively by

$$\kappa_r = -\frac{d^2 w}{dr^2} \quad , \quad \kappa_\theta = -\frac{1}{r}\frac{dw}{dr} \quad . \tag{2}$$

The usual assumptions of plane cross-sections under cylindrical bending will be adopted, so that the strains in the bottom and top sheets of the sandwich are taken to be

$$\varepsilon_r = \overset{+}{\underset{-}{}} \kappa_r h \quad , \quad \varepsilon_\theta = \overset{+}{\underset{-}{}} \kappa_\theta h \quad . \tag{3}$$

These strains will produce stresses in the two sheets. Since the stress components in the direction perpendicular to the plane of the sheet will be zero, the result is a state of plane stress. Furthermore, if the material is isotropic in both the elastic and plastic ranges the principal strains given in equation (3) will produce principal stresses which we shall term σ_r and σ_θ. The net force per unit length in the r and θ directions which is associated with this stress distribution in the top and bottom plates will be zero, but there will be resultant moments per unit length M_r and M_θ given in terms of the stresses in the lower sheet as

$$M_r = 2th\sigma_r \quad , \quad M_\theta = 2th\sigma_\theta \quad . \tag{4}$$

The appropriate form of the stress-strain relations for an isotropic material in the elastic range will be (Section 1.4, equation 7)

$$\varepsilon_r = \frac{\sigma_r}{E} - \frac{\nu\sigma_\theta}{E} \quad , \quad \varepsilon_\theta = \frac{\sigma_\theta}{E} - \frac{\nu\sigma_r}{E} \quad . \tag{5}$$

Substituting these equations into (3), using equations (4) and putting

$$2th^2 E = D \quad , \tag{6}$$

we obtain the elastic constitutive equations for the plate;

$$\kappa_r = \frac{1}{D}(M_r - \nu M_\theta) \quad , \quad \kappa_\theta = \frac{1}{D}(M_\theta - \nu Mr) \quad . \tag{7}$$

We shall assume that the plates are composed of an *elastic, perfectly plastic material* with an isotropic yield surface. Since the principal stress directions are known at each point in the plate, it is particularly convenient to use the Tresca yield condition in plane stress (Chapter 4, equation 44). With the third principal stress equal to zero, the yield condition is the figure enclosed by the six independent lines

$$\pm\sigma_r - \sigma_o = 0 \quad ,$$

$$\pm\sigma_\theta - \sigma_o = 0 \quad , \tag{8}$$

$$\pm(\sigma_r - \sigma_\theta) - \sigma_o = 0 \quad .$$

Yielding will occur in the sandwich cross-section when σ_r, σ_θ satisfy any one of equations (8) and the remaining yield functions are non-positive. Since the stresses in the upper and lower sheets are equal in magnitude but opposite in sign, we shall consider without loss in generality the stresses in the lower sheet which are given in terms of the moments in equations 4. Consider first the part of the yield surface given by $\sigma_r = \sigma_o$, $0 < \sigma_\theta < \sigma_o$. From equations (4), putting

$$2th\sigma_o = M_o \quad , \tag{9}$$

we see that yield occurs for

$$M_r = M_o \quad , \quad 0 < M_\theta < M_o \quad . \tag{10}$$

This condition is plotted as the line AB in Figure 2. Alternatively, consider the part of the yield surface given by

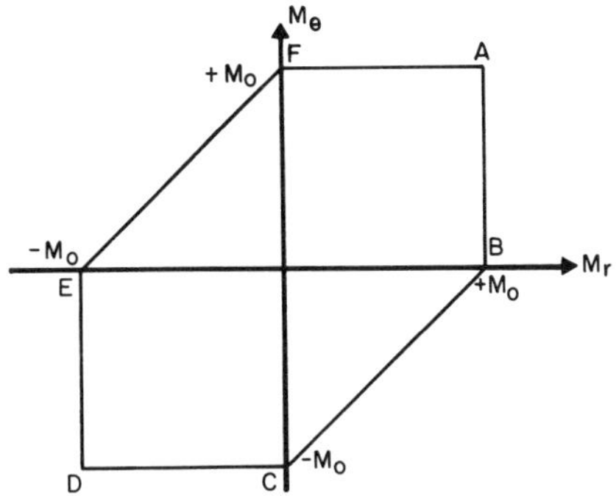

Figure 2. Tresca yield surface for the sandwich plate

$(\sigma_r - \sigma_\theta) = \sigma_o$, with $0 < \sigma_r < \sigma_o$ and $-\sigma_o < \sigma_\theta < 0$. Substitution of equation (4) gives yielding of the cross-section for

$$M_r - M_\theta = M_o \quad \text{for} \quad 0 < M_r < M_o \text{ and } -M_o < M_\theta < 0 \ . \tag{11}$$

This condition appears as the line BC in Figure 2. We may continue in this manner with the four other sides of the yield surface of equations (8). However, it is evident that because M_r, M_θ are directly proportional to σ_r, σ_θ respectively the yield surface for the plate will be identical to the Tresca yield surface in biaxial tension with M_r, M_θ, M_o substituted in turn for σ_r, σ_θ, σ_o.

The yield surface plotted in Figure 2 is the initial yield surface. It is also the limit surface, as we can establish by showing that flow occurs whenever the generalized stress M_r, M_θ lies on the yield surface. Suppose, for example, that the generalized stress point lies on AB in Figure 2. The stress point σ_r, σ_θ satisfies the yield condition $\sigma_r - \sigma_o = 0$, and

$$d\varepsilon_r^p = \lambda \ , \quad d\varepsilon_\theta^p = 0 \ , \quad \text{with} \quad \lambda \geq 0 \ , \tag{12}$$

as a result of the flow rule. Hence, from equation (3),

$$d\kappa_r = \frac{\lambda}{h} \ , \quad d\kappa_\theta = 0 \ . \tag{13}$$

If we define

$$\kappa_r = \kappa_r^e + \kappa_r^p \,, \qquad \kappa_\theta = \kappa_\theta^e + \kappa_\theta^p \,, \tag{14}$$

with κ_r^e, κ_θ^e given respectively by the expressions in equation (7), it follows that $d\kappa_r^e$, $d\kappa_\theta^e$ are zero when M_r, M_θ are constant. Hence equation (13) gives

$$d\kappa_r^p = \frac{\lambda}{h} \,, \qquad d\kappa_\theta^p = 0 \tag{15}$$

when M_r, M_θ lie on AB. Plastic flow is thus taking place under constant moment, and the restriction $\lambda \geq 0$ shows that the generalized plastic strain increment vector $d\kappa_r^p$, $d\kappa_\theta^p$ has the direction of the outward normal on AB. In a similar manner when M_r, M_θ lie on BC we find that

$$d\kappa_r^p = \frac{\lambda}{h} \,, \qquad d\kappa_\theta^p = -\frac{\lambda}{h} \,, \tag{16a}$$

or that

$$d\kappa_r^p = -d\kappa_\theta^p \,. \tag{16b}$$

This generalized plastic strain increment vector again has the direction of the appropriate outward normal to the yield surface. Similarly, flow may be shown to occur whenever M_r, M_θ lie on the yield surface. At the corner B the plastic generalized strain increments of equations (15) and (16) must be combined independently, giving

$$d\kappa_r^p = \frac{\lambda_1 + \lambda_2}{h} \,, \qquad d\kappa_\theta^p = -\frac{\lambda_2}{h} \,; \quad \lambda_1 \geq 0 \,, \ \lambda_2 \geq 0 \,. \tag{17}$$

All combinations of permissible values of λ_1, λ_2 will lead to plastic generalized strain increment vectors which lie in the fan defined by the outward normals adjacent to B i.e. by the normals to AB and BC. Thus the conventional flow rule holds at

B, and may be similarly shown to hold at the other corners.

8.2 Uniformly Loaded Simply Supported Plate

We proceed now with the elastic solution of the plate problem
shown in Figure 3, since clearly for sufficiently small values
of the load p, as p increased from zero and $\kappa_r^p = \kappa_\theta^p = 0$, the
behavior will be entirely elastic. The appropriate equilibrium
equations for the plate (Section 1.3, equations 1 and 2) are

$$\frac{d}{dr}\,(rM_r) \;=\; M_\theta - rQ_r \;\;,\tag{18a}$$

$$\frac{d}{dr}\,(rQ_r) \;=\; pr \;\;,\tag{18b}$$

where Q_r is the shear force per unit length acting across a circumferential line on the middle surface of the plate. Q_r may
be eliminated immediately by integrating equation (18b) between
the limits 0 and r. With p constant, we therefore obtain the
single equation

$$\frac{d}{dr}\,(rM_r) \;=\; M_\theta - \frac{pr^2}{2} \;\;.\tag{19}$$

The curvatures may be obtained in terms of M_r, M_θ for the elastic case by inverting equations (7). The inversion gives

$$M_r \;=\; \frac{D}{(1-\nu^2)}\,(\kappa_r + \nu\kappa_\theta) \;\;,$$

$$\tag{20}$$

$$M_\theta \;=\; \frac{D}{(1-\nu^2)}\,(\kappa_\theta + \nu\kappa_r) \;\;.$$

The curvatures may be expressed in terms of the displacements
w(r) by means of equations (2); hence by substituting equations
(20) and then equations (2) into equation (19) we obtain a
third order ordinary differential equation in w;

$$\frac{d}{dr}\,[r(\frac{d^2w}{dr^2} + \frac{\nu}{r}\frac{dw}{dr})] \;=\; \frac{1}{r}\frac{dw}{dr} + \nu\frac{d^2w}{dr^2} + \frac{(1-\nu^2)}{2D}\,pr^2 \;\;.\tag{21}$$

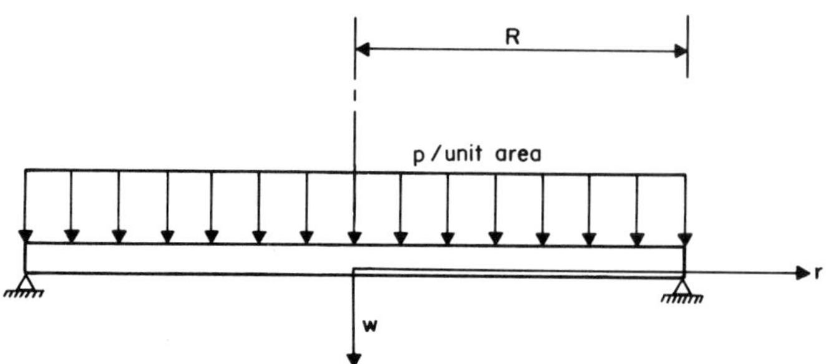

Figure 3. Uniformly loaded simply supported circular plate

Only terms involving derivatives in w appear in this equation, hence we proceed with the solution by considering the second order differential equation in w' = dw/dr;

$$\frac{d}{dr}\left[r\frac{dw'}{dr} + \nu w'\right] = \frac{1}{r}w' + \nu\frac{dw'}{dr} + \frac{(1-\nu^2)}{2D}pr^2 \ . \tag{22}$$

The solution of this equation is

$$w' = \frac{(1+\nu)}{r}B + (1-\nu)Cr + \frac{(1-\nu^2)}{16D}pr^3 \ . \tag{23}$$

After determining dw'/dr by differentiation, the moments may now be found by substituting equation (23) into equations (2) and (20);

$$M_r = \frac{DB}{r^2} - DC - \frac{(3+\nu)}{16}pr^2 \ , \tag{24a}$$

$$M_\theta = -\frac{DB}{r^2} - DC - \frac{(1+3\nu)}{16}pr^2 \ . \tag{24b}$$

There are two constants of integration which are determined by substituting known values of w', M_r, M_θ into equations (23) and (24); at the boundary of the plate, from the simple support condition,

$$M_r(R) = 0 \ . \tag{25}$$

Furthermore, because it is not possible to distinguish between the r and θ directions at the center of the plate,

$$M_r(0) = M_\theta(0) \; . \tag{26}$$

The constants B and C can thus be evaluated. The expression for w' becomes

$$w' = \frac{p}{16D} (1 - \nu^2) \{r^3 - \frac{(3+\nu)}{(1+\nu)} R^2 r\} \; . \tag{27}$$

The function w' is continuous, as expected, and w'(0) = 0. Equation (27) may now be integrated to give the displacement $w(r)$, with the additional constant of integration determined by means of the support condition

$$w(R) = 0 \; . \tag{28}$$

The displacements are

$$w = \frac{p(1-\nu)}{64D} \{(1+\nu)r^4 - 2(3+\nu)R^2 r^2 + (5+\nu)R^4\} \; . \tag{29}$$

After substituting the appropriate values of B and C into the expressions for the moments (equation 24), we find that

$$M_r = \frac{(3+\nu)}{16} p [R^2 - r^2] \; , \tag{30a}$$

$$M_\theta = \frac{p}{16} [(3+\nu)R^2 - (1+3\nu)r^2] \; . \tag{30b}$$

It is convenient to plot the *generalized stress profile* in the M_r, M_θ space in order to determine when first yield will occur. When the parameter r^2 is eliminated it is seen that a linear relation exists between M_r and M_θ; in the M_r M_θ space a straight line may be drawn between the points $M_r(R) = 0$, $M_\theta(R)$ and $M_r(0)$, $M_\theta(0)$. It is evident that as p is increased monotonically, first yield will occur at the corner A of the yield surface, as shown diagrammatically in Figure 4. First yield thus

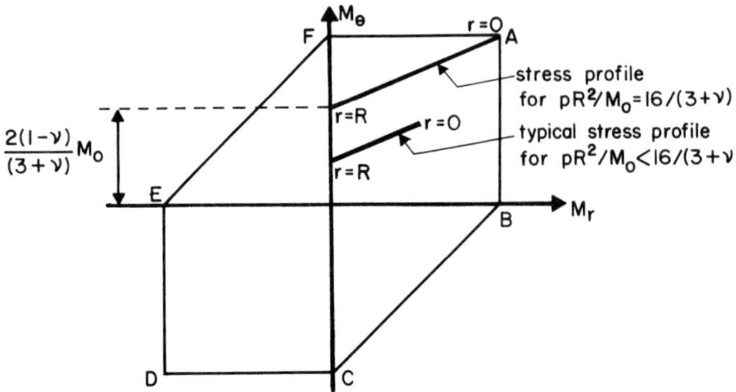

Figure 4. Generalized stress profile for the elastic plate

occurs for $M_r(0) = M_\theta(0) = M_0$, i.e. for

$$\frac{pR^2}{M_0} = \frac{16}{(3+\nu)} \, .$$

(31)

The solution for the elastic range is now complete. We have
a choice of procedures in extending the analysis into the elastic-plastic range. On the one hand we can follow the more direct procedure and consider a series of incremental analyses
until the limit load is reached and flow takes place in the
plate. However, it is extremely difficult and time consuming
to obtain an analytical solution (if one can be found) by this
procedure. If the problem were to be solved numerically this
would be the logical method of continuing the analysis.

On the other hand, we can assume that a comparatively simple
analytical solution exists, make certain further assumptions regarding the form of the solution, and then attempt to find an
analytical solution. Then, in an important final step, we must
confirm that the assumptions made were justified, i.e. that the
final solution satisfies in all respects the static and kinematic requirements and the constitutive equations.

In this particular problem we note first that additional material must become plastic if the plate is to deform plastically. At least part of the stress profile must therefore lie

on the yield surface. From the general form of the elastic solution we expect that M_r will vary continuously from a maximum
value at $r = 0$ to zero at $r = R$. Hence the stress profile cannot lie on AB (or be concentrated at A) since this would give
M_r constant for a range of values of r. Consequently we assume
that for p greater than the value which first causes yield the
stress profile has the form shown in Figure 5. We define a
parameter s, which we term the elastic-plastic boundary, such
that for

$$r \;=\; 0 \qquad M_r = M_\theta = M_o \;\; ,$$

$$0 < r \le s \qquad M_\theta = M_o \;\; , \quad M_r < M_o \;\; , \tag{32}$$

$$s < r \le R \qquad M_\theta < M_o \;\; , \quad M_r < M_o \;\; .$$

The moments M_r, M_θ are both taken to be positive, so that it is
evident that for $r < s$ plastic deformation may occur, while for
$r > s$ the plate must deform elastically. Furthermore, for
$0 < r < s$ we have from the flow rule that $d\kappa_r^p = 0$ with $d\kappa_\theta^p$ non-
negative but otherwise unspecified by the constitutive rela-
tions. For $r = 0$ both $d\kappa_r^p$ and $d\kappa_\theta^p$ are non-negative but not
necessarily zero.

   If this assumption is to lead us to the correct solution, s
must be determined by the load p which has been achieved during

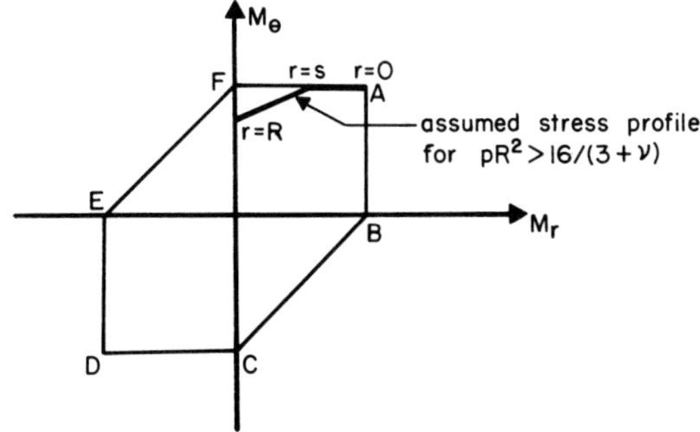

Figure 5. Generalized stress profile for the plastic plate

the monotonic increase in load. Thus any section of the plate
which lies currently in the range $0 < r < s$ has been subjected
to a history of plastic strain. Let us further *assume* that no
unloading has occurred during this process. This enables us to
say that M_r, M_θ for any point currently in the plastic range
first lay inside the yield surface and then subsequently on the
side AF as the load was increased. Thus, independent of the
precise history of stress at any point in the plastic region
$0 < r < s$, the generalized plastic strain increment vector always has the same direction (excepting the point $r = 0$) because AF is a straight line. This permits us to put

$$\kappa_\theta^p = \gamma(r), \quad \kappa_r^p = 0 \quad \text{for} \quad 0 < r < s \quad , \tag{33}$$

i.e. to integrate the plastic strain increments and write an
expression for the *total* generalized plastic strain in the plastic region. The function $\gamma(r)$ will be determined when the solution is found; however, the solution can be correct only if
$\gamma(r)$ is non-negative and is a monotonically increasing function
of p, since it has in reality been obtained by integrating over
the loading program a succession of non-negative increments λ.

We are now in a position to continue the analysis in terms
of total deformations, and if our assumptions are found to be
correct, we have obviated the need for an incremental analysis.
In the plastic region, $0 < r < s$ we follow a procedure similar
to that used in the elastic case. In order to obtain a differential equation in terms of $w(r)$ from the strain-displacement
relations (2) and the equilibrium equation (19), we must ex-
press M_r, M_θ in terms of κ_r, κ_θ. Noting that for $0 < r < s$ we
have assumed $M_\theta = M_o$, $\kappa_r^p = 0$, we obtain from equations (14), (7)
and (33)

$$\kappa_r = \kappa_r^e = \frac{1}{D}(M_r - \nu M_o) \quad , \tag{34}$$

$$\kappa_\theta = \kappa_\theta^e + \kappa_\theta^p = \frac{1}{D}(M_o - \nu M_r) + \gamma \quad .$$

Inverting these equations, expressions for M_r and γ can be obtained;

$$M_r = D\kappa_r + \nu M_o \quad , \tag{35a}$$

$$\gamma = (\kappa_\theta + \nu\kappa_r) - \frac{(1-\nu^2)}{D} M_o \quad . \tag{35b}$$

Thus, equations (35a), (2) and (19) give

$$-\frac{d}{dr}\left(r\,\frac{d^2w}{dr^2}\right) = \frac{(1-\nu)}{D} M_o - \frac{pr^2}{2D} \quad . \tag{36}$$

As in the elastic case, we consider this as a second order differential equation in $w' = dw/dr$. The solution is readily found to be

$$w' = \frac{pr^3}{18D} - \frac{(1-\nu)M_o r}{D} - \frac{G}{r^2} + H, \quad 0 < r < s \quad . \tag{37}$$

Substituting this solution into the strain-displacement relations (2) and then into (35a), we obtain an expression for M_r;

$$M_r = \frac{-pr^2}{6} + M_o - \frac{G}{r} \quad , \quad \text{for } 0 < r < s$$

$$\tag{38}$$

$$M_\theta = M_o$$

In the elastic region the equilibrium relation (19), the strain displacement relation (2) and the elastic moment-curvature relations still hold. Hence the differential equations (21) and (22) remain valid for $s < r < R$. Repeating the solution to this equation we have for $r < s < R$

$$w' = \frac{(1+\nu)}{r}B + (1-\nu)Cr + \frac{(1-\nu^2)}{16D}pr^3 \quad , \tag{39a}$$

$$M_r = \frac{DB}{r^2} - DC - \frac{(3+\nu)}{16}pr^2 \quad , \tag{39b}$$

$$M_\theta = -\frac{DB}{2} - DC - \frac{(1+3\nu)}{16}pr^2 \quad . \tag{39c}$$

The two solutions involve five unknowns, B, C, G, H and s. Although these terms are constants in the sense that they are independent of r, they must be expected to depend on p. This, in physical terms, expresses the dependence of the solution in the elastic-plastic range on the history of loading. For each value of the load p we can invoke five conditions from which these constants can be determined;

$$M_r(0) = M_\theta(0) = M_o \quad , \tag{40a}$$

$$M_r(R) = 0 \quad , \tag{40b}$$

$$M_o = M_\theta(s^-) = M_\theta(s^+) \quad , \tag{40c}$$

$$M_r(s^-) = M_r(s^+) \quad , \tag{40d}$$

$$w'(s^-) = w'(s^+) \quad . \tag{40e}$$

The first of these equations involves our assumption that the stress point at the center of the plate lies at A in Figure 5. The second derives from the simple support condition, and the remaining three from the continuity of M_r, M_θ and w' at the elastic-plastic boundary. While this continuity is expected we have already assumed that it exists by postulating a continuous stress profile (see Figure 5) in which $\kappa_r^p = 0$ for r = s.

The first boundary condition (40a) gives immediately from equation (38a) that

$$G = 0 \quad . \tag{41}$$

Conditions (40b), (40c) and (40d) lead to three equations in B,

C and s. If we first eliminate B and C, we obtain after some manipulation a relation between p and s which has the form

$$\frac{pR^2}{M_o} \left\{ 3(3+\nu) - 2 \left(\frac{s}{R}\right)^2 (1+3\nu) + (1+3\nu) \left(\frac{s}{R}\right)^4 \right\} = 48 \ . \qquad (42)$$

This is a one to one relation between (pR^2/M_o) and (s/R), but we cannot conveniently express s in terms of p. We see, however, that when s = 0 we have

$$\frac{pR^2}{M_o} = \frac{16}{(3+\nu)} \qquad (43)$$

which agrees with the load parameter at first yield given in equation (31). At the other limit, s = R, on the other hand, we have

$$\frac{pr^2}{M_o} = 6 \ . \qquad (44)$$

This value is independent of ν, and, since we might expect that when plastic deformation occurs over the whole plate flow is possible, this result confirms that the limit load is independent of the elastic constants. Let us now introduce dimensionless variables

$$\bar{p} = \frac{pR^2}{M_o} \ , \quad \xi = \frac{s}{R} \ , \quad \rho = \frac{r}{R} \ ,$$

$$\qquad (45)$$

$$m_r = \frac{M_r}{M_o} \ , \quad m_\theta = \frac{M_\theta}{M_o} \ .$$

The relation between $\bar{p}$ and ξ for the special case $\nu = 1/3$ is plotted in Figure 6. It is now convenient to express the remaining constants in terms of $\bar{p}$ and ξ, always remembering that $\xi = s/R$ is determined from equation (42). Thus we find that

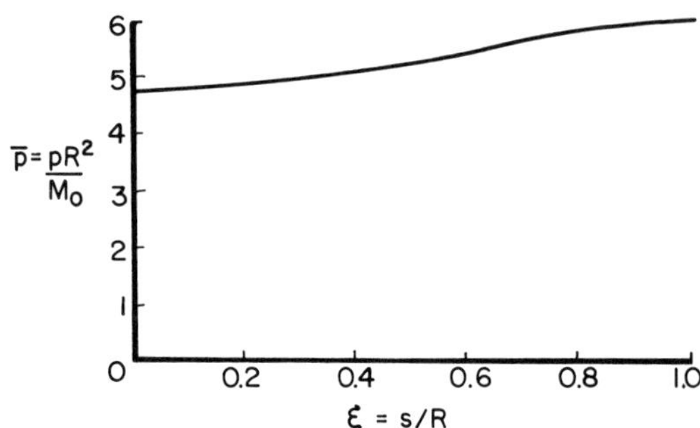

Figure 6.Position of the elastic-plastic boundary

$$\frac{DC}{M_o} = -1 - \frac{\bar{p}\xi^2}{24}(1+3\nu) \quad , \tag{46a}$$

$$\frac{DB}{M_o R^2} = \frac{(3+\nu)}{16}\bar{p} - \frac{(1+3\nu)}{24}\bar{p}\xi^2 - 1 \quad . \tag{46b}$$

Substituting equations (46a) and (46b) into equations (38a), (39a) and (39b), expressions for the moments can be written;

$$m_r = \frac{M_r}{M_o} = 1 - \frac{\bar{p}}{6}\rho^2 , \qquad 0 < \rho < \xi \quad , \tag{47a}$$

$$m_r = \left(1 - \frac{1}{\rho^2}\right) + \frac{(3+\nu)}{16}\bar{p}\left\{\frac{1}{\rho^2} - \rho^2\right\}$$

$$+ \frac{(1+3\nu)}{24}\bar{p}\xi^2\left(1 - \frac{1}{\rho^2}\right), \quad \xi < \rho < 1 \quad , \tag{47b}$$

$$m_\theta = 1 \quad 0 < \rho < \xi \quad . \tag{47c}$$

Stress profiles for various values of $\bar{p}$ are plotted in Figure 7 for the particular case $\nu = 1/3$. These confirm the validity of the general form of the stress profile assumed in Figure 5.

The remaining constant H is evaluated by means of the continuity condition (40e). This gives a complex expression

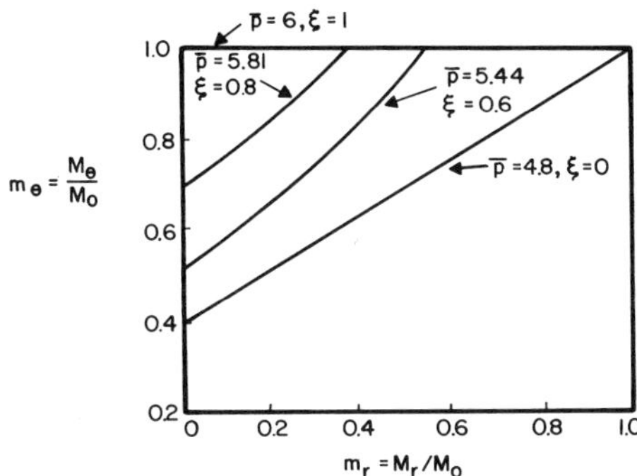

Figure 7. Stress profiles for circular plate

$$\frac{DH}{M_o R} = \frac{(1+\nu)}{\xi}\left\{\frac{\bar{p}}{16}(3+\nu) - 1\right\} - \frac{\xi\bar{p}}{24}(1+3\nu)(1+\nu)$$

$$- \frac{\xi^3\bar{p}}{144}(5-3\nu)(1+3\nu) \quad . \tag{48}$$

The constant H can be readily interpreted in physical terms.
From equation (37), we see that

$$H = w'(0) \quad . \tag{49}$$

Since H is in general non-zero, the slope of the plate is discontinuous at the center. This pointed vertex can arise only
from generalized plastic strains at the center of the plate. A
discontinuity in the slope must be associated with a delta function in curvature, implying that κ_r is infinite at the center.
This is admissible, and corresponds to the hinges found in beam
problems (c.f. Chapter 6). However, the sign of the curvature
must nevertheless be of the sense required by the flow rule at
point A in Figure 5. If we integrate the first of equations (2)
across the interval $0^- < r < 0^+$, the finite rotation α will be
equal to the area under the delta function κ_r and must be positive from the flow rule. It can be related to the jump in w'
by means of the relation

$$\alpha = \int_{0^-}^{0^+} \kappa_r\, dr = \left[-w'\right]_{0^-}^{0^+} = -2w'(0^+) \quad , \tag{50}$$

where $w'(0^+)$ is in fact the value given in equation (49). Hence, if our assumptions about the solution are to be confirmed, H must be a non-positive monotonically decreasing function of $\bar{p}$. The slope $w'(0)$ is plotted in Figure 8 for the special case $\nu = 1/3$ as a function of ξ, and our suppositions are confirmed. The values of H are also central in determining whether $\gamma(r)$ is for each value of r a monotonically increasing function of p. From the expression for w' in the range $0 < r < s$ (equation 37 with G = 0) and equations (2) and (35b), we find

$$\frac{D\gamma}{M_o} = \frac{\bar{p}}{18}\rho^2(1+3\nu) - \frac{1}{\rho}\left(\frac{DH}{M_o R}\right) . \tag{51}$$

Thus γ is infinite and positive for $\rho = 0$ (since $H < 0$). This corresponds again to the discontinuity in slope or pointed vertex at the center of the plate. $D\gamma/M_o$ is plotted for representative values of $\bar{p}$ in Figure 9 for the special case $\nu = 1/3$,

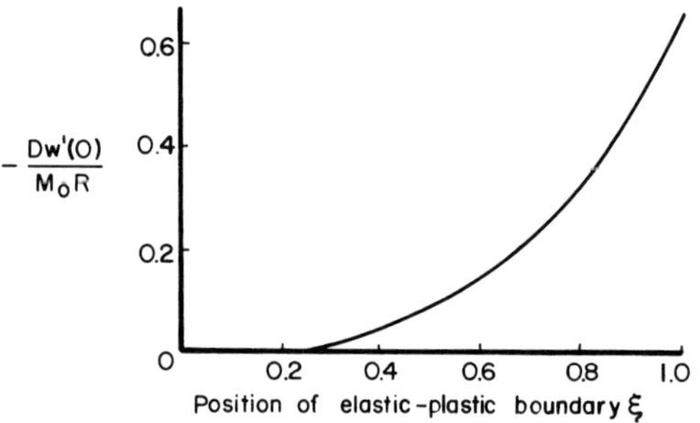

Figure 8.Rotation at the center of the plate

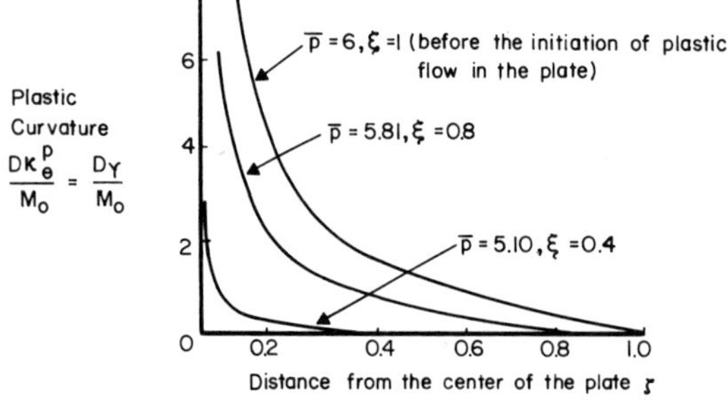

Figure 9. Plastic curvature distribution

and again our assumptions are confirmed.

The displacements of the plate are found by integrating the expressions for w' (equations 37 and 39a). Without substituting the values of the constants which have been hitherto determined (except for G = 0) we then obtain;

$$0 < r < s \; : \; w' = \frac{pr^4}{72D} - (1-\nu) \frac{M_o r^2}{2D} + Hr + J \quad , \tag{52a}$$

$$s < r < R \quad w' = (1+\nu)B \log r + \frac{(1-\nu)Cr^2}{2} + \frac{(1-\nu^2)}{64D} pr^4 + K \; . \tag{52b}$$

The two additional constants of integration are obtained from the simple support condition and the continuity of displacements at the elastic-plastic boundary,

$$w(R) \;\; = \;\; 0 \quad , \tag{53a}$$

$$w(s^-) \;\; = \;\; w(s^+) \quad . \tag{53b}$$

From equation (52a) it can be seen that the displacement at the center of the plate where r = 0 is given simply by the constant J. Thus, solving for J from equations (53a) and (53b),

$$\frac{Dw(0)}{M_o R^2} = \frac{-DH\xi}{M_o R} + \frac{(1-\nu)}{2} \xi^2 + \frac{(1+\nu)}{M_o R^2} DB \log \xi - \frac{(1-\nu)DC}{2M_o} (1-\xi^2)$$

$$- \frac{(1-\nu^2)}{64} \bar{p} \{1 - \frac{(1-9\nu^2)}{9(1-\nu^2)} \xi^4\} \quad . \tag{54}$$

Because s is defined between the values 0 and R, the solution we have developed for the elastic-plastic range holds only between the load values

$$\frac{16}{(3+\nu)} \leq \frac{pR^2}{M_o} = \bar{p} < 6 \quad . \tag{55}$$

As we approach the load parameter $\bar{p}$ = 6 as p is increased

monotonically from zero the displacements w and slopes w' remain finite and determined, and thus do not indicate that flow
has been initiated in range of loads. Nevertheless, we expect
that $\bar{p} = 6$ is in fact the limit load, since the stress profile
coincides with side AF of the yield surface (Figure 7) independent of the value of ν. We must now confirm that it is indeed
possible for flow of the plate to take place. With $\bar{p} = 6$ the
moments are given by

$$m_r = \frac{M_r}{M_o} = 1 - \frac{\bar{p}\rho}{6} , \qquad \text{for} \quad 0 < \rho < 1 \tag{56}$$

$$m_\theta = 1 ,$$

and do not change if the load is held constant. We seek an incremental displacement field $dw(r) = \bar{w}(r)$ such that the genera-
lized plastic strain increments $d\kappa_r^p = \bar{\kappa}_r^p$, $d\kappa_\theta^p = \bar{\kappa}_\theta^p$ and are admissible in terms of the flow rule. It is evident that we have

$$\bar{\kappa}_\theta^p > 0 \quad 0 \le \rho \le 1 , \tag{57a}$$

$$\bar{\kappa}_r^p = 0 \quad 0 < \rho \le 1 , \tag{57b}$$

$$\bar{\kappa}_r^p \ge 0 \quad \rho = 0 . \tag{57c}$$

Since, from the incremental form of equation (2),

$$\bar{\kappa}_r^p = -\frac{d^2\bar{w}}{dr^2} , \tag{58}$$

we must have $d\bar{w}/dr$ independent of r i.e. the slope of the incremental displacement field is constant. This suggests the
incremental displacement field

$$\bar{w} = a(1 - \rho) , \tag{59}$$

where a is non-negative but unspecified constant. Equations
(57a) and (57b) are then satisfied. There will be a discontinuity in the slope of the incremental velocity field given by
$\bar{a}$, say, as shown in Figure 10 and $\bar{\kappa}_r^p$ will be infinite. However,
this discontinuity in slope is of the same sign as the discon-
tinuity in the elastic-plastic case (i.e. $\bar{w}'(0) = -a < 0$) and
the increments in κ_r^p are positive as required by the equation
(57c).

The solution is now complete, since we have established that
flow of the plate can occur under the constant load $\bar{p} = 6$. The
central displacement w(0) is plotted in Figure 11 as a function
of the monotonically increasing load.

It should be noted how easy it is to determine the limit load
solution alone provided that an intelligent guess can be made
regarding the stress profile while flow occurs. If the correct
stress profile given in Figure 7 for $\bar{p} = 6$ is arrived at, integration of the equilibrium equations gives equation (56) immediately, and the flow displacement increment field would be
found in the same manner as given above.

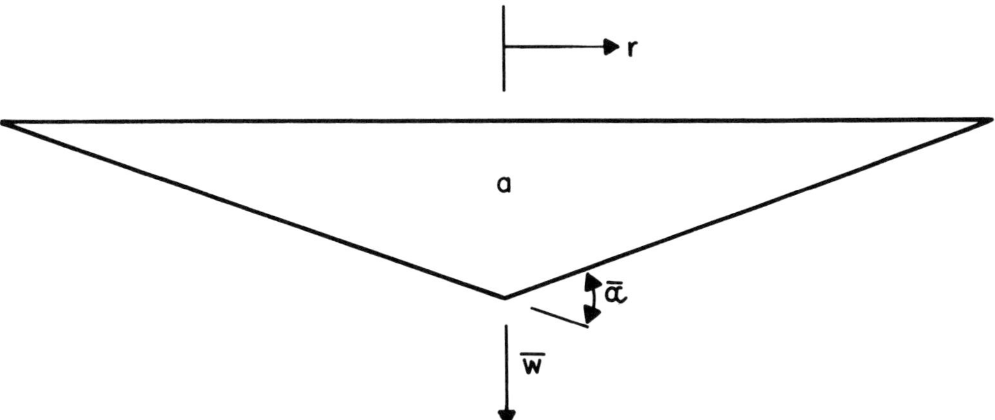

Figure 10. Incremental displacement field for plate during flow

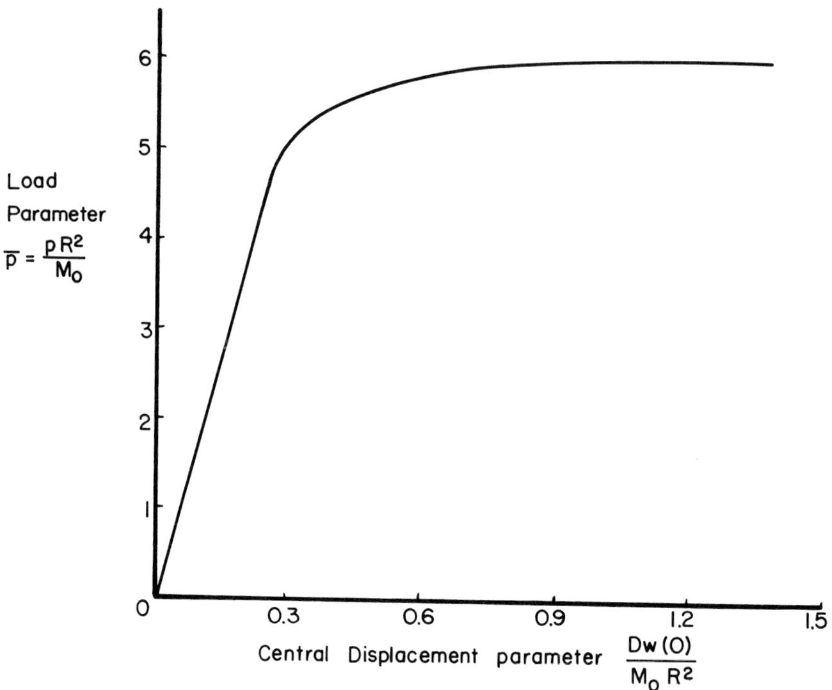

Figure 11. Central displacement of elastic, perfectly plastic circular plate for a monotonically increasing uniform load

8.3 Bibliographical Remarks

The problem of a simply supported elastic, perfectly plastic plate subjected to a monotonically increasing uniform load is one of the few elastic, plastic plate problems for which an analytic solution can be found. Such problems were first studied by Haythornthwaite [1955], Tekinalp [1955], Prager [1956], Boyce [1956] and Perrone and Hodge [1959]. Reviews of problems in this case, including discussions of the effects of membrane stresses which are clearly of great importance in a realistic analysis, have been given by Goodier and Hodge [1958] and Hodge [1960], [1963].

References

W.Boyce   1956   "The bending of a work-hardening cir-
                                 cular plate by a uniform transverse
                                 load", Quart.Appl.Math.,14, 277.

R.M.Haythornthwaite 1956   "The deflection of plates in the
                            elastic-plastic range", Proc.2nd U.S.
                            Nat.Congr.Appl.Mech. (ASME), 521.

J.N.Goodier and   1958   *Elasticity-Plasticity*,Wiley (N.Y.).
P.G.Hodge,Jr.

P.G.Hodge,Jr.   1960   "Boundary value problems in plasti-
                            city", *Plasticity*, edited by E.H.Lee
                            and P.S.Symonds, Pergamon Press
                            (N.Y.), 297.

P.G.Hodge,Jr.   1963   *Limit Analysis of Rotationally Sym-
                            metric Plates and Shells*, Prentice-
                            Hall (Englewood Cliffs,N.J.).

N.Perrone and   1959   "Strain hardening solutions to plate
P.G.Hodge,Jr.   problems", J.Appl.Mech., $\underline{26}$, 276.

W.Prager   1956   "A new method of analyzing stresses
                            and strains in work-hardening plas-
                            tic solids", J.Appl.Mech., $\underline{23}$, 493.

B.Tekinalp   1955   "Elastic, plastic bending of a simply
                            supported circular plate under a uni-
                            formly distributed load", DAM Rep.
                            C11-6, Brown University, Providence,
                            R.I.

PART III
____

LIMIT ANALYSIS AND DESIGN
____

Chapter 9

THE THEOREMS OF LIMIT ANALYSIS

9.1 Introductory Remarks

In Part II some examples of the analysis of elastic-plastic
structures have been presented, and it has been noted that the
solution of such problems can be expected to be extremely difficult except in certain special cases. These difficulties
arise from the necessity of carrying out an incremental analysis in most problems. There are clear advantages to be gained,
therefore, from exploring those aspects of an elastic-plastic
problem about which information can be obtained *without* embark-
ing on an incremental analysis. This information can be regarded as a partial solution to the problem. It may well be sufficient for the purposes for which the solution is attempted,
and in such cases a great deal of tedious analysis will have
been circumvented. It may be necessary to carry out a detailed
analytical or numerical solution of the problem, in which case
an independently obtained partial solution can be of great value
as a check on more complex analysis. Finally, a detailed exact
solution may not be possible (either through lack of time or
inability to carry out the analysis), and here the partial solutions are invaluable in constructing an approximate analysis of
the problem.

The partial solution we shall consider in this chapter is the
determination of the *limit surface for elastic, perfectly plastic structures* i.e. the determinations of loads which cause flow
in a structure composed of an elastic, perfectly plastic material. We have already noted (Chapter 3) that the limit surface
for the structure is *path independent*, and it is this feature
of the limit surface which can be considered as the element
which permits a direct determination of the limit surface. The
direct determination of the limit surface is termed the *limit*

analysis problem. In Part III we shall be primarily concerned with establishing the *upper and lower bound theorems of limit analysis* and applying these theorems to the exact and approximate solution of the limit analysis problem.

The results of Chapter 3 form the starting point for the discussion; the most important features will be reviewed briefly. We are concerned exclusively with materials which exhibit flow. There is thus associated with the material a limit function $\psi(Q_j)$ which is independent of history. The material may be elastic, perfectly plastic, in which case the initial yield function $\phi(Q_j)$ coincides with the limit surface, or it may exhibit hardening within the limit surface. The limit surface in load space (i.e. for the structure rather than for the material) is not affected by the behavior when $\psi(Q_j) < 0$, and so without loss in generality we shall assume that the material is elastic, perfectly plastic with

$$\psi(Q_j) \;=\; \phi(Q_j) \;. \tag{1}$$

This has the slight advantage that it is not necessary to distinguish between plastic strain increments which occur during flow and those that occur during hardening.

It is also notationally convenient at this point to introduce the concept of a *generalized strain rate* $\dot{q}_j$ in place of a generalized strain increment dq_j. The materials which we are considering are time independent, and time does not enter explicitly into the constitutive relations. However, because the materials are path dependent, the order in which mechanical events take place is clearly important. We can therefore introduce a parameter t *which increases monotonically with time, but at an unspecified rate.* Thus if event A occurs at 'time' t_1, and event B occurs at time t_2, and $t_2 > t_1$, we know that event B took place after event A, but the real time which elapsed

between the events is not specified. In our present context,
let the strain at some instant which we denote by t be q_j, and
let the strain be changed to $q_j + dq_j$ at some later 'time'
t + dt. Because t increases monotonically, dt > 0. We can
then define the *strain rate* at time t as

$$\dot{q}_j = \frac{dq_j}{dt} \ .$$ (2)

Because t is not explicitly defined in terms of real time, the
only information conveyed in any one strain rate component, say
$\dot{q}_1$, is whether it is positive, negative or zero. It is, however, meaningful to compare two or more strain rates, so that
if, for example, $\dot{q}_1 > \dot{q}_2$ we know that $dq_1 > dq_2$ over any increment in t. In a similar way we define stress rates $\dot{Q}_j$, load
rates $\dot{\underline{p}}$ and displacement rate $\dot{\underline{u}}$.

   The constitutive relations for an elastic, perfectly plastic
material with a continuously differentiable yield function take
the form

$$\dot{q}_j = \dot{q}_j^e + \dot{q}_j^p \ ,$$

$$\dot{q}_j^e = C_{jk}\dot{Q}_k \ ,$$

$$\dot{q}_j^p = \lambda\frac{\partial\phi}{\partial Q_j} \ ,$$ (3)

$$\lambda > 0 \quad \text{for} \quad \phi = 0 \quad \text{and} \quad \frac{\partial\phi}{\partial Q_j}dQ_j = 0 \ ,$$

$$\lambda = 0 \quad \text{for} \quad \phi < 0$$

$$\text{or} \quad \phi = 0 \quad \text{and} \quad \frac{\partial\phi}{\partial Q_j}dQ_j < 0 \ .$$

The strain rates $\dot{q}_j^e$ and $\dot{q}_j^p$ are respectively the elastic and
plastic components of the total strain rate. Equations (3) can

be returned to their incremental form by multiplying through by dt.

The yield function $\phi(Q_j)$ must be convex, and stress states such that $\phi(Q_j) > 0$ are not accessible. We also admit the possibility of a number of independent yield surfaces ϕ_1, ϕ_2, ..., ϕ_m (see Chapter 2) but it is not necessary for us explicitly to write the complete equations for this case. We note only that at a corner in the yield surface the direction of the plastic strain rate vector $\dot{q}_j^p$ is not uniquely defined, but is bounded by the adjacent normals to the yield surface.

The most important feature of the constitutive relations for the purposes of limit analysis is in fact the inequality which leads to convexity and the flow rule (Section 2.6, equation 129). If Q_j is a state of stress which is such that $\phi(Q_j) = 0$ and which has associated with it a plastic strain increment dq_j^p or a plastic strain rate $\dot{q}_j^p$, and Q_j^a is any state of stress such that $\phi(Q_j^a) \leq 0$, then

$$(Q_j - Q_j^a)\dot{q}_j^p \geq 0 \ . \tag{4}$$

As in Chapter 3, it is convenient to represent the given loads on the structure (acting on S_p) in terms of the load parameters Γ_α ($\alpha = 1$, ..., μ) and the body forces, if present, by the single parameter $\Gamma_{\mu+1}$. Thus (Section 3.2, equations 33 and 34) we have

$$\underline{p} = \Gamma_1\underline{\pi}_1(s) + \Gamma_2\underline{\pi}_2(s) + \ldots + \Gamma_\mu\underline{\pi}_\mu(s) \ ,$$
$$\underline{F} = \Gamma_{\mu+1}\underline{f} \ , \tag{5}$$

where $\underline{\pi}_\alpha(s)$ and $\underline{f}$ are dimensionless vector functions. Similarly we introduce generalized displacement rates $\dot{\xi}_\alpha$ ($\alpha = 1$, ..., $\mu + 1$) defined by (Section 3, equations 36)

$$\dot{\xi}_\alpha = \int_S \underline{\pi}_\alpha \cdot \underline{\dot{u}} \, ds \qquad \alpha = 1, \, \ldots, \, \mu \quad ,$$

$$\xi_{\mu+1} = \int_V \underline{f} \cdot \underline{\dot{u}} \, dV \quad . \tag{6}$$

In Chapter 3 the limit function in load space $\Psi(\Gamma_\alpha)$ was introduced, and we shall review briefly the most important results obtained. $\Psi(\Gamma_\alpha)$ is defined in such a way that flow of the structure will occur under constant loads when

$$\Psi(\Gamma_\alpha) = 0 \quad , \tag{7}$$

and flow will not occur when

$$\Psi(\Gamma_\alpha) < 0 \quad . \tag{8}$$

It was shown that the limit function $\Psi(\Gamma_\alpha)$ is unique, which implies that it is not a function of the path of loading in load space. Using inequality (4) and the principle of virtual work, we have for any load parameters Γ_α such that $\Psi(\Gamma_\alpha) = 0$, with associated generalized displacement increments $d\xi_\alpha$ or generalized displacements rates $\dot{\xi}_\alpha$, and any load parameters Γ_α^a such that $\Psi(\Gamma_\alpha^a) \leq 0$,

$$(\Gamma_\alpha - \Gamma_\alpha^a) \, \dot{\xi}_\alpha \geq 0 \quad . \tag{9}$$

This inequality implies convexity of the limit surface in load space, and normality of the generalized displacement rate $\dot{\xi}_\alpha$ at points on the limit surface where the outward normal is uniquely defined. It may in fact be necessary to express $\Psi(\Gamma_\alpha)$ in terms of a number of independent limit functions $\Psi_1(\Gamma_\alpha)$, $\Psi_2(\Gamma_\alpha)$, $\ldots$, but it is again not necessary to write out these functions explicitly. It can be assumed that $\Psi(\Gamma_\alpha)$ is a continuous function with the limit surface $\Psi(\Gamma_\alpha) = 0$ piecewise continuously differentiable. At a corner in the limit surface, $\dot{\xi}_\alpha$ must lie between the adjacent outward normals to the limit surface.

In the development of these results it was also shown that when flow occurs under constant loads the stress increments dQ_j or the stress rates $\dot{Q}_j$ are zero throughout the structure. This in turn implies that *the strain rates $\dot{q}_j$ which occur during flow are composed of plastic strain rates* and the elastic strain rates are zero everywhere in the structure. Throughout Part III, in which we discuss conditions during flow exclusively, a strain rate field $\dot{q}_j$ is interpreted as consisting entirely of *plastic strain rates*; we shall not use the notation $\dot{q}_j^p$ to distinguish plastic strain rates because there is no possibility of confusion.

9.2 The Theorems of Limit Analysis

The theorems of limit analysis are complementary principles which permit us to compute load parameters Γ_α which lie respectively within or outside the limit surface in load space. They are established by arguments which are very closely related to those used to show uniqueness of the limit surface in Chapter 3, and indeed uniqueness of the limit surface can be alternatively demonstrated as a consequence of the limit theorems. Like other dual principles in structural mechanics, the two theorems are obtained by comparing first the constraints imposed on the solution by equilibrium requirements and the constitutive equations, and second the constraints imposed by kinematic requirements and the constitutive equations with the complete solution, which must satisfy the requirements of equilibrium, kinematics and the constitutive relations.

Let us first suppose that for load parameters Γ_α we can find a stress field $Q_j(s)$ in the structure which satisfies all the equilibrium requirements imposed on the structure. In terms of the phrase introduced in Section 1.6 $Q_j(s)$ is a *pertinent statically admissible field* for loads Γ_α. In arriving at this field we are not concerned in any way with the strains or strain rates associated with $Q_j(s)$. We require *only* that $Q_j(s)$ should satisfy the internal equilibrium equations appropriate for the structure, and should be in external equilibrium with the loads $\hat{\underline{p}}$ defined on S_p by the load parameters Γ_α (equation 5). Further, let us suppose that we can find for load parameters Γ_α a pertinent statically admissible field $Q_j(s)$ such that at every point in the structure $\phi(Q_j) \leq 0$. Thus the stress point does not fall outside of the yield surface in stress space at any point in the structure. We shall term $Q_j(s)$ which satisfied both the requirements $\phi(Q_j) \leq 0$ and equilibrium a *safe, pertinent statically admissible field*.

The static theorem of limit analysis may now be stated.

Load parameters Γ_α for which a safe, pertinent statically admissible field can be found are such that $\Psi(\Gamma_\alpha) \leq 0$.

Alternatively, Γ_α cannot lie outside of the limit surface in load space if it is possible to associate with the loads a stress field Q_j which satisfies all equilibrium requirements and for which $\phi(Q_j) \leq 0$.

In order to prove the theorem we introduce loads Γ_α^* for which $\Psi(\Gamma_\alpha^*) = 0$. Flow may occur under these loads, with displacement rates $\dot{\xi}_\alpha^*$, stresses Q_j^* and plastic strain rates $\dot{q}_j^*$. A balance of external work rates and internal energy dissipation rates for this set gives

$$\Gamma_\alpha^* \, \dot{\xi}_\alpha^* \;=\; \int_V Q_j^* \, \dot{q}_j^* \, dV \; . \tag{10}$$

If Γ_α is a set of load parameters for which a safe pertinent statically admissible field $Q_j(s)$ is known, we can use the kinematically admissible velocity field $\dot{\xi}_\alpha^*$, $\dot{q}_j^*$ and the principle of virtual work to write

$$\Gamma_\alpha \, \dot{\xi}^*_\alpha \;=\; \int_V Q_j \, \dot{q}^*_j \, dV \quad . \tag{11}$$

Subtracting equation (11) from equation (10),

$$(\Gamma^*_\alpha - \Gamma_\alpha)\dot{\xi}^*_\alpha \;=\; \int_V (Q^*_j - Q_j) \, \dot{q}^*_j \, dV \quad . \tag{12}$$

Since $\dot{q}^*_j$ is a *plastic* strain rate $\phi(Q^*_j) = 0$ when $\dot{q}^*_j \neq 0$; further, $\phi(Q_j) \leq 0$, and hence from the fundamental inequality
(equation 4)

$$(Q^*_j - Q_j) \, \dot{q}^*_j \;\geq\; 0 \quad . \tag{13}$$

Hence

$$(\Gamma^*_\alpha - \Gamma_\alpha)\dot{\xi}^*_\alpha \;\geq\; 0 \quad . \tag{14}$$

If we now suppose that the theorem is not true and that $\Psi(\Gamma_\alpha) > 0$,
we can easily show that inequality (14) will be violated for
certain choices of Γ^*_α. This can be shown diagrammatically
(Figure 1); let Γ_α lie outside of the limit surface. Γ^*_α must
lie on the limit surface, but it can be arbitrarily chosen.
Figure 1 shows a choice of Γ^*_α for which the scalar product
$(\Gamma^*_\alpha - \Gamma_\alpha)\dot{\xi}^*_\alpha$ is negative. Thus, in order that inequality (14)
should hold for any Γ^*_α for which $\Psi(\Gamma^*_\alpha) = 0$, we must have Γ_α inside or on the limit surface, i.e., $\Psi(\Gamma_\alpha) \leq 0$, and the theorem

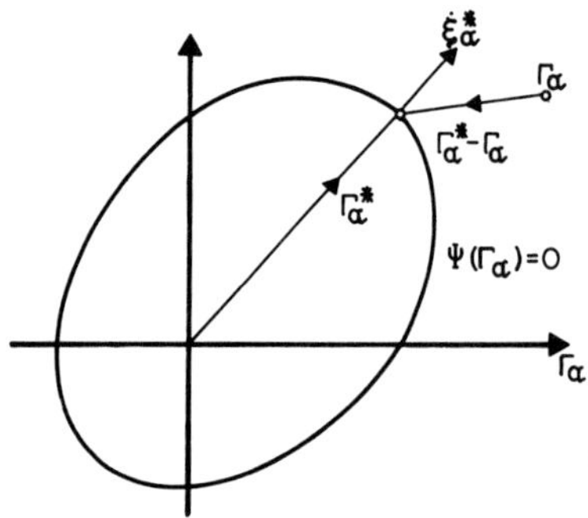

Figure 1. Proof of the static theorem

is true.

In order to formulate the second theorem we consider a set of displacement rates $\underline{\dot{u}}$ and strain rates $\dot{q}_j$ which satisfy the kinematic relations (i.e., the appropriate strain rate, displacement rate and compatibility relations) and the conditions imposed on S_u for the given problem. In the phrase introduced in Section 1.6, $\underline{\dot{u}}$, $\dot{q}_j$ constitute a pertinent kinematically admissible set. In choosing this velocity field we are not concerned with the equilibrium requirements. We require only that the strain rates should be derived from the velocities by means of the appropriate relations for the structure, and that the displacement boundary conditions should be met.

We assume that the strain rates $\dot{q}_j$ obtained from the velocity field are *plastic strain rates*, and thus that the stress rates are zero. Associated with the velocity field $\underline{\dot{u}}$, therefore, will be an internal plastic energy dissipation rate D_{int} given by

$$D_{int} = \int_V Q_j \dot{q}_j \, dV \; , \tag{15}$$

where Q_j are the stresses associated with the plastic strain rates $\dot{q}_j$ by means of the constitutive equation. We note that, given $\dot{q}_j$, the specific dissipation rate $Q_j \dot{q}_j$ is uniquely determined as a result of convexity and the flow rule. This may be seen diagrammatically in Figure 2. If $\phi(Q_j) = 0$ is strictly

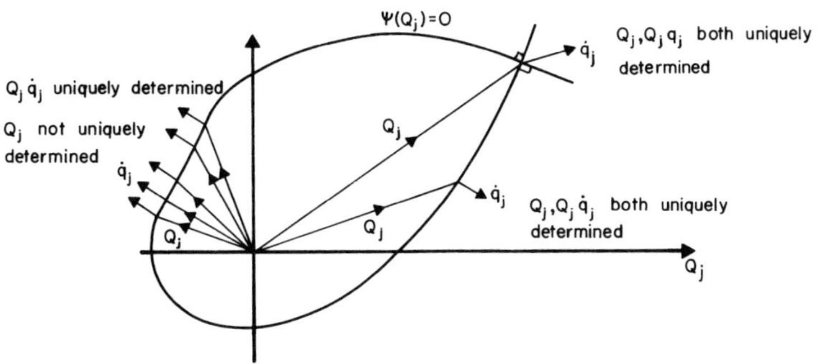

Figure 2. Determination of Q_j, $Q_j \dot{q}_j$ when $\dot{q}_j$ is given

convex (i.e. if it has a continuously turning outward normal
vector), Q_j is uniquely determined when $\dot{q}_j$ is known; there is
only one stress vector which has associated with it a plastic
strain rate of a given direction. Consequently the specific
dissipation rate $Q_j \dot{q}_j$ is uniquely determined. If the yield su -
face is not strictly convex there will be flat regions where a
number of stress states have associated with them plastic strain
rate vectors which have the same direction. Thus, given $\dot{q}_j$, Q_j
is not uniquely defined. However, the scalar product $Q_j \dot{q}_j$ *is*
uniquely determined, since the stress vectors are non-unique
only in the sense that the component *perpendicular* to $\dot{q}_j$ is not
unique; the component of Q_j parallel to $\dot{q}_j$ is uniquely determined. Finally, if $\dot{q}_j$ is zero, the specific dissipation rate
$Q_j \dot{q}_j$ is also zero. Thus the internal energy dissipation rate
D_{int} given in equation (15) can be unambiguously determined.
This is an extremely important feature of perfectly plastic materials.

Also associated with the velocity field $\dot{\underline{u}}$ are generalized
velocities $\dot{\xi}_\alpha$ defined by the vector functions $\underline{\pi}_\alpha(s)$ and $\underline{f}(s)$
which control the load distribution (equation 6). If the struc-
ture is subjected to loads defined by the load parameters Γ_α the
rate of work of the loads D_{ext} would then be

$$D_{ext} \;=\; \Gamma_\alpha \, \dot{\xi}_\alpha \; . \tag{16}$$

We can now *compute* load parameters by equating the external
work rate and the internal energy dissipation rate.

From

$$D_{ext} \;=\; D_{int} \; , \tag{17}$$

we have

$$\Gamma_\alpha \, \dot{\xi}_\alpha \;=\; \int_V Q_j \, \dot{q}_j \, dV \; . \tag{18}$$

It must be emphasized that equation (18) is a work rate balance; it does not contain the implication that Γ_α and Q_j are statically admissible, and in general they will not be. The resemblance of equation (18) to the principle of virtual velocities is superficial. Equation (18) is work rate balance and it is *solved* for Γ_α. The solution is not unique; equation (18) is in fact, a *single linear equation* in the components of Γ_α, and can be interpreted as the equation of a *hyperplane* in the load space, as shown diagrammatically in Figure 3. All load parameters Γ_α which lie on this plane will satisfy equation (18).

The kinematic theorem of limit analysis can now be stated.

Load parameters Γ_α determined by equating the external work rate and the internal energy dissipation rate associated with a pertinent, kinematically admissible field are such that

$$\Psi(\Gamma_\alpha) \geq 0.$$

Alternatively, any set of load parameters computed from a pertinent kinematically admissible field $\underline{\dot{u}}$, $\dot{q}_j$ by means of equation (18) cannot lie *inside* the limit surface.

To prove the theorem we introduce again any set of load parameters Γ_α^* such that $\Psi(\Gamma_\alpha^*) = 0$, and assume that during flow under these loads the stresses are Q_j^*. Since Γ_α^*, Q_j^* are certainly statically admissible, we may use the kinematically admissible velocity field $\dot{\xi}_\alpha$, $\dot{q}_j$ and the principle of virtual velocities to

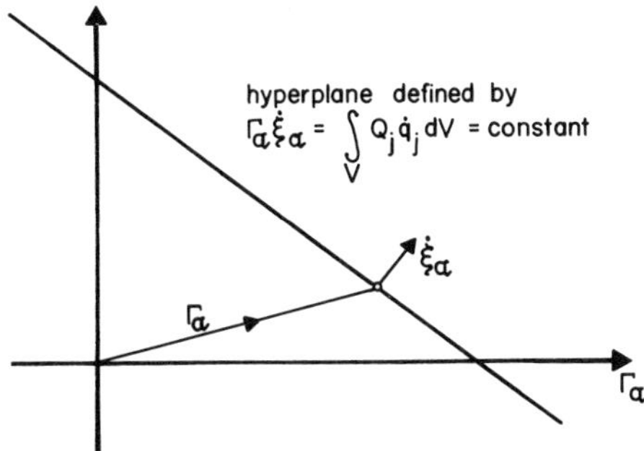

Figure 3. Hyperplane defined by equation (18)

write

$$\Gamma_\alpha^* \, \dot{\xi}_\alpha \;=\; \int_V Q_j^* \, \dot{q}_j \, dV \; . \tag{19}$$

Subtracting equation (19) from equation (18), we have

$$(\Gamma_\alpha - \Gamma_\alpha^*)\dot{\xi}_\alpha \;=\; \int (Q_j - Q_j^*)\dot{q}_j \, dV \; . \tag{20}$$

Since $\dot{q}_j$ is a plastic strain rate, $\phi(Q_j) = 0$ when $\dot{q}_j \neq 0$; further, $\phi(Q_j^*) \leq 0$ since it is associated with a structure in which
flow occurs. The fundamental inequality (equation 4) then gives

$$(Q_j - Q_j^*)\dot{q}_j \;\geq\; 0 \; . \tag{21}$$

Consequently, equation (21) gives

$$(\Gamma_\alpha - \Gamma_\alpha^*)\dot{\xi}_\alpha \;\geq\; 0 \; . \tag{22}$$

We note further that

$$\Gamma_\alpha \, \dot{\xi}_\alpha \;\geq\; 0 \; . \tag{23}$$

This follows again from the fundamental inequality (equation 4
with $Q_j^a = 0$) and equation 18.

    Now suppose the theorem is not true. The hyperplane defined
by equation (18) may then pass through the limit surface, as
shown diagrammatically in Figure 4. $\Gamma_\alpha \, \dot{\xi}_\alpha$ has the same value
for all load parameters on the hyperplane, and consequently $\dot{\xi}_\alpha$

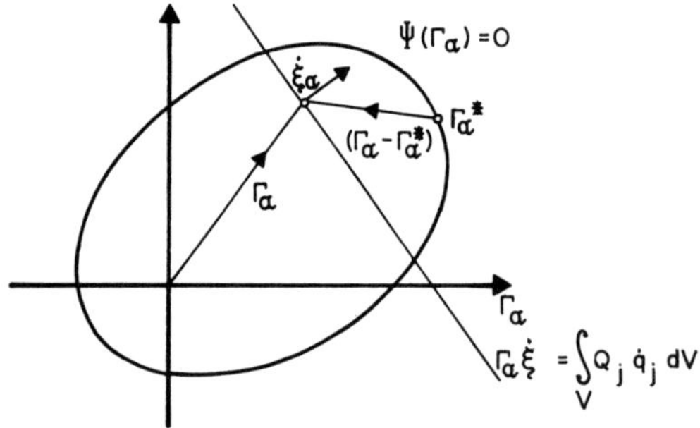

Figure 4. Proof of the kinematic theorem

is *normal to the hyperplane.* Further, from equation (23) $\dot{\xi}_\alpha$
has the normal direction away from the origin. Γ_α^* may be
chosen to lie anywhere on the limit surface; let it lie on the
side of the hyperplane remote from the origin. It is evident
then that the scalar product $(\Gamma_\alpha - \Gamma_\alpha^*)\dot{\xi}_\alpha$ is negative, which
contravenes the requirement given in inequality (22). Inequa-
lity (22) will be true for all Γ_α^* only if the hyperplane which
contains Γ_α lies outside of the limit surface or is tangential
to it, and the theorem holds.

In many problems to which limit analysis is applied we are
concerned with *proportional loading,* i.e., all the loads are
increased in proportion. The state of load may then always be
described by

$$\Gamma_\alpha = \Gamma\, C_\alpha\ , \tag{24}$$

where C_α is a fixed vector in the load space, and Γ is *non-
negative.* This is shown diagrammatically in Figure 5; Γ^* may
be defined as the value of Γ such that

$$\Psi(\Gamma^*\, C_\alpha) = 0\ , \tag{25}$$

and this is the *limit load parameter* for proportional loading.
In most cases it is convenient to arrange the generalized loads
in such a way that C_α coincides with one of the coordinate axes
in the load space. In either case, we are no longer concerned

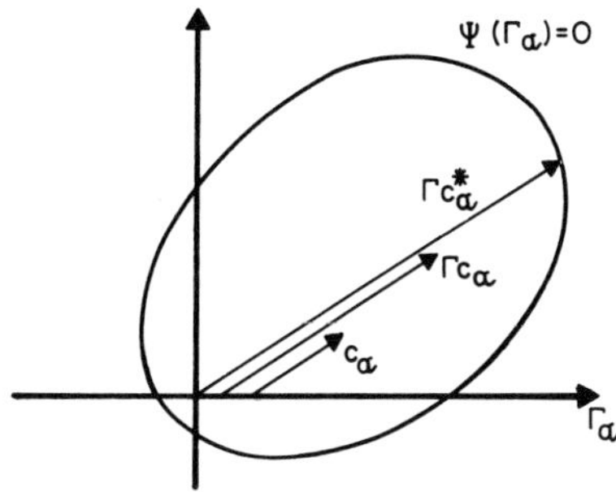

Figure 5. Proportional loading

with the entire limit surface $\Psi = 0$ but only with the intersection of the limit surface and the radial line in the load space defined by equation (24).

The limit theorems now take the form of *upper and lower bound theorems*. Since C_α in equation (24) is a fixed vector, Γ is proportional to the magnitude of the vector ΓC_α. It follows from this and the fact that the limit surface is convex and contains the origin that if

$$\Psi(\Gamma C_\alpha) \leq 0 \quad , \quad \Psi(\Gamma^* C_\alpha) = 0 \quad , \quad \Gamma \leq \Gamma^* \quad , \tag{26a}$$

and if

$$\Psi(\Gamma C_\alpha) \geq 0 \quad , \quad \Psi(\Gamma^* C_\alpha) = 0 \quad , \quad \Gamma \geq \Gamma^* \quad . \tag{26b}$$

The limit theorems can now be restated. The *lower bound theorem* takes the form:

A proportional loading parameter Γ for which a safe, pertinent statically admissible field can be found is not greater than the limit load parameter Γ^* (i.e. $\Gamma \leq \Gamma^*$).

The second theorem is again based on a pertinent kinematically admissible field. The energy dissipation relation (equation 18) is written, and equation (24) is used to substitute for Γ_α. Thus the work rate equation now gives

$$\Gamma C_\alpha \, \dot{\xi}_\alpha = \int_V Q_j \, \dot{q}_j \, dV \quad ,$$

or

$$\Gamma = \frac{\int_V Q_j \, \dot{q}_j \, dV}{C_\alpha \, \dot{\xi}_\alpha} \quad . \tag{27}$$

In this process we are simply finding the point where the radial line of equation (24) cuts the hyperplane defined by equation

(18) and shown in Figure 3.

A restatement of the second theorem, bearing in mind equation (27), gives the upper bound theorem.

A proportional loading parameter Γ determined by equating the external work rate and the internal energy dissipation rate associated with a pertinent, kinematically admissible field is not less than the limit load parameter Γ^* (i.e., $\Gamma \geq \Gamma^*$).

Both statements of the limit theorems (i.e. for both general loading and proportional loading) are given in a form appropriate for a limit analysis problem where the limit surface or limiting parameter for proportional loads is required for a given structure. Some alternative statements and corollaries of the limit theorems will be considered shortly.

The limit theorems can be used in two related ways. First, they are a tool by which the exact limit surface can be found. If a given set of loads Γ_α can be shown to be such that $\psi(\Gamma_\alpha) \leq 0$ and $\psi(\Gamma_\alpha) \geq 0$, then clearly $\psi(\Gamma_\alpha) = 0$ and we have determined a point on the limit surface. On the other hand, if the limit surface cannot be found exactly, we can attempt to find bounds on the limit surface by determining a series of points within the limit surface and a series of points outside of the limit surface. In many problems such bounds may suffice for the purposes at hand. In Chapters 10, 11, and 12 we shall treat briefly some examples of the application of the limit theorems to a number of particular examples. These examples are all comparatively simple, and are characterized by the fact that an *ad hoc* approach may be taken in each case to determine the exact limit loads or close bounds. Such approaches are certainly admissible, but they draw heavily on insight and experience in particular problems if they are to be efficient.   In Chapter 13 we shall consider briefly attempts to set up systematic (or automatic)

methods of determining the limit load by exploiting the limit
theorems for various classes of problems. In Chapter 15 we
shall touch briefly also on an important class of continuum
problems, and in particular problems in plane strain, where we
solve directly for the velocity field during flow by showing
that the governing equations are hyperbolic and making use of
the method of characteristics.

9.3 Alternative Statements of the Limit Theorems

It is essentially trivial to establish the converses of the two
limit theorems, although a slight restatement of the upper
bound theorem is necessary.

The static theorem of limit analysis states that load parameters Γ_α for which a safe, pertinent statically admissible
field can be found are such that $\Psi(\Gamma_\alpha) \geq 0$. The converse of
this theorem states that:

*Load parameters Γ_α for which no safe, pertinent statically ad-
missible field can be found are such that $\Psi(\Gamma_\alpha) > 0$.*

Alternatively, Γ_α cannot lie inside the limit surface in load
space if it is not possible to associate with the loads at least
one stress field Q_j which satisfies all the equilibrium requirements and for which $\psi(Q_j) \leq 0$ everywhere. This statement follows directly from the lower bound theorem of limit analysis.

In the statement of the kinematic theorem of limit analysis
a pertinent kinematically admissible velocity field $\dot{\xi}_\alpha$, $\dot{q}_j$ was
introduced, and load states Γ_α were defined by the equation

$$\Gamma_\alpha \dot{\xi}_\alpha = D_{ext} = D_{int} = \int_V Q_j \dot{q}_j \, dV \; . \tag{28}$$

It was then shown that these load states do not lie within the
limit surface in load space.

It is sometimes convenient to begin with a given load state

Γ_α, and ask whether $\Psi(\Gamma_\alpha)$ is positive or negative. Remembering that Γ_α is now considered to be a *fixed* set of loads, the upper bound theorem can be written as follows:

If for a load parameter Γ_α there exists at least one pertinent kinematically admissible field $\dot\xi_\alpha$, $\dot q_j$ such that $D_{ext} \geq D_{int}$, then $\Psi(\Gamma_\alpha) \geq 0$.

In order to prove this result, we note that since

$$\Gamma_\alpha \dot\xi_\alpha \; \geq \; D_{int} \tag{29}$$

for at least one pertinent kinematically admissible field, it will be possible to associate with this field a multiplier α, $0 < \alpha \leq 1$, such that

$$\alpha\Gamma_\alpha \dot\xi_\alpha \; = \; D_{int} \; . \tag{30}$$

It follows then from the kinematic theorem as stated in Section 9.2 that $\Psi(\alpha\Gamma_\alpha) \geq 0$; since Γ_α is fixed, this implies that $\Psi(\Gamma_\alpha) \geq 0$.

Having an alternative statement of the kinematic theorem, we may now give the converse.

If for a load parameter Γ_α there does not exist a pertinent kinematically admissible velocity field for which $D_{ext} > D_{int}$, then $\Psi(\Gamma_\alpha) \leq 0$.

Alternatively, if for *all* pertinent kinematically admissible fields $D_{ext} < D_{int}$, the load state Γ_α lies within or on the limit surface in load space. This statement of the converse of the limit theorem will be used in Chapter 14 in the discussion of limit design.

It has been established that the limit surface in load space,

$$\Psi(\Gamma_\alpha) \; = \; 0 \tag{31}$$

is convex, and that the flow rule

$$\dot{\xi}_\alpha \;=\; \Lambda \frac{\partial \Psi}{\partial \Gamma_\alpha} \;\;,\;\; \Lambda \geq 0 \tag{32}$$

relates the generalized displacement rates and the generalized
loads. These results can be exploited in the limit analysis
problem, particularly in situations where the limit load parameters for certain proportional loading states are known and
the entire limit surface in load space is required.

We may formally state two results which follow from the convexity of the limit surface and the flow rule.

If $\Psi(\Gamma_\alpha^a) \leq 0$, $\quad \Psi(\Gamma_\alpha^b) < 0$,

then $\Psi\{\beta\Gamma_\alpha^a + (1-\beta)\Gamma^b\} < 0$, $\quad 0 \leq \beta \leq 1.$

Alternatively, if two load states are known to lie within or
on the limit surface, all load states on the line joining these
two load states also lie within or on the limit surface. This
statement can serve as a definition of convexity, and in geometric terms the result is immediately apparent.

If flow occurs under loads Γ_α^a, *with generalized displacement
rates* $\dot{\xi}_\alpha^a$ *and an internal energy dissipation rate* D_{int}^a, *then all
load states satisfying the equation*

$$\Gamma_\alpha \dot{\xi}_\alpha^a \;=\; D_{int}^a \tag{33}$$

are such that $\Psi(\Gamma_\alpha) \geq 0.$

Alternatively, if a load state lying on the limit surface
and the associated generalized displacement rates are known,
the *tangent hyperplane* at that point can be constructed. This
tangent hyperplane cannot cut the convex limit surface, and
consequently all load states on the hyperplane are such that
$\Psi \geq 0.$

Consider a structure which may be subjected to generalized loads Γ_α. Let the limit surface in generalized stress space be given at each point in the structure by $\psi(Q_j) = 0$, where $\psi(Q_j)$ may depend on the space variables. Suppose that the limit surface in load space for the structure is $\Psi(\Gamma_\alpha) = 0$.

Consider now a second structure, identical in all respects except that the limit surface in stress space is given by $\bar\psi(Q_j) = 0$. Further, we require that at *each point* in the structure $\bar\psi(Q_j)$ is such that for any state of stress for which $\bar\psi(Q_j) \leq 0$ it is also true that $\psi(Q_j) \leq 0$. This condition can also be interpreted diagrammatically (Figure 6a). It requires that the convex surface $\psi(Q_j) = 0$ should entirely contain or be coincident with the convex surface $\bar\psi(Q_j) = 0$. Suppose that with this modified limit surface in stress space, the limit surface in load space becomes $\bar\Psi(\Gamma_\alpha) = 0$.

It then follows that for any state of load Γ_α for which $\bar\Psi(\Gamma_\alpha) \leq 0$, it is also true that $\Psi(\Gamma_\alpha) \leq 0$. Alternatively, in the load space the convex limit surface $\Psi(\Gamma_\alpha) = 0$ completely contains, or is coincident with, the limit surface $\bar\Psi(\Gamma_\alpha) = 0$, as shown in Figure 6(b). The result is a direct consequence of the lower bound theorem of limit analysis. For a load state Γ_α for which $\bar\Psi(\Gamma_\alpha) \leq 0$ it is certainly possible to find at least one safe pertinent statically admissible stress distribution

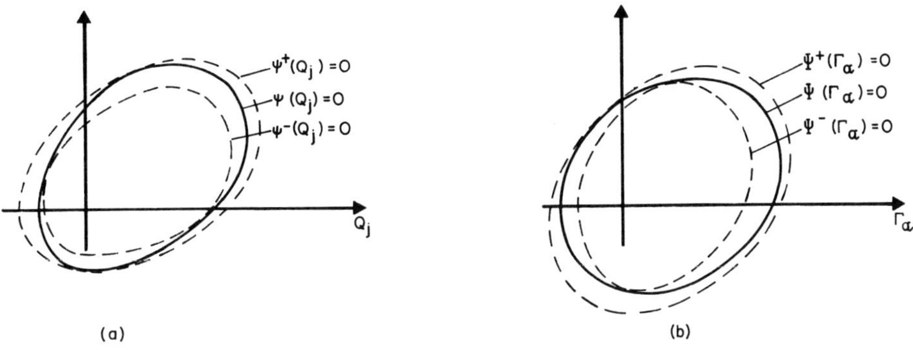

Figure 6. Circumscribed and inscribed limit surfaces

$Q_j(s)$ for which $\psi^-(Q_j) \leq 0$ at each point in the structure. However, if $\psi^-(Q_j) \leq 0$ it follows from our restriction that $\psi(Q_j) \leq 0$, and $Q_j(s)$ is also a safe pertinent statically admissible field for the original problem. Hence $\Psi(\Gamma_\alpha) \leq 0$.

A second result can now be stated without proof. Let a third structure be identical to the original problem except that the limit surface in generalized stress space $\psi^+(Q_j) = 0$ is coincident with or contains the limit surface $\psi(Q_j) = 0$. Then the associated limit surface $\Psi^+(\Gamma_\alpha) = 0$ in load space will be coincident with or contain the limit surface $\Psi(\Gamma_\alpha) = 0$. These results are shown in Figures 6(a) and 6(b).

The implications of this result when the limit surface in stress space is idealized are evident. It is preferable to choose two idealized limit surfaces, one which *inscribes* the actual limit surface and one which *circumscribes* it. If the structural limit analysis problem can be solved using each of these limit surfaces, for proportional loading $\Gamma_\alpha = \Gamma C_\alpha$ say, the two limit load parameters obtained bound the actual limit load parameter. Further, if the structural problem cannot be solved exactly a lower bound is computed using the inscribed limit surface and an upper bound is computed using the circumscribed limit surface. These two results again bound the actual limit load parameter. Finally, since it is always possible to choose the inscribing and circumscribing idealizations to have the same *shape*, in the case when the limit surface is uniform throughout the structure the process described above will provide bounds for only one computation of the limit load parameter. This follows because when a homogeneous limit surface in stress space is expanded without change in shape the limit surface in load space will also expand in the same way.

9.4 The Specific Dissipation Function

In applications of the upper bound theorem we are required to determine the specific dissipation rate $Q_j \dot{q}_j$ when the plastic strain rate $\dot{q}_j$ is given. It has been noted that the specific dissipation rate is a unique function of the strain rate $\dot{q}_j$. This function, which we shall denote by $D(\dot{q}_j)$, has several interesting properties which merit further investigation.

Let us first assume that the limit function $\psi(Q_j)$ is *strictly convex* (i.e. it has no flats) and is continuously differentiable. In order to evaluate

$$D(\dot{q}_j) \;=\; Q_j \dot{q}_j \tag{34}$$

for all $\dot{q}_j$ we must solve the equations

$$\dot{q}_j \;=\; \lambda \frac{\partial \psi}{\partial Q_j} \;, \quad \psi(Q_j) \;=\; 0 \;, \tag{35}$$

for λ and Q_j and substitute Q_j back into equation (34). This inversion cannot be carried out in general terms. Examples of the inversion will be given, but the properties of $D(\dot{q}_j)$ will first be studied.

It is convenient to plot $D(\dot{q}_j)$ in the *generalized strain rate* space. The specific dissipation rate $D(\dot{q}_j)$ is a non-negative scalar function, and we may plot *level surfaces* in the strain rate space.

We note first that $D(\dot{q}_j)$ is *homogeneous* and of *degree one* in the components of $\dot{q}_j$. This follows directly from equation (34) in that a change in the *magnitude* of the strain rate $\dot{q}_j$ does not lead to a change in the stresses Q_j associated with $\dot{q}_j$, but only a change in λ. Consequently $D(\dot{q}_j)$ depends *linearly* on the magnitude of the plastic strain rate, and vanishes when $\dot{q}_j = 0$.

Secondly, we note that $D(\dot{q}_j)$ is a potential function which provides Q_j on differentiation. From equation (34),

$$\frac{\partial D}{\partial \dot{q}_j} = Q_j + \frac{\partial Q_k}{\partial \dot{q}_j}\dot{q}_k \ . \tag{36}$$

Since Q_j is regarded as a function of $\dot{q}_j$, $(\partial Q_k/\partial \dot{q}_j)d\dot{q}_j$ is the change in stress caused by a change in the strain rate $d\dot{q}_j$. However, the change in stress must lie in the yield surface in stress space, and hence, for any fixed j, $\partial Q_k/\partial \dot{q}_j$ is a *vector* lying in the *tangent hyperplane* to the limit surface in stress space. The strain rate $\dot{q}_k$ is *normal* to this tangent hyperplane, and thus

$$\frac{\partial Q_k}{\partial \dot{q}_j}\dot{q}_k = 0 \ . \tag{37}$$

It follows then that

$$Q_j = \frac{\partial D}{\partial \dot{q}_j} \ . \tag{38}$$

Thirdly, we note that $D(\dot{q}_j)$ is a *convex function*. For two associated states $Q_j^a, \dot{q}_j^a$ and $Q_j^b, \dot{q}_j^b$ the fundamental inequality gives

$$(Q_j^b - Q_j^a)\dot{q}_j^b \geq 0 \ . \tag{39}$$

Subtracting $Q_j^a\dot{q}_j^a$ from each side of this inequality,

$$Q_j^b\dot{q}_j^b - Q_j^a\dot{q}_j^a \geq Q_j^a\dot{q}_j^b - Q_j^a\dot{q}_j^a = (\dot{q}_j^b - \dot{q}_j^a)Q_j^a \ . \tag{40}$$

Using equations (34) and (38), inequality (40) may be written as

$$D(\dot{q}_j^b) - D(\dot{q}_j^a) \geq (\dot{q}_j^b - \dot{q}_j^a)\frac{\partial D}{\partial \dot{q}_j}\bigg|_{\dot{q}_j^a} \ . \tag{41}$$

This inequality is essentially a definition of convexity.

These results may be interpreted as stating that *level sur-faces* of $D(\dot{q}_j)$ in strain rate space are similar, that the magnitude of D increases linearly with distance from the origin, that a level surface of D is a convex figure in the same sense that the limit surface in stress space is convex, and that the stress vector Q_j plotted with its origin on the level surface of D has the direction of the outward normal. This is shown diagrammatically in Figure 7.

As an example of the construction of the specific dissipation function, consider a plane stress state in which the principal stresses σ_x, σ_y are required. The principal plastic strain rates are $\dot{\varepsilon}_x$, $\dot{\varepsilon}_y$. Let the limit function be the von Mises yield function,

$$\psi = \sigma_x^2 - \sigma_x \sigma_y + \sigma_y^2 - \sigma_o^2 . \tag{42}$$

The flow rule gives

$$\dot{\varepsilon}_x = \lambda(2\sigma_x - \sigma_y) \quad , \quad \dot{\varepsilon}_y = \lambda(2\sigma_y - \sigma_x) . \tag{43}$$

Setting $\psi = 0$, we have in addition

$$\sigma_x^2 - \sigma_x \sigma_y + \sigma_y^2 = \sigma_o^2 . \tag{44}$$

Equations (43) give directly that

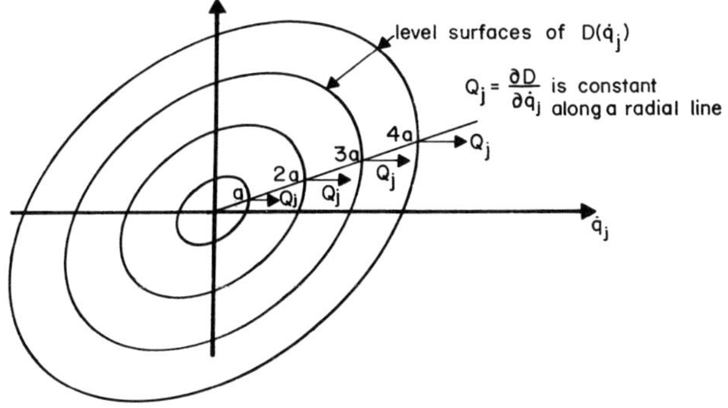

Figure 7. Level surfaces of $D(\dot{q}_j)$

$$\sigma_x \;=\; \frac{1}{3\lambda}\,(2\dot{\varepsilon}_x + \dot{\varepsilon}_y) \quad , \qquad \sigma_y \;=\; \frac{1}{3\lambda}\,(\dot{\varepsilon}_x + 2\dot{\varepsilon}_y) \;. \tag{45}$$

Substituting equations (45) into equation (44), we find that

$$\lambda^2 \;=\; \frac{1}{3\sigma_o^2}\,(\dot{\varepsilon}_x^2 + \dot{\varepsilon}_x\dot{\varepsilon}_y + \dot{\varepsilon}_y^2) \;. \tag{46}$$

Hence, from equations (45) and (46),

$$D(\dot{\varepsilon}_x,\dot{\varepsilon}_y) \;=\; \sigma_x\dot{\varepsilon}_x + \sigma_y\dot{\varepsilon}_y$$

$$\;=\; \frac{\sigma_o}{\sqrt{3}}\,\{\dot{\varepsilon}_x^2 + \dot{\varepsilon}_x\dot{\varepsilon}_y + \dot{\varepsilon}_y^2\}^{1/2} \;. \tag{47}$$

It may readily be confirmed that

$$\sigma_x \;=\; \frac{\partial D}{\partial \dot{\varepsilon}_x} \quad , \qquad \sigma_y \;=\; \frac{\partial D}{\partial \dot{\varepsilon}_y} \;. \tag{48}$$

The arguments given above can readily be extended to limit
surfaces in stress space which contain flats and corners. A
detailed proof will not be given, but two interesting features
will be noted. First, a *flat* on the surface $\psi = 0$ becomes a
corner on the level surfaces of D. It will be remembered that
for given $\dot{q}_j$ normal to a flat on $\psi = 0$, the stress Q_j is not
unique but D is uniquely determined. This implies one point on
a level surface of D where $\partial D/\partial \dot{q}_j$ is not unique. Secondly, a
corner in the limit surface becomes a *flat* on a level surface
of D. This follows because the same stress $\partial D/\partial \dot{q}_j$ is associated
with a range of strain rates, and that for D constant the com-
ponent of $\dot{q}_j$ in the direction of Q_j (or $\partial D/\partial \dot{q}_j$) is constant.
Figure 8(a) shows the Tresca yield condition for principal
stresses σ_x,σ_y in plane stress, and Figure 8(b) shows the level
surface of D in the $\dot{\varepsilon}_x,\dot{\varepsilon}_y$. Confirmation of the details is left
to the reader.

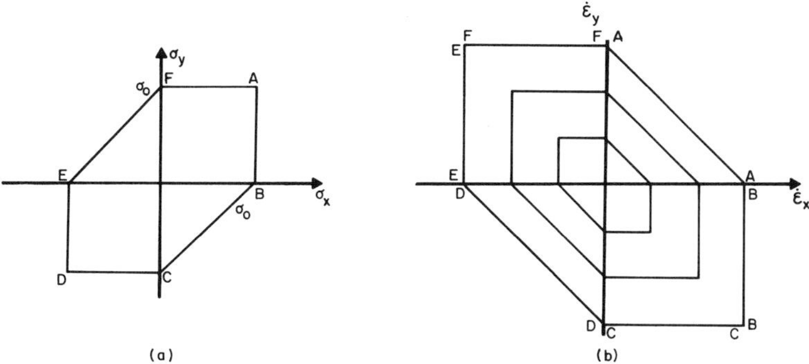

Figure 8. ψ and D surfaces for the Tresca yield condition

9.5 Historical and Bibliographical Remarks

Attempts to formulate the limit theorems precisely accompanied
attempts to formulate the limit analysis problem and to establish the uniqueness of the limit surface in load space. Early
contributions were those of Kazincky [1914] and Kist [1917];
although an account of the development of the concepts of limit
analysis in much earlier work has been given by Prager [1972].
Further progress was made by Gvozdev [1938], Feinberg [1949] and
Horne [1950]. The first complete formulation of both theorems
in terms equivalent to those given in this section were given in
the papers of Greenberg and Prager [1951] (for beams and frames)
and Drucker, Greenberg and Prager [1951], [1952] (for plane and
general continuum problems respectively). Hill [1951], [1952]
arrived at an alternative formulation which can be shown to lead
to the theorems presented in this section.

It can be pointed out again that the papers of Drucker,Green-
berg and Prager approach the limit analysis problem from the
point of view of elastic-plastic materials with the conventional
small displacement assumption, while Hill approaches it from the
point of view of rigid-plastic materials. There is no essential
difference between these approaches, as we showed in Chapter 3,
provided that we limit ourselves to problems in which the geometry of the body or structure during flow is, or is assumed to

be, known *a priori*. We shall impose this limitation throughout
this chapter.

The solution of the limit analysis problem does not of course
depend entirely on the limit theorems. Various direct methods
can be used; it is for example possible to determine conditions
under which flow occurs simply by carrying out an incremental
analysis. Early examples of the direct approach to limit analysis can be found in the volumes by van den Broek [1948], Hill
[1950] and Prager and Hodge [1951]. The importance of the limit
theorems lies in that they show us in the general case that
whether or not a given load state lies within or without the
limit surface in load space does not depend on the loading path
but on much simpler conditions.

Finally, it may be noted briefly that Melan [1938] obtained
a general theorem for shakedown in elastic, perfectly plastic
structures (see Part IV) which actually contains the first limit
theorem as a special case; this result was overlooked during the
intensive period of study which led to the formulation of the
limit theorems in the early 1950's.

References

J.A.van den Broek	1948	*Theory of Limit Design*, Wiley (N.Y.).
D.C.Drucker, H.H.Greenberg, and W.Prager	1951	"The Safety Factory of an Elastic-Plastic Body in Plane Strain", J. Appl.Mech., $\underline{18}$, 371.
D.C.Drucker, H.J.Greenberg and W.Prager	1952	"Extended Limit Design Theorems for Continuous Media", Quart.Appl.Math., $\underline{9}$, 381.
S.M.Feinberg	1948	"The Principle of Limiting Stress", Prik.Mat.Mekh., $\underline{7}$, 85.
H.J.Greenberg and W.Prager	1951	"Limit Design of Beams and Frames", Proc.ASCE, $\underline{77}$, (Separate 59).

A.A.Gvozdev 1938 "The Determination of the Value of the Collapse Load for Statically Indeterminate Systems undergoing Plastic Deformations", Akad.Nauk. (Moscow-Leningrad), 19. Translation by R.M.Haythornthwaite, Int.J.Mech. Sci., $\underline{1}$, 322, (1960).

R.Hill 1951 "On the State of Stress in a Plastic-Rigid Body at the Yield Point", Phil.Mag., $\underline{42}$.

R.Hill 1952 "A Note on Estimating the Yield Points in a Plastic-Rigid Body", Phil.Mag., $\underline{43}$ (7), 353.

R.Hill 1950 *Mathematical Theory of Plasticity,* Oxford University Press, Ch.III.

M.R.Horne 1950 "Fundamental Propositions in the Plastic Theory of Structures", J. Instn.Civil Engrs., $\underline{34}$, 174.

G.von Kazincky 1914 "Tests with Fixed End Beams", Betonszemle, Vol.2.

N.C.Kist 1917 *Inaugural Dissertation,* Technical University, Delft.

E.Melan 1938 "Zur Plastizitat des Raumlichen Kontinuums", Ing.Arch., $\underline{9}$, 116.

W.Prager 1972 "Limit Analysis: The Development of a Concept", *Foundation of Plasticity* (edited by A.Sawczuk), Noordhoff, $\underline{2}$.

W.Prager and P.G.Hodge,Jr. 1951 *Theory of Perfectly Plastic Solids,* Wiley (N.Y.) 1951. Republished by Dover (N.Y.) 1968.

Chapter 10

LIMIT ANALYSIS OF BEAMS AND FRAMES

10.1 General Features

One class of structural problems to which the limit theorems
have been applied with great success is that of beams and plane
frames loaded in the plane of the structure. These structures
are of great importance in civil engineering, and limit analysis has become accepted as a preliminary basis for design.

We consider beams and frames composed of members whose cross-
section dimensions are small compared with their length. It is
assumed that the structure lies entirely in one plane, and that
the applied loads also lie in this plane. The generalized
stresses which act on any element are bending moment M, axial
force N and shear force V. It is assumed that bending deformation predominates, and in order to simplify the analysis we
consequently assume that the generalized strains associated with
shear and axial force are zero. This assumption is justifiable
if the shear and axial forces are comparatively small in the
sense that whenever flow occurs at a section the bending moment
M is close to the limit moment for the case $V = N = 0$. The
assumption is implemented by adopting a limit function ψ which
is independent of V and N; the derivatives of ψ with respect to
N and V are then always zero. For simplicity we shall assume
that the limit bending moment has the same magnitude for positive and negative moment, and, since we are concerned only with
flow, that the material is rigid, perfectly plastic.

The limit function can then be replaced by a limit condition
on the bending moment M,

$$- M_o \leq M \leq M_o \qquad \text{or} \qquad |M| \leq M_o \ . \tag{1}$$

The curvature rate $\dot{\kappa}$ during flow is then given by

$$\dot{\kappa} \;=\; \lambda \,\langle M\rangle \;, \tag{2}$$

where

$$\langle M\rangle \;=\; +1 \quad\text{when}\quad M = +M_{o} \;,$$
$$\langle M\rangle \;=\; \;\;0 \quad\text{when}\quad -M_{o} < M < M_{o} \;, \tag{3}$$
$$\langle M\rangle \;=\; -1 \quad\text{when}\quad M = M_{o} \;,$$

and λ is a non-negative but otherwise unspecified scalar. $\dot{\kappa}$ is of course the plastic curvature rate, but since the elastic curvature rates are by definition always zero the superscript can be dropped without ambiguity.

The application of the theorems of limit analysis to beams and frames under these assumptions is particularly simple because the choice of possible safe statically admissible bending moment distributions $M(s)$ and of kinematically admissible curvature rate distributions can be restricted.   Let us consider each of these factors in turn.

The bending moment distribution $M(s)$ is of course a linear function of the generalized loads Γ_{α} which are applied to the structure.   It is also a linear function of a discrete number of additional parameters which we may refer to as the *redundant forces*.   The redundant forces cannot be determined by statical considerations alone, but their number is limited and equal to the degree of indeterminacy of the problem.   In general the bending moment distribution can be written as

$$M(s) \;=\; M^{*}(s) \;+\; \sum_{j=1}^{m} X_{j}M_{j}(s) \;. \tag{4}$$

The moment $M^{*}(s)$ is any pertinent statically admissible bending moment distribution, and is a linear function of the generalized loads Γ_{α}. m is the degree of indeterminacy, and X_{j} are the *redundant forces*. $M_{j}(s)$ are linearly independent self-equilibrating moment distributions (i.e. they are each in equilibrium for zero loads on S_p, or in equilibrium for $\Gamma_\alpha = 0$) and are independent of the generalized loads.

The lower bound theorem can thus be stated as follows: any generalized loads Γ_α for which we can find a set of redundant forces X_j such that $-M_o \leq M(s) \leq M_o$ lies within or on the limit surface in load space. The total number of parameters which must be handled in any application of the lower bound theorem is the sum of the number of load parameters and the number of redundant forces; this is a comparatively small number, and considerably simplifies the problem of constructing the limit surface.

The application of the upper bound theorem is also simplified by the observation that plastic deformation during flow is generally confined to *plastic hinges*. This localization of the plastic deformation was discussed in Chapter 6 and occurs because the bending moment distribution, which must have the form given in equation (4), is generally such that $M = \pm M_o$ only at a discrete number of cross-sections, with $|M| < M_o$ at adjacent cross-sections. If the curvature rate were finite at these discrete points flow could not occur. We are thus led to the conclusion that the curvature rate is infinite at these points, and that there is a discontinuity or jump in the gradient of the velocity field. Such discontinuities are termed *plastic hinges,* and the velocity field consists of a number of regions within which the gradient is constant. These regions are rigid during flow, and the velocity field is referred to as a *mechanism* by virtue of its analogy with the motion of a mechanism composed of rigid bars connected by hinges. The position of the positions of peak moments may all be known *a priori*; this will occur when the peak values of $M(s)$ are independent of the redundant forces. More generally the positions of all peak moments may not be known **exactly**, or variations in $M_o(s)$ may mean that plastic deformation does not necessarily occur at positions of peak moment. In any event the class of possible velocity fields which must be investigated in applying the upper bound theorem is considerably restricted: it consists of all one degree of freedom mechanisms which can be constructed from the possible locations of the plastic hinges. It may of course happen that the discontinuity in the gradient of the velocity field at a position of peak moment is zero. This would occur if the peak moment has an absolute value less than M_o. Thus the rotation rate of any potential plastic hinge may be zero.

The sign of the discontinuity in slope or the rotation rate of a plastic hinge is governed by the sign of the bending moment. In the application of the upper bound theorem we are concerned with the dissipation rate at a plastic hinge, and this is always positive. Thus if the rotation rate at the k-th hinge is given by $\dot{\theta}_k$, the dissipation rate is $M_o|\dot{\theta}_k|$. In applying the upper bound theorem we construct a velocity field or mechanism with generalized displacements rates $\dot{\xi}_\alpha$ and hinge rotation rates $\dot{\theta}_k$ (k=1, ..., ℓ) at the potential hinge positions. The upper bound theorem states that generalized loads Γ_α which satisfy the equation

$$\Gamma_\alpha \dot{\xi}_\alpha = \sum_{k=1}^{\ell} M_o|\dot{\theta}_k| \tag{5}$$

do not be within the limit surface in load space.

10.2 Examples of the Limit Analysis of Beams and Frames

Let us now consider a number of examples of the application of the limit theorems to beams and frames in which bending predominates. We shall first treat some specific cases with the object of illustrating certain facets of the general problem. Following this, we shall consider briefly the foundations of one

method of systematically determining the limit load. Such
systematic methods are important in dealing with complex beam
and frame problems.

Our first observation is that in *statically determinate* problems the application of the lower bound theorem becomes extremely simple. This occurs because, by definition, the generalized stresses in a statically determinate problem are determined by statical or equilibrium considerations alone. Thus
in equation (4) the redundant forces X_j (which express our inability to compute the moment distribution $M(s)$ by statics)
vanish; $M(s)$ is linear and homogeneous in the generalized loads
Γ_α. This in turn implies that for any *fixed* ratio of the generalized loads, or for proportional loading,

$$\Gamma_\alpha = \Gamma c_\alpha \, , \tag{6}$$

the bending moment at any point is a linear and homogeneous
function of a single parameter Γ. The position or cross-
section at which $M(s)$ has its greatest or least value, say $s=s^*$,
is *independent* of Γ. Thus those values of Γ for which

$$-M_o \leq M(s^*) \leq M_o \tag{7}$$

indicate load states which lie within or on the limit surface
in load space. Alternatively, points lying on the limit surface in load space can be obtained directly from the equations

$$M(s^*) = \pm M_o \, . \tag{8}$$

As an example of a statically determinate structure consider
the uniform simply supported beam shown in Figure 1(a). The
beam is subjected to two point loads P_1 and P_2, and our object
is to find the limit surface in the P_1, P_2 plane. The reactions
at the ends of the beam A and D can readily be computed (Figure
1a), and the bending moment diagram is shown in Figure 1(b).

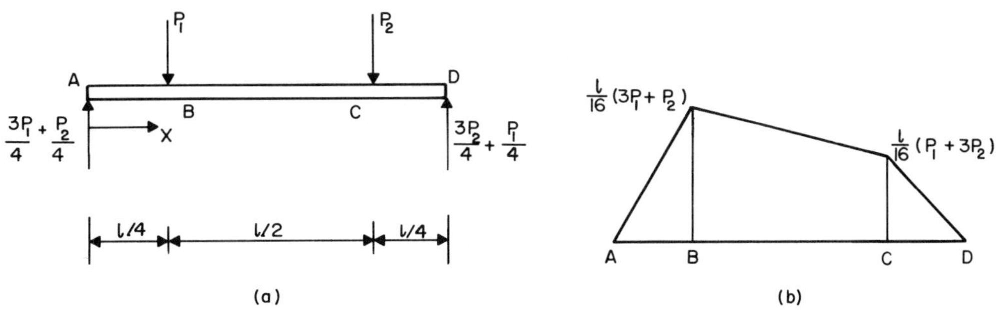

Figure 1. Simply supported beam

Letting x be the distance along the beam measured from A, it is evident that the point where the bending moment has its greatest magnitude is either B or C. Since the beam is uniform (i.e. M_o is independent of X), flow must occur when

either $M(\frac{\ell}{4})$ = M_o

$$\text{for}\quad M(\tfrac{\ell}{4}) \geq M(\tfrac{3\ell}{4})\quad , \tag{8a}$$

or $M(\frac{3\ell}{4})$ = $-M_o$

and when

either $M(\frac{\ell}{4})$ = $-M_o$

$$\text{for}\quad M(\tfrac{\ell}{4}) \leq M(\tfrac{3\ell}{4})\quad . \tag{8b}$$

or $M(\frac{3\ell}{4})$ = M_o

If we now substitute the values of $M(\ell/4)$ and $M(3\ell/4)$ into equations (8) we obtain piecewise equations for the limit surface. Since

$$M(\tfrac{\ell}{4}) \;=\; \tfrac{\ell}{16}\,(3P_1 + P_2)\quad , \qquad \text{and} \tag{9a}$$

$$M(\tfrac{3\ell}{4}) \;=\; \tfrac{\ell}{16}\,(P_1 + 3P_2)\quad , \tag{9b}$$

equations (8) become

either $3P_1 + P_2 = + \dfrac{16M_o}{\ell}$

$$\text{for } P_1 \geq P_2 \,, \qquad (10a)$$

or $P_1 + 3P_2 = - \dfrac{16M_o}{\ell}$

and

either $3P_1 + P_2 = - \dfrac{16M_o}{\ell}$

$$\text{for } P_1 \leq P_2 \,. \qquad (10b)$$

or $P_1 + 3P_2 = + \dfrac{16M_o}{\ell}$

These are the conditions under which flow may occur. Plotting the equations in the P_1, P_2 plane (Figure 2) provides the limit surface in load space. Note that $P_1 \geq P_2$ requires that the load point lie on the fourth quadrant side of the dotted line, and that $P_1 \leq P_2$ requires that the load point lie on the second quadrant side of the same line.

We can confirm that the surface shown in Figure 2 is indeed the limit surface by showing that for each load point on the surface there exists a velocity field such that the internal dissipation rate is equal to the external dissipation rate. It can be noted that, since the beam is uniform and since the moment diagram has its maximum or minimum values at B and C (Figure 1b), plastic hinges can only form at points B and C. Let

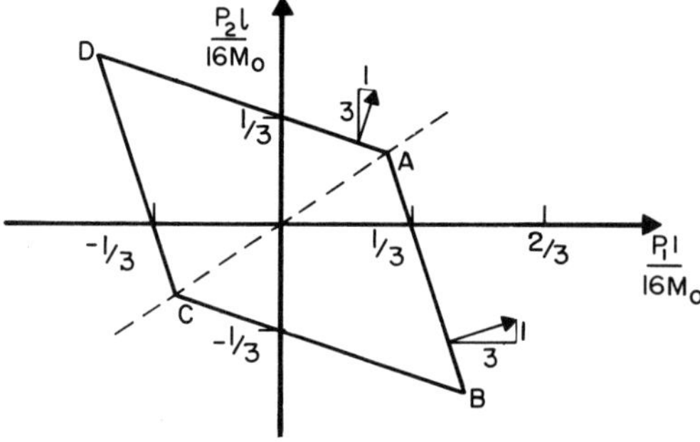

Figure 2. Limit surface for simple beam

us consider two possibilities: that hinges form at either B or
C and that the velocity is positive (i.e. downwards) everywhere
along the beam. The two velocity fields are shown in Figures
3(a) and 3(b) respectively. The generalized displacement rates
appropriate for this problem are the vertical velocities of
points B and C, say $\dot{u}_1$ and $\dot{u}_2$ respectively. It is a simple
geometric problem to compute the hinge rotation rates for the
two velocity fields; the results are shown in Figure 3. Equating internal and external work rates (in the form of equation
5), we obtain for the velocity field or mechanism of Figure 3
(a);

$$P_1\dot{u}_1 + P_2\dot{u}_2 = M_o|\dot{\theta}_1| \quad , \quad \text{or}$$

$$3P_1 + P_2 = \frac{16M_o}{\ell} \quad .$$

(11)

This equation is the first of equations (10a), and confirms that
side AB of the surface in Figure 2 is part of the limit surface.
It may be noted that the generalized displacement rates for this
velocity field $\dot{u}_1, \dot{u}_2$ bear to each other the ratio 3:1. This is
predicted by the direction of the outward normal to the limit
surface on the side AB.

Similarly, for the mechanism of Figure 3(b) we obtain

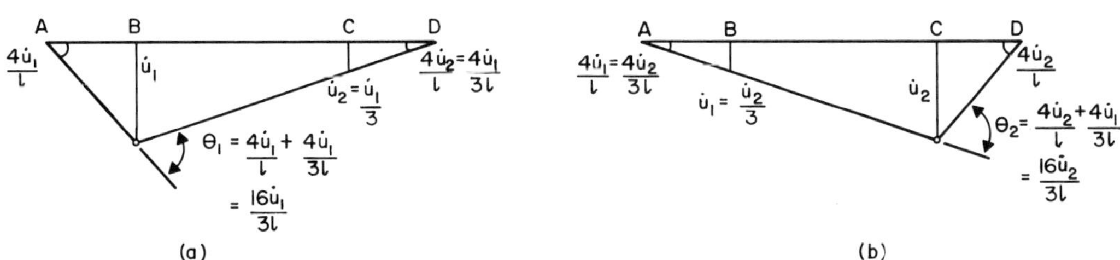

Figure 3. Velocity fields for simple beam

$$P_1 + 3P_2 = \frac{16M_o}{\ell} \, . \tag{12}$$

This equation is the same as the second of equations (10b), and confirms that side DA of the surface in Figure 2 is part of the limit surface. Note also that the ratio of $\dot{u}_1$ to $\dot{u}_2$ is 1:3 in Figure 3(b), and this is evident in the direction of the normal vector on side DA.

Equations (11) and (12) intersect at A, and for flow with $P_1 = P_2 = 4M_o/\ell$ *the velocity field is not unique.* Hinges appear at both B and C in the beam, as shown in Figure 4, and provided that $\dot{\theta}_1$ and $\dot{\theta}_2$ are each non-negative, the relative magnitudes of $\dot{\theta}_1$ and $\dot{\theta}_2$ cannot be predicted. This non-uniqueness is expressed in the non-unique direction of the normal vector at A on the limit surface. The adjacent normal vectors provide the constraint that $\dot{\theta}_1$ and $\dot{\theta}_2$ are each non-negative. It should be emphasized that the work rate balance equation

$$P_1 \dot{u}_1 + P_2 \dot{u}_2 = M_o |\dot{\theta}_1| + M_o |\dot{\theta}_2| \tag{13}$$

is not affected by this non-uniqueness. Geometrical considerations in Figure 4 show that

$$\dot{\theta}_1 = \frac{4\dot{u}_1}{\ell} + \frac{2(\dot{u}_1 - \dot{u}_2)}{\ell} \, , \tag{14a}$$

$$\dot{\theta}_2 = \frac{4\dot{u}_2}{\ell} - \frac{2(\dot{u}_1 - \dot{u}_2)}{\ell} \, . \tag{14b}$$

Putting $P_1 = P_2$ and substituting from equations (14), equation (13) gives

$$P_1(\dot{u}_1 + \dot{u}_2) = M_o\{|\dot{\theta}_1| + |\dot{\theta}_2|\} = \frac{4M_o(\dot{u}_1 + \dot{u}_2)}{\ell} \, , \quad \text{or}$$

$$P_1 = P_2 = \frac{4M_o}{\ell} \, . \tag{15}$$

The upper bound theorem can be employed to show that sides DC and CB are also part of the limit surface. Non-unique velocity fields occur for flow when $P_1 = P_2$ or $P_1 = -P_2$, i.e., at corners A,B,C and D of the limit surface. As we have pointed out, however, upper bound calculations are not essential for statically determinate structures, provided of course that the cross-section at which the bending moment reaches it maximum value can be predicted with confidence.

In statically indeterminate beam and frame structures it is frequently easier to start the limit analysis problem by first applying the upper bound theorem. As a second example consider the uniform fixed end beam subjected to a single point load at a third point, as shown in Figure 5(a). It is evident from inspection that peak bending moments occur at the built-in ends and under the load; hence the potential plastic hinge positions are limited to sections A,B and C. It is then evident that only

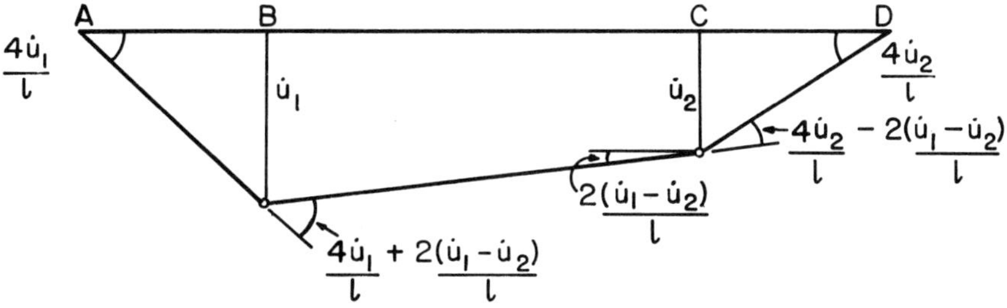

Figure 4. Flow in simple beam with two hinges

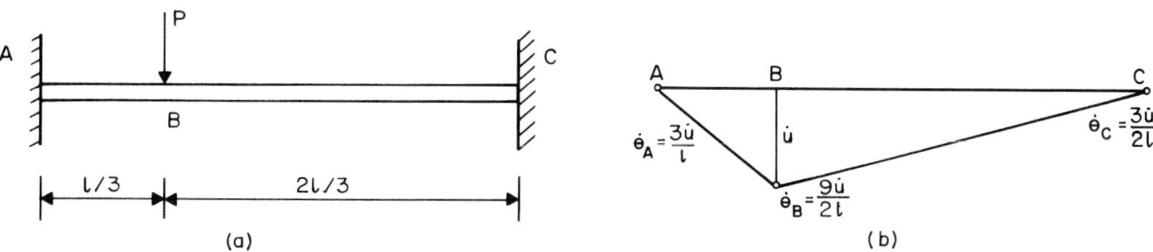

Figure 5. Fixed end beam

one possible velocity field can be constructed; it is that
shown in Figure 5(b). The hinge rotation rates are obtained in
terms of the generalized displacement $\dot{u}$ by geometry, and are
shown in Figure 5(b). Equating the internal and external work
rate we see that

$$P\dot{u} \;=\; M_o\left|\dot{\theta}_A\right| + M_o\left|\dot{\theta}_B\right| + M_o\left|\dot{\theta}_C\right| \tag{16}$$

$$\;=\; M_o\left(\frac{3\dot{u}}{\ell} + \frac{9\dot{u}}{2\ell} + \frac{3\dot{u}}{2\ell}\right) = \frac{9M_o\dot{u}}{\ell} \;.$$

Eliminating $\dot{u}$,

$$P \;=\; \frac{9M_o}{\ell} \tag{17}$$

is certainly not less than the limit load.

   If we now draw the free body diagrams for the rigid segments
AB and BC of the beam (Figure 6), we can immediately draw in
the terminal moments. Since the bending moment must vary lin-
early between the terminal values, it is clear that $\left|M(x)\right| \leq M_o$
at all points. Further, we can compute the terminal shear forces
(equal and opposite in each unloaded segment) required to maintain each segment in equilibrium. If we now rejoin the segments
the unbalanced force $P = 9M_o/\ell$ cannot be greater than the limit
load, and the analysis is complete.

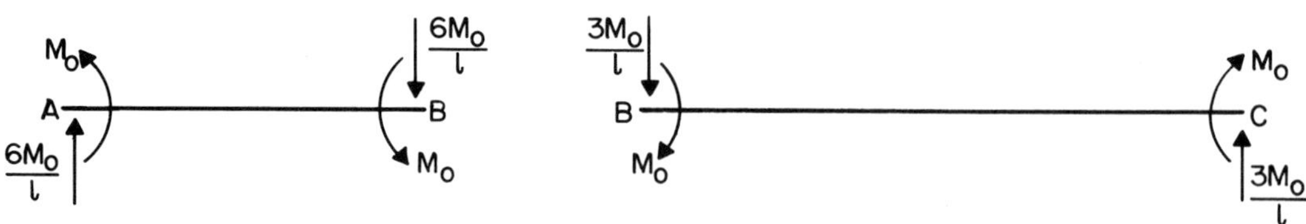

Figure 6.Moment field for fixed beam

Consider now the example shown in Figure 7(a). A uniform fixed end beam is subjected to a uniformly distributed load p per unit length along half of its length. It is clear that there will be three moment peaks, at A, C and somewhere in the region AB. These will be the positions of potential hinges. Since only one mechanism or velocity field can be constructed with three hinges, let us assume that the third hinge falls a distance $\rho\ell$ from A. The velocity field is shown in Figure 7(b) with the hinge rotation rates marked. For convenience, we assume that the velocity at the point where the third hinge forms is unity. A work rate balance gives

$$\int_{0}^{\ell/2} p\dot{u}(x)dx \;=\; M_0|\dot{\theta}_A| + M_0|\dot{\theta}| + M_0|\dot{\theta}_C| \quad . \tag{18}$$

We find that

$$\int_{0}^{\ell/2} p\dot{u}(x)dx \;=\; \int_{0}^{\rho\ell} p\left(\frac{x}{\rho\ell}\right)dx + \int_{\rho\ell}^{\ell/2} p\left\{1 - \frac{(x-\rho\ell)}{(1-\rho)\ell}\right\}dx \tag{19}$$

$$= \;\frac{p\ell}{8(1-\rho)}(3-4\rho) \quad .$$

Further

$$M_0|\dot{\theta}_A| + M_0|\dot{\theta}| + M_0|\dot{\theta}_C| \;=\; M_0\left\{2\left(\frac{1}{\rho\ell}\right) + 2\left(\frac{1}{(1-\rho)\ell}\right)\right\} \tag{20}$$

$$= \;\frac{2M_0}{\ell}\left\{\frac{1}{\rho(1-\rho)}\right\} \quad .$$

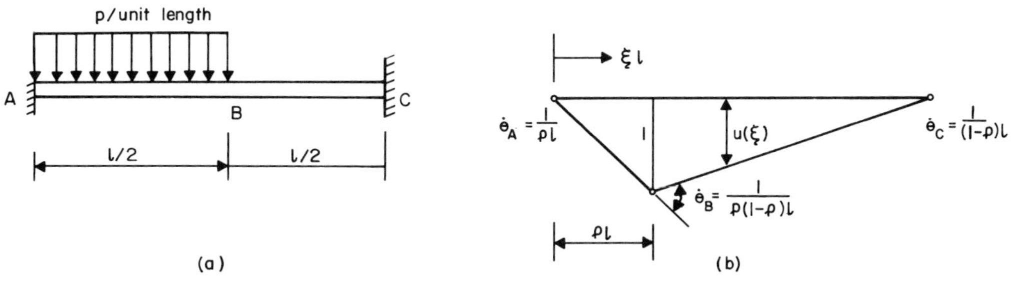

Figure 7. Fixed beam with distributed load

Equating (19) and (20) it is seen that

$$p = \frac{16}{\rho(3-4\rho)} \frac{M_o}{\ell^2} \quad . \tag{21}$$

The values of p given by equation (21) cannot be less than the limit value. The best estimate of the limit load obtainable from this mechanism is readily shown to occur when $\rho = 3/8$. This value is

$$p = \frac{256}{9} \frac{M_o}{\ell^2} \quad . \tag{22}$$

While it is expected that this is the limit load, a check must be carried out to see whether a uniform load of $(256M_o/9\ell^2)$ per unit length can be equilibrated with $|M(x)| \leq M_o$. The check can be most readily carried out by the same method used in the previous example, and is left as an exercise for the reader.

In these two fixed end beam examples the application of the upper bound theorem has been particularly simple because there is in effect only one possible mechanism or velocity field. Let us progress to a slightly more complex problem. Consider the frame shown in Figure 8(a), which is subjected to a horizontal load H and a vertical load V. Let us suppose that the cross-section is uniform for the entire frame, and let us determine the limit surface in the first quadrant of the H-V plane.

The potential hinge positions occur at the ends of members and under the vertical load; they are shown in Figure 8(a) by heavy dots at points A, B, C, D and E. It should be noted that in general a hinge at the intersection of two members of uniform but unequal section will occur in the weaker member. However, since we have assumed that the entire frame is uniform, it does not matter whether the hinge forms in the beam or the column at points B and D.

We now attempt to construct all the possible one degree of freedom mechanisms which will be relevant for H, V both positive. Two such mechanisms, the beam flow mechanism shown in Figure 8(b) and the sway flow mechanism shown in Figure 8(c), are readily apparent. Not quite so obvious is the third mechanism shown in Figure 8(d) in which no plastic deformation takes place in the hinge at B. We shall return to discuss ways in which this mechanism can be recognised after completing the analysis.

The magnitudes of the hinge rotation rates and the displacement rates u, v in the lines of action of H and V respectively (i.e. the generalized velocities) are shown in Figure 8. Each field is expressed in terms of a single parameter, in these cases a rotation rate. Equating external and internal work rates, we obtain for the three velocity fields respectively:

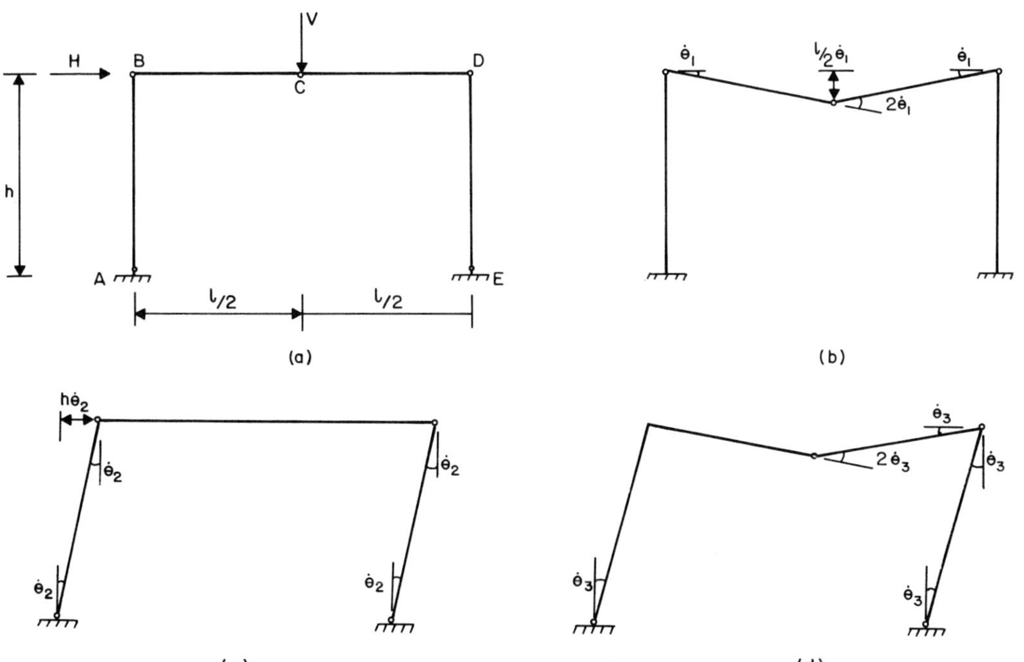

Figure 8. Rectangular portal frame

$$\frac{V\ell}{2}\,\dot{\theta}_1 = 4M_o\dot{\theta}_1 \qquad \text{or} \qquad \frac{V\ell}{M_o} = 8 \quad , \tag{23a}$$

$$Hh\dot{\theta}_2 = 4M_o\dot{\theta}_2 \qquad \text{or} \qquad \frac{Hh}{M_o} = 4 \quad , \tag{23b}$$

$$\frac{V\ell}{2}\,\dot{\theta}_3 + Hh\dot{\theta}_3 = 6M_o\dot{\theta}_3 \quad \text{or} \quad \frac{V\ell}{M_o} + 2\frac{Hh}{M_o} = 12 \quad . \tag{23c}$$

We may now plot these three equations in the first quadrant of
the H-V plane, as shown in Figure 9. The upper bound theorem
assures us that the *inner envelope* of these three lines does not
lie within the limit surface. If we have correctly identified
the only feasible velocity fields this inner envelope will be
the limit surface. We now proceed to check whether this is true
by applying the lower bound theorem.

Let us consider in detail the mechanism shown in Figure 8(d).
If flow occurs in this mechanism the moments at points A, C, D
and E are fixed by virtue of the fact that plastic flow occurs
at these cross-sections, and the moments have magnitude M_o. We
draw the free body diagrams for the rigid segments AB, BC, CD
and DE (Figure 10) and start by drawing the moments at points A,
C, D and E. The *signs* of these moments must agree with the
signs of the hinge rotation rates. The moment at B is not known;
let it be M_B with the sign arbitrarily chosen. We can now compute the equal and opposite shear forces acting at the ends of

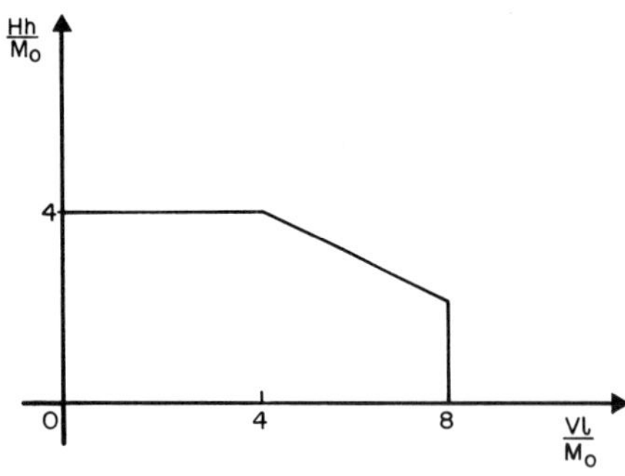

Figure 9. Load space for frame

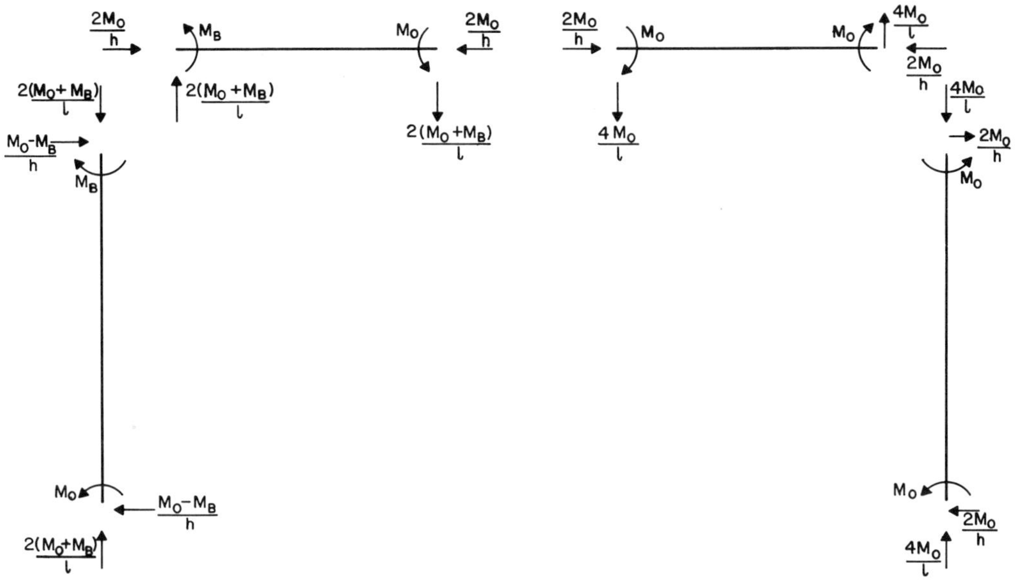

Figure 10. Moments for portal frame

each segment. Joint equilibrium then allows us to find the
axial forces acting in each segment. We now have an equilibrated generalized stress field (the moments varying linearly and
the shear and axial forces constant along each segment). The
unbalanced horizontal and vertical forces at B and C respectively give H and V;

$$H = \frac{3M_o}{h} - \frac{M_B}{h} , \tag{24a}$$

$$V = \frac{6M_o}{\ell} + \frac{2M_B}{\ell} . \tag{24b}$$

These loads will lie inside or on the limit surface in load
space provided that $|M(s)| \leq M_o$ at all points on the frame. This
condition will be met in this case if

$$- M_o \leq M_B \leq M_o \tag{25}$$

Note first that equations (24) are expressed in terms of the parameter M_B. Eliminating M_B from the two equations we see that

$$\frac{V\ell}{M_o} + 2\frac{Hh}{M_o} = 12 \; . \tag{26}$$

This equation is identical to equation (24b). It indicates that at least part of this line must coincide with the limit surface. The limits imposed on M_B (equation 25) provide the additional information. At one extreme $M_B = M_o$; substituting this into equation (24a) we see that

$$\frac{V\ell}{M_o} = 8 \; . \tag{27a}$$

The other extreme is $M_B = -M_o$; substituting this into equation (24b) we find

$$\frac{Hh}{M_o} = 4. \tag{27b}$$

These are precisely the limits which appear on equation (23c) or (26) in Figure 9. This portion of the envelope thus coincides with the limit surface. Similar computations will confirm that the remainder of the inner envelope obtained by the application of the upper bound theorem also coincides with the limit surface.

Two observations can simplify the execution of the process which has just been described. First, let us consider the three mechanisms of Figure 8. Notice that, taking into account the signs of the hinge rotations and putting $\dot{\theta}_1 = \dot{\theta}_2$, the mechanism or velocity field of Figure 8(a) is simply the sum of the velocity fields of mechanisms of Figures 8(b) and 8(c). When the two are added the equal and opposite hinge rotations at B ($\dot{\theta}_1 = \dot{\theta}_2$) cancel each other, and that hinge disappears. The hinge rotations at the other hinges in the mechanism of Figure 8(a) are

the sums of the appropriate hinge rotations in the mechanisms of
8(b) and 8(c) if we put $\dot{\theta}_3 = \dot{\theta}_2 = \dot{\theta}_1$. More generally, any one
mechanism in Figure 8 can be generated as a suitable linear com-
bination of the other two (involving the disappearance of one or
more hinges). This suggests that, at least for simple frames,
it is possible to generate all the one degree of freedom mechanisms from a set of simply derived basic mechanisms in a systematic way. We shall return to this point shortly.

The second observation is that the relation given in equation
(26) is an inevitable consequence of the assumption that the
bending moment M has the magnitude M_o and the sign dictated by
the sign of the hinge rotation at points A, C, D and E where
plastic deformation takes place. This occurs because *under
these circumstances* the work rate balance equation which is used
in the application of the upper bound theorem can be interpreted
as a statement of the principle of virtual velocities. Reconsider briefly the steps in the lower bound calculation to confirm that the mechanism of Figure 8(d) provides part of the
limit surface. Our assumption is that a statically admissible
bending moment distribution, $M^c(s)$ say, can be found and that
$M^c(s) = \pm M_o$ at points A, C, D and E with the sign controlled by
the sign of the hinge rotation. Now apply the principle of virtual velocities to the statically admissible set H, V and $M^c(s)$
and the velocity field of Figure 8(d). Since the generalized
strain rates are zero at all sections other than A, C, D and E
and since we have assumed that $M^c(s)$ has the magnitude M_o and
the same sign as the hinge rotation, the principle of virtual
velocities gives equation (23c) i.e., it is identical to the
work rate balance. Our only object in carrying out the statical
analysis is to find the conditions under which the unspecified
bending moments (in this case only M_B) do not violate the yield

condition. M_B can be found directly by another application of
the principle of virtual velocities, this time using the velocity fields of either Figures 8(b) or 8(c). Choosing that of
Figure 8(b) first, we see that $M^c(s)$ is known at the three
points where the plastic strain rate is non-zero, Hence

$$\frac{V\ell\dot{\theta}_1}{2} = M_B\dot{\theta}_1 + M_o(2\dot{\theta}_1) + M_o(\dot{\theta}_1)$$

(28a)

$$\text{or} \quad \frac{V\ell}{M_o} = 6 + \frac{2M_B}{M_o}$$

This is identical to equation (24b). In this case the signs of
the moments and hinge rotations are such that the terms on the
right hand side are positive. This is not true at B when we
use the mechanism of Figure 8(c), since the rotation rates at B
in the mechanisms Figures 8(b) and 8(c) have opposite signs. The
other terms will have positive signs, however; this can be seen
by noting that the other hinge rotation rates in the mechanism
of Figure 8(d), which fixes the sign of the moment, and the mechanism Figure 8(d) are the same. Hence using the mechanism of
Figure 8(c) in the principle of virtual velocities with the
statically admissible set N, V, $M^c(s)$,

$$Hh\dot{\theta}_2 = M_o\dot{\theta}_2 - M_B\dot{\theta}_2 + M_o\dot{\theta}_2 + M_o\dot{\theta}_2$$

(28b)

$$\text{or} \quad \frac{Hh}{M_o} = 3 - \frac{M_B}{M_o}$$

This equation is identical to equation (24a).

We see now that by starting with the easily recognizable mechanism or velocity fields of Figure 8(b) and 8(c) it is possible to proceed with the determination of the limit surface
without any further physical insight; the application of both

the upper and the lower bound theorems follows more or less automatically. This suggests that it should be possible to set up for more complex frames systematic methods of determining limit loads. One such systematic approach is known as the method of *combination of mechanisms*; we shall describe it briefly in concluding this Chapter.

10.3 The Method of Combination of Mechanisms

Let us limit ourselves, for illustrative purposes, to multibay, multistorey frames which are made up of (vertical) columns and (horizontal) beams. Such structures are encountered in the study of steel framed buildings. Let us further assume that the members are uniform between joints or beam-column intersections. The joints are assumed to be sufficiently strong that the limit moment in any member can be developed. To a first approximation let the floor loads be represented by point loads acting at the center of each beam and wind loads by point loads acting horizontally at the level of the beams. It is very seldom that the complete limit surface in generalized load space is required for such problems; generally we limit ourselves to certain combinations of loads which are assumed to act proportionally. The limit load factor is required for each set of proportional loads.

The success of the method of combination of mechanisms in these problems is founded on the easy recognition of the basic mechanism. The basic mechanisms are of three types, two of which we have encountered in the previous example. First,there are the *beam mechanisms*, in which each beam is assumed to flow with plastic hinges at the ends (i.e. at the nearest joints)and under the load. The mechanism of Figure 8(b) is simply applied to each loaded beam. Secondly, there are the *sway mechanisms* in which each storey is assumed to shear under the action of the horizontal loads with plastic hinges at the tops and bottoms of

the columns. The third set of basic mechanisms arises because
we must take into account joints at which three or four members
intersect. We have previously noted that at a joint where two
members intersect the hinge will always form in the weaker member. The situation is considerably more complex when more than
two members intersect. To handle this case systematically we
introduce basic mechanisms which consist of *joint rotations*: the
rigid joint is assumed to rotate with hinges occurring in all
the intersecting members at the joint in question.

The basic mechanisms and the subsequent application of the
method are best illustrated by reference to a specific example.
Consider the frame shown in Figure 11. The heights of the sto-
reys and the span of the beams are indicated by arbitrary length
units. The loads, given in arbitrary force units, which act on
the structure are shown. The absolute magnitudes of the loads
are not important; they merely indicate the *ratios* of the

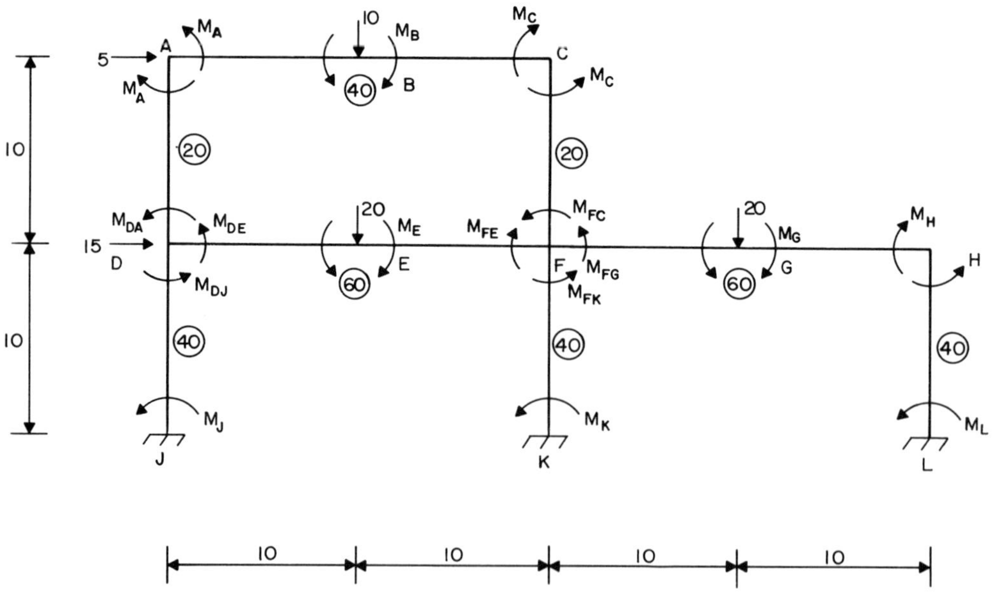

Figure 11. Multi-storey, multi-bay frame

magnitudes of the loads. We are required to find the limit load
parameter Γ by which each load must be multiplied in order that
quasi-static flow will occur in the structure. The limit moment
of each beam and column is indicated by a figure in a circle adjacent to the member; the units of the limit moments are consistent with the force and length units. A sign convention for moments at joints and under loads is also adopted and is shown in
Figure 11. At joint where two members meet the double subscript
is unnecessary and is omitted.

The basic mechanisms are shown in Figures 12(a)-12(f). A
calculation of Γ for each of the beam and sway mechanisms is
shown under each diagram. In the absence of concentrated couples
acting at the joints the joint rotation mechanisms do not provide a value of Γ (or, alternatively, they give an infinitely
large value of Γ). The two joint rotations are shown together
in Figure 12(f), but they are independent. It is rather difficult to illustrate the joint rotation mechanisms clearly; it
should be remembered that these mechanisms are velocity fields
and that the hinges form *at* the joint although they are shown in
the figure as occurring a short distance from the joint for
clarity.

It is unnecessary to give complete details of each calculation in Figure 12; for illustrative purposes consider the bound
on Γ obtained in Figure 12(e). The hinge rotation rates are
seen by geometry to be each equal to θ. The limit moment of
each column is 40; hence the internal work rate is $(40).(6\theta)$.
The horizontal forces 5Γ and 15Γ are each doing work; the horizontal velocity of both A and D is (10θ); hence the external
work rate is $(5\Gamma+15\Gamma).(10\theta)$. Equating work rates, we find Γ =
1.2. The values of Γ found in Figure 12 are each larger than or
equal to the value of Γ at which flow occurs. The lowest value
of Γ so far obtained occurs for mechanisms (a), (b), (c) and (e).

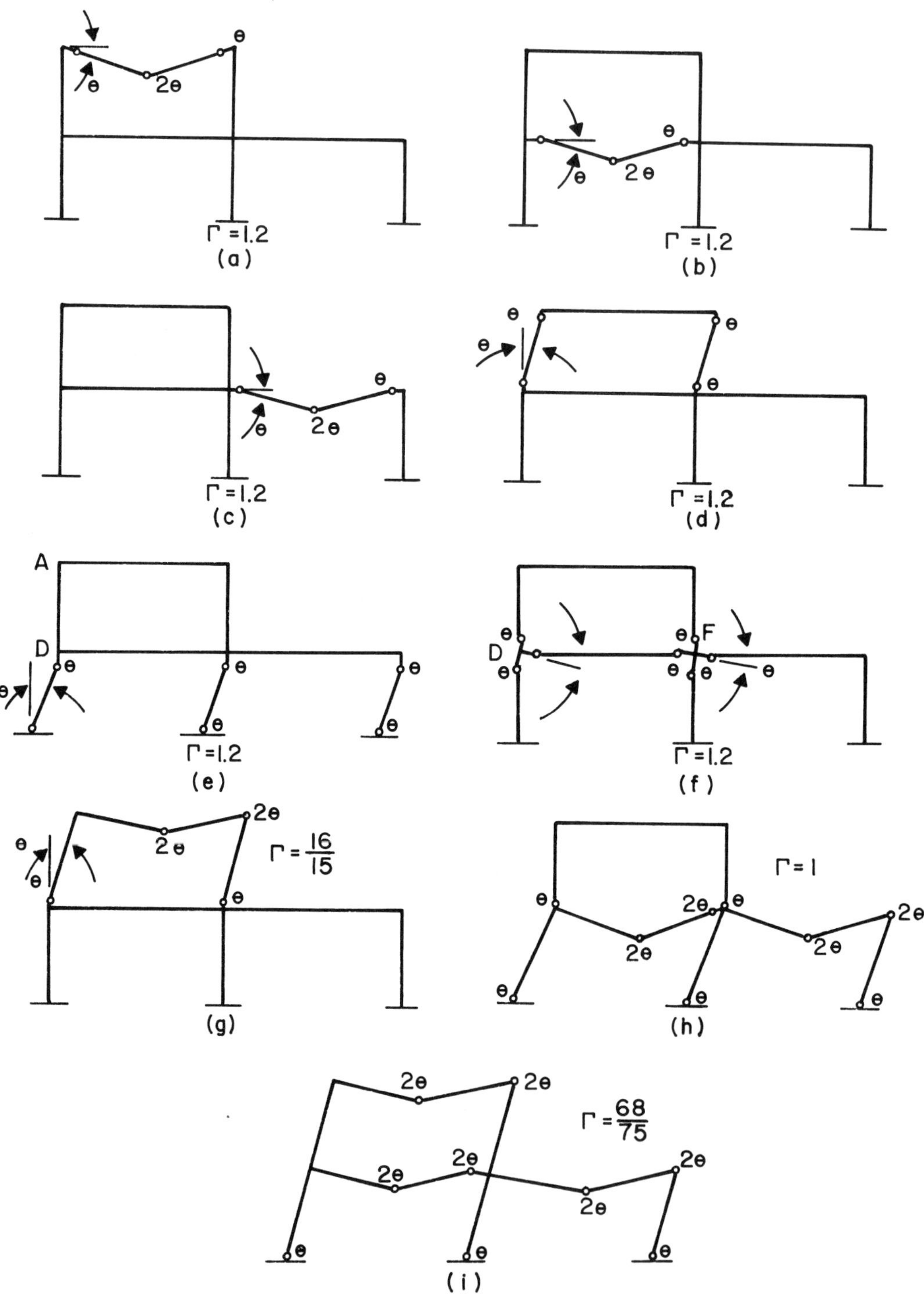

Figure 12. Mechanisms for multi-storey frame

Our next object is to find improved (i.e. lower) values of Γ using mechanisms formed by combining the basic mechanisms.

Let us begin with one of the basic mechanisms which gives a low value of the limit load parameter, say that of Figure 12(a), and modify it to see whether the load parameter can be reduced. We might first enquire whether sway of the top storey (for which $\Gamma = 1.6$, Figure 12d) combined with the beam mechanism is favorable for our purpose. Combination of mechanism (a) and (d) is accomplished by eliminating the hinge at A, and leads to the mechanism of Figure 12(g). For this mechanism the rate of work of the external loads is 150 units, and the internal work rate is 160 units. Thus $\Gamma = 16/15$. This represents an improvement over the previous least value of Γ.

Let us now investigate another basic mechanism which leads to a low load factor (Figure 12e) and check whether the combination of the beam mechanisms of Figures 12(b) and (c) with the sway of the lower storey leads to a reduction in the load factor. The combined mechanism is shown in Figure 12(h). Rotations of joints D and F have also been introduced in order to reduce the internal work rate at those joints as much as possible; this is equivalent to allow the hinges to form in the weaker members. The external work rate is 600 units, and the internal work rate is 600 units. Hence for this mechanism $\Gamma = 1.0$, an improvement on our previous best value.

The improvements achieved by using the mechanisms of Figures 12(g) and 12(h) suggest a combination of these two mechanisms. The new combination is shown in Figure 12(i). No further joint rotations are required at either D or F. For this mechanism the external work rate is 750 units, and the internal work rate is 680 units. Hence $\Gamma = 68/75$, a further improvement. Further attempts could be made to find a lower value of Γ. Let us proceed however, to attempt to associate a safe, statically admissible bending moment distribution with the mechanism of Figure 12(i).

We assume that the moments at the hinges have the magnitude
of the respective limit moment; the sign must be deduced from
the sign of the hinge rotation rate. Our task is to determine
equilibrium values of the moments at the other potential hinge
positions. As occurs frequently in problems of this type, there
is not a unique equilibrating moment field. Notice that the
frame is indeterminate to the ninth degree; if the loads are
given, nine independent internal forces or reactions must be
specified if the moment distribution is to be uniquely determined. In limit analysis, however, the load magnitude is not
known; it is in fact computed by means of an equilibrium relation. This means that we must specify one more independent internal force to permit us to compute the additional unknown
quantity. A necessary condition for a unique moment field is
that we should have one more hinge than the degree of indeterminacy (it is not a sufficient condition because the moment values at the hinges may not be independent). In this case we
have only nine hinges, and consequently a non-unique moment
field. This is simply a reflection of the general result which
we have discussed previously; the generalized stress field during flow is not necessarily completely determined by the conditions of flow alone, but may depend on the path of loading.

A possible moment distribution is shown in Figure 13. All
equilibrium requirements have been satisfied with external loads
which have magnitudes which are 68/75 of those shown in Figure
11. The moments do not exceed the respective limit values at
any point, and consequently Γ = 68/75 is the true limit load
parameter.

Had it transpired that some moment or moments exceeded the
limit values, it is a straightforward matter to compute a lower

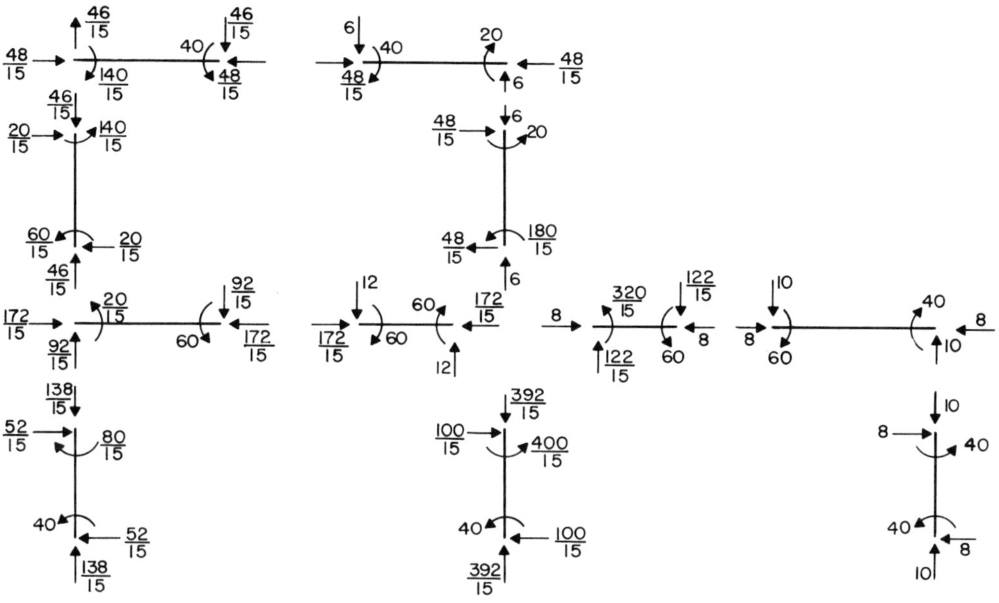

Figure 13. Moment distribution for multi-storey frame

bound on the limit load parameter. This is achieved by multiplying the loads and moments of Figure 13 by a factor, which
will be less than unity,so that the moment which exceeds its
limiting value by the largest amount is reduced to the limiting
value. The resulting loads and moments then satisfy the requirements of the lower bound theorem.

As an example of the application of this procedure, let us
suppose that we wish to compute the limit load parameter for
the frame of Figure 11 with the lower storey columns replaced
by members whose limit moment is 30 units. Using the mechanism
of Figure 12(i), we find an upper bound given by $\Gamma = 13/15$. To
compute a lower bound we multiply the moment field of Figure 13
by 0.75, observing that the moments now satisfy the modified
limit values in the lower columns. The load parameter is now
$\Gamma = (68/75) \times 0.75 = 17/25$. Thus immediately we have established

that the limit load parameter Γ is such that $0.68 \leq \Gamma \leq 0.87$, and can now continue to refine the calculation.

10.4 Bibliographical and Historical Remarks

The limit analysis of framed structures is extremely important in civil engineering, and a number of texts have appeared which discuss this problem in great detail. These texts include van den Broek [1948], Baker, Horne and Heyman [1956], Neal [1956], Heyman [1957], Beedle [1958], Hodge [1959], Massonet and Save [1965], Baker and Heyman [1969] and Horne [1971]. The limit analysis problem described in this Section is of course only the first step in applying the concepts of plasticity to structural design; consideration must be given to more complex factors such as the effect of axial force and shear, the computation of deflections and local and overall instability, each of which may mean that the limit load of the structure cannot be realized. It is remarkable, however, that in many cases correction to the limit analysis solution can be made to allow for the other factors; a complete analysis based on a more sophisticated theory is not always necessary.

The method of combination of mechanisms to which we have referred is due to Neal and Symonds [1951], [1952]. It is of interest to note that attempts have been made to formulate corresponding methods which systematically improve lower bound calculations in a simple manner. Such methods are not easily carried out by hand (e.g. Horne [1954]), but have been used successfully with programming approaches and other numerical methods. Heyman [1968] has recently described systematic methods of determining the moment field, particularly when the moments are not uniquely determined, which can be used in conjunction with the method of combination of mechanisms.

References

J.F.Baker and J.Heyman	1969	*Plastic Analysis of Frames,* Cambridge.
J.F.Baker, M.R.Horne and J.Heyman	1956	*The Steel Skeleton,* Vol.2, Cambridge.
L.S.Beedle	1958	*Plastic Design of Steel Frames,* Wiley (N.Y.).
J.A.van den Broek	1948	*Theory of Limit Design,* Wiley,(N.Y.).
J.Heyman	1957	*Plastic Design of Portal Frames,* Cambridge.
J.Heyman	1968	"Bending moment distributions in collapsing frames", in *Engineering Plasticity* (edited by J.Heyman and F.A. Leckie), Cambridge University Press, 219.
P.G.Hodge,Jr.	1959	*Plastic Analysis of Structures,* McGraw-Hill (N.Y.).
M.R.Horne	1954	"A moment-distribution method for the analysis and design of structures by the plastic theory", Proc.I.C.E., $\underline{3}$.
M.R.Horne	1971	*The Plastic Theory of Structures,* Nelson.
C.Massonet and M.Save	1965	*Plastic Analysis and Design,* Blaisdell (Boston).
B.G.Neal	1956	*The Plastic Methods of Structural Analysis,* Chapman and Hall (London).
B.G.Neal and P.S.Symonds	1951	"The calculation of collapse loads for framed structures", Proc.I.C.E., $\underline{35}$, 21.
B.G.Neal and P.S.Symonds	1952	"The rapid calculation of plastic collapse loads for a framed structure", Proc.I.C.E., $\underline{1}$ (3), 58.

Chapter 11

LIMIT ANALYSIS OF PLATES

11.1 Basic Equations

In this section we shall consider two problems in the determination of the limit load for rotationally symmetric plates. The fundamental equations for this class of problems are given in Section 1.4. The circumferential and radial bending moments M_θ, M_r must satisfy the equilibrium equations

$$\frac{d}{dr}\left(rM_r\right) = M_\theta - rQ_r \quad , \tag{1a}$$

$$\frac{d}{dr}\left(rQ_r\right) = pr \quad , \tag{1b}$$

where Q_r is the radial shear force per unit length and $p(r)$ is the applied load. Q_r may be eliminated from these equations by integrating equation (1b) between the limits $[0,r]$ and substituting into equation (1a). This gives a single equilibrium relation

$$\frac{d}{dr}\left(rM_r\right) = M_\theta - \int_0^r prdr \quad . \tag{2}$$

In the rate form, the circumferential and radial curvatures $\dot{\kappa}_\theta$, $\dot{\kappa}_r$ are given in terms of the downward displacement rate $\dot{w}(r)$ by means of the kinematic relations

$$\dot{\kappa}_\theta = -\frac{1}{r}\frac{d\dot{w}}{dr} \quad , \qquad \dot{\kappa}_r = -\frac{d^2\dot{w}}{dr^2} \quad . \tag{3}$$

The yield or limit surface in the M_r, M_θ space will be taken to be identical with that given in Chapter 7. This yield surface, shown in Figure 1, was derived for a sandwich plate assuming that the material of which the face sheets of the sandwich are composed satisfied a Tresca yield condition. It is an

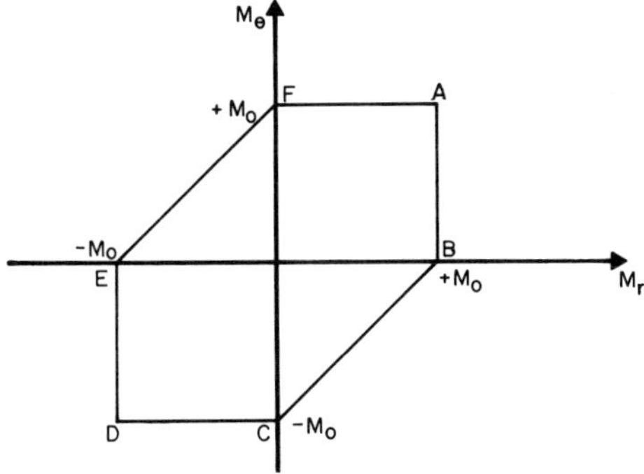

Figure 1. Yield surface for axisymmetric plate

exact surface for a sandwich plate, but, with a suitable rein-
terpretation of the value of M_o, it is also the yield condition
for a perfectly plastic uniform (or solid) plate under pure
bending. The flow rule requires that the generalized strain
rate vector $\dot{\kappa}_r$, $\dot{\kappa}_\theta$ should be normal to the yield surface at the
stress point during flow, with the appropriate restrictions
given by adjacent normals when the normal is not uniquely defined at a corner.

11.2 A Simply Supported Plate

Consider, as a first example, the simply supported circular
plate subjected to a uniform symmetric load p which extends a
distance a from the center of the plate, as shown in Figure 2(a).
For a given value of a, we are required to determine the magnitude of the load p for which flow will take place.

The equilibrium equation for the plate can be written in two
parts, after substituting the loads of Figure 2 into equation
(2);

$$\frac{d}{dr}(rM_r) = M_\theta - \frac{pr^2}{2} \quad , \quad 0 \leq r \leq a \quad , \tag{3a}$$

$$\frac{d}{dr}(rM_r) = M_\theta - \frac{pa^2}{2} \quad , \quad a \leq r \leq R \quad . \tag{3b}$$

Using our experience with the uniformly loaded plate for which an

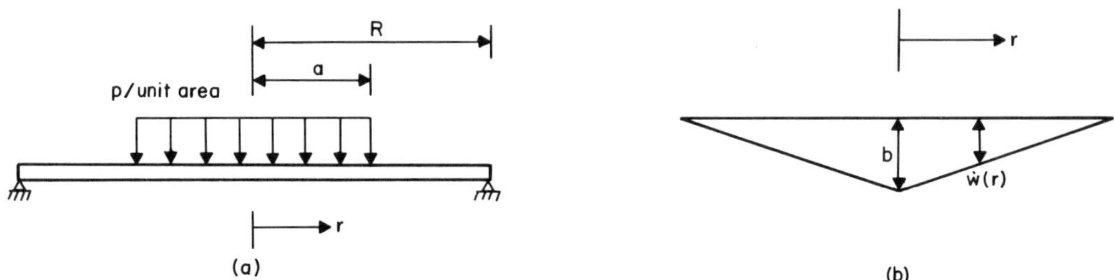

Figure 2. Simply supported plate

elastic-plastic analysis was carried out (Chapter 7), we might suppose that M_θ is constant and equal to M_o over the whole plate. The stress profile will then be constrained to lie on the side AF of the yield condition in Figure 1. Using this assumption, we can construct a safe pertinent statically admissible field by integrating equations (3), and ensuring that the conditions imposed on the moment field are satisfied and that no stress point lies outside the yield surface. Integrating equations (3a) and (3b), we obtain

$$M_r = M_o - \frac{pr^2}{6} + \frac{A}{r} \, , \quad 0 \leq r \leq a \, , \tag{4a}$$

$$M_r = M_o - \frac{pa^2}{2} + \frac{B}{r} \, , \quad a \leq r \leq R \, . \tag{4b}$$

From the isotropy condition at the center of the plate we require that

$$M_r(0) = M_\theta(0) = M_o \, . \tag{5}$$

The simply supported boundary condition gives

$$M_r(R) = 0 \, . \tag{6}$$

Finally, we require that M_r should be continuous at r=a. These three conditions enable us to evaluate the constants of integration A, B *and the load* p. This gives

$$A = 0 \ , \tag{7a}$$

$$B = \frac{2M_o a}{(3 - 2 \frac{a}{R})} \ , \tag{7b}$$

$$\frac{pa^2}{6M_o} = \frac{1}{(3 - \frac{2a}{R})} \ . \tag{7c}$$

Substituting these values into the expressions for M_r (equations 4(a) and 4(b)) the complete moment field can be obtained:

$$\frac{M_r}{M_o} = 1 - \frac{6r^2}{a^2(3 - \frac{2a}{R})} \ , \quad 0 \leq r \leq a \ , \tag{8a}$$

$$\frac{M_r}{M_o} = 1 - \frac{(3 - \frac{2a}{r})}{(3 - \frac{2a}{R})} \ , \quad r \leq a \leq R \ , \tag{8b}$$

$$\frac{M_\theta}{M_o} = 1 \ . \tag{8c}$$

It can readily be confirmed from these equations that

$$0 \leq M_r \leq M_o \tag{9}$$

for the range of values $0 \leq r \leq R$. This in turn confirms that the stress profile lies entirely on the side AF of the yield surface, and does not lie outside the yield surface. The load value given in equation (7c) is consequently a lower bound on the load which causes flow of the plate.

A suitable velocity field to determine an upper bound would be that found in the case of a uniformly loaded plate in Chapter 7. This velocity field is shown in Figure 2(b), and is expressed by

$$\dot{w} = b(1 - \frac{r}{R}) \ . \tag{10}$$

The external work rate is

$$D_{ext} = \int_o^a p\dot{w}(2\pi r)\,dr$$

$$= \frac{pb}{3}\,a^2\left(3 - \frac{2a}{R}\right)\ .$$

(11)

The curvature rates are obtained by substituting the velocity field of equation (10) into equations (3),

$$\dot{\kappa}_\theta = -\frac{1}{r}\frac{d\dot{w}}{dr} = \frac{b}{rR}\ ,$$

(12a)

$$\dot{\kappa}_r = -\frac{d^2\dot{w}}{dr^2} = 0\ .$$

(12b)

The curvature rate $\dot{\kappa}_\theta$ is infinite at the center of the plate, and the discontinuity in $d\dot{w}/dr$ at $r = 0$ also indicates that $\dot{\kappa}_r$ is infinite and positive at the center. Thus the moments associated with this velocity field at $r = 0$ are $M_r = M_\theta = M_o$ from the flow rule. However, because this discontinuity in slope is confined to a *single point*, the contribution to the total internal energy dissipation rate will be zero. It is only along a *line* of discontinuity that a finite contribution to the internal dissipation rate in a plate will be obtained. Over the remainder of the plate $\dot{\kappa}_\theta$ is positive and $\dot{\kappa}_r$ is zero; hence, from the flow rate, the energy dissipation rate per unit area is given by $M_o\dot{\kappa}_\theta$. The total internal energy dissipation rate is thus

$$D_{int} = \int_L^R M_o\kappa_\theta(2\pi r)\,dr$$

$$= 2\pi M_o b\ .$$

(13)

Equating the internal energy dissipation rate (equation 13) and

the external work rate (equation 11), an upper bound on the
limit load is

$$\frac{pa^2}{6M_o} = \frac{1}{(3 - \frac{2a}{R})} \quad . \tag{14}$$

This coincides with the lower bound given in equation (7c), and
hence this is in fact the limit load. We may note that for the
case a = R, this limit load corresponds to the calculation of
the limit load given for the uniform load in Chapter 7. It is
also of interest to express the limit load in terms of the total load on the plate,

$$P = \pi pa^2 \quad . \tag{15}$$

Substituting this total load into equation (14), the limit load
is

$$P = \frac{6\pi M_o}{(3 - 2a/R)} \tag{16}$$

If we now let a go to zero while P remains finite (achieved by
allowing p to become infinitely large in equation 15), equation
(16) gives the limit load for a circular plate subjected to a
transverse concentrated load at the center;

$$P = 2\pi M_o \quad . \tag{17}$$

Although it is beyond the scope of this monograph, it is of interest to note that the limiting value of a transverse point
load on a plate of any size and shape is given by $2\pi M_o$.

11.3 A Clamped Plate

As a second example let us consider a clamped plate subjected
to a uniform load p per unit area over the whole plate (Figure
3). In view of our experience with the simply supported plate,
we expect that for the clamped plate the velocity field $\dot{w}$ will

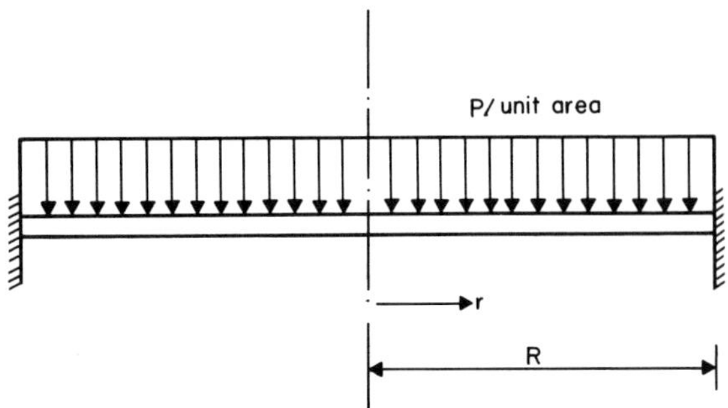

Figure 3. Clamped plate

be a decreasing function of r. Thus, assuming that plastic deformation occurs everywhere in the plate during flow, $\dot{k}_\theta$ (from
equation 3) must be positive everywhere and the stress profile
must be on the sides AFE of the yield surface (Figure 1) as a
consequence of the flow rule. Moreover, $M_r = M_\theta$ will still be
expected to be positive at the center of the plate, and we may
assume that the corresponding stress point is at A. The clamped
boundary at $r = R$ implies that M_r is negative at the boundary
and thus the stress point on the boundary lies on FE. Let us
assume that $M_r(R) = -M_o$ (i.e. its smallest possible value) and
further let $M_r = 0$ when $r = s$. The guessed form of the stress
profile thus gives

For $0 \leq r \leq s$ $M_\theta = M_o$,   (18a)

$$M_r(0) = M_o , \qquad (18b)$$

$$M_r(s) = 0 , \qquad (18c)$$

For $s \leq r \leq R$ $M_\theta - M_r = M_o$,   (19a)

$$M_r(s) = 0 , \qquad (19b)$$

$$M_r(R) = -M_o . \qquad (19c)$$

Equations (18) and (19) include boundary conditions and

continuity requirements.

With a uniform load over the entire plate the equilibrium equation (2) becomes

$$\frac{d}{dr}(rM_r) = M_\theta - \frac{pr^2}{2} \quad . \tag{20}$$

Substituting for M_θ from equations (18a) and (19a), we may integrate equation (20) over the two ranges of r;

$$M_r = M_o - \frac{pr^2}{6} + \frac{A}{r} \qquad \text{for} \quad 0 \leq r \leq s \quad , \tag{21a}$$

$$M_r = M_o \log r - \frac{1}{4} pr^2 + B \quad \text{for} \quad r \leq s \leq R \quad . \tag{21b}$$

Equations (18b), (18c), (19b) and (19c) provide four conditions which can be used to evaluate the constants of integration A, B and the other parameters p, s. Substitution of these conditions leads to a transcendental equation for s,

$$1 + \log \frac{R}{s} = \frac{3}{2} \left(\frac{R^2}{s^2} - 1 \right) \quad . \tag{22}$$

This equation must be solved numerically for (s/R). The load parameter may be expressed in terms of s, given by

$$\frac{pR^2}{6M_o} = \frac{R^2}{s^2} \quad . \tag{23}$$

The moment expressions (21a) and (21b) become

$$\frac{M_r}{M_o} = 1 - \frac{pR^2}{M_o} \left(\frac{r}{R} \right)^2 \quad , \tag{24a}$$

$$\frac{M_r}{M_o} = \log\left(\frac{r}{s} \right) - \frac{3}{2} \left(\frac{r^2}{s^2} - 1 \right) \quad . \tag{24b}$$

The solution of equation (22) gives R/s = 1.369, so that

$$\frac{pR^2}{6M_o} = 1.876 \ . \tag{25}$$

It can readily be confirmed that the stress profile, which can be determined completely from equations (24), (18a) and (19a), lies entirely on the yield surface. Thus we have a safe pertinent statically admissible stress field for the load parameter given in equation (25), and this is a lower bound on the limit load.

In order to illustrate an upper bound calculation which does not in fact provide the limit load but an upper bound, let us assume a velocity field identical to that used in the simply supported plate,

$$\dot{w} = b(1 - \frac{r}{R}) \ . \tag{26}$$

The external work rate may be obtained from equation (11), with a = R,

$$D_{ext} = \frac{\pi p b R^2}{3} \ . \tag{27}$$

In order that the clamped boundary condition should be met we must take cognizance of the discontinuity of slope at r = R. $\dot{\kappa}_r$ will be infinite and negative at this line hinge, while $\dot{\kappa}_\theta$ is zero. Hence the associated moment is $-M_o$ (corresponding to DE in Figure 1) and the energy dissipation rate per unit length of the hinge circle will be $M_o b/R$. Over the remainder of the plate the energy dissipation rate will be identical to that in the simply supported case (equation 13). Hence for the clamped plate

$$\begin{aligned} D_{int} &= 2\pi M_o b + 2\pi R(M_o b/R) \\ &= 4\pi M_o b \ . \end{aligned} \tag{28}$$

Equating the work rates, we find that

$$\frac{pR^2}{6M_o} = 2 \quad . \tag{29}$$

This value is an upper bound, and cannot be less than the limit load. The difference between this upper bound and the lower bound of equation (25) is about 6%, and thus a fairly good approximation to the limit load has been obtained.

If we consider that the velocity field of equation (26) has $\dot{\kappa}_\theta = 0$, $\dot{\kappa}_r < 0$ for $r = R$ and $\dot{\kappa}_r = 0$, $\dot{\kappa}_\theta > 0$ for $0 < r < R$, it is evident that it corresponds to a stress profile which is on DE for $r = R$ and on AF for the remainder of the plate. This could not be the correct stress profile, and thus we are led to attempt to improve the upper bound.

We expect, in fact, that part of the stress profile will be on FE. In this region we have $\dot{\kappa}_\theta = -\dot{\kappa}_r$ from the flow rule. From equations (3) this would require that

$$\frac{d^2\dot{w}}{dr^2} + \frac{1}{r}\frac{dw}{dr} = 0 \quad . \tag{30}$$

Integration of this equation gives

$$\dot{w} = c \log r + d \quad , \tag{31}$$

which does not agree with the velocity field of equation (26). However, the linear relation between $\dot{w}$ and r is associated with a stress profile on AF, which is expected in the inner part of the plate. Let us then define a circle on the plate distance ξ from the center, such that within this circle the velocity field is conical ($\dot{w}$ varies linearly with r) and outside of the circle $\dot{w}$ has the form of equation (31). Taking into account the requirement that $\dot{w}$ and $d\dot{w}/dr$ should be continuous for $r = \xi$ and that $\dot{w}(R) = 0$, this velocity field can be expressed in terms of one unspecified parameter;

$$\dot{w} = b \left\{ 1 - \frac{(r/\xi)}{1-\log(r/\xi)} \right\} \quad , \quad 0 \le r \le \xi \, , \tag{32a}$$

$$\dot{w} = - \frac{b}{1-\log(r/\xi)} \, \log \frac{r}{R} \quad , \quad \xi \le r \le R \, . \tag{32b}$$

The external work rate associated with this velocity field can readily be found by integration. After some manipulation, we obtain

$$D_{ext} = \int_{o}^{R} 2\pi p\dot{w} \, rdr$$

$$= \frac{\pi p b R^2}{6(1-\log \, \xi/R)} \, (3 - \frac{\xi^2}{R^2}) \quad . \tag{33}$$

The curvature rates are determined from the velocity field by means of equation (3). They are

$$\dot{\kappa}_\theta = \frac{b}{r\xi(1-\log \frac{\xi}{R})} \, , \, \dot{\kappa}_r = 0 \quad \text{for} \quad 0 \le r \le \xi \, , \tag{34a}$$

$$\dot{\kappa}_\theta = -\dot{\kappa}_r = \frac{b}{r^2(1-\log \frac{\xi}{R})} \quad \text{for} \quad \xi \le r \le R \, . \tag{34b}$$

There are discontinuities in $d\dot{w}/dr$ at the center (which is of no concern because the contribution to the internal energy dissipation rate is again zero) and at the edge of the plate. Differentiating equation (32b) we find that

$$\dot{w}'(R) = - \frac{b}{R(1-\log \frac{\xi}{R})} \, , \tag{35}$$

and since the clamped edge condition requires $\dot{w}'(R) = 0$, the rotation rate at the hinge circle is given by equation (35). Referring to Figure 1, we may now locate the stresses associated with the plastic curvature rates given by equations (34a), (34b). First, for $0 < r < \xi$, $\dot{\kappa}_\theta > 0$ and $\dot{\kappa}_r = 0$. The associated stress

points lie consequently on AF. As before, the precise stress
profile is not required since the internal energy dissipation
rate per unit area of plate is uniquely given by $M_o \dot{\kappa}_\theta$. For
$\xi \leq r \leq R$, $\dot{\kappa}_\theta = -\dot{\kappa}_r > 0$, so that the stress profile lies on FE.
Making use of the uniqueness of the internal energy dissipation
rate per unit area, we can assume that the stress point is at
(say) F, so that

$$M_\theta \dot{\kappa}_\theta + M_r \dot{\kappa}_r = M_o \dot{\kappa}_\theta \quad . \tag{36}$$

Alternatively, we could choose the stress point $M_\theta = 0$, $M_r = -M_o$,
or any other point, to obtain the same result. At the hinge
circle at the edge of the plate $\dot{\kappa}_r$ is infinite and negative, and
hence the stress point is at E, with $M_r = -M_o$. The total internal energy dissipation rate is then given by

$$D_{int} = \int_0^\xi 2\pi M_o \dot{\kappa}_\theta r\,dr + \int_\xi^R 2\pi M_o \dot{\kappa}_\theta r\,dr + 2\pi R M_o |\dot{w}(R)|$$

$$\tag{37}$$

$$= \frac{2\pi M_o b}{(1-\log \frac{\xi}{R})} (2 - \log \frac{\xi}{R}) \quad .$$

Finally, we obtain a value of the load parameter by equating the
internal and external work rates,

$$\frac{pR^2}{6M_o} = \frac{2(2-\log \frac{\xi}{R})}{(3 - \frac{\xi^2}{R^2})} \quad . \tag{38}$$

This value of p, by the upper bound theorem of plasticity, must
be larger than the load required to cause flow in the plate for
any choice of ξ/R. If, for example, we put $\xi/R = 1$ we recover
the entirely conical velocity field and the upper bound given in
equation (29). The best upper bound given by equation (38) is

found by putting $dp/d\xi = 0$, which gives

$$(3 - \frac{\xi^2}{R^2}) = 2(2-\log \frac{\xi}{R}) \frac{\xi^2}{R^2} \quad . \tag{39}$$

With some rearrangement, the equation in ξ is identical to that in s given in equation (22). This is to be expected if we have the correct velocity field, since $r = \xi$ and $r = s$ are in each case the points where $M_r = 0$. The solution of equation (39) gives $R/\xi = 1.369$, and substituting equation (39) into equation (38) the best upper bound is

$$\frac{pR^2}{6M_o} = \frac{R^2}{\xi^2} = 1.876 \quad . \tag{40}$$

This coincides exactly with the lower bound obtained earlier and hence the load given in equation (40) lies on the limit surface.

11.4 Further Plate Examples

As an example of the use which can be made of the corollaries of the limit theorems given in Chapter 9, consider the problem shown in Figure 4. A uniform simply supported circular plate of radius R is subject to a concentrated load P at the center and a uniformly distributed load p. The Tresca yield condition for a sandwich plate will be adopted. Let the generalized loads be

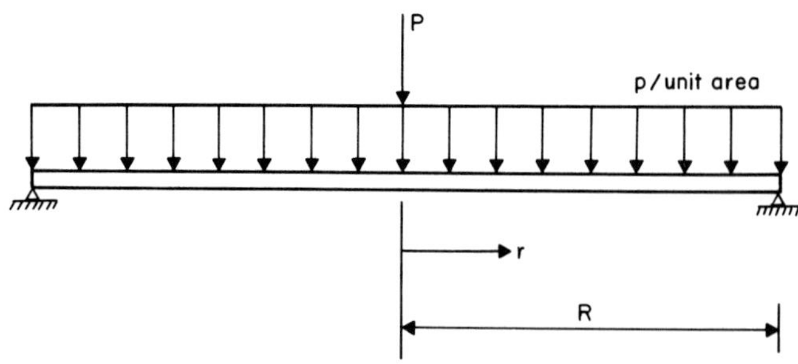

Figure 4. Combined loading on plate

$$\Gamma_1 = P , \quad \Gamma_2 = \pi p R^2 . \tag{41}$$

The transverse velocity field will be axisymmetric, and hence a function of the radius r of a generic point. Denoting the velocities by $\dot{w}(r)$,

$$\dot{\xi}_1 = \dot{w}(0), \quad \dot{\xi}_2 = \frac{1}{\pi R^2} \int_A \dot{w}(r)dA = \frac{2}{R^2} \int_o^R \dot{w}(r)rdr . \tag{42}$$

Let us suppose that the complete limit surface in load space is required. It is our present intention to show that some information about this limit surface can be obtained, starting with the results given in Section 11.2 for a point load and a uniform load acting alone.

We know that (equation 17)

$$\Psi(\pm 2\pi M_o, 0) = 0 , \tag{43}$$

and that the velocity field for flow under a concentrated central load is

$$\dot{w} = \pm b_1 (1 - \frac{r}{R}) . \tag{44}$$

Thus, using equation (42),

$$\frac{\dot{\xi}_1}{\dot{\xi}_2} = 3 . \tag{45}$$

Further, we know that (equation 14)

$$\Psi(0, \pm 6\pi M_o) = 0 , \tag{46}$$

and that the velocity field is

$$\dot{w} = \pm b_2 (1 - \frac{r}{R}) . \tag{47}$$

Hence we again have

$$\frac{\dot{\xi}_1}{\dot{\xi}_2} = 3 \qquad\qquad (48)$$

at this point on the limit surface .

Equations (43) and (46) permit us to plot four points in the Γ_1, Γ_2 space which lie on the limit surface. It follows then that the quadrilateral ABCD shown in Figure 5(a) contains a set of load states which lie within the limit surface. The tangent lines at points A, B, C and D can be drawn, making use of equations (45) and (48), and are shown in Figure 5(b). All load states on these lines lie without or on the limit surface. It may be noted that lines AB and CD appear in both Figure 5(a) and Figure 5(b); consequently these lines form part of the limit surface and provide a result which is not immediately obvious. Very little information is obtained about the second and fourth quadrants; further analysis is required to complete the surface.

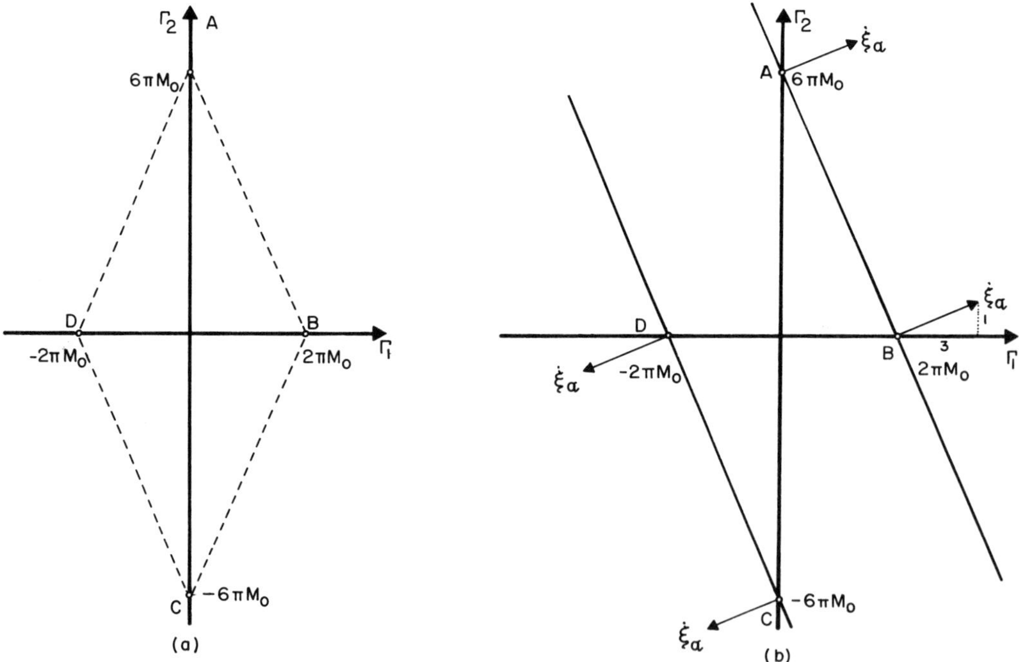

Figure 5.Results for combined loading

It may be appreciated, however, that one further point in each of these quadrants (say $\Gamma_1 = -\Gamma_2$) will permit a very good approximation of the limit surface to be drawn.

As a second example, consider again the limit analysis of an axisymmetric, uniform circular plate subjected to a uniform pressure p. Let us assume that we wish to find the limiting value of p for a von Mises yield surface,

$$M_r^2 - M_r M_\theta + M_\theta^2 = M_o^2 \ . \tag{49}$$

This problem was solved for a Tresca yield condition in Section 11.2. Flow occurs for $pR^2/M_o = 6$, where R is the radius of the plate. The von Mises yield surface of equation (49) circumscribes the Tresca surface of equation (50), as shown in Figure 6. Furthermore, the Tresca yield surface

$$|M_r| \quad = \quad 2M_o/\sqrt{3}$$

$$|M_\theta| \quad = \quad 2M_o/\sqrt{3} \tag{50}$$

$$|M_r - M_\theta| \quad = \quad 2M_o/\sqrt{3}$$

circumscribes the von Mises surface. It follows immediately that the limit load parameter for the plate with the yield surface of equation (49) can be bounded from above and below:

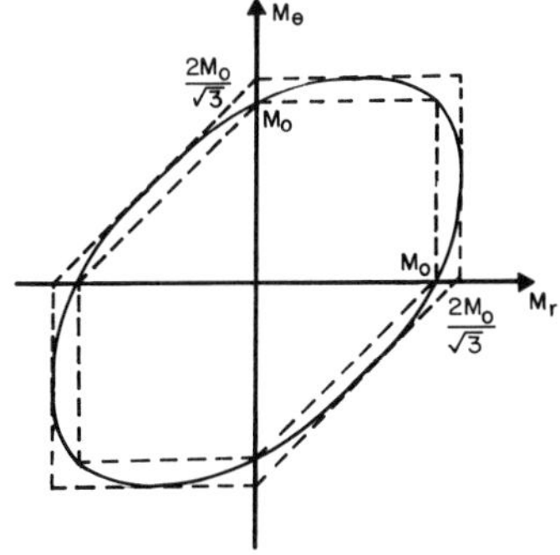

Figure 6. Inscribed and circumscribed yield surface

$$6 \le \frac{pR^2}{M_0} \le 6\,\frac{2}{\sqrt{3}} = 4\sqrt{3} \tag{51}$$

11.5 Bibliographical and Historical Remarks

The solutions presented in this Chapter were first given by Hopkins and Prager [1953]. Hopkins and Wang [1955] generalized the results to arbitrary yield conditions. Zaid [1958] and Schumann [1958] demonstrated that the limiting value of a concentrated load on a plate of arbitrary shape is $2\pi M_o$. The solution of the similar, but more complex, problem of an annular plate has been discussed by Hodge [1959]. Further examples of the limit analysis of plates and shells may be found in the monograph by Hodge [1963].

Mention should also be made of *yield line theory*, a method developed for determining the load carrying capacity of reinforced concrete slabs by Johansen [1943]. Accounts are also given by Jones and Wood [1967] and Johnson [1967]. Yield line theory bears a close resemblance to the kinematic approach in limit analysis, and it was a matter of contention for some time as to whether it could in fact be based on limit analysis. This question has been answered affirmatively (see, for example, Braestrup [1970]); yield line theory provides an upper bound on the limit load, but cannot always predict the actual limit load.

References

M.N.Braestrup   1970   "Yield line theory and limit analysis of plates and slabs", Magazine of Concrete Research, <u>22</u>, 99.

P.G.Hodge,Jr.   1959   "Yield point load of an annular plate", J.Appl.Mech., <u>26</u>, 454.

P.G.Hodge,Jr.   1963   *Limit Analysis of Rotationally Symmetric Plates and Shells*, Prentice-Hall (Englewood Cliffs, N.J.).

H.G.Hopkins and W.Prager | 1953 | "The load carrying capacities of circular plates", J.Mech.Phys.Sol., $\underline{2}$, 1.

H.G.Hopkins and A.J.Wang | 1955 | "Load carrying capacities for circular plates of perfectly plastic material with an arbitrary yield condition", J.Mech.Phys.Sol., $\underline{3}$, 117.

K.W.Johansen | 1943 | *Brudlinieteorier*, Gjellerup (Copenhagen). (English Edition: *Yield Line Theory*, Cement and Concrete Association, London, 1962).

R.P.Johnson | 1967 | *Structural Concrete*, McGraw-Hill.

L.L.Jones and R.H.Wood | 1967 | *Yield Line Analysis of Slabs*, Thames and Hudson, Chatto and Windus (London).

W.Schumann | 1958 | "On limit analysis of plates", Q.Appl.Math., $\underline{16}$, 61.

M.Zaid | 1958 | "On the carrying capacity of plates of arbitrary shape and variable fixity under a concentrated load", J.Appl.Mech., $\underline{25}$, 598.

Chapter 12

LIMIT ANALYSIS IN PLANE STRESS AND PLANE STRAIN PROBLEMS

12.1 Discontinuities in Stress and Velocity Fields

As an introduction to the use of limit analysis in three dimensional continua, we shall consider briefly in this Chapter the
application of the theorems of limit analysis to problems in
which either the plane stress or plane strain assumptions are
valid. We shall also attempt to illustrate the way in which
the limit theorems can be used as an approximate method to obtain bounds on the limit load in proportional loading problems
with the expenditure of comparatively little effort. Such applications of the limit theorems are of great value when a
reasonably accurate estimate of the limit load is the primary
purpose of the calculation: it will be seen as we proceed that
close bounds on the limit load can be found without necessarily
determining the exact, or even close estimates of, stress and
strain rate fields in the deforming regions during flow.

We shall consider first the nature of discontinuities in the
stress and velocity fields. Since for the most part we shall be
considering simple stress and strain rate fields it will be
particularly convenient to work in terms of principal stresses
and principal strain rates: we can then use the Tresca criterion
for the limit surface in stress space. The discussion of discontinuities will be followed by a study of the Tresca condition
in the special cases of plane stress and plane strain. Thereafter, we shall consider a variety of particular examples.

It has been seen that in limit analysis problems in beams
and plates discontinuities in the generalized velocity field (in
the form of discontinuities in the rotation rate) can be an important feature during flow. Discontinuities in the generalized
stresses also occur frequently in beam problems, notably in the

shear force at the point of application of a concentrated load.
These indicators are sufficient to cause us to suspect that
similar discontinuities either occur in continuum problems during flow or may be exploited in the applications of limit analysis to these cases. It will consequently be convenient to investigate at this point the conditions under which discontinuous
stress and velocity fields may be found without violating the
conditions of equilibrium or compatibility.

Consider first the existence of a surface of discontinuity
S_D in the stress field, σ_{ij}, in a three-dimensional continuum,
as shown in Figure 1(a). Let ν_j be a unit vector which is normal to the surface of discontinuity. The body may be separated
into two parts by dividing it by a surface which contains the
discontinuity. The free body diagrams of the two separate
pieces are shown in Figure 1(b). The two surfaces created in
this manner are acted upon by surface tractions

$$T_i^+ = \sigma_{ij}^+ \nu_j \quad , \quad T_i^- = -\sigma_{ij}^- \nu_j \quad , \tag{1}$$

which represent respectively the mechanical effect of one part
of the body on the other. The stresses σ_{ij}^+ and σ_{ij}^- are the
stresses at corresponding points in the two bodies; thus
$(\sigma_{ij}^+ - \sigma_{ij}^-)$ is the discontinuity in the stress field as we pass
across the surface S_D. The requirements of equilibrium, however,

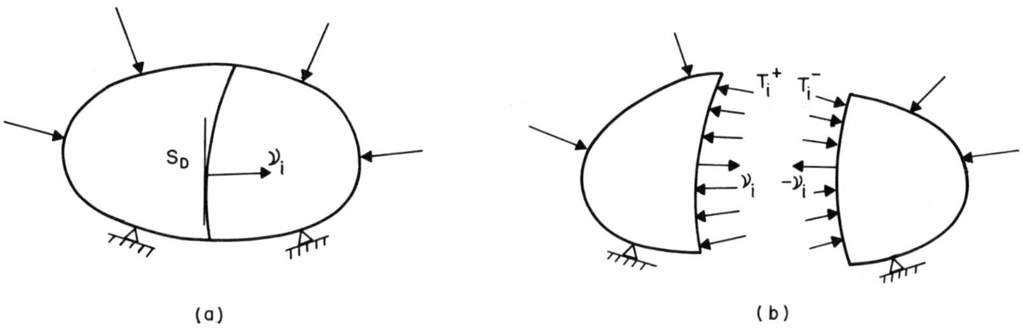

Figure 1. Surface of discontinuity in the stress field

demand that T_i^+ and T_i^- should be equal and opposite. Thus

$$(\sigma_{ij}^+ - \sigma_{ij}^-)\nu_j = 0 \tag{2}$$

at any point on the surface S_D in the original undivided body.
This condition is more usually expressed as the condition that
the jump in $\sigma_{ij}\nu_j$ across the discontinuity should be zero, and
is written in the form

$$[\sigma_{ij}\nu_j]_{S_D} = 0 \ . \tag{3}$$

The import of this restriction can readily be appreciated if
we consider a plane of discontinuity and choose a cartesian coordinate system in such a way that the y-axis, say, is normal to
the plane S_D. Figure 2 shows a cubic element of material, one
of whose faces lies in the plane S_D. The unit vector ν_j has in
this case components $(0,1,0)$, and consequently equation (3) becomes

$$[\sigma_{yx}]_{S_D} = [\sigma_{yy}]_{S_D} = [\sigma_{yz}]_{S_D} = 0 \ . \tag{4}$$

The stress components σ_{yx}, σ_{yy} and σ_{yz}, which are the components
acting on the face of the cube which *coincides with the plane*
S_D, must thus vary continuously with y. The remaining components $\sigma_{xx}, \sigma_{xz}, \sigma_{zz}$ need not be continuous functions of y.

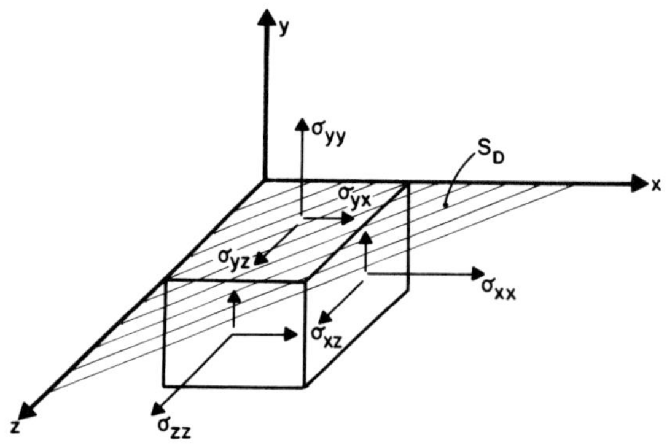

Figure 2. Element of material on plane of discontinuity

The differential equations of equilibrium,

$$\sigma_{ij}\nu_j + F_i = 0 \tag{5}$$

must be satisfied on each side of the discontinuity. Note that these equations, which are given fully as

$$\frac{\partial\sigma_{xx}}{\partial x} + \frac{\partial\sigma_{xy}}{\partial y} + \frac{\partial\sigma_{yz}}{\partial z} + F_x = 0 \ ,$$

$$\frac{\partial\sigma_{xy}}{\partial x} + \frac{\partial\sigma_{yy}}{\partial y} + \frac{\partial\sigma_{yz}}{\partial z} + F_y = 0 \ , \tag{6}$$

$$\frac{\partial\sigma_{xz}}{\partial x} + \frac{\partial\sigma_{yz}}{\partial y} + \frac{\partial\sigma_{zz}}{\partial z} + F_z = 0 \ .$$

do not contain the partial derivatives $\partial\sigma_{xx}/\partial y$, $\partial\sigma_{xz}/\partial y$, $\partial\sigma_{zz}/\partial y$, which become delta functions if the stress conponents are discontinuous. The partial derivatives $\partial\sigma_{xy}/\partial y$, $\partial\sigma_{yy}/\partial y$, $\partial\sigma_{yz}/\partial y$ may or may not be continuous functions of y; in any event they are not delta functions on the plane S_D.

It may also be remarked that the free surface of any body is a surface of discontinuity of the stress field. The usual boundary condition

$$T_i = \sigma_{ij}\nu_j \ , \tag{7}$$

where T_i is the surface traction, imposes similar conditions on the stress field at the free surface to those imposed on the stress field on an interior surface on which the stress field is not continuous.

In a similar manner, consider a body in which the velocity field is denoted by $\dot{u}_i$. Let the body contain a surface S_D along which the velocity field is discontinuous. Clearly if the adjacent sides of the body move away from each other, creating vacancies along the surface, or move into each other, causing one point in space to be occupied by more than one particle, the requirements of compatibility will be violated. However, it is admissible that one part of the body should slide along S_D with respect to the other. Thus we require only that the component of the velocity field normal to the surface S_D should be continuous. If ν_j is a unit vector normal to S_D, this is expressed as condition that

$$[\dot{u}_i \nu_i]_{S_D} = 0 \ . \tag{8}$$

The component of $\dot{u}_i$ lying in the surface S_D need not be continuous across the surface.

This condition can also be better appreciated by studying a simple situation. Consider a plane of discontinuity with the y-axis perpendicular to the plane (Figure 3a). The normal unit vector has components $(0,1,0)$. Let the components of $\dot{u}_i$ be $\dot{u}$, $\dot{v}$, $\dot{w}$ respectively. Equation (8) then requires that

$$[\dot{v}]_{S_D} = 0 \tag{9}$$

i.e. that $\dot{v}$ should be a continuous function of y. The velocity components $\dot{u}$ and $\dot{w}$ need not be continuous, indicating that $\partial \dot{u}/\partial y$ and $\partial \dot{w}/\partial y$ may be delta functions on the plane S_D. Note that the shear strain rates

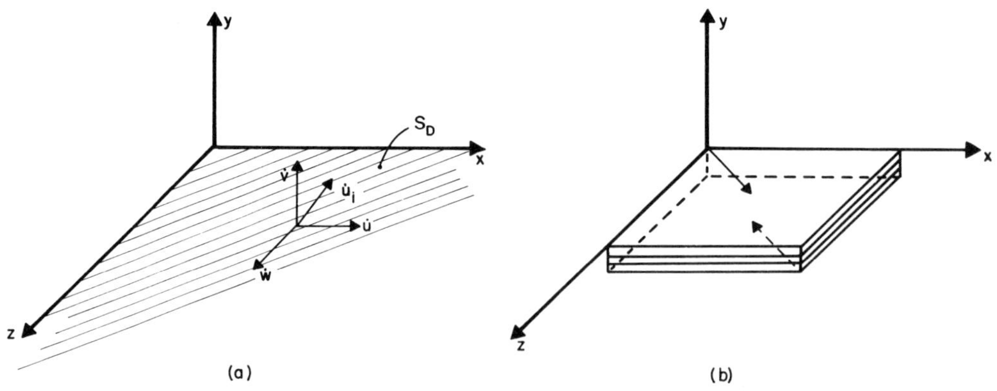

Figure 3. Discontinuity in the velocity field

$$\dot{\varepsilon}_{yz} = \frac{1}{2} \left(\frac{\partial \dot{w}}{\partial y} + \frac{\partial \dot{v}}{\partial z} \right)$$

$$\dot{\varepsilon}_{yx} = \frac{1}{2} \left(\frac{\partial \dot{u}}{\partial y} + \frac{\partial \dot{v}}{\partial x} \right)$$

(10)

contain these derivatives and thus become infinitely large on S_D. The plane of discontinuity is thus a plane of intense shearing. Figure 3(b) shows a thin sandwich in which the shear strain rates $\dot{\varepsilon}_{yz}$ and $\dot{\varepsilon}_{yx}$ are very large. The limiting case obtained when the thickness of the sandwich goes to zero is the case of a discontinuous velocity field which satisfies the compatibility relations.

Finally we may remark that a surface S_D may be a surface of discontinuity on both stress and velocity. However, the rate of dissipation along the discontinuity is always finite, just as it is in a plastic hinge on a beam or a hinge line in a plate. Considering our simple illustrative example, note that while $\dot{\varepsilon}_{xy}$ and $\dot{\varepsilon}_{yz}$ are delta functions, σ_{xy} and σ_{xz} are continuous (equation 4). Hence the rate of dissipation of energy per unit area of the plane of the discontinuity is readily seen to be

$$\int_{S_D} \sigma_{ij} \dot{\varepsilon}_{ij} ds = 2(\sigma_{xy}[\dot{u}]_{S_D} + \sigma_{xz}[\dot{w}]_{S_D}) \ . \tag{11}$$

The general form of this result can be obtained by considering the surface of discontinuity S_D as a film of finite thickness, as shown in Figure 4. Denoting values of field quantities on either side of the film by superscripts + and − we use the principle of virtual velocities to write

$$\int_{S_D} \sigma_{ij} \dot{\varepsilon}_{ij} ds = \int_{S_D} T_i^+ \dot{u}_i^+ ds + \int_{S_D} T_i^- \dot{u}_i^- ds \ , \tag{12}$$

where T_i^+ and T_i^- are the tractions on either side of the surface.

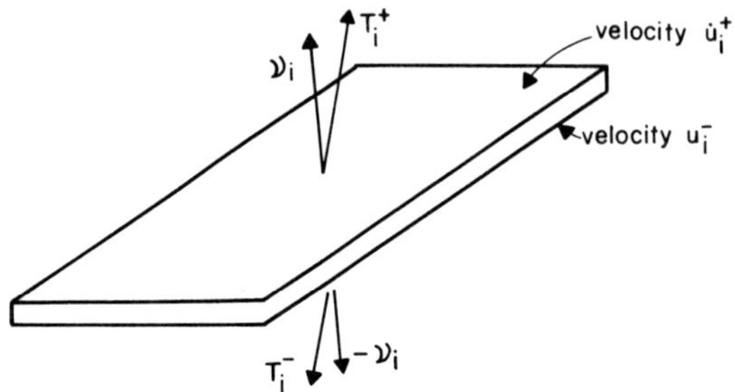

Figure 4. Work rate on a surface of discontinuity

Equation (12) remains valid when the film thickness tends to zero. However, from equations (1), (2) and (3),

$$T_i^+ \;=\; -T_i^- \;=\; \sigma_{ij}\nu_j \;\;,\tag{13}$$

so that

$$\int_{S_D} \sigma_{ij}\dot{\varepsilon}_{ij}\,ds \;=\; \int_{S_D} \sigma_{ij}\nu_j[\dot{u}_i]_{S_D}\,ds \;\;.\tag{14}$$

It may now be noted that $\sigma_{ij}\nu_j$ is a vector which can be resolved into a component σ_N normal to S_D, in the direction ν_i, and a component σ_s lying in S_D in the direction of a unit vector μ_i. σ_N is the normal stress on the surface of the discontinuity and σ_s is the *resolved shear* stress on the surface S_D. It is evident that (see section 3.2)

$$\sigma_N \;=\; \sigma_{ij}\nu_j\nu_i \;\;,\tag{15}$$

and that

$$\sigma_s^2 \;=\; \sigma_{ij}\nu_j\sigma_{ik}\nu_k - \sigma_N^2 \;\;.\tag{16}$$

The vector u_i can be obtained from the equation

$$\sigma_{ij}\nu_j \;=\; \sigma_N\nu_i + \sigma_s\mu_i \;\;.\tag{17}$$

Further, on substituting (17) into (14), and using equation (8), we see that

$$\int_{S_D} \sigma_{ij}\dot{\varepsilon}_{ij}\,ds \;=\; \int_{S_D} (\sigma_N \nu_i + \sigma_s \mu_i)[\dot{u}_i]_{S_D}\,ds$$

$$=\; \int_{S_D} \{\sigma_N [\nu_i \dot{u}_i]_{S_D} + \sigma_s [\mu_i \dot{u}_i]_{S_D}\}\,ds \qquad\qquad (18)$$

$$=\; \int_{S_D} \sigma_s [\mu_i \dot{u}_i]_{S_D}\,ds \quad .$$

Thus the dissipation rate per unit area on the surface of discontinuity S_D is the product of the resolved shear stress and
the component of the jump in the velocity field $\dot{u}_i$ in the direction of the resolved shear stress. We shall make extensive
use of this result in the application of the upper bound theorem.

12.2 The Tresca Yield Condition in Plane Stress and Plane Strain

Throughout this Chapter we shall be dealing with simple assumed
stress and velocity fields. It will be convenient to carry out
the discussion in terms of *principal stresses* and velocity discontinuities which involve sliding along surfaces of discontinuity. These factors contribute to lead us to choose the
Tresca condition for the limit surface (or yield surface of the
material is taken to be elastic, perfectly plastic or rigid,
perfectly plastic) rather than the von Mises condition.

The general form of the Tresca condition was developed in
Chapter 4. In our present context, flow is assumed to be possible when the maximum value of the resolved shear stress on
any plane passing through a point reaches a critical value, say
k. The maximum value of the resolved shear stress is one half
the difference between the greatest and the least principal
stress. Alternatively we introduce six yield functions

$$\phi_{1,2} = {}^{+}_{-}(\sigma_I - \sigma_{II}) - 2k \quad , \tag{19a}$$

$$\phi_{3,4} = {}^{+}_{-}(\sigma_{II} - \sigma_{III}) - 2k \quad , \tag{19b}$$

$$\phi_{5,6} = {}^{+}_{-}(\sigma_I - \sigma_{III}) - 2k \quad . \tag{19c}$$

The yield surface, it will be recalled, is a hexagonal prism in the principal stress space with its axis coinciding with the line $\sigma_I + \sigma_{II} + \sigma_{III} = 0$. The principal strain rates are given by

$$\dot{\varepsilon}_I = +\lambda_1 \quad \text{where} \quad \lambda_1 \geq 0 \quad \text{if} \quad \phi_1 = 0 \quad ,$$

$$\dot{\varepsilon}_{II} = -\lambda_1 \qquad\qquad \lambda_1 \leq 0 \quad \text{if} \quad \phi_2 = 0 \quad , \tag{20a}$$

$$\dot{\varepsilon}_{III} = 0 \qquad\qquad \lambda_1 = 0 \quad \text{if} \quad \phi_1 < 0 \quad , \quad \phi_2 < 0 \quad .$$

$$\dot{\varepsilon}_{II} = +\lambda_2 \quad \text{where} \quad \lambda_2 \geq 0 \quad \text{if} \quad \phi_3 = 0 \quad ,$$

$$\dot{\varepsilon}_{III} = -\lambda_2 \qquad\qquad \lambda_2 \leq 0 \quad \text{if} \quad \phi_4 = 0 \quad , \tag{20b}$$

$$\dot{\varepsilon}_I = 0 \qquad\qquad \lambda_2 = 0 \quad \text{if} \quad \phi_3 < 0 \quad , \quad \phi_4 < 0 \quad .$$

$$\dot{\varepsilon}_I = +\lambda_3 \quad \text{where} \quad \lambda_3 \geq 0 \quad \text{if} \quad \phi_5 = 0 \quad ,$$

$$\dot{\varepsilon}_{III} = -\lambda_3 \qquad\qquad \lambda_3 \leq 0 \quad \text{if} \quad \phi_6 = 0 \quad , \tag{20c}$$

$$\dot{\varepsilon}_{II} = 0 \qquad\qquad \lambda_3 = 0 \quad \text{if} \quad \phi_5 < 0 \quad , \quad \phi_6 < 0 \quad .$$

In this formulation we take advantage of the fact that ϕ_I and ϕ_2 cannot be simultaneously equal to zero, and so on. The multipliers $\lambda_1, \lambda_2, \lambda_3$ are otherwise unspecified, and no yield function is permitted to be greater than zero.

Each of the principal strain states in equations (20) represents pure shear, with shearing occurring along planes parallel to the plane of maximum shear stress. These planes (cf. Chapter 4) are shown in Figure 5 together with the principal stresses and the appropriate yield conditions. In each case the resolved

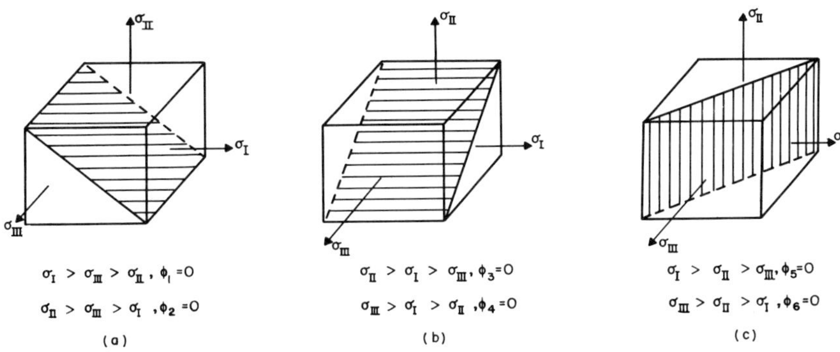

Figure 5. Slip directions for Tresca yield condition

shear stress on the plane of shearing in the direction of sliding is k. The specific rate of dissipation of energy in each
case can be expressed as $k\dot{\gamma}$, where $\dot{\gamma}$ is the *engineering* shear
strain rate (i.e. twice the tensorial shear strain rate) in
each case. To demonstrate this, consider flow with $\phi_1 = 0$. The
specific rate of dissipation is

$$
\begin{aligned}
\sigma_{I}\dot{\varepsilon}_{I} + \sigma_{II}\dot{\varepsilon}_{II} + \sigma_{III}\dot{\varepsilon}_{III} &= \sigma_{I}\lambda_1 - \sigma_{II}\lambda_1 \\
&= (\sigma_{I} - \sigma_{II})\lambda_1 \\
&= 2k\lambda_1 \\
&= k(\dot{\varepsilon}_{I} - \dot{\varepsilon}_{II}) \\
&= k\dot{\gamma} \quad ,
\end{aligned}
\tag{21}
$$

where $\dot{\gamma} = \dot{\varepsilon}_{I} - \dot{\varepsilon}_{II}$.

Consider now the specific case of *plane stress*, which has
been discussed previously in Chapter 5. Let us assume that the
thin sheet of material lies in the x-y plane, with the z direction normal to the plane of the sheet. The plane stress assumption requires that

$$
\sigma_{zz} = \sigma_{xz} = \sigma_{yz} = 0 \quad .
\tag{22}
$$

It follows that the z-axis is always a principal stress

direction, and that the other principal stress directions lie in the plane of the sheet. Putting $\sigma_{III} = \sigma_{zz} = 0$, the yield functions of equation (19) become

$$\phi_{1,2} = \pm(\sigma_I - \sigma_{II}) - 2k \quad , \tag{23a}$$

$$\phi_{3,4} = \pm\sigma_{II} - 2k \quad , \tag{23b}$$

$$\phi_{5,6} = \pm\sigma_I - 2k \quad . \tag{23c}$$

The yield surface is then the hexagonal figure shown in Figure 6.

It should be noted that the restriction that $\sigma_{III} = 0$ still permits shearing along any one of the planes shown in Figure 5. The situation shown in Figure 5(a), which will occur when σ_I and σ_{II} have *opposite* signs, represents shearing along planes which are perpendicular to the sheet. The other cases, when σ_I and σ_{II} have the same sign, represent shearing along planes inclined at 45° to the sheet: the velocity field will have a component in the z-direction.

Consider next the specific case of plane strain. The body is assumed to be infinitely large in, say, the z-direction, so that the component of the velocity field in that direction is identically zero from symmetry considerations. Furthermore, the remaining field quantities are constant in the z-direction.

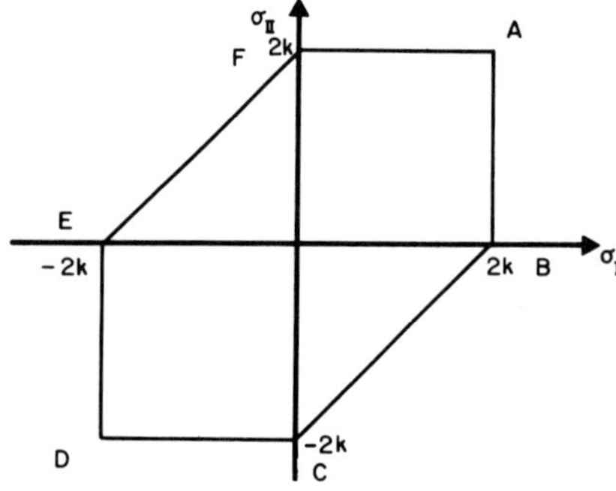

Figure 6. Tresca condition in plane stress

These restrictions lead to the conclusion that

$$\dot{\varepsilon}_{zz} = \dot{\varepsilon}_{zx} = \dot{\varepsilon}_{zy} = 0 \quad . \tag{24}$$

Again, it follows that $\dot{\varepsilon}_{zz}$ is always a principal strain rate and that the z axis is always a principal direction. Let us assume, therefore, that

$$\dot{\varepsilon}_{III} = \dot{\varepsilon}_{zz} = 0 \quad . \tag{25}$$

If this result is compared with the strain rates predicted by the flow rule (equations 20) and shown in Figure 5, we are led to the conclusion that the plane strain restrictions can only be met if *flow does not occur* with either ϕ_3, ϕ_4, ϕ_5 or $\phi_6 = 0$. This in turn can only be achieved if $\sigma_{III} = \sigma_{zz}$ always takes a value such that it is the intermediate principal stress; thus we find that either $\sigma_I \geq \sigma_{zz} \geq \sigma_{II}$ or $\sigma_{II} \geq \sigma_{zz} \geq \sigma_I$. Apart from this restriction, σ_{zz} is not uniquely specified by the Tresca condition. Since σ_{zz} is always a principal stress, it also follows that $\sigma_{xz} = \sigma_{yz} = 0$. The analysis can consequently be carried out in terms of the stress components σ_{xx}, σ_{xy} and σ_{yy} and the strain rate components $\dot{\varepsilon}_{xx}$, $\dot{\varepsilon}_{xy}$ and $\dot{\varepsilon}_{yy}$, or in terms of the principal stresses and strains σ_I, σ_{II}, $\dot{\varepsilon}_I$, $\dot{\varepsilon}_{II}$. The yield condition and the flow rule are given by equations (19a) and (20a). The remaining stress component σ_{zz} then adopts a suitable value. It should be noted that if we put

$$\sigma_{zz} = \sigma_{III} = \frac{\sigma_I + \sigma_{II}}{2} \tag{26}$$

σ_{III} will always be the intermediate principal stress. The von Mises yield condition, in the case of plane strain, leads to equation (26) directly.

12.3 Symmetrical Internal and External Notches in a Rectangular Bar

In studying the fracture of metals, experiments are frequently
carried out on notched rectangular bars. If the material is
ductile and can with reasonable accuracy be approximated by an
elastic, perfectly plastic material, it is extremely important
in the interpretation of the experimental results to compare
the load at which failure occurs with the limit load of the bar.
If failure occurs significantly below the limit load the problem is essentially one of *brittle fracture*; if, however, it
occurs at or above the limit load considerable plastic deformation will take place. In the latter case failure will be primarily due to instability (necking) or other phenomena associated with large inelastic deformation. In problems of this type
limit analysis provides an important criterion to distinguish
between two very distinct physical mechanisms.

As an example of the limit analysis problem in this context,
consider a bar of width b and thickness t subjected to a tensile force. We shall not concern ourselves with the end conditions; it is assumed that the axial force is applied in such
a way that the limit load of the unnotched bar (which is given
by 2kbt) can be sustained and such that bending moments are not
induced in the bar. If the thickness t is very small compared
with the width b the plane stress assumptions will be applicable, while if t is very large compared to b the plane strain
assumptions may be used. In order to compare the two extremes
we shall compute the tensile force per unit thickness P.

Consider first the case where we have a symmetrical internal
slit or crack of width a which is perpendicular to the axial
force and extends uniformly through the thickness. The bar is
shown in Figure 7(a) together with a coordinate system.

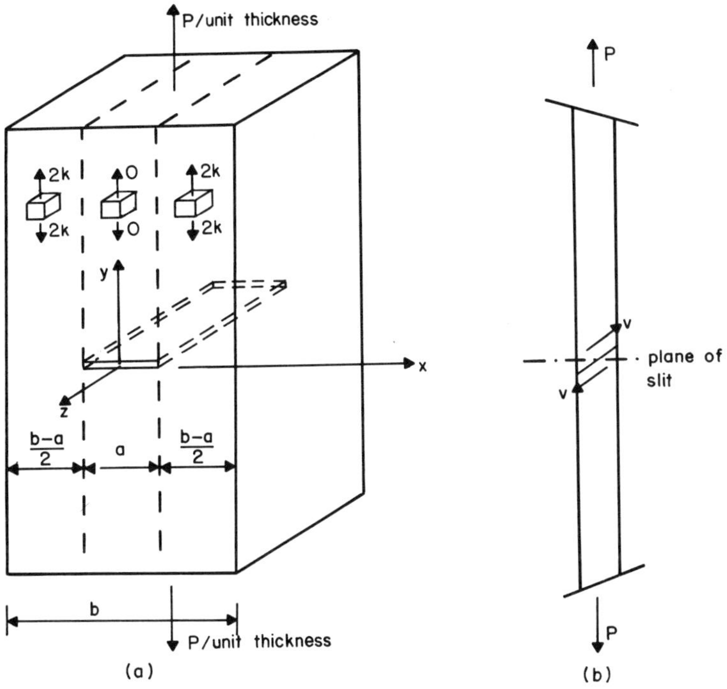

Figure 7. Bar with an internal notch

In the application of the lower bound theorem we are concerned only with equilibrium requirements. For t small, there-
fore, it is essential that we have σ_{zz}, σ_{zx} and σ_{zy} equal to
zero, but no further requirements need be introduced for the
plane strain case. The simplest stress field which meets the
requirements of the lower bound theorem is

$$\sigma_{yy} = 2k \quad \text{for} \quad a/2 \leq x \leq (b-a)2$$
$$\text{and} \quad -(b-a)/2 \leq x \leq -a/2 \quad , \qquad (27)$$

$$\sigma_{yy} = 0 \quad \text{for} \quad -a/2 \leq x \leq a/2 \quad ,$$

with all other stress components zero. This stress field is
discontinuous, the bar being divided, as shown in Figure 7(a),
into three regions in which the stress field is homogeneous.
However, the discontinuity in the stress field satisfies the
restriction of equation (3), and the equilibrium equations are

satisfied within each region where the stress is homogeneous.
The yield condition is satisfied in general (and certainly for
both plane stress and plane strain) and the plane stress conditions are also satisfied. For this case

$$P \;=\; (2k)\,\frac{2(b-a)}{2} \;=\; 2k(b-a) \quad , \tag{28}$$

and this will be a lower bound on the load which causes flow
for all values of t. This calculation simply assumes that the
yield stress in tension 2k can be achieved over the net area of
material at the cross section which contains the slit.

Consider an upper bound calculation for the case of $t \ll b$,
i.e. the case of plane stress. Rather than attempt to construct a continuous strain rate field, let us think in terms of
rigid blocks sliding with respect to each other along planes of
discontinuity of the velocity field. Any one of the shearing
deformations shown in Figure 5 is admissible in plane stress:
take first the case corresponding to flow when $\sigma_{yy} = 2k$ and
$\sigma_{zz} = 0$ is the smallest principal stress. As shown in Figure
7(b) assume that the plane along which the velocity is discontinuous is inclined at 45° to the plane of the thin sheet. The
velocity of the rigid blocks on either side of the sheet is
also shown. The component of the velocity field perpendicular
to the plane of sliding is zero, satisfying equation (8), and
the discontinuity in velocity parallel to the sheet in 2v. The
resolved shear stress on the plane of sliding is 2k, as given
by the Tresca yield condition. Hence the internal energy dissipation rate is

$$D_{int} \;=\; k(2v)\{\,\frac{2(b-a)}{2}\,t\sqrt{2}\,\}$$
$$\;=\; 2\sqrt{2}\,k(b-a)tv \quad . \tag{29}$$

The external work rate is

$$D_{ext} = 2 \frac{Ptv}{\sqrt{2}} = \sqrt{2}\,Ptv \quad . \tag{30}$$

Equating these quantities we find that

$$P = 2k(b-a) \quad . \tag{31}$$

The limit load is thus not less than $2k(b-a)$. However, we have found (equation 28) that this quantity is a lower bound on the limit load. Consequently flow will occur when $P = 2k(b-a)$.

Two points should be noted. First, the fact that we have determined the limit load exactly does not imply that the stress or strain rate fields are necessarily those which will in fact be found during flow in an elastic, perfectly plastic material. Discontinuous stress fields would very seldom occur in plane stress when elastic deformations are present, and we shall see shortly that alternate velocity fields also give the exact limit load. Secondly, the upper bound computation of equation (31) is strictly true only for plane stress. The plane of sliding is shown in greater detail in Figure 8. It can be seen that sliding also occurs on the four triangular shaded areas adjacent to the ends of the crack, and the contribution of these areas was not taken into account in equation (29). The velocity discontinuity parallel to these planes is v, and consequently for arbitrary t

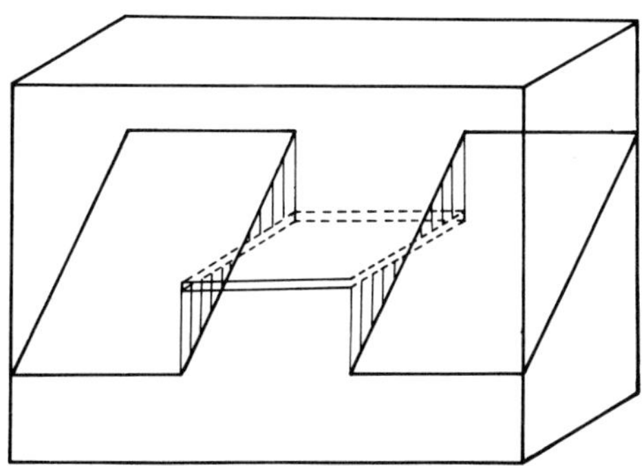

Figure 8. Velocity field for internal notch

$$D_{int} = 2\sqrt{2}\ k(b-a)tv + 4kv\ \frac{t^2}{8}$$
$$= \{2\sqrt{2}(b-a)t + \frac{t^2}{2}\}\ kv\ . \tag{32}$$

Equating this to the external work rate of equation (30), which remains unchanged for arbitrary t, we see that

$$P = \{2(b-a) + \frac{t}{2\sqrt{2}}\}\ k$$
$$= 2k(b-a)\ \{1 + \frac{1}{4\sqrt{2}}\ \frac{t}{(b-a)}\}\ . \tag{33}$$

This expression reduces to equation (31) when t is small com-pared to (b-a), a statement which more properly expresses the conditions under which the plane stress assumptions are appli-cable in the vicinity of the crack tip. It is clear that as t becomes large compared to (b-a) this upper bound becomes large, and hence it cannot be expected that the bound will be of any value in the plane strain case.

We also have the opportunity of using the shearing deforma-tion shown in Figure 5(a), corresponding to the case of flow when $\sigma_{yy} = 2k$ and $\sigma_{xx} \leq 0$ is the smallest principal stress. As shown in Figure 9, introduce two parallel planes normal to the x-y plane and inclined at 45° to the x and y axes, each emanting from the root of the crack. This velocity field is admissible

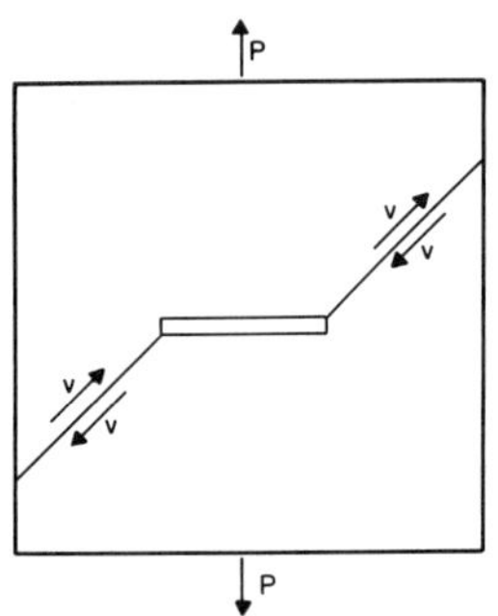

Figure 9. Velocity field for internal notch

for plane stress and plain strain, and for arbitrary t. Let
the velocity discontinuity be as shown in Figure 9. It is
apparent that

$$D_{ext} = 2Pt \frac{v}{\sqrt{2}} , \qquad (34)$$

and

$$D_{int} = 2k(2v) \{\frac{(b-a)}{\sqrt{2}} t\}$$

$$= 2\sqrt{2} (b-a) tkv . \qquad (35)$$

Equating these work rates, we see that

$$P = 2k(b-a) \qquad (36)$$

is not less than the limit load for all t. Comparing this to
equation (28), which is also valid for all t, we find that
flow occurs when P = 2k(b-a) for plates of any thickness with a
symmetrical internal crack.

Let us now compare these results to the case of a similar
bar, with width b and thickness t, with symmetric external slits
or cracks each of width a/2 and uniform through the thickness,
as shown in Figure 10. A little study will reveal that the
lower bound for the internal crack based on the stress field
shown in Figure 7 can be applied to the symmetrical external
crack problem. In essence we assume that yield stress in

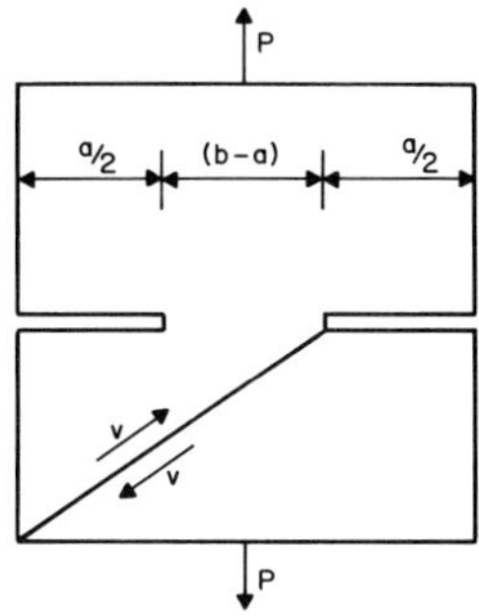

Figure 10. Bar with external notches

tension is developed over the net area of the cross section
containing the crack, and continue the stress field over the
entire bar.   Thus

$$P = 2k(b-a) \tag{37}$$

remains a lower bound on the limit load. Similarly, the velocity field shown in Figures 7(b) and 8 can be used with a geometrical rearrangement.   The upper bound is not affected: hence
flow will certainly occur for

$$P = 2k(b-a) \left\{1 + \frac{1}{4\sqrt{2}}\frac{t}{(b-a)}\right\} \tag{38}$$

However, when sliding is assumed to occur along a plane perpendicular to the x-y plane and inclined at $45°$ to the x and
y axes, we find a significant difference from the velocity
field shown in Figure 9.   Because of the rearrangement of the
slits, a plane of sliding of the form shown in Figure 10 must
be adopted. This plane has a different area from that of
Figure 9. The internal work rate is

$$D_{int} = k(2v)\sqrt{2}\left(b - \frac{a}{2}\right) t \; . \tag{39}$$

The external work rate is unchanged from that given in equation
(34), and consequently an upper bound on the limit oad is

$$P = 2k(b-a/2) = 2k(b-a) \left\{1 + \frac{a}{2(b-a)}\right\} \; . \tag{40}$$

While this result is valid for all t it does not coincide with
the lower bound of equation (37). The difference between the
bounds depends on the ratio a/b, being small when a/b is small
and becoming progressively larger as a/b increases. Figure 11
shows a comparison of the lower bound and the upper bounds of
equations (38) and (40) for some ratios of a/b. For small
cracks, at least, the bounds are very close. Further analysis

would be required for deep cracks; in the next section we shall
obtain improved bounds for the plane strain case (i.e. t large
compared to b-a) for large ratios of a/b.

12.4 The Punch Problem in Plane Strain

In the classical and frequently studied punch problem we attempt
to find the load required to indent a rigid punch into a block
of metal. If the material is perfectly plastic the limit load
provides an estimate of the load required to initiate the indentation. The simplest formulation of the problem treats the
indentation of an infinitely long punch of width h into a half
space of perfectly plastic material. Considering the plane
perpendicular to the punch, the problem is one of plane strain.
Let P be the load per unit length in the direction perpendicular
to the x-y plane. We shall assume that the contact between the
punch and the half space is smooth, so that the punch may apply

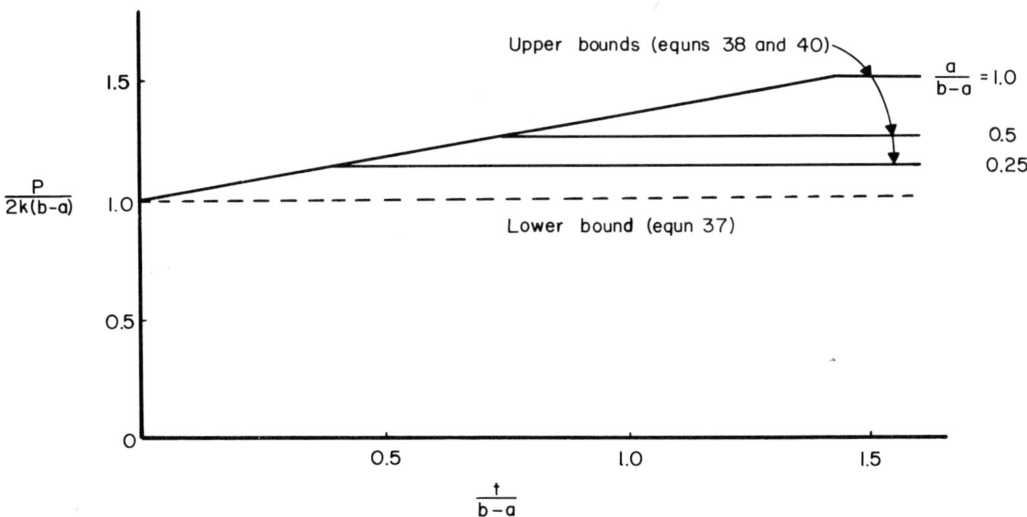

Figure 11. Bounds for the externally notched bar

only normal tractions to the surface over the width h.

A first attempt at the computation of a lower bound on the limit load may exploit the result we obtained in the previous section: consider the discontinuous stress field shown in Figure 12(a). The stress field is homogeneous in each of the three regions, the equilibrium equations are satisfied in each region, and the restrictions along the lines of discontinuity are satisfied. Clearly flow will occur for a load which is larger than or equal to

$$P = 2kh \quad . \tag{41}$$

In this computation it is imagined that the load is carried by a single vertical strip of material. Let us attempt to improve the lower bound by considering additional strips, each of which carries the yield stress in compression. These strips may then be superposed, and the stress state in the overlapping regions must be found.

Figure 12(b) shows the superposition of two strips each inclined at 30° to the x-axis. The stress state in the triangular area ABC is the sum of the stress states in the two inclined strips. We thus have six regions in which the stress field is homogeneous, and consequently the equilibrium equations are satisfied. The restrictions on discontinuities in the stress are met, from our previous experience, on all lines of discontinuity except AC and BC. We are required, therefore, to consider two questions; whether the stress discontinuities along AC and BC are admissible and whether the stress state within the triangle ABC is admissible.

We may assure ourselves that the stress discontinuity is admissible by recalling that the restriction on the jump in stress takes the form

$$[\sigma_{ij}\nu_j]_{S_D} = 0 \quad . \tag{42}$$

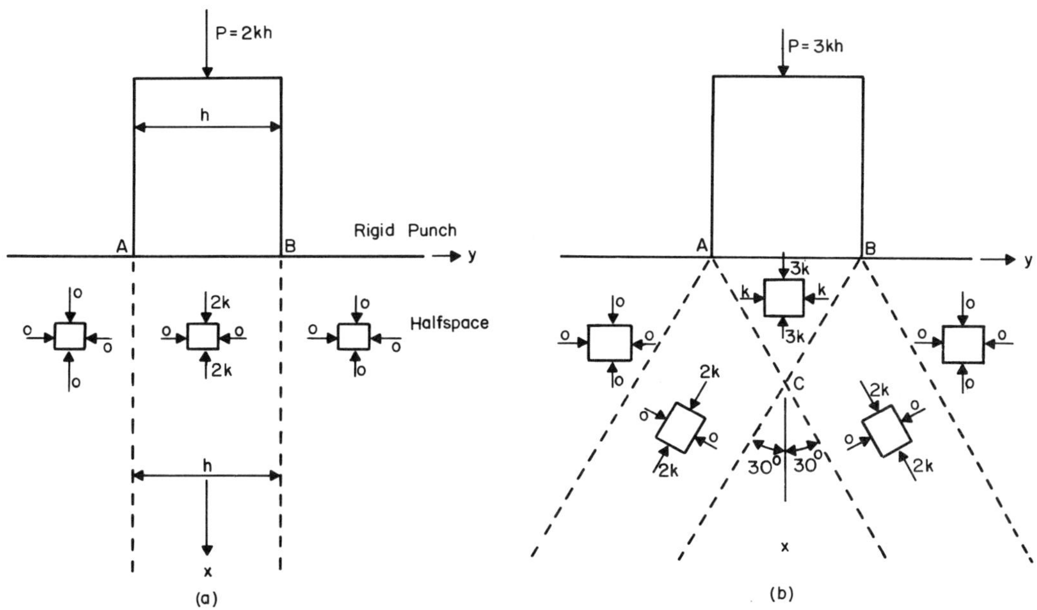

Figure 12. Stress fields for punch indentation

This expression is linear in σ_{ij}. Consequently if σ_{ij} is the sum of two stress fields, $\sigma_{ij}^{(1)}$ and $\sigma_{ij}^{(2)}$, equation (42) becomes

$$[\sigma_{ij}^{(1)}\nu_j + \sigma_{ij}^{(2)}\nu_j]_{S_D} = 0 \ . \tag{43}$$

A sufficient condition that equation (43) should hold is that

$$[\sigma_{ij}^{(1)}\nu_j]_{S_D} = 0 \ , \quad [\sigma_{ij}^{(2)}\nu_j]_{S_D} = 0 \ . \tag{44}$$

This is precisely the situation which we find in this example. The stress field in ABC is the sum of the stress field in ACED extended into ABC, and the stress field in BCFG extended into ABC. The first of these stress fields is continuous in all respects across AC, while the jump in the second across AC is admissible. Thus the discontinuities in stress across AC, and similarly across BC, are admissible.

In order to sum the stress fields of regions ACED and BCFG
it is necessary to express the two stress states in the same
coordinate system, add, and then redetermine the principal directions. Taking advantage of symmetry, let us express each
stress field in terms of a coordinate system coinciding with the
x-y system of Figure 12. A facility in the use of the Mohr
circle is a great advantage in calculations of this type. However, in this simple example we may carry out the transformation
of axes directly. Figures 13(a) and 13(b) show the force diagrams on the respective elements; these lead directly to the
stress states of Figures 13(c) and 13(d) respectively. It will
be remembered that the third principal stress, σ_{zz}, remains unchanged during a rotation of the coordinate system about the z
axis. Further, since the sum of the three direct stresses (the
diagonal terms on the stress tensor) is invariant, the sum of
the two direct stresses in the x-y plane remains unchanged. The

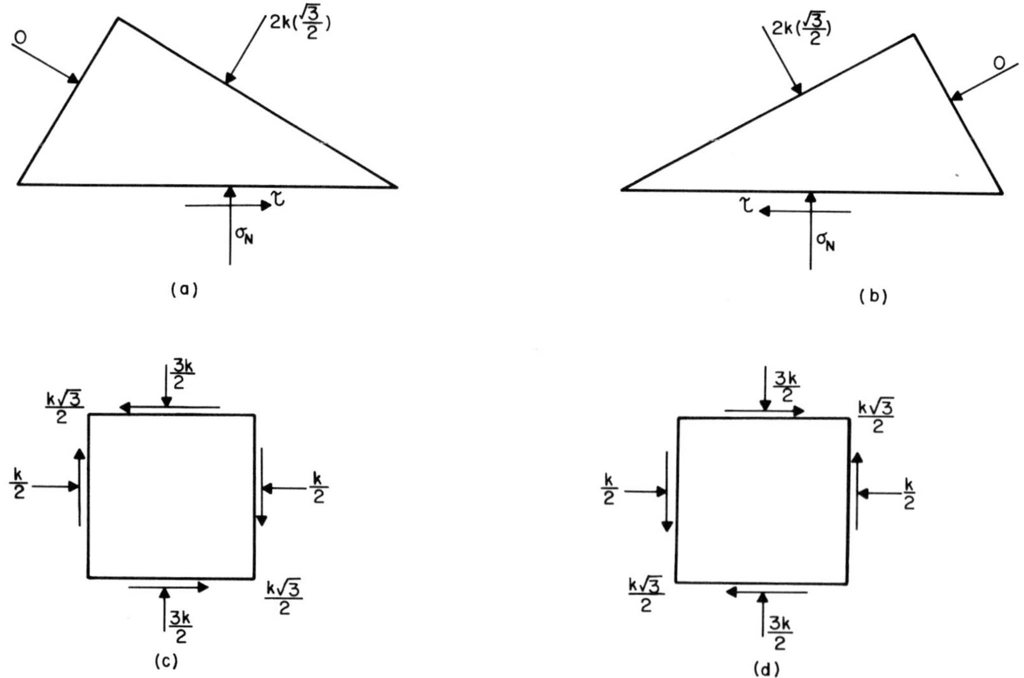

Figure 13. Transformed stress fields

sum of the stress states of Figures 13(c) and 13(d) leads fortuitously in this case to a principal stress state which is
shown in Figure 12(b). The difference between the principal
stresses is 2k, and consequently the yield condition is satis-
field. Thus

$$P = 3kh \tag{45}$$

is a lower bound on the limit load.

A third computation of a lower bound may be obtained by
superposing the stress fields of Figures 12(a) and 12(b). That
the stress discontinuities are admissible we accept from extensions of our argument that it is sufficient that each field
which makes up the sum should be admissible. We may note immediately, however, that the yield condition will not be satisfied in region ABC, since the difference in the greatest and
least principal stress is now 4k. This yield violation can be
accommodated by introducing a horizontal strip at the free
surface in which there is a horizontal compressive stress 2k.
If the width of this strip is as shown in Figure 14, the regions of homogeneous stress states can be seen. A calculation
involving transformations of axes need be carried out only for
triangle ADE; each of the other stress states is obtained
straighforwardly (note that in triangle ACD we add a state of
hydrostatic compression to the stress state in the inclined leg).
Thus

$$P = 5kh \tag{46}$$

is a lower bound on the limit load.

We have thus far obtained a dramatic improvement in the
lower bound by two further calculations. Further improvement
will necessitate dealing with more complex superposed stress
states, and it is advisable at this point to attempt upper bound

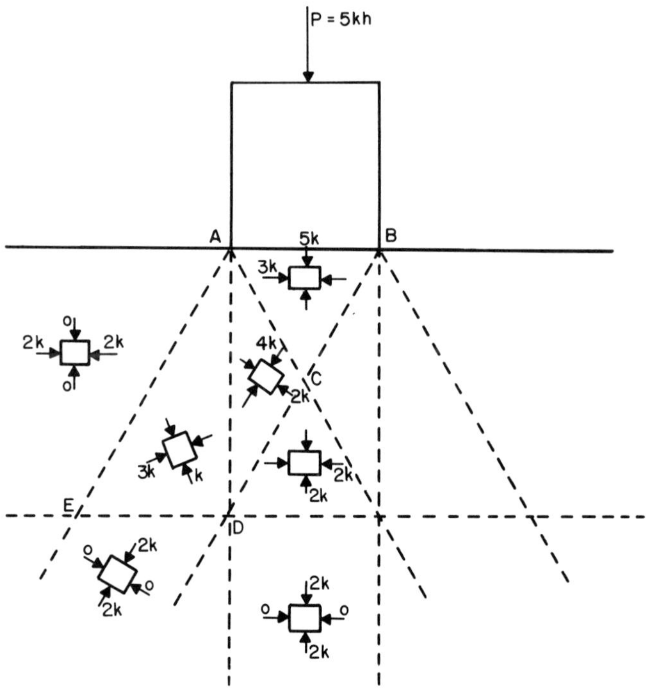

Figure 14. Combined stress field

calculations to determine whether any improvement is worthwhile.

In computing upper bounds we shall again make use of discontinuities in the velocity field, thinking at first in terms of rigid blocks sliding with respect to each other. A particularly simple velocity field is then shown in Figure 15(a). The punch rotates about the point B, and sliding takes place along a circular arc of radius h. If the mean velocity of the punch is v, the discontinuity in the tangential component of velocity along the circular arc is 2v. The velocity field is continuous (and zero) in the direction normal to the arc. Equating internal and external work rates we find that

$$Pv = k(2v)(\pi h) \quad,$$

or that

$$P = 2\pi kh \tag{47}$$

is not less than the limit load.

It is to be expected that the velocity field during flow will
be symmetric about the vertical plane through the center of the
punch, and that flow will lead to the extrusion of material on
either side of the punch. Figure 15(b) shows a discontinuous
velocity field which meets these requirements with sliding rigid
blocks. Letting the vertical velocity of the punch be v, the
velocities of the blocks are obtained by using the requirement
that the normal component of velocity across the lines AC and
AD should be continuous. The tangential component of velocity
is discontinuous across lines AC, AD, CD and DE. Equating external and internal work rates,

$$\frac{P}{2}\,v \;=\; k(\frac{h}{\sqrt{2}})(\sqrt{2}v) \;+\; k(\frac{v}{\sqrt{2}}) \;+\; khv \;+\; k(\frac{h}{\sqrt{2}})(\frac{v}{\sqrt{2}})\;.$$

Thus

$$P \;=\; 6kh \qquad\qquad\qquad\qquad (48)$$

is an upper bound on the limit load.

An improved bound can be found by eliminating the velocity
discontinuity across AD and reducing the dissipation rate along
CD or AC. This is accomplished in the velocity field shown in
Figure 15(c). We replace the triangle ACD by a quadrant of a
circle and assume that the tangential velocity at any point in

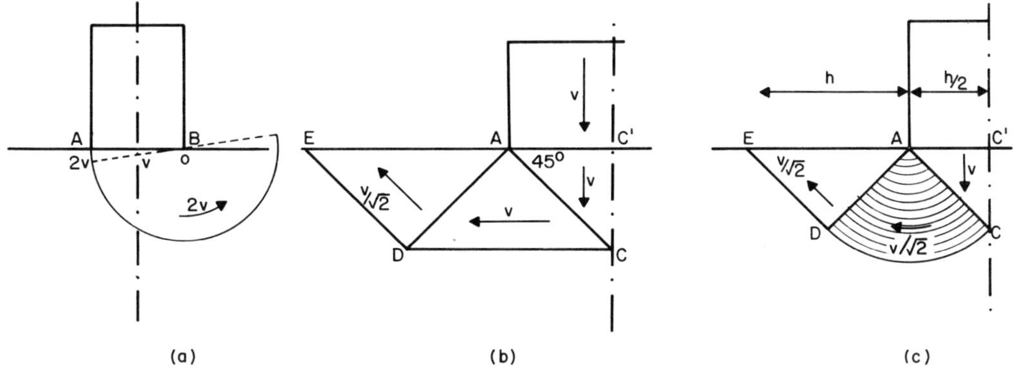

Figure 15. Velocity fields for punch indentation

the quadrant is $v/\sqrt{2}$ while the radial velocity is zero. This
leads to a state of pure shearing strain rate in ACD, with the
engineering shear strain $\dot{\gamma}$ given by

$$\dot{\gamma} \;=\; \frac{v}{r\sqrt{2}} \tag{49}$$

where r is the radial distance from A. The dissipation rate
within ACD is thus $kv/r\sqrt{2}$ per unit volume. The tangential velocity discontinuity along the arc CD and the line AC is now
$v/\sqrt{2}$, and there is no discontinuity along AD. Equating the external and internal work rates we find that

$$\frac{Pv}{2} \;=\; k\left(\frac{h}{\sqrt{2}}\right)\left(\frac{v}{\sqrt{2}}\right) \;+\; k\left(\frac{\pi h}{2\sqrt{2}}\right)\left(\frac{v}{\sqrt{2}}\right) \;+\; \int_{o}^{h/\sqrt{2}} \frac{kv}{r\sqrt{2}} \cdot \frac{\pi}{2}\, r\,dr \;+\; k\left(\frac{h}{\sqrt{2}}\right)\left(\frac{v}{\sqrt{2}}\right) \;.$$

Thus

$$P \;=\; (2+\pi)kh \tag{50}$$

is not less than the limit load.

A comparison of the lower bound of equation (46) and the
upper bound of equation (50) shows that we have obtained narrow
limits on the value of the limit load. To continue to improve
the bounds would obscure the major point in analyses of this
type; good estimates of the limit load can be obtained with
relatively little work. In fact the result given in equation
(50) is the correct value of the limit load.

The problem we have considered treats the indentation of a
punch of width h into a half space. More frequently the finite
dimensions of the body on which the punch acts are comparable
with the width of the punch h. Let us briefly study some results for two limiting cases: that where the body has finite
width and that where it has finite depth.

Suppose that the punch is symmetrically located on a body of

width b. The stress field of Figure 14 is no longer acceptable,
and hence the result P = 5kb is not necessarily a lower bound.
It can in fact be shown that the result of equation (50) is the
correct limit load for b $\geq$ 8.7h, but we shall not concern ourselves with stress fields for this case.

It may be noted that the upper bound derived from the velocity field of Figure 15(c) is valid for b $\geq$ 3h. Let us attempt
to find a lower bound for the base b = 3h. We can do so by modifying the stress field of Figure 12(b), restricting the width
of the regions in which the stress is non-zero to 3h. Such an
attempt is shown in Figure 16. The stress field is cut off by
the line DE, and we introduce additional lines of discontinuity
in stress CD and CE. The homogeneous stress states in the triangles ABC, ACD and BCE correspond to similar regions in Figure
12(b). Furthermore, in DEFG we assume that the stress is simply
compressive with a magnitude dictated by the magnitude of P.

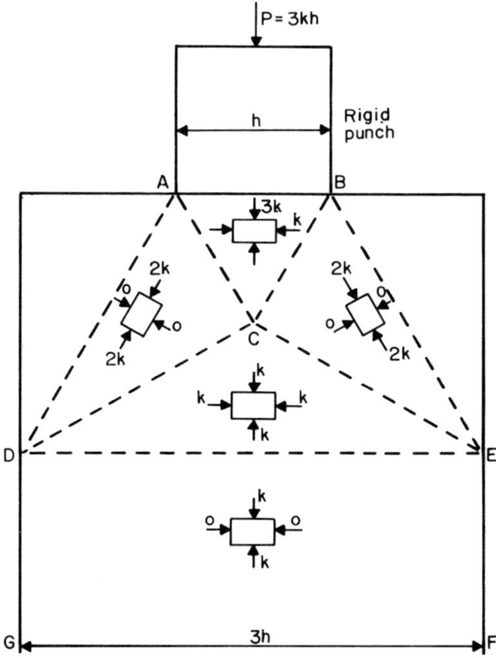

Figure 16. Punch on bar of finite width

If we can now find a homogeneous stress state in CDE such that
the restrictions on stress discontinuities are met along CD, CE
and DE and such that the yield condition is satisfied, P = 3kh
will be a lower bound. By symmetry consideration it can be expected that the principal stress directions in CDE will be the
vertical and horizontal directions. Moreover, the continuity
of tractions on DE requires that the principal stress in the
vertical direction should be compressive and equal to k. To
determine whether the horizontal principal stress in CDE, σ_{xx},
can be given a value such that the stress discontinuity along
CD is admissible, we consider the tractions acting on CD determined respectively from the stress states in ACD and CDE. The
force diagrams are shown in Figure 17; it can be seen that if
σ_{xx} = k the tractions will be continuous across CD. It follows
then that P = 3kh is a lower bound for this problem.

The punch problem in the case where the body is of finite
width is mathematically identical to the case of the bar with
symmetrical external notches under conditions of plane strain.
Comparing Figure 16 with Figure 10 it can be seen that if the
notched bar is subjected to a compressive load rather than a
tensile load we have through symmetry about the plane of the
notch precisely the same problem as that of the punch indented
into a block of finite width. The smooth contact between the

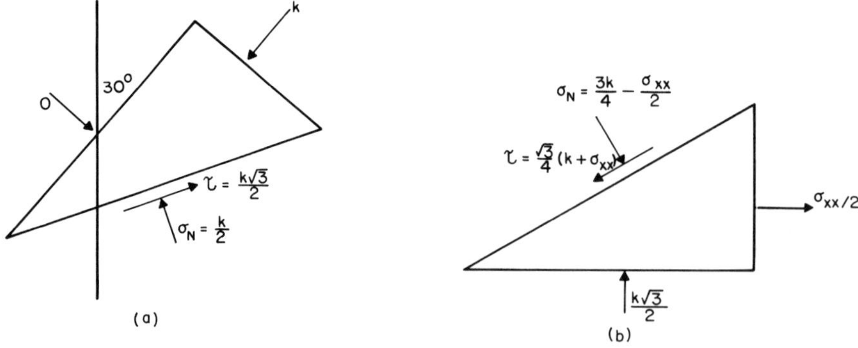

Figure 17. Stress transformations

punch and the half-space is acceptable because the shear stresses
along the plane of symmetry through the notch must vanish. The
notation used to describe the geometry in the two cases can be
reconciled by putting

$$h = b-a \ . \tag{51}$$

In comparing results for the case $b/h = 3$ or $a = 2b/3$ it will
be seen that the lower bound $P = 3kh$ or $P = 3k(b-a)$ is an improvement over that found in the external notch problem. On the
other hand, the upper bound of equation (40), obtained using the
velocity field of Figure 10, is an improvement over the upper
bound $P = (2+\pi)kh$ for the case $b/h = 3$. Substituting from
equation (51), the upper bound of equation (40) is

$$P = 4kh \ . \tag{52}$$

This is to be expected, since clearly when the block is of finite
width the velocity field of Figure 10 will lead to good upper
bounds when b/h is small. A short calculation will show that
the velocity field of Figure 10 is preferable to that of Figure
15(c) for $b/h \leq (1+\pi)$; equation (4) gives, on substituting
equation (51),

$$P = kh(1+b/h) \tag{53}$$

and thus an upper bound for any value of b/h.

Let us make one further attempt to improve the upper bound
for the punch indented into the finite block of width b. Consider the velocity field of Figure 18; this field is made up of
sliding rigid blocks with tangential velocity discontinuities
along AC, BC, CD and CE. The horizontal velocity u can easily
be calculated by the incompressibility condition: the rate at
which the punch indents into the material must be equal to the
rate at which material is extruded from the original configuration of the block. Thus

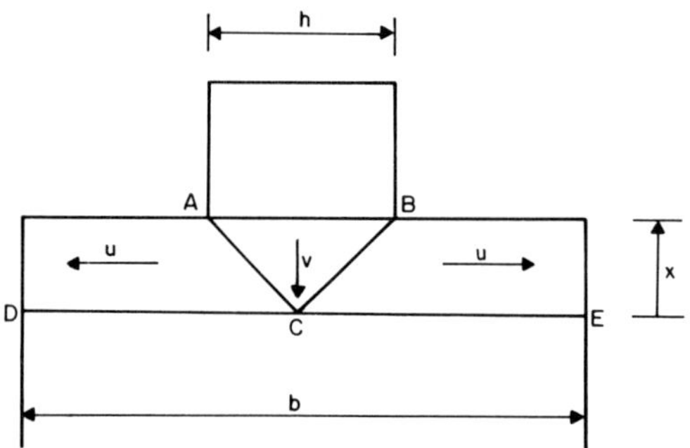

Figure 18. Velocity field for indentation of a bar of finite width

$$vh = 2ux ,$$
or
$$u = \frac{hv}{2x} ,$$
(54)

where x is the depth of the blocks which move horizontally. In terms of the angle θ, the tangential velocity discontinuity along AC or BC is (v cos θ+u sin θ). The length of the plane of discontinuity AC is x/cos θ. Thus the work rate equation gives

$$Pv = k(v \cos \theta + u \sin \theta) \frac{2x}{\cos \theta} + kbu$$
$$2kx(v + u \tan \theta) + kbu .$$
(55)

Now

$$\tan \theta = \frac{h}{2x} ,$$
(56)

and consequently, using equation (54) and putting x/h = ξ

$$\frac{P}{kh} = (2\xi + \frac{1}{2\xi}) + \frac{1}{2\xi} \frac{b}{h} .$$
(57)

This value of P is an upper bound for any choice of ξ and b/h. For any value of b/h we wish to choose that value of ξ which gives the least value of P. Hence, putting $\partial P/\partial \xi = 0$, the optimum choice of ξ is

$$\xi = \frac{\sqrt{1 + b/h}}{2} .$$
(58)

Substituting this expression into equation (57), an upper bound
on P is given by

$$\frac{P}{kh} = 2\sqrt{1+b/h} \quad . \tag{59}$$

The results we have so far obtained for the punch indented
into a block of finite width are plotted in Figure 19. The region in which the correct solution must lie is within the hatch
marks. Our efforts to obtain lower bounds have been limited,
and would require further attention, particularly for larger
values of b/h. Experience indicates, however, that it is easier
to obtain good upper bounds than good lower bounds in problems
of this type.

As a final example in this section, consider the case of two
aligned punches indented into a strip of finite thickness H.
The problem is again taken to be one of plane strain, requiring

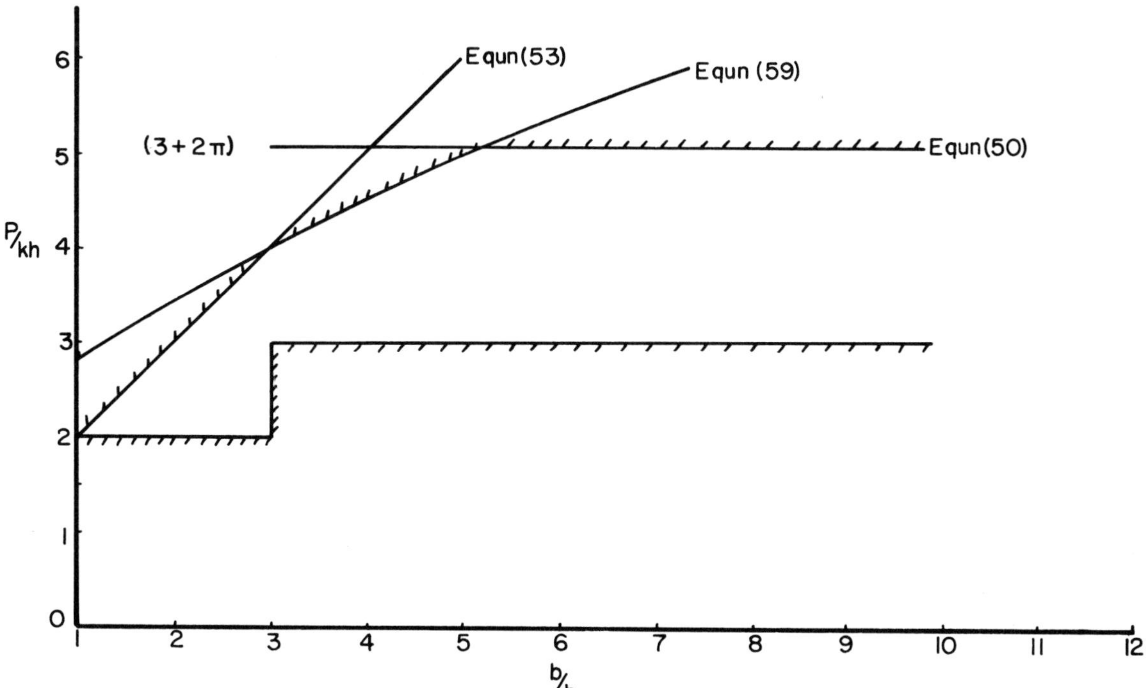

Figure 19. Results for the punch problem

that the strip and the punch extend a very large distance in
in the z direction. We also assume that the length of the strip
in the x direction is large compared to the punch width h. We
shall limit ourselves to upper bound calculations, initially
exploiting the results we have already obtained.

Returning to the problem of the punch indented into a half-
space, the velocity field of Figure 15(c), which provided the
best upper bound for the punch, extends into the half-space a
distance of $h/\sqrt{2}$. This velocity field may be used in our present problem provided that $h/\sqrt{2}$ is less than the half width H/2
of the strip. Thus, from the computations leading to equation
(50), an upper bound is given by

$$P = (2+\pi)kh \quad \text{for} \quad H/h > \sqrt{2} . \tag{60}$$

The velocity field of Figure 18 may also be modified to
apply to the present problem, as shown in Figure 20(a). Symmetry about the x axis (Figure 20a) removes the velocity discontinuity along DE in Figure 18 if we put x = H/2. In all
other respects the major characteristics of the two velocity
fields are identical. Thus, from equation (54)

$$u = \frac{h}{H} v , \tag{61}$$

and from equations (55) and (56), dropping the term describing
the internal work rate along DE in Figure 18, an upper bound on

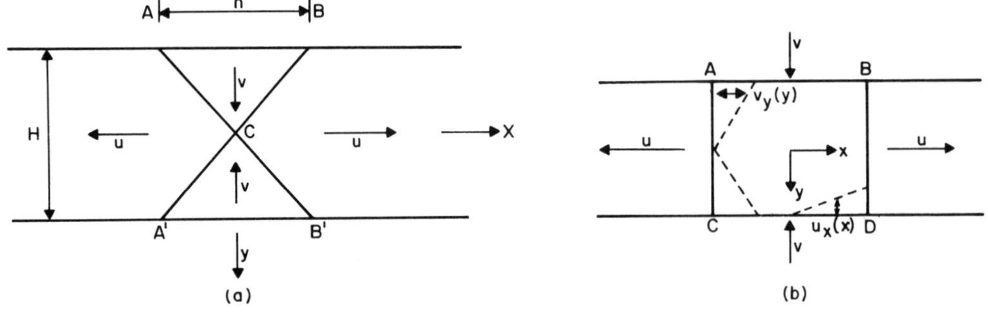

Figure 20. Velocity fields for strip of finite thickness

the limit load is given by

$$P = kh \left\{ \frac{H}{h} + \frac{h}{H} \right\} \ . \tag{62}$$

For some values of the width of the punch (i.e. for some values of H/h) it might be expected that a sliding block velocity field would be less appropriate than a velocity field which involves a continuous and largely compressive strain rate field under the punch. Figure 20(b) shows such a velocity field. The region ABCD is assumed to be in a homogeneous strain rate state. The horizontal velocity of the lines BC and AD is, by the compressibility condition, given by equation (61), u = hv/H. Within ABCD let u_x be the velocity component in the x direction and v_y be the velocity in the y direction. We choose

$$v_y = -2yv/H \tag{63}$$

$$u_x = 2xu/h = 2xv/H \ .$$

The velocity field is then continuous in the direction normal to the lines AB, BC, CD and DA. The tangential velocity discontinuity across BD and AC is v_y. There is also a tangential velocity discontinuity across AB and CD; however, we assume that the punch is smooth and this discontinuity consequently does not contribute to the dissipation rate. Within the block ABCD

$$\dot{\varepsilon}_{xx} = -\dot{\varepsilon}_{yy} = 2v/H \ , \tag{64}$$

and consequently the dissipation rate per unit volume is 4kv/H. Equating the internal and external work rates, an upper bound is given by

$$2Pv = 4 \int_o^{H/2} k\left(\frac{2yv}{H}\right) \, dy + hH \left(\frac{4kv}{H}\right) \ , \qquad \text{or}$$

$$\tag{65}$$

$$P = kh \left\{ 2 + \frac{1}{2} \frac{H}{h} \right\} \ .$$

The upper bounds of equations (60), (62) and (65) are plotted in Figure 21 as a function of H/h. The exact solution is shown by a dotted line, and it is remarkable that in this example three simple velocity fields provide a very good estimate of the limit load. Lower bound calculations would of course be required in the normal course of events, since an unsupported upper bound calculation cannot be accepted.

It has been our intention in this section to demonstrate how limit analysis can be used as an approximate and rapid method of obtaining information about the limit load. This approach exploits the intuitive understanding of a problem which an inexperienced analyst may possess, and can be extremely fruitful. It is possible, however, to formulate a mathematically systematic method of determining the exact solution for the stress and velocity fields in the deforming regions of bodies in plane strain. This method, known as slip line theory, exploits the

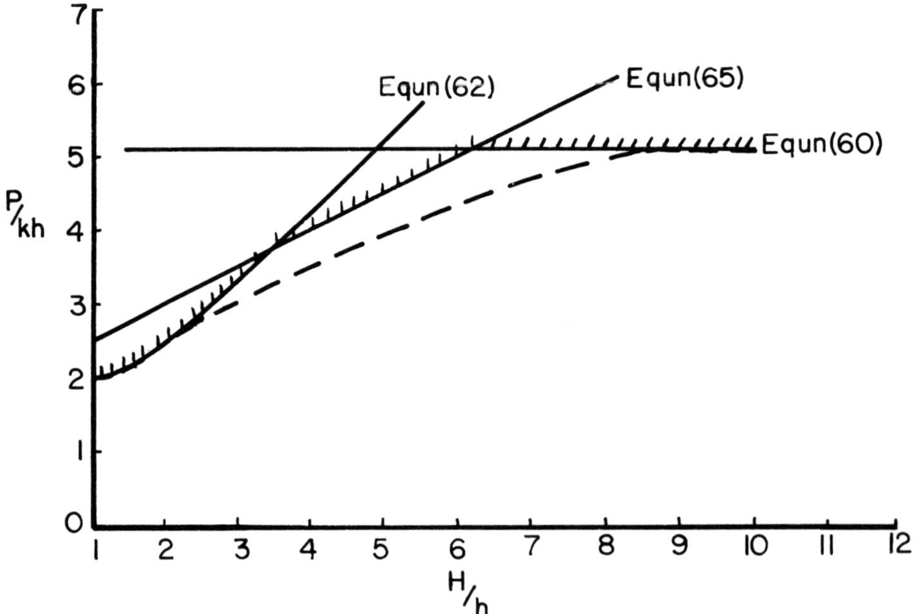

Figure 21.Results for punching of a strip of finite thickness

fact that the governing equations in the deforming region are
hyperbolic, and is essentially the method of characteristics
applied to the plane strain flow problem in plasticity. Complete solutions for the limit load are not obtained directly,
since the stress field in the rigid regions of the body are not
provided. However, the method is more informative than the use
of the upper bound theorem of limit analysis in that the stress
field in parts of the body is found. To complete the solution
this stress field must be extended into the rigid region without
violating the yield condition. While slip line theory is not a
general method of attacking limit analysis in plasticity, confined as it is to plane strain problems, it is of sufficient
general interest and technological importance that a brief outline of its major features will be given. This discussion will
be found in Chapter 15.

12.5 Some Remarks on Friction

In many problems in which limit analysis may be applied we find
metal surfaces in contact with each other with compressive
forces holding them together. The punch problem is one example
of this situation. Quite generally we expect that the traction
transmitted across the interface may have a normal compressive
component and a tangential component governed to a first approximation by the idealized linear friction relation between the
normal component of traction and the maximum permissible magnitude of the tangential component. That the maximum value of
the tangential component of the traction is limited suggests a
limit surface which determines whether or not sliding will occur. Moreover, when sliding does occur it is at least superfi-
cially similar to sliding which occurs along a discontinuity in
a velocity field in a rigid, perfectly plastic material; energy
is dissipated during the sliding, and the motion could be treated

in the same way as plastic flow.

We may ask, therefore, whether a problem containing inter-
faces of this type can be treated within the framework of limit
analysis. Unfortunately this is not possible, except in two
special cases; frictional behavior is, in general, not included
in the theory of plasticity. The reasons why this is so can
best be explained in terms of a simple model. If we consider a
structure or body subjected to two generalized loads we have at
this point a good understanding of the general features of the
behavior if the material of which the structure is composed is
rigid, perfectly plastic. We would expect to find a convex
limit surface in load space, such that flow will not occur when
the load point lies within the limit surface. When the load
point lies on the limit surface unaccelerated or time independent flow is possible, with the generalized displacement rate
vector normal to the limit surface, or lying between adjacent
normals at a singular point. When the stress point lies outside
of limit surface accelerating motion must occur, or alternatively,
if we exclude the possibility of accelerated motion, the load
point may not be outside of the limit surface. In order to test
whether frictional behavior falls within the framework of plasticity as we have defined it in this volume, we may consider a
simple "structure" which involves friction and compare its behavior to the general features of plastic behavior we have just
described.

The simplest "structure" which we can study is that shown in
Figure 22. It consists of a block resting on a plane surface,
subjected to a vertical "generalized load" P and a horizontal
"generalized load" Q. It is sufficient that we confine ourselves
to a horizontal load Q which always acts in a given vertical
plane. Let us introduce a coefficient of friction μ, and further
assume that the static and dynamic coefficients of friction are

the same. The idealized friction law tells us that the block
will not slide when

$$|Q| < \mu P \; . \tag{66}$$

When $|Q| > \mu P$ the block will accelerate, and we exclude this pos-
sibility. When $|Q| = \mu P$ time independent motion may take place,
and is acceptable. This suggests that we may introduce a limit
function

$$\Psi(P,Q) \;=\; |Q| - \mu P \; . \tag{67}$$

The function Ψ will be negative when inequality (66) is satisfied, and zero when "flow" occurs. The limit surface is plotted
in Figure 23; it is convex and consequently acceptable within
the theory of plasticity.

However, when the generalized displacement rate vector is
considered, it can be seen that the flow rule is not satisfied.

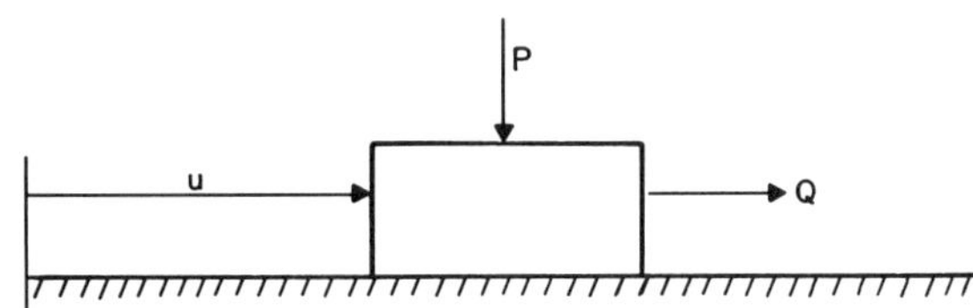

Figure 22. Frictional model

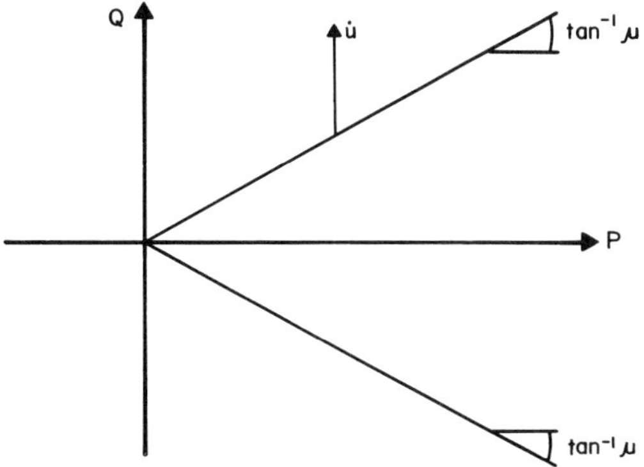

Figure 23. Limit surface for frictional model

The block may slide along the plane; let its distance from some
fixed origin be u, so that the generalized displacement rate
associated with Q is $\dot{u}$. The generalized displacement rate associated with P is always zero, so that the displacement rate
vector plotted with it origin at the load point on the yield
surface (shown at a typical point in Figure 23) is not normal
to the limit surface under conditions of flow. It is this that
demonstrates the distinction between frictional behavior and
limit behavior as we have defined it in this volume.

The special cases in which frictional behavior can be treated
by limit analysis are those where the flow rule is satisfied.
While they are somewhat trivial, they have important implications
in some of the problems we have already considered. First, when
$\mu = 0$, the limit surface coincides with the positive P axis, and
the flow rate is satisfied even though the rate of dissipation
of energy is zero during flow and no horizontal force can be
supported by the block. This case is referred to as a *smooth*
contact. Secondly, when μ tends to infinity, the displacement
rate vector vanishes and the limit surface coincides with the Q
axis. Any horizontal force can be supported and in the absence
of motion the flow rule is not violated. Finally if P is a constant force (i.e. it cannot be altered in the problem we are
considering) the limit surface may be treated as a function of
Q alone, $\Psi = \Psi(Q)$. It is no longer necessary to use the P-Q
plane. The limiting values of Q are $\pm\mu P$, and the flow rate demands only that $\dot{u}$ be positive when $Q = \mu P$ and negative when
$Q = -\mu P$. Since this is satisfied the block behaves like a
rigid-plastic structure subjected to a single generalized load.

In considering an interface between surfaces in a plane
strain problem (as in the punch problem, for example) the limit
surface of Figure 23 can be interpreted as a limit surface in
stress space, P becoming the normal component of the traction

σ_N transmitted across the surface and Q the tengantial component σ_S. The generalized displacement $\dot{u}$ is interpreted as the tangential velocity discontinuity across the interface. The normal component of the velocity discontinuity is zero, corresponding to the zero displacement rate associated with P. That frictional behavior cannot be accepted with the context of perfect plasticity is most succinctly stated by observing that the fundamental inequality

$$(Q_j - Q_j^a)\, \dot{q}_j^p \geq 0 \tag{68}$$

(Chapter 9, equation (4)) is not satisfied. Q_j is any state of stress for which $\psi(Q_j) = 0$, $\dot{q}_j^p$ is the associated strain rate and Q_j^a is any state of stress for which $\psi(Q_j) \leq 0$. In our present context the left hand side of inequality (68) becomes

$$(\sigma_S - \sigma_S^a)\dot{u} \quad ,$$

where $|\sigma_S| = \mu\sigma_N$ and $|\sigma_S^a| \leq \mu\sigma_N^a$. Clearly for any given σ_N, σ_S it is possible to choose σ_N^a, σ_S^a so that this expression is negative.

It must be noted that σ_N is generally determined by the magnitude of the external load. Since the magnitude of the external load is unknown (it is in fact sought in the analysis) and the distribution of σ_N along the interface is also unknown, σ_N can certainly not be treated as a constant.

The proof of the limit theorems is based on inequality (68), and it follows that in general the limit theorems cannot be applied when frictional sliding is present. The exceptions are the cases where $\mu = 0$, or a smooth contact, and a rough contact. A rough contact can be defined as a case in which μ is sufficiently large that plastic flow will occur in layers of material adjacent to the contact surface before slipping occurs along the surface. In the case of smooth and rough contacts the limit theorems are applicable because inequality (68) holds.

Another perspective on the failure of the limit theorems in the general case can be obtained by noting that when flow occurs in perfectly plastic materials the specific rate of dissipation is uniquely determined by the generalized strain rate (or tangential velocity discontinuity) and the properties of the material, since the associated generalized stresses are determined from the limit surface. This is not possible when the flow rule does not hold; further information about the stress field would be required before the rate of dissipation could be computed.

12.6 Bibliographical and Historical Remarks

The nature of discontinuities in the stress and velocity fields during flow have been discussed by Winzer and Carrier [1948] and Prager [1955].

The presentation of the material in the remainder of this Chapter is based largely on the philosophy expressed by Drucker and Chen [1968] and Calladine [1969]. The plane stress and plane strain problems involving notched bars were discussed by Drucker [1954], and extensions in the same spirit were given, for example, by Brady and Drucker [1955]. The implication of these results in fracture problems have been discussed by Drucker [1963] and McClintock and Irwin [1965]. Early discussions of the applications of limit analysis to indentation problems in plane strain were given by Shield and Drucker [1953], Bishop [1953] and Bishop, Green and Hill [1956]. The volumes by Hill [1950] and Johnson and Mellor [1962] summarize results which are available in this area.

References

J.F.W.Bishop   1953   "On the complete solution to problems of deformation of a plastic-rigid material", J.Mech.Phys.Sol.,$\underline{2}$, 43.

J.F.W.Bishop,   1956   "A note on the deformable region in
A.P.Green and   a rigid-plastic body", J.Mech.Phys.
R.Hill   Sol., $\underline{4}$, 256.

W.G.Brady and   1955   "Investigation and limit analysis of
D.C.Drucker   net area in tension", Trans.ASCE,
                            $\underline{120}$, 1133.

C.R.Calladine   1969   *Engineering Plasticity*(Chapter VII),
                            Pergamon Press (London).

D.C.Drucker   1954   "On obtaining plane strain or plane
                            stress conditions in plasticity",
                            Proc. 2nd U.S.Nat.Congr.Appl.Mech.
                            (ASME, N.Y.), 485.

D.C.Drucker   1963   *Fracture of Solids* (edited by D.C.
                            Drucker and J.J.Gilman) Wiley (N.Y.)
                            3.

D.C.Drucker and   1968   "On the use of simple discontinuous
W.F.Chen   fields to bound limit loads", *Eng-
                            gineering Plasticity* (edited by J.
                            Heyman and F.A.Leckie), Cambridge
                            University Press, 129.

R.Hill   1950   *The Mathematical Theory of Plasticity*,
                            Oxford.

W.Johnson and   1962   *Plasticity for Mechanical Engineers*,
P.B.Mellor   Van Nostrand (London, N.Y.).

F.A.McClintock and   1965   "Fracture toughness testing and its
G.R.Irwin   applications", STP 381 (ASTM), 84.

W.Prager   1955   "Discontinuous fields of plastic
                            stress and flow", Proc. 2nd U.S. Nt.
                            Congr.Appl.Mech. (ASME,N.Y.), 21.

R.T.Shield and   1953   "The application of limit analysis
D.C.Drucker   to punch indentation problems", J.
                            Appl.Mech., $\underline{20}$, 453.

A.Winzer and   1948   "The interaction of discontinuity
G.F.Carrier   surfaces in plastic fields of stress",
                            J.Appl.Mech., $\underline{15}$, 261.

Chapter 13

LIMIT ANALYSIS AS A PROGRAMMING PROBLEM

13.1 Restatement of the Limit Theorems

The applications of limit analysis which are described in Chapters 10-12 are analytic in nature and with some exceptions are applied in an *ad hoc* manner in each specific problem. While a degree of proficiency in these applications of limit analysis can lead to remarkable results, either in terms of exact solutions or bounds, there nevertheless remain many problems which can be solved only by a systematic numerical approach. In this section we shall discuss formulations of the limit analysis problem which can be adapted to numerical computation.

Two important points should be noted. A numerical approach requires that the problem should be formulated in a discrete form. The most suitable discretization will frequently depend on the particular problem under consideration. We shall consequently defer a discussion of this step until the general form of the numerical problem is evident. Most commonly the numerical formulation of limit analysis will lead to a *programming problem* i.e. a problem in which a function of several variables must be maximized or minimized subject to certain equality and inequality constraints. It is possible to study the general nature of this problem and derive results which apply independently of the precise way in which the discretization is carried out: this we shall now proceed to do. Consider a body subjected to proportional generalized loads Γc_α. The discussion will be limited to the case of a continuously differentiable limit function $\psi(Q_j)$ and a continuously differentiable specific dissipation rate function $D(\dot{q}_j)$; generalization will be readily apparent.

To state the upper bound theorem in the required form we introduce the class of pertinent kinematically admissible fields, denoting velocities by $\underline{u}(s)$, strain rates by $\dot{q}_j(s)$ and generalized velocities by $\dot{\xi}_\alpha$. For each member of this class, we may equate the external work rate and the internal energy dissipation rate,

$$\Gamma c_\alpha \dot{\xi}_\alpha \;=\; \int_V Q_j \dot{q}_j dV \;=\; \int_V D(\dot{q}_j) dV \;. \tag{1}$$

Thus

$$\Gamma \;=\; \frac{\int_V D(\dot{q}_j) dV}{c_\alpha \dot{\xi}_\alpha} \;. \tag{2}$$

The upper bound theorem of limit analysis (Chapter 9) states that the limit load parameter Γ^* is not greater than Γ, i.e.

$$\Gamma^* \;\leq\; \Gamma \;. \tag{3}$$

The limit analysis problem becomes a programming problem if we rephrase inequality (3) to state that

$$\Gamma^* \;=\; \min\{\Gamma\} \;, \tag{4}$$

since we may determine Γ^* by minimizing Γ (given by equation (2)) subject to the constraint that $\dot{\xi}_\alpha, \dot{q}_j$ should constitute a pertinent kinematically admissible field.

The statement given in equation (4) is acceptable if flow will occur in the body for some value of Γ. The limit load parameter Γ^* is then unique, and is distinguished in that we can associate it with a safe, pertinent statically admissible field $Q_j(s)$ and a pertinent kinematically admissible field $\dot{q}_j(s)$ such that the constitutive equations are satisfied at each point. That Γ^* is the least value of Γ under these conditions can also be demonstrated by variational methods by considering the conditions under which Γ takes its minimum value.

A sufficient condition that Γ (equation (2)) is a local minimum is that we can associate with $\dot{q}_j(s)$ a safe pertinent statically

admissible field $Q_j(s)$, *with* $Q_j = \partial D/\partial \dot{q}_j$ *where* $\dot{q}_j \neq 0$ *and*
$\psi(Q_j) \leq 0$ *where* $\dot{q}_j = 0$.

In order to establish this result, we may first without loss in
generality limit the kinematically admissible velocity fields
to the normalized set which satisfy the equation

$$c_\alpha \dot{\xi}_\alpha = 1 \quad . \tag{5}$$

Consider a member of this class, and let V_1 denote those parts
of the body where $\dot{q}_j \neq 0$ and V_2 those parts of the body where
$\dot{q}_j = 0$. Note that $D(\dot{q}_j) = 0$ in V_2.
   Now consider small variations $\delta\xi_\alpha, \delta\dot{q}_j$ in the kinematically
admissible field. From constraint (5),

$$c_\alpha \delta\dot{\xi}_\alpha = 0 \quad . \tag{6}$$

The variation in Γ is given by

$$\delta\Gamma = \int_{V_1} \frac{\partial D}{\partial \dot{q}_j} \delta\dot{q}_j \, dV + \int_{V_2} D(\delta\dot{q}_j) dV \quad . \tag{7}$$

In V_1, $\partial D/\partial \dot{q}_j$ gives the stress Q_j associated with $\dot{q}_j$ through the
constitutive equation. Since $D(\dot{q}_j)$ is convex, in V_2 we observe
that

$$D(\delta\dot{q}_j) \geq \frac{\partial D}{\partial \dot{q}_j}\Big|_{\dot{q}_j=0} \delta\dot{q}_j \quad . \tag{8}$$

The gradient of D at the origin of the strain space may be interpreted as any stress state Q_j which satisfies the condition
$\psi(Q_j) \leq 0$. Consequently equation (7) becomes

$$\delta\Gamma \geq \int_V Q_j \delta\dot{q}_j dV \quad , \tag{9}$$

with Q_j interpreted in the manner described in V_1 and V_2. A
necessary and sufficient condition that Q_j should be a pertinent

statically admissible field is that

$$\Gamma c_\alpha \delta \dot{\xi}_\alpha \;=\; \int_V Q_j \delta \dot{q}_j \, dV \;\; , \tag{10}$$

where $\delta \dot{q}_j$ is an arbitrary variation. Noting equation (6), it is then evident that a sufficient condition that

$$\delta \Gamma \;\geq\; 0 \tag{11}$$

is that $Q_j(s)$ in equation (9) should be statically admissible.

In order to reformulate the lower bound theorem we introduce the class of pertinent statically admissible fields $Q_j(s)$. It is useful to note that we can put

$$Q_j(s) \;=\; \Gamma Q_j^S(s) + \rho_j(s) \;\; , \tag{12}$$

where $Q_j^S(s)$ is any stress field which is statically admissible with generalized loads c_α (i.e. $\Gamma = 1$) and $\rho_j(s)$ is the class of self-equilibrating stress fields, i.e. the class of statically admissible stress fields for $\Gamma = 0$.

If for any load parameter Γ, a pertinent statically admissible stress field can be found for which

$$\psi(Q_j) \;\leq\; 0 \tag{13}$$

at each point in the body, the lower bound theorem assures us that

$$\Gamma \;\leq\; \Gamma^* \;\; , \tag{14}$$

where Γ^* is again the limit load parameter. We rephrase this inequality to state that

$$\Gamma^* \;=\; \max\{\Gamma\} \;\; . \tag{15}$$

We may then determine Γ^* by finding the maximum value of Γ subject to the constraints of inequality (13), and that $Q_j(s)$ should be statically admissible.

In order to establish that we can associate a kinematically admissible velocity field with the greatest value of Γ, we state the problem in yet another form. Inequality (13) may be written equivalently as the requirement that

$$\max_V \{\psi(Q_j)\} \leq 0 \; , \tag{16}$$

where we indicate that the largest value of $\psi(Q_j)$ in V must be non-positive. Consider now the sub-class of fields $Q_j(s)$, defined by Γ and ρ_j as in equation (12), for which

$$\max_V \{\psi(Q_j)\} = 0 \; . \tag{17}$$

Since Q_j is linear in the load parameter Γ,

$$\max_V \{\psi(Q_j)\}$$

increases monotonically with Γ if $\rho_j(s)$ is held constant. It follows then that the set Γ, ρ_j for which $\delta\Gamma \leq 0$ for all $\delta\Gamma$, $\delta\rho_j$ in the subclass defined by equation (17) is characterized by the condition

$$\delta[\max_V \{\psi(Q_j)\}]_{\Gamma=\text{constant}} \geq 0 \tag{18}$$

for arbitrary variations $\delta\rho_j$.

A sufficient condition that Γ is a local maximum is that we can associate with Q_j a kinematically admissible velocity field $\dot{q}_j$, with $\dot{q}_j = \lambda\partial\psi/\partial Q_j$ where $\psi(Q_j) = 0$, $\dot{q}_j = 0$ where $\psi(Q_j) < 0$.

To establish this result, we show that the conditions are sufficient for inequality (18) to hold. Now

$$\delta[\max_V \{\psi(Q_j)\}]_{\Gamma=\text{constant}} = \max_V \{\psi(Q_j+\delta\rho_j)\}$$

$$- \max_V \{\psi(Q_j)\} \; . \tag{19}$$

Let the points in the body at which $\psi(Q_j) = 0$ be denoted by V_1, and those point for which $\psi(Q_j) < 0$ by V_2. Certainly

$$\max_{V}\{\psi(Q_j)\}$$

occurs in V_1, and

$$\max_{V}\{\psi(Q_j)\} = 0 \quad . \tag{20}$$

It is not evident whether

$$\max_{V}\{\psi(Q_j+\delta\rho_j)\}$$

occurs in V_1 or in V_2. However, we can say that

$$\max_{V_1}\{\psi(Q_j+\delta\rho_j)\} = \max_{V_1}\{\frac{\partial\psi}{\partial Q_j}\,\delta\rho_j\} \quad , \tag{21}$$

and that

$$\max_{V}\{\psi(Q_j+\delta\rho_j)\} \geq \max_{V_1}\{\psi(Q_j+\delta\rho_j)\} \quad . \tag{22}$$

Combining equations (19)-(22), we now have

$$\delta[\max_{V}\{\psi(Q_j)\}]_{\Gamma=\text{constant}} \geq \max_{V_1}\{\frac{\partial\psi}{\partial Q_j}\,\delta\rho_j\} \quad . \tag{23}$$

We now introduce a scalar function $\lambda(s)$ such that

$$\lambda \geq 0 \quad \text{in} \quad V_1 \,, \quad \lambda = 0 \quad \text{in} \quad V_2 \quad . \tag{24}$$

Multiplying inequality (23) by the positive quantity

$$\int_{V_1} \lambda(s)ds \quad ,$$

we see that

$$\delta\left[\max_{V}\{\psi(Q_j)\}\right]_{\Gamma=\text{constant}} \int_{V_1} \lambda\, ds \geq \max_{V_1}\left\{\frac{\partial\psi}{\partial Q_j}\,\delta\rho_j\right\} \int_{V_1} \lambda\, ds$$

$$\geq \int_{V_1} \lambda\left\{\frac{\partial\psi}{\partial Q_j}\,\delta\rho_j\right\} dV \tag{25}$$

$$= \int_{V_1} \left(\lambda\frac{\partial\psi}{\partial Q_j}\right)\delta\rho_j\, dV \quad.$$

This now suggests that we may introduce a strain rate field $\dot{q}_j(s)$ given by

$$\dot{q}_j(s) = \lambda\frac{\partial\psi}{\partial Q_j} \quad \text{in} \quad V_1 \quad,$$

$$\tag{26}$$

$$\dot{q}_j(s) = 0 \quad \text{in} \quad V_2 \quad.$$

If $\dot{q}_j(s)$ is a pertinent kinematically admissible field, it follows that

$$\int_V \dot{q}_j\delta\rho_j\, dV = \int_{V_1}\left(\lambda\frac{\partial\psi}{\partial Q_j}\right)\delta\rho_j\, dV = 0 \quad. \tag{27}$$

On substituting equation (27) into inequality (25), we see that
if $\dot{q}_j(s)$ as given in equations (26)(which satisfies the constitutive relations) is kinematically admissible, inequality (18)
holds, and hence $\delta\Gamma \leq 0$ for arbitrary variations in the stress
field.

    In the following sections we shall illustrate the application
of these forms of the limit theorems to some selected problems.
The problems are chosen to illustrate ways in which the discretization necessary for numerical exploitation of the theorems
may be carried out.

13.2 Application to Trusses and Beams

The simplest application of programming techniques to limit
analysis occurs in problems where discretization of continuous

fields is not formally necessary. Two such cases will be treated briefly: pinjointed trusses and beams subjected to point loads.

In the pinjointed truss each member is in a homogeneous state of tension or compression. Hence whether or not the limit function is non-positive in any one member can be determined by substituting the bar force into the appropriate limit function. Since the number of bars is finite, the limitations imposed on the generalized stress field in the lower bound approach are in discrete form. Similarly, the dissipation rate in each bar may be written in terms of the extension rate, and the conditions of kinematic admissibility of the velocity field are a discrete set involving the extension rate.

As an example, consider the truss shown in Figure 1. The bars are uniform and identical in cross section. A horizontal load Γ and a vertical load 2Γ are applied at node D. Denoting the tension forces in AD, BD and CD by N_A, N_B and N_C respectively, the equilibrium equations are (cf. Chapter 6)

$$N_A + 2N_B + N_C = 4\Gamma \ ,$$

$$N_A - N_C = \frac{2}{\sqrt{3}}\Gamma \ . \tag{28}$$

Assuming that the strain rates in AD, BC and CD are $\dot{\varepsilon}_A, \dot{\varepsilon}_B, \dot{\varepsilon}_C$

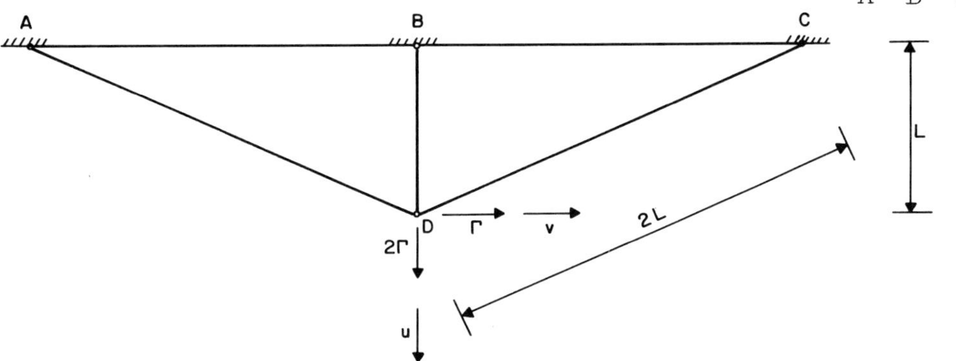

Figure 1. Truss problem

respectively, and that the vertical and horizontal velocities of D and $\dot{u}$ and $\dot{v}$ respectively, the strain rate, displacement rate relations are

$$\dot{\varepsilon}_A = \frac{1}{2L} \{\frac{\dot{u}}{2} + \frac{\sqrt{3}}{2} \dot{v}\} \quad,$$

$$\dot{\varepsilon}_B = \frac{\dot{u}}{L} \quad, \tag{29}$$

$$\dot{\varepsilon}_C = \frac{1}{2L} \{\frac{\dot{u}}{2} - \frac{\sqrt{3}}{2} \dot{v}\} \quad.$$

Eliminating $\dot{u}$, $\dot{v}$ from these equations, we have the requirement that

$$\dot{\varepsilon}_A + \dot{\varepsilon}_C = \frac{1}{2} \dot{\varepsilon}_B \quad. \tag{30}$$

In applying the lower bound theorem, we designate N_B, say, as the redundant or hyperstatic force, and solve the equilibrium equations for the remaining bar forces. Thus,

$$N_A = (2 + \frac{1}{\sqrt{3}})\Gamma - N_B \quad,$$

$$\tag{31}$$

$$N_C = (2 - \frac{1}{\sqrt{3}})\Gamma - N_B \quad.$$

Assuming that the limit function is the same in each bar and is given by

$$\psi = |N| - N_o \quad,$$

the limit load is given by the solution of the following problem;

$$\text{maximize } \Gamma \tag{32a}$$

subject to

$$- N_o \leq (2 + \frac{1}{\sqrt{3}})\Gamma - N_B \leq N_o \quad ,$$

$$- N_o \leq N_B \leq N_o \quad , \tag{32b}$$

$$- N_o \leq (2 - \frac{1}{3})\Gamma - N_B \leq N_o \quad .$$

A graphical solution can be found for this problem. A domain can be plotted in the Γ, N_B space within which inequalities (32) are satisfied. The lines bounding this domain are the six equations

$$(2 + \frac{1}{\sqrt{3}})\Gamma - N_B = \pm N_o \quad ,$$

$$N_B = \pm N_o \quad , \tag{33}$$

$$(2 - \frac{1}{\sqrt{3}})\Gamma - N_B = \pm N_o \quad .$$

We limit Γ to be non-negative, but N_B is unrestricted in sign. The domain is plotted in Figure 2 in dimensionless form. Level lines of Γ/N_o are parallel to the N_B/N_o axis; clearly the maximum value of Γ contained within the domain occurs at point F, where

$$\frac{\Gamma}{N_o} = \frac{2\sqrt{3}}{2\sqrt{3}+1} = 0.776 \quad , \qquad \frac{N_B}{N_o} = 1 \quad . \tag{34}$$

Applying the upper bound theorem, we see from an energy dissipation balance that

$$2\Gamma\dot{u} + \Gamma\dot{v} = 2LN_o|\dot{\varepsilon}_A| + LN_o|\dot{\varepsilon}_B| + 2LN_o|\dot{\varepsilon}_C| . \tag{35}$$

Adopting $\dot{\varepsilon}_A$, $\dot{\varepsilon}_B$ and $\dot{\varepsilon}_C$ as the variables, the compatibility equation (equation (30)) applies as a constraint. In addition, we may require that

$$2\dot{u} + \dot{v} = 2L \quad . \tag{36}$$

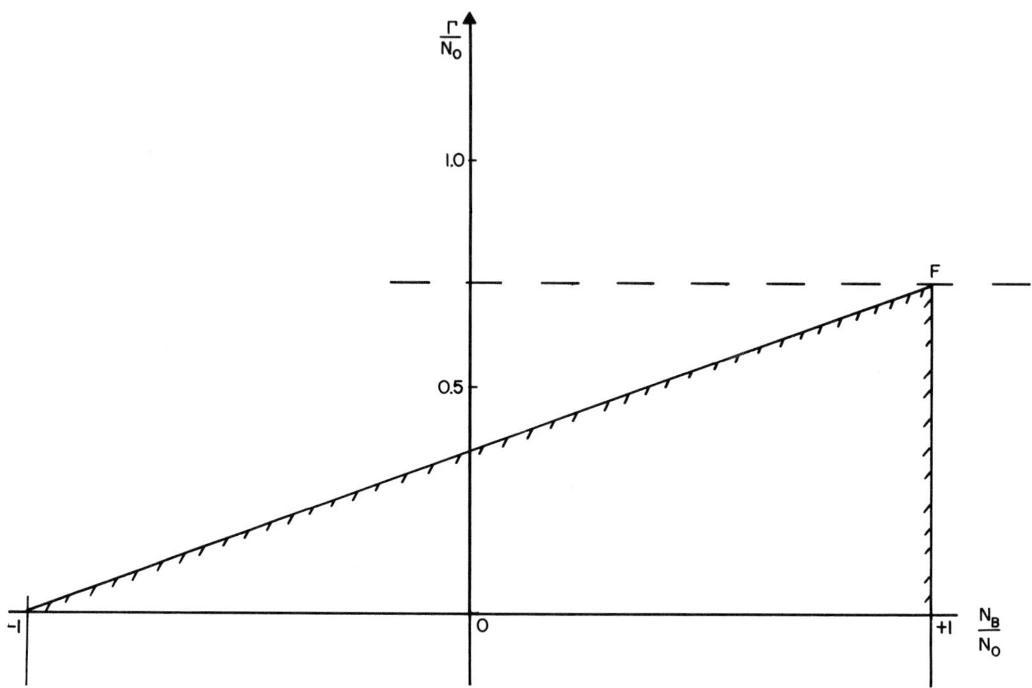

Figure 2. Limit load of truss

Thus, after using equations (29), to determine the limit load we must find the solution of the following problem;

$$\text{minimize } \frac{1}{2} N_o (2|\dot{\varepsilon}_A| + |\dot{\varepsilon}_B| + 2|\dot{\varepsilon}_C|) \tag{37a}$$

subject to

$$2\dot{\varepsilon}_A - \dot{\varepsilon}_B + 2\dot{\varepsilon}_C = 0 , \tag{37b}$$

$$\frac{1}{\sqrt{3}} \dot{\varepsilon}_A + \dot{\varepsilon}_B - \frac{1}{\sqrt{3}} \dot{\varepsilon}_C = 1 . \tag{37c}$$

The strain rates $\dot{\varepsilon}_A$, $\dot{\varepsilon}_B$, $\dot{\varepsilon}_C$ are unrestricted in sign.

Using equations (37b) and (37c) to eliminate, say $\dot{\varepsilon}_B$ and $\dot{\varepsilon}_C$, the objective function can be written in terms of a single un-constrained variable

$$\frac{\Gamma}{N_o} = \frac{1}{2}\{2|\dot{\varepsilon}_A| + \frac{2}{(2\sqrt{3}-1)} |\sqrt{3} - 2\dot{\varepsilon}_A| + \frac{2}{(2\sqrt{3}-1)} |\sqrt{3}-(2\sqrt{3}+1)\dot{\varepsilon}_A|\} . \tag{38}$$

This function is plotted in Figure 3, and is seen to be piece-wise linear. Its least value occurs for

$$\dot{\varepsilon}_A = \frac{\sqrt{3}}{2\sqrt{3}+1} \quad , \quad \dot{\varepsilon}_B = \frac{1}{2}\dot{\varepsilon}_A \quad , \quad \dot{\varepsilon}_C = 0 \quad ,$$

$$\frac{\Gamma}{N_o} = \frac{2\sqrt{3}}{2\sqrt{3}+1} = 0.776 \quad .$$

(39)

We see then that the programming problems yield the same limit load; they are in fact dual problems. This formulation offers advantages because the problems of equations (32) and (37) can be written as standard *linear programming problems* which can be solved numerically by routine methods. Before considering this transformation, however, we shall briefly study an example of another class of problems which also lead to linear programming problems.

In beam and frame problems in which shear and axial deformation

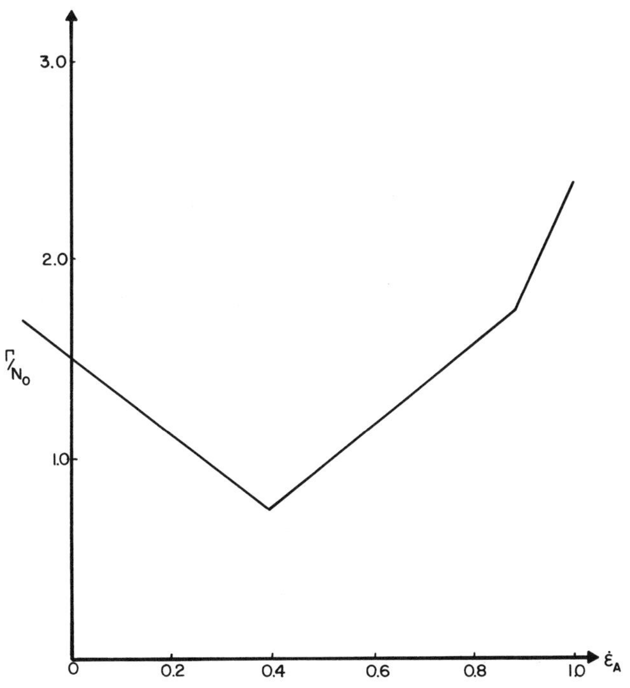

Figure 3. Limit load for truss

can be neglected the bending moment diagram will be piecewise
linear if the loads consist of concentrated loads at various
points on the structure. As we have seen in Chapter 10, this
implies that a discrete number of points can be located at which
the bending moment may reach its limiting value or at which
hinges can potentially occur. The regions between these critical points are thus known *a priori* to be rigid during flow.
Since in the application of the lower bound theorem the bending
moment can be expressed in terms of a discrete number of hyperstatic forces and the limit condition need be imposed only at a
discrete number of points, the programming problem is discrete.
Similarly, a discrete number of compatibility relations are imposed on potential hinge rotations, making the upper bound
problem discrete.

As an example, consider the frame shown in Figure 4. The
discrete points at which flow may occur at labelled 1-7 on the
assumption that members AB, BC, CD and DE are uniform. We shall
assume that AB and DE have limit moment M_o, and that BD has limit
moment $3M_o/2$. With this information points 3 and 5 can be eliminated, since flow cannot occur at these points in preference
to points 2 and 6 respectively.

In applying the lower bound theorem, let us give the bending
moments in terms of the three redundants M_1, V_1 and H_1, as shown

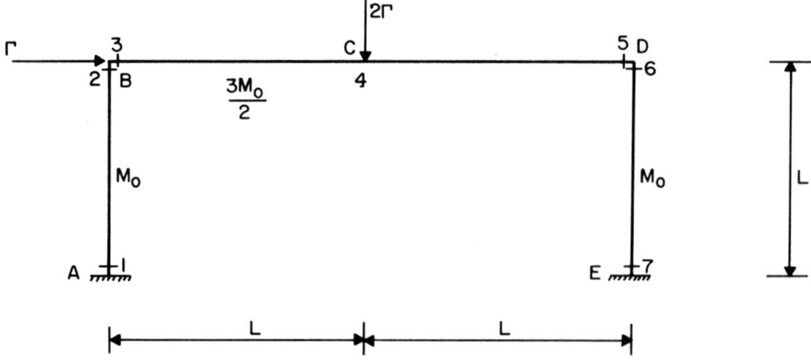

Figure 4. Frame example

in Figure 5. Then, using the sign convention shown, we find
that

$$M_2 = -M_1 + H_1 L \ ,$$

$$M_4 = -M_1 + H_1 L + V_1 L \ ,$$

$$M_6 = M_1 - H_1 L - 2V_1 L + 2\Gamma L \ , \tag{40}$$

$$M_7 = -M_1 + 2V_1 L - \Gamma L \ .$$

Thus the limit load is given by the solution of the following
problem;

maximize $\qquad \Gamma$

subject to
$$\begin{aligned}
-M_0 &\le M_1 &&\le M_0 \ , \\[4pt]
-M_0 &\le -M_1 + H_1 L &&\le M_0 \ , \\[4pt]
\tfrac{-3}{2} M_0 &\le -M_1 + H_1 L + V_1 L &&\le \tfrac{3}{2} M_0 \ , \\[4pt]
-M_0 &\le M_1 - H_1 L - 2V_1 L + 2\Gamma L &&\le M_0 \ , \\[4pt]
-M_0 &\le -M_1 + 2V_1 L - \Gamma &&\le M_0 \ .
\end{aligned} \tag{41}$$

In applying the upper bound theorem, we permit hinges with
rotation rates $\dot\theta_1$, $\dot\theta_2$, $\dot\theta_4$, $\dot\theta_6$, $\dot\theta_7$ at the points 1,2,4,6 and 7.
Assume that the signs of the hinge rotation rates are positive

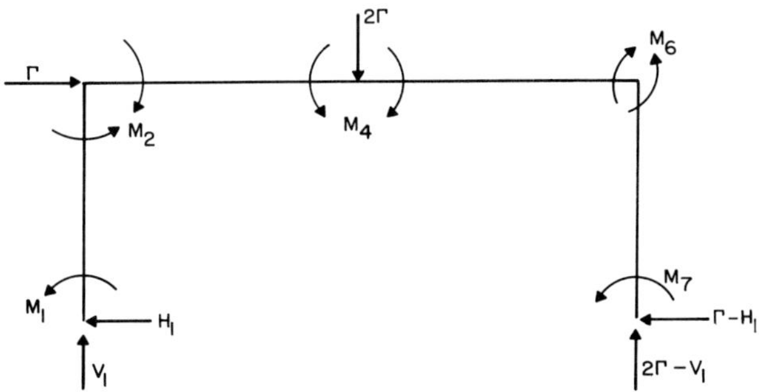

Figure 5. Sign convention for frame

when they have the same sense as the moments shown in Figure 5.
The compatibility equations may be obtained by using the self-
equilibrating moment distributions for the frame. There will
be three such distributions, and their values at the critical
points can be obtained by putting $\Gamma=0$ in equations (40) and
assigning any three independent sets of values to the triplet
(M_1,V_1,H_1). Choosing $(1,0,0)$, $(0,1,0)$ and $(0,0,1)$ and equating
the sum of the products of moments and hinge rotation rates to
zero, the compatibility equations are

$$\dot{\theta}_1-\dot{\theta}_2-\dot{\theta}_4+\dot{\theta}_6-\dot{\theta}_7 = 0 \quad ,$$

$$\dot{\theta}_4-2\dot{\theta}_6+2\dot{\theta}_7 = 0 \quad , \tag{42}$$

$$\dot{\theta}_2+\dot{\theta}_4-\dot{\theta}_6 = 0 \quad .$$

Equating the external work rate and the internal dissipation
rate, and denoting the horizontal velocity at B by $\dot{u}$ and the
vertical velocity at C by $\dot{v}$, we find that

$$\Gamma(2\dot{v}+\dot{u}) = M_o|\dot{\theta}_1| + M_o|\dot{\theta}_2| + \frac{3}{2}M_o|\dot{\theta}_4| + M_o|\dot{\theta}_6| + M_o|\dot{\theta}_7| \quad . \tag{43}$$

We can express $(2\dot{v}+\dot{u})$ in terms of the hinge rotation rates by
again using equations (40), with $\Gamma=1$ and $M_1=V_1=H_1=0$. Thus

$$1.(2\dot{v}+\dot{u}) = 2L\dot{\theta}_6 - L\dot{\theta}_7 \quad . \tag{44}$$

Normalizing by requiring that

$$2\dot{v}+\dot{u} = L \quad , \tag{45}$$

the limit load is then given by the solution of the following
problem;

$$\text{minimize } \{ \frac{M_o}{L} (|\dot{\theta}_1|+|\dot{\theta}_2| + \frac{3}{2}|\dot{\theta}_4| + |\dot{\theta}_6| + |\dot{\theta}_7|\}$$

subject to $2\dot{\theta}_6 - \dot{\theta}_7 = 1$,

$$\dot{\theta}_1 - \dot{\theta}_2 - \dot{\theta}_4 + \dot{\theta}_6 - \dot{\theta}_7 = 0 \ ,$$

$$\dot{\theta}_2 + \dot{\theta}_4 - \dot{\theta}_6 = 0 \ ,$$

$$\dot{\theta}_4 - 2\dot{\theta}_6 + 2\dot{\theta}_7 = 0 \ . \tag{46}$$

The two problems considered suffice to show that the general form of the programming problems for trusses and frames is as follows. For the lower bound approach, we are required to solve the following problem;

maximize R_1

subject to $-M_{ok} \leq b_{k\alpha} R_\alpha \leq M_{ok}$, $\tag{47}$

where $R_1 = \Gamma$ and $R_\alpha (\alpha=2,3,\ldots)$ are hyperstatic forces and k=1, 2,... represent critical points. M_{ok} is the limit value of the appropriate generalized stress.

In the upper bound approach, we are required to

minimize $M_{ok}|\dot{\theta}_k|$

subject to $\hat{b}_{\alpha k}\dot{\theta}_k = a_\alpha$ $(a_1=1, a=0 \text{ for } \alpha > 1)$, $\tag{48}$

where $\dot{\theta}_k$ (k=1,2,...) are kinematic variables.

The *general form of the linear programming problem*, in similar notation, is given as

maximize $A_\alpha X_\alpha$

subject to $B_{k\alpha} X_\alpha \leq C_k$ $\tag{49a}$

$$X_\alpha \geq 0$$

and its *dual*,

minimize $C_k Y_k$

subject to $B_{\alpha k} Y_k \geq A_\alpha$, (49b)

$$Y_k \geq 0 .$$

The lower bound approach (equation (47)) can be written in the form of equation (49a) if the unrestricted variables R_α can be written in terms of non-negative variables. Noting that any number, positive or negative, can be written as the difference of two non-negative numbers, we introduce R^+, R^- such that

$$R_\alpha = R_\alpha^+ - R_\alpha^-$$

$$R_\alpha^+ \geq 0 , \quad R_\alpha^- \geq 0 .$$ (50)

Thus, equation (47) can be written as

maximize $\{R_1^+ - R_1^-\}$

subject to $b_{k\alpha}(R_\alpha^+ - R_\alpha^-) \geq M_o$,

$$-b_{k\alpha}(R_\alpha^+ - R_\alpha^-) \geq M_o ,$$ (51)

$$R_\alpha^+ \geq 0 , \quad R_\alpha^- \geq 0 .$$

In the upper bound approach, we again replace the unrestricted variables $\dot\theta_k$ by

$$\dot\theta_k = \dot\theta_k^+ - \dot\theta_k^- , \quad \dot\theta_k^+ \geq 0 , \quad \dot\theta_k^- \geq 0 .$$ (52)

The objective function involves $|\dot\theta_k|$, and to express this in terms of the new restricted variables we put

$$|\dot\theta_k| = \dot\theta_k^+ + \dot\theta_k^- .$$ (53)

Note that on solving equations (52) and (53) simultaneously,

$$\dot{\theta}_k = \dot{\theta}_k^+ \quad , \quad \dot{\theta}_k^- = 0 \quad \text{if} \quad \dot{\theta}_k > 0 \quad ,$$

$$\dot{\theta}_k = -\dot{\theta}_k^- \quad , \quad \dot{\theta}_k^+ = 0 \quad \text{if} \quad \dot{\theta}_k < 0 \quad . \tag{54}$$

We must further express the equality constraints of equation (48) as simultaneous inequalities. Hence (48) becomes

$$\text{minimize} \quad \{M_{ok}(\dot{\theta}_k^+ + \dot{\theta}_k^-)\}$$

$$\text{subject to} \quad \hat{b}_{\alpha k}(\dot{\theta}_k^+ - \dot{\theta}_k^-) \geq a_\alpha \quad ,$$

$$-b_{\alpha k}(\dot{\theta}_k^+ - \dot{\theta}_k^-) \leq a_\alpha \quad , \tag{55}$$

$$\dot{\theta}_k^+ \geq 0 \quad , \quad \dot{\theta}_k^- \leq 0 \quad .$$

The form of the objective function ensures that the solution will be such that either $\dot{\theta}_k^+$ or $\dot{\theta}_k^-$ is zero, satisfying equation (54).

It remains to demonstrate that equations (51) and (55) are indeed dual problems in the sense of (49a) and (49b). We shall not provide a general proof in programming notation, but we can illustrate the result by reformulating the frame problem considered above. Introducing the dimensionless variables

$$\gamma = \frac{\Gamma L}{M_o} \quad , \quad m = \frac{M_1}{M_o} \quad , \quad h = \frac{H_1 L}{M_o} \quad , \quad v = \frac{V_1 L}{M_o} \tag{56}$$

and using matrix notation, problem (41) becomes

$$\text{maximize} \left\{ (1 \ 0 \ 0 \ 0) \begin{bmatrix} \gamma^+ - \gamma^- \\ m^+ - m^- \\ h^+ - h^- \\ v^+ - v^- \end{bmatrix} \right\}$$

$$\tag{57}$$

subject to

$$\pm \begin{bmatrix} 0 & 1 & 0 & 0 \\ 0 & -1 & 1 & 0 \\ 0 & -1 & 1 & 1 \\ 2 & 1 & -1 & -2 \\ -1 & -1 & 0 & 2 \end{bmatrix} \begin{bmatrix} \gamma^+ - \gamma^- \\ m^+ - m^- \\ h^+ - h^- \\ v^+ - v^- \end{bmatrix} \leq \begin{bmatrix} 1 \\ 1 \\ 3/2 \\ 1 \\ 1 \end{bmatrix}$$

with non-negative variables.

Problem (46) becomes

$$\text{minimize} \quad \{ (1 \ \ 1 \ \ 3/2 \ \ 1 \ \ 1) \begin{bmatrix} \dot\theta_1^+ + \dot\theta_1^- \\ \dot\theta_2^+ + \dot\theta_2^- \\ \dot\theta_4^+ + \dot\theta_4^- \\ \dot\theta_6^+ + \dot\theta_6^- \\ \dot\theta_7^+ + \dot\theta_7^- \end{bmatrix} \}$$

$$(58)$$

subject to

$$\pm \begin{bmatrix} 0 & 0 & 0 & 2 & -1 \\ 1 & -1 & -1 & 1 & -1 \\ 0 & 0 & 1 & -1 & 0 \\ 0 & 0 & 1 & -2 & 2 \end{bmatrix} \begin{bmatrix} \dot\theta_1^+ - \dot\theta_1^- \\ \dot\theta_2^+ - \dot\theta_2^- \\ \dot\theta_4^+ - \dot\theta_4^- \\ \dot\theta_6^+ - \dot\theta_6^- \\ \dot\theta_7^+ - \dot\theta_7^- \end{bmatrix} \leq \begin{bmatrix} 1 \\ 0 \\ 0 \\ 0 \end{bmatrix}$$

with non-negative variables.

Problems (57) and (58) are duals in the sense of (49a) and
(49b), as can be seen by comparing the formulations. Note that
the (4x5) matrix in (58) is the transpose of the (5x4) matrix
in (57). The solution of the problem given in equations (58) is

$$\gamma = 2.33, \quad \dot\theta_1^+ = 0.333, \quad \dot\theta_4^+ = 0.667, \quad \dot\theta_6^+ = 0.667, \quad \dot\theta_7^+ = 0.333$$

$$(59)$$

with all other variables zero. This answer may be confirmed by direct analysis.

13.3 Application to Plates

As an example of the discretization of a continuous problem by finite differences, consider an axisymmetric circular sandwich plate subjected to uniform pressure p. The boundary conditions at the boundary of the plate r=R will not be immediately specified.

A lower bound approach will be formulated. Using the notation of Chapter 11, the equilibrium equation is

$$\frac{d}{dr}(rM_r) - M_\theta = -\frac{1}{2}pr^2 \ . \tag{60}$$

Introducing the dimensionless variables,

$$s = \frac{r}{R}, \quad \bar{p} = \frac{pR^2}{M_o}, \quad m_r = \frac{M_r}{M_o}, \quad m = \frac{M_\theta}{M_o}, \tag{61}$$

the equilibrium equation becomes

$$\frac{d}{ds}(sm_r) - m_\theta = -\frac{1}{2}\bar{p}s^2 \ . \tag{62}$$

The Tresca yield condition, in dimensionless form, is given by

$$-1 \leq m_r \leq 1 \ ,$$

$$-1 \leq m_\theta \leq 1 \ , \tag{63}$$

$$-1 \leq m_r - m_\theta \leq 1 \ .$$

In continuous form, the limit load $\bar{p}^*$ is given by the largest value of $\bar{p}$ for which we can find $\bar{p}$, $m_r(s)$, $m_\theta(s)$ which satisfy constraints (62) and (63).

In order to discretize the problem, we divide the interval (0,1) of the variable s into n sub-intervals each of length

$\delta = 1/n$. We now consider only the values of m_r, m_θ at the discrete points $s = k\delta$ $(k=0,1,\ldots,n)$. Putting

$$m_{rk} = m_r(k\delta) \quad , \quad m_{\theta k} = m_\theta(k\delta) \quad , \tag{64}$$

we write the equilibrium equation in finite difference form, here using a central difference approximation of order δ^2. For the k-th mesh point,

$$k\delta \, \frac{\{m_{r(k+1)} - m_{r(k-1)}\}}{2\delta} + m_{rk} - m_{\theta k} = \frac{1}{2} \, \bar{p} \, (k\delta)^2 \quad . \tag{65}$$

The yield condition is applied only at the mesh points $s = k$; hence equation (63) becomes

$$-1 \leq \quad m_{rk} \leq 1 \quad ,$$

$$-1 \leq \quad m_{\theta k} \leq 1 \quad , \qquad k=0,1,\ldots,n \tag{66}$$

$$-1 \leq m_{rk} - m_{\theta k} \leq 1 \quad .$$

The boundary conditions for the plate must now be imposed. In all cases we require that $m_r(0) = m_\theta(0)$, and this becomes

$$m_{ro} = m_{\theta o} \quad . \tag{67}$$

No further conditions need be imposed for a clamped plate. If the plate is simply supported, we require that in addition

$$m_{rn} = 0 \quad . \tag{68}$$

The variables in the programming problem are now $\bar{p}, m_{rk}, m_{\theta k}$ $(k=0,1,\ldots,n)$. We are required to maximize $\{\bar{p}\}$ subject to constraints (65), (66), (67) and, if applicable, (68). Since the objective function and the constraints are linear, this problem can be transformed into a linear programming problem.

The solution for n=10 is compared with the exact solution (cf. Chapter 11) in Table 1 for the simply supported and clamped

Table 1. Bounds for Collapse Loads and Associated Stress Fields for Circular Plates (after Koopman and Lance, 1965)

	Simply supported plate				Clamped plate			
Critical load	Exact 6.00	Approx. 5.97			Exact 11.26	Approx. 11.28		
Stress field	m_r		m_θ		m_r		m_θ	
x	E	A	E	A	E	A	E	A
0	1.000	1.000	1.0	1.0	1.000	1.000	1.000	1.000
0.1	0.990	0.991	1.0	1.0	0.981	0.982	1.000	1.000
0.2	0.960	0.959	1.0	1.0	0.925	0.924	1.000	1.000
0.3	0.910	0.912	1.0	1.0	0.831	0.832	1.000	1.000
0.4	0.840	0.839	1.0	1.0	0.699	0.697	1.000	1.000
0.5	0.750	0.754	1.0	1.0	0.530	0.533	1.000	1.000
0.6	0.640	0.639	1.0	1.0	0.323	0.320	1.000	1.000
0.7	0.510	0.516	1.0	1.0	0.79	0.082	1.000	1.000
0.8	0.360	0.359	1.0	1.0	-0.208	-0.207	0.792	0.793
0.9	0.190	0.199	1.0	1.0	-0.571	-0.570	0.429	0.430
1.0	0	0	1.0	1.0	-1.000	-1.000	0	0

E Exact   A Approximate

cases. The numerical solution is of adequate accuracy for most purposes. An estimate of errors due to round-off by the computer and due to the finite difference approximation can be found.

In general it should be noted that in lower bound approaches of this type the yield condition is satisfied only at discrete points, and may be violated at other points. The degree of error which may be introduced will depend on the manner in which the stresses or generalized stresses vary with the space

variables. This does mean that the computed limit load may not
necessarily be less than the true limit load.

13.4 Use of Finite Elements in the Programming Problem

The finite element approach to the discretization of problems
in continuum mechanics has a number of advantages. The body is
divided into elements of finite size and predetermined generic
shape; the simplest elements are triangles in plane problems
and tetrahedra in three-dimensional bodies. In the case of
either displacement fields or stress fields a suitable number
of parameters are chosen so that statically admissible stresses
or kinematically admissible strains are defined within each
element and the appropriate continuity requirements between adjacent elements are satisfied. This process can be carried out
without fixing the exact shape or size of the element,permitting
the choice of elements of different size and facilitating the
processing of assembling relations for the whole body.

As an example of the application of the approach to programming problems in limit analysis, consider a thin sheet in a
state of plane stress. The kinematic approach is conceptually
simpler since we deal with a discretized velocity field, so
that we shall first consider the application of the upper bound
theorem. Triangular elements will be used in both cases.

Figure 6 shows a typical triangular element. The apices of

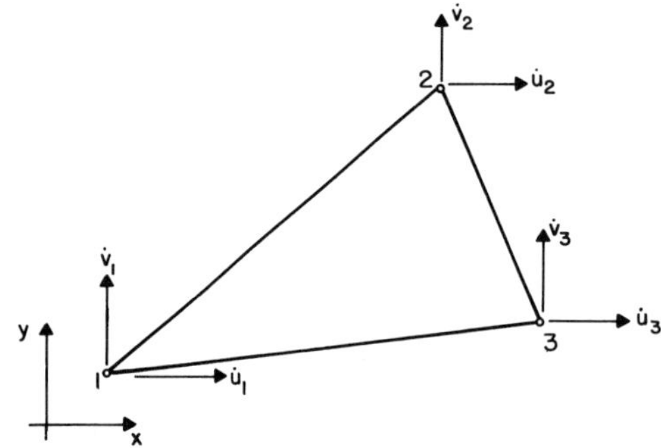

Figure 6. Triangular element

of the triangle, referred to as the *nodes*, may have arbitrary positions and thus the triangle may have arbitrary size and shape. The nodes are numbered 1,2,3 in the generic element. The velocity field will be required to be continuous throughout the body, and consequently the velocity field in any typical element must be chosen to satisfy this restriction. The simplest discretization is constructed so that the strain rate is constant in any one element. Let the velocities of the nodes be $(\dot{u}_1,\dot{v}_1)$, $(\dot{u}_2,\dot{v}_2)$, $(\dot{u}_3,\dot{v}_3)$ respectively, and let the coordinates of the nodes be (x_1,y_1), (x_2,y_2),(x_3,y_3). The velocities within the elements are taken to be linear to provide constant strain rate, and are written as

$$\dot{u} = C_1x + C_2y + C_3 \;,$$

$$\dot{v} = D_1x + D_2y + D_3 \;. \tag{69}$$

The coefficients in equation (69) are then functions of the node velocities and the node coordinates, and may be determined explicitly:

$$
\begin{bmatrix} C_1 & D_1 \\ C_2 & D_2 \\ C_3 & D_3 \end{bmatrix}
=
\begin{bmatrix} x_1 & y_1 & 1 \\ x_2 & y_2 & 1 \\ x_3 & y_3 & 1 \end{bmatrix}^{-1}
\begin{bmatrix} u_1 & v_1 \\ u_2 & v_2 \\ u_3 & v_3 \end{bmatrix}
\tag{70}
$$

The strain rates within the element are then

$$\dot{\varepsilon}_{xx} = \frac{\partial \dot{u}}{\partial x} = C_1 \;,$$

$$\dot{\varepsilon}_{yy} = \frac{\partial \dot{v}}{\partial y} = D_2 \;, \tag{71}$$

$$\dot{\varepsilon}_{xy} = \frac{1}{2}(\frac{\partial \dot{u}}{\partial x} + \frac{\partial \dot{v}}{\partial y}) = \frac{1}{2}(C_2 + D_1) \;.$$

If the entire sheet is divided into triangular elements, with

the restriction that node points occur only at the apices of
any triangle, the velocity field is defined over the entire
sheet in terms of the node velocities. The velocity field is
continuous; this can be demonstrated by noting that the velocity
along any side of an element depends only on the coordinates
and velocities of the nodes defining that side and not on the
coordinates or velocity of the third node. Hence the velocities
along abutting sides of adjacent elements are identical.

The von Mises yield condition is a convenient choice since
the principal stress or strain rate directions are not immediately known. The specific dissipation function (a generalization
of the result given in Chapter 9) is

$$D(\dot{\varepsilon}_x, \dot{\varepsilon}_y, \dot{\varepsilon}_{xy}) \;=\; \frac{2\sigma_o}{\sqrt{3}} \, [\dot{\varepsilon}_x^2 + \dot{\varepsilon}_x \dot{\varepsilon}_y + \dot{\varepsilon}_y^2 + (\frac{\dot{\varepsilon}_{xy}}{2})^2]^{1/2} . \tag{72}$$

It follows then that the energy dissipation rate in a typical
element is

$$D_{int}^e \;=\; \frac{2\sigma_o}{\sqrt{3}} \, [C_1^2 + C_1 D_2 + D_2^2 + (\frac{C_2 + D_1}{2})^2]^{1/2} \, V_e \; , \tag{73}$$

where V_e is the volume of the element, i.e. its area multiplied
by unit thickness. This expression can be written in terms of
the node velocities by means of equation (70), and the total
internal energy dissipation rate is found by summing over all
the elements.

Curved boundaries of the sheet must clearly be replaced by
polygonal boundaries which become the sides of triangular elements. Where the velocities are prescribed to be zero the appropriate node velocities are set equal to zero. The rate of
work done by boundary tractions and by body forces D_{ext} can
then be computed and expressed in terms of the node velocities

by means of equations (69) and (70). Proportional loading is
considered, so that D_{ext} is linear and homogeneous in a single
scalar parameter Γ. The programming problem then requires that
we minimize Γ, where Γ is expressed in terms of the node velocities. This is an unconstrained programming problem.

In using the finite element approach in conjunction with the
lower bound theorem, consider again a triangular element. Within the element the equilibrium equations

$$\frac{\partial \sigma_x}{\partial x} + \frac{\partial \sigma_{xy}}{\partial y} = 0 \ ,$$

$$\frac{\partial \sigma_{xy}}{\partial x} + \frac{\partial \sigma_y}{\partial y} = 0 \ , \tag{74}$$

must be satisfied identically; the case of zero body force is
taken for simplicity. This can be achieved by assuming that
the stress distribution within each element has the form

$$\sigma_x = A_1 x^2 + A_2 xy + A_3 y^2 + A_4 x + A_5 y + A_6 \ ,$$

$$\sigma_y = A_7 x^2 + A_8 xy + A_1 y^2 + A_9 x + A_{10} y + A_{11} \ , \tag{75}$$

$$\sigma_{xy} = -\frac{1}{2} A_8 x^2 - 2A_1 xy - \frac{1}{2} A_2 y^2 - A_{10} x - A_4 y + A_{12} \ .$$

This field depends on twelve parameters $A_1, \ldots, A_{12}$, and is
derivable from a general fourth-order Airy stress function.

At abutting edges of adjacent elements the normal and tangential stresses σ_n, σ_t must be continuous. These components
are given by

$$\sigma_n = \sigma_x n_x + \sigma_{xy} n_y$$

$$\sigma_t = \sigma_{xy} n_x + \sigma_y n_y \ . \tag{76}$$

Along boundaries of the plate (which must again be polygonal)
the normal and tangential components must be equal to the prescribed normal and tangential tractions if the tractions are
given, but are unconstrained if the velocity is prescribed to
be zero. While complete details of the manner in which these
conditions are satisfied will not be given, it may be noted that
the normal stress distribution along any edge will be quadratic.
This implies that if the normal and tangential components of
stress at three discrete points on the abutting edges of two
adjacent elements are continuous, the continuity requirements
will be satisfied at all points on that edge.

It is convenient to replace the coefficients of equation (75)
with physically meaningful parameters: this may be accomplished
by introducing the twelve independent normal and tangential
stress values shown in Figure 7. These may be labelled $\sigma_{\alpha k}$,
where $\alpha = 1, \ldots, \nu$ identifies the element and $k = 1, \ldots, 12$.
It follows then that by using equations (75) and (76) a relation
between the A_k and the $\sigma_{\alpha k}$ may be written

$$\sigma_{\alpha k} = \sum_{\ell=1}^{12} B_{\alpha k \ell} A_\ell \ . \tag{77}$$

The continuity requirements and the stress boundary conditions
then lead to a set of equality constraints involving the $\sigma_{\alpha k}$.

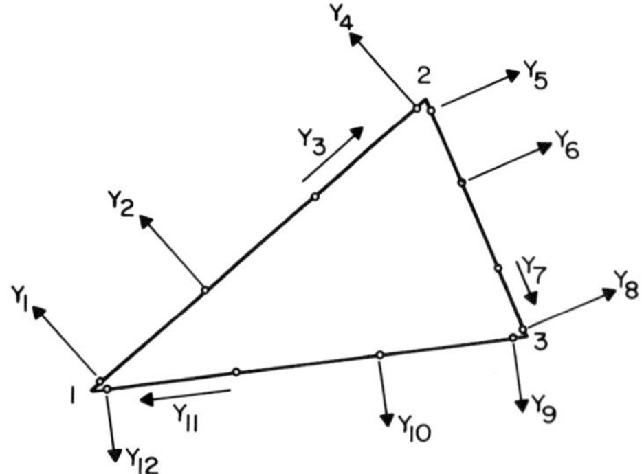

Figure 7. Normal and tangential stresses on element

The formulation of these constraints involves consideration of all elements, and through the boundary conditions will include the load parameter Γ. They will have the general form

$$\sum_{\alpha=1}^{\nu} \sum_{k=1}^{12} B_{p\alpha k}\sigma_{\alpha k} = \Gamma c_p \tag{78}$$

where $p=1,\ldots,m$. The total number of constraints is thus given by m.

The von Mises yield function is

$$\phi = \sigma_x^2 + \sigma_x\sigma_y + \sigma_y^2 + \left(\frac{\sigma_{xy}}{2}\right)^2 - \sigma_o^2 \tag{79}$$

and we require that $\phi \leq 0$ at all points in the body. A set of discrete points x_j, y_j in the body which include all relative maxima of ϕ can be identified, and by means of equations (75) and (77) a set of inequality constraints is formulated:

$$\phi(\sigma_{\alpha k}, x_j, y_j) \leq 0 \quad . \tag{80}$$

The programming problem thus requires that we maximize Γ subject to constraints (78) and (80).

The two programming problems described will provide upper and lower bounds on the limit load parameter since it is not likely that the discretized fields contain the actual solution. The accuracy of the bounds will depend on the number of elements and their arrangement: smaller elements should be used in areas of the sheet where the gradients of the actual fields are expected to be large. It should be noted that the constant strain rate elements in the upper bound approach are simpler than the quadratic stress elements in the lower bound approach. Consequently more elements in the kinematic formulation given here should be used than in the static approach for comparable accuracy.

Figure 8(a) shows a sheet with a square opening subjected to uniaxial tension. Figures 8(b) and 8(c) show the division of the plate into triangular elements for the kinematic and static approaches respectively. The upper and lower bounds obtained in computations for this problem are

$$\sigma^+ = 0.764 \, \sigma_0 \quad ,$$
$$\sigma^- = 0.693 \, \sigma_0 \quad . \tag{81}$$

These results would provide acceptable accuracy for many applications. Improvements in the results would require refinements in the finite element net.

13.5 Incremental Methods of Determining the Limit Load

Before concluding this discussion of numerical methods of determining the limit load, it must be pointed out that approaches which do not exploit the limit theorems directly can be used.

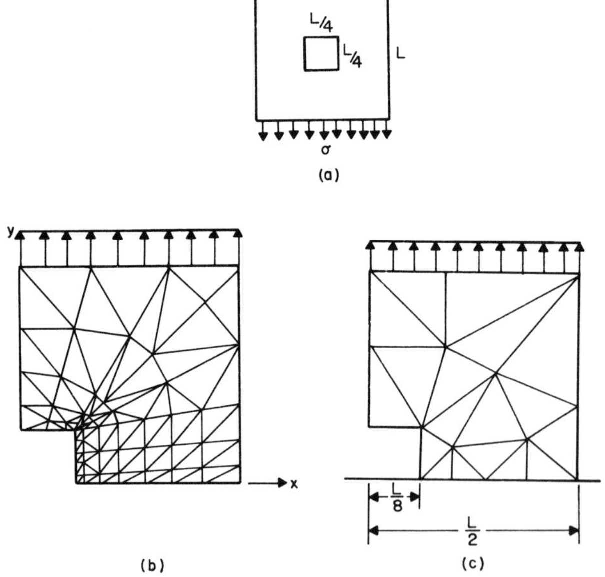

Figure 8. Sheet with square opening
(after Belytschko and Hodge 1970)

The most important alternative is *incremental analysis*.   The
structure is assumed to be elastic, perfectly plastic and the
complete solution for a monotonically increasing proportional
loading state is determined. The load at which flow occurs
identifies the limit load parameter. Incremental analysis will
be discussed in Part V, and a detailed account will be deferred.
It may be noted, however, that the incremental analysis involves a series of linear problems, a process which is comparable with step by step solutions of the programming problem. It
may occur that incremental analysis is more efficient, and it
provides considerably more information about the behavior of
the structure.

13.6 Bibliographical Remarks

Programming methods for solutions of the limit analysis problem
were first studied for trusses, beam and frames: these are problems in linear programming. Significant papers are those of
Dorn and Greenberg [1957], Charnes, Lemke and Zienkiewicz [1959]
and Gavarini [1966]. The development of more powerful compu-
ters has more recently made feasible the solution of complex
programming problems. Limit analysis of arches was studied by
Belytschko and Hodge [1968], of plates by Koopman and Lance
[1965], Hodge [1966] and Hodge and Belytschko [1968] and of
shells by Biron and Hodge [1967]. Examples of the solution of
plane stress problems by finite elements and finite differences
are given by Hayes and Marcal [1967], Neal [1968], and Belytschko and Hodge [1970]. Dynamic programming methods have been applied to shells by Palmer [1969].

References

T.Belytschko and   1968   "Program for the yield-point load of
P.G.Hodge,Jr.   arches",Proc.ACSE, 94 (ST6), 1383.

T.Belytschko and   1970   "Plane stress limit analysis by finite
P.G.Hodge,Jr.   elements",Proc.ASCE,96 (EM6), 931.

A.Biron and   1967   "Limit analysis of rotationally sym-
P.G.Hodge,Jr.   metric shells under central boss
                              loadings by numerical methods", J.
                              Appl.Mech., $\underline{34}$, 640.

A.Charnes,   1959   "Virtual work, linear programming and
C.E.Lemke and   plastic limit analysis", Proc.Roy.
O.C.Zienkiewicz   Soc., $\underline{A251}$, 110.

W.S.Dorn and   1967   "Linear programming and plastic limit
H.J.Greenberg   analysis of structures", Q.Appl.Math.,
                              $\underline{15}$, 155.

C.Gavarini   1966   "The plastic theorems of limit analy-
                              sis and the duality in linear pro-
                              gramming", Ingegneria Civile, $\underline{18}$.

D.J.Hayes and   1967   "Determination of upper bounds for
P.V.Marcal   problems in plane stress using
                              finite element techniques", Int.J.
                              Mech.Sci., $\underline{9}$, 245.

D.C.A.Koopman and   1965   "On linear programming in limit
R.H.Lance   analysis", J.Mech.Phys.Sol.,$\underline{13}$, 77.

P.G.Hodge,Jr.   1966   "Yield-point determination by non-
                              linear programming", Proc.Int.Congr.
                              Appl.Mech.(1965), Springer, 554.

P.G.Hodge,Jr. and   1968   "Numerical methods for the limit
T.Belytschko   analysis of plates", J. Appl.Mech.,
                              $\underline{35}$, 796.

B.G.Neal   1968   "Limit Load of a Cantilever in Plane
                              Stress", *Engineering Plasticity* (ed.
                              by J.Heyman and F.A.Leckie),
                              Cambridge, 473.

A.C.Palmer   1969   "Limit analysis of cylindrical shells
                              by dynamic programming", Int.J.Solids
                              Structures, $\underline{5}$, 289.

THE LIMIT DESIGN PROBLEM

14.1 Introductory Remarks

In many applications of structural mechanics we are required to design a structure to fulfill a given purpose. In general the class of acceptable designs is quite large, and we usually attempt to select from this class a design which adequately satisfies some design criterion. There will exist among the class of acceptable designs at least one design which best meets this objective, and this design is termed the *optimal design*. Optimization, or near-optimization, of the design has traditionally been accomplished by trial and error, involving analysis and re-analysis, and guided by previous experience with similar design problems. In a growing class of problems, however, we may explicitly formulate the design problem and devise methods of directly generating the optimal design.

A properly formulated problem in *structural design*, as opposed to structural analysis, requires a mathematical characterization of the *function* of the structure, *geometrical* and *behavioral constraints* imposed on the problem, and the *design objective*. An *acceptable design* is any design which fulfills the function of the structure and satisfies the constraints. The *optimal design* is that member of the class of acceptable designs which minimizes (or maximizes) the design objective.

In structures composed of an elastic, perfectly plastic material, one of the primary functions of the structure is that it should carry a given set of loads without flow occurring in the structure. If this is the *only* function of the structure, the structural design problem is referred to as the *limit design problem*. The limit design problem is fairly well developed, and in this section we shall discuss some representative problems in limit design. This will be accomplished by choosing a simple fundamental problem in limit design which will be discussed in detail. Generalizations of this problem will then be considered in a more concise form.

14.2 A Fundamental Problem in Limit Design

Consider a one or two-dimensional structure whose *configuration is fixed*. Thus in a bar structure the center lines of the members are given, and in a plate or shell the middle surface is fixed. Suppose that the structure is subjected to a *single* set of proportional generalized loads ΓC_α, where $0 \leq \Gamma < \Gamma^*$. Since the limit surface in load space is convex for any structure composed of a conventional elastic-plastic material, and contains the origin, we can ensure that the loads ΓC_α can be supported by the structure without flow occurring by requiring that for all acceptable designs

$$\Psi(\Gamma^* C_\alpha) \leq 0 \quad . \tag{1}$$

Thus if inequality (1) is satisfied the structure will fulfill its function.

The cross-section of the structure may be chosen at the discretion of the designer. We shall limit ourselves throughout this section to cases in which only one cross-sectional parameter can be varied in the design process. Further, we shall assume that this cross-section parameter is such that it can be expressed as a monotone function of some *design parameter* Q_o which has the dimensions of a generalized stress component. A *design* is then given by a choice of the function $Q_o(s)$, where s represents the spatial variables in the structure.

The limit surface in generalized stress space will then be a function of the appropriate generalized stresses Q_j and the design parameter Q_o. Thus

$$\psi = \psi(Q_j, Q_o) \quad . \tag{2}$$

We shall further assume that $\psi(Q_j, Q_o)$ can be expressed in the form

$$\psi(Q_j, Q_o) = \zeta(Q_j) - Q_o \quad , \tag{3}$$

where $\zeta(Q_j)$ is *convex, homogeneous, and of degree one* in the components of Q_j, and is *independent* of Q_o. The limit surface in stress space will thus be given by

$$\zeta(Q_j) = Q_o \quad . \tag{4}$$

This equation represents a family of similar convex surfaces in stress space which expand as Q_o increases. It is evident that, in order that equations (3) and (4) should be meaningful, Q_o must be non-negative. Since Q_o is related to some cross-section dimension, this is consistent with the obvious statement that the cross-section dimensions must be non-negative. Thus we have a geometrical constraint

$$Q_o(s) \geq 0 \quad . \tag{5}$$

As the design objective we shall choose the mass (or weight) of material used in the structure. The material is assumed to be homogeneous, and since Q_o is a function of the cross-section dimensions, the *specific weight* $w(s)$ is taken to be a monotonically increasing function of Q_o. Thus

$$w = w(Q_o) \quad . \tag{6}$$

The total weight of the structure is then

$$C = \int_V w\,dV \quad . \tag{7}$$

The total weight C will be the *design objective*. An *acceptable design* is any choice of $Q_o(s)$ for which inequalities (1) and (5) are satisfied. The *limit design* or *optimal design* is the acceptable design for which the design objective C has its *least*

value. A design will be said to be a *global optimum* if its
weight C is not greater than all other *acceptable designs*. The
design is a local optimum if its weight C is not greater than
all *neighboring acceptable designs*.

Before proceeding to consider the limit design, let us briefly
consider one possible way in which equations (3)-(6) are generated. Consider a truss composed of uniform pin-jointed circular
bars. At each cross-section the axial force N is the only generalized stress component which need be considered. For a circular bar subjected to an axial force, the limiting magnitude
of the axial force will be

$$N_o = \pi \sigma_o R^2 , \tag{8}$$

where R is the radius of the bar. N_o increases monotonically
with R, and may be chosen as the design parameter Q_o. The limit
function is

$$\psi(N,N_o) = |N| - N_c , \tag{9}$$

which has the form of equation (3). The specific weight of the
bar is

$$w = \rho \pi R^2 , \tag{10}$$

where ρ is the density of the material. From equations (8) and
(10),

$$w = \frac{\rho}{\sigma_o} N_o . \tag{11}$$

The specific weight is a linear function of the design parameter. Note that if ρ, σ_o are independent of the space parameters
s, as is usually the case, we may without loss of generality
simply put

$$w = N_o . \tag{12}$$

The design objective C will be proportional to the total weight, and to minimize the design objective will be equivalent to minimizing the total weight.

Returning to the general problem, we seek $Q_o(s)$ such that constraints (1) and (5) are satisfied and the design objective C (equation (7)) has its least value. This is a *programming problem*, and we could proceed to formulate it as such. It is more instructive, however, to seek conditions which identify the optimal design, i.e. which determine how the optimal design may be recognized among other members of the class of acceptable designs.

In formulating conditions for optimality it is convenient to work with the class of pertinent statically admissible fields for the structure under generalized loads $\Gamma^* C_\alpha$. Let this class of statically admissible fields be denoted by $\bar{Q}_j(s)$. The fields can be unambiguously defined because the configuration is fixed; they do not depend on $Q_o(s)$. We may now proceed to state conditions for optimality. The proofs of these propositions are based on the limit theorems.

Our first statement provides a sub-class of the class of acceptable designs which contains the optimal design.

The optimal design is contained within the sub-class of acceptable designs given by

$$Q_o(s) = \zeta\{\bar{Q}_j(s)\} \ . \tag{13}$$

This proposition is proved by appealing to the lower bound theorem of limit analysis. Consider an acceptable design $Q_o^{(1)}(s)$ which does not fall into the sub-class defined by equation (13). This implies that there exists at least one statically admissible field $\bar{Q}_j^{(1)}(s)$ such that

$$Q_o^{(1)}(s) = \zeta(\bar{Q}_j^{(1)}) \tag{14}$$

over part of the structure, V_1 say, and

$$Q_o^{(1)}(s) > \zeta(\bar{Q}_j^{(1)}) \tag{15}$$

over the remainder of the structure V_2. Consequently, given that $w(Q_o)$ is a monotonically increasing function of Q_o, there exists another design $Q_o^{(2)}(s)$,

$$Q_o^{(2)}(s) = \zeta(\bar{Q}_j^{(1)}) \quad \text{over} \quad V_1 \text{ and } V_2 \tag{16}$$

which satisfies the constraints (1) and (5) and whose weight is less than that of the design $Q_o^{(1)}$. This follows because, from equations (15) and (16)

$$Q_o^{(2)}(s) = Q_o^{(1)} \quad \text{in} \quad V_1 \ ,$$
$$Q_o^{(2)}(s) < Q_o^{(1)} \quad \text{in} \quad V_2 \ . \tag{17}$$

Thus for every acceptable design which does not fall into the sub-class of equation (13), there exists a design of less weight in the sub-class. Alternatively, equation (13) can be considered as a *necessary condition for optimality*, both in the local and the global sense.

In the general case we do not know which member of the class of pertinent statically admissible fields is associated with the optimal design, and hence equation (13) is not *sufficient* for optimality. An important standing exception, however, is found in *statically determinate structures*, where $\bar{Q}_j(s)$ contains only one member. The optimal design is uniquely determined by equation (13), and the condition is sufficient for optimality in this case.

In statically indeterminate structures we may limit ourselves to the sub-class of acceptable designs given by equation (13), but we must formulate conditions which allow us to identify the

generalized stress field associated with optimal design. To do
this we make use of the calculus of variations. Consider a design which is locally optimal. Working within the sub-class of
acceptable designs given by equation (13), we vary the genera-
lized stress field $\bar{Q}_j(s)$ by $\delta\bar{Q}_j(s)$. Note that since $\bar{Q}_j(s)$ are
statically admissible, $\delta\bar{Q}_j(s)$ is a self-equilibrating stress
field. Since the cost C is given by

$$C = \int w(Q_0)dV \quad , \tag{18}$$

a variation about a locally optimal design must give a variation
in the cost which is non-negative,

$$\delta C = \int_V \frac{dw}{dQ_0}\bigg|_{Q_0=\zeta(\bar{Q}_j)} \delta\zeta \, dV \geq 0 \quad . \tag{19}$$

The inequality sign is employed because we have no reason to
expect that the local minimum will be analytic. In many cases
it is not, as we shall see in the illustrative examples which
follow.

Provided that $\zeta(\bar{Q}_j) > 0$, the variation $\delta\zeta$ can be written as

$$\delta\zeta = \frac{\partial\zeta}{\partial Q_j}\bigg|_{\bar{Q}_j} \delta\bar{Q}_j \quad . \tag{20}$$

However, this expression does not hold when $\zeta(\bar{Q}_j) = 0$. In this
case

$$\frac{\partial\zeta}{\partial Q_j}\bigg|_{Q_j=0}$$

is not uniquely defined, and indeed

$$\delta\zeta = \zeta(\delta\bar{Q}_j) \quad . \tag{21}$$

Notice that because $\zeta(Q_j)$ is homogeneous and of order one in
the components of Q_j, $\partial\zeta/\partial Q_j$ is homogeneous and of order zero.

This permits us to obtain an expression for $\delta\zeta$ in equation (21).

The function $\zeta(Q_j)$ is required to be convex. This implies that, if we choose arbitrarily two states of stress Q_j^a, Q_j^b, we can write

$$\zeta(Q_j^b) - \zeta(Q_j^a) \geq \left.\frac{\partial\zeta}{\partial Q_j}\right|_{Q_j^a} (Q_j^b - Q_j^a) \quad . \tag{22}$$

Now suppose that we allow Q_j^a to approach zero along some radial path in stress space. The function $\zeta(Q_j^a)$ approaches zero, but since $\partial\zeta/\partial Q_j$ does not depend on the magnitude of Q_j^a (being of order zero) but only its direction in stress space it remains constant. Hence we can write

$$\zeta(Q_j^b) \geq \frac{\partial\zeta}{\partial Q_j} \, Q_j^b \quad , \tag{23}$$

where, since we can allow Q_j^a to approach zero from any direction in stress space, $\partial\zeta/\partial Q_j$ may be evaluated for *any arbitrarily chosen non-zero state of stress*. Putting $Q_j^b = \delta\bar{Q}_j$, we have

$$\delta\zeta \; = \; \zeta(\delta\bar{Q}_j) \geq \frac{\partial\zeta}{\partial Q_j} \, \delta\bar{Q}_j \quad . \tag{24}$$

Further, if $0 \leq \beta \leq 1$, it is evident that

$$\delta\zeta \geq \beta \, \frac{\partial\zeta}{\partial Q_j} \, \delta\bar{Q}_j \quad . \tag{25}$$

Returning to the expression for δC (equation (19)) it is convenient to denote collectively all points on the structure at which $\zeta(\bar{Q}_j) = 0$ in the optimal design by V_o. All points at which $\zeta(\bar{Q}_j) > 0$ are denoted by $V-V_o$. From equations (19) and (25), with $\beta = \beta(s)$,

$$\delta C = \int_{V-V_o} \left.\frac{dw}{dQ_o}\right|_{\zeta(\bar{Q}_j)} \left.\frac{\partial \zeta}{\partial Q_j}\right|_{\bar{Q}_j} \delta\bar{Q}_j dV + \int_{V_o} \left.\frac{dw}{dQ_o}\right|_{Q_o=0} \delta\zeta dV$$

$$\geq \int_{V-V_o} \left.\frac{dw}{dQ_o}\right|_{\zeta(\bar{Q}_j)} \left.\frac{d\zeta}{dQ_j}\right|_{\bar{Q}_j} \delta\bar{Q}_j dV + \int_{V_o} \beta \left.\frac{dw}{dQ_o}\right|_{Q_o=0} \frac{d\zeta}{dQ_j} \delta\bar{Q}_j dV \quad .$$

$$(26)$$

Comparing equation (19) with equation (26), it is evident that a *sufficient condition for optimality* is that

$$\int_{V-V_o} \left.\frac{dw}{dQ_o}\right|_{\zeta(\bar{Q}_j)} \left.\frac{d\zeta}{dQ_j}\right|_{\bar{Q}_j} \delta\bar{Q}_j dV + \int_{V_o} \beta \left.\frac{dw}{dQ_o}\right|_{Q_o=0} \frac{d\zeta}{dQ_j} \delta\bar{Q}_j dV = 0 \quad . \quad (27)$$

This condition admits a simple physical interpretation. Noting that $\delta\bar{Q}_j(s)$ is an *arbitrary* self-equilibrating field, the condition

$$\int_V \dot{q}_j \, \delta\bar{Q}_j dV = 0 \tag{28}$$

implies that $\dot{q}_j(s)$ is a pertinent kinematically admissible strain rate field. Hence equation (27) will hold if

$$\dot{q}_j(s) = \left.\frac{dw}{dQ_o}\right|_{\zeta(\bar{Q}_j)} \left.\frac{\partial \zeta}{\partial Q_j}\right|_{\bar{Q}_j} \quad \text{in} \quad V-V_o \quad , \tag{29a}$$

$$\dot{q}_j(s) = \beta(s) \left.\frac{dw}{dQ_o}\right|_{Q_o=0} \frac{\partial \zeta}{\partial Q_j} \quad \text{in} \quad V_o \quad , \tag{29b}$$

can be interpreted as a kinematically admissible field, with $\dot{q}_j(s)$ derivable from a velocity field $\dot{\underline{u}}(s)$ which satisfies the kinematic boundary conditions.

Flow will occur in the optimally designed structure under loads $\Gamma^* C_\alpha$, in view of the necessary condition for optimality (equation (13)) which requires that each cross-section yields.

At a cross-section in $V-V_o$, the strain rate has the form

$$\dot{q}_j \;=\; \lambda\,\frac{\partial\psi}{\partial Q_j} \;=\; \lambda\,\frac{\partial\zeta}{\partial Q_j}\;. \tag{30}$$

Associating λ in equation (30) with dw/dQ_o in equation (29a), the strain rates of equations (29) can be interpreted as a *possible* strain rate field during flow of the optimally designed structure. The structure need not *necessarily* flow with this strain rate field (or any scalar multiple of it) since the velocity field is not necessarily unique and seldom is in optimally designed structures.

The *sufficient conditions for optimality* are that the design parameter Q_o and the generalized stress field $\bar{Q}_j(s)$ under loads $\Gamma^* C_\alpha$ be related through equation (13), and that equations (29), with $\beta(s)$, $0 \leq \beta \leq 1$ and $\partial\zeta/\partial Q_j$ in V_o chosen at the discretion of the designer, should represent a pertinent kinematically admissible field for the structure and a possible velocity field during flow.

We have thus far limited attention to the case where the limit function (equation (3)) is continuously differentiable. If the limit surface contains corners, we assume that there exist two or more continuously differentiable limit functions. It will suffice that we treat the case where there exist two limit functions. These are assumed to have the form

$$\psi_1(Q_j,Q_o) \;=\; \zeta_1(Q_j) - Q_o\;,$$
$$\psi_2(Q_j,Q_o) \;=\; \zeta_2(Q_j) - Q_o\;, \tag{31}$$

where $\zeta_1(Q_j)$, $\zeta_2(Q_j)$ are subject to the same limitations which were placed on $\zeta(Q_j)$ in equation (3).

The first result concerning conditions for optimality (equation (13)) is readily generalized. Since both yield functions must be non-positive in the optimal design, it is evident that the optimal design is contained within the sub-class of acceptable designs given by

$$Q_o(s) = \zeta_1\{\bar{Q}_j(s)\} \quad \text{if} \quad \zeta_1\{\bar{Q}_j(s)\} \geq \zeta_2\{\bar{Q}_j(s)\} \quad ,$$

$$Q_o(s) = \zeta_2\{\bar{Q}_j(s)\} \quad \text{if} \quad \zeta_2\{\bar{Q}_j(s)\} \geq \zeta_1\{\bar{Q}_j(s)\} \quad . \tag{32}$$

The generalization of equations (29) is not quite as obvious. We expect that quantities which can be interpreted as strain rates should be required to be kinematically admissible, but these quantities can be more freely interpreted. Consider first equation (29a). It might be expected that this will be replaced by

$$\dot{q}_j(s) = \frac{dw}{dQ_o}\bigg|_{\zeta_1(\bar{Q}_j)} \frac{\partial\zeta_1}{\partial Q_j}\bigg|_{\bar{Q}_j} \quad \text{in } V\text{-}V_o \text{ if } \zeta_1(\bar{Q}_j) > \zeta_2(\bar{Q}_j) \quad , \tag{33a}$$

$$\dot{q}_j(s) = \frac{dw}{dQ_o}\bigg|_{\zeta_2(\bar{Q}_j)} \frac{\partial\zeta_2}{\partial Q_j}\bigg|_{\bar{Q}_j} \quad \text{in } V\text{-}V_o \text{ if } \zeta_2(\bar{Q}_j) > \zeta_1(\bar{Q}_j) \quad . \tag{33b}$$

Further consideration must be given to the case $\zeta_1(\bar{Q}_j) = \zeta_2(\bar{Q}_j)$ when the stress point lies at a corner on the limit surface. A fairly simple way in which the required form of the result can be seen is to consider the product $\bar{Q}_j\dot{q}_j$ at any point in the structure where equation (29a) (or equation 33a or equation 33b) hold. Since $\zeta(Q_j)$ is homogeneous and of degree one, and in view of equation (13),

$$\bar{Q}_j\dot{q}_j = \bar{Q}_j \frac{dw}{dQ_o} \frac{\partial\zeta}{\partial Q_j} = \frac{dw}{dQ_o} \zeta(\bar{Q}_j)$$

$$= \frac{dw}{dQ_o} Q_o \quad . \tag{34}$$

If this relation is to hold when $\zeta_1(\bar{Q}_j) = \zeta_2(\bar{Q}_j) = Q_o$ when there

exist two limit functions, equations (33) must be supplemented
by

$$\dot{q}_j(s) = \frac{dw}{dQ_o} \{\alpha \frac{\partial \zeta_1}{\partial Q_j} + (1-\alpha) \frac{\partial \zeta_2}{\partial Q_j}\} \quad \text{in } V-V_o$$

$$\text{when} \quad \zeta_1(\bar{Q}_j) = \zeta_2(\bar{Q}_j) \quad , \tag{35}$$

where $0 \leq \alpha \leq 1$, but α is otherwise arbitrary and may be a
function of the space parameters s. This further suggests that
equation (29b) is replaced by

$$\dot{q}_j(s) = \beta \frac{dw}{dQ_o}\bigg|_{Q_c=0} \{\alpha \frac{\partial \zeta_1}{\partial Q_j} + (1-\alpha) \frac{\partial \zeta_2}{\partial Q_j}\} \quad \text{in } V_o \quad , \tag{36}$$

where $\partial \zeta_1/\partial Q_j$ and $\partial \zeta_2/\partial Q_j$ may be evaluated for any one arbitrarily chosen state of stress.

In order to demonstrate that equation (31) together with the
requirement that the strain rates of equations (33), (35) and
(36) should be kinematically admissible are sufficient for optimality, we shall present an alternative proof which gives
stronger results for the case where $w(Q_o)$ is a convex function
than we have hitherto obtained.

Let us denote an acceptable design which also satisfies the
sufficient conditions for optimality by $Q_o^{(1)}(s)$. The statically admissible stress field $\bar{Q}_j^{(1)}(s)$ will satisfy equation
(31), and equations (32)-(36) give a kinematically admissible
strain rate field $\dot{q}_j^{(1)}$. Since the constitutive equations are
satisfied in $V-V_o$, this strain rate field is also a potential
flow field. Specifically, during quasi-static flow under loads
Γ^*C_α, using equation (34)

$$\Gamma^*C_\alpha \dot{\xi}_\alpha = \int_V \bar{Q}_j^{(1)} \dot{q}_j^{(1)} dV = \int_{V-V_o} \frac{dw}{dQ_o} {}_{Q_o^{(1)}} Q_o^{(1)} dV \quad , \tag{37}$$

where $\dot{\xi}_\alpha$ is the velocity field from which $\dot{q}_j^{(1)}$ is derived.

Consider also an admissible design $Q_o^{(2)}(s)$. With this design
we can associate a statically admissible stress field $\overline{Q}_j^{(2)}(s)$
such that

$$\zeta_1(\overline{Q}_j^{(2)}) \leq Q_o^{(2)} \quad ,$$

$$\zeta_2(\overline{Q}_j^{(2)}) \leq Q_o^{(2)} \quad . \tag{38}$$

The difference in weight in the two designs is

$$c^{(2)} - c^{(1)} = \int_V \{w(Q_o^{(2)}) - w(Q_o^{(1)}))\} \, dV \quad . \tag{39}$$

If $w(Q_o)$ is a *convex function* of Q_o, we permit $Q_o^{(2)}$ to be *any*
acceptable design. Using the convexity of w,

$$w(Q_o^{(2)}) - q(Q_o^{(1)}) \geq \frac{dw}{dQ_o}\Big|_{Q_o^{(1)}} (Q_o^{(2)} - Q_o^{(1)}) \quad . \tag{40}$$

If, on the other hand, $w(Q_o)$ is not convex, we limit the choice
of $Q_o^{(2)}$ to a design which neighbors $Q_o^{(1)}$, so that

$$\Delta Q_o = Q_o^{(2)} - Q_o^{(1)} \tag{41}$$

is small. Then, to first order,

$$w(Q_o^{(2)}) - w(Q_o^{(1)}) = \frac{dw}{dQ_o}\Big|_{Q_o^{(1)}} (Q_o^{(2)} - Q_o^{(1)}) \quad . \tag{42}$$

We now consider the term

$$\int_V \frac{dw}{dQ_o}\Big|_{Q_o^{(1)}} (Q_o^{(2)} - Q_o^{(1)}) dV = \int_V \frac{dw}{dQ_o}\Big|_{Q_o^{(1)}} Q_o^{(2)} dV$$

$$- \int_V \frac{dw}{dQ_o}\Big|_{Q_o^{(1)}} Q_o^{(1)} dV \quad , \tag{43}$$

which, according to equations (39)-(42) does not exceed the cost difference in either case.

Since the design $Q_o^{(2)}$ is safe, in that $\Psi(\Gamma^*C_\alpha) \leq 0$, the internal energy dissipation rate for any kinematically admissible velocity field must exceed the work rate of the external loads, according to the upper bound theorem of limit analysis. Employing the kinematically admissible field $\dot{\xi}_\alpha, \dot{q}_j^{(1)}$, we see that

$$\int_{V-V_o} \left.\frac{dw}{dQ_o}\right|_{Q_o^{(1)}} Q_o^{(2)} dV + \int_{V_o} \beta \left.\frac{dw}{dQ_o}\right|_{Q_o^{(1)}} Q_o^{(2)} dV \geq \Gamma^*C_\alpha \dot{\xi}_\alpha \ . \tag{44}$$

Since $0 \leq \beta \leq 1$, it follows that

$$\int_V \left.\frac{dw}{dQ_o}\right|_{Q_o^{(1)}} Q_o^{(2)} dV \geq \Gamma^*C_\alpha \dot{\xi}_\alpha \ . \tag{45}$$

Hence, from equations (37), (45), (43) and (39),

$$C^{(2)} \geq C^{(1)} \ . \tag{46}$$

This shows that the conditions we have set down are *sufficient* for *local optimality* when $w(Q_o)$ is any monotonically increasing function of Q_o, and *sufficient for global optimality when* $w(Q_o)$ *is convex.*

As an example of the exploitation of the sufficient conditions for optimality for this class of problems, consider first the limit design of transversely loaded sandwich beams. The cross-section of the beam, shown in Figure 1, consists of a core of depth h and width b and two thin cover plates each of width b and thickness t. It is assumed that b and h are fixed, and that the core resists shear but does not contribute to the bending strength. The thickness t may be varied at the discretion of the designer, and the weight of material to be minimized is the weight of the cover plates. If the cover plates flow under direct stress σ_o, the limit function may be written as

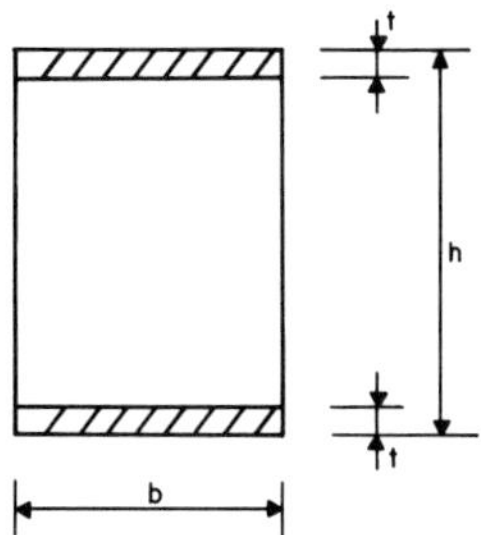

Figure 1. Cross-section of a sandwich beam

$$\psi(M,M_o) \;=\; |M| - M_o \;, \tag{47}$$

where M is the bending moment and

$$M_o \;=\; \sigma_o bth \;. \tag{48}$$

The area of the cover plates is 2bt, so that M_o is a linear
function of the specific weight. Without loss in generality,
we may put

$$w \;=\; M_o \;. \tag{49}$$

This linear function is convex, and consequently we shall be
able to find global optimal designs. Equation (47) satisfies
the restrictions imposed on equation (3), with

$$\zeta(M) \;=\; |M| \;. \tag{50}$$

Noting that

$$\frac{d\zeta}{dM} \;=\; \text{sgm } M \;,$$

where

$$\text{sgm } M = +1 \quad \text{if} \quad M > 0$$
$$\text{sgm } M = -1 \quad \text{if} \quad M < 0 \;, \tag{51}$$

equations (13), (29a) and (29b) become

$$M(x) = M_o \;,\tag{52a}$$

$$\dot{\kappa}(x) = \text{sgm } M(x) \quad \text{where} \quad |M| > 0 \;,\tag{52b}$$

$$-1 \leq \dot{\kappa}(x) \leq 1 \quad \text{where} \quad |M| = 0 \;,\tag{52c}$$

The last of these expressions is obtained by eliminating β from the requirement that $\dot{\kappa} = \beta \text{ sgm } M,\; 0 \leq \beta \leq 1$.

Consider the simply supported beam shown in Figure 2. The span is ℓ, and a concentrated load P is applied at a distance $a\ell$ from the left hand support. This beam is statically determinate, and the bending moment $M(x)$ is

$$M(x) = P(1-a)x \quad \text{for} \quad 0 \leq x \leq a\ell \;,$$
$$M(x) = Pa(\ell-x) \quad \text{for} \quad a\ell \leq x \leq \ell \;.\tag{53}$$

Since $M(x) > 0$ for all x, we obtain immediately from equation (52a) that

$$M_o(x) = P(1-a)x \quad \text{for} \quad 0 \leq x \leq a\ell \;,$$
$$M_o(x) = Pa(\ell-x) \qquad a\ell \leq x \leq \ell \;.\tag{54}$$

The thickness of the cover plates $t(x)$ can now be obtained from equation (48). It is not necessary that the kinematic conditions (equations (52b) and (52c)) should be checked, but it is instructive to show that an admissible velocity field can be

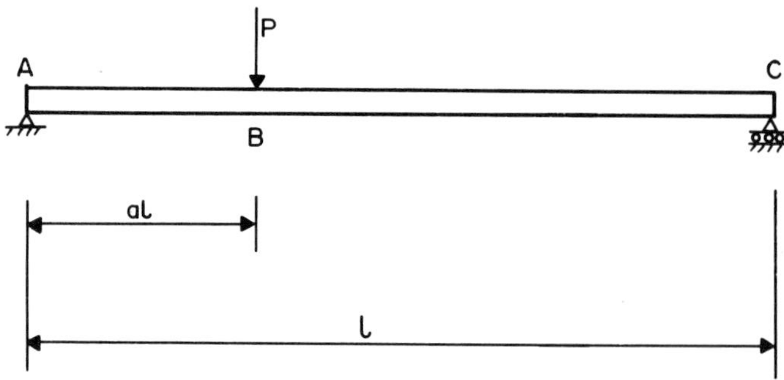

Figure 2. Simply supported beam example

constructed. From equation (52b), we seek a velocity field
$\dot{u}(x)$ such that

$$- \frac{d^2 \dot{u}}{dx^2} = \dot{\kappa}(x) = 1 \quad . \tag{55}$$

This equation can be integrated with boundary conditions $\dot{u}(0)$
$= \dot{u}(\ell) = 0$, giving

$$\dot{u}(x) = \frac{1}{2} x(\ell - x) \quad . \tag{56}$$

Note again that this velocity field is *a possible velocity
field* during flow, but not the only possible one.

Now consider the fixed end beam shown in Figure 3. Ths
span is again ℓ, and a concentrated load P is applied at a
distance a from the left support. This beam is statically
indeterminate. Choosing as the redundant forces the moments
M_A and M_C at the ends of the beam, the class of pertinent statically admissible bending moments is shown in Figure 4 and is
given by

$$M(x) = P(1-a)x - \frac{xM_A}{\ell} - (\ell-x) \frac{M_C}{\ell} \quad \text{for} \quad 0 \leq x \leq a\ell \quad ,$$

$$M(x) = Pa(\ell-x) - \frac{xM_A}{\ell} - (\ell-x) \frac{M_C}{\ell} \quad \text{for} \quad a \leq x \leq \ell \quad . \tag{57}$$

The bending moment may change sign over the length of the
beam. Thus, from equations (52b) and (52c), we seek a velocity
field in which $\dot{\kappa} = +1$ for $M > 0$ and $\dot{\kappa} = -1$ for $M < 0$, with the
restriction that $-1 \leq \dot{\kappa} \leq +1$ when $M=0$. Assume first that $M_A >$
0, $M_C > 0$, and that the bending moment does indeed change sign.
From equations (57), we find that $M(x) = 0$ when

$$x = x_1 = \frac{M_C \ell}{P\ell(1-a) - M_A + M_C} \quad , \tag{58}$$

and

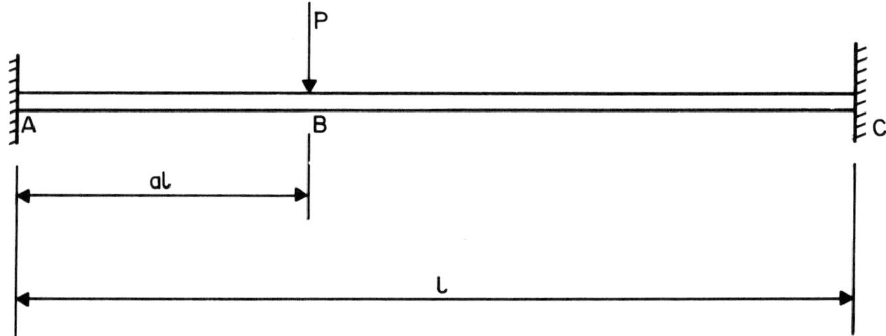

Figure 3. Fixed beam example

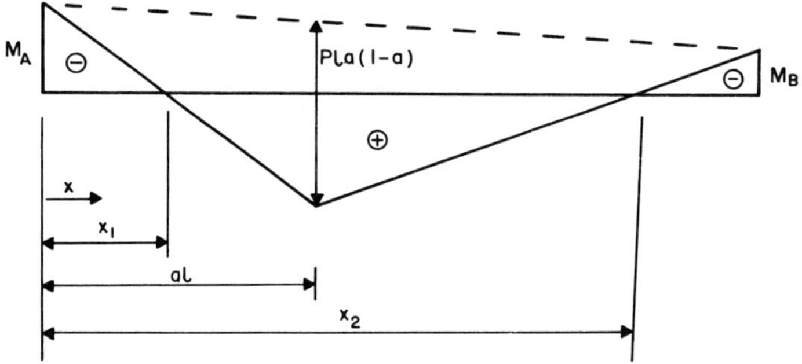

Figure 4. Bending moments for fixed beam

$$x = x_2 = \frac{Pa\ell^2 - M_C\ell}{P\ell a + M_A - M_C} \quad . \tag{59}$$

Hence

$$\dot{\kappa}(x) = -1 \quad \text{for} \quad 0 \le x \le x_1$$

$$\text{and} \quad x_2 \le x \le \ell \quad , \tag{60}$$

$$\dot{\kappa}(x) = +1 \quad \text{for} \quad x_1 \le x \le x_2 \quad .$$

In general these curvatures will not lead to a velocity field which satisfies the boundary conditions $\dot{u}(0) = \dot{u}(\ell) = \dot{u}'(0) = u'(\ell) = 0$. We must select those values of M_A and M_C for which the velocity field is a pertinent kinematically admissible field. Two necessary conditions for the satisfaction of the boundary conditions can be written using the *self-equilibrating* moment fields

$$m_A = x \quad , \quad m_B = (\ell - x) \quad . \tag{61}$$

From the principle of virtual velocities, we see that

$$\int_0^\ell m_A \dot{k}(x)\,dx \;=\; \int_0^\ell m_B \dot{k}(x)\,dx \;=\; 0 \tag{62}$$

if $\dot{u}(x)$, $\dot{k}(x)$ are kinematically admissible and satisfy the boundary conditions at $x=0$ and $x=\ell$.

Substituting from equations (60) and (61) into equation (62), we require that

$$\int_0^{x_1} (-1)x\,dx + \int_{x_1}^{x_2} (+1)x\,dx + \int_{x_2}^\ell (-1)x\,dx \;=\; 0 \quad ,$$

$$\int_0^{x_1} (-1)(\ell-x)\,dx + \int_{x_1}^{x_2} (+1)(\ell-x)\,dx + \int_{x_2}^\ell (-1)(\ell-x)\,dx \;=\; 0 \quad ,$$

or that

$$-x_1^2 + x_2^2 \;=\; \frac{\ell^2}{2} \quad ,$$

$$\text{and} \quad x_1(x_1-2\ell) + x_2(2\ell-x_2) \;=\; \frac{\ell^2}{2} \quad . \tag{63}$$

The solution of these equations can be found by inspection: it is

$$x_1 \;=\; \frac{\ell}{4} \quad , \quad x_2 \;=\; \frac{3\ell}{4} \quad . \tag{64}$$

Substitution of equation (64) into equations (58) and (59) provides two algebraic equations which can be solved for M_A and M_C, giving

$$M_A \;=\; \frac{P\ell}{8}(3-4a) \quad , \quad M_C \;=\; \frac{P\ell}{8}(4a-1) \quad . \tag{65}$$

Substitution of these values into equations (57) gives $M(x)$,

and $M_o(x)$ follows from equation (52a). The solution applies,
however, only for M_A and M_C non-negative. The design has thus
been obtained for the range

$$1/4 \leq a \leq 3/4 \quad . \tag{66}$$

The optimal designs are shown for the cases a = 1/2, a = 3/8
and a = 1/4 in Figure 5. Note that in each case the curvature
rate distribution which satisfies the sufficient conditions for
optimality is identical; it is given by equations (60) and
equations (64). This field is shown in Figure 6(a). The condition which applies for M=0 (equation (52c)) is satisfied at
the points $x=x_1$ and $x=x_2$, although it is not really operative
in the design.

The optimal design for values of a which lie outside the
range given by inequalities (66) is suggested by the optimal
design for the cases a = 1/4 and a = 3/4. Choosing the range

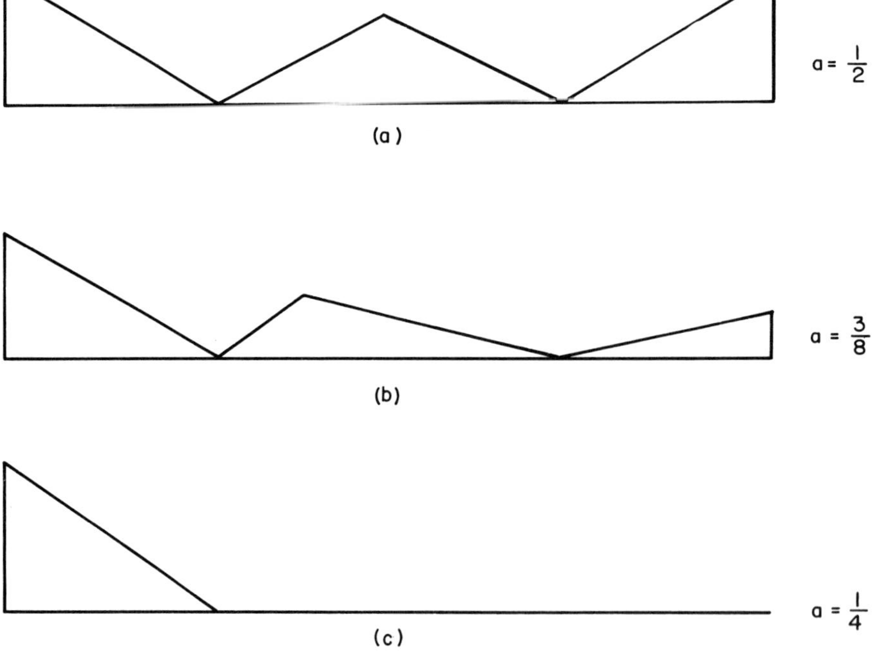

Figure 5. Optimal designs

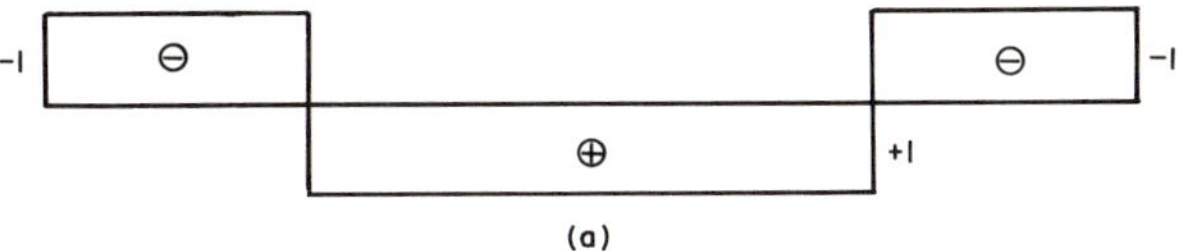

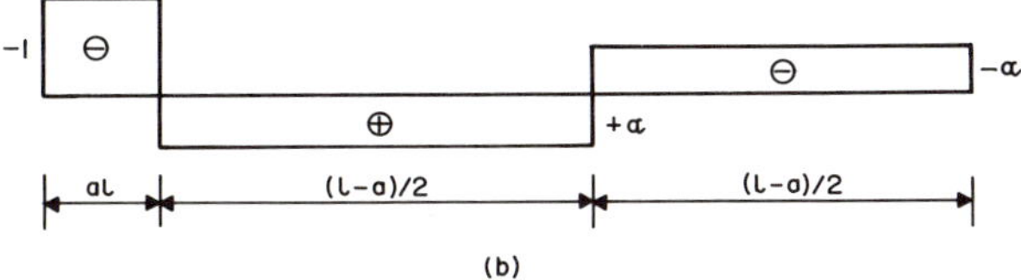

Figure 6. Curvature rates for fixed beam

$0 \leq a \leq 1/4$ as an example, we may seek a design in which $M(x) = 0$ for $x > a$. The fixed end beam then essentially degenerates into a cantilever AB, with $M_0 = 0$ over BC (Figure 3). Evidently in this case

$$M(x) = -P(a\ell-x) \text{ for } 0 \leq x \leq a\ell \ , \tag{67a}$$

$$M_0(x) = P(a\ell-x) \text{ for } 0 \leq x \leq a\ell \ , \tag{67b}$$

$$M(x) = M_0(x) = 0 \text{ for } a\ell \leq x \leq \ell \ . \tag{67c}$$

This design is optimal if, from equations (52b) and (42c), we can find a kinematically admissible velocity field $\dot{u}(x)$ which satisfies the boundary conditions $\dot{u}(0) = \dot{u}(\ell) = 0$, $\dot{u}'(0) = \dot{u}'(\ell) = 0$, and such that

$$-\frac{d^2\dot{u}}{dx^2} = \dot{\kappa}(x) = -1 \qquad 0 \leq x \leq a\ell \ , \tag{68a}$$

$$-1 \leq -\frac{d^2\dot{u}}{dx^2} = \dot{\kappa}(x) \leq 1 \qquad a\ell \leq x \leq \ell \ . \tag{68b}$$

To demonstrate that this is indeed possible suppose that the curvature rate distribution is as shown in Figure 6(b). It can be quickly checked that this distribution satisfies equations (62) when $\alpha = 2a/(1-2a)$. Equation (68b) is then satisfied.

The importance of the condition which applies in regions

when $M(x) = 0$ can be seen by considering some alternative designs which, in the absence of equation (52c), may plausibly be the optimal design. As an example, for the case $a = 1/2$, we might suppose that the bending moment distribution in the optimal design is

$$M(x) = - P(\frac{\ell}{2} - x) \quad \text{for} \quad 0 \leq x \leq \ell/2 ,$$
$$M(x) = 0 \quad \text{for} \quad \ell/2 \leq x \leq \ell . \tag{69}$$

This bending moment distribution is statically admissible, and the design is a cantilever of span $\ell/2$, with $M_o(x)$ given by

$$M_o(x) = P(\frac{\ell}{2} - x) \quad \text{for} \quad 0 \leq x \leq \ell/2 ,$$
$$M_o(x) = 0 \quad \text{for} \quad \ell/2 \leq x \leq \ell . \tag{70}$$

The necessary condition for optimality (equation (52a)) is satisfied, but it is not possible to find a velocity field $\dot{u}(x)$ which satisfies the boundary conditions for the *fixed end beam* and equations (52b) and (52c). A simple calculation will confirm that the design given by equations (70) is indeed of greater weight than that shown in Figure 5 for $a = 1/2$. If the kinematic boundary conditions are relaxed, however, and we do *not* require that $\dot{u}(\ell) = \dot{u}'(\ell) = 0$, the design of equation (70) *will* satisfy all of the sufficient conditions for optimality. It is the optimal design for a *cantilever* of span $\ell/2$ with a concentrated load at its tip.

As a second example consider an axi-symmetric sandwich plate in which the thickness of the cover plates may be varied as a function of the radius of the plate. Adopting the Tresca yield criterion, the limit surface in the M_r, M_θ space (cf. Chapter 11) is shown in Figure 7, where

$$M_o = \sigma_o th , \tag{71}$$

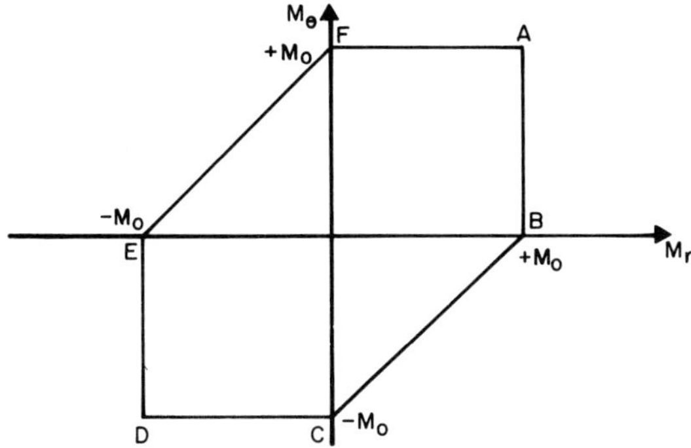

Figure 7. Yield condition for a sandwich plate

with t denoting the thickness of the plate, h the depth of the core and σ_0 the yield stress of the plate material in simple tension. It is evident that the weight per unit area is proportional to t, so that we may put

$$w(M_0) = M_0 \ . \tag{72}$$

Let us treat the limit design of a simply supported circular plate of radius R subjected to a uniform load p. From our experience in the limit analysis of this case (Chapter 11) we expect that $M_r(r) > 0$ and $M_\theta(r) > 0$. The conditions $M_r(0) = M_\theta(0)$ and $M_r(R) = 0$ must also be satisfied. This suggests that the limit design may have the form $M_r(r) = M_\theta(r) = M_0(r)$, so that each point on the plate lies at the corner A of the limit surface. The equilibrium equation is

$$\frac{d}{dr} (rM_r) = M_\theta - \frac{pr^2}{2} \tag{73}$$

With $M_r = M_\theta$ and the boundary condition $M_r(R) = 0$, we then have

$$M_0 = M_r = M_\theta = \frac{p}{4} (R^2 - r^2) \ . \tag{74}$$

Since the stress point lies at a corner, equation (35) applies in giving the associated strain rates which must be kinematically admissible. Let us put

$$\zeta_1 = M_r - M_o \tag{75}$$

for side AB of the limit surface, and

$$\zeta_2 = M_\theta - M_o \tag{76}$$

for side AF. Equation (35) then gives, with $dw/dM_o = 1$ from equation (72),

$$\dot{k}_r = \alpha + 0 \ , \tag{77}$$
$$\dot{k}_\theta = 0 + (1-\alpha) \ .$$

Alternatively, these equations may be written as

$$\dot{k}_r + \dot{k}_\theta = 1 \ . \tag{78}$$

If it is possible to find $\dot{k}_r(r)$, $\dot{k}_\theta(r)$, each non-negative, so that they are derived from a velocity field which satisfies the boundary conditions and equation (78), the design of equation (74) will be a global optimum.

If w is the transverse velocity of the plate, we have

$$\dot{k}_r = - \frac{d^2 w}{dr^2} \ , \qquad \dot{k}_0 = - \frac{1}{r}\frac{dw}{dr} \ . \tag{79}$$

Substituting these expressions into equation (78),

$$\frac{d^2 w}{dr^2} + \frac{1}{r}\frac{dw}{dr} = -1 \ . \tag{80}$$

Since in view of equations (35) and (36) no infinite strain rates and hence no discontinuities in dw/dr are permitted, the boundary conditions are

$$w(R) = 0 \ , \qquad \frac{dw}{dr}(0) = 0 \ . \tag{81}$$

The solution of equations (80) and (81) is

$$w(r) = \frac{1}{4}(R^2 - r^2) \ . \tag{82}$$

Thus we have a velocity field which satisfies the boundary conditions. On determining $\dot{\kappa}_r$ and $\dot{\kappa}_\theta$ from equation (79), it will be seen that equation (80) is satisfied. Hence equation (74) gives a global least weight design.

The basic limit design problem we have thus far considered is easily formulated, and conditions have been derived which permit us to determine the optimal design for a variety of specific problems. Nevertheless, the basic formulation is limited in several practical respects. We shall now consider briefly one elaboration of the basic limit design problem

14.3 Piecewise Uniform Designs

One practical difficulty in the basic limit design problem formulated in Section 14.2 is that the cost of constructing structures of a continuously varying cross-section may be very high. Thus the design of least weight may not be the design of least cost. Material for construction is generally available as elements of standardized dimensions, and the designer is frequently constrained to choose from these standard elements. It is consequently of importance to consider the optimal design of structures which are made up of elements whose size and cross-section may be assigned by the designer but whose cross-section is uniform. Such designs will be referred to as *piecewise uniform designs*.

In this section we shall treat briefly a special case of the piecewise uniform design problem. We shall limit ourselves to *bar, beam or frame structures* where the *lengths* over which the cross-section is to be uniform are fixed *a priori*, and to the case where the design parameter Q_o may be chosen to have any non-negative value rather than any one of a series of discrete values. While this problem is not the most general in this class it is nevertheless of sufficiently wide potential application to be of interest.

The basic problem of Section 14.2 is modified in the follow-
ing way. Let the structure, which has total generalized volume
V, be divided into segments each of volume V_α ($\alpha=1,2,\ldots,\nu$).
Let a space parameter s_α be defined for each segment, and let
the design parameter be constrained to have the value $Q_{o\alpha}$ in V_α.
The total weight of the structure is now given by

$$C = \sum_{\alpha=1}^{\nu} V_\alpha w(Q_{o\alpha}) \ . \tag{83}$$

Our object is to minimize C subject to the same constraints as
formerly applied viz. equations (1) and (5). As before, an
acceptable design is any choice of $Q_{o\alpha}$ which satisfies constraints (1) and (5).

The class of pertinent statically admissible generalized
stress fields $\bar{Q}_j(s)$ can be written in terms of the new space
variables s_α, so that $\bar{Q}_j(s)$ in V becomes $\bar{Q}_j(s_\alpha)$ in V_α. It is
evident that we can limit our search for the optimal design to
a subclass of acceptable designs given by

$$Q_{o\alpha} = \max_{s_\alpha} \ [\zeta\{\bar{Q}_j(s_\alpha)\}] \ , \tag{84}$$

the right hand side indicating that we choose the maximum value
of $\zeta(\bar{Q}_j)$ in s_α. The proof of this statement follows the same
lines as that used to establish equation (13), and will not be
repeated.

In order to establish sufficient conditions for optimality
we consider a small variation in the generalized stress field
and hence in the total weight. From equation (83),

$$\delta C = \sum_{\alpha=1}^{\nu} V_\alpha \left. \frac{dw}{dQ_o}\right|_{Q_o} \delta Q_o \ . \tag{85}$$

Equation (84) may then be used to express $\delta Q_{o\alpha}$ in terms of $\delta\bar{Q}_j$.

In general

$$\delta Q_{o\alpha} = \max_{s_\alpha}[\zeta\{\bar{Q}_j + \delta\bar{Q}_j\}] - \max_{s_\alpha}[\zeta(\bar{Q}_j)] \quad . \tag{86}$$

This expression cannot be conveniently evaluated, but a comparatively simple formulation results if we impose some further restrictions.

Let us assume that

$$\max_{s_\alpha}[\zeta\{\bar{Q}_j(s)\}]$$

occurs at a number of discrete locations $s_{\alpha\beta}$, $\beta=1,\ldots,\mu_\alpha$. (Parenthetically we may note that it is not enough to assume that ζ has only one maximum value in V_α. The optimal design will frequently be such that $\mu_\alpha \geq 2$ in at least one segment.) We further assume that

$$\max_{s_\alpha}[\zeta(\bar{Q}_j + \bar{Q}_j)]$$

occurs at *one of the locations* $s_{\alpha\beta}$. This assumption is not unreasonable if the number of segments α is large, or the generalized stresses vary linearly with s_α. With these limitations,

$$\delta Q_{o\alpha} = \max_{s_{\alpha\beta}} \left\{ \left. \frac{\partial\zeta}{\partial Q_j} \right|_{\bar{Q}_j} \delta\bar{Q}_j \right\} \tag{87}$$

provided that $\zeta(\bar{Q}_j) \neq 0$ in V_α. For brevity we shall omit consideration of the zero cross-section constraint, assuming that the optimal design does not contain any elements in which $Q_{o\alpha} = 0$. Equation (99) indicates that $\delta Q_{o\alpha}$ is the largest value of

$$\left. \frac{\partial\zeta}{\partial Q_j} \right|_{\bar{Q}_j} \delta\bar{Q}_j$$

evaluated at the discrete points $s_{\alpha\beta}$. Clearly this implies that

$$\delta Q_{o\alpha} = \max \left\{ \frac{\partial \zeta}{\partial Q_j}\bigg|_{\bar{Q}_j} \delta \bar{Q}_j \right\}_{S_{\alpha\beta}} \geq \left\{ \frac{\partial \zeta}{\partial Q_j}\bigg|_{\bar{Q}_j} \delta \bar{Q}_j \right\}_{S_{\alpha\beta}} \quad . \tag{88}$$

Let us now introduce for each segment V_α a set of non-negative parameters $\Theta_{\alpha\beta}$ such that

$$\sum_{\beta=1}^{\mu_\alpha} \Theta_{\alpha\beta} = V_\alpha \frac{dw}{dQ_o}\bigg|_{Q_{o\alpha}} \quad . \tag{89}$$

Making use of equation (88), it follows that

$$V_\alpha \frac{dw}{dQ_o}\bigg|_{Q_{o\alpha}} \delta Q_{o\alpha} = \max \left\{ \frac{\partial \zeta}{\partial Q_j}\bigg|_{\bar{Q}_j} \delta \bar{Q}_j \right\}_{S_{\alpha\beta}} \sum_{\beta=1}^{\mu_\alpha} \Theta_{\alpha\beta}$$

$$\geq \sum_{\beta=1}^{\mu_\alpha} \Theta_{\alpha\beta} \left\{ \frac{\partial \zeta}{\partial Q_j}\bigg|_{\bar{Q}_j} \delta \bar{Q}_j \right\}_{S_{\alpha\beta}} \quad . \tag{90}$$

Note that the summation convention is not used for the subscripts α and β. Substituting equation (90) into equation (85), we see that

$$\delta C \geq \sum_{\alpha=1}^{\nu} \sum_{\beta=1}^{\mu_\alpha} \Theta_{\alpha\beta} \left\{ \frac{\partial \zeta}{\partial Q_j}\bigg|_{\bar{Q}_j} \delta \bar{Q}_j \right\}_{S_{\alpha\beta}} \quad . \tag{91}$$

It follows then that a *sufficient condition for local optimality* ($\delta C \geq 0$) is that there exist non-negative parameters $\Theta_{\alpha\beta}$ satisfying equation (89) and such that

$$\sum_{\alpha=1}^{\nu} \sum_{\beta=1}^{\mu_\alpha} \Theta_{\alpha\beta} \left\{ \frac{\partial \zeta}{\partial Q_j}\bigg|_{\bar{Q}_j} \delta \bar{Q}_j \right\}_{S_{\alpha\beta}} = 0 \quad . \tag{92}$$

In order to interpret this requirement we regroup the terms in the expression and write equation (92) as

$$\sum_{\alpha=1}^{\nu} \sum_{\beta=1}^{\mu_{\alpha}} \left\{ \Theta_{\alpha\beta} \left. \frac{\partial \zeta}{\partial Q_j} \right|_{\bar{Q}_j(s_{\alpha\beta})} \right\} \delta \bar{Q}_j(s_{\alpha\beta}) = 0 \quad . \tag{93}$$

This expression now has a form similar to equation (27). Whereas equation (27) involves the product of a continuous function and δQ_j integrated over the volume, equation (93) is the sum of the product of a quantity and the variation δQ_j at the discrete sections $s_{\alpha\beta}$. This suggests that we can interpret

$$\Theta_{\alpha\beta} \left. \frac{\partial \zeta}{\partial \bar{Q}_j} \right|_{\bar{Q}_j(s_{\alpha\beta})}$$

as a discontinuity in the generalized velocity field, or delta function in strain rate, at the points $s_{\alpha\beta}$. Equation (93) expresses the requirement that this set should be derivable from a pertinent kinematically admissible generalized velocity field. Further, since $\Theta_{\alpha\beta}$ are non-negative, these discontinuities in the imagined generalized velocity field satisfy the constitutive equations for the material. Consequently the imagined velocity field can also be regarded as a possible velocity field during flow for the optimal structure.

As an example of the application of the optimality conditions for piecewise uniform optimal design, consider the beam shown in Figure 8(a). Assume that the beam has a sandwich cross-section, and that shear forces are carried by the core. The weight of the face plates per unit length is

$$w(M_o) = M_o \tag{94}$$

and the yield condition is

$$\psi(M,M_o) = \zeta(M) - M_o$$
$$= |M| - M_o \quad . \tag{95}$$

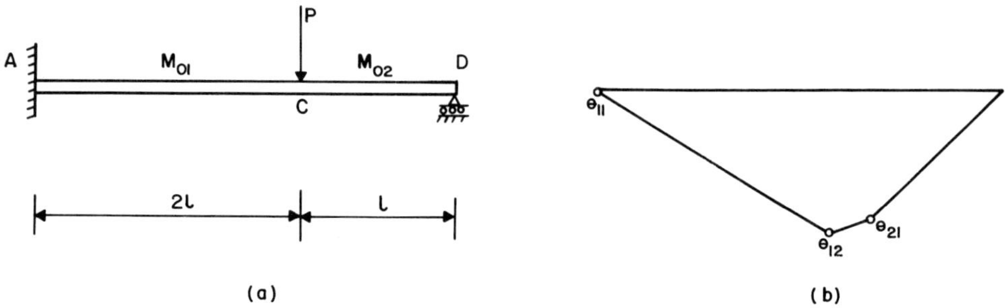

Figure 8. Piecewise uniform propped cantilever

Let us assume that the segments AC and CD each have a uniform cross-section, the design parameters being M_{o1} and M_{o2} respectively.

Letting the reaction at the free end be R, it is clear that peak values in the bending moment occur at A and C. We have that

$$M_C = R \quad , \quad M_A = 3R - 2P \quad . \tag{96}$$

Hence

$$M_{o1} = \max \{|M_A|, \ |M_C|\} = \max \{|R\ell|, \ |3R\ell - 2P\ell|\} \ , \tag{97a}$$

$$M_{o2} = |M_C| = |R\ell| \quad . \tag{97b}$$

The total weight is

$$C = 2M_{o1}\ell + M_{o2}\ell \ , \tag{98}$$

and hence depends on the single parameter R.

Some insight into the form of the optimal design can be obtained by noting that the optimal design must in some way permit us to choose the reaction R. In this simple example this can only occur if

$$|M_A| = |M_C| \tag{99}$$

in the optimal design. Substituting equation (96) into equation

(99) we see that R is positive and

$$R = P/2 \ .$$ (100)

This in turn gives

$$M_{o1} = M_{o2} = P\ell/2$$ (101)

$$\text{and} \quad C = \frac{3P\ell^2}{2} \ .$$ (102)

In order to check whether this is indeed the optimal design we introduce non-negative parameters Θ_{11}, Θ_{12}, Θ_{21} at each of the cross-sections in each segment where the moment has a maximum value. Noting that

$$\frac{\partial \zeta}{\partial M} = \text{sgm } M \ ,$$ (103)

the parameters Θ_{11}, Θ_{12}, Θ_{21} can be regarded as the magnitudes of hinge rotation rates as shown in Figure 9(b). The actual hinge rotation rate at A is $-\Theta_{11}$ since M_A is negative, while the remaining hinge rotation rates are positive.

The condition that the velocity field of Figure 8(b) should be kinematically admissible can most easily be written by using the self-equilibrating moment distribution for the once indeterminate beam of Figure 8(a). This moment field is obtained from Figure 8(a) with P=0 and R=1. Using the principle of virtual velocities, kinematic admissibility requires that

$$-3\Theta_{11} + \Theta_{12} + \Theta_{21} = 0 \ .$$ (104)

The sufficient conditions for optimality (equation (89)) also require that

$$\Theta_{11} + \Theta_{12} = 2\ell \ ,$$ (105)

$$\Theta_{21} = \ell \ .$$ (106)

We may solve these equations for Θ_{11}, Θ_{12}, Θ_{21}. If the solution shows that Θ_{11}, Θ_{12}, Θ_{21} are each non-negative, the design will be optimal. Indeed, we find that

$$\Theta_{11} = \frac{3\ell}{4} \quad , \quad \Theta_{12} = \frac{5\ell}{4} \quad , \quad \Theta_{21} = \ell \quad , \tag{107}$$

which satisfies the final restriction.

As a more complex example of the optimal design of a sandwich beam, consider the fixed end beam of Figure 9, with uniform segments AB, BC, CD denoted by M_{o1}, M_{o2}, M_{o3} respectively. The beam is twice indeterminate, and hence we might guess that two segments are such that two equal peak moments occur. The most likely choice is that

$$M_A = M_B \quad , \quad M_C = M_D \quad . \tag{108}$$

Working out the equilibrium equations, we find the bending moment distribution of Figure 10(a). Hence we suspect that the

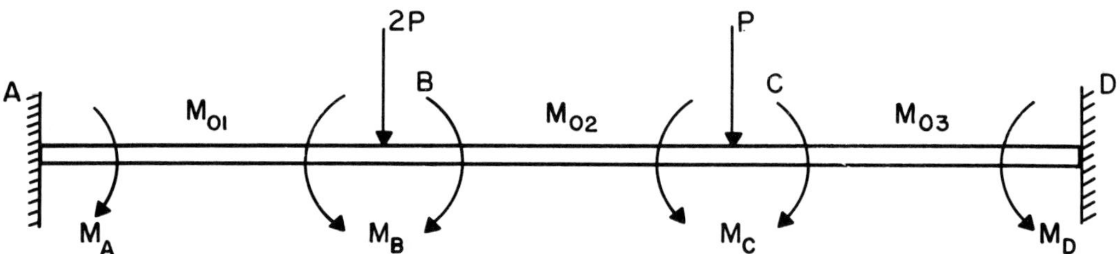

Figure 9.   Fixed beam example

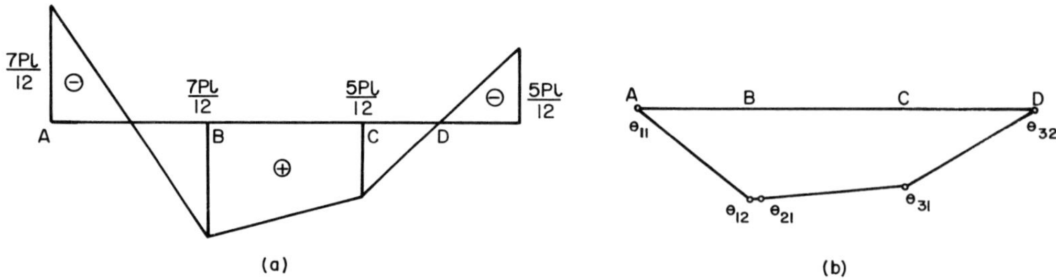

Figure 10.   Bending moment and velocity field for fixed beam

optimal design is

$$M_{o1} = 7P\ell/12 , \quad M_{o2} = 7P\ell/12 , \quad M_{o3} = 5P\ell/12 . \tag{109}$$

To determine whether this is indeed optimal, we consider the velocity field of Figure 10(b) with the rotation rates Θ_{11}, Θ_{32} negative and the others positive. The optimality conditions (equations 89) require that

$$\Theta_{11} + \Theta_{12} = \ell ,$$

$$\Theta_{21} = \ell , \tag{110}$$

$$\Theta_{31} + \Theta_{32} = \ell .$$

Using the self-equilibrating moment field adopted for the previous example, and its mirror image, the conditions for kinematic admissibility are satisfied if

$$-3\Theta_{11} + 2\Theta_{12} + 2\Theta_{21} + \Theta_{31} = 0 ,$$

$$\Theta_{12} + \Theta_{21} + 2\Theta_{31} - 3\Theta_{32} = 0 . \tag{111}$$

The solution of equations (110) and (111) gives

$$\Theta_{11} = 7\ell/8, \ \Theta_{12} = \ell/8, \ \Theta_{21} = \ell, \ \Theta_{31} = 3\ell/8, \ \Theta_{32} = 5\ell/8 . \tag{112}$$

Each of these values is non-negative, and hence the design is optimal.

Problems of the general class discussed above can be formulated as *programming problems*. Let $s_{\alpha\beta}$ define all points in each segment at which the function $\zeta(\bar{Q}_j)$ could reach a peak value. The class of *acceptable designs* is then given by those choices of $Q_{o\alpha}$ for which

$$\zeta\{\bar{Q}_j(s_{\alpha\beta})\} \leq Q_{o\alpha} . \tag{113}$$

Since the weight function is given by equation (95), the optimal design problem is

$$\text{minimize} \quad \sum_{\alpha=1}^{\nu} V_\alpha w(Q_{o\alpha}) \tag{114}$$

$$\text{subject to} \quad Q_{o\alpha} \geq \zeta\{\bar{Q}_j(s_{\alpha\beta})\} \quad \alpha=1, \ldots, \nu$$

$$\beta=1, \ldots, \mu_\alpha \quad .$$

As an example of this formulation, consider the problem shown in Figure 9. The points A, B, C, D are the points where the bending moments reach peak values. The moments M_A, M_B, M_C and M_D do not constitute a statically admissible set; we must impose two independent equality constraints which ensure equilibrium. These may be written as

$$-2M_A + 3M_B - M_D = 5P\ell \quad ,$$

$$- M_A + 3M_C - 2M_D = 4P\ell \quad . \tag{115}$$

The optimal design problem is then

$$\text{minimize} \quad (M_{o1}\ell + M_{o2}\ell + M_{o3}\ell) \tag{116}$$

subject to

$$-2M_A + 3M_B - M_D = 5P\ell \quad ,$$

$$- M_A + 3M_C - 2M_D = 4P\ell \quad ,$$

$$-M_{o1} \leq M_A \leq M_{o1} \quad ,$$

$$-M_{o1} \leq M_B \leq M_{o1} \quad ,$$

$$-M_{o2} \leq M_B \leq M_{o2} \quad ,$$

$$-M_{o2} \leq M_C \leq M_{o2} \quad ,$$

$$-M_{o3} \leq M_C \leq M_{o3} \quad ,$$

$$-M_{o3} \leq M_D \leq M_{o3} \quad .$$

This is a *linear programming problem*.

In beam and frame problems the programming problem can be formulated in an alternative way. Because there exist a limited number of positions at which the moment can reach a peak value, there exist a limited number of potential plastic hinge locations. It is thus possible to identify all potential one degree of freedom flow mechanisms. If the M_o are chosen such that

$$D_{ext} \overset{<}{=} D_{int} \tag{117}$$

for every potential mechanism the design will be acceptable: this follows from the alternative statement of the limit theorem (Chapter 9). This set of inequalities then replaces the constraints of the programming problem (116).

As an example, the set of possible mechanisms for the beam of Figure 9 is shown in Figure 11. The optimal design problem then becomes

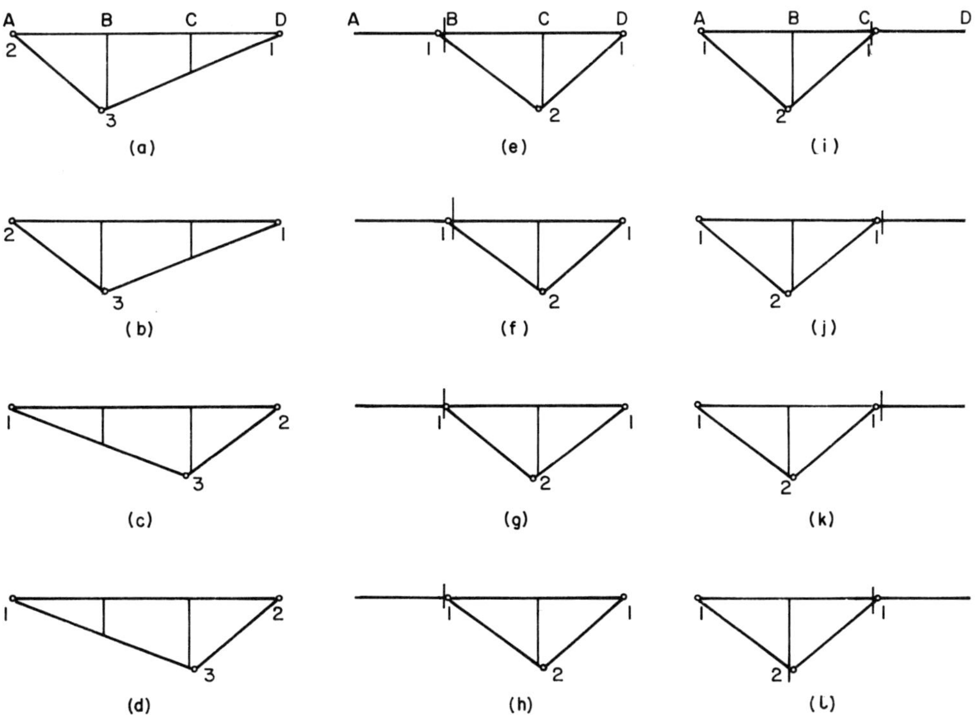

Figure 11. Mechanisms for the fixed beam example

minimize $(M_{o1}\ell + M_{o2}\ell + Mo_3\ell)$ (118)

subject to

$$5M_{o1} \qquad\qquad + M_{o3} \geq 5P\ell \ ,$$

$$2M_{o1} + 3M_{o2} + M_{o3} \geq 5P\ell \ ,$$

$$M_{o1} + 3M_{o2} + 2M_{o3} \geq 4P\ell \ ,$$

$$M_{o1} \qquad\qquad + 5M_{o3} \geq 4P\ell \ ,$$

$$M_{o1} + 2M_{o2} + M_{o3} \geq P\ell \ ,$$

$$M_{o1} \qquad\qquad + 3M_{o3} \geq P\ell \ ,$$

$$3M_{o2} + M_{o3} \geq P\ell \ ,$$

$$M_{o2} + 3M_{o3} \geq P\ell \ ,$$

$$3M_{o1} \qquad\qquad + M_{o3} \geq 2P\ell \ ,$$

$$3M_{o1} + M_{o2} \qquad\qquad \geq 2P\ell \ ,$$

$$M_{o1} + 3M_{o2} \qquad\qquad \geq 2P\ell \ ,$$

$$M_{o1} + 2M_{o2} + M_{o3} \geq 2P\ell \ .$$

This is a second linear programming problem.

It should be noted that problem (118) is not the dual of problem (116), even though static and kinematic approaches are apparently used. In fact, problem (118) may be formally obtained from problem (116) by eliminating the variables M_A, M_B, M_C and M_D. Such elimination is possible since the moment variables do not appear in the objective function. Thus the two problems are equivalent. Each admits a dual, but we shall not treat these formulations for lack of space.

The methods discussed in this section may be extended to the piecewise uniform design of plate and shell structures. Indeed, an approach of this kind is necessary for a numerical solution of the optimal design problem.

14.4 Historical and Bibliographical Remarks

Limit design was first studied in detail in the context of
beams and plane frames. Representative papers are those of
Heyman [1951] and Foulkes [1952], [1954] (the latter first recognizing that problems of this type may be treated as linear
programming problems). Beam and frame problems are treated in
detail in the volumes by Baker, Horne and Heyman [1956], Neal
[1956] and Hodge [1959].

The formulation of the more general problem followed this
initial work. Drucker and Shield [1956], [1957] provided important advances, and further studies of interest are those of
Shield [1960], Marcal and Prager [1964], Prager and Shield
[1967], Prager and Taylor [1968], Save [1968] and Charret and
Rozvany [1972]. Axisymmetric circular plates were studied by
Freiberger and Tekinalp [1956] and Marcal [1967]. The systematic solutions of discrete or discretized problems has been
intensively studied; examples are given in the papers by Heyman [1959], Dorn, Gomory and Greenberg [1964], Megarefs [1968]
and Sheu and Prager [1969].

Multi-purpose structures were studied more recently; repre-
sentative papers are by Shield [1963], Save and Prager [1963],
Save and Shield [1964], Prager [1967], Mayeda and Prager [1967],
Chan [1969] and Prager [1971].

The relation between limit analysis and limit design is discussed by Cohn,Ghosh and Parimi [1972] and Rozvany [1973].

References

J.F.Baker,   1956   *The Steel Skeleton*, Vol.2,Cambridge.
M.R.Horne and
J.Heyman

H.S.Y.Chan   1969   "On Foulkes' mechanism in portal
                           frame design for alternative loads",
                           J.Appl.Mech., $\underline{36}$, 73.

D.E.Charrett and G.I.N.Rozvany — 1972 — "Extensions of the Prager-Shield theory of optimal plastic design", Int.J.Non-Linear Mech., $\underline{7}$, 51.

M.Z.Cohn, S.K.Ghosh and S.R.Parimi — 1972 — "A unified approach to the theory of plastic structures", Rept.No.78, Solid Mechanics Div., Univ. of Waterloo, Proc.ASCE $\underline{98}$ (EM5), 1133.

N.S.Dorn, R.E.Gomory and H.J.Greenberg — 1964 — "Automatic design of optimal structures", J.de Mécanique, $\underline{3}$, 25.

D.C.Drucker and R.T.Shield — 1956 — "Design for minimum-weight", Proc. 9th Int.Congr.Appl.Mech. (Brussels), 212.

D.C.Drucker and R.T.Shield — 1957 — "Bounds on minimum-weight design", Q.Appl.Math., $\underline{15}$, 269.

J.Foulkes — 1954 — "The minimum-weight design of structural frames", Proc.Roy.Soc, $\underline{223A}$, 482.

J.Foulkes — 1952 — "Minimum weight design and the theory of plastic collapse", Q.Appl.Math., $\underline{10}$, 347.

W.Freiberger and B.Tekinalp — 1956 — "Minimum weight design of circular plates", J.Mech.Phys.Sol., $\underline{4}$, 294.

J.Heyman — 1951 — "Plastic design of beams and plane frames for minimum material consumption", Q.Appl.Math., $\underline{8}$, 373.

J.Heyman — 1959 — "On the absolute minimum weight design of framed structures", Q.J. Mech.Appl.Math., $\underline{12}$, 314.

P.G.Hodge,Jr. — 1959 — *Plastic Analysis of Structures*, McGraw-Hill (N.Y.).

P.V.Marcal — 1967 — "Optimal design of circular plates", Int.J.Solids and Structures, $\underline{3}$, 427.

P.V.Marcal and W.Prager — 1964 — "A method of optimal plastic design", J.de Mécanique, 3, 509.

R.Mayeda and W.Prager — 1967 — "Minimum weight design of beams for multiple loading", Int.J.Solids and Structures, $\underline{3}$, 1001.

G.F.Megarefs   1968   "Minimal design of sandwich axisym-
                            metric plates",Proc.ASCE $\underline{94}$,(EM1),177.

B.G.Neal   1956   *The Plastic Methods of Structural
                            Analysis*, Chapman and Hall (London).

W.Prager   1967   "Optimum plastic design of a portal
                            frame for alternative loads", J.
                            Appl.Mech., $\underline{34}$, 772.

W.Prager   1971   "Foulkes' mechanism in optimal plas-
                            tic design for alternative loads",
                            Int.J.Mech.Sci., $\underline{13}$, 971.

W.Prager and   1967   "A general theory of optimal plastic
R.T.Shield   design", J.Appl.Mech., $\underline{34}$, 184.

W.Prager and   1968   "Problems of optimal structural de-
J.E.Taylor   sign",J.Appl.Mech., $\underline{35}$, 102.

G.I.N.Rozvany   1973   "Optimal plastic design for partially
                           preassigned strength distribution",
                           J.O.T.A., $\underline{11}$, 421.

M.A.Save   1968   "Some Aspects of Minimum Weight De-
                           sign", *Engineering Plasticity* (edi-
                           ted by J.Heyman and F.A.Leckie),
                           Cambridge University Press, 611.

M.A.Save and   1963   "Minimum weight design of beams sub-
W.Prager   ject to fixed and moving loads",
                           J.Mech.Phys.Sol., $\underline{11}$, 255.

M.A.Save and   1964   "Minimum weight design of beams sub-
R.T.Shield   jected to fixed and moving loads",
                           Proc.11th Int.Congr.Appl.Mech.
                           (Munich), 341.

C.Y.Sheu and   1969   "Optimal plastic design of circular
W.Prager   and annular sandwich plates with
                           piecewise constant cross-section",
                           J.Mech.Phys.Sol., $\underline{17}$, 11.

R.T.Shield   1960   "Optimum design methods for struc-
                           tures", *Plasticity* (Proc.2nd Symp.
                           on Naval Structural Mechanics) Per-
                           gamon Press, 580.

R.T.Shield   1963   "Optimum design methods for multiple
                           loading", Z.angew.Math.Phys., $\underline{14}$,
                           38.

SLIP LINE THEORY

15.1 Partial Differential Equations Governing Plastic Flow in Plane Strain

Slip line theory is a highly developed technique which may be used to determine limit solutions for rigid, perfectly plastic bodies in *plane strain*. It may in fact be applied to a wider class of problems than those considered in this volume, since it may be used to investigate steady or pseudo-steady flow of rigid-plastic materials. Such problems include drawing, extrusion, rolling and other processes of technological importance. A complete development of the technique is beyond the scope of this monograph in that a detailed discussion of the extensive literature in the field would consume space to a degree which would not be appropriate in a work devoted primarily to general concepts and theorems that can be applied to a variety of configurations. Nevertheless, the technique is of such importance that a brief introduction to the major concepts will be given.

A variety of approaches to the basic ideas in slip line theory may be given. Essentially, however, slip line theory exploits the fact that the partial differential equations for stress and velocity in the plastic region of a body deforming in plane strain are *hyperbolic*. Slip line theory is an application of the *method of characteristics* developed for the solution of hyperbolic partial differential equations. The method of characteristics finds many other applications in continuum mechanics, particularly in wave propagation, and we shall develop our discussion from the basic mathematical ideas.

Conditions governing plastic flow in plane strain have been previously discussed in Chapters 7 and 12. In a fixed cartesian coordinate system, suppose that the body extends an infinite distance in the z-direction. The velocity component w in the z-direction is zero, and the *velocity components* u,v in the x, y directions respectively are

$$u = u(x,y) \quad , \quad v = v(x,y) \quad . \tag{1}$$

The strain rates are then

$$\dot{\varepsilon}_x = \frac{\partial u}{\partial x} \quad , \quad \dot{\varepsilon}_y = \frac{\partial v}{\partial y} \quad ,$$

$$\dot{\varepsilon}_{xy} = \frac{1}{2} \left(\frac{\partial u}{\partial y} + \frac{\partial v}{\partial x} \right) \quad . \tag{2}$$

We shall limit ourselves to isotropic materials, so that we may immediately write the stress tensor as

$$\sigma_{ij} = \begin{bmatrix} \sigma_x & \sigma_{xy} & 0 \\ \sigma_{xy} & \sigma_y & 0 \\ 0 & 0 & \sigma_z \end{bmatrix} \tag{3}$$

Assuming that body forces are zero, and that the motion is quasi-static, the appropriate equilibrium equations are

$$\frac{\partial \sigma_x}{\partial x} + \frac{\partial \sigma_{xy}}{\partial y} = 0 \quad , \quad \frac{\partial \sigma_{xy}}{\partial x} + \frac{\partial \sigma_y}{\partial y} = 0 \quad . \tag{4}$$

Boundary conditions must of course be given in a form consistent with plane strain.

We restrict our attention to the *plastic region* of the body. *Hence the yield condition and the normality rule must be satisfied at each point under consideration.* The von Mises condition is chosen, its advantage over the Tresca condition being that the stress component σ_z is given unambiguously as

$$\sigma_z = \frac{\sigma_x + \sigma_y}{2} \quad . \tag{5}$$

Thus we require at each point that

$$\left(\frac{\sigma_x - \sigma_y}{2}\right)^2 + \sigma_{xy}^2 = k^2 \ . \tag{6}$$

The strain rates are given by

$$\dot{\varepsilon}_x = \lambda \frac{\partial \phi}{\partial \sigma_x} = \frac{\lambda}{2}(\sigma_x - \sigma_y) \ ,$$

$$\dot{\varepsilon}_y = \lambda \frac{\partial \phi}{\partial \sigma_y} = -\frac{\lambda}{2}(\sigma_x - \sigma_y) \ , \tag{7}$$

$$\dot{\varepsilon}_{xy} = \lambda \frac{\partial \phi}{\partial \sigma_{xy}} = \lambda \, \sigma_{xy} \ .$$

We shall also make explicit use of the incompressibility condi_
tion in terms of the velocity components; from equations (2)
and (7),

$$\dot{\varepsilon}_x + \dot{\varepsilon}_y = \frac{\partial u}{\partial x} + \frac{\partial v}{\partial y} = 0 \ . \tag{8}$$

Equations (1)-(8) thus govern the stress and velocity fields
in the plastic region. These equations may be reduced to four
quasi-linear hyperbolic partial differential equations by trans-
forming the stresses to a set of axes in which the direct
stresses are equal. In these axes an element is subject to hydrostatic tension and shear in the x,y plane. It is evident
that in view of the yield condition (6) the magnitude of the
shear stresses in these axes is k. Figure 1 shows an element
of material referred to the x,y axes, to the principal axes and
to the axes in which the direct stresses are equal and the shear
stress reaches its greatest value. Let the latter set of axes
be referred to as the α,β axes. The α,β axes are right handed,
and the α axis is chosen so that the maximum principal stress
lies in the first quadrant of the α,β plane. Let the counter-
clockwise rotation from the x-axis to the α-axis be ϕ.

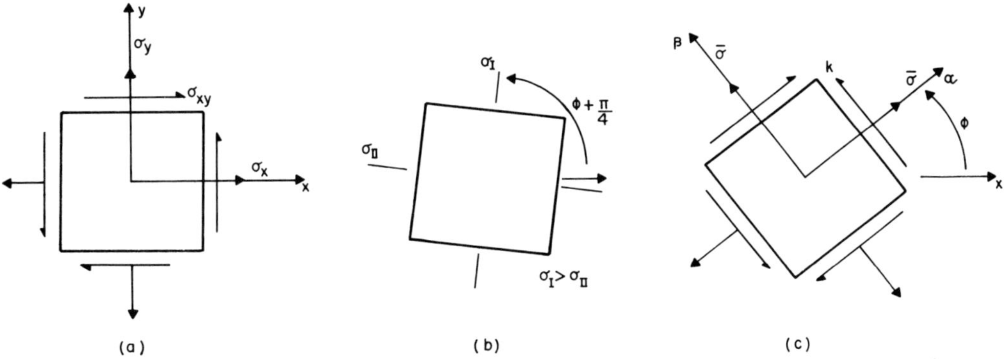

Figure 1. Stress transformations

By direct transformation (see Chapter 4) or by use of the Mohr circle, we see that

$$\sigma_x = \bar{\sigma} - k\sin 2\phi \quad ,$$

$$\sigma_y = \bar{\sigma} + k\sin 2\phi \quad , \tag{9}$$

$$\sigma_{xy} = k\cos 2\phi \quad .$$

where $\bar{\sigma}$ is the hydrostatic tension.

In the plastic region of the body under consideration

$$\bar{\sigma} = \bar{\sigma}(x,y) \quad , \quad \phi = \phi(x,y) \quad . \tag{10}$$

However, k is taken to be constant, implying that the body is composed of a uniform material. We now substitute equations (9) into equations (4), thus expressing the equilibrium equations in terms of the transformed stresses. This leads to

$$\frac{\partial \bar{\sigma}}{\partial x} - (2k\cos 2\phi)\frac{\partial \phi}{\partial x} - (2k\sin 2\phi)\frac{\partial \phi}{\partial y} = 0 \quad , \tag{11a}$$

$$\frac{\partial \bar{\sigma}}{\partial y} - (2k\sin 2\phi)\frac{\partial \phi}{\partial x} + (2k\cos 2\phi)\frac{\partial \phi}{\partial y} = 0 \quad . \tag{11b}$$

Furthermore, from equations (7) and (9), we may eliminate λ by noting that

$$(\dot{\varepsilon}_x - \dot{\varepsilon}_y)\cos 2\phi + 2\dot{\varepsilon}_{xy}\sin 2\phi$$

$$= \lambda\{-k\sin2\phi\cos2\phi - k\sin2\phi\cos2\phi + 2k\cos2\phi\sin2\phi\} \qquad (12)$$

$$= 0 \ .$$

Thus, by substituting equations (2) into equation (12), and by using the incompressibility condition (8) directly, we may generate two further equations involving the velocity components;

$$- \cos2\phi\frac{\partial u}{\partial x} - \sin2\phi\frac{\partial v}{\partial x} - \sin2\phi\frac{\partial u}{\partial y} + \cos2\phi\frac{\partial v}{\partial y} = 0 \ , \qquad (13a)$$

$$\frac{\partial u}{\partial x} + \frac{\partial v}{\partial y} = 0 \ . \qquad (13b)$$

Equations (11) and (13) are quasi-linear partial differential equations, the coefficients depending only on $\phi(x,y)$. We may expect that these four equations will permit us to determine $\bar{\sigma}$, ϕ, u, v in the body when the appropriate boundary conditions are given. We may consider the two sets of equations separately, or alternatively, combine them. Let the column vector $\underset{\sim}{w}$ be defined as

$$\underset{\sim}{w} = \begin{bmatrix} \phi \\ u \\ v \\ \bar{\sigma} \end{bmatrix} . \qquad (14)$$

There is no possibility of confusion between $\underset{\sim}{w}$ and the velocity components in the z-direction, w, which is zero and will not enter our discussion. Multiplying equation (13a) by 2k, and rearranging the order of the equations, equations (11) and (13) may be written as

$$\underset{\sim}{A} \frac{\partial \underset{\sim}{w}}{\partial x} + \underset{\sim}{B} \frac{\partial \underset{\sim}{w}}{\partial y} = 0 \ , \qquad (15)$$

where

$$
\underset{\sim}{A} =
\begin{bmatrix}
0 & -2k\cos2\phi & -2k\sin2\phi & 0 \\
-2k\cos2\phi & 0 & 0 & 1 \\
-2k\sin2\phi & 0 & 0 & 0 \\
0 & 1 & 0 & 0
\end{bmatrix} ,
\tag{16a}
$$

$$
\underset{\sim}{B} =
\begin{bmatrix}
0 & -2k\sin2\phi & 2k\cos2\phi & 0 \\
-2k\sin2\phi & 0 & 0 & 0 \\
+2k\cos2\phi & 0 & 0 & 1 \\
0 & 0 & 1 & 0
\end{bmatrix} .
\tag{16b}
$$

Note that in this form $\underset{\sim}{A}$, $\underset{\sim}{B}$ are symmetric, non-singular matrices.

Our task is now to solve these equations. In the next section we shall inquire into this problem, establishing that equations (15) are hyperbolic and introducing the elements of the method of characteristics as it applies in this problem.

15.2 The Nature of the Partial Differential Equations

Consider in general terms the four first order quasi-linear partial differential equations defined by

$$
\underset{\sim}{A} \frac{\partial \underset{\sim}{w}}{\partial x} + \underset{\sim}{B} \frac{\partial \underset{\sim}{w}}{\partial y} = 0 ,
\tag{17}
$$

where $\underset{\sim}{A}$, $\underset{\sim}{B}$ may depend on x,y and $\underset{\sim}{w}(x,y)$. Let us suppose that along a line $f(x,y) = 0$ in the x,y plane values of $\underset{\sim}{w}$ are prescribed. Thus

$$
\underset{\sim}{w}(x,y) = \underset{\sim}{w}^{o}(x,y) \quad \text{on} \quad f(x,y) = 0 .
\tag{18}
$$

The line $f=0$ may be a boundary of the body, and $\underset{\sim}{w}^{o}$ may be obtained from the boundary conditions. We now ask whether it is possible to compute values of $\underset{\sim}{w}$ in the neighborhood of the line $f=0$. Consider a point A on the line $f=0$, and a neighboring point B which lies off the line (Figure 2). It is evident that to first order

$$\underset{\sim}{w}^B = \underset{\sim}{w}^A + \left.\frac{\partial\underset{\sim}{w}}{\partial x}\right|_{\underset{\sim}{w}^A} dx + \left.\frac{\partial\underset{\sim}{w}}{\partial y}\right|_{\underset{\sim}{w}^A} dy \quad . \tag{19}$$

Thus with $\underset{\sim}{w}^A$ given, since it is contained in $\underset{\sim}{w}^o$, we can compute $\underset{\sim}{w}^B$ if we can determine $\partial\underset{\sim}{w}/\partial x$ and $\partial\underset{\sim}{w}/\partial y$ at A. The partial differential equation (17) provides 4 equations for the eight partial derivatives, and hence 4 further equations are needed. The additional information is provided by the $\underset{\sim}{w}^o(x,y)$ on $f=0$; if $\underset{\sim}{w}^o$ is given the derivatives of $\underset{\sim}{w}$ along the line $f=0$ can be computed. Let this derivative be represented by $\underset{\sim}{D}$. Let us assume for the present that the given data is smooth, so that $\underset{\sim}{D}$ is continuous along $f=0$.

In order to make use of this information, we must find an expression for the derivative of $\underset{\sim}{w}$ along the line $f=0$. It is evident that

$$df = \frac{\partial f}{\partial x} dx + \frac{\partial f}{\partial y} dy = 0 \tag{20}$$

along $f=0$, and hence along this line

$$\frac{dy}{dx} = -\frac{\partial f/\partial x}{\partial f/\partial y} = -\mu(x,y), \quad \text{say.} \tag{21}$$

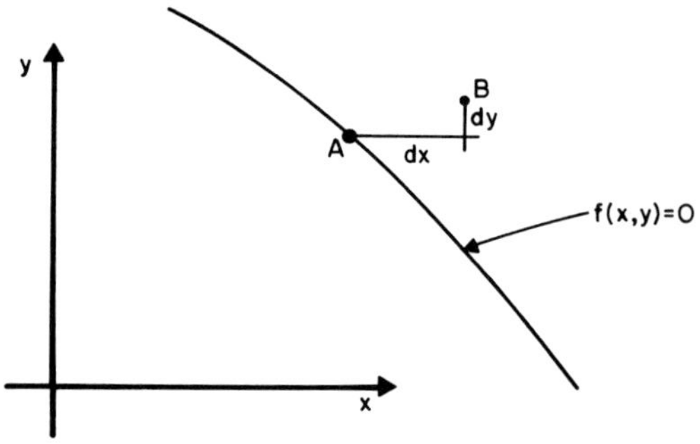

Figure 2. Line on which $\underset{\sim}{w}$ is given

It follows then that we may put

$$
\underset{\sim}{D} \;=\; \frac{d\underset{\sim}{w}}{dx}\bigg|_{\text{along } f=0} \;=\; \frac{\partial \underset{\sim}{w}}{\partial x} + \frac{\partial \underset{\sim}{w}}{\partial y}\,\frac{dy}{dx} \tag{22}
$$

$$
=\; \frac{\partial \underset{\sim}{w}}{\partial x} - \mu(x,y)\,\frac{\partial \underset{\sim}{w}}{\partial y} \;.
$$

Four additional equations for the derivatives $\partial \underset{\sim}{w}/\partial x$, $\partial \underset{\sim}{w}/\partial y$ are
provided by evaluating equation (22) at A.

Eliminating $\partial \underset{\sim}{w}/\partial x$ from equations (17) and (22), we see that

$$
(\mu \underset{\sim}{A} + \underset{\sim}{B})\,\frac{\partial \underset{\sim}{w}}{\partial y} \;=\; -\,\underset{\sim}{A}\,\underset{\sim}{D} \;. \tag{23}
$$

This equation will provide a solution for $\partial \underset{\sim}{w}/\partial y$, and hence for
$\partial \underset{\sim}{w}/\partial x$ from equations (17) (note that $\underset{\sim}{A}$ and $\underset{\sim}{B}$ are both non-
singular) provided that

$$
\det(\mu \underset{\sim}{A} + \underset{\sim}{B}) \;\neq\; 0 \;. \tag{24}
$$

If A is taken to be a generic point on the line f=0, it is evident that the solution can be extended to the neighborhood of
f=0 as long as condition (24) holds along f=0. Whether or not
condition (24) holds depends of course on $\mu(x,y)$ and hence on
the slope of the line f=0.

An alternative but equivalent view can be obtained by considering the derivative of $\underset{\sim}{w}$ across the line f=0. Let this derivative be denoted by $\underset{\sim}{D}^{*}$. Since the slope of a line crossing
f=0 at right angles is

$$
\frac{dy}{dx} \;=\; \mu^{*}(x,y) \;=\; +\,\frac{1}{\mu(x,y)} \;, \tag{25}
$$

we see from the same argument that led to equation (22) that

$$
\underset{\sim}{D}^{*} \;=\; \frac{d\underset{\sim}{w}}{dx}\bigg|_{\text{across } f=0}
$$

$$= \frac{\partial \underset{\sim}{w}}{\partial x} + \frac{1}{\mu(x,y)} \frac{\partial \underset{\sim}{w}}{\partial y} \quad . \tag{26}$$

If we now eliminate $\partial \underset{\sim}{w}/\partial x$ and $\partial \underset{\sim}{w}/\partial y$ from equations (17), (22) and (26), we find after some manipulation that

$$(\mu \underset{\sim}{A} + \underset{\sim}{B}) \underset{\sim}{D}^* = -(\frac{1}{\mu} \underset{\sim}{A} - \underset{\sim}{B}) \underset{\sim}{D} . \tag{27}$$

$$= (\mu^* \underset{\sim}{A} + \underset{\sim}{B}) \underset{\sim}{D} \quad .$$

Thus, given $\underset{\sim}{D}$, we can compute $\underset{\sim}{D}^*$ provided that condition (24) is met. In this case the derivatives across the line f=0 can be found from the derivative along the line f=0, and clearly the solution can be extended to the neighborhood of f=0.

It is of interest to consider the conditions under which (24) does *not* hold. To this end we seek $\gamma(x,y)$ such that

$$\det (\gamma \underset{\sim}{A} + \underset{\sim}{B}) = 0 \quad . \tag{28}$$

In general (i.e. without further specifying the matrices $\underset{\sim}{A}$ and $\underset{\sim}{B}$) equation (28) is a fourth order algebraic equation. If all the roots are *real*, the original partial differential equations are said to be *hyperbolic*. For each root we solve the differential equation

$$\frac{dy}{dx} = - \psi(x,y) \quad , \tag{29}$$

determining four *families of curves* which we shall denote by

$$\psi(x,y) = c \quad , \tag{30}$$

where c is a constant. The families of curves are said to be *characteristic curves*, or simply *characteristics*.

Some important observations about the behavior of the solution can be drawn from consideration of the characteristics. First, let us define the left null vector $\underset{\sim}{\ell}(x,y)$ such that

$$\underset{\sim}{\ell}^T (\gamma \underset{\sim}{A} + \underset{\sim}{B}) = 0 \quad . \tag{31}$$

The rate of change of $\underset{\sim}{w}$ along a characteristic $\psi = c$ may be determined by considering equation (23); evidently

$$\underset{\sim}{A}\ \underset{\sim}{D}\ =\ -(\gamma\underset{\sim}{A}+\underset{\sim}{B})\ \frac{\partial \underset{\sim}{w}}{\partial y}\ . \tag{32}$$

Multiplying equation (32) by $\underset{\sim}{\ell}^{T}$, we see that

$$\underset{\sim}{\ell}^{T}\underset{\sim}{A}\ \underset{\sim}{D}\ =\ 0 \tag{33}$$

along a characteristic. This is usually written as

$$(\underset{\sim}{\ell}^{T}\underset{\sim}{A})\ d\underset{\sim}{w}\ =\ 0\ \ \text{along}\ \psi = c\ . \tag{34}$$

This equation is known as the *characteristic relation*. The solution $\underset{\sim}{w}(x,y)$ must satisfy the appropriate characteristic relation along each family of characteristics. Alternatively, $\underset{\sim}{w}(x,y)$ *cannot be arbitrarily prescribed along a characteristic*.

Secondly, we see from equation (27) that if $\underset{\sim}{D}$ is known along a characteristic $\psi = c$, the derivative across the characteristic $\underset{\sim}{D}^{*}$ cannot be computed from $\underset{\sim}{D}$ in view of equation (28). Consequently the variation of the solution $\underset{\sim}{w}$ along a characteristic does not influence the variation of $\underset{\sim}{w}$ across a characteristic. Most importantly, it does not constrain $\underset{\sim}{D}^{*}$ to be *continuous* across a characteristic. This implies that *the derivatives of* $\underset{\sim}{w}(x,y)$ *may be discontinuous across characteristics*.

Because there are several special features which distinguish the equations governing plane plastic flow (15 and 16) from the general case of four simultaneous quasi-linear partial differential equations, we shall return to determine the characteristics and characteristic relations for plane plastic flow before discussing how this knowledge can be exploited in solving boundary value problems.

15.3 Characteristics and Characteristic Relations in Plane Plastic Flow

From equation (28) the slopes of the characteristics for the partial differential equations of equation (17) are found by solving the algebraic equation

$$\det\left(\gamma \underset{\sim}{A} + \underset{\sim}{B}\right) = 0 \ . \tag{35}$$

In the case of plane plastic flow, the matrices A and B are given by equations (16). Substituting these values into equation (35), we have

$$\begin{vmatrix} 0 & \gamma(-2k\cos2\phi)-2k\sin2\phi & \gamma(-2k\sin2\phi)+2k\cos2\phi & 0 \\ \gamma(-2k\cos2\phi)-2k\sin2\phi & 0 & 0 & \gamma \\ \gamma(-2k\sin2\phi)+2k\cos2\phi & 0 & 0 & 1 \\ 0 & \gamma & 1 & 0 \end{vmatrix} = 0. \tag{36}$$

Evaluating this determinant, we find after some manipulation that

$$- \left[\gamma^2(2k\sin2\phi)-\gamma(4k\cos2\phi)-2k\sin2\phi\right]^2 = 0 \ . \tag{37}$$

It is evident that there are only two distinct roots; these are

$$\gamma = \frac{4k\cos2\phi \overset{+}{-} \sqrt{16k^2\cos^2 2\phi+16k^2\sin^2 2\phi}}{4k\sin2\phi}$$
$$= \frac{\cos2\phi\overset{+}{-}1}{\sin2\phi} \ . \tag{38}$$

Further simplification is obtained on considering the individual values; either

$$\gamma = \frac{\cos2\phi-1}{\sin2\phi} = -\tan\phi \ , \tag{39a}$$

$$\gamma = \frac{\cos2\phi+1}{\sin2\phi} = \cot\phi \ . \tag{39b}$$

From equation (29), we thus see that we have two distinct families of characteristics defined by the equations

$$\frac{dy}{dx} = \tan \phi , \tag{40a}$$

$$\frac{dy}{dx} = -\cot \phi . \tag{40b}$$

An immediate physical interpretation of these families of curves is available; consider the transformation of axes from the x,y directions to the α,β directions which led us to transform the stress components (Figure 1). It is seen that the characteristic directions coincide with the axes in which the shear stress reaches its maximum value, and that the characteristic directions are *orthogonal*. Alternatively, a material surface perpendicular to the x-y plane and coinciding with a characteristic is subjected to tractions $\bar{\sigma}$ and k in the normal and tangential directions respectively, as shown in Figure 3.

It is convenient to refer to the family of characteristics which coincides with the α-direction as the α-*lines*, and the family which coincides with the β-direction as the β-*lines*. It should be noted that the characteristic directions depend only on ϕ. We shall see that once $\phi(x,y)$ is known the entire solution can be found by means of the characteristic relations and the boundary conditions. That there exist only two families of

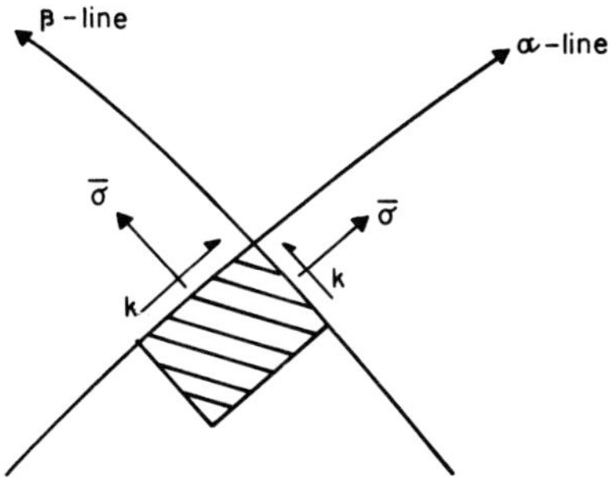

Figure 3. Element in the α,β axes

orthogonal characteristics which depend on only one of the four unknown functions in the problem constitutes a considerable simplification by comparison with the most general case.

In order to write the characteristic relations we must first determine the null vector $\underset{\sim}{\ell}^T = (\ell_1, \ell_2, \ell_3, \ell_4)$. Taking the root of equation (39a), we seek a vector $\underset{\sim}{\ell}$ such that

$$(\ell_1, \ell_2 . \ell_3, \ell_4)$$

$$\begin{bmatrix} 0 & 2k\cos2\phi\tan\phi - 2k\sin2\phi & 2k\sin2\phi\tan\phi + 2k\cos2\phi & 0 \\ 2k\cos2\phi\tan\phi - 2k\sin2\phi & 0 & 0 & -\tan\phi \\ 2k\sin2\phi\tan\phi + 2k\cos2\phi & 0 & 0 & 1 \\ 0 & -\tan\phi & 1 & 0 \end{bmatrix}$$

$$= 0 \ . \tag{41}$$

It can be seen that we have two independent sets of homogeneous equations in ℓ_1, ℓ_4 and ℓ_2, ℓ_3 respectively. This means that we can find two linearly independent null vectors, resolving the apparent difficulty that with only two families of characteristics we may not have sufficient characteristic relations for the four unknown functions. We choose

$$\underset{\sim}{\ell} = \begin{bmatrix} 1 \\ 0 \\ 0 \\ -2k \end{bmatrix} \quad \text{and} \quad \underset{\sim}{\ell} = \begin{bmatrix} 0 \\ 1 \\ \tan\phi \\ 0 \end{bmatrix} \quad \text{for the } \alpha\text{-lines.} \tag{42}$$

Replacing $(-\tan\phi)$ by $\cot\phi$ in equation (41), by a similar process we choose

$$\underset{\sim}{\ell} = \begin{bmatrix} 1 \\ 0 \\ 0 \\ 2k \end{bmatrix} \quad \text{and} \quad \underset{\sim}{\ell} = \begin{bmatrix} 0 \\ 1 \\ -\cot\phi \\ 0 \end{bmatrix} \quad \text{for the } \beta\text{-lines.} \tag{43}$$

Our next step is to compute $(\underset{\sim}{\ell}^T \underset{\sim}{A})$ for each of these null vectors. The characteristic relations are then given by

$$(\underset{\sim}{\ell}^T \underset{\sim}{A}) \, d\underset{\sim}{w} \;=\; 0 \tag{44}$$

from equation (34), where, from equation (14),

$$d\underset{\sim}{w} \;=\; \begin{bmatrix} d\phi \\ du \\ dv \\ d\bar{\sigma} \end{bmatrix} . \tag{45}$$

We find that along α-lines

$$(\underset{\sim}{\ell}^T \underset{\sim}{A}) \;=\; \begin{bmatrix} -2k \\ 0 \\ 0 \\ 1 \end{bmatrix}^T \quad \text{and} \quad (\underset{\sim}{\ell}^T \underset{\sim}{A}) \;=\; \begin{bmatrix} 0 \\ -2k(1+\cos2\phi) \\ -2k\sin2\phi \\ 0 \end{bmatrix}^T , \tag{46a}$$

while along β-lines

$$(\underset{\sim}{\ell}^T \underset{\sim}{A}) \;=\; \begin{bmatrix} 2k \\ 0 \\ 0 \\ 1 \end{bmatrix}^T \quad \text{and} \quad (\underset{\sim}{\ell}^T \underset{\sim}{A}) \;=\; \begin{bmatrix} 0 \\ 2k(\cos2\phi-1) \\ 2k\sin2\phi \\ 0 \end{bmatrix}^T . \tag{46b}$$

The four characteristic relations are obtained by substituting equations (45) and (46) into (44). It is convenient to divide these equations into two groups. Using the first of equations (46a) and (46b), we obtain two relations involving only the quantities describing the stress field:

$$-2k \, d\phi + d\bar{\sigma} \;=\; 0 \quad \text{along an } \alpha\text{-line} \;, \tag{47a}$$

$$2k \, d\phi + d\bar{\sigma} \;=\; 0 \quad \text{along a } \beta\text{-line} \;. \tag{47b}$$

The coefficients in these equations are constants, and hence the equations are integrable. Thus

$$\bar{\sigma} - 2k\phi = \text{constant} \quad \text{along an } \alpha\text{-line} \quad, \tag{48a}$$

$$\bar{\sigma} + 2k\phi = \text{constant} \quad \text{along a } \beta\text{-line} \quad. \tag{48b}$$

The constants in these equations may vary from one member of the family of characteristics to another. However, $(\bar{\sigma}-2k\phi)$ will be constant along any one member of the family of α-lines, and $(\bar{\sigma}+2k\phi)$ will be constant along any one member of the family of β-lines.

Taking the second vectors of equations (46a) and (46b), we find that

$$\begin{aligned}
(1+\cos2\phi)du + (\sin2\phi)dv &= 0 \quad, \quad \alpha\text{-line} \\
(1-\cos2\phi)du - (\sin2\phi)dv &= 0 \quad, \quad \beta\text{-line} \quad.
\end{aligned} \tag{49}$$

These relations may be simplified to give

$$du + (\tan\phi)dv = 0 \quad, \quad \alpha\text{-line} \tag{50a}$$

$$du - (\cot\phi)dv = 0 \quad, \quad \beta\text{-line} \quad. \tag{50b}$$

The coefficients of these equations involve $\phi(x,y)$, and consequently cannot be integrated without knowledge of the characteristics.

It can be appreciated at this point that in our present approach to plane plastic flow the characteristics, or the α- and β-lines, can be regarded as the primary unknowns in the problem. Once the α- and β- lines are known, ϕ follows directly from equations (40) and $\bar{\sigma}$, u and v may then be obtained by use of the characteristic relations (48) and (50) and the boundary conditions.

This observation suggests a means of constructing the solution. If a plausible net of α- and β-lines can be constructed, the stress and velocity fields can be found by integration of the characteristic relations and the use of some of the boundary conditions. If the remaining boundary conditions are satisfied, the suggested net of characteristics is correct; if not it must be modified and the process repeated.

The domain in the x-y plane occupied by the characteristic net will generally consist of regions in which the solution and its derivatives are continuous. The boundaries of these regions will coincide with members of the families of characteristics with one possible exception discussed in the following section. It is frequently possible to construct a plausible net of characteristics by inferring or guessing the position of the particular characteristics across which discontinuities occur, and then completing the net in the several regions where the solution and its derivatives are continuous. In order to apply these ideas efficiently, it is necessary that we consider in greater detail the physical and mathematical implications of the characteristic relations, the nature of discontinuities across characteristics and the utilization of the boundary conditions. These studies will provide us with information which will make the task of constructing plausible characteristic nets considerably simpler. To this end we shall consider first the stress field and then the velocity field in the following sections.

Before beginning these studies, however, it is instructive to consider the physical implications of the characteristic relations (50) for the velocity field. These relations simply express the condition that the *extension rate along an* α- *or* β-*line is zero*. This result is not unexpected; an element of material (Figure 1c) oriented in the α,β axes is subjected to hydrostatic tension and shear. Since the perfectly plastic material is incompressible and isotropic, the element must deform in pure shear and the extension rates along the α,β axes must be zero.

To confirm the result kinematically, consider two adjacent

points in the x-y plane, (x,y) and $(x+dx,y+dy)$. The distance between these two points, ds, may be written as

$$ds^2 = dx^2 + dy^2 \ . \tag{51}$$

The material derivative of the quantity ds^2 is

$$\frac{d}{dt}(ds^2) = \frac{d}{dt}(dx)^2 + \frac{d}{dt}(dy^2)$$
$$= 2\{dx\frac{d}{dt}(dx) + dy\,\frac{d}{dt}(dy)\} \ . \tag{52}$$

Since

$$dx = (x+dx) - x \ ,$$

we see that

$$\frac{d}{dt}(dx) = \frac{d}{dt}(x+dx) - \frac{d}{dt}(x)$$
$$= \frac{\partial u}{\partial x}\,dx + \frac{\partial u}{\partial y}\,dy \ . \tag{53}$$

Similarly

$$\frac{d}{dt}(dy) = \frac{\partial v}{\partial x}\,dx + \frac{\partial v}{\partial y}\,dy \ . \tag{54}$$

Hence

$$\frac{d}{dt}(ds^2) = 2\{\frac{\partial u}{\partial x}\,dx^2 + (\frac{\partial u}{\partial y} + \frac{\partial v}{\partial x})dxdy + \frac{\partial u}{\partial y}\,dy^2\} \ . \tag{55}$$

If the two adjacent points both lie on an α-line,

$$\frac{dy}{dx} = \tan\phi \ . \tag{56}$$

Thus for a line element lying in an α-line

$$\frac{d}{dt}(ds^2) = 2\{\frac{\partial u}{\partial x} + (\frac{\partial u}{\partial x} + \frac{\partial v}{\partial x})\tan\phi + \frac{\partial v}{\partial y}\,\tan^2\phi\}\,dx^2 \ . \tag{57}$$

Now, along an α-line

$$du = (\frac{\partial u}{\partial x} + \tan\phi \, \frac{\partial u}{\partial y})dx \quad ,$$

$$dv = (\frac{\partial v}{\partial x} + \tan\phi \, \frac{\partial v}{\partial y})dx \quad . \tag{58}$$

On comparing equations (57) and (58), we see that

$$\frac{d}{dt}(ds^2) = (du + dv\tan\phi)dx^2 \tag{59}$$

The coefficient of the right hand side of this equation is constrained to be zero by the characteristic relation (50a); hence a line element coinciding with an α-line does not extend. A similar argument shows that a line element in the β-direction does not extend.

A thin strip of material enclosed between adjacent members of the same family of characteristics is thus seen to undergo simple shear deformation. The characteristics are for this reason generally referred to as *slip lines*, and the net of characteristics is termed the *slip line field*.

15.4 Considerations Involving the Stress Field

We place ourselves in the position of attempting to construct an orthogonal net of slip lines which is plausible for any given boundary value problem. Limitations are placed on the slip lines by the characteristic relations (48), by discontinuities in the stresses and their derivatives, and by the stress boundary conditions which occur in any problem.

Consider first the characteristic relations

$$\bar{\sigma} - 2k\phi = \text{const. along an } \alpha\text{-line} \quad , \tag{60a}$$

$$\bar{\sigma} + 2k\phi = \text{const. along a } \beta\text{-line} \quad . \tag{60b}$$

The constant has the same numerical value along any one member of a family of slip lines; it may vary from one member of a family to another. These relations are completely equivalent

to the equilibrium relations in the deforming region, and impose restrictions on $\phi(x,y)$ and hence on the slip line net.

One way in which these restrictions can be stated is to consider two points A,B in the x-y plane, as shown in Figure 4,
and the α- and β-lines through each point. Let the α-line
through A and the β-line through B intersect at P, and the β-
line through A and the α-line through B intersect at Q. Let
the values of $\bar{\sigma}(x,y)$ and $\phi(x,y)$ at the points A,B,P,Q be $\bar{\sigma}_A$,
$\bar{\sigma}_B$, $\bar{\sigma}_P$, $\bar{\sigma}_Q$ and ϕ_A, ϕ_B, ϕ_P, ϕ_Q respectively. Applying equation
(60a) along the two α-lines, we see that

$$\bar{\sigma}_A - 2k\phi_A = \bar{\sigma}_P - 2k\phi_P \ ,$$

$$\bar{\sigma}_B - 2k\phi_B = \bar{\sigma}_Q - 2k\phi_Q \ . \tag{61}$$

Similarly, applying equation (60b) to the two β-lines,

$$\bar{\sigma}_Q + 2k\phi_Q = \bar{\sigma}_A + 2k\phi_A \ ,$$

$$\bar{\sigma}_P + 2k\phi_P = \bar{\sigma}_B + 2k\phi_B \ . \tag{62}$$

Adding all four equations, we find that the hydrostatic tensions
are eliminated, and that

$$\phi_P + \phi_Q = \phi_A + \phi_B \ . \tag{63}$$

This result is generally known as Hencky's theorem.

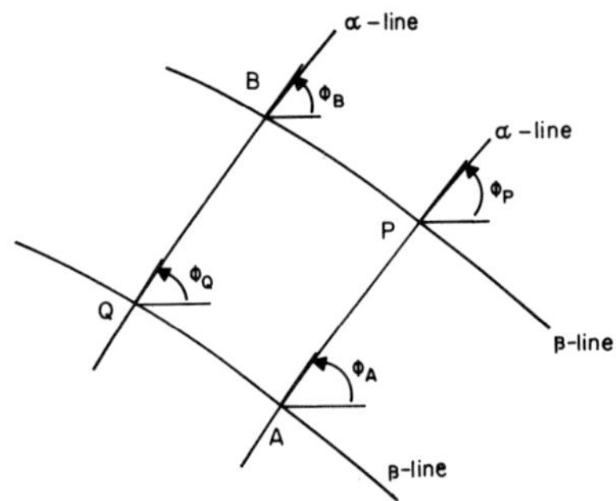

Figure 4. Pairs of α- and β-lines

If we rewrite equation (63) as

$$\phi_B - \phi_P = \phi_Q - \phi_A \quad , \tag{64}$$

and recall that the points A,B are arbitrarily chosen, we may interpret the result as follows: *the change in $\phi(x,y)$ as we pass along any β-line between two particular α-lines is constant.* Similarly, if equation (63) is written as

$$\phi_B - \phi_Q = \phi_P - \phi_A \quad , \tag{65}$$

we see that *the change in $\phi(x,y)$ as we pass along any α-line between two particular β-lines is constant.*

These statements are evidently constraints on the geometry of the net. From a physical point of view, the constraints arise by virtue of the fact that the change in hydrostatic tension between any two points in the x-y plane can be computed along alternative paths; clearly the result must be the same for all paths if the hydrostatic pressure is single-valued at the terminal points.

An important corollary of Hencky's theorem is the result that if any one α-line between two β-lines is straight, all α-lines are straight between the same two β-lines. Again, α and β can be interchanged in this argument.

The restrictions on geometry of the slip line field imposed by the characteristic relations (60) or by Hencky's theorem can be expressed in a variety of different ways. We do not have sufficient space to develop this discussion in greater detail; the topic is of importance in constructing plausible slip line fields for given problems and the reader is referred to the references given in the bibliography following this Chapter.

Consider now some simple slip line fields which do satisfy Hencky's theorem. These simple fields occur very frequently in some regions of the solution of many boundary value problems.

First, suppose that the slip line field consists of two orthogonal famiies of *straight lines*, as shown in Figure 5. It
is seen that ϕ remains constant in both the α- and β-directions.
This field satisfies the requirements of Hencky's theorem. From
equations (60a) and (60b), it is evident that $\bar{\sigma}$ is constant
along both α-lines and β-lines; $\bar{\sigma}$ is thus constant over the entire region covered by the straight line net. This in turn im-
plies that the region is one in which *the stresses are constant.*

Next, consider a slip line field in which the α-lines, say,
are straight, but the β-lines are curved. Such a field is known
as a *fan*. In the general case the α-lines will not be concurrent but may be considered to be tangents to a curve A-B (Figure 6) which is the *base curve* of the fan. Hencky's theorem
then requires that the β-lines should be *involutes* of the base
curve (it will be recalled that involutes are the loci of
points on a straight line which rolls on the base curve without

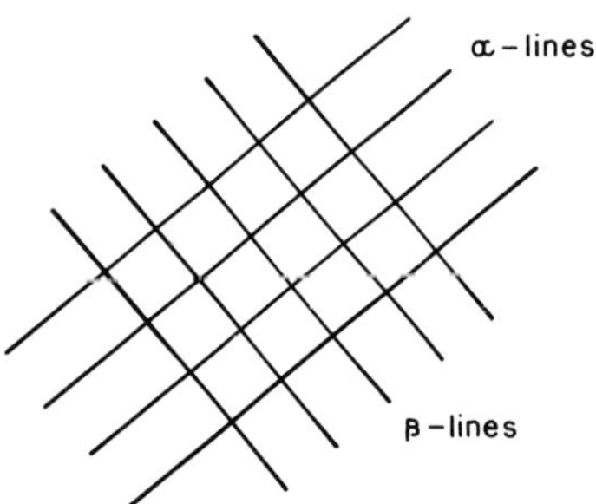

Figure 5. Straight line net

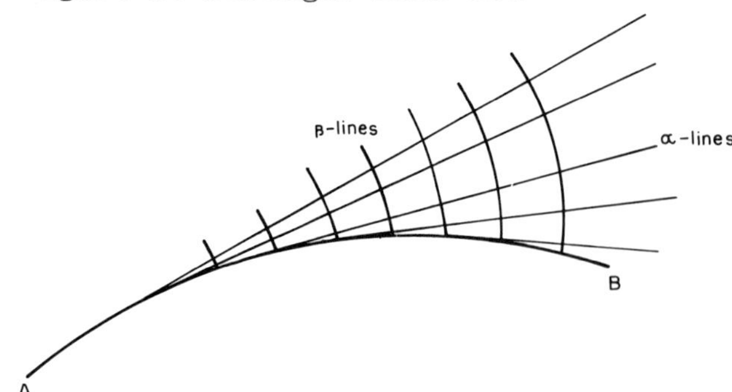

Figure 6. Fan

sliding).

It should be noticed that the base curve can be considered as a line on which one family of characteristics coincides; the second family cannot be extended beyond this line. For this reason the base curve is referred to as a *limiting line*. It must coincide with a rigid-plastic boundary or in some cases an external surface of the body under discussion. Limiting lines also occur in slip line fields in which both families of characteristics are curved.

If the family of straight lines in a fan is concurrent, the slip line field becomes a *centered fan* (Figure 7). The β-lines will now be the arcs of circles whose centers coincide with the center of the fan. Using the angle ϕ together with the distance r from the center of the fan to construct a polar coordinate system, it is evident from equations (60) that $\bar{\sigma} = \bar{\sigma}(\phi)$, and that we may write

$$\bar{\sigma}(\phi) - \bar{\sigma}(\phi_o) = 2k(\phi-\phi_o) \quad . \tag{66}$$

The limiting line degenerates to a point, the center of the fan. Since $\bar{\sigma}$ and ϕ are not single-valued at the center of the fan, this point represents a singularity in the stress field.

The characteristic relations (60) also show us that if $\phi(x,y)$ is continuous and continuously differentiable along a slip line,

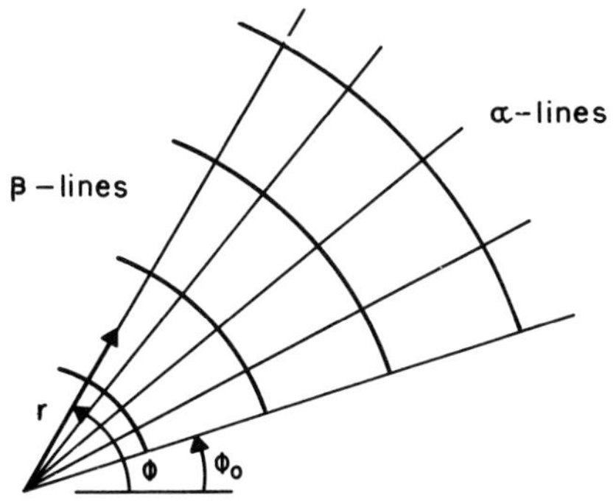

Figure 7. Centered fan

$\bar{\sigma}(x,y)$ will also be continuous and continuously differentiable along the same line. If ϕ is continuous, but its derivative is discontinuous, $\bar{\sigma}$ will be continuous but have a discontinuous derivative. Discontinuities in the derivatives of ϕ along α-lines, say, must be such that the line passing through the points on each α-line where the derivative of ϕ is discontinuous must be a β-line. It can be shown that the jump in the radius of curvature of the α-lines will be constant along the β-line which forms the surface of discontinuity. Note, however, that this condition is meaningless when the α-lines are straight on one side of the discontinuity, since the radius of curvature is infinite.

In Chapter 12 we have exploited the possibility of discontinuities in stress. It may be expected that such discontinuities could be encountered in plane strain problems: the commonest example is the discontinuity in direct stress which occurs across the neutral axis of a sheet subjected to cylindrical bending. Consider a surface across which the stresses are discontinuous. We may impose the physical restriction, arising from the equilibrium equations, that the traction normal to the surface of discontinuity must be continuous. Furthermore, the yield condition must be satisfied on each side of the discontinuity. Thus, as shown in Figure 8(a), let the normal direct

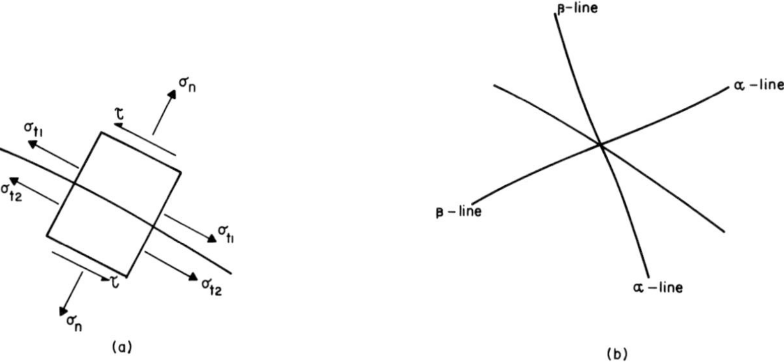

Figure 8. Stress discontinuities

stress σ_n and the shear stress τ acting on elements adjacent to the surface of discontinuity be continuous, and let the tangential direct stress have values σ_{t1}, σ_{t2} on either side of the discontinuity. From the yield condition (6), σ_{t1} and σ_{t2} are roots of the equation

$$(\sigma - \sigma_n)^2 + 4\tau^2 = 4k^2 \quad . \tag{67}$$

It follows then that

$$\sigma_{t1} = \sigma_n + 2\sqrt{(k^2 - \tau^2)} \quad , \tag{68a}$$

$$\sigma_{t2} = \sigma_n - 2\sqrt{(k^2 - \tau^2)} \quad . \tag{68b}$$

The jump in the tangential direct stress is thus $4\sqrt{(k^2 - \tau^2)}$, and the jump in $\bar{\sigma} = (\sigma_n + \sigma_t)/2$ is $2\sqrt{(k^2 - \tau^2)}$. Along the α-lines, from equation (60a), the jump in ϕ must be $\sqrt{(k^2 - \tau^2)}/k$.

Without loss in generality we may suppose that the y-axis is tangential to the surface of discontinuity of the stresses at the point under consideration. It follows then that we may put $\sigma_n = \sigma_x$, $\sigma_t = \sigma_y$, $\tau = \sigma_{xy}$ in equations (9). Inverting these equations, ϕ_1 and ϕ_2, the values of ϕ on each side of the discontinuity, must satisfy the equations

$$\sin 2\phi_1 = \sqrt{(k^2 - \tau^2)}/k, \quad \sin 2\phi_2 = -\sqrt{(k^2 - \tau^2)}/k \quad , \tag{69}$$

$$\cos 2\phi_1 = \cos 2\phi_2 = \tau/k \quad .$$

This requires that

$$\phi_2 = \pi - \phi_1 \quad , \tag{70}$$

and shows that the α-lines on one side of the discontinuity are, locally, reflections of the lines on the other side. More generally, the α or β directions on either side of a surface of discontinuity in the stresses are reflections in the surface, as

shown in Figure 8(b). This result indicates that a surface
across which the stresses are discontinuous cannot coincide
with a slip line.

Since the derivatives of the stresses will generally be discontinuous across a line of discontinuity of the stresses, it
is expected that the derivatives of ϕ along α- and β-lines, or
the curvatures of the slip lines, will change abruptly across
the line of discontinuity.

The same kind of ambiguity that permits jumps in the tangential component of direct stresses across a surface also appears
in the exploitation of stress boundary conditions. Suppose
that on an external boundary of the body the y-axis is tangential to the boundary at the point under consideration. If the
tractions on the boundary are known at this point, $\sigma_x = \sigma_n$ and
$\sigma_{xy} = \tau$ are given. This is not sufficient to determine $\bar{\sigma}$ and ϕ
from equations (9). The yield condition must be satisfied, but
because of the quadratic nature of the condition, we find that

$$\sigma_y = \sigma_n \pm 2\sqrt{(k^2 - \tau^2)} \ . \tag{71}$$

From equations (9), we see that

$$\bar{\sigma} = \sigma_n \pm \sqrt{(k^2 - \tau^2)} \ ,$$
$$\tan 2\phi = \pm \frac{\sqrt{(k^2 - \tau^2)}}{\tau} \ . \tag{72}$$

It may be possible from the nature of the specific boundary value problem under consideration to choose the correct root in
equation (71). If not, the solution must be attempted with
each root, the correct choice becoming clear only when further
considerations are taken into account.

Finally in our discussion of the stress field, let us consider the problem of constructing the slip line field by

approximate methods which can be adapted to numerical computation. Consider first the case where $\bar{\sigma}(x,y)$, $\phi(x,y)$ are given
along the arc AB of a curve which is not a characteristic. The
curve will usually be a boundary, and the slip line field is to
be constructed on one side of the curve.

Choose two neighboring points (x_1,y_1), (x_2,y_2) on the arc AB.
Let the values of $\bar{\sigma}(x,y),\phi(x,y)$ at these points be $\bar{\sigma}_1,\phi_1$ and $\bar{\sigma}_2$,
ϕ_2 respectively. From ϕ_1 and ϕ_2 we know the slopes of the slip
lines through the two points. Suppose, as shown in Figure 9,
that the β-line through (x_1,y_1) and the α-line through (x_2,y_2)
intersect at the point (x_3,y_3) and that $\bar{\sigma}(x_3,y_3) = \bar{\sigma}_3$, $\phi(x_3,y_3)$
$= \phi_3$. The characteristic relations provide $\bar{\sigma}_3,\phi_3$, for on applying equations (60),

$$\bar{\sigma}_3 - 2k\phi_3 \;=\; \bar{\sigma}_2 - 2k\phi_2 \quad,$$

$$\bar{\sigma}_3 + 2k\phi_3 \;=\; \bar{\sigma}_1 + 2k\phi_1 \quad. \tag{73}$$

Solving for $\bar{\sigma}_3,\phi_3$, we find

$$\bar{\sigma}_3 \;=\; \tfrac{1}{2}\{(\sigma_1+\bar{\sigma}_2) + 2k(\phi_1-\phi_2)\} \quad,$$

$$\phi_3 \;=\; \frac{1}{4k}\{(\bar{\sigma}_1-\bar{\sigma}_2) + 2k(\phi_1+\phi_2)\} \quad. \tag{74}$$

It remains to determine the coordinates (x_3,y_3). If the points

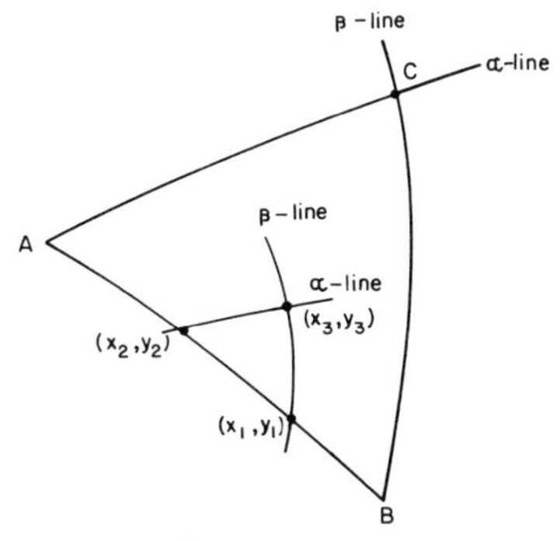

Figure 9. Construction of net from a boundary

are sufficiently close together, we can replace the slip line
arcs by straight lines through (x_1,y_1) and (x_2,y_2) inclined at
angles $\phi_1 + \pi/2$, ϕ_2 to the x-axis. The point (x_3,y_3) is then
approximately given by the intersection of these segments. A
better approximation is to replace the arcs by straight lines
having the mean slope of the terminal points of the arcs; thus
the line through (x_1,y_1) is inclined at $(\phi_1 + \phi_3 + \pi)/2$ to the
x-axis, and the straight line through (x_2,y_2) at $(\phi_2 + \phi_3)/2$,
ϕ_3 being obtained from equations (75). A still better approximation is to replace the arcs by circular arcs. The slip line
arc through (x_1,y_1) would be replaced by the arc of a circle
inclined at ϕ_1 to the x-axis at (x_1,y_1) and at ϕ_3 at (x_3,y_3),
for example.

It is evident that this process can be used to find $\bar{\sigma}_3$, ϕ_3,
x_3,y_3, and that it can be reapplied to find additional information at other points on the net. The solution can be extended over the area ABC in Figure 9, where C lies at the intersection of the α-line through A and the β-line through B. This
area is the *domain of influence* of the arc AB.

It is also possible to extend the solution when $\bar{\sigma},\phi$ are given
along two intersecting slip lines. Figure 10 shows a possible
situation. Application of the characteristic relations to the
four discrete sets of values shown gives

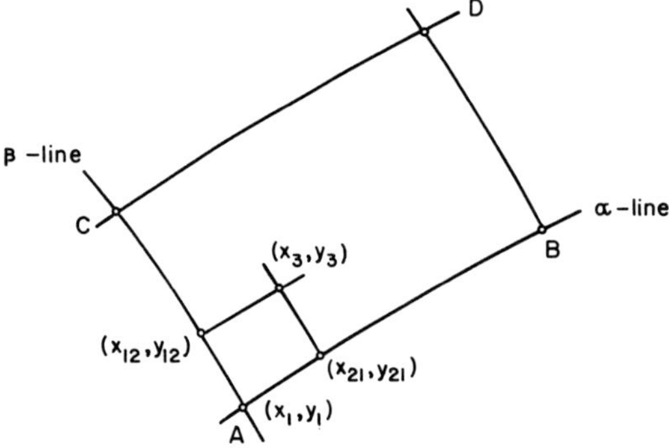

Figure 10. Construction of net from two slip lines

$$\bar{\sigma}_3 - 2k\phi_3 = \bar{\sigma}_{21} - 2k\phi_{21} \quad ,$$
$$\bar{\sigma}_3 + 2k\phi_3 = \bar{\sigma}_{12} + 2k\phi_{12} \quad .$$
$$(75)$$

These equations can be solved for $\bar{\sigma}_3, \phi_3$, and x_3, y_3 can be found in the manner described above. By a combination of the methods of both these problems, the solution can be found within the area ABDC if $\bar{\sigma}(x,y), \phi(x,y)$ are given along AB, AC in Figure 10.

Thirdly, we can also extend the field if $\bar{\sigma}(x,y), \phi(x,y)$ are given along a slip line and either $\bar{\sigma}(x,y), \phi(x,y)$ is given along a curve AB which is not a slip line. In this case it is possible to compute the missing value along the line AB in the process of extending the field by the two methods given above. Figure 11 shows the first step. We have from equations (60),

$$\bar{\sigma}_3 + 2k\phi_3 = \bar{\sigma}_2 + 2k\phi_2 \quad . \qquad (76)$$

If ϕ_3 is given, $\bar{\sigma}_3$ can be found, and *vice versa*. By a combination of all three methods, the solution can be found over ABC if $\bar{\sigma}(x,y), \phi(x,y)$ are given along AC, and either $\bar{\sigma}(x,y)$ or $\phi(x,y)$ is given along AB.

15.5 Considerations Involving the Velocity Field

In treating the velocity field we assume that the slip line field, and hence $\phi(x,y)$, is known *a priori*. The slip line field may be known exactly, or may be hypothetical. The velocity field

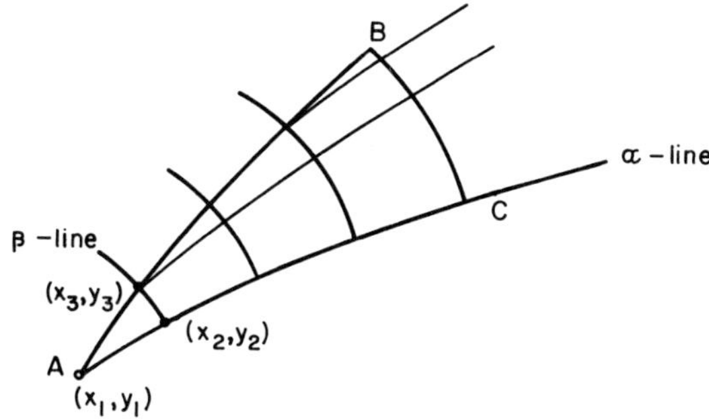

Figure 11. Construction of net from one slip line

is then found by integrating the characteristic relations (50);
this is a comparatively simple problem.

Before proceeding, however, it is convenient to refer the
characteristic equations to velocity components in the slip line
directions. At any point in the x-y plane let v_α, v_β be the components of the velocity vector in the α- and β-directions respectively. The components u,v in the x,y directions may be
written in terms of v_α, v_β by means of the transformation (cf.
Figure 4)

$$u = v_\alpha\cos\phi - v_\beta\sin\phi \quad ,$$

$$v = v_\alpha\sin\phi + v_\beta\cos\phi \quad . \tag{77}$$

It follows that

$$du = dv_\alpha\cos\phi - dv_\beta\sin\phi - v_\alpha d\phi\sin\phi - v_\beta d\phi\cos\phi \quad ,$$

$$dv = dv_\alpha\sin\phi + dv_\beta\cos\phi + v_\alpha d\phi\cos\phi - v_\beta d\phi\sin\phi \quad . \tag{78}$$

Substituting equations (78) into equations (50), the characteristic relations become

$$dv_\alpha - v_\beta d\psi = 0 \quad \text{along an } \alpha\text{-line} \quad , \tag{79a}$$

$$dv_\beta + v_\alpha d\phi = 0 \quad \text{along a } \beta\text{-line} \quad . \tag{79b}$$

A special case of interest occurs, when, say, $v_\beta = 0$ along
an α-line. It follows that along the α-line v_α = constant,
$v_\beta = 0$, and the α-line is a *stream line*.

With the characteristic equations in the form given in equation (79), it can be seen that if the slip line field is known
and v_α, v_β are given at any one point, the velocity field can be
extended by integrating along the characteristics. In the general case this can be done numerically, replacing equations (79)
by appropriate difference equations. In some simple cases the
integration can be carried out analytically.

Suppose for example that the slip line field consists of two orthogonal families of straight lines, so that $\phi(x,y)$ is constant. From equations (79), we see that v_α and v_β are constant along α- and β-lines respectively (note that in general the velocity component along a straight slip line will always be constant). This does not imply that v_α and v_β are necessarily constant over the entire straight line field, but only that v_α varies only in the β-direction and that v_β varies only in the α-direction. Thus if α,β are coordinates in the α- and β-directions,

$$v_\alpha = v_\alpha(\beta) \ , \quad v_\beta = v_\beta(\alpha) \ . \tag{80}$$

The velocity field thus consists of two *independent* shear flows in orthogonal directions.

A second simple example is a centered fan in which, say, the α-lines are straight and concurrent, and the β-lines are circular arcs. Using a polar coordinate system r,ϕ (Figure 7), we see that from equation (79a)

$$v_\alpha = v_\alpha(\phi) \ , \tag{81}$$

and that from equation (79b)

$$\frac{\partial}{\partial\phi} v_\beta(r,\phi) = - v_\alpha(\phi) \ . \tag{82}$$

On integrating, we may write

$$v_\beta(r,\phi) = v_\beta(r,\phi_o) - \int_{\phi_o}^{\phi} v_\alpha(\phi')d\phi'$$
$$= g(r) + h(\phi) \ . \tag{83}$$

It is thus seen that v_β consists of the sum of independent functions of r and ϕ. Note that if the velocity field is continuous at $r=0$, v_α must be independent of ϕ and hence constant over the fan.

Discontinuities in the derivatives of ϕ, or discontinuities in the radii of curvature of slip lines, imply discontinuities in the derivatives of the velocity components. If the velocity field is continuous, the jumps in the derivatives of the velocity along a slip line can be obtained readily from the jump in the curvature of the slip line by means of the characteristic relations (79). Discontinuities in the velocity field may occur, as we have seen in Chapter 12. We may impose the physical restriction that the normal component of velocity across a surface of discontinuity must be continuous; only the tangential component may be discontinuous. We may also extend this physical reasoning to argue that a surface of discontinuity of the velocity field must coincide with a slip line. In Section 12.1 we treated the discontinuity in the tangential velocity across a surface as the limit of a thin sheet in which the shear strain rate in axes tangential and normal to the sheet is very large; clearly the shear stress in these axes is k, and the axes must coincide with the α- and β-directions.

Finally, we may combine this argument and an earlier result to show that a boundary between the deforming region of a body and a region which is at rest must coincide with a slip line. Either the tangential velocity is continuous across this boundary or it is not. If it is not continuous, the argument of the previous paragraph requires the boundary to be a slip line. If it is continuous, the fact that one side of the boundary is at rest requires that both v_α and v_β must be zero along the boundary. If the boundary is not a slip line, this in turn implies that both v_α and v_β are zero throughout the domain of influence of the boundary (see Figure 9). This leads to a contradiction; it follows that the boundary must be a slip line.

15.6 The Solution of Specific Problems

The information we have obtained in Sections 15.4 and 15.5 may
now be exploited to construct solutions of problems. It must
be emphasized that we are by no means in a position to systematically construct the complete solution in a straightforward or
automatic process. We have discussed the construction of the
slip line field within the domain of influence of stress boundary conditions, but this will generally not be sufficient to
construct the solution. Even if all the boundary conditions
are given in terms of stresses, some stress components may be
unknown (remembering that we are attempting to determine the
load parameter for which flow occurs) or the extent of the solution may not be known. In mixed boundary value problems the
velocity boundary conditions must be satisfied; again on parts
of the boundary where the velocity is given the tractions are
not known.

In general, therefore we must approach the construction of
the solution on a trial and error basis. The slip line field
is extended from parts of the boundary where the tractions are
known and the material is known to be deforming. It is hypo-
thetically completed, and checks must be made to see whether the
remaining boundary conditions are satisfied. If not, a new hypothesis is made, and the process is repeated. Finally, it is
advisable to ensure that the stress and velocity fields are such
that the rate of work is non-negative at each point in the body.

As a first example, let us reconsider the problem (see Section 12.4) of a rigid punch of width h indented into a half-
space of rigid-plastic material under the assumption that the
contact between the punch and the half-space is smooth. The
problem is shown diagrammatically in Figure 12. We are concerned with the load P per unit length of the punch required to
initiate plastic flow. At the instant flow begins, the punch

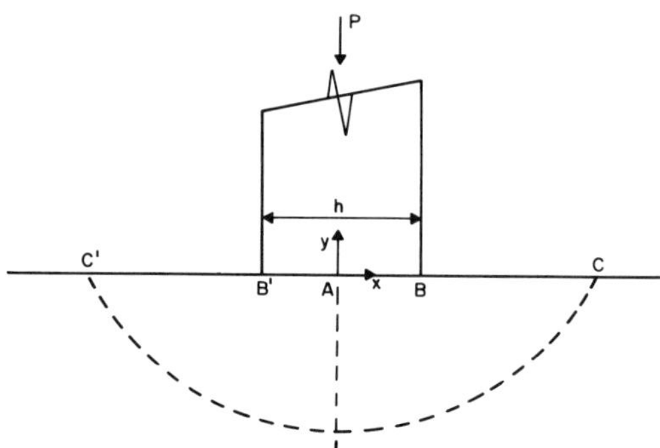

Figure 12. Punch problem

will move vertically with velocity v_o.

The extent of the plastic region under the punch can be expected to be as shown by the dotted line in Figure 12. Along
BC and B'C' the tractions are zero. Along BAB' the tangential
component of traction vanishes, and the component of velocity
in the y-direction is $(-v_o)$. We must now construct a plausible
slip line field.

First, in the neighborhood of a straight, free boundary along
which the tangential component of traction vanishes, the simplest slip line field consists of orthogonal straight lines intersecting the boundary at $45°$. Assuming this to be correct,
we can construct the domain of influence of BC, which of course
depends on the length BC which is so far unknown. Next, we observe that a singularity in stress must occur at B. This suggests that a centered fan abuts the domain of influence of BC.
Under the punch itself we might assume that the unknown normal
tractions are constant. We could conceivably then have a second region of slip lines meeting the surface at $45°$.

Even with the assumption of a straight line field under the
punch, its extent is important in fixing the length of BC. Let
us assume that AB and AB' have separate domains of influence.
Then, considering only one half of the deforming field, the slip
lines take the form shown in Figure 13(a). This field consists

of three regions ABD, BDE and BCE in which the solution is continuous. The regions abut along BD and BE, and ADEC is a rigid-plastic boundary.

We may start to construct the stress field in BCE, since the tractions along the boundary are known. At a point on BC, therefore, $\sigma_y = \sigma_{xy} = 0$. It follows from the yield condition that

$$(\frac{\sigma_x}{2})^2 = k^2 \quad , \quad \text{or} \quad \sigma_x = \pm 2k \quad . \tag{84}$$

It can be expected that the region BCE will be in hydrostatic pressure rather than hydrostatic tension, and hence we choose $\sigma_x = -2k$, and $\bar{\sigma} = -k$. From equations (9),

$$\sin 2\phi = 1 \quad , \quad \cos 2\phi = 0 \quad , \tag{85}$$

and hence $2\phi = \pi/2 + 2\pi n$, where n is an integer. The simplest choice is n=0, giving $\phi = \pi/4$. This immediately identifies the α-direction and the α-lines. The α,β axes are shown in the region BCE in Figure 13(b). Since $\bar{\sigma} = k$, $\phi = \pi/4$ along the line BC, it follows that $\bar{\sigma} = -k$, $\phi = \pi/4$ throughout the region BCE.

The centered fan is shown in Figure 13(c). The stresses will be taken to be continuous across BE, and the α- and β-lines are readily identified. It is evident that ϕ changes from $\pi/4$ along BE to $-\pi/4$ along BD. The hydrostatic tension $\bar{\sigma}$ will be constant along the β-lines. If F is a generic point along DE, the change in ϕ along any α-line between BE and BF is constant. From the characteristic relation along an α-line,

$$\bar{\sigma}_{BF} - 2k\phi_{BF} = (-k) - 2k\pi/4 \quad , \quad \text{or}$$
$$\bar{\sigma}_{BF} = (-k) + 2k(\phi_{BF} - \pi/4) \quad . \tag{86}$$

In particular

$$\bar{\sigma}_{BD} = -k(1+\pi) \quad . \tag{87}$$

The region ABD is shown in Figure 13(d), with the α- and

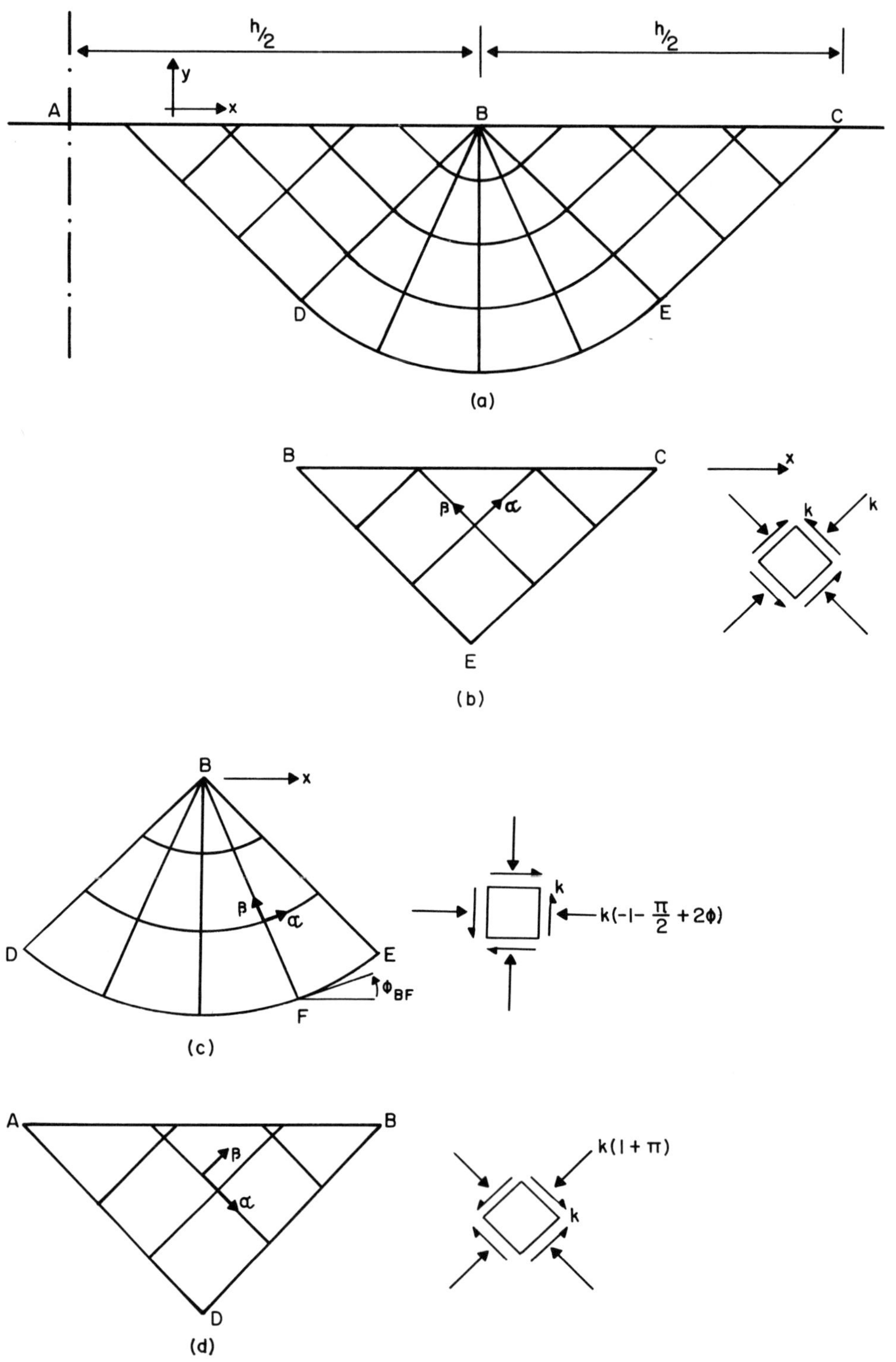

Figure 13. Slip line field for punch

β-lines identified. From continuity of the stresses along BD, $\bar{\sigma} = -k(1+\pi)$, $\phi = -\pi/4$ throughout ABD. Using equations (9), the stress field in ABD can be written as

$$\sigma_x = \bar{\sigma} - k\sin2\phi = -k\pi \quad ,$$

$$\sigma_y = \bar{\sigma} + k\sin2\phi = -k(2+\pi) \quad , \tag{88}$$

$$\sigma_{xy} = k\cos2\phi = 0 \quad .$$

These stresses satisfy the tangential traction boundary condition along AB. The normal traction is constant, and we can put

$$\frac{P}{h} = |\sigma_y| \quad , \quad \text{or}$$

$$P = (2+\pi)kh \quad . \tag{89}$$

These calculations show that the slip line field is acceptable as far as the stresses are concerned. We must now consider the velocity field. Here we start with region ABD, since the velocities are known along AB.

Considering Figure 13(d), along AB we must have

$$v_\alpha - v_\beta = \sqrt{2}\, v_0 \quad . \tag{90}$$

Along AD, which is a rigid-plastic boundary, continuity of the normal component of velocity requires that $v_\beta = 0$. Since $dv_\beta = 0$ along any straight β-line, the conditions on AB and AD force us to put $v_\beta = 0$ everywhere in ABD. The α-lines are thus stream lines, and $v_\alpha = \sqrt{2}\, v_0$ throughout ABD. There is a discontinuity in the tangential component of velocity along AD.

Now consider the centered fan BDE (Figure 13c). Continuity of the normal component of velocity across BD requires that $v_\alpha = \sqrt{2}\, v_0$ along BD. Assume that v_β is also continuous across BD; this gives $v_\beta = 0$ along BD. Continuity of normal component of velocity across DE requires that $v_\beta = 0$ along DE; hence v_β

is independent of ϕ and equal to zero throughout the fan. The α-lines are again streamlines, and $v_\alpha = \sqrt{2}\,v_o$ throughout BDE. There will be a discontinuity in the tangential component of velocity across DE.

Assuming no discontinuity in the tangential component of velocity across BE, the velocity is easily extended into BEC (Figure 13b). We find that $v_\alpha = \sqrt{2}\,v_o$, $v_\beta = 0$, so that the α-lines are again streamlines, and that there is a discontinuity in the tangential component of velocity along DC.

These computations show that the velocity field implied by the suggested slip line field is acceptable. It remains to check whether the stresses and strain rates are such that the work rate is non-negative. Figures 13(b), (c) and (d) have inserts showing the stresses in the α,β axes. It is evident that these stresses are consistent with the strain rates implied by the velocity field in BDE (the strain rates are zero in ABD and BCE) and with the velocity discontinuity along ADEC. A more detailed computation will show that the internal and external work rates are equal. The slip line field of Figure 13 is thus acceptable.

The solution of the problem is not unique. In Section 12.4 (Figure 15c) we treated a velocity field for the same problem which can be identified with the slip line field of Figure 14.

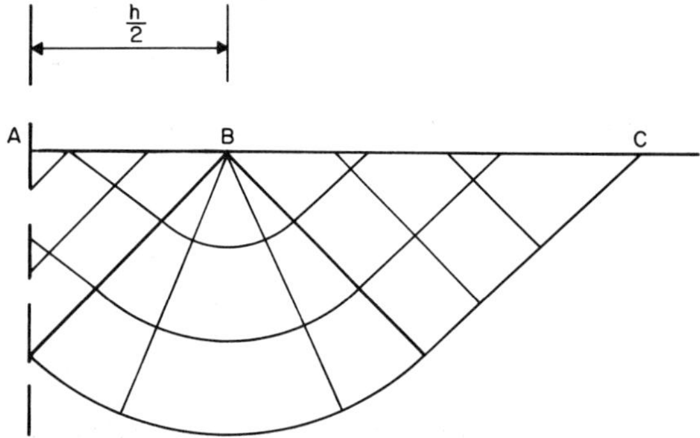

Figure 14. Alternative slip line field for punch

In this field the velocity is discontinuous at B, and the tangential component of velocity is discontinuous across BD and DEC. The limiting value of P is identical to that given in equation (89). Alternative slip line fields, representing combinations of Figures 13 and 14, can also be found.

It has been argued that for an elastic, perfectly plastic solid the solution of Figure 13 is more informative. As the load P is increased monotonically from zero yield will first occur in the neighborhood of B and B'. As soon as the two plastic regions join at the center A, flow will commence. It is unlikely that the more extensive deforming region of Figure 14 will be established. Once flow begins, of course, the free boundary will deform and our solution based on the original geometry will not longer be valid.

As a second example, consider the deeply notched strip shown in Figure 15(a). The notches are symmetric about the x and y axes, and the bar is subjected to a tension force P. It is evident that when flow begins the plastic regions will be confined to the area around the notches: let the rigid regions remote from the notch move away from the x-axis with velocity v_o at the instant flow begins.

This problem is in many respects similar to the punch problem, and the slip line field can be considered a modification of that shown in Figure 13. The part of the field which appears in the fourth quadrant in the x-y plane is shown in Figure 15(b). The complete field is symmetric about the x and y axes. We again treat two regions ABD and BCE in which both families of slip lines are straight, and a centered fan BDE.

Using the traction free boundary conditions along BC, and assuming that BCE is in hydrostatic tension, we find that $\bar{\sigma} = k$, $\phi = -(\theta+\pi/4)$ in BCE. Assuming that the stress field is continuous across BD and BE, and noting that $\bar{\sigma} = \bar{\sigma}(\phi)$ in BDE, we find from the characteristic relation along the β-lines of the fan that $\bar{\sigma} = k(1+\pi-2\theta)$ in ABD. Since $\phi = -3\pi/4$ in ABD, the stresses in ABD are

$$\sigma_x = k(\pi-2\theta) \quad,$$

$$\sigma_y = k(2+\pi-2\theta) \quad, \tag{91}$$

$$\sigma_{xy} = 0 \quad.$$

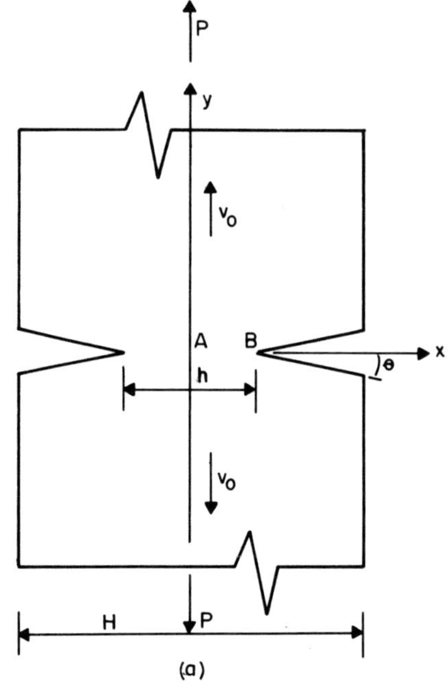

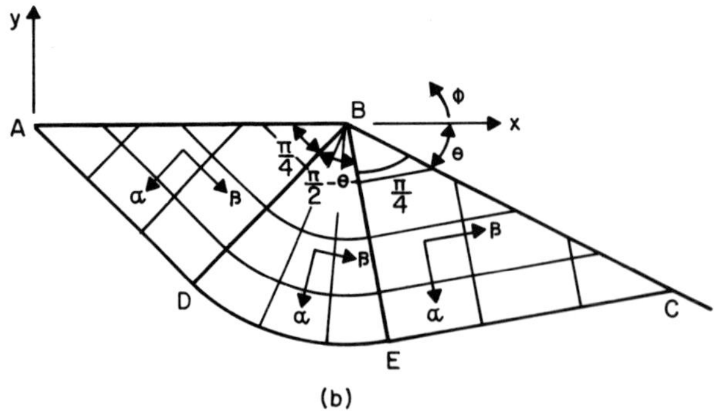

Figure 15. Notched strip

These stresses are acceptable along AB, and we thus find that

$$P = kh(2+\pi-2\theta) \quad . \tag{92}$$

In constructing the velocity field, we begin with ABD. Symmetry and continuity require that the normal component of velocity is zero and $v_o/\sqrt{2}$ on AB and AD respectively. These conditions require that $v_\alpha = -v_\beta = v_o/\sqrt{2}$ in ABD. There will be a tangential velocity discontinuity of $v_o\sqrt{2}$ along AD, and the velocity at A is discontinuous.

Across BD, v_β must be continuous, and along DC continuity requires that $v_\alpha = v_o \sin(-\phi)$. These conditions are sufficient to give the velocities in BDE;

$$v_\alpha = v_o \sin(-\phi) \quad ,$$

$$v_\beta = v_\beta(\phi = -\frac{3\pi}{4}) - \int_{\frac{5\pi}{4}}^{\phi} v_\alpha(\phi')d\phi' \tag{93}$$

$$= -v_o(\sqrt{2} + \cos\phi) \quad .$$

Across BE v_β must be continuous, and hence along BE

$$v_\beta = -\frac{v_o}{\sqrt{2}} (2 + \cos\theta - \sin\theta) \quad . \tag{94}$$

Continuity across CE in turn gives

$$v_\alpha = \frac{v_o}{\sqrt{2}} (\sin\theta + \cos\theta) \tag{95}$$

along CE. These conditions imply that v_α and v_β are constant throughout BCE.

Tangential velocity discontinuities appear along DE and CE, and the velocity at B is singular. The velocity field is acceptable and it remains only to check that the dissipation rate is non-negative at each point. This is left as an exercise for the reader.

15.7 Historical and Bibliographical Remarks

In the context of limit analysis slip line theory is a kinematic method. In problems of incipient plastic flow which fall into the domain of limit analysis, as do the examples presented in the previous section, the load parameters obtained by slip line theory will be not less than,but not necessarily equal to, the limit load parameters. A correct slip line field gives a kinematically admissible velocity field, and a safe statically admissible stress field within the deforming region. However, it provides no information about the stress field in the rigid regions. It must be possible to extend the stress field into the rigid region, satisfying both the equilibrium equations (4) and the yield condition,

$$\left(\frac{\sigma_x-\sigma_y}{2}\right)^2 + \sigma_{xy}^2 \le k^2 \, , \tag{96}$$

before the slip line solution can be accepted as giving the true limit load parameter.

Thus, while the solution for the punch problem can be shown to be correct, the stress field for the notched bar of Figure 15(b) can be extended only for H/h > 10.

The illustrative examples presented in the previous section, while consistent with the class of problems discussed in this volume, do not accurately reflect the power of slip line theory in plane strain problems. A number of problems in steady and pseudo-steady plastic flow, important in metal processing operations, have been successfully investigated by this method. The reader is referred to the bibliography for more extensive discussions of the theory.

The development of slip line theory in metal plasticity is based on the work of Hencky [1923], Prandtl [1923], and Caratheodory and Schmidt [1923]. Early studies were devoted almost

exclusively to statically determinate problems (i.e. problems
in which the slip line field can be obtained from stress boundary conditions alone). A considerable amount of work was done
on statically determinate problems (see,for example,Sokolovskij
[1946])despite the limited applicability of an approach which
ignores the velocity field.

The characteristic relations for velocities were obtained by
Geiringer [1930], leading the way to consideration of more realistic problems. Early work in slip line theory is described
in detail by Hill [1950] and Prager and Hodge [1951]; these
discussions include problems of incipient plastic flow and problems of steady and pseudo-steady flow. Approaches to slip line
theory which are geometric in nature have been given by Prager
[1953], [1959].

The particular solutions given as examples are due to Hill
[1949], while the alternative solution to the punch problem is
due to Prandtl [1923]. Recent accounts of the application of
slip line theory, particularly in metal forming problems, have
been given by Ford [1963], Thomsen, Yang and Kobayashi [1965],
and Johnson and Sowerby [1970].

Slip line theory also finds applications in soil mechanics,
and more generally in materials in which plastic volume changes
may occur. The applications to soil mechanics in fact predate
applications in metals; some of the results given by Hencky
[1923] are, for example, implied in the work of Kötter [1903].
Examples of the use of slip line theory in soil mechanics are
given by Wu [1966]. Berg [1971] has discussed the general problem of determining conditions under which the plane strain
problem in dilational plasticity is hyperbolic.

References

C.A.Berg	1971	"On Plane Dilational Plasticity", *Studies in Applied Mathematics*, <u>50</u> (MIT).
C.Caratheodory and E.Schmidt	1923	"Ueber die Hencky-Prandtlschen Kurrenscharen", Z.angew.Math.Mech., <u>3</u>, 468.
H.Ford	1963	*Advanced Mechanics of Materials*, Wiley (N.Y.).
H.Geiringer	1930	"Beit zum Vollständigen ebenen Plastizitäts-problem", Proc.3rd Intern. Congr.Appl.Mech., <u>2</u>, 185.
H.Hencky	1923	"Ueber einige statisch bestimmte Faelle des Gleichgewichts in plastischen Koerpern", Z.angew.Math.Mech., <u>3</u>, 241.
R.Hill	1949	"The plastic yielding of notched bars under tension", Quart.J.Mech.Appl. Math., <u>2</u>, 40.
R.Hill	1950	*The Mathematical Theory of Plasticity*, Oxford University Press.
W.Johnson and R.Sowerby	1970	*Plane Strain Slip Line Fields* (Edward Arnold).
F.Kötter	1903	"Die Bestimmung des Druckes an Gekrümmten Gleitflächen", Ber.Akad.der Wiss. (Berlin).
W.Prager	1953	"A geometrical discussion of the slip line field in plane plastic flow", Trans.Roy.Inst.Tech.(Stockholm),<u>65</u>,1.
W.Prager	1959	*An Introduction to Plasticity*, Addison-Wesley (Reading,Mass.).
W.Prager and P.G.Hodge,Jr.	1951	*Theory of Perfectly Plastic Solids*, Wiley, N.Y. (Dover, 1968).
L.Prandtl	1923	"Anwendungsbeispiele zu einem Henchyschen Satz ueber das plastische Gleichgewicht", Z.angew.Math.Mech., <u>3</u>, 468.
V.V.Sokolovskij	1946	*Theory of Plasticity*, Moscow.
E.G.Thomsen, C.T.Yang and S.Kobayashi	1965	*Mechanics of Plastic Deformation in Metal Processing*, Macmillan (N.Y.).
T.H.Wu.	1966	*Soil Mechanics*, Allyn and Bacon (Boston).

PART IV

ELASTIC, PERFECTLY PLASTIC PROBLEMS

PROPERTIES OF ELASTIC-PLASTIC SOLUTIONS

16.1 Convergence of Solutions

In this chapter we shall continue to investigate properties of
the solutions of elastic, perfectly plastic structural problems
which, to some extent at least, do not depend on a complete incremental solution of the problem. We shall be concerned primarily with histories of loading on the structure (or paths of
loading in the load space) which do *not* cause flow. Such histories can be characterized by generalized loads $\Gamma_\alpha(t)$ which are
such that

$$\Psi[\Gamma_\alpha(t)] < 0 \tag{1}$$

for all t. Time t is introduced in the same sense as in Chapter
9; it measures the order of events and increases monotonically
with real time. However, the loads are applied sufficiently
slowly that inertia effects are negligible.

The material is again assumed to be elastic, perfectly plastic in the sense that the initial yield surface $\phi(Q_j)$ and the
limit surface in generalized stress space $\psi(Q_j)$ are identical.
To emphasize that we are dealing with an elastic, perfectly plastic material rather than a rigid-perfectly plastic material we
shall use $\phi(Q_j)$ to denote the yield and limit functions throughout this chapter. The constitutive equations are consequently
given by

$$\dot{q}_j = \dot{q}_j^e + \dot{q}_j^p \ ,$$

$$\dot{q}_j^e = C_{jk}\dot{Q}_k \ ,$$

$$\dot{q}_j^p = \lambda \frac{\partial \phi}{\partial Q_j} \quad \text{for} \quad \phi(Q_j) = 0 \quad \text{and} \quad \frac{\partial \phi}{\partial Q_j}\dot{Q}_j = 0 \ , \tag{2}$$

$$\dot{q}_j^p = 0 \quad \text{for} \quad \phi(Q_j) < 0$$

$$\text{or} \quad \phi(Q_j) = 0 \quad \text{and} \quad \frac{\partial \phi}{\partial Q_j} \dot{Q}_j < 0 \quad ,$$

$$\lambda \geq 0 \quad .$$

As before, $\dot{q}_j^e$ and $\dot{q}_j^p$ are respectively the elastic and plastic components of strain rate, $\phi(Q_j)$ is a convex function, and stresses such that $\phi(Q_j) > 0$ are not permitted.

The most convenient starting point in this discussion is the notion of the convergence of solutions of problems which differ only in their initial conditions. The complete solution of any problem in which generalized loads $\Gamma_\alpha(t)$ are given, with the displacements and rotations given zero at an appropriate number of supports, will depend on the initial stresses $Q_j(s,t=0)$. We thus admit the possibility that $\Gamma_\alpha(0)$ may not coincide with the origin of the load space, and that the structure may have been subjected to some previous history of loading which has caused plastic deformation of the structure. Once $Q_j(s,0)$ is given, the response of the structure for the given loads and displacements can always be determined using an incremental analysis.

Suppose that in one instance the initial stress field in the structure is given as $Q_j^{(1)}(s,0)$. Let the solution be characterized by stresses $Q_j^{(1)}(s,t)$, plastic strain rates $\dot{q}_j^{p(1)}(s,t)$ and displacement rates $\dot{\xi}_\alpha^{(1)}(s,t)$. Alternatively, let the solution with initial stresses $Q_j^{(2)}(s,0)$ be characterized by $Q_j^{(2)}(s,t)$, $\dot{q}_j^{p(2)}(s,t)$ and $\dot{\xi}_\alpha^{(2)}(s,t)$. We note first that the difference between the generalized stress fields $(Q_j^{(1)} - Q_j^{(2)})$ is self-equilibrating (i.e. in equilibrium with zero generalized loads) since each stress field is in equilibrium with the same loads. Further, the displacement rates $(\dot{\xi}_\alpha^{(1)} - \dot{\xi}_\alpha^{(2)})$ and the strain rates $(\dot{q}_j^{(1)} - \dot{q}_j^{(2)})$ are kinematically admissible, since $\dot{\xi}_\alpha^{(1)}$, $\dot{q}_j^{(1)}$ and $\dot{\xi}_\alpha^{(2)}$, $\dot{q}_j^{(2)}$ are each kinematically admissible.   The principle of virtual velocities thus gives

$$0 = \int_V (Q_j^{(1)}-Q_j^{(2)})(\dot{q}_j^{(1)}-\dot{q}_j^{(2)})dV$$

$$= \int_V (Q_j^{(1)}-Q_j^{(2)})(\dot{q}_j^{e(1)}-\dot{q}_j^{e(2)})dV \tag{3}$$

$$+ \int_V (Q_j^{(1)}-Q_j^{(2)})(\dot{q}_j^{p(1)}-\dot{q}_j^{p(2)})dV \quad .$$

Making use of the relation between stress and the elastic component of strain (equations 2),

$$\int_V (Q_j^{(1)}-Q_j^{(2)})(\dot{q}_j^{e(1)}-\dot{q}_j^{e(2)})dV = \int_V C_{jk}(Q_j^{(1)}-Q_j^{(2)})(\dot{Q}_k^{(1)}-\dot{Q}_k^{(2)})dV$$

$$= \frac{d}{dt} \int_V C_{jk}(Q_j^{(1)}-Q_j^{(2)})(Q_k^{(1)}-Q_k^{(2)})dV \quad . \tag{4}$$

Since $\phi(Q_j^{(1)}) \leq 0$, $\phi(Q_j^{(2)}) \leq 0$, the fundamental inequality on
which convexity of the yield and limit surfaces and the norma-
lity rule (Chapter 2, equation 129 and Chapter 9, equation 4)
gives

$$(Q_j^{(1)}-Q_j^{(2)})\dot{q}_j^{p(1)} \geq 0 \quad , \tag{5a}$$

$$(Q_j^{(2)}-Q_j^{(1)})\dot{q}_j^{p(2)} \geq 0 \quad . \tag{5b}$$

Note that these statements hold also when $\dot{q}_j^{p(1)}$ or $\dot{q}_j^{p(2)}$ are
identically zero. Adding (5a) and (5b),

$$(Q_j^{(1)}-Q_j^{(2)})(\dot{q}_j^{p(1)}-\dot{q}_j^{p(2)}) \geq 0 \quad . \tag{6}$$

Substitution of equation (4) and (6) into (3) gives finally

$$\frac{d}{dt} \int_V C_{jk}(Q_j^{(1)}-Q_j^{(2)})(Q_k^{(1)}-Q_k^{(2)})dV$$

$$= - \int_V (Q_j^{(1)}-Q_j^{(2)})(\dot{q}_j^{p(1)}-\dot{q}_j^{p(2)})dV \tag{7}$$

$$\leq 0 \quad .$$

The matrix of the elastic coefficients C_{jk} is positive definite, and consequently the expression

$$A(Q_j^{(1)}, Q_j^{(2)}) = \int_V C_{jk}(Q_j^{(1)} - Q_j^{(2)})(Q_k^{(1)} - Q_k^{(2)})\,dV \tag{8}$$

is a quadratic form. It is non-negative, and zero only when $Q_j^{(1)}(s,t) \equiv Q_j^{(2)}(s,t)$. The function $A(Q_j^{(1)}, Q_j^{(2)})$ may be thought of as a scalar quantity which measures, in the sense defined by equation (8), the *difference between the stress fields* $Q_j^{(1)}$ and $Q_j^{(2)}$. Equation (7) shows that

$$\frac{dA}{dt} \leq 0 \tag{9}$$

i.e. that the difference between the stress fields can only decrease with time.

It can immediately be seen that inequalities (7) and (9) imply uniqueness of the response of the elastic, perfectly plastic structure. Let us assume that for given initial conditions there are in fact two distinct solutions to a structure subjected to loads $\Gamma_\alpha(t)$. $Q_j^{(1)}, \dot{q}_j^{p(1)}, \dot{\xi}_\alpha^{(1)}$ and $Q_j^{(2)}, \dot{q}_j^{p(2)}, \dot{\xi}_\alpha^{(2)}$ can be taken to be these two solutions. However, since the initial conditions are the same, we put $Q_j^{(1)}(s,0) = Q_j^{(2)}(s,0)$. Thus, from equation (8),

$$A(t=0) = 0 \quad . \tag{10}$$

However, from inequality (9), A is a non-increasing function. Since A is non-negative, we must have

$$A(t) = 0 \quad \text{and} \quad Q_j^{(1)} = Q_j^{(2)} \quad . \tag{11}$$

Our supposition that two distinct solutions exist is therefore contradicted, and the solution is unique.

If we return to the case where the two initial stress fields are different, so that $A(t=0) > 0$, inequality (9) indicates that

the stress fields can only converge with time in the sense that
A can only get smaller and closer to zero. In fact, A will decrease (i.e. dA/dt < 0), so that convergence of the stress fields
occurs, unless, from equations (6) and (7),

$$(Q_j^{(1)}-Q_j^{(2)})(\dot{q}_j^{p(1)}-\dot{q}_j^{p(2)}) \ = \ 0 \tag{12}$$

at each point in the body. There are three non-trivial situations in which equation (12) can be satisfied for any point in
the body. The first is that

$$Q_j^{(1)} \ = \ Q_j^{(2)} \ , \tag{13}$$

i.e. the stresses are identical at that point in the body. The
second case is that the vectors $(Q_j^{(1)}-Q_j^{(2)})$ and $(\dot{q}_j^{p(1)}-\dot{q}_j^{p(2)})$
are orthogonal so that the scalar product is zero. Bearing in
mind that the quantity given in equation (12) is the sum of two
non-negative terms (equations 5a and 5b), this can occur only
when $Q_j^{(1)}$ and $Q_j^{(2)}$ both lie on the same flat region of the yield
surface, so that $\dot{q}_j^{p(1)}$ and $\dot{q}_j^{p(2)}$ both have the same direction in
stress space. This case is shown diagrammatically in Figure 1,
and also includes the possibility that $\dot{q}_j^{p(1)} = \dot{q}_j^{p(2)} \neq 0$.
Thirdly, equation (12) may be satisfied at a point if

$$\dot{q}_j^{p(1)} \ = \ \dot{q}_j^{p(2)} \ = \ 0 \ . \tag{14}$$

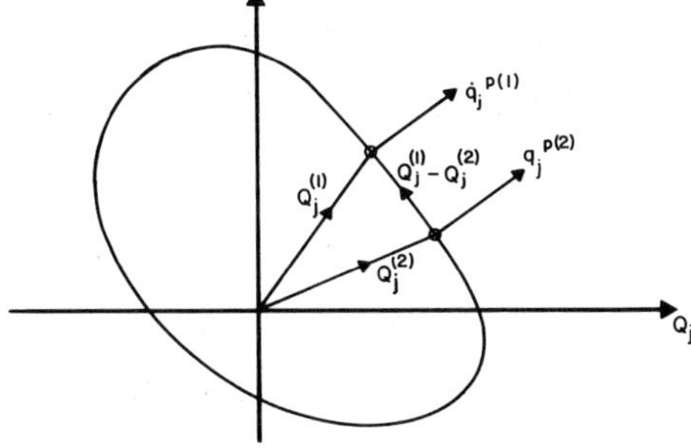

Figure 1. Flats on the yield surface

This means that the behavior at the point under consideration is *elastic*.

The significance of the convergence phenomenon in elastic, perfectly plastic structures lies in the uncertainty about the initial stresses which always occurs in practical structures. Inelastic strains will usually result from the process by which the structure and its components are manufactured (i.e. rolling, welding, etc.) and these will lead to initial stresses in the structure. In analyzing the structure it will generally be impossible to determine this initial stress distribution - indeed the structure is generally analyzed before it is built. It is necessary, therefore, to *assume* initial conditions in order to estimate the response of the structure to additional loads. The structure may be assumed to be in its virgin state, with the initial stresses zero, or some suitable initial stress distribution may be chosen. In either case the designer must be concerned about the effect of his initial assumptions on the stresses in the structure. The convergence phenomenon ensures that whenever plastic deformation takes place such that

$$(Q_j^{(1)} - Q_j^{(2)})(\dot{q}_j^{p(1)} - \dot{q}_j^{p(2)}) \neq 0 \tag{15}$$

at any point in the structure, where $Q_j^{(1)}$, $\dot{q}_j^{p(1)}$ are the stresses and plastic strains in the actual structure with its uncertain initial conditions and $Q_j^{(2)}$, $\dot{q}_j^{p(2)}$ are the stresses and plastic strains in the analysis with assumed initial conditions, the stress fields will converge in the sense that $\dot{A}$ (equations 8 and 9) will be negative. Thus the adoption of erroneous initial stresses in an analysis is mitigated whenever (15) holds. *Complete* convergence cannot be assumed, since the circumstances under which equation (12) holds may pertain; however, the designer is assured that erroneous initial conditions cannot lead to a *divergence* of the stresses determined in the analysis and those found in the actual structure. If the loading is such that flow takes place ($\Psi\{\Gamma_\alpha\} = 0$) the initial conditions do not affect the limit surface and the possible differences in the stress field are reduced to the extent to which the stress field during flow may be non-unique in any particular case.

16.2 Superposition of Elastic and Residual Stress Distributions

A convenient device by means of which the discussion of convergence can be carried out is that of expressing the stress field $Q_j(s,t)$ as the sum of two parts. Suppose that the problem is solved on the assumption that the material behaves *elastically* at every point, so that the constitutive equation is simply

$$q_j = C_{jk} Q_k \tag{16}$$

rather than the constitutive equations given in equation (2). The resulting stress field, which we will denote by $Q_j^E(s,t)$, is both path and time independent and consequently depends on the instantaneous value of the applied loads $\Gamma_\alpha(t)$. The strains associated with this stress field

$$q_j^E = C_{jk} Q_k^E \tag{17}$$

are of course kinematically admissible.

Let the stresses in the elastic, perfectly plastic structure $Q_j(s,t)$ be written as

$$Q_j(s,t) = Q_j^E(s,t) + \rho_j(s,t) \ . \tag{18}$$

Because the equilibrium equations are linear, and because $Q_j(s,t)$ and $Q_j^E(s,t)$ are each statically admissible and in equilibrium with the same external loads $\Gamma_\alpha(t)$, it follows that $\rho_j(s,t)$ is statically admissible and in equilibrium with *zero generalized loads*. ρ_j will be termed a *residual stress field* or

a *self-equilibrating stress field.* It can also be thought of as
the stress field which would be present in the structure if the
loads $\Gamma_\alpha(t)$ were removed *provided that no further plastic deformation took place as the structure is unloaded.* The residual
stress field is of course a function of the path of loading in
load space.

In any particular problem the initial stress field is given
by

$$Q_j(s,0) \;=\; Q_j^E(s,0) + \rho_j(s,0) \;\; . \tag{19}$$

Since $Q_j^E(s,0)$ is a function only of $\Gamma_\alpha(0)$ and can always be computed, any effects of previous history of deformation are contained in the initial residual stress field. Furthermore, if
we consider two problems identical except for their initial residual stress fields $\rho_j^{(1)}(s,0)$ and $\rho_j^{(2)}(s,0)$, we see that on
substituting equation (19) into equation (8),

$$\begin{aligned}
A(Q_j^{(1)},Q_j^{(2)}) &= \int_V C_{jk}(Q_j^{(1)}-Q_j^{(2)})(Q_k^{(1)}-Q_k^{(2)})dV \\[2mm]
&= \int_V C_{jk}(\rho_j^{(1)}-\rho_j^{(2)})(\rho_k^{(1)}-\rho_k^{(2)})dV \\[2mm]
&= A(\rho_j^{(1)}, \rho_j^{(2)}) \;\; .
\end{aligned} \tag{20}$$

We still have, from equation (19),

$$\dot{A} \;\leq\; 0 \;\; . \tag{21}$$

Hence whenever plastic deformation takes place such that (15)
holds at any point in the body, the residual stress fields converge in the sense that $A(\rho_j^{(1)},\rho_j^{(2)})$ decreases. In addition,
following the uniqueness argument given above, if $\rho_j(s,0)$ is
specified, $\rho_j(s,t)$ is uniquely determined for a given program of
loading $\Gamma_\alpha(t)$. It can thus be considered that $\rho_j(s,t)$ stores
the effect of the previous history of deformation in the structure .

As such, $\rho_j(s,t)$ is intimately connected with the plastic strain distribution $q_j^p(s,t)$. In the next section we shall consider this relation in greater detail.

16.3 Elastic Solutions for Given Plastic Strains

It is convenient in this section to return to the general formulation of the structural problem. Thus we consider a body subjected to loads $\hat{\underline{p}}(s)$ on part of the surface S_p, body forces $\hat{\underline{F}}(s)$ on V and displacements $\hat{\underline{u}}(s)$ on the remainder of the surface S_u, whereas in the previous section we dealt with generalized loads Γ_α derived from the loading on S_p and the body forces with the implicit assumption that the displacements on S_u are zero. The loading given is the instantaneous value at some time in the loading program; the previous history is not specified.

We shall make use of the elastic solution of this problem (i.e. the solution obtained when the material is assumed to behave elastically for all stress states). The elastic solution will be characterized by stresses $Q_j^E(s)$, strains $q_j^E(s)$, displacements $\underline{u}^E(s)$ and reactions $\underline{p}^E(s)$ on S_u. Clearly

$$q_j^E = C_{jk} Q_k^E \ , \tag{22}$$

and the sets $\hat{\underline{p}}$ on S_p, $\underline{p}^E$ on S_u, $\hat{\underline{F}}$ on V, Q_j^E and $\underline{u}^E$ on V and S_p, $\hat{\underline{u}}$ on S_u, q_j^E are respectively statically and kinematically admissible.

The problem we wish to consider is the following: in addition to the loading described above, we wish to impose on the structure *inelastic strains* $q_j^p(s)$. *Assuming that no further plastic deformation occurs* as $q_j^p(s)$ is imposed, we wish to determine the stresses $Q_j(s)$, strains $q_j(s)$ and displacements $\underline{u}(s)$ which characterize the new solution. Note that this is an *elastic* problem, since we continue to suspend the yield condition and the plastic part of the constitutive relations.

Because we are dealing with a linear elastic material, we can approach the solution in two ways. We could find the solution directly, in which case the stresses $Q_j(s)$ must be in equilibrium with forces $\hat{\underline{p}}(s)$ on S_p and body forces $\hat{\underline{F}}(s)$ on V. The strains $q_j(s)$ which are the result of elastic deformation,

$$q_j = C_{jk}Q_k \quad , \tag{23}$$

must be such that the *total* strains $(q_j+q_j^p)$ are kinematically admissible, i.e. $(q_j+q_j^p)$ may be derived from the displacement field $\underline{u}(s)$ which satisfies the condition $\underline{u} = \hat{\underline{u}}$ on S_u.

Alternatively, we can make use of the principle of superposition. The solution for the loading $\hat{\underline{p}}$ on S_p, $\hat{F}$ on V and $\hat{\underline{u}}$ on S_u has already been obtained; it is Q_j^E, q_j^E, $\underline{u}^E$, $\underline{p}^E$. We seek a solution to the second problem in which the loads on S_p, the body forces and the displacements on S_u are all zero. The in-elastic strains $q_j^p(s)$ are given. The solution to this problem may be characterized by stresses $\rho_j(s)$, strains $q_j^r(s)$ and dis-placements $\underline{u}^r(s)$. The stress field $\rho_j(s)$ is in equilibrium with zero forces on S_p and zero body forces (or zero generalized loads), and consequently it is a *self-equilibrating or residual stress field*. The strains $q_j^r(s)$ are the elastic strains asso-ciated with the residual stresses,

$$q_j^r(s) = C_{jk}\rho_k(s) \quad , \tag{24}$$

and the total strains $(q_j^r+q_j^p)$ are kinematically admissible and may be derived from displacements $\underline{u}^r$ $(\underline{u}^r = 0$ on $S_u)$.

Now consider the superposed solutions. The superposed stress and strain fields are statically and kinematically admissible respectively, because of the linearity of the equilibrium and kinematic equations. In addition, the boundary conditions on S_p and S_u are satisfied. Because of the linearity of the stress

strain relation, the superposed stresses and strains satisfy the
constitutive relations. Thus we may put

$$Q_j(s) = Q_j^E(s) + \rho_j(s) \ ,$$

$$q_j(s) = q_j^E(s) + q_j^r(s) \ , \tag{25}$$

$$\underline{u}(s) = \underline{u}^E(s) + \underline{u}^r(s) \ .$$

relating the direct solution and that found by superposition.

The uniqueness of the elastic solution for the given loading
has already been established as part of the consideration of the
plastic case (Chapter 3). We shall concern ourselves here only
with the question of whether the solution to the second of the
two elastic sub-problems is unique. The contrary is assumed,
so that we consider two solutions $\rho_j^{(1)}$, $q_j^{r(1)}$, $\underline{u}^{r(1)}$ and $\rho_j^{(2)}$,
$q_j^{r(2)}$, $\underline{u}^{r(2)}$ for given q_j^p, with $\rho_j^{(1)}$, $\rho_j^{(2)}$ both residual stress
fields in order to satisfy the boundary conditions on S_p, and
$\underline{u}^{r(1)} = \underline{u}^{r(2)} = 0$ on S_u. Using the linearity of the equilibrium
and the strain-displacement relations, the stress field $(\rho_j^{(1)} -$
$\rho_j^{(2)})$ is statically admissible (and in equilibrium with external
forces which are non-zero only on S_u), and the strain field
$\{(q_j^{r(1)}+q_j^p) - (q_j^{r(2)}+q_j^p)\}$ is kinematically admissible (and associated with displacements which are zero on S_u). Hence, by the
principle of virtual work,

$$\int_V (\rho_j^{(1)}-\rho_j^{(2)})(q_j^{r(1)}-q_j^{r(2)})dV = 0 \ . \tag{26}$$

Further,

$$q_j^{r(1)}-q_j^{r(2)} = C_{jk}\rho_k^{(1)} - C_{jk}\rho_k^{(2)}$$

$$= C_{jk}(\rho_k^{(1)} - \rho_k^{(2)}) \ , \tag{27}$$

so that equation (26) becomes

$$\int_V C_{jk}(\rho_j^{(1)}-\rho_j^{(2)})(\rho_k^{(1)}-\rho_k^{(2)})dV = 0 \; . \tag{28}$$

The matrix C_{jk} is positive definite, and thus the integrand of
this expression is a non-negative quadratic form. The integral
can be zero only if the integrand vanishes for all s, i.e. if

$$\rho_j^{(1)}(s) = \rho_j^{(2)}(s) \; . \tag{29}$$

Hence there cannot exist two distinct solutions to the problem,
and the uniqueness of the solution is ensured.

<u>16.4 Minimum Principles for the Stresses and Elastic Strains</u>

Two minimum principles can be established for this problem.
These may be regarded as generalizations of the classical minimum complementary and minimum potential energy theorems to the
case of imposed inelastic strains. Each of these minimum principles can be framed in terms of the complete problem (external
loads and inelastic strains imposed together) or in terms of
the second sub-problem (inelastic strains only). We shall state
and prove the minimum principles for the complete problem, and
then reduce them to the more restricted problem.

The minimum principle for the stresses is framed in terms of
the class of pertinent statically admissible stress fields
$Q_j^*(s)$. $Q_j^*(s)$ is required to equilibrate the loads $\hat{p}(s)$ on S_p
and $\underline{\hat{F}}$ on V. It is evident that the class Q_j^* contains the particular stress distribution Q_j which characterizes the solution
of the problem.

*The minimum principle for the stresses then states that the
expression*

$$\int_V \Omega(Q_j^*)dV + \int_V Q_j^* q_j^p \, dV - \int_{S_u} \underline{p}^* \cdot \underline{\hat{u}}dS \tag{30}$$

attains an absolute minimum for $Q_j^* = Q_j$.

In this expression $\Omega(Q_j^*)$ is the complementary energy density associated with the stresses Q_j^*, and for a linear elastic material is given by

$$\Omega(Q_j^*) = \int_0^{Q_j^*} q_j^e dQ_j = \frac{1}{2} C_{jk} Q_j^* Q_k^* \quad . \tag{31}$$

The reactions or forces on S_u, $\underline{p}^*(s,t)$, are associated with the pertinent statically admissible stress field $Q_j^*(s,t)$.

The minimum principle for elastic strains is phrased in terms of a class of strain fields q_j^*, defined in such a way that $(q_j^*+q_j^p)$, $\underline{u}^*$ constitute a pertinent kinematically admissible field, with $\underline{u}^* = \underline{\hat{u}}$ on S_u. It is evident that the class q_j^*, $\underline{u}^*$ contains the particular strain and displacement field q_j, $\underline{u}$ which characterizes the solution of the problem.

The minimum principle for the elastic strains states that the expression

$$\int_V W(q_j^*)dV - \int_{S_p} \underline{\hat{p}} \cdot \underline{u}^* dS - \int_V \underline{\hat{F}} \cdot \underline{u}^* dS \tag{32}$$

attains an absolute minimum value when $q_j^* = q_j$, $\underline{u}^* = \underline{u}$.

In this expression $W(q_j^*)$ is the strain energy density given by

$$W(q_j^*) = \int_0^{q_j^*} Q_j dq_j^e = \frac{1}{2} D_{jk} q_j^* q_k^* \quad , \tag{33}$$

where $D_{jk} = C_{jk}^{-1}$ is the inverse of the matrix of elastic coefficients which appears in equation (22).

The proofs for both minimum principles parallel those of the classical elastic energy theorems (Section 1.6), and may be obtained from inequalities expressing the convexity of the complementary and strain energy. Consider two arbitrary stress states Q_j^a, Q_j^b and their associated elastic strains q_j^{ea}, q_j^{eb}.

The convexity of the complementary energy $\Omega(Q_j)$ permits us to write (Chapter 1, equation 78),

$$\Omega(Q_j^a) - \Omega(Q_j^b) \geq (Q_j^a - Q_j^b)q_j^{eb} \quad . \tag{34}$$

For our purposes, this inequality is more conveniently expressed as

$$\Omega(Q_j^a) - Q_j^a q_j^{eb} \geq \Omega(Q_j^b) - Q_j^b q_j^{eb} \quad . \tag{35}$$

Similarly, the convexity of the elastic strain energy function leads to (Chapter 1, equation 74),

$$W(q_j^{ea}) - W(q_j^{eb}) \geq (q_j^{ea} - q_j^{eb})Q_j^b \quad . \tag{36}$$

This inequality may be rearranged to give

$$W(q_j^{ea}) - Q_j^b q_j^{ea} \geq W(q_j^{eb}) - Q_j^b q_j^{eb} \quad . \tag{37}$$

Inequality (35) holds for any pair of stresses Q_j^a, Q_j^b so that in order to establish the minimum principle for stresses stated above we may put $Q_j^a = Q_j^*$ and $Q_j^b = Q_j$ at each point in the body. The inequality may then be integrated over the whole body, giving

$$\int_V \Omega(Q_j^*)\,dV - \int_V Q_j^* q_j\,dV \geq \int_V \Omega(Q_j)\,dV - \int_V Q_j q_j\,dV \quad . \tag{38}$$

Note that the elastic strains $q_j = C_{jk}Q_k$ are such that $q_j + q_j^p$, $\underline{u}$ are kinematically admissible and $\underline{u} = \hat{\underline{u}}$ on S_u. Both Q_j^*, Q_j are pertinent statically admissible fields, so that by the principle of virtual work

$$\int_{S_p} \hat{\underline{p}}\cdot\underline{u}\,dS + \int_{S_u} \underline{p}^*\cdot\hat{\underline{u}}\,dS + \int_V \hat{\underline{F}}\cdot\underline{u}\,dV = \int_V Q_j^*(q_j + q_j^p)\,dV \quad , \tag{39a}$$

$$\int_{S_p} \hat{\underline{p}}\cdot\underline{u}\,dS + \int_{S_u} \underline{p}\cdot\hat{\underline{u}}\,dS + \int_V \hat{\underline{F}}\cdot\underline{u}\,dV = \int_V Q_j(q_j + q_j^p)\,dV \quad . \tag{39b}$$

Finally, we substitute (39a) and (39b) into inequality (38) and eliminate terms which appear on both sides; we find that

$$\int_V \Omega(Q_j^*)dV + \int_V Q_j^* q_j^P dV - \int_{S_u} \underline{p}^* \cdot \hat{\underline{u}} dS$$

$$\geq \int_V \Omega(Q_j)dV + \int_V Q_j q_j^P dV - \int_{S_u} \underline{p} \cdot \hat{\underline{u}} dS \quad , \tag{40}$$

and the minimum principle is established.

In order to establish the minimum principles for elastic strains, we use inequality (16) and put $q_j^{ea} = q_j^*$, $q_j^{eb} = q_j$. $(q_j^* + q_j^P)$, $\underline{u}^*$ and $(q_j + q_j^P)$, $\underline{u}$ are both kinematically admissible, and $\underline{u}^* = \underline{u} = \hat{\underline{u}}$ on S_u. Hence by the principle of virtual work

$$\int_{S_p} \hat{\underline{p}} \cdot \underline{u}^* dS + \int_{S_u} \underline{p} \cdot \hat{\underline{u}} dS + \int_V \hat{\underline{F}} \cdot \underline{u}^* dV = \int_V Q_j(q_j^* + q_j^P)dV \quad , \tag{41a}$$

$$\int_{S_p} \hat{\underline{p}} \cdot \underline{u} dS + \int_{S_u} \underline{p} \cdot \hat{\underline{u}} dS + \int_V \hat{\underline{F}} \cdot \underline{u} dV \doteq \int_V Q_j(q_j + q_j^P)dV \quad . \tag{41b}$$

Further, inequality (37) may be integrated over the volume to give

$$\int_V W(q_j^*)dV - \int_V Q_j q_j^* dV \geq \int_V W(q_j)dV - \int_V Q_j q_j dV \quad . \tag{42}$$

Substituting into (42) from equations (41a) and (41b), we find finally that

$$\int_V W(q_j^*)dV - \int_{S_p} \hat{\underline{p}} \cdot \underline{u}^* dS - \int_V \hat{\underline{F}} \cdot \underline{u}^* dV$$

$$\geq \int_V W(q_j)dV - \int_{S_p} \hat{\underline{p}} \cdot \underline{u} dS - \int_V \hat{\underline{F}} \cdot \underline{u} dV \quad , \tag{43}$$

and the minimum principle for elastic strains is established.

In the more restricted form of the problem where the generalized loads and displacements are given by Γ_α, ξ_α, with $\hat{\underline{u}}(s) = 0$

on S_u, the expressions (30) and (32) which are minimized by the correct solution take respectively the forms

$$\int_V \Omega(Q_j^*)dV + \int_V Q_j^* q_j^p dV \tag{44}$$

and

$$\int_V W(q_j^*)dV - \Gamma_\alpha \xi_\alpha^* \quad . \tag{45}$$

Following Chapter 3, equations (28), ξ_α^* are defined by the expressions

$$\xi_\alpha^* = \int_S \underline{\pi}_{-\alpha} \cdot \underline{u}^* dS \quad \alpha=1,\ldots,\mu \quad ,$$

$$\xi_{\mu+1}^* = \int_V \underline{F} \cdot \underline{u}^* dV \quad . \tag{46}$$

16.5 Minimum Principles for Residual Stresses and Residual Elastic Strains

The minimum principles can be written for the sub-problem involving zero external forces and imposed inelastic strains q_j^p either by a direct process similar to that already carried out, or by reducing the expressions already obtained. In following the second course, we note that Q_j^* may be written without loss in generality as

$$Q_j^* = Q_j^E + \rho_j^* \quad , \tag{47}$$

where ρ_j^* is any residual stress distribution. Substituting (47) into (30), we now consider the expression

$$\int_V \Omega(Q_j^E + \rho_j^*)dV + \int_V (Q_j^E + \rho_j^*)q_j^p dV - \int_{S_u} \underline{p}^* \cdot \underline{\hat{u}} dS \quad . \tag{48}$$

To eliminate the last term in this expression, we make use of the principle of virtual work with the statically admissible field $\underline{\hat{p}}$ on S_p, $\underline{\hat{F}}$ on V, $\underline{p}^*$ on S_u, Q_j^* and the kinematically admissible field q_j^E, $\underline{u}^E$ on S_p and V, $\underline{\hat{u}}$ on S_u:

$$\int\limits_{S_p} \hat{\underline{p}}.u^E dS + \int\limits_{S_u} \underline{p}^*.\hat{\underline{u}} dS + \int\limits_{V} \hat{\underline{F}}.\underline{u}^E dV = \int\limits_{V} Q_j^* q_j^E dV \tag{49}$$

$$= \int\limits_{V} Q_j^E q_j^E dV + \int\limits_{V} \rho_j^* q_j^E dV \ .$$

The first term in (48) may be written as

$$\int\limits_{V} \Omega(Q_j^E + \rho_j^*) dV = \frac{1}{2} \int\limits_{V} C_{jk} Q_j^E Q_k^E dV + \int\limits_{V} C_{jk} Q_j^E \rho_k^* dV + \int\limits_{V} \frac{1}{2} C_{jk} \rho_j^* \rho_k^* dV \tag{50}$$

$$= \frac{1}{2} \int\limits_{V} Q_j^E q_j^E dV + \int\limits_{V} \rho_j^* q_j^E dV + \int\limits_{V} \Omega(\rho_j^*) dV$$

If we substitute equations (49) and (50) into (48), and eliminate all terms which do *not* depend on ρ_j^*, we can restate the minimum principle for the stresses in the following terms:

The expression

$$\int\limits_{V} \Omega(\rho_j^*) dV + \int\limits_{V} \rho_j^* q_j^P dV \tag{51}$$

attains an absolute minimum when $\rho_j^* = \rho_j$.

As before, ρ_j is the residual stress field given by the solution to the problem.

In a similar way, without loss in generality we may put

$$q_j^* = q_j^E + q_j^{r*} \ , \tag{52}$$

where q_j^{r*} is a strain field such that $(q_j^{r*} + q_j^P)$, $\underline{u}^{r*}$ ($\underline{u}^{r*} = 0$ on S_u) is a pertinent kinematically admissible field. Expression (32) then becomes

$$\int\limits_{V} W(q_j^E + q_j^{r*}) dV - \int\limits_{S_p} \hat{\underline{p}}.(\underline{u}^E + \underline{u}^{r*}) dS - \int\limits_{V} \hat{\underline{F}}.(\underline{u}^E + \underline{u}^{r*}) dV \ . \tag{53}$$

We make use of the principle of virtual work, with the statically admissible set $\hat{\underline{p}}$ on S_p, $\hat{\underline{F}}$ on V, $\underline{p}^E$ on S_u, Q_j^E and the kinematically admissible set $\underline{u}^*$ on S_p and V, $\hat{\underline{u}}$ on S_u, q_j^*;

$$\int_{S_p} \hat{\underline{p}} \cdot \underline{u}^* dS + \int_{S_u} \underline{p}^E \cdot \hat{\underline{u}} dS + \int_V \hat{\underline{F}} \cdot \underline{u}^* dV = \int_V Q_j^E q_j^* dV \tag{54}$$

$$= \int_V Q_j^E q_j^E dV + \int_V Q_j^E q_j^{r*} dV \quad .$$

The first term in (53) may be written as

$$\int_V W(q_j^E + q_j^{r*}) dV = \frac{1}{2} \int_V Q_j^E q_j^E dV + \int_V Q_j^E q_j^{r*} dV + \int_V W(q_j^{r*}) dV \quad . \tag{55}$$

Again, substituting equations (54) and (55) into (53), and eliminating all terms which are independent of q_j^{r*}, we may restate the minimum principle for elastic strains as follows:

The expression

$$\int_V W(q_j^{r*}) dV \tag{56}$$

attains an absolute minimum when $q_j^{r*} = q_j^r$.

As before, q_j^r is the elastic strain field associated with the residual stress field given by the solution.

16.6 Conclusions

There are a number of important comments which can be made regarding the uniqueness proof and the minimum principles for the elastic problem when inelastic strains are imposed on the structure. First, it should be appreciated that while the inelastic strains have been denoted by plastic strains q_j^p, their origin plays no part in the determination of the stress, residual stress and elastic strain fields. They can equally well be inelastic strains caused by thermal effects, viscous flow or any other source of inelastic deformation. The minimum principles consequently apply to a much wider class of problems than those of plasticity. Further, q_j^p may of course be caused by plastic strains in a hardening material. We have no assurance that if we impose an arbitrary $q_j^p(s)$ the resulting stresses $Q_j = Q_j^E + \rho_j$ will be such that

$$\phi(Q_j) \leq 0 \ . \tag{57}$$

However, if the plastic strains q_j^p are the *result of a program of loading* on an elastic-plastic structure we are assured that equation (57) will be satisfied and, further, that the stress and elastic strain distribution obtained from an incremental analysis which leads to $q_j^p(s)$ at an instant of time will be identical to the stress and elastic strain distribution obtained by solving the elastic problem with $q_j^p(s)$ and the *instantaneous* external loads given. This distinction between an arbitrary distribution of inelastic strain and a distribution of plastic strain resulting from a loading program must be borne in mind when the results obtained in this section are applied to elastic -plastic or elastic, perfectly plastic materials.

Secondly, it should be noted that the results presented in this section apply equally well in cases where *increments in plastic strain* dq_j^p are given, together with increments in the external loading. In such situations we are concerned with increments in the stress and residual stress field and in the elastic strain field. If these increments occur in time dt we can refer to *rates* rather than increments, in the sense introduced in Chapter 9.

Thus, for given plastic strain rates $\dot{q}_j^p(s)$ and given $\dot{\underline{p}}(s)$ on S_p, $\dot{\hat{\underline{u}}}(s)$ on S_u, $\dot{\hat{\underline{F}}}(s)$ on V we are assured that the stress rates $\dot{Q}_j^p$ and the residual stress rates $\dot{\rho}_j$ connected by the relation

$$\dot{Q}_j(s) = \dot{Q}_j^E(s) + \dot{\rho}_j(s) \tag{58}$$

are unique.

If $\dot{Q}_j^*(s)$ is any pertinent statically admissible stress rate field, the expression

$$\int_V \Omega(\dot{Q}_j^*)dV + \int_V \dot{Q}_j^* \dot{q}_j^P dV - \int_{S_u} \underline{\dot{p}}^* \cdot \underline{\hat{u}} dV \tag{59}$$

attains an absolute minimum when $\dot{Q}_j^* = \dot{Q}_j$.

If $(\dot{q}_j^* + \dot{q}_j^P), \underline{u}^*$ is any pertinent kinematically admissible field, the expression

$$\int_V W(\dot{q}_j^*)dV - \int_{S_p} \underline{\hat{\dot{p}}} \cdot \underline{\dot{u}}^* dS - \int_V \underline{\hat{\dot{F}}} \cdot \underline{\dot{u}}^* dV \tag{60}$$

attains an absolute minimum when $\dot{q}_j^* = \dot{q}_j$.

Finally, it should be noted that while a given plastic strain or plastic strain rate distribution implies a unique residual stress field or residual stress rate field, the converse is not true. Thus a given residual stress distribution $\rho_j(s)$ does not imply a unique plastic strain distribution $q_j^P(s)$. The most important reason for this can be seen by considering expression (51)

$$\int_V \Omega(\rho_j^*)dV + \int_V \rho_j^* q_j^P dV \quad,$$

which is minimized when $\rho_j^* = \rho_j$. Suppose that $q_j^P = \bar{q}_j^P$, which is distinguished because it is kinematically admissible, i.e. $\bar{q}_j^P$ can be derived from a displacement field $\underline{\bar{u}}$, with $\underline{\bar{u}} = 0$ on S_u. Then

$$\int_V \rho_j^* \bar{q}_j^P dV = 0 \tag{61}$$

by the principle of virtual work, since ρ_j^* is in equilibrium with zero external forces on S_p. Expression (51) then becomes

$$\int_V \Omega(\rho_j^*)dV \quad, \tag{62}$$

which is minimized when $\rho_j^* = 0$ ($\rho_j^* = 0$ being an admissible residual stress distribution).

This consequence can be stated in a number of different ways which it is useful to enumerate.

(a) If $\bar{q}_j^p$ is kinematically admissible in the sense defined above, the plastic strains do not cause stresses in the body.

(b) If $\dot{q}_j^p$ is kinematically admissible, $\dot{\rho}(s) = 0$ and the residual stress field remains constant.

(c) If the plastic strains undergo changes $\Delta q_j^p(s)$ which are kinematically admissible in the time interval $0 < t < T$, the change in the *residual stress field* $[\rho_j(s,T) - \rho_j(s,0)]$ will be zero.

(d) Any residual stress field $\rho_j(s)$ may be associated with an infinite number of plastic strain fields which differ from each other by an inelastic strain field which is kinematically admissible.

Some of these results will be used in the next Chapter when we consider periodic loading.

The principles given in this section find application in a number of problems. Let us briefly review, however, the significance of the results given in this and the previous section for the problem under discussion; that of an elastic-perfectly plastic structure subjected to a loading program $\Gamma_\alpha(t)$.

First, in order to obtain the solution for $t > t_1$ we need to know only the boundary conditions for $t > t_1$ and $\rho_j(t_1), q_j^p(t_1)$. $\rho_j(t_1)$ depends, as we have just seen, on the kinematically inadmissible part of q_j^p. Thus the entire effects of history of loading are stored in the current plastic strain field q_j^p. This is consistent with the observation that the yield surface in load space depends on the stresses in the structure, which may be considered as the sum of Q_j^E (functions of Γ_α) and ρ_j (a function of history). Hence

$$\Phi[\Gamma_\alpha(t),\rho_j(s,t)] \;=\; \Phi[\Gamma_\alpha(t),q_j^p(s,t)] \;\;. \tag{63}$$

Further, $q_j^p(s,t)$ and $\rho_j(s,t)$ can be used interchangeably provided we are concerned with the determination of the stresses; it
is only when displacements are concerned that $\rho_j(s,t)$ is not
sufficient to store all the effects of previous history. In the
next Chapter we shall make use of this conception of the behavior in order to study the response of a structure to a periodic
loading program.

16.7 Historical and Bibliographical Remarks

The minimum principles discussed in this Chapter are due to
Colonetti [1918]. It should be noted again that the inelastic
strains are not necessarily derived from plastic strains, and
that the theorems have wider application than the theory of
plasticity. The principles were further discussed by Colonetti
[1950], [1955], and were independently established by Reissner
[1931] and Melan [1938a], [1938b].

References

G.Colonetti   1918   "Sul problema delle coazioni elas-
                           tiche", Rend.Accad.Lincei, 27, Serie
                           5a, 2 sem.

G.Colonetti   1950   "Elastic equilibrium in the presence
                           of permanent set", Quart.Appl.Math.,
                           7, 353.

G.Colonetti   1955   L'équilibre des corps déformables,
                           Dunod (Paris).

H.Reissrer   1931   "Eigenspannungen und Eigenspannungs-
                           quellen", ZAMM, 11, 1.

E.Melan   1938a   "Der Spannungszustand eines Hencky-
                           Mises'schen Kontinuums bei verander-
                           lichen Belastung", Sitz.Ber.Ak.Wiss.
                           Wien, IIa, 147, 73.

E.Melan   1938b   "Zur Plastizitat des raumlichen Kon-
                           tinuums, Ing.Arch., 9, 116.

Chapter 17

CYCLIC LOADING AND SHAKEDOWN

17.1 Cyclic Loading

In many problems of technological interest structures or bodies
are subjected to repeated programs of loading. For this reason
it is of interest to consider in greater detail the response of
elastic, perfectly-plastic structures which are subjected to
cyclic loading. We shall carry out this discussion in terms of
generalized loads $\Gamma_\alpha(t)$ with the implication that displacements
on S_u are specified to be zero. A cyclic loading program with
period T must be such that

$$\Gamma_\alpha(t) = \Gamma_\alpha(t+T) \tag{1}$$

for all t. Since t is any parameter which increases monotonically with real time, equation (1) is a statement about the order in which the loading program is carried out, and the period
T need not be strictly related to real time.

If we write the stress distribution as the sum of the elastic
solution and a residual stress distribution,

$$Q_j(s,t) = Q_j^E(s,t) + \rho_j(s,t) \quad , \tag{2}$$

as in Chapter 16 (equation 18), we note that $Q_j^E(s,t)$ is a cyclic
function since it depends on the instantaneous value of the external loads. Thus

$$Q_j^E(s,t) = Q_j^E(s,t+T) \quad . \tag{3}$$

Further, because of the cyclic nature of the loads, the stress
distributions $Q_j(s,t)$ and $Q_j(s,t+T)$ are in equilibrium with the
same external loads, and the stress field $\{Q_j(s,t) - Q_j(s,t+T)\}$
is self-equilibrating or in equilibrium with zero external loads.
In consequence we can write, by the principle of virtual velocities,

$$0 = \int_V \{Q_j(t) - Q_j(t+T)\}\{\dot{q}_j(t) - (t+T)\}dV \quad , \tag{4}$$

where $\dot{q}_j(t)$ are the total strain rates. This equation is similar in form to that used as the basis of the convergence statement (Chapter 16, equation 3). By dividing the total strain
rates into elastic and plastic parts and making use of the elastic part of the constitutive relation, we can write equation (4)
in the form (see Chapter 16, equation 7)

$$\frac{d}{dt} \int_V C_{jk}\{Q_j(t) - Q_j(t+T)\}\{Q_k(t) - Q_k(t+T)\}dV$$

$$= -\int_V \{Q_j(t) - Q_j(t+T)\}\{\dot{q}_j^p(t) - \dot{q}_j^p(t+T)\}dV \tag{5}$$

$$\leq 0 \quad .$$

The expression

$$A\{Q_j(t), Q_j(t+T)\} = \int_V C_{jk}\{Q_j(t) - Q_j(t+T)\}\{Q_k(t) - Q_k(t+T)\}dV \tag{6}$$

is positive definite, and equation (5) shows that its rate of
change is never positive. We note also from equations (2) and
(3) that

$$Q_j(t) - Q_j(t+T) = \{Q_j^E(t) - Q_j^E(t+T)\} + \{\rho_j(t) - \rho_j(t+T)\}$$

$$= \rho_j(t) - \rho_j(t+T) \quad , \tag{7}$$

$$\text{and} \quad A\{Q_j(t), Q_j(t+T)\} = A\{\rho_j(t), \rho_j(t+T)\} \quad . \tag{8}$$

As the structure responds to the cyclic load we expect that
in general A will decrease (not necessarily continuously) whenever

$$\{Q_j(t) - Q_j(t+T)\}\{\dot{q}_j^p(t) - \dot{q}_j^p(t+T)\} \neq 0 \tag{9}$$

at any point in the structure. Since A cannot become negative
and hence cannot continue to decrease in every cycle of loads,

we expect that there will be a time, denoted say by the passage of n cycles, such that for $t > nT$

$$\{Q_j(s,t)-Q_j(s,t+T)\}\{\dot{q}_j^p(s,t)-\dot{q}_j^p(s,t+T)\} \;=\; 0 \tag{10}$$

Thus for $t > nT$, $\dot{A} = 0$ and A becomes constant. The circumstances in which $\dot{A} = 0$ have been laid out in Section 16.1 and two of these conditions are of interest in the present context. First, equation (10) may be satisfied because

$$Q_j(s,t) \;=\; Q_j(s,t+T) \quad \text{for} \quad t > nT \;, \tag{11}$$

everywhere in the body. The stress field has thus also become cyclic. Alternatively, equation (10) may be satisfied because

$$\dot{q}_j^p(t) = \dot{q}_j^p(t+T) \;=\; 0 \quad \text{for} \quad t > nT \;. \tag{12}$$

This condition implies that the structure is behaving elastically for $t > nT$ and that $q_j^p(s,t)$ is constant for $t > nT$. Under these circumstances the stress field will also be cyclic, since $\rho_j(s,t)$ is uniquely determined by $q_j^p(t)$, as shown in Section 16.3. If $q_j^p(s,t)$ is constant for $t > nT$, it follows that $\rho_j(s,t)$ is also constant for $t > nT$, and

$$Q_j(s,t) \;=\; Q_j^E(s,t) + \rho_j(s,t) \tag{13}$$

is cyclic because $Q_j^E(s,t)$ is cyclic.

It can be seen from equation (6) that $A = 0$ implies that the stress field becomes cyclic, and the two cases discussed above are sufficient conditions for $A=0$ for $t > nT$. (It does not appear at present that necessary conditions for A eventually to become zero can be given. There remains a possibility that A can reach a steady non-zero value, particularly if the yield surface contains flat regions so that the vectors $\{Q_j(t)-Q_j(t+T)\}$ and $\{\dot{q}_j^p(t)-\dot{q}_j^p(t+T)\}$ may be orthogonal. On the other hand

it can be expected that most structures will exhibit a cyclic
response eventually, with A=0, and that the situations in which
A reaches a constant non-zero value and the response is never
cyclic are very special cases).

We shall continue our discussion on the assumption that the
stress field in a structure subjected to cyclic loading will become cyclic after a sufficiently large number of cycles. As we
further narrow our range of interest it will be seen that no
loss in generality is implied in this assumption.

17.2 Shakedown

The response of the structure for $t > nT$ when it has become cyclic and A=0 can be conveniently divided into two categories.
First, suppose that

$$Q_j(s,t) \;=\; Q_j(s,t+T) \;, \tag{14}$$

but that the plastic strain rates are not zero for all points in
the body for all $t > nT$. We can consider the behavior of the
structure in terms of the response to one cycle of loads, such
that $nT \leq t \leq \overline{n+1}\,T$. At the beginning of the cycle the residual
stress field is $\rho_j(s,t+T)$; the response of the structure over
the loading period is uniquely determined by this residual stress
field. The cyclic nature of the response is made evident in the
observation that $\rho_j(s,t)$ is such that $\rho_j(s,nT) = \rho_j(s,\overline{n+1}\,T)$.
The elastic strain rates are given by

$$\dot{q}^e_j(s,t) \;=\; C_{jk}\dot{Q}^E_k(s,t) + C_{jk}\dot{\rho}_k(s,t) \;. \tag{15}$$

The change in the elastic strains over a cycle is zero, since

$$\Delta q^e_j(s) = q^e_j(s,\overline{n+1}\,T) - q^e_j(s,nT)$$

$$= \int_{nT}^{\overline{n+1}\,T} C_{jk}\dot{Q}^E_k(s,t)dt + \int_{nT}^{\overline{n+1}\,T} C_{jk}\dot{\rho}_k(s,t)dt \tag{16}$$

$$= 0 \;.$$

However, the change in the total strains

$$\Delta q_j(s) = q_j(s, \overline{n+1}\ T) - q_j(s, nT) \tag{17}$$

must be kinematically admissible, since it represents the difference between two kinematically admissible fields. In consequence the *change in the plastic strains* over a cycle, defined by

$$\Delta q_j^P(s) = \int_{nT}^{\overline{n+1}\ T} \dot{q}_j^P(s,t)\, dt \tag{18}$$

must be kinematically admissible when the structure reaches a cyclic state. The changes in the plastic strains can thus be associated with a change in the generalized displacements $\Delta\xi_\alpha$. The changes $\Delta q_j^P(s)$, $\Delta\xi_\alpha$ will recur in each cycle when the response is periodic.

It is convenient further to subdivide the response into two classes. First, consider the case where $\Delta q_j^P(s) = 0$ (remembering that we are considering a category in which $\dot{q}_j^P(s,t)$ is *not* identically zero for all $t > nT$). This condition implies that the *plastic strains* $q_j^P(s,t)$ and the *generalized displacements* $\xi_\alpha(t)$ *are cyclic* as well as the plastic strain rates $\dot{q}_j^P(s,t)$ and the displacements rates $\dot{\xi}_\alpha(t)$. This condition has been denoted as one of *alternating plastic deformation*, since it is most commonly met when points in the structure undergo plastic strains which change sign in the cycle. A tensile specimen subjected to alternating plastic flow in tension and compression will fracture after a small number of cycles, and consequently cyclic loading on a structure which leads eventually to alternating plastic deformation can be expected to lead to failure of the structure through the appearance of localized fractures.

The second subcategory is that when $\Delta q_j^P(s) \neq 0$. In this case the structure undergoes increments in displacement $\Delta\xi_\alpha$ over each cycle. The work done by the external loads in each cycle is positive, for

$$
\overline{\int_{nT}^{n+1\;T}} \Gamma_\alpha \dot{\xi}_\alpha dt = \overline{\int_{nT}^{n+1\;T}} dt \int_V Q_j \dot{q}_j^e dV + \overline{\int_{nT}^{n+1\;T}} dt \int_V Q_j \dot{q}_j^p dV
$$

(19)

$$
= \overline{\int_{nT}^{n+1\;T}} dt \int_V Q_j \dot{q}_j^p dV > 0 \quad .
$$

The elastic term vanishes because the stresses are cyclic, and $Q_j \dot{q}_j^p > 0$. Thus the structure will undergo increasing deformations in each cycle. If the cyclic loading continues indefinitely the structure may fail, or may no longer be able to perform its function, as a result of the unbounded deformations. This type of behavior has been termed *incremental collapse.*

The second major category of interest is that where the plastic strain rates are identically zero after the structure attains a cyclic state, i.e. $\dot{q}_j^p(s,t) = 0$ for $t > nT$. In this case the behavior is elastic for $t > nT$, and the generalized displacements $\xi_\alpha(t)$ are cyclic. The residual stress field $\rho_j(s,t)$ is uniquely determined by the plastic strains $q_j^p(s,t)$ which have taken place in earlier cycles, but since $\dot{q}_j^p(s,t) = 0$ we have $\dot{\rho}_j(s,t) = 0$. The stresses are given by

$$
Q_j(s,t) \;=\; Q_j^E(s,t) + \rho_j^o(s) \quad .
$$

(20)

This type of behavior is termed *shakedown*, and $\rho_j^o(s)$ is termed the *shakedown residual stress field.*

From the point of view of the designer, a structure subjected to cyclic loads *in which shakedown does not occur* represents a situation in which failure is almost certain. The mechanism of failure may be that of alternating plastic flow, incremental collapse or a combination of these mechanisms represented by the cases in which the response of the structure does not become strictly cyclic. In consequence, our attention will be devoted to the problem of determining *whether or not shakedown will occur* for a given structure and a given cyclic loading. In this sense our neglect of the cases where a cyclic response does not necessarily ensue does not limit the discussion; it will be taken that all cases where shakedown does not occur are potentially dangerous, and we shall not enter a detailed study of the mechanical behavior in these cases. In the following section we shall consider some general results which assist in determining whether or not shakedown will occur in a given structure. The remainder of this section will deal with some simple illustrative examples which are intended to assist in understanding the phenomenon of shakedown.

In a general way, shakedown can be understood by considering the behavior of the yield surface in load space. Consider, as shown in Figure 1, a load cycle $\Gamma_\alpha(t)$. This cycle lies within

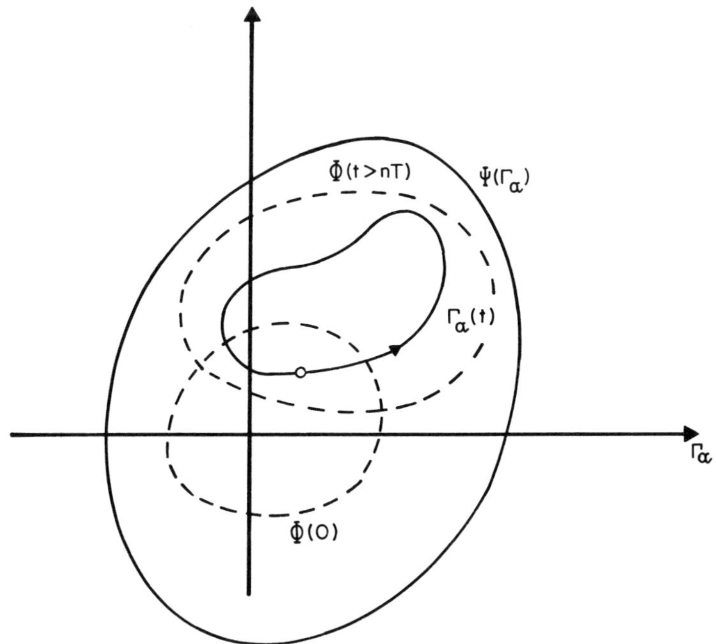

Figure 1. Behavior of yield surface when shakedown occurs

the limit surface $\Psi(\Gamma_\alpha)$. Assuming some initial stress distribution in the structure, generally the virgin state, the initial
yield surface in load space $\Phi(t=0)$ can be plotted. When the
structure is subjected to loading which passes beyond the initial
surface, the yield surface will move, forming subsequent yield
surfaces which change with time. Whether or not shakedown will
eventually occur depends entirely on *whether or not a subsequent
yield surface can be set up in the structure which contains the
load cycle* $\Gamma_\alpha(t)$. If this does occur, the behavior thereafter
will be elastic. If not, the yield surface will continue to
move every time the cycle is traversed, implying continuing
plastic deformation.

The mechanism of failure of a structure due to alternating
plastic deformation was recognized early in the development of
plasticity. Recognition of incremental collapse as a factor to
be taken into account in design, however, proved more elusive.
This is no doubt due to complexity of the calculations which
must be carried out to analyse the response of a realistic structure subjected to cyclic loading. For this same reason we shall
confine ourselves to presenting a very simple problem in order
to illustrate the mechanisms of alternating plastic deformation,
incremental collapse and shakedown.

17.3 Shakedown in a Truss

In the case of a truss composed of pin jointed bars which is
indeterminate to degree one and is subjected to a one parameter
loading system a very convenient model of the behavior of the
structure can be given in the sense that it avoids the necessity
of carrying out a complete analysis. Consider the truss which
was discussed in detail in Chapter 6 and which is reproduced in
Figure 2. The bars are uniform along their length and have
identical cross-sectional properties. If the tensile force in

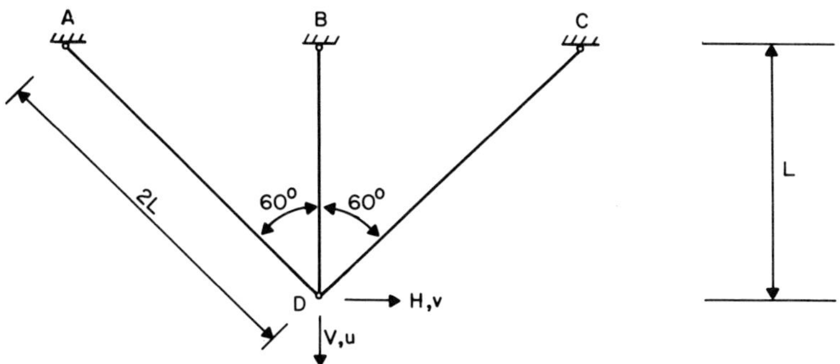

Figure 2. Truss example

the bar is N and the extension per unit length is ε, the constitutive equations are (cf. Chapter 6, equations 23)

$$d\varepsilon^e = \frac{dN}{AE} ,$$

$$d\varepsilon^p = \lambda <N> ,$$

(21)

where $<N> = +1$ if $N = N_o$,

$\qquad <N> = 0$ if $N_o > N > -N_o$,

$\qquad <N> = -1$ if $N = -N_o$,

$\qquad \lambda \geq 0$ but otherwise unspecified.

Bar forces such that $N > N_o$ or $N < -N_o$ are not admitted. The
bars are thus elastic, perfectly plastic with limit values of N
given by N_o and $-N_o$.

For the two parameter loading system V,H shown in Figure 2,
the tensile forces in the bars AD, BD, CD denoted respectively
by N_A, N_B, N_C must satisfy an equilibrium relation at joint D
(Chapter 6, equations 3)

$$N_A + 2N_B + N_C = 2V ,$$

(22a)

$$N_A - N_C = \frac{2H}{\sqrt{3}} .$$

(22b)

The strains in the bars, denoted by $\varepsilon_A, \varepsilon_B, \varepsilon_C$ are related to the

displacements u,v of the joint D by means of the strain displacement relations

$$\varepsilon_A = \frac{1}{2L} \{\frac{u}{2} + \frac{\sqrt{3}}{2} v\} \quad , \tag{23a}$$

$$\varepsilon_B = \frac{u}{L} \quad , \tag{23b}$$

$$\varepsilon_C = \frac{1}{2L} \{\frac{u}{2} - \frac{\sqrt{3}}{2} v\} \quad . \tag{23c}$$

Elimination of u,v from these equations leads to a compatibility
equation,

$$\varepsilon_A + \varepsilon_C = \frac{1}{2} \varepsilon_B \quad . \tag{24}$$

The elastic solution to this problem (neglecting the possibility
of plastic deformation) is obtained by substituting the bar
forces into equation (24) by means of the elastic stress-strain
relation, and solving the simultaneous equations (22a), (22b)
and (24). This gives (Chapter 6, equation 7)

$$N_A^E = \frac{V}{5} + \frac{H}{\sqrt{3}} \quad , \qquad N_B^E = \frac{4V}{5} \quad , \qquad N_C^E = \frac{V}{5} - \frac{H}{\sqrt{3}} \quad . \tag{25}$$

After plastic deformation takes place the bar forces will
change; however, as in Chapter 16, they can always be written
in terms of the sum of the elastic solution and a residual
stress field. Thus

$$N_A = N_A^E + N_A^R \quad , \qquad N_B = N_B^E + N_B^R \quad , \qquad N_C = N_C^E + N_C^R \quad . \tag{26}$$

This particular structure is *indeterminate to degree one;* this
implies that there is only one parameter set of residual bar
forces N_A^R, N_B^R, N_C^R which will satisfy the equilibrium equations
(22) with the external loads V,H set equal to zero. These forces
are

$$N_A^R = -a \quad , \quad N_B^R = +a \quad , \quad N_C^R = -a \quad . \tag{27}$$

Thus, combining equations (25), (26) and (27), the bar forces are given by

$$N_A = \frac{V}{5} + \frac{H}{\sqrt{3}} - a \quad , \tag{28a}$$

$$N_B = \frac{4V}{5} + a \quad , \tag{28b}$$

$$N_C = \frac{V}{5} - \frac{H}{\sqrt{3}} - a \quad . \tag{28c}$$

As demonstrated in Chapter 16, the parameter a which defines the residual force distribution depends only on the plastic strains in the bars. Suppose that the plastic strains are ε_A^p, ε_B^p, ε_C^p. The total strains must satisfy the compatibility equation (24); however the elastic strains associated with the elastic solution (25) already satisfy this condition by definition. Hence a is determined by the condition that the strains composed of the sum of the elastic strains associated with the residual force distribution ($-a/AE$, a/AE, $-a/AE$) and the plastic strains ($\varepsilon_A^p, \varepsilon_B^p, \varepsilon_C^P$) should satisfy equation (25). This leads to

$$a = \frac{2AE}{5} (\varepsilon_A^p - \frac{1}{2} \varepsilon_B^p + \varepsilon_C^p) \quad . \tag{29}$$

When the structure is in its virgin state, $\varepsilon_A^p = \varepsilon_B^p = \varepsilon_C^p = 0$, it is clear that a=0. Further, from the rate form of equation (29),

$$\dot{a} = \frac{2AE}{5} (\dot{\varepsilon}_A^p - \frac{1}{2} \dot{\varepsilon}_B^p + \dot{\varepsilon}_C^p) \quad , \tag{30}$$

it is evident that a=0 or a remains constant when the plastic strains remain constant. Alternatively, a can change only when the plastic strains change, and such changes can occur only when one or more of the bars yields, with $N = \pm N_o$.

Making use of these statements it is possible to determine the way in which a changes with the loads without working out the detailed plastic strain distribution. This is most easily done with a one parameter loading system. As a first example, consider the structure shown in Figure 2 with H=0. We can then plot on a plane diagram (Figure 3) a *state point* for the structure with coordinates (V,a). Clearly if V and a are known, i.e. for a given state point, the bar forces can be determined from equations (28). The domain of achievable state points is limited by the yield condition $-N_o \leq N \leq N_o$. The boundary of this domain is the inner envelope of the six lines

$$N_A = \pm N_o \, , \qquad N_B = \pm N_o \, , \qquad N_C = \pm N_o \, . \qquad (31)$$

Substituting from equation (28), with H=0, these six lines have been plotted on Figure 3. Because of the symmetry introduced

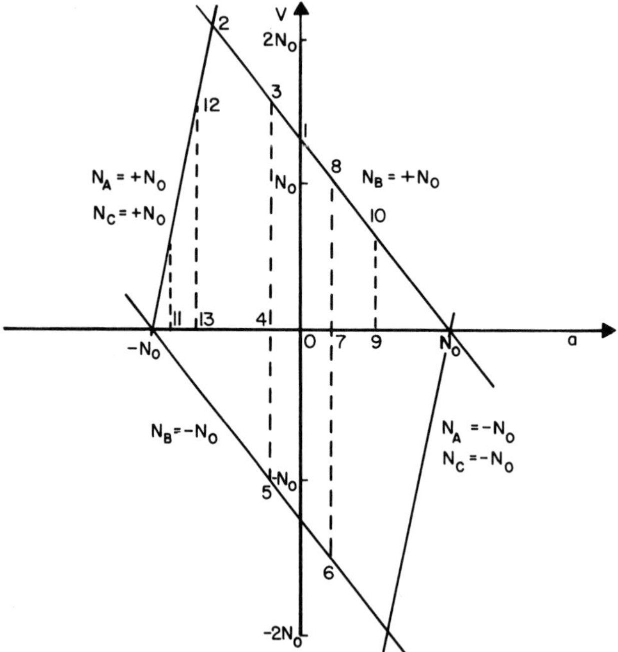

Figure 3. State diagram for H=0

when H=0, the lines $N_A = N_o$, $N_C = N_o$ and $N_A = -N_o$, $N_C = -N_o$ coincide.

Before we consider a cyclic loading program, let us treat the simple case of a monotonic increase in V. Assume the structure is initially in its virgin state, so that a=0 at time t=0. Since the load is also zero, the state point is initially at the origin. V is now increased monotonically; the behavior will be initially elastic, so that a remains zero and the state point moves up the V axis. When the state point reaches point 1, bar BD yields in tension. The state point cannot move outside of the admissible domain, and hence as V is increased further it will move up the line $N_B = +N_o$. The parameter a changes during this process, but this is acceptable since at least one bar is plastic. The state point will continue to move up the line $N_B = +N_o$ until it reaches point 2. Thereafter V cannot be increased further without the state point leaving the admissible domain. From this diagram we can thus determine a at each instant in the loading history. First yield in the structure occurs for $V = 1.25\ N_o$. Thereafter the relation between V and a can be read from the line $N_B = +N_o$. Point 2 represents the limit state, with $V = 2N_o$, $a = -0.6N_o$. The forces in the bars, from equation (28), are $N_A = N_B = N_C = N_o$, and flow takes place in the structure. Any loading program can be dealt with in this way. Figure 3 will give a at each instant in the loading program, and also indicate which bars are yielding at any instant. The only thing it does not provide are the plastic strains ε_A^p, ε_B^p, ε_C^p which are needed to determine the displacements.

Now consider the cyclic loading which consists of the application of the load $V = +1.5N_0$, the removal of this load, the application of the load $V = -1.5N_o$ and the removal of this second load. This process is repeated continuously, and the load cycle is sketched in Figure 4(a). It is clear that these loads

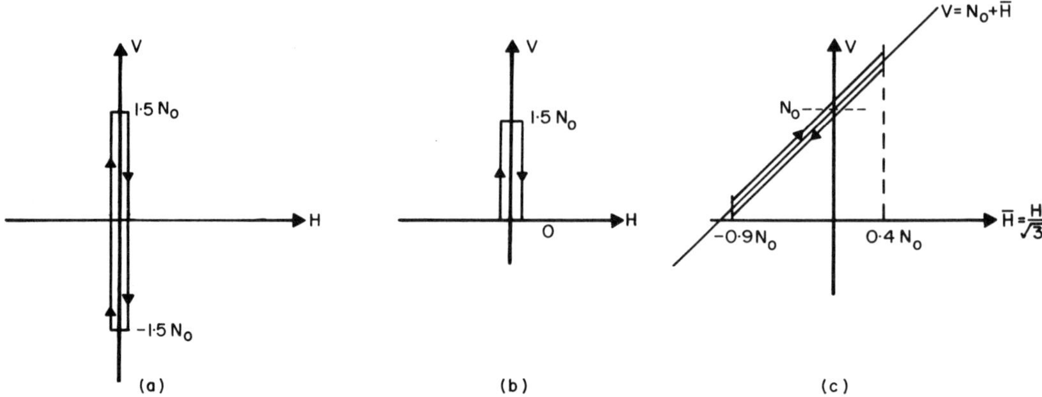

Figure 4. Loading programs

do not pass outside the limit surface in load space. Assuming
that the structure is initially in its virgin state, the application of the load will cause the state point to follow the path
0-1-3, causing a tensile plastic strain in bar BD. The removal
of the load (i.e. a decrease in V) will take place elastically,
with a constant and the state point following the path 3-4. The
imposition of the load $V = -1.5N_o$ will cause the state point to
move along the path 4-5-6, causing compressive plastic strain
in bar BD. The removal of the load, and the completion of the
load cycle, will again take place elastically with the state
point following the path 6-7. The response of the structure to
this first cycle of loads has not been cyclic, since a has different values at the beginning and end of the cycle.

However, if we reapply the cycle of loads the state point
will follow the path 7-8-3-4-5-6-7. Thus the *response* of the
structure has become cyclic after the first cycle, and it will
exhibit the same response for further load cycles. In each cycle after the first bar BD will undergo plastic strain in tension (8-3) and in compression (5-6). The magnitude of the plastic strains in tension and compression will be equal (since a
returns to the same value and bar BD is the only bar in which
plastic strain can take place). Alternating plastic flow will

thus take place, and early fatigue failure can be expected.

To demonstrate the independence of the cyclic solution on the initial conditions, suppose that the initial state point is at point 9 as a consequence of some previous plastic deformation. The response of the structure to the first cycle of loads will cause the state point to follow the path 9-10-3-4-5-6-7. The second cycle of loads will produce the same cyclic response as was obtained when the initial value of a was zero.

Consider as an alternative cycle of loads simply the addition and removal of the load $V = +1.5N_o$, as shown in Figure 4(b). The stress point will follow the path 0-1-3-4 during the first cycle if the structure is in its virgin state. Thereafter, for subsequent cycles, it will follow the path 4-3-4. The parameter a will remain constant, no further plastic strain will take place, and the structure is said to have *shaken down*. Note that the magnitude of a in the shakedown state is dependent on the initial value of a; if, as a result of previous plastic deformation, the initial state point is at 11, the first cycle of loading will take the state point along the path 11-12-13; for subsequent cycles it will follow the path 13-12-13. This type of *dependence of the shakedown residual stress field on the initial residual stress field* will generally occur. We shall show in the next section, however, that if shakedown will take place for one initial stress field it will take place for all initial stress fields. The reader is invited to confirm this for the load cycles of Figure 4(b) and others he may devise.

The structure under discussion will not exhibit incremental collapse when H=0. To illustrate this alternative type of behavior under a one parameter loading system consider the load cycle shown in Figure 4(c). A load $V=N_o$ is first applied monotonically. The loads are then altered in such a way that the load point always lies on the line in load space defined by

$$V = N_o + \frac{H}{\sqrt{3}} . \qquad (32)$$

In order to avoid the radical it is convenient to introduce

$$\bar{H} = \frac{H}{\sqrt{3}} , \qquad (33)$$

so that the loads lie on the line

$$V = N_o + \bar{H} \qquad (34)$$

in the $V\text{-}\bar{H}$ space. Then, as shown in Figure 4(c), the loads are increased ($\dot{\bar{H}}, \dot{V} > 0$) until $\bar{H} = 0.4N_o$. The loads are then decreased ($\dot{\bar{H}}, \dot{V} < 0$), with the load point still confined to the line given in equation (34), until $\bar{H} = -0.9N_o$. The loads are then again increased until $\bar{H}=0$, completing the cycle. Equation (34) permits us to specify both V and $\bar{H}$ in terms of one parameter $\bar{H}$. As in the previous example, we introduce a state point with coordinates $\bar{H}$ and a (Figure 5). The domain of admissible states is obtained by plotting the inner envelope of the six lines

$$N_A = \overset{+}{-}N_o, \qquad N_B = \overset{+}{-}N_o, \qquad N_C = \overset{+}{-}N_o \qquad (35)$$

where N_A, N_B, N_C are obtained by substituting equation (34) into equations (8). The admissible domain is plotted in Figure 5; the lines $N_C = -N_o$ and $N_B = -N_o$ do not contribute to the envelope. We see also from the first example that if the structure is in its virgin state the monotonic addition of the load $V = N_o$ does not cause plastic deformation. Consequently the state point will be at the origin at the beginning of the first cycle for a virgin structure.

Consider now the response of the structure on the diagram in Figure 5. As $\bar{H}$ is increased the behavior will first be elastic, with a=0. At point 1 bar BD yields in tension, and the state

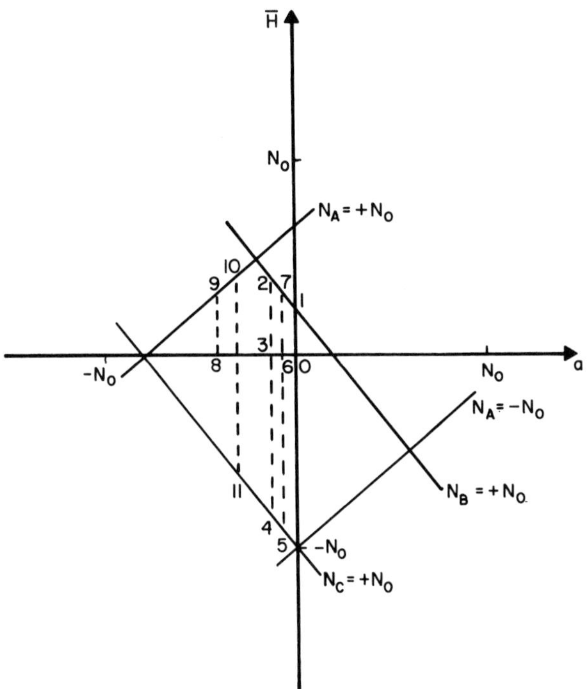

Figure 5. State diagram showing incremental collapse

point follows the line $N_B = \pm N_o$. When $\bar{H} = 0.4N_o$, the state
point will be at 2. $\bar{H}$ is then decreased, with a remaining constant as the state point passes along the line 2-3-4. Bar CD
then yields in tension, with the state point moving along the
line $N_C = +N_o$. When $\bar{H} = -0.9N_o$, the state point will be at 5.
$\bar{H}$ is then again increased, a remains constant, and the first
cycle is completed when the state point is at 6. In the second
cycle of loading the state point will follow the path 6-7-2-3-
4-5-6; the response of the structure is thus cyclic and subsequent cycles will elicit an identical response.

Thus, after the first cycle, the loading produces a tensile
plastic strain $\Delta\varepsilon_B^p$ in bar BD and a tensile plastic strain $\Delta\varepsilon_C^p$
in bar CD. While we cannot determine the magnitude of these
plastic strain increments from our restricted approach, they
must be kinematically admissible (i.e. they must satisfy equation

24) since they do not produce a change over a cycle in the residual stress field. Hence, from equation (24),

$$\Delta\varepsilon^p_C = \frac{1}{2} \Delta\varepsilon^p_B \ . \tag{36}$$

Inverting equations (23), the plastic strain increments over a cycle produce displacement increments Δu, Δv over a cycle given by

$$\Delta u = L\Delta\varepsilon^p_B \ , \qquad \Delta v = \frac{-L}{\sqrt{3}} \Delta\varepsilon^p_B \ . \tag{37}$$

These displacement increments will occur for each cycle of load after the first; as the number of cycles builds up the displacements will also increase. The structure is undergoing incremental collapse.

In order further to illustrate the behavior of the structure the displacements u,v at the end of each cycle have been plotted in Figure 6. In order to compute these displacements a more detailed analysis is required than has so far been given. The additional calculations are straightforward, however, and the details will not be presented. It should be appreciated that the displacements which take place during the cycle (which are not plotted in Figure 6) may be very much larger than the increments per cycle when the number of cycles is small. However, after a large number of cycles the displacement changes during the cycle will become relatively less important than the accumulated displacements caused by incremental collapse.

If the initial state point lies away from the origin as a result of previous plastic deformation, it can be readily seen that the cyclic response is unchanged. Suppose that the state point is initially at 8 (Figure 5); it will follow the path 8-9-10-11-5-6 during the first cycle of loads. Thereafter it will follow the cyclic path 6-7-2-3-4-5-6 which is found when

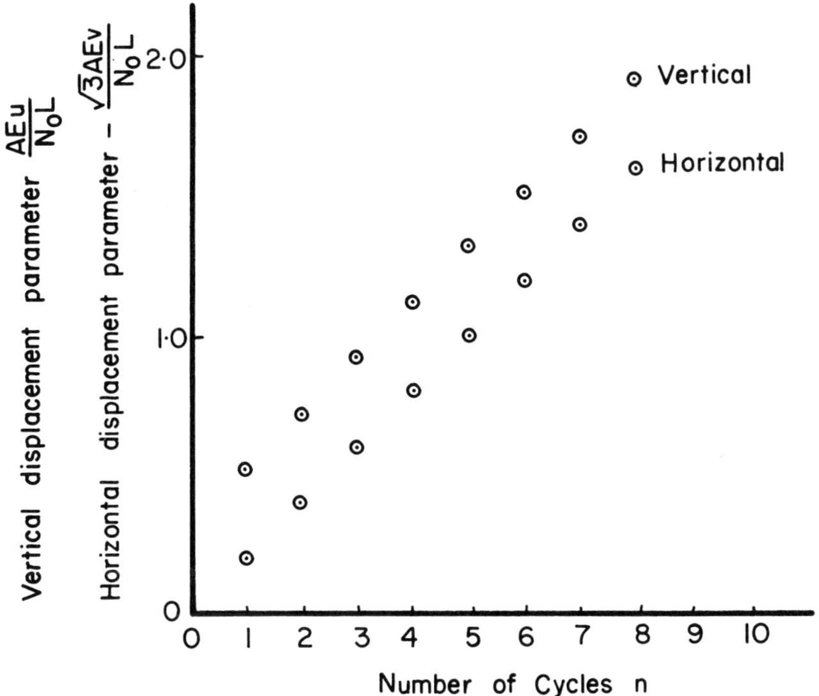

Figure 6. Displacements during incremental collapse

the initial state point is at the origin.

It is simple to devise alternative cycles for which shakedown will occur. If, for example the maximum value of $\bar{H}$ is
$0.4N_o$, as before, and the minimum value of $\bar{H}$ is $-0.8N_o$ shakedown
occurs. The reader is invited to confirm that if shakedown occurs for a given cycle with any one initial state it will occur
for all initial states.

In the next Chapter we shall consider general approaches to
the problem of determining whether or not shakedown will occur
for a given structure under a given cyclic loading.

17.4 Historical and Bibliographical Remarks

While a recognition of the importance of determining the nature
of the response of an elastic-plastic structure to repeated or
cyclic loading occurred fairly early in the development of the
theory of plasticity, much of the terminology is more modern.

The term *shakedown* was introduced by Prager [1948], and the phenomena of *alternating plasticity* and *incremental collapse* were
isolated and named by Horne [1949] and Neal and Symonds [1950]
respectively. Also of interest in the study of shakedown in
simple structures are the experiments of Symonds [1952]. A detailed bibliography follows Chapter 18.

References

M.R.Horne   1949   "Effect of variable repeated loads in
                          the plastic theory of structures",
                          Engineering Structures (Academic
                          Press,N.Y.), 141.

B.G.Neal and   1950   "A method for calculating the failure
P.S.Symonds   load for a framed structure subjected
                          to fluctuating loads", Proc. I.C.E.,
                          35, 186.

W.Prager   1948   "Problem types in the theory of per-
                          fectly plastic materials", J.Aero.
                          Sci., 15, 337.

P.S.Symonds   1952   "Cyclic loading tests on small-scale
                          portal frames", Final Report 4th
                          Congr.Intern.Assoc. Bridge and Struc-
                          tural Eng., Cambridge, 109.

Chapter 18

SHAKEDOWN THEOREMS

18.1 The Static Shakedown Theorem

The shakedown theorems are counterparts (or in fact generalizations) of the limit theorems in the sense that they provide static and kinematic approaches to the question of whether or not shakedown will occur for a given structure under a given cyclic loading, whereas the limit theorems provide static and kinematic approaches to the question of whether or not flow will occur in a given structure under given loads. Although they are not as easy to apply in any given situation as the limit theorems, they do permit a consideration of the problem of shakedown in terms which are *independent of history*. Thus, like the limit theorems they provide a partial solution of the problem (which nevertheless contains important information) at a cost in computational time and effort which is significantly less than that required to complete a full analysis.

The static theorem, which is generally referred to as *Melan's theorem*, can be stated as follows:

If any time-independent distribution of residual stresses $\bar{\rho}_j(s)$ can be found such that

$$\phi\{Q_j^E(s,t) + \bar{\rho}_j(s)\} < 0 \tag{1}$$

the structure will shakedown.

As before, $Q_j^E(s,t)$ is the elastic solution to the given structure for the given cyclic loads. If shakedown does occur for any initial residual stress distribution $\rho_j(s,0)$, the shakedown residual stress distribution will of course become time-independent after a sufficiently large number of cycles. If we denote the shakedown residual stress field by $\rho_j^o(s)$ (Chapter 17, equation 20), we have, by definition, that

$$\phi(Q_j^E + \rho_j^o) \leq 0 \ . \tag{2}$$

The stress field $\rho_j^o(s)$ in equation (2) is not necessarily identical to $\bar{\rho}_j$ in the statement of Melan's theorem and in equation (1). The theorem states that if any $\bar{\rho}_j(s)$ can be found which satisfies equation (1), shakedown will occur with *some* shakedown residual stress field $\rho_j^o(s)$.

An alternative statement of the theorem can consequently be given as follows: if shakedown occurs under one initial residual stress field (leading to a shakedown residual stress field $\bar{\rho}_j$), shakedown will occur for all admissible initial residual stress fields. From the point of view of the designer this is an extremely important result for, as we have repeatedly pointed out, uncertainty always exists regarding the initial conditions in practical situations. A correct assessment of the initial conditions, therefore, is not necessary in determining whether or not shakedown will occur.

The proof of the theorem makes use of the convergence criterion as set out in Chapter 16. If $Q_j^{(1)}(s,t), \dot{q}_j^{(1)}(s,t)$ and $Q_j^{(2)}(s,t), \dot{q}_j^{(2)}(s,t)$ represent solutions to two problems which are identical (i.e. same structure and loading) except that the initial stress fields $Q_j^{(1)}(s,0), Q_j^{(2)}(s,0)$ may be different, we have (Chapter 16, equations 7,8)

$$\frac{dA}{dt} = -\int_V (Q_j^{(1)} - Q_j^{(2)})(\dot{q}_j^{p(1)} - \dot{q}_j^{p(2)}) dV \leq 0 \ , \tag{3}$$

where

$$A = \int_V C_{jk}(Q_j^{(1)} - Q_j^{(2)})(Q_k^{(1)} - Q_k^{(2)}) dV$$

$$= \int_V C_{jk}(\rho_j^{(1)} - \rho_j^{(2)})(\rho_k^{(1)} - \rho_k^{(2)}) dV \geq 0 \ . \tag{4}$$

Let the first solution $Q_j^{(1)}(s,t)$ be identified with given initial conditions $\rho_j(s,0)$ in the structure subjected to cyclic loading. We wish to determine whether or not shakedown will occur in this solution. We assume that there exists a residual stress field $\bar{\rho}_j(s)$ which is time independent and such that

$$\phi(Q_j^E + \bar{\rho}_j) < 0 \quad . \tag{5}$$

Let the second solution be such that the initial residual stress field is given by

$$\rho_j^{(2)}(s,0) = \bar{\rho}_j(s) \quad . \tag{6}$$

Since the yield condition (equation 5) is always satisfied the behavior will be entirely elastic, i.e.

$$\dot{q}_j^{p(2)} = 0 \quad . \tag{7}$$

Substituting these two solutions into equations (3) and (4) we have

$$\frac{dA}{dt} = - \int_V (Q_j^{(1)} - Q_j^{(2)})\dot{q}_j^{p(1)} dV \leq 0 \quad , \tag{8}$$

where

$$A = \int_V C_{jk}[\rho_j(s,t) - \bar{\rho}_j(s)][\rho_k(s,t) - \bar{\rho}_k(s)] dV \quad . \tag{9}$$

Thus A will decrease whenever $\dot{q}_j^{p(1)} \neq 0$. Since A is non-negative it cannot decrease indefinitely; eventually after a sufficiently large number of cycles, $t > nT$

$$\frac{dA}{dt} = - \int_V (Q_j^{(1)} - Q_j^{(2)})\dot{q}_j^{p(1)} dV = 0 \quad , \tag{10}$$

and A becomes constant. Equation (1) can be satisfied only when

$$Q_j^{(1)}(s,t) = Q_j^{(2)}(s,t), \quad \rho_j(s,t) = \bar{\rho}(s) \quad , \quad \text{or} \tag{11a}$$

$$\dot{q}_j^{p(1)}(s,t) = 0 \quad \text{for} \quad t > nT \tag{11b}$$

at each point in the body. In this case we do not take into account the possibility that the vectors $(Q_j^{(1)}-Q_j^{(2)})$ and $\dot{q}_j^{p(1)}$ may be orthogonal on a flat region of the yield surface since $\phi(Q_j^{(2)}) < 0$ and $Q_j^{(2)}$ always lies within the yield surface. If equation (11a) is satisfied at any point in the body $\phi(Q_j^{(1)}) < 0$ and hence no plastic strain changes can take place. Hence both (11a) and (11b) imply that the first solution is behaving elastically for $t > nT$, and shakedown has occurred. Melan's theorem is thus proved.

The converse of Melan's theorem can be established very easily. It can be stated as follows;

If no time-independent distribution of residual stresses $\bar{\rho}_j(s)$ can be found such that

$$\phi(Q_j^E + \bar{\rho}_j) < 0 \ , \tag{12}$$

shakedown will not occur.

This statement is self-evident, since by definition shakedown requires that a time-independent residual stress field $\rho_j^o(s)$ be established in the structure such that $\phi(Q_j^E + \rho_j^o) < 0$ as a consequence (cf. Chapter 16) of the requirement that $\dot{q}_j^p(s,t)$ = 0 for $t > nT$. If no such residual stress fields can be found, clearly shakedown can never occur.

18.2 The Kinematic Shakedown Theorem

The second shakedown theorem, generally referred to as *Koiter's theorem*, is framed in terms of an *admissible plastic strain rate cycle* $\dot{q}_j^{*p}(s,t)$ for $0 < t < T$. The admissible plastic strain rate cycle is thus characterized by plastic strain rates defined over *one cycle of loading*; it is further limited by the requirement that

$$\Delta q_j^{*p}(s) \ = \ \int_o^T \dot{q}_j^{*p}(s,t)dt \tag{13}$$

should constitute a pertinent kinematically admissible strain distribution. We thus require that Δq_j^{*p} can be derived from an incremental displacement field $\Delta\xi_\alpha^*$ which is such that the displacement boundary conditions are satisfied. The energy dissipated in plastic work associated with this admissible plastic strain rate cycle is

$$W_{int}^* = \int_0^T dt \int_V Q_j^*(s,t)\dot{q}_j^{*p}(s,t)dV = \int_0^T dt \int_V D(\dot{q}_j^{*p})dV \quad . \tag{14}$$

The field $Q_j^*(s,t)$ is the stress associated with $\dot{q}_j^{*p}(s,t)$ through the plastic part of the constitutive relation. As noted in the proof of the limit theorems (Chapter 9), Q_j^* is not uniquely determined if the yield surface has flat regions or if $\dot{q}_j^{*p} = 0$; however, $D(\dot{q}_j^{*p}) = Q_j^*\dot{q}_j^{*p}$ is always uniquely determined in elastic, perfectly plastic materials.

In addition, making use of the results of Chapter 16, we associate with the admissible plastic strain rate cycle a pertinent kinematically admissible velocity field $\dot{\xi}_\alpha^*(t)$, $0 \leq t \leq T$. To do so we consider the structure subjected to *zero external loads* and *plastic strains* $\dot{q}_j^{*p}(s,t)$, and *assume that it behaves elastically*. At each instant the solution of this elastic problem will yield a residual stress rate distribution $\dot{\rho}_j^*(s,t)$, which is in equilibrium with zero loads. The condition that the elastic strain rates

$$\dot{q}_j^{*e} = C_{jk}\dot{\rho}_k^* \quad , \tag{15}$$

which together with the imposed admissible plastic strain rate cycle $\dot{q}_j^{*p}$ will provide total strain rates

$$\dot{q}_j^* = \dot{q}_j^{*e} + \dot{q}_j^{*p} \tag{16}$$

that constitute a pertinent kinematically admissible field,will be sufficient to determine $\rho_j^*(s,t)$ uniquely. The total strain

rates can thus be derived from a velocity field $\dot{\xi}^*_\alpha$ which satisfies the displacement boundary conditions.

It may be noted that in view of equation (13) the change in
ρ^*_j over a cycle will be zero;

$$\rho^*_j(s,T) - \rho^*_j(s,0) = \int_0^T \dot{\rho}^*_j(s,t)dt = 0 \ . \tag{17}$$

Consequently the change in the elastic strain field over a cycle
will also be zero.

$$\Delta q^{*e}_j = \int_0^T \dot{q}^{*e}_j(s,t)dt = \int_0^T C_{jk}\dot{\rho}^*_k(s,t)dt = 0 \ . \tag{18}$$

A further consequence of equations (17) and (18) is that the in-
cremental displacement field $\Delta\xi^*_\alpha$ and the velocity field $\dot{\xi}^*_\alpha(t)$
are related by

$$\Delta\xi^*_\alpha = \int_0^T \dot{\xi}^*_\alpha(t)dt \ . \tag{19}$$

The displacement rates $\dot{\xi}^*_\alpha$ can be used to define an expression
which gives the external work done over a cycle by the loads. We
write

$$W^*_{ext} = \int_0^T \Gamma_\alpha(t)\dot{\xi}^*_\alpha(t)dt \ . \tag{20}$$

The statement of Koiter's theorem involves the expressions
defined in equations (14) and (20);

*If shakedown occurs in a given structure under given cyclic
loading,*

$$W^*_{ext} \leq W^*_{int} \tag{21}$$

for all admissible plastic strain rate cycles $\dot{q}^{*p}_j(s,t)$.

If shakedown occurs in a structure, we know from Melan's
theorem that there exists a time independent residual stress
distribution $\bar{\rho}_j(s)$ such that the stresses

$$\bar{Q}_j(s,t) \;=\; Q_j^E(s,t) + \bar{\rho}_j(s) \tag{22}$$

satisfy the condition

$$\phi\{\bar{Q}_j(s,t)\} < 0 \quad . \tag{23}$$

Choose *any* admissible strain rate cycle $\dot{q}_j^{*P}(s,t)$. From (23) and
the fundamental inequality for perfectly plastic materials
(Chapter 9, equation 4)

$$(Q_j^* - \bar{Q}_j)\dot{q}_j^{*P} \geq 0 \tag{24}$$

at every point in the structure for all $0 \leq t \leq T$. As above,
Q_j^* is associated with $\dot{q}_j^{*P}$ through the flow rule, and $\bar{Q}_j$ lies ss
within the limit surface for the material. Inequality (24) may
be integrated over the structure and over the cycle, giving

$$\int_o^T dt \int_V \bar{Q}_j\dot{q}_j^{*P}dV \leq \int_o^T dt \int_V Q_j^*\dot{q}_j^{*P}dV \;=\; W_{int}^* \quad . \tag{25}$$

Making use of equation (16), and noting that Γ_α, $\bar{Q}_j$ are statically admissible and $\dot{q}_j^*$, ξ_α^* are kinematically admissible, the
principle of virtual velocities integrated over the cycle
gives

$$
\begin{aligned}
W_{ext}^* \;&=\; \int_o^T \Gamma_\alpha\dot{\xi}_\alpha^* dt \;=\; \int_o^T dt \int_V \bar{Q}_j\dot{q}_j^* dV \\[2mm]
&=\; \int_o^T dt \int_V \bar{Q}_j\dot{q}_j^{*e} dV + \int_o^T dt \int_V \bar{Q}_j\dot{q}_j^{*P} dV \quad .
\end{aligned}
\tag{26}
$$

The first term on the right hand side of equation (26) can be
shown to be identically zero. First, we use equations (15) and
(22) to write

$$\int_o^T dt \int_V \bar{Q}_j\dot{q}_j^{*e} dV = \int_o^T dt \int_V C_{jk}Q_j^E\dot{\rho}_k^* dV + \int_o^T dt \int_V C_{jk}\bar{\rho}_j\dot{\rho}_k^* dV \quad . \tag{27}$$

The first of these terms may be rewritten (using the reciprocal theorem) as

$$\int_o^T dt \int_V C_{jk} Q_j^E \dot{\rho}_k^* dV = \int_o^T dt \int_V \dot{\rho}_j^* q_j^E dV = 0 \quad . \tag{28}$$

This follows because $\dot{\rho}_j^*$ is a self-equilibrating stress field in equilibrium with zero external load and q_j^E is the strain field obtained from the elastic solution to the problem and is kinematically admissible. The second term in equation (27) may be integrated with respect to time because $\bar{\rho}_j$ is time independent, and using equation (17) we see that

$$\int_o^T dt \int_V C_{jk} \bar{\rho}_j \dot{\rho}_k^* dV = \int_V C_{jk} \bar{\rho}_j \{ \int_o^T \dot{\rho}_k^* dt \} \, dV = 0 \quad . \tag{29}$$

Substituting equations (27), (28) and (29) into (26), it follows that

$$W_{ext}^* = \int_o^T \Gamma_\alpha \dot{\xi}_\alpha^* dt = \int_o^T dt \int_V \bar{Q}_j \dot{q}_j^{*P} dV \quad . \tag{30}$$

Consequently, substituting equation (30) into inequality (25),

$$W_{ext}^* \; \leq \; W_{int}^* \tag{31}$$

for *all* admissible plastic strain rate cycles if the structure shakes down. Koiter's theorem is thus established.

An alternative form of Koiter's theorem can be stated as follows:

If any admissible plastic strain rate cycle $\dot{q}_j^{*P}(s,t)$ *can be found such that*

$$W_{ext}^* \; > \; W_{int}^* \tag{32}$$

shakedown will not occur.

This statement is self-evident, for if it were not true the

first statement of Koiter's theorem would be contradicted.

18.3 Alternative Statements of the Shakedown Theorems

It is evident from the definition of an admissible plastic strain
rate cycle that if a structure undergoes alternating plastic deformation or incremental collapse after achieving a cyclic response to cyclic loads, the actual plastic strain rates over a
cycle must be a member of the class of admissible plastic strain
rate cycles. Further, if $\dot{\xi}_\alpha(t)$ is the cyclic velocity field in
the response, and $\dot{\xi}_\alpha^E(t)$ is the velocity field obtained in the
elastic solution of the problem, $(\dot{\xi}_\alpha - \dot{\xi}_\alpha^E)$ must be a member of the
class $\dot{\xi}_\alpha^*$. The work done by the external loads over a cycle in
the elastic solution will be zero,

$$\int_0^T \Gamma_\alpha \dot{\xi}^E dt \;=\; 0 \; . \tag{33}$$

Thus, when we substitute the actual plastic strain rate cycle
into equations (14) and (20) we find that $W^*_{ext} = W^*_{int}$. A knowledge of a cyclic solution in which continued plastic strain
can occur is, by the convergence result, a certain indication
that shakedown will not occur, but the second statement of
Koiter's theorem would not assure us of this on the basis of the
actual cyclic solution alone.

A similar ambiguity exists in the case of Melan's theorem,
since a shakedown residual stress field ρ_j^o is set up after some
cycles in which plastic deformation occurs will generally be
such that $\phi(Q_j^E + \rho_j) = 0$ at some points in the structure at some
instants during the cycle although no plastic deformation will
occur at these points.

These ambiguities can be traced in part to loading cycles
which lie on the borderline of shakedown and failure by continued plastic deformation. They can be clarified by thinking
in terms of a loading cycle whose amplitude can be varied. Let

$c_\alpha(t)$ be a given set of cyclic loads, and let the structure be subjected to loads

$$\Gamma_\alpha(t) = \Gamma c_\alpha(t) , \tag{34}$$

where Γ is a non-negative constant scalar multiplier. When Γ is very small it can be expected that the structure will behave elastically. As Γ is increased the structure can be expected to deform plastically at first, yet eventually shakedown. If Γ is sufficiently large we may expect that a shakedown limit Γ^S may be exceeded, and shakedown will no longer occur. Γ can not be increased indefinitely, since the loads must remain within the limit surface in load space. We may define a limiting value Γ^L such that

$$\Psi(\Gamma^L c_\alpha) = 0 \quad \text{for some } t, \ 0 < t < T ,$$
$$\Psi(\Gamma^L c_\alpha) < 0 \quad \text{otherwise.} \tag{35}$$

It is clear then that the shakedown limit Γ^S is such that

$$\Gamma^S \leq \Gamma^L . \tag{36}$$

In some structures, including all statically determinate structures, $\Gamma^S = \Gamma^L$, and we gain no further information from shakedown analysis than we obtain from limit analysis.

The concept of a shakedown limit Γ^S is supported by the following result, which may not at first sight be obvious; if the
structure shakes down for a multiplier Γ', it will also shake
down for a multiplier $\Gamma'' < \Gamma'$. To demonstrate this result we
note first that if the elastic solution for the problem with
loads $c_\alpha(t)$ gives a stress field $Q_j^{OE}(s,t)$, by virtue of the
linearity of the elastic solution loads $\Gamma c_\alpha(t)$ will give a
stress field $\Gamma Q_j^{OE}(s,t)$. If shakedown occurs with a multiplier
Γ', there exists a shakedown residual stress field $\bar{\rho}_j(s)$ such
that

$$\phi(\Gamma'Q_j^{OE}+\bar{\rho}_j) \; < \; 0 \tag{37}$$

from Melan's theorem. In diagrammatic terms, the vector $(\Gamma'Q_j^{OE}+\bar{\rho}_j)$ does not penetrate the yield or limit surface at any point in the structure at any instant in the cycle. It is evident then, as shown diagrammatically in Figure 1, that the vector

$$\frac{\Gamma''}{\Gamma'} \; (\Gamma'Q_j^{OE}+\bar{\rho}_j) \; = \; (\Gamma''Q_j^{OE} + \frac{\Gamma''}{\Gamma'} \bar{\rho}_j) \tag{38}$$

does not penetrate the yield surface provided that $\Gamma'' < \Gamma'$, and provided that the yield surface is *convex* and contains the origin. Since this last requirement is satisfied, there exists a time independent residual stress field $\Gamma''\bar{\rho}_j/\Gamma'$ which satisfies Melan's theorem for loads $\Gamma''c_\alpha$, and shakedown will not occur.

The behavior when $\Gamma = \Gamma^S$ is transitional and is not clearly defined. If we approach Γ^S from below, Γ^S is the limiting case for which shakedown will not occur. However, it will not be possible to find a residual stress field $\bar{\rho}_j(s)$ such that $\phi(\Gamma^S Q_j^{OE}+\bar{\rho}_j) < 0$ everywhere during the cycle. It will in general be possible to find a $\bar{\rho}_j$ such that $\phi(\Gamma^S Q_j^{OE}+\bar{\rho}_j) \leq 0$ everywhere; thus yielding will take place but in fact no plastic strain will occur during the cycle. Alternatively, approaching Γ^S from above, the behavior is such that Γ^S is the limiting case for which shakedown will not occur; alternating plastic deformation,

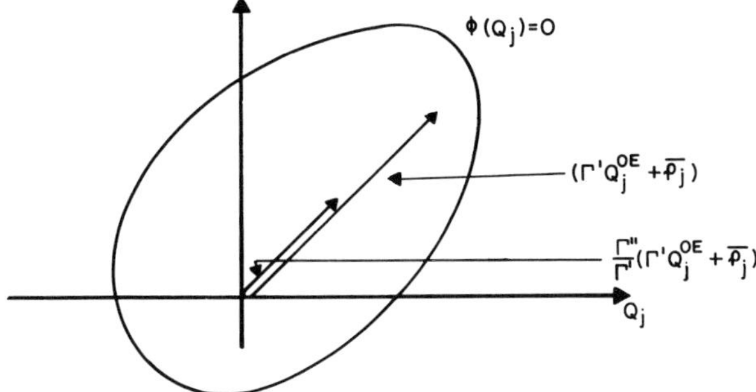

Figure 1. Stress space

incremental collapse or some form of non-cyclic behavior is evident, but the plastic strain which occurs during each cycle will be vanishingly small, and an infinite number of cycles is required to produce noticeable damage.

Understanding that Γ^S is a limit which may be approached from above or below, we can restate Melan's and Koiter's theorems in such a way that bounds on Γ^S are obtained. Melan's theorem can in fact be restated directly;

Γ^S is the largest multiplier Γ for which a time dependent residual stress field $\bar{\rho}_j(s)$ can be found such that

$$\phi(\Gamma Q_j^{oE} + \rho_j) \leq 0 \tag{39}$$

at all points in the body for $0 \leq t \leq T$.

The strict inequality of the original statement of Melan's theorem (equation 1) is replaced by a not greater than sign because we are assured that if Γ is *reduced* by an infinitesimal amount, $\delta\Gamma$, there exists a residual stress field $(\Gamma-\delta\Gamma)\bar{\rho}_j/\Gamma$ for which the strict inequality is satisfied (cf. Figure 1). This implies that shakedown will occur for all $\Gamma < \Gamma^S$ without involving the ambiguous case $\Gamma = \Gamma^S$.

In the case of Koiter's theorem we define Γ by equating W^*_{ext}, given by

$$W^*_{ext} = \int_0^T \Gamma_\alpha \dot{\xi}^*_\alpha dt = \Gamma \int_0^T c_\alpha \dot{\xi}^*_\alpha dt \quad , \tag{40}$$

and W^*_{int} for any admissible strain rate cycle $\dot{q}_j^{*P}(s,t)$. Since c_α and $\dot{q}_j^{*P}$ are independent of Γ,

$$W^*_{ext} = W^*_{int} \tag{41}$$

gives a linear equation for Γ. From the statements of Koiter's

theorem we do not know whether or not shakedown will occur for
the value of Γ so obtained; however, if we *increase* Γ by an infinitesimal amount $\delta\Gamma$ we are assured that shakedown will not
occur for $\Gamma+\delta\Gamma$, since

$$(\Gamma+\delta\Gamma) \int_0^T c_\alpha \dot{\xi}^*_\alpha dt > W^*_{int} \quad . \tag{42}$$

It follows then that Koiter's theorem can be rewritten as follows:

Γ^S *is the smallest multiplier* Γ *obtained from the expression*

$$\Gamma = \frac{\int_0^T dt \int_V D(\dot{q}^{*P}_j) dV}{\int_0^T c_\alpha \dot{\xi}^*_\alpha dt} \tag{43}$$

for any choice of an admissible plastic strain rate cycle
$\dot{q}^{*P}_j(s,t), \dot{\xi}^*_\alpha(t)$.

Any value of Γ obtained from equating W^*_{ext} and W^*_{int} in Koiter's
theorem is thus an upper bound on Γ^S. The result also implies
that shakedown will not occur for any $\Gamma > \Gamma^S$, again without involving the ambiguous behavior when $\Gamma = \Gamma^S$.

18.4 Relation between the Shakedown and Limit Theorems

The analogy between the shakedown theorems and the limit theorems can be clearly seen when both sets of theorems are stated
in terms of bounds on particular load multipliers. The shakedown theorems are in fact generalizations of the limit theorems
to more complex loading. To show this, consider the *special cycle* in which the loads $c_\alpha(t)$ remain *constant*, i.e. do not vary
with time. This is an admissible cyclic load, since it is

certainly true that $c_\alpha(t) = c_\alpha(t+T)$. However, the period T may be arbitrarily chosen when the loads are constant. The elastic solution for the loads c_α will lead to a *time-independent* stress field Q_j^{oE}.

Melan's theorem states that Γ^S is the largest multiplier for which a residual stress field $\bar\rho_j$ can be found such that

$$\phi(\Gamma Q_j^{oE} + \bar\rho_j) \leq 0 \ . \tag{44}$$

When Q_j^{oE} is constant, the stress field $(\Gamma Q_j^{oE}+\bar\rho_j)$ becomes any time-independent stress field in equilibrium with loads Γc_α. From the lower bound theorem of limit analysis (Chapters 9 and 13) the largest multiplier for which (44) holds is the limit load multiplier Γ^L. Hence for *constant loads* $\Gamma^S = \Gamma^L$, and the concept of shakedown becomes identical with that of loading which does not produce flow in the structure.

In the application of Koiter's theorem to the case of constant loads we again make use of the observation that the period T is undefined. The admissible plastic strain rate cycle $\dot q_j^{*P}$ (s,t) is defined in such a way that

$$\Delta q_j^{*P}(s) \ = \ \int_o^T \dot q_j^{*P}(s,t)\,dt \tag{45}$$

must be a pertinent kinematically admissible field. If T is undefined this can be true only when $\dot q_j^{*P}(s,t)$ is itself a pertinent kinematically admissible field. Hence $\dot\rho_j^*(s,t)$ (equation 15) will be zero, and $\dot\xi_\alpha^*$ becomes a velocity field from which the plastic strains are derived. The equation $W_{ext}^* = W_{int}^*$ reduces to

$$D_{ext} \ = \ \Gamma c_\alpha \dot\xi_\alpha^* \ = \ \int_V D(\dot q_j^{*P})\,dV = D_{int} \ , \tag{46}$$

because T is undefined. From the upper bound theorem of limit analysis (Chapters 9 and 13) the smallest multiplier Γ which can be obtained from equation (46) is the limit load multiplier Γ^L. The concept of failure as a result of continued plastic deformation for cyclic loading is thus identical to the concept of the inability of the structure to equilibrate the loads when the loads are constant.

The shakedown limit Γ^S can thus be regarded as replacing the limit load parameter Γ^L in structures subjected to cyclic loading in the sense that for $\Gamma > \Gamma^S$ failure of the structure will certainly occur, either as a result of accumulated plastic deformation or fatigue, provided that the assumptions in the analysis remain valid. Unfortunately the shakedown theorems are considerably more difficult to apply than the limit theorems. An elastic analysis must first be carried out, and in addition, direct applications of the kinematic theorem require additional computations to evaluate the denominator in equation (43). Not surprisingly, few complete solutions for Γ^S have appeared in the literature for any but the simplest classes of problems. In the next section we shall give some examples of the applications of the shakedown theorems to simple structures.

In practice the situation is redeemed to some extent by the observation that frequently Γ^S is not much smaller than Γ^L, often in the range $\Gamma^S = 0.8\Gamma^L$ to $0.9\Gamma^L$ (with Γ^L defined as in equation 35). If a structure is designed with an ample load factor against flow, it may be unnecessary to take shakedown explicitly into account. Thus if a lower bound calculation on Γ^S can show that Γ^S is close to Γ^L, an exact determination of Γ^S, or a calculation of an upper bound on Γ^S, can be waived.

18.5 Non-cyclic Loading Programs

Another important practical consideration is that a structure may be subjected to a number of loading states which are not necessarily applied in a strictly cyclic order. As an example,

a building may be subjected to floor loads, wind loads and roof
loads due to snow. These loads may be applied singly or in combination, in a random order. The concept of shakedown is unal-
tered in this situation; shakedown is said to occur if eventually the response of the structure becomes elastic. The behavior
when shakedown does not occur becomes a great deal more complicated, but the dangers of accumulating plastic deformation and
fatigue remain.

The shakedown theorems are applicable to non-cyclic repeated
loading provided that we construct a *circumscribing cycle* in a
suitable manner. This can be demonstrated quite effectively in
the case of a structure which is subjected to two generalized
loads $\Gamma_1(t)$ and $\Gamma_2(t)$. Suppose that we know only that

$$\Gamma_1^- \leq \Gamma_1(t) \leq \Gamma_1^+ \ ,$$

$$\Gamma_2^- \leq \Gamma_2(t) \leq \Gamma_2^+ \ . \tag{47}$$

At any instant the load point in load space may lie within a
domain given by inequalities (47), as shown in Figure 2. Let us
arbitrarily suppose that the loads follow the cycle which passes
along the outer boundary of this domain (ABCDA in Figure 2). If
shakedown occurs for this cycle, a yield surface in load space
can be found which contains the domain, or alternatively a field
$\bar{\rho}_j(s)$ can be found such that $\phi(Q_j^E + \bar{\rho}_j) < 0$ when Q_j^E is the elastic

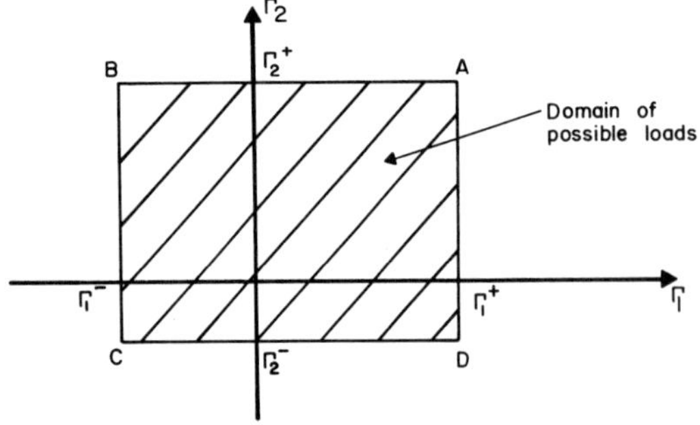

Figure 2. Arbitrary loading within prescribed limits

solution for any load state in the domain. Shakedown is thus possible for any load path which remains within the domain. The convergence concept tells us that if shakedown is possible for any loading program it will always occur; we consider one solution with arbitrary initial conditions and one with initial residual stresses $\bar{\rho}_j(s)$ and substitute them into equation (3). dA/dt must eventually become zero, and for times larger than this both solutions will behave elastically.

Melan's theorem can thus be broadened. Let a structure or body be subjected to an arbitrary loading program $\Gamma_\alpha(t)$ such that $\Gamma_\alpha^- \leq \Gamma_\alpha(t) \leq \Gamma_\alpha^+$. The elastic solution to this problem will yield stresses Q_j^E which are linear functions of the generalized loads, i.e.

$$Q_j^E = Q_j^E(\Gamma_\alpha) \quad . \tag{48}$$

Shakedown will occur if it is possible to find any time independent residual stress field $\bar{\rho}_j(s)$ such that

$$\phi[Q_j^E(\Gamma_\alpha) + \bar{\rho}_j] < 0 \tag{49}$$

for all admissible Γ_α.

A formal proof of this statement may be obtained on the lines indicated above, and is left to the reader.

18.6 Historical and Bibliographical Remarks

Shakedown was first discussed in general terms by Bleich [1932], who gave a proof of the static shakedown theorem for a truss with a single redundancy. A general proof for trusses was given by Melan [1936a], [1936b]. Independent proofs of the static theorem for trusses and framed structures were given by Symonds and Prager [1950] and Neal [1951] respectively. The generalization of the static shakedown theorem to general bodies was achieved by Melan [1938a], [1938b]. While it is remarkable

that Melan formulated these results so early in the development
of the theory of plasticity, his arguments are difficult to follow. Considerable simplifications of the proof of the static
theorem were given by Symonds [1951] and Koiter [1952]. The kinematic shakedown theorem was given by Koiter [1956].

References

H.Bleich	1932	"Über die Bemessung statish unbes- timmter Stahl tragwerke unter der Berücksichtigung des elastisch- plastischen Verhaltens des Baustof- fes", Bauingenieur, $\underline{13}$, 261.
W.T.Koiter	1952	"Some remarks on plastic shakedown theorems", Proc.8th Int.Congr.Appl. Mech.(Istanbul). $\underline{1}$. 220.
W.T.Koiter	1956	"A new general theorem on shakedown of elastic-plastic structures", Proc. Koninkl.Ned.Akad.Wetenschap, $\underline{B59}$, 24.
E.Melan	1936a	"Theorie statisch unbestimmter Sys- teme", Prelim.Publ.2nd Congr.Intern. Assoc.Bridge and Struc.Eng., Berlin, 43.
E.Melan	1936b	"Theorie statisch unbestimmter Sys- teme aus ideal plastischem Baustoff", Sitz.Ber.Akad.Wiss.Wien, Abt. 11a, $\underline{145}$, 195.
E.Melan	1938a	"Der Spanning zustand eines "Mises- Henckyschen" Kontinuums bei veränd- lichen Belastung", Sitz.Ber.Akad. Wiss.Wien, Abt. 11a, $\underline{147}$, 73.
E.Melan	1938b	"Zur Plastizitat des räumlichen Kon- tinuums", Ing.Arch., $\underline{9}$, 116.
B.G.Neal	1951	"The Behavior of Framed Structures under Repeated Loading", Quart.J. Mech.Appl.Math., $\underline{4}$, 78.
P.S.Symonds	1951	"Shakedown in continuous media", J. Appl.Mech., $\underline{18}$, 85.
P.S.Symonds and W.Prager	1950	"Elastic-plastic analysis of struc- tures subjected to loads varying be- tween prescribed limits", J.Appl. Mech., $\underline{17}$, 315.

Chapter 19

APPLICATION OF THE SHAKEDOWN THEOREMS

19.1 The Static Shakedown Theorem as a Programming Problem

In discussing applications of the shakedown theorems we shall consider systematic methods of exploiting the theorems to determine the shakedown load parameter. While some of the results may be applied quite generally, we shall direct our attention primarily to beams and plane frames in which bending alone is considered. Shakedown analysis has been studied most extensively in this class of problems.

For a loading cycle $\Gamma_\alpha(t) = \Gamma c_\alpha(t)$, $0 \leq t \leq T$, the shakedown load factor Γ^S, according to the statement of the static shakedown theorem given in Section 18.3, is given by the solution of the following problem;

$$\text{maximize} \quad \Gamma$$
$$\text{subject to} \quad \phi(\Gamma Q_j^{oE} + \bar{\rho}_j) \leq 0 \quad \text{in} \quad V \quad \text{for} \quad 0 \leq t \leq T . \tag{1}$$

The stress field $Q_j^{oE}(s,t)$ is the elastic solution for loads $c_\alpha(t)$, and $\bar{\rho}_j(s)$ is a self-equilibrating or residual stress field. The determination of Γ^S is thus a programming problem. However, in order to arrange the problem in a form suitable for numerical computation, the constraints given in (1) must be placed in discrete form. Since the yield condition is applied to quantities which vary in both space and time, the discretization must make it possible to consider only a finite number of instants in the loading program and a finite number of points in the body.

Discretization of the time parameter t is most conveniently achieved by replacing the loading program $c_\alpha(t)$ by a program $\bar{c}_\alpha(t)$ which is piecewise linear in the load space. This can be effectively demonstrated in the case $\alpha = 2$; Figure 1 shows the

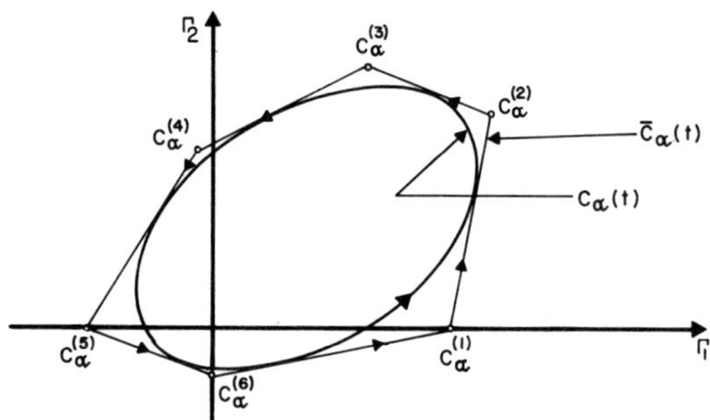

Figure 1. Segmental loading program

replacement of a loading program which traces out a curved line
in load space by one which traces out a series of straight line
segments.

Let the vertices of the locus of the loading program $\bar{c}_\alpha(t)$ in
load space be denoted by $c_\alpha^{(k)}$, k=1,2,...,m, and let the elastic
stress field for load $c_\alpha^{(k)}$ be denoted by $Q_j^{oE(k)}(s)$. We now make
use of the following result. If

$$\phi(\Gamma Q_j^{oE(k)} + \bar{\rho}_j) \leq 0 \quad \text{in } V \quad , \tag{2a}$$

and $\quad \phi(\Gamma Q_j^{oE(k+1)} + \bar{\rho}_j) \leq 0 \quad \text{in } V \quad ,$ \hfill (2b)

then $\quad \phi\{\beta\Gamma Q_j^{oE(k)} + (1-\beta)\Gamma Q_j^{oE(k+1)} + \bar{\rho}_j\} \leq 0 \quad \text{in } V$ \hfill (2c)

for $\quad 0 \leq \beta \leq 1$.

Alternatively, if the yield condition is satisfied everywhere in
the body for loads at two adjacent vertices in the loading program in load space, the yield condition will be satisfied at
all points on the straight line in load space joining the two
vertices.

This result follows trivially from consideration of any one
point in the body. The stress states

$$\Gamma\{\beta Q_j^{oE(k)} + (1-\beta)Q_j^{oE(k+1)}\}, \quad 0 \leq \beta \leq 1$$

lie on the straight line in stress space joining the states obtained by putting $\beta = 0$ and $\beta = 1$. If these extreme states lie within or on the yield surface, convexity of the yield surface in stress space assures us that stress states on the line joining the extreme states do not lie outside the yield surface.

The programming problem (1) then becomes:

maximize Γ

$$\text{subject to} \quad \phi\{\Gamma Q_j^{oE(k)}(s) + \bar{\rho}_j(s)\} \leq 0 \quad \text{in } V \quad , \tag{3}$$

$$k = 1,2,\ldots,m.$$

We now further modify the problem by discretizing the stress fields. It may be necessary to use such a discretization to compute $Q_j^{oE(k)}$. It is certainly necessary in the programming problem to express $\bar{\rho}_j(s)$ in terms of a discrete number of parameters. Finally, we apply the yield condition at a discrete number of points in the body. The manner in which these steps are carried out will depend on the configuration of the body. It is particularly simple in trusses, and in beams and frames subjected to concentrated loads, since we can express the generalized stress field in these cases as linear functions of a finite number of hyperstatic forces. In addition, we can readily identify points at which the yield function must be checked in order to ensure that $\phi \leq 0$ everywhere in the body. In more complex problems, methods such as finite differences or finite elements must be used. A variety of means of discretizing the stress field was discussed in Chapter 13.

The solution of the final discrete programming problem will not in general be identical to Γ^S. If the piecewise linear loading cycle $\bar{c}_\alpha(t)$ is identical to $c_\alpha(t)$, or circumscribes $c_\alpha(t)$, the discrete stress field is statically admissible and the yield condition is satisfied everywhere in V, the programming

problem will give a value of Γ which does not exceed Γ^S. While
the determination of a conservative estimate of Γ^S is preferable,
it is not always possible

In the next section we shall consider as an example the shakedown analysis of a portal frame subjected to two varying generalized loads. The problem will suffice to illustrate application of the concepts we have just discussed.

19.2 The Static Shakedown Theorem Applied to a Portal Frame

Consider the portal frame shown in Figure 2. The cross-section
is taken to be uniform, with limit moment M_o. Hence the yield
condition requires that the bending moment $M(s)$ should satisfy
the inequalities

$$- M_o \leq M(s) \leq M_o \tag{4}$$

at all points on the frame.

The frame is subjected to independent loads H and V. We shall
consider a variety of loading conditions, and shall defer placing limits on H and V for the present. However, because the
bending moment diagram will be piecewise linear, we see immediately that it is necessary only that we check the yield condition
at points 1-5 marked on Figure 2. If the yield condition is
satisfied at each of these points, it will be satisfied everywhere on the frame.

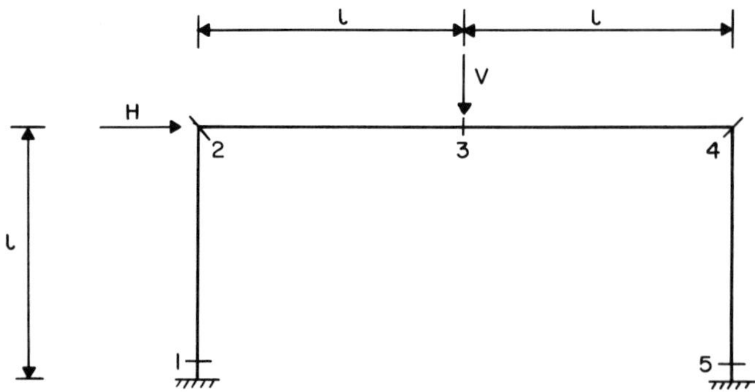

Figure 2. Portal frame example

The frame is indeterminate to the third degree, and consequently we can find three independent self-equilibrating moment distributions. The three shown in Figures 3(a), 3(b) and 3(c) will be adopted; moments which cause tension on the inside of the frame are chosen to be positive. The most general residual moment field will be the sum of these three independent fields, and hence may be expressed in terms of the three parameters h, m and v. Denoting by $M_1^r, M_2^r, \ldots, M_5^r$ the residual moments at points 1-5 respectively, we see that

$$
\begin{aligned}
M_1^r &= m\ell + v\ell \ , \\
M_2^r &= -h\ell + m\ell + v\ell \ , \\
M_3^r &= -h\ell + m\ell \ , \\
M_4^r &= -h\ell + m\ell - v\ell \ , \\
M_5^r &= m\ell - v\ell \ .
\end{aligned}
\tag{5}
$$

The elastic analysis of the frame is carried out by conventional methods. Assuming that the flexural rigidity EI is

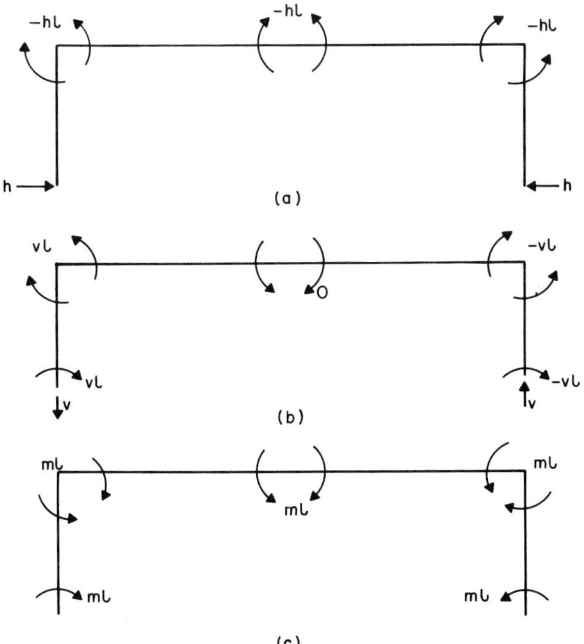

Figure 3. Independent self-stress systems

uniform, we consider the two loading conditions H=1, V=0, and H=0, V=1. These we denote by (1,0) and (0,1). The bending moments at the critical sections 1-5 are labelled $M_1^{HE} - M_5^{HE}$ for loading (1,0) and $M_1^{VE} - M_5^{VE}$ for loading (0,1). These moments are

$$M_1^{HE} = -0.3125\ell \quad , \qquad\qquad M_1^{VE} = +0.1\ell \quad ,$$

$$M_2^{HE} = +0.1875\ell \quad , \qquad\qquad M_2^{VE} = -0.2\ell \quad ,$$

$$M_3^{HE} = 0 \quad , \qquad\qquad M_3^{VE} = +0.3\ell \quad , \qquad\qquad (6)$$

$$M_4^{HE} = -0.1875\ell \quad , \qquad\qquad M_4^{VE} = -0.2\ell \quad ,$$

$$M_5^{HE} = +0.3125\ell \quad . \qquad\qquad M_5^{VE} = +0.1\ell \quad .$$

Consider first the case where the loads may take the values

$$0 \leq V \leq \Gamma \quad , \qquad -\Gamma \leq H \leq \Gamma \quad . \qquad\qquad (7)$$

A circumscribing piecewise linear cycle in load space is a program which proceeds linearly between the states $(\Gamma,0)$, (Γ,Γ), $(-\Gamma,\Gamma)$ and $(-\Gamma,0)$. It is necessary to apply the yield condition only for these four loading states. Let us introduce the dimensionless parameters

$$\bar{M} = \frac{M}{M_o} \quad , \qquad \bar{h} = \frac{h\ell}{M_o} \quad , \qquad \bar{m} = \frac{m\ell}{M_o} \quad , \qquad \bar{v} = \frac{v\ell}{M_o} \quad , \qquad \bar{\Gamma} = \frac{\Gamma\ell}{M_o} \quad . \qquad\qquad (8)$$

The yield condition is then

$$- 1 \leq \bar{M} \leq 1 \quad . \qquad\qquad (9)$$

We can limit the number of inequalities which must be treated in the programming problem by noting that if the largest moment at any section is not greater than 1 and the smallest moment is not less than -1, all intermediate values will fall between 1 and -1. For example, for the load states $(\Gamma,0)$, (Γ,Γ), $(-\Gamma,\Gamma)$ and

$(-\Gamma, 0)$ the largest moment at section 1 is, from equations (5) and (6),

$$0.4125\overline{\Gamma} + \overline{m} + \overline{v} \ ,$$

corresponding to loading $(-\Gamma, \Gamma)$, and the smallest moment is

$$-0.3125\overline{\Gamma} + \overline{m} + \overline{v} \ ,$$

corresponding to loading $(\Gamma, 0)$. Applying this to all 5 critical sections, the shakedown load parameter $\overline{\Gamma}^{s}$ is given by the solution of the following problem:

maximize $\overline{\Gamma}$

subject to

$$
\begin{array}{lll}
(\text{section 1}) & 0.4125\overline{\Gamma} + \overline{m} + \overline{v} \leq 1 & \\
& -0.3125\overline{\Gamma} + \overline{m} + \overline{v} \geq -1 & \\
(\text{section 2}) & 0.1875\overline{\Gamma} - \overline{h} + \overline{m} + \overline{v} \leq 1 & \\
& -0.3875\overline{\Gamma} - \overline{h} + \overline{m} + \overline{v} \geq 1 & \\
(\text{section 3}) & 0.3\overline{\Gamma} - \overline{h} + \overline{m} \leq 1 & (10) \\
& -\overline{h} + \overline{m} \geq -1 & \\
(\text{section 4}) & 0.1875\overline{\Gamma} - \overline{h} + \overline{m} - \overline{v} \leq 1 & \\
& -0.3875\overline{\Gamma} - \overline{h} + \overline{m} - \overline{v} \geq -1 & \\
(\text{section 5}) & 0.4125\overline{\Gamma} + \overline{m} - \overline{v} \leq 1 & \\
& -0.3125\overline{\Gamma} + \overline{m} - \overline{v} \geq -1 \ . &
\end{array}
$$

This is a linear programming problem which can be solved by standard techniques. The solution is

$$
\begin{aligned}
\overline{\Gamma} &= 2.759 \ , \\
\overline{h} &= -0.207 \ , \\
\overline{m} &= -0.137 \ , \\
\overline{v} &= 0 \ .
\end{aligned}
\qquad (11)
$$

On substituting the solution back into the constraints of problem (10), it can be seen that we have yield under both positive
and negative moment at sections 1 and 5, and yield under negative moments at sections 2 and 4 for the cycle with $\bar{\Gamma} = 2.759$.
This indicates that alternating plastic deformation will occur
for $\bar{\Gamma} = 2.759$.

   Consider now the case where

$$0 \leq V \leq \Gamma \quad , \quad 0 \leq H \leq \Gamma \quad . \tag{12}$$

A circumscribing piecewise linear cycle has vertices at load
states $(\Gamma,0)$, (Γ,Γ), $(0,\Gamma)$ and $(0,0)$. The shakedown load parameter is thus given by the modified problem:

maximize $\bar{\Gamma}$

subject to

$$
\begin{aligned}
0.1\bar{\Gamma} && + \bar{m} + \bar{v} &\leq 1 \\
-0.3125\bar{\Gamma} && - \bar{m} - \bar{v} &\geq -1 \\
0.1875\bar{\Gamma} - \bar{h} &+ \bar{m} + \bar{v} &\leq 1 \\
-0.2\bar{\Gamma} && - \bar{h} + \bar{m} + \bar{v} &\geq -1 \\
0.3\bar{\Gamma} && - \bar{h} + \bar{m} &\leq 1 \\
&& - \bar{h} + \bar{m} &\geq -1 \\
&& - \bar{h} + \bar{m} - \bar{v} &\leq 1 \\
-0.3875\bar{\Gamma} && - \bar{h} + \bar{m} - \bar{v} &\geq -1 \\
0.4125\bar{\Gamma} && + \bar{m} - \bar{v} &\leq 1 \\
&& + \bar{m} - \bar{v} &\geq -1
\end{aligned}
\tag{13}
$$

The solution of this problem is

$$\bar{\Gamma} = 2.857 \quad,$$
$$\bar{h} = -0.286 \quad,$$
$$\bar{m} = -0.143 \quad,$$
$$\bar{v} = 0.035 \quad. \tag{14}$$

Substitution of these results into the inequalities shows that yield occurs for positive moments at Sections 3 and 5, and for negative moments at Sections 1 and 4. This indicates incremental collapse for the cycle with $\Gamma > 2.857$, as we shall see on reconsideration of this example by the kinematic method.

Finally, consider the single loading state (Γ,Γ). This may be interpreted as a degenerate cycle. The shakedown load parameter is given by the solution of the following problem:

maximize $\bar{\Gamma}$

subject to

$$-0.2125\bar{\Gamma} \quad\quad + \bar{m} + \bar{v} \leq 1$$
$$-0.2125\bar{\Gamma} \quad\quad + \bar{m} + \bar{v} \geq -1$$
$$-0.0125\bar{\Gamma} - \bar{h} + \bar{m} + \bar{v} \leq 1$$
$$-0.0125\bar{\Gamma} - \bar{h} + \bar{m} + \bar{v} \geq -1$$
$$0.3\bar{\Gamma} \quad - \bar{h} + \bar{m} \quad\quad \leq 1$$
$$0.3\bar{\Gamma} \quad - \bar{h} + \bar{m} \quad\quad \geq -1 \tag{15}$$
$$-0.3875\bar{\Gamma} - \bar{h} + \bar{m} - \bar{v} \leq 1$$
$$-0.3875\bar{\Gamma} - \bar{h} + \bar{m} - \bar{v} \geq -1$$
$$0.4125\bar{\Gamma} \quad\quad + \bar{m} - \bar{v} \leq 1$$
$$0.4125\bar{\Gamma} \quad\quad + \bar{m} - \bar{v} \geq -1$$

The solution to this problem is

$$\bar{\Gamma} \;=\; 3.000 \;\;,$$

$$\bar{h} \;=\; -0.4 \;\;,$$

$$\bar{m} \;=\; -0.3 \;\;,$$

$$\bar{v} \;=\; -0.0625 \;\;.$$

It may be confirmed directly that this is the limit load parameter for the proportional loading (Γ,Γ), confirming that limit analysis may be regarded as a particular case of shakedown analysis.

19.3 Some Remarks on Alternating Plastic Deformation in Beams

The application of the static shakedown theorem discussed in the preceding sections does not distinguish between the mechanisms of incremental collapse or alternating plastic flow. We may remark, however, that the load parameter at which alternating plastic flow will first occur can frequently be determined rather more easily than the shakedown load parameter. This is particularly true for beams and frames, or any structure in which only one generalized stress component is considered. We shall limit our discussion to the case of beams and frames, using the example of Section 19.2 to illustrate the concepts.

Alternating plastic flow in a beam or frame composed of members which have an elastic, perfectly plastic moment curvature relation requires that one or more sections should yield under both positive moment and negative moment during the cycle. At a particular section s_i of the beam or frame, let the largest elastic moment at any instant during the cycle be $M^E(s_i)_{max}$, and let the smallest moment be $M^E(s_i)_{min}$. Let the residual moment be $M^r(s_i)$. Assuming that the limit moment has the same magnitude for positive and negative bending moment, alternating plastic flow will be incipient at a section if

$$M^E(s_i)_{max} + M^r(s_i) = + M_o \ , \tag{16a}$$

$$M^E(s_i)_{min} + M^r(s_i) = - M_o \ . \tag{16b}$$

Subtracting equation (16b) from equation (16a), we can eliminate the residual moment $M^r(s_i)$:

$$M^E(s_i)_{max} - M^E(s_i)_{min} = 2M_o \ . \tag{17}$$

It follows then that we can check each critical point on the frame on the basis of the elastic solution alone, determining for each section the load parameter which will ensure that alternating plastic deformation will not occur at that section. Any load parameter determined in this way must be greater than or equal to the shakedown load parameter.

In order to illustrate these calculations, consider equations (10) in Section 19.2 which contain the greatest and least moments at the sections 1-5 for the frame of Figure 2 for the loading cycle $(\Gamma,0)$, (Γ,Γ), $(-\Gamma,\Gamma)$, $(-\Gamma,0)$. Replacing the inequalities by equalities, subtracting the second equation from the first for each section, and solving for Γ in each case, we find:

(section 1) $0.725\Gamma = 2$, $\Gamma = 2.759$

(section 2) $0.575\Gamma = 2$, $\Gamma = 3.478$

(section 3) $0.3\Gamma \ = 2$, $\Gamma = 6.667$ $\tag{18}$

(section 4) $0.575\Gamma = 2$, $\Gamma = 3.478$

(section 5) $0.725\Gamma = 2$, $\Gamma = 2.759$.

Since the least value of Γ obtained in this way is 2.759, it follows that alternating plastic flow will not take place for $\Gamma < 2.759$, and that $\Gamma^S \leq 2.759$. We know in fact that $\Gamma^S = 2.759$,

and this indicates that for $\Gamma > 2.759$ alternating plastic flow
will occur for this cycle.

Consider now the second cycle, $(\Gamma,0)$, (Γ,Γ), $(0,\Gamma)$, $(0,0)$ for
the same frame. The largest and smallest moments appear in the
constraints of equations (13). Repeating the same process, we
find;

$$(\text{section 1}) \quad 0.4125\Gamma = 2 \quad , \quad \Gamma = 4.849$$

$$(\text{section 2}) \quad 0.3875\Gamma = 2 \quad , \quad \Gamma = 5.161$$

$$(\text{section 3}) \quad 0.3\Gamma \quad = 2 \quad , \quad \Gamma = 6.667 \tag{19}$$

$$(\text{section 4}) \quad 0.3875\Gamma = 2 \quad , \quad \Gamma = 5.161$$

$$(\text{section 5}) \quad 0.4125\Gamma = 2 \quad , \quad \Gamma = 4.849 \ .$$

These calculations tell us that alternating plastic flow cannot
occur for $\Gamma < 4.849$. Notice that this value of Γ exceeds the
limit load parameter $\Gamma^L = 3$. Alternating plastic flow cannot
consequently occur for this cycle.

This approach to alternating plastic deformation is of practical importance because it can be extended to cover *kinematically hardening* beams. Suppose first that the yield function
has the form

$$-M_y \leq M - b\kappa_p \leq M_y \quad , \tag{20}$$

where M_y is the yield moment, b is a hardening parameter, and
κ_p is the plastic curvature at the cross-section. The moment~
curvature relation has a bi-linear form for monotonic loading.
The yield range at any instant is $2M_y$; alternatively we can rewrite equation (20) as

$$-M_y + b\kappa_p \leq M \leq M_y + b\kappa_p \quad . \tag{21}$$

If a structure composed of a material of this type is subjected
to cyclic loading, plastic deformation will lead to the

establishment of a residual moment distribution M^r. Hence

$$M = M^E + M^r \quad . \tag{22}$$

The major differences between the hardening case and the perfectly plastic cases are that hinges will not occur, and that the
limit moment M_o is replaced by $(M_y + b\kappa_p)$ which depends on the extent of plastic deformation. We shall not enter a detailed and
rigorous discussion of this problem. However, we note that for
alternating plastic deformation to be incipient at a section we
find that when the cyclic response is established,

$$M^E_{max} + M^r = M_y + b\kappa_p \quad , \tag{23a}$$

$$M^E_{min} + M^r = -M_y + b\kappa_p \quad . \tag{23b}$$

Subtracting equation (23b) from equation (23a), we eliminate
both M^r and κ^p;

$$M^E_{max} - M^E_{min} = 2M_y \quad . \tag{24}$$

This rather limited discussion is of practical importance in
the case of beams and frames because the *moment~curvature relation* is not elastic, perfectly plastic even when the material
is elastic, perfectly plastic. For a monotonic increase or decrease in the bending moment M, the moment~curvature relation
will have a curved knee, with first yield when $M = \pm M_y$, and
flow occurring for $M = \pm M_o$. Plastic deformation will take place
for $M > M_y$ or $M < -M_y$, but this will be limited to parts of the
section. On unloading it will be found that the yield range is
always $2M_y$.

Plastic deformation which takes place over part of the cross-
section cannot contribute greatly to the overall deformation of
the frame. Hence the elastic, perfectly plastic idealization is
adequate when incremental collapse is discussed. However,

alternating plastic deformation of part of the cross-section is
of concern, since local failure due to fatigue could eventually
lead to failure of the entire section. It is advisable, therefore to use the first yield moment M_y rather than the limit moment M_o in considering alternating plastic flow.

In carrying out these calculations it is common to introduce
a shape factor α defined by

$$M_o = \alpha M_y \quad . \tag{25}$$

For a rectangular section $\alpha = 1.5$, while for an I-beam $\alpha \simeq 1.15$.
Equation (24) is replaced by

$$M^E_{max} - M^E_{min} = \frac{2M_o}{\alpha} \quad . \tag{26}$$

Let us now reconsider the frame example which we discussed
above. The load parameter for which the structure is safe
against alternating plastic flow throughout the cross-section
may be denoted by Γ^{ap}. For the cycle $(\Gamma,0)$, (Γ,Γ), $(-\Gamma,\Gamma)$,
$(-\Gamma,0)$ we see from equations (23) that

$$\Gamma^{ap} = 2.758/\alpha \quad , \tag{27}$$

while for the cycle $(\Gamma,0)$, (Γ,Γ), $(0,\Gamma)$, $(0,0)$, from equations
(24),

$$\Gamma^{ap} = 4.849/\alpha \quad .$$

If the frame is constructed of an I-section, with $\alpha = 1.15$, we
see that Γ^{ap} is 2.399 and 4.217 respectively. In the first
case this represents a significant reduction in the shakedown
load parameter. In the second case it is seen that alternating
plastic deformation is still not a factor.

19.4 A Modified Statement of the Kinematic Theorem

The kinematic shakedown theorem is stated as a minimization

problem in Section 18.3. The shakedown load parameter Γ^s is the least value of

$$\Gamma = \frac{\displaystyle\int_0^T dt \int_V D(\dot{q}_j^{*p})\,dV}{\displaystyle\int_0^T c_\alpha \dot{\xi}_\alpha^*\,dt} \; . \tag{28}$$

In this expression $\dot{q}_j^{*p}(s,t)$, $0 \leq t \leq T$, is an admissible strain rate cycle as defined in Section 18.2. The velocities $\dot{\xi}_\alpha^*(t)$ represent the response of the structure to the admissible strain rate cycle with the external loads set equal to zero. For any chosen admissible strain rate cycle the numerator of equation (28) can be readily evaluated. However, the denominator is considerably more difficult to evaluate for all possible plastic strain rate cycles, and as a result few direct applications of the kinematic theorem have been given.

We seek in this Section to modify the denominator of equation (28) so that it can be more readily evaluated. It will be seen that this process will lead us to a conservative estimate Γ^s, except in a few special cases where we can obtain Γ^s exactly. The special cases include beams and frames in which only bending is considered, and are thus of practical importance.

Using the notation of Chapter 18, we may use the principle of virtual velocities to write

$$c_\alpha \dot{\xi}_\alpha^* = \int_V Q_j^{oE}(\dot{q}_j^{*r} + \dot{q}_j^{*p})\,dV \; . \tag{29}$$

The stress field Q_j^{oE} is certainly in equilibrium with loads c_α, since it is the elastic solution for these loads, and $(\dot{q}_j^{*r} + \dot{q}_j^{*p})$ and $\dot{\xi}_\alpha^*$ are kinematically admissible. Further

$$\dot{q}_j^{*r} = C_{jk}\dot{\rho}_k^* \; , \tag{30}$$

where ρ_k^* is the residual stress field induced by $\dot{q}_j^{*p}$. Hence

$$\int_V Q_j^{oE}\dot{q}_j^{*r} = \int_V C_{jk}Q_j^{oE}\dot{\rho}_k^* \, dV = \int_V \dot{\rho}_k^* q_k^{oE} dV = 0 \quad , \qquad (31)$$

since $\dot{\rho}_K^*$ equilibrates zero loads and q_k^{oE} is kinematically admissible. Hence on substituting equation (31), we find that

$$\int_o^T c_\alpha \dot{\xi}_\alpha^* dt = \int_o^T dt \int_V Q_j^{oE}(s,t)\dot{q}_j^{*p}(s,t) dV \qquad (32)$$

This expression can be evaluated without an elastic analysis for the residual stress and elastic strain rates associated with $\dot{q}_j^{*p}(s,t)$, but is nevertheless still a complex problem.

Suppose, however, that at any *one point* in the body, $\dot{q}_j^{*p}(s,t)$ is a vector which has a *fixed direction* in strain space. At the *same point* in the body, let $\{Q_j^{oE}\}_{max}$ indicate the value of $Q_j^{oE}(s,t)$ which has the *greatest component in the direction* of $\dot{q}_j^{*p}(s,t)$. It is evident then that

$$\{Q_j^{\ E}\}_{max} \dot{q}_j^{*p}(s,t) \geq Q_j^{oE}(s,t)\dot{q}_j^{*p}(s,t). \qquad (33)$$

On substituting inequality (33) into equation (32), and recalling that

$$\Delta q_j^{*p} = \int_o^T \dot{q}_j^{*p} \, dt \qquad (34)$$

is the kinematically admissible plastic strain cycle, we see that

$$\int_o^T c_\alpha \dot{\xi}_\alpha^* dt \leq \int_V \{Q_j^{oE}\}_{max} \Delta q_j^{*p} dV \quad . \qquad (35)$$

The right hand side of this expression is now comparatively simple to evaluate. Suppose we now define a new problem, that of minimizing

$$\Gamma = \frac{\int\limits_{o}^{T} dt \int\limits_{V} D(\dot{q}_j^{*p})dV}{\int \{Q_j^{oE}\}_{max} \Delta q_j^{*p}dV} \quad . \tag{36}$$

The least value of Γ, as given by equation (36), must as a consequence of inequality (35) be *not greater than* Γ^s.

Since both the numerator and denominator of equation (36) are homogeneous and of degree one in the components of Δq_j^{*p}, remembering that $\dot{q}_j^{*p}$ has a fixed direction at each point in the body, we can without loss in generality restate the problem of equation (36) in the programming form;

$$\text{minimize} \quad \int\limits_{V} D(\Delta q_j^{*p})dV$$

$$\tag{37}$$

$$\text{subject to} \quad \int\limits_{V} \{Q_j^{oE}\}_{max} \Delta q_j^{*p}dV = 1 \quad ,$$

for all kinematically admissible plastic strain cycles Δq_j^{*p}.

One circumstance in which the replacement of problem (28) by problem (37) is reasonable is that of a frame or beam which fails by incremental collapse. We shall limit ourselves to a detailed discussion of this particular problem.

First, the frame problem is spatially discrete in the sense that the plastic deformation can occur only in hinges at a discrete number of sections. If alternating plastic deformation is not considered, the rotations of the hinges during plastic deformation can be taken to have a *fixed sign* at any one particular hinge. This fulfills the assumption made in the previous paragraphs, and limits consideration to incremental collapse. Finally, and perhaps most importantly, the complete class of kinematically admissible plastic strain cycles leading to incremental collapse can be identified. Since in incremental collapse

the displacements of the beam or frame which occur during any
one cycle appear as if a limited amount of flow has taken place,
the class of kinematically admissible plastic strain cycles will
be qualitatively identical to the class of kinematically admissible velocity fields for *flow* under constant loads. The generation of this latter class was discussed in Chapters 10 and 13.

The application of the method will be illustrated by recon-
sidering the examples treated in Section 19.2 and shown in
Figure 2. We shall discuss only the load cycle $(0,\Gamma)$, (Γ,Γ),
$(\Gamma,0)$, $(0,0)$. The kinematically admissible plastic strain cycles are shown in Figure 4, and the hinge rotations are tabulated in terms of a single parameter in Table 1. The signs are
consistent with the sign convention for moments used in Section
19.2; moments causing tension on the inside of the frame do positive work on a positive hinge rotation.

We shall first carry out the calculation by hand, computing

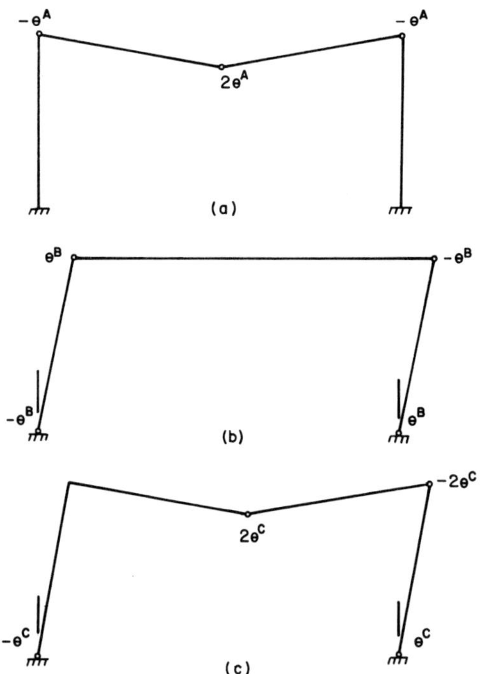

Figure 4. Mechanisms of incremental collapse

the shakedown load parameter for each of the incremental modes shown in Figure 4 independently. To do so, we select from the elastic solution (given in equation 6) the largest moment in the cycle if the hinge rotation is positive and the least moment if the hinge rotation is negative. These moments are given in Table 2.

Table 1. Hinge Rotations

Section	Cycle(a)	Cycle(b)	Cycle(c)
1	0	$-\theta^B$	$-\theta^C$
2	$-\theta^A$	$+\theta^B$	0
3	$+2\theta^A$	0	$+2\theta^C$
4	$-\theta^A$	$-\theta^B$	$-2\theta^C$
5	0	$+\theta^B$	θ^C

Table 2. Maximum Moment Values

$\{M^{OE}\}_{max}$			
Section	Cycle(a)	Cycle(b)	Cycle(c)
1	–	-0.3125ℓ	-0.3125ℓ
2	-0.2ℓ	$+0.1875\ell$	–
3	$+0.3\ell$	–	$+0.3\ell$
4	-0.3875ℓ	-0.3875ℓ	-0.3875ℓ
5	–	$+0.4125\ell$	$+0.4125\ell$

Consider cycle (a) in Figure 4. From Tables 1 and 2, Θ^A is chosen to satisfy the constraint of problem (37). We thus put

$$0.2\ell\Theta^A + 0.6\ell\Theta^A + 0.3875\ell\Theta^A = 1 \quad , \quad \text{or}$$
$$\Theta^A = 1/1.875 \quad . \tag{38}$$

The shakedown parameter is then

$$\Gamma = M_o|\Theta_2| + M_o|\Theta_3| + M_o|\Theta_4|$$

$$= 4M_o\Theta^A \tag{39}$$

$$= \frac{3.368M_o}{\ell} \quad .$$

Incremental collapse will not occur in this mode for $\Gamma\ell/M_o < 3.368$.

For cycle (b) in Figure 4 we require that

$$0.3125\ell\Theta^B + 0.1875\ell\Theta^B + 0.3875\ell\Theta^B + 0.4125\ell\Theta^B = 1 \quad , \quad \text{or}$$
$$\Theta^B = 1/1.3 \quad . \tag{40}$$

Then

$$\Gamma = M_o|\Theta_1| + M_o|\Theta_2| + M_o|\Theta_4| + M_o|\Theta_5|$$

$$= 4M_o\Theta^B \tag{41}$$

$$= \frac{3.078M_o}{\ell} \quad .$$

For cycle (c) in Figure 4,

$$0.3125\Theta^C + 0.6\Theta^C + 0.775\Theta^C + 0.4125\Theta^C = 1 \quad , \tag{42}$$
$$\Theta^C = 1/2.1 \quad .$$

Then

$$\Gamma = M_o \left| \Theta_1 \right| + M_o \left| \Theta_3 \right| + M_o \left| \Theta_4 \right| + M_o \left| \Theta_5 \right|$$

$$= 6 M_o \Theta^C \tag{43}$$

$$= \frac{2.857 M_o}{\ell} \ .$$

In considering all three possible modes, the lowest shakedown parameter is given by cycle (c), with

$$\frac{\Gamma \ell}{M_o} = 2.857 \ . \tag{44}$$

This result cannot exceed the correct shakedown load parameter for incremental collapse; in fact it coincides with the result obtained in the static approach given in Section 18.2 for this loading cycle.

It is pertinent to enquire why the exact result is obtained in view of the use of inequality (33) in the kinematic approach. When the structure undergoes incremental collapse plastic deformation, appearing as hinge rotations, takes place as the load point moves along the straight line segments of the load cycle. The actual moment $M(s,t)$ at a point undergoing plastic deformation must be constant and equal to M_o. For the *correct* shakedown load parameter and the correct incremental mode, however, the *residual moments are also constant* since we are considering the limiting case between shakedown and incremental collapse. It follows then that $M^E(s,t)$, and hence $M^{oE}(s,t)$, is also constant while plastic deformation occurs and equal to their maximum or minimum values depending on the sign of the hinge rotation. Consequently for the correct incremental mode inequality (33) is in fact an equality and no errors are introduced. The shakedown parameters for the incorrect modes, on the other hand, may be conservative.

The method of computation we have just outlined is not particularly suitable for use on the computer. We may avoid identifying the incremental collapse modes directly by making use of the compatibility equations for the plastic hinge rotations. Using the self-equilibrating moments given in Figure 3, these equations are

$$\Theta_2 + \Theta_3 + \Theta_4 = 0 \quad,$$

$$\Theta_1 + \Theta_2 - \Theta_4 - \Theta_5 = 0 \quad, \tag{45}$$

$$\Theta_1 + \Theta_2 + \Theta_3 + \Theta_4 + \Theta_5 = 0 \quad.$$

These equations appear as constraints. Reverting to dimensionless variables, the objective function of problem (37) becomes

$$\frac{\Gamma\ell}{M_o} = |\Theta_1| + |\Theta_2| + |\Theta_3| + |\Theta_4| + |\Theta_5| \quad. \tag{46}$$

In order to eliminate the moduli which appear in equation (46) and to use variables of restricted sign, we introduce new variables Θ_i^+, Θ_i^- (i=1,...,5)

$$\Theta_i = \Theta_i^+ - \Theta_i^- \quad,$$

$$|\Theta_i| = \Theta_i^+ + \Theta_i^- \quad, \tag{47}$$

$$\Theta_i^+ \geq 0 \quad, \quad \Theta_i^- \geq 0 \quad.$$

The objective function is then

$$\frac{\Gamma\ell}{M_o} = \Theta_1^+ + \Theta_1^- + \Theta_2^+ + \Theta_2^- + \Theta_3^+ + \Theta_3^- + \Theta_4^+ + \Theta_4^- + \Theta_5^+ + \Theta_5^- \tag{48}$$

The constraint of problem (37) is introduced by first writing the greatest and smallest elastic moments M^{oE} for each point where a hinge may occur. These are again taken from equations (6), and are tabulated in Table 3. The greatest value is used when Θ is positive, and the smallest when Θ is negative. We can incorporate this requirement by noting that, from equations (47), $\Theta_i^+ = 0$ when $\Theta_i < 0$ and $\Theta_i^- = 0$ when $\Theta_i > 0$. Hence the constraint can be written as

$$0.1\Theta_1^+ + 0.3125\Theta_1^- + 0.1875\Theta_2^+ + 0.2\Theta_2^- + 0.3\Theta_3^+ + 0.3875\Theta_4^-$$

$$+ 0.4125\Theta_5^+ = 1 \quad . \tag{49}$$

The compatibility equations (45) become

$$\Theta_2^+ - \Theta_2^- + \Theta_3^+ - \Theta_3^- + \Theta_4^+ - \Theta_4^- = 0 \quad ,$$

$$\Theta_1^+ - \Theta_1^- + \Theta_2^+ - \Theta_2^- \qquad - \Theta_4^+ + \Theta_4^- - \Theta_5^+ + \Theta_5^- = 0 \quad , \tag{50}$$

$$\Theta_1^+ - \Theta_1^- + \Theta_2^+ - \Theta_2^- + \Theta_3^+ - \Theta_3^- + \Theta_4^+ - \Theta_4^- + \Theta_5^+ - \Theta_5^- = 0 \quad .$$

We thus minimize the objective function (48) subject to the

Table 3. Values of $\{M^{oE}\}_{max}$

Section	$\{M^{oE}\}_{max}$	
	+ve	−ve
1	0.1ℓ	-0.3125ℓ
2	0.1875ℓ	-0.2ℓ
3	0.3ℓ	0
4	0	-0.3875ℓ
5	0.4125ℓ	0

equality constraints of equations (49) and (50). This is a
linear programming problem, and the solution is

$$\frac{\Gamma\ell}{M_o} = 2.857, \quad \theta_1^- = 0.476, \quad \theta_3^+ = 0.952, \quad \theta_4^- = 0.952, \quad \theta_5^+ = 0.476, \quad (51)$$

with all other variables being zero.

We can in fact show that this linear programming problem is
the dual of the appropriate static problem in Section 19.2. The
reader is invited to study the relation between the kinematic
problem in shakedown and the limit analysis problem discussed
in Chapter 13. It will be seen that if we limit the load cycle
to a single point in load space, say (Γ,Γ), the limit analysis
problem is recovered.

It might be remarked that the method described in this section can be modified to include alternating plastic deformation
by using inequality (33) over successive parts of the period T,
thus modifying inequality (35). In beams and frames, however,
alternating plastic deformation can be adequately checked by
the method described in Section 19.3, and the application of
the kinematic method can be restricted to consideration of incremental collapse.

19.5 Historical and Bibliographical Remarks

By far the largest number of applications of the shakedown theorems have been carried out for framed structures. Detailed
descriptions of this work can be found in the volumes by Neal
[1956], Baker, Horne and Heyman [1956] and Hodge [1959]. The
most significant early advance in methods of computing shakedown loads was the kinematic method for frames described in
Section 19.4, given by Neal and Symonds [1951] and Symonds and
Neal [1951]. This work predated the development of the general
kinematic theorem of which it is a special case. Methods of
computing shakedown load parameters for framed structures have

been developed considerably, largely in the context of linear programming. Representative papers are those of Hodge [1954], Heyman [1959], Tocher and Popov [1962], Ceradini and Gavarini [1965], Gavarini [1956], Davies [1967] and Cohn, Ghosh and Parimi [1971].

A more general programming formulation of the shakedown problem has been given by Maier [1969]. Examples of the application of the theorems to more complex structures are given in the work of Leckie [1956] and Hodge and Kalinowski [1967].

An important area in which shakedown analysis is applied is in problems involving thermal cycles. While such problems fall beyond the scope of this volume, they have been discussed by Prager [1957], Rosenblum [1957], [1958] and de Donato [1970].

Recent reviews of work in shakedown have been given by Eyre and Galambos [1967] and Leckie [1971].

References

J.F.Baker,M.R.Horne and J.Heyman	1956	*The Steel Skeleton*, Vol.2, Cambridge Univ.Press.
G.Ceradini and C.Gavarini	1967	"Calcolo a volture e programazione lineare",Giornale del Genio Civile, $\underline{1}$.
M.Z.Cohn,S.K.Ghosh and S.R.Parimi	1971	"A unified approach to the theory of plastic structures", Proc.ASCE. $\underline{98}$, (EM5), 1133.
J.M.Davies	1967	"Collapse and shakedown loads of plane frames", Proc.Struc. Div.ASCE, $\underline{93}$, (ST3), 35.
O.de Donato	1970	"Second shakedown theorem allowing for cycles of both load and temperature", Instituto Lombardo-Academia di Scienze e Lettere,Rendiconti $\underline{A104}$, 265.
D.G.Eyre and T.V.Galambos	1967	"Variable repeated loading – a literature survey", Civil and Environmental Eng.Dept., Washington Univ.(St.Louis), Struc.Div.Rept.No.3.

C.Gavarini   1966   "Plastic analysis of structures and
                              duality in linear programming",
                              Meccanica, 1.

J.Heyman   1959   "Automatic analysis of steel-framed
                              structures under fixed and varying
                              loads", Proc.I.C.E., 12, 39.

P.G.Hodge,Jr.   1954   "Shakedown of elastic-plastic struc-
                              tures", *Residual Stresses in Metals
                              and Metal Construction* (edited by W.
                              R.Osgood), Reinhold, N.Y.

P.G.Hodge,Jr.   1959   *Plastic Analysis of Structures*,
                              McGraw-Hill, N.Y.

P.G.Hodge,Jr. and   1967   "Shakedown interaction curve for a
A.J.Kalinowski   circular arch", Illinois Inst. of
                              Tech. Report DOMMIT, 1.

F.A.Leckie   1965   "Shakedown pressures for flush cy-
                              linder-sphere shell intersections",
                              J.Mech.Eng.Sci., 7, 367.

F.A.Leckie   1971   "Interpretative Report on Shakedown
                              and Ratchetting at Elevated Tempera-
                              ture", Report to the Pressure Vessel
                              Research Committee.

G.Maier   1969   "Shakedown theory in perfect elasto-
                              plasticity with associated and non-
                              associated flow laws: a finite ele-
                              ment, linear programming approach",
                              Meccanica, 4, 1.

B.G.Neal   1956   *The Plastic Methods of Structural
                              Analysis*, Chapman and Hall (London).

B.G.Neal and   1951   "A method for calculating the failure
P.S.Symonds   load for a framed structure subjected
                              to fluctuating loads", J.Inst.Civ.
                              Engrs., 35, 186.

W.Prager   1957   "Shakedown in elastic-plastic media
                              subjected to cycles of load and tem-
                              perature", Symp. sulla Plasticità
                              nelle Scienze delle Costruzioni,
                              Bologna.

W.J.Rosenblum   1957   "On the shakedown theorem for elastic-
                              plastic bodies", Izv.Akad.Nauk.USSR,
                              OTN 6, 47.

W.J.Rosenblum   1958   "On the shakedown problem of unevenly heated elastic-plastic bodies", Izv. Akad.Nauk.USSR, OTN $\underline{7}$, 136.

P.S.Symonds and   1951   "Recent progress in the plastic methods of structural analysis", J. Franklin Inst., $\underline{252}$, 383.
B.G.Neal

J.L.Tocher and   1962   "Incremental collapse analysis of rigid frames", Proc.4th U.S.Nat. Congr.Appl.Mech., ASME, N.Y., 727.
E.P.Popov

Chapter 20

DEFORMATIONS OF ELASTIC, PERFECTLY PLASTIC STRUCTURES

20.1 Bounds on Plastic Work

Limit analysis and shakedown analysis provide an indication of the limit of serviceability of an elastic, perfectly plastic structure or body, since deformations of the structure will increase markedly when flow begins or when the shakedown limit is achieved. However, if deformations of the structure are important, it is of concern to know whether large displacements will occur during a loading program other than as a result of flow or incremental collapse. Large deformations, particularly those that occur as a result of the combination of elastic and plastic strains, may render a structure unserviceable, or may invalidate the assumption of linear equilibrium equations and strain-displacement relations.

It is frequently essential, therefore, to compute or estimate displacements of the structure at critical instants in the loading program. This can be done by carrying out a complete analysis, a problem we shall discuss further in Part V. Alternatively attempts can be made to estimate displacements to determine whether or not a complete analysis is necessary. A variety of methods of estimating displacements is available. Some of these methods will also be discussed in Part V, but it is convenient at this point to introduce some special results of interest in elastic, perfectly plastic structures.

As a starting point, let us consider the *total plastic work* done during a loading program. If the total plastic work is small, we can expect that the total contribution of the *plastic strains* to the *gross deformation* of the structure will also be small. Of course, this does not preclude the possibility of very large localized plastic strains, or the possibility of large displacements resulting from elastic deformation.

Nevertheless, the total plastic work serves as a first indication of the extent of the contribution of the plastic strains to the behavior of the structure.

Consider a structure composed of an elastic, perfectly plastic material and subjected to a generalized loading program $\Gamma_\alpha(t)$. Let the actual generalized stresses and total strain rates in the structure be $Q_j(s,t)$ and $\dot{q}_j(s,t)$ respectively. The loading program is taken to be such that $\Psi\{\Gamma_\alpha(t)\} < 0$.

It is convenient to rewrite the loading program as

$$\Gamma_\alpha(t) \;=\; \Gamma(t)c_\alpha(t) \;\;, \tag{1}$$

where $c_\alpha(t)$ is a unit vector in the load space and $\Gamma(t)$ is a non-negative load parameter. We may then introduce $\Gamma^L(t)$ such that the load point

$$\Gamma^L(t) \;=\; \Gamma^L(t)\,c_\alpha(t) \tag{2}$$

always lies on the limit surface. This permits us to introduce a safety factor against flow, $k^L(t)$, defined at each instant in the loading program and given by

$$k^L(t) = \frac{\Gamma^L(t)}{\Gamma(t)} \;\;. \tag{3}$$

Note that $k^L(t) > 1$, with $k=1$ indicating that flow occurs.

The lower bound theorem of limit analysis asserts that it is possible to find a pertinent statically admissible stress field for the loading program $\Gamma^L(t)$ which, at each instant, everywhere satisfies the yield condition. This field we can write as $k^L(t)Q_j^S(s,t)$. This then implies, by virtue of the linearity of the equilibrium equations, that $\Gamma_\alpha(t), Q_j^S(t)$ are statically admissible, and that

$$\phi\{k^L Q_j^S(s,t)\} \leq 0 \;\;. \tag{4}$$

Since $Q_j(s,t)$ and $Q_j^S(s,t)$ equilibrate the same loads, the principle of virtual velocities may be used to show that

$$\int_V (Q_j - Q_j^S)\dot{q}_j \, dV = 0 \quad . \tag{5}$$

From inequality (4) and the fundamental inequality for perfectly plastic materials, we have that

$$(Q_j - k^L Q_j^S)\dot{q}_j^p \geq 0 \quad , \tag{6}$$

where $\dot{q}_j$ is divided into elastic and plastic parts, $\dot{q}_j^e$ and $\dot{q}_j^p$. Dividing inequality (6) by k^L (noting again that $k^L > 1$), integrating over the volume, and subtracting the result from equation (5), we find that

$$\frac{k^L-1}{k^L} \int_V Q_j \dot{q}_j^p \, dV \leq \int_V (Q_j^S - Q_j)\dot{q}_j^e \, dV \quad . \tag{7}$$

Assuming that the loading program begins at time t=o with the structure in its virgin state, inequality (7) may be integrated over the interval (o,t) to bound the total plastic work, W^p, from below;

$$W^p = \int_o^t d\tau \int_V Q_j \dot{q}_j^p \, dV$$

$$\leq \frac{k^L}{k^L-1} [\int_o^t d\tau \int_V \{Q_j^S(s,t) - Q_j(s,t)\} \, \dot{q}_j^e(s,t) \, dV] \quad . \tag{8}$$

If t is small, the expression on the right hand side will be finite for $k > 1$, and must be quite limited in magnitude for values of k which are not very close to unity. The restriction to small t must be considered carefully; t does not measure real time but merely the order of events. Thus t will be small for a *simple* loading program, where the number of events is

small, and large for a complex loading program, when the number
of events is large.

An example of a simple loading program is a proportional
loading program, where

$$\Gamma_\alpha(t) = \Gamma(t)c_\alpha \quad , \quad \dot{\Gamma} > 0 \quad , \tag{9}$$

and c_α is constant. In this case the total plastic work, and
hence the effect of plastic strain on the gross deformation,
will be limited until the load parameter $\Gamma(t)$ becomes close to
the limit load parameter Γ^L.

Provided that shakedown occurs, this result can be extended
to complex loading programs. We introduce a factor of safety
against failure by alternating plastic deformation or incremental collapse, k^S, by assuming that the loading program

$$\Gamma_\alpha^S(t) = k^S \Gamma(t) c_\alpha(t) \quad , \quad k^S > 1 \tag{10}$$

lies on the borderline between shakedown and failure.

Let $Q_j(s,t)$, $\dot{q}_j(s,t)$ characterize the solution for the loading program $\Gamma_\alpha(t)$ with stress free initial conditions, $\rho_j(s,o)=0$.
Let $Q_j^E(s,t)$ be the elastic solution for this same loading program. Since there must exist a shakedown residual stress field
$\bar{\rho}_j(s)$ for the loading program $\Gamma_\alpha^S(t)$, it follows that

$$\phi\{\bar{Q}_j(s,t)\} < 0 \quad , \quad \text{where} \tag{11}$$

$$\bar{Q}_j(s,t) = k^S Q_j^E(s,t) + \bar{\rho}_j(s) \quad .$$

Finally, let $\rho_j(s,t)$ be defined by

$$Q_j(s,t) = Q_j^E(s,t) + \rho_j(s,t) \quad . \tag{12}$$

It follows now that the stress fields $Q_j(s,t)$, with which there
are associated plastic strains $\dot{q}_j^p(s,t)$, and $(1/k^S)\bar{Q}_j(s,t)$, with

which there are no plastic strains associated, are both solutions
for the loading program $\Gamma_\alpha(t)$. Substituting these two solutions
into the convergence criterion given in Chapter 16 (cf. equations 7, 8 and 20), we find that

$$\frac{dA}{dt} = - \int_V (Q_j - \frac{1}{k^s} \bar{Q}_j) \dot{q}_j^p dV \quad , \tag{12a}$$

where

$$A = \frac{1}{2} \int_V C_{jk} (\rho_j - \frac{1}{k^s} \bar{\rho}_j)(\rho_k - \frac{1}{k^s} \bar{\rho}_k) dV \tag{12b}$$

After multiplying through by k^s, equation (12a) may be rewritten
as

$$k^s \frac{dA}{dt} + (k^s - 1) \int_V Q_j \dot{q}_j^p dV \leq - \int_V (Q_j - \bar{Q}_j) \dot{q}_j^p dV \quad . \tag{13}$$

In view of inequality (11),

$$(Q_j - \bar{Q}_j) \dot{q}_j^p \geq 0 \quad , \tag{14}$$

and hence (13) becomes

$$\int_V Q_j \dot{q}_j^p dV \leq - \frac{k^s}{(k^s - 1)} \frac{dA}{dt} \quad . \tag{15}$$

If we now integrate inequality (15) over the internal (o,t),
the total plastic work W^p is given by

$$W^p = \int_o^t d\tau \int_V Q_j \dot{q}_j^p dV$$

$$\leq \frac{k^s}{(k^s - 1)} \{A(o) - A(t)\} \quad . \tag{16}$$

We cannot evaluate $A(t)$, but from equation (12b) we note
that it is non-negative. Hence (16) may be finally written as

$$W^p \leq \frac{k^s}{k^s - 1} A(o)$$

$$= \frac{1}{2k^s(k^s-1)} \int_V C_{jk} \bar{\rho}_j \bar{\rho}_k dV \quad .$$

$$(17)$$

Thus W^p is bounded from above by a quantity which is finite, and which is limited in magnitude if k^s does not approach unity. In this case the contributions of the plastic strains to the gross deformations during the complex loading program will be limited in extent.

20.2 Upper Bounds on Displacements

The most severe limitation of the work bounds is that they give information about overall behavior rather than localized deformations. In this section we shall consider a means of bounding more specific properties of the displacement field. While these bounds are not particularly accurate, they do give a conservative estimate with a comparatively simple calculation.

Consider a structure or body subjected to a loading program $\hat{\underline{p}}(s,t)$ on S_p, $\hat{\underline{u}} = 0$ on S_u and zero body forces. The body is elastic, perfectly plastic, and flow does not occur during the program. We shall further assume that the body is in its virgin state at time $t = 0$ and that $\hat{\underline{p}}(s,o) = 0$. Let the solution be characterized by $Q_j(s,t)$, $q_j(s,t)$, $\underline{u}(s,t)$ on V and S_p. Further, let the elastic solution be characterized by $Q_j^E(s,t)$, $q_j^E(s,t)$, $\underline{u}^E(s,t)$. The residual stress field $\rho_j(s,t)$ is again defined as

$$\rho_j = Q_j - Q_j^E \quad .$$

$$(18)$$

The bound is obtained by introducing constant dummy loads $\hat{\underline{p}}^*(s)$ on S_p. Let the _elastic_ solution associated with the problem $\hat{\underline{p}}^*(s)$ on S_p, $\hat{\underline{u}} = 0$ on S_u, and zero body forces be $Q_j^{*E}(s)$,

$q_j^{*E}(s), \; \underline{u}^{*E}(s).$

The quantity we shall bound is

$$\int_{S_p} \hat{\underline{p}}^*(s) \cdot \underline{u}(s,T) dS \quad . \tag{19}$$

This permits us to choose $\hat{\underline{p}}^*(s)$ in such a way that useful information regarding the displacement field $\underline{u}(s,t)$ at time $t = T$ is obtained.

We introduce an arbitrary self-equilibrating field $\bar{\rho}_j(s)$, and we impose a constraint on the choice of $\underline{p}^*(s)$ by requiring that it must be possible to choose $\bar{\rho}_j(s)$ such that

$$\phi\{Q_j^E(s,t) + Q_j^{*E}(s,t) + \bar{\rho}_j(s)\} \leq 0 \quad . \tag{20}$$

We note first that, by the principle of virtual velocities,

$$\int_V \{\bar{\rho}_j(s) - \rho_j(s,t)\} \, \dot{q}_j(s,t) dV = 0 \quad . \tag{21}$$

Again using the principles of virtual velocities, the quantity we wish to bound is given by

$$\int_{S_p} \hat{\underline{p}}^*(s) \cdot \underline{u}(s,T) dS = \int_o^T dt \int_V Q_j^{*E}(s) \dot{q}_j(s,t) dV \quad . \tag{22}$$

Integrating equation (21) over the interval $(0,T)$, adding and subtracting $Q_j^E(s,t)$ to the integrand, and using equation (18), we see that equation (22) can be written as

$$\int_{S_p} \hat{\underline{p}}^*(s) \cdot \underline{u}(s,T) dS = \int_o^T dt \int_V (Q_j^E + Q_j^{*E} + \bar{\rho}_j - Q_j) \dot{q}_j dV$$

$$\leq \int_o^T dt \int_V (Q_j^E + Q_j^{*E} + \bar{\rho}_j - Q_j) \dot{q}_j^e dV \tag{23}$$

$$= \int_o^T dt \int_V (\bar{\rho}_j - \rho_j) \dot{q}_j^e dV + \int_o^T dt \int_V Q_j^{*E} \dot{q}_j^e dV \quad .$$

The inequality follows from the observation that, in view of
inequality (20),

$$\{Q_j - (Q_j^E + Q_j^{*E} + \bar{\rho}_j)\}\dot{q}_j^p \geq 0 \ .\tag{24}$$

Using the relation between stress and elastic strain, the first
terms on the right hand side of (23) may be rearranged:

$$\int_o^T dt \int_V (\bar{\rho}_j - \rho_j)\dot{q}_j^e dV = \int_o^T dt \int_V C_{jk}(\bar{\rho}_j - \rho_j)\dot{\bar{Q}}_k dV$$

$$= - \int_o^T dt \int_V C_{jk}(\bar{\rho}_j - \rho_j) \frac{d}{dt}(\bar{\rho}_k - \rho_k)dV\tag{25}$$

$$\leq \int_V \frac{1}{2} C_{jk}\bar{\rho}_j\bar{\rho}_k dV \ ,$$

in view of observation that

$$\int_V \frac{1}{2} C_{jk}\{\bar{\rho}_j(s) - \rho_j(s,T)\}\{\bar{\rho}_k(s) - \rho_k(s,T)\}dV$$

is non-negative.

Further, using the reciprocal theorem and the principle of
virtual work, the second term on the right hand side of equation (23) may also be rearranged. Noting first that

$$\int_V \dot{\rho}_j q_j^{*E} dV = 0 \ ,\tag{26}$$

we see that

$$\int_o^T dt \int_V Q_j^{*E}\dot{q}_j^e dV = \int_o^T dt \int_V \dot{Q}_j(s,t)q_j^{*E}(s)dV$$

$$= \int_o^T dt \int_V \dot{Q}_j^E q_j^{*E} dV + \int_o^T dt \int_V \dot{\rho}_j q_j^{*E} dV$$

$$= \int_o^T dt \int_V Q_j^{*E}\dot{q}_j^E dV\tag{27}$$

$$= \int_{S_p} \hat{\underline{p}}^*(s)\cdot\underline{u}^E(s,T)dS \ .$$

Finally, combining (23),(25) and (27), the bound is established:

$$\int_{S_p} \hat{\underline{p}}*(s).\underline{u}(s,T)dS \;\leq\; \int_V \tfrac{1}{2} C_{jk}\bar{\rho}_j\bar{\rho}_k\,dV + \int_{S_p} \hat{\underline{p}}*(s).\underline{u}^E(s,T)dS \quad . \tag{28}$$

As an example of the application of this result, consider an elastic, perfectly plastic cylinder of radius R subjected to a torque T about its axis. This problem has been considered in detail in Chapter 7, and the results given here are taken from Section 7.4.

The elastic solution to this problem is given by

$$T = \frac{\pi E R^4 \theta^E}{2(1+\nu)} \quad , \tag{29}$$

where θ is the twist per unit length, and E, ν are Young's modulus and Poisson's ratio respectively. The stress distribution is most conveniently given in cylindrical coordinates; the only non-zero stress component is

$$\sigma_{rz}^E = \frac{E\theta^E}{(1+\nu)}\, r = \frac{2T}{\pi R^3}\,\frac{r}{R} \tag{30}$$

where r is the distance from the axis. Assuming that yield occurs when $\sigma_{rz} = \sigma_o$, first yield will occur for monotonic loading when the torque has the value

$$T_y = \frac{\pi \sigma_o R^3}{2} \quad . \tag{31}$$

We may also introduce

$$\theta_y = \frac{\sigma_o(1+\nu)}{ER} \quad . \tag{32}$$

The elastic relation, equation (29), now becomes

$$\frac{\theta^E}{\theta_y} = \frac{T}{T_y} \quad , \tag{33}$$

and the elastic stress distribution is

$$\frac{\sigma_{rz}^{E}}{\sigma_{o}} = \frac{T}{T_{y}} \frac{r}{R} \quad . \tag{34}$$

Let us also assume that the limit solution is known. This is

$$\sigma_{rz} = \sigma_{o} \quad ,$$

$$T_{L} = \frac{2\pi\sigma_{o}R^{3}}{3} = \frac{4}{3}T_{y} \quad . \tag{35}$$

It is now evident that

$$\frac{\sigma_{rz}}{\sigma_{o}} = 1 = \frac{3T_{L}}{4T_{y}} \tag{36}$$

is a statically admissible stress distribution. Subtracting
equation (34) from equation (36), and replacing T_{L}/T_{y} by an
arbitrary factor γ,

$$\frac{\bar{\sigma}_{rz}}{\sigma_{o}} = \gamma(\frac{3}{4} - \frac{r}{R}) \tag{37}$$

is a self stress system for this problem.

Let us put

$$T = \alpha T_{y} \quad , \tag{38a}$$

and choose a dummy torque

$$T^{*} = \beta T_{y} \quad . \tag{38b}$$

If θ is the twist per unit length resulting from a history of
torque T which is zero at t=0 and has a final value of αT_{y}, the
form of the bound (28) becomes

$$\beta T_{y}\theta \leq A(\bar{\sigma}_{rz}) + \beta T_{y}\theta^{E} \quad , \tag{39}$$

where

$$A(\bar{\sigma}_{rz}) = \int_{o}^{R} \frac{1}{2} \frac{(1+\nu)}{E} \bar{\sigma}_{rz}^{\,2} (2\pi r)dr \quad . \tag{40}$$

Since

$$\frac{\sigma*_{rz}^{E}}{\sigma_{o}} = \beta \frac{r}{R} \quad , \tag{41}$$

we see that

$$\frac{(\sigma_{rz}^{E} + \sigma*_{rz}^{E} + \bar{\sigma}_{rz})}{\sigma_{o}} = (\alpha+\beta) \frac{r}{R} - \gamma (\frac{r}{R} - \frac{3}{4}) \quad . \tag{42}$$

This stress distribution must satisfy the yield condition
for $0 < r < R$. Putting $(\sigma_{rz}^{E} + \sigma*_{rz}^{E} + \bar{\sigma}_{rz}) = \sigma_{o}$ at $r = R$, we see
that

$$\alpha + \beta - \frac{\gamma}{4} = 1 \quad ,$$

or

$$\gamma = 4(\alpha+\beta-1) \quad . \tag{43}$$

At $r = 0$, the condition that $(\sigma_{rz}^{E} + \sigma*_{rz}^{E} + \bar{\sigma}_{rz}) \leq \sigma_{o}$ gives

$$\frac{3\gamma}{4} \leq 1 \quad ,$$

or

$$\alpha + \beta \leq \frac{4}{3} \quad . \tag{44}$$

We may now evaluate $A(\bar{\sigma}_{rz})$, using the value of γ given in equation (43). It is found that

$$A(\bar{\sigma}_{rz}) = \gamma^{2}\pi \frac{(1+\nu)}{32E} \sigma_{o}^{2} R^{2} = (\alpha+\beta-1)^{2} T_{y}\theta_{y} \quad . \tag{45}$$

The form of the bound (39) is thus now

$$\beta T_y \theta \leq (\alpha+\beta-1)^2 \, T_y \theta_y + \alpha\beta T_y \theta_y \quad ,$$

or

$$\frac{\theta}{\theta_y} \leq \frac{(\alpha+\beta-1)^2}{\beta} + \alpha \quad . \tag{46}$$

One choice we may make is that of (cf. equation 43)

$$\alpha + \beta = \frac{4}{3} \quad , \quad \beta = \frac{4}{3} - \alpha \quad . \tag{47}$$

The bound then becomes

$$\frac{\theta}{\theta_y} \leq \frac{1}{12-9\alpha} + \alpha \quad . \tag{48}$$

This bound will apply for $0 \leq \alpha \leq 4/3$, and for a loading program which begins with $T=0$, terminates at $T = \alpha T_y$, and is such that $-2T_y/3 \leq T \leq \alpha T_y$ for intermediate times.

Certainly a monotonic loading program satisfies these constraints. For this case, from Section 7.4, we know that

$$\frac{T}{T_y} = \frac{4}{3} \left\{ 1 - \frac{1}{4}\left(\frac{\theta_y}{\theta}\right)^3 \right\} \quad . \tag{49}$$

In Figure 1 the bound is compared with the exact solution for the monotonic loading case.

This simple example illustrates the application of the method. Bounds over ranges of α can be improved by optimizing the right hand side of (46). While this case is essentially trivial, a construction of a bound which uses only the elastic and limit stress fields offers a possibility of obtaining useful results even when the incremental analysis is more difficult than the torsion problem.

20.3 Deflections in Frames at Incipient Flow

One further case of practical importance in which it is possible

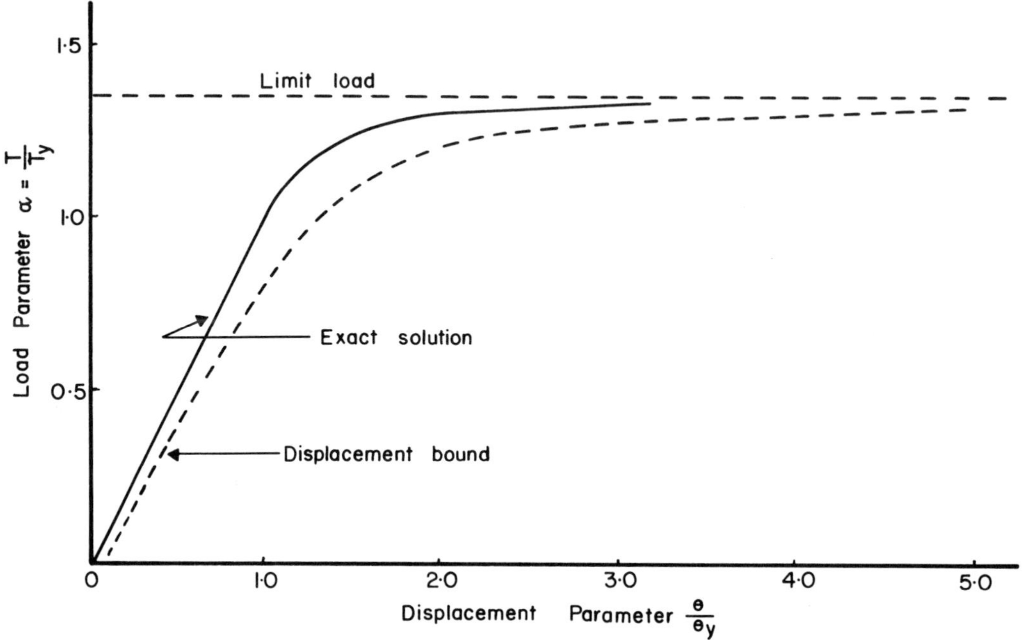

Figure 1. Bounds on the twist of a circular bar

to make direct calculations of displacements occurs in frame
problems in which only the deformations due to bending are considered. If the limit solution is known, it is possible to
compute the displacements at the point of incipient flow directly on the basis of a number of frequently satisfied assumptions.
We deal only with frames of piecewise uniform section, so that
a limited number of potential hinge positions can be identified
in advance. The frame is assumed to be subjected to monotonically increasing proportional loading and it is further assumed
that *unloading does not occur* at any hinge.

It follows then that if the limit load parameter and the velocity or mechanism of flow is known, we know at the instant at
which flow commences the bending moment at a number of points
in the frame and that plastic deformation may have occurred at
these points during loading. At one such point no plastic

deformation can have occurred, since when the load parameter
reaches its limiting value the moment at at least one section
must have just reached the value M_o, permitting flow to commence. Precisely which section has not deformed is in general
not known.

If we proceed to formulate the compatibility conditions for
the frame using the moment values derived from the limit solution and assuming that deformation has occurred at all sections
which deform during flow, we will generate one equation less
than the number of independent hyperstatic unknowns and independent hinge rotations. We may arbitrarily assign any one
hinge rotation the value zero, and solve for the remaining rotations. The assignation will be correct only if the signs of
the non-zero hinge rotations correspond to the sign of the bending moment at that point. If this is not so, we may proceed to
assume that each hinge rotation in turn is non-zero until conditions at the remaining hinges are met.

Alternatively, we may note that deformations of the frame
corresponding to the flow mechanism satisfy the compatibility
equations. Hence we may superimpose on our first and incorrect
solution a set of hinge rotations bearing a relation to each
other corresponding to the flow mechanism and calculated to
bring to zero that hinge rotation which to the greatest extent
has a sign opposite to the bending moment. This process leaves
us with one hinge undeformed and all remaining hinge rotations
in the correct sense. All hyperstatic unknowns can also be
found, and displacements at any point on the frame can readily
be calculated.

This process can be illustrated by a simple example. Consider
the frame shown in Figure 2. The frame is taken to be uniform,
with limit moment M_o and flexural rigidity EI. The limit solution, corresponding to $\Gamma \ell / M_o = 3$, is shown in Figure 3. Flow

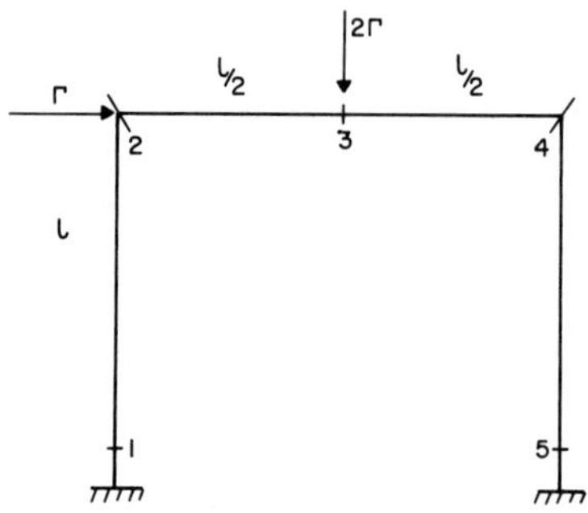

Figure 2. Frame example

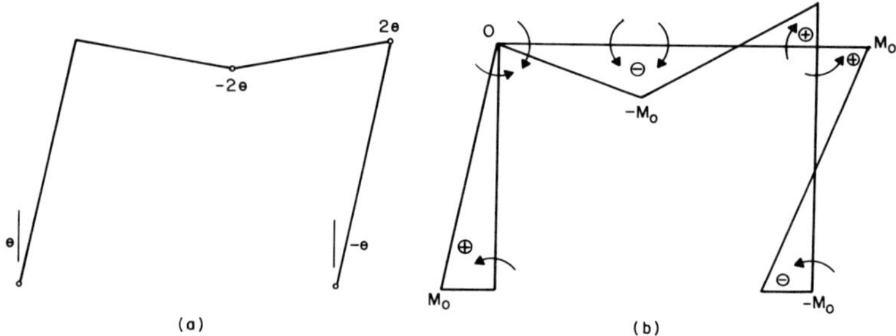

(a)   (b)

Figure 3. Limit solution for frame

occurs at sections 1, 3, 4 and 5, and the complete bending moment distribution M(s) can be found. Bending moments causing
tension on the outside of the frame are taken to be positive.

We assume now that at the instant of incipient flow, when
$\Gamma \ell/Mo$ has just reached the value 3, rotations $\theta_1, \theta_3, \theta_4, \theta_5$ have
taken place at sections 1, 3, 4, 5 respectively. Three inde-
pendent self stress systems, $m_1(s)$, $m_2(s)$, $m_3(s)$ are shown in
Figure 4. From the principle of virtual work, we see that

$$\int m_1 \frac{M}{EI} \, dS + \theta_1 \ell + \theta_3 \ell + \theta_4 \ell + \theta_5 \ell = 0 \quad ,$$

$$\int m_2 \frac{M}{EI} \, dS \qquad + \theta_3 \ell + \theta_4 \ell \qquad = 0 \quad , \qquad (50)$$

$$\int m_3 \frac{M}{EI} \, dS + \frac{1}{2} \theta_1 \ell - \frac{1}{2} \theta_4 \ell - \frac{1}{2} \theta_5 \ell = 0 \quad .$$

After evaluating the integrals, which can be done rapidly for

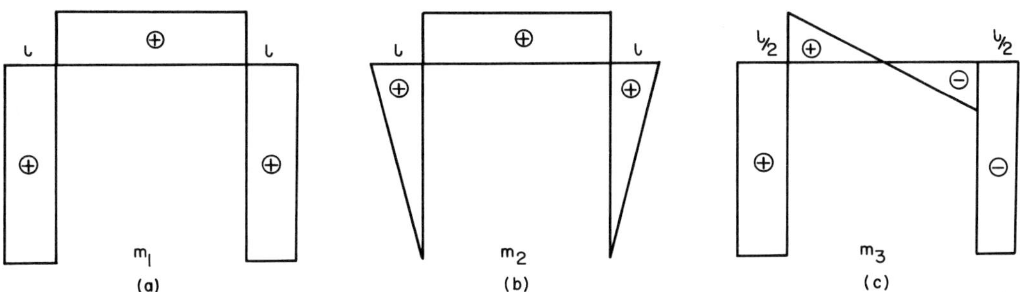

Figure 4. Self stress systems

products of piecewise linear functions, we find

$$
\begin{bmatrix}
1 & 1 & 1 & 1 \\
0 & 1 & 1 & 0 \\
1 & 0 & -1 & -1
\end{bmatrix}
\begin{bmatrix}
\theta_1 \\
\theta_3 \\
\theta_4 \\
\theta_5
\end{bmatrix}
= \frac{M_o\ell}{12EI}
\begin{bmatrix}
3 \\
1 \\
4
\end{bmatrix} .
\tag{51}
$$

Now let us set, say, $\theta_4 = 0$. Solving the three equations we find

$$
\begin{bmatrix}
\theta_1 \\
\theta_3 \\
\theta_4 \\
\theta_5
\end{bmatrix}
= \frac{M_o\ell}{12EI}
\begin{bmatrix}
3 \\
1 \\
0 \\
-1
\end{bmatrix} .
\tag{52}
$$

On comparison with the signs of the moments in Figure 3(b), we see that θ_3 is positive while the moment at that point is negative. The other signs correspond to those of the moments. A set of hinge rotations, taken from Figure 3(a)

$$
\begin{bmatrix}
\theta_1 \\
\theta_3 \\
\theta_4 \\
\theta_5
\end{bmatrix}
= \Theta
\begin{bmatrix}
+1 \\
-2 \\
+2 \\
-1
\end{bmatrix} ,
\tag{53}
$$

will satisfy equations (51). Hence if we choose

$$\theta = \frac{M_o\ell}{24EI} \quad ,$$

and superimpose the rotations of equations (52) and (53), we obtain the rotations

$$\begin{bmatrix} \theta_1 \\ \theta_3 \\ \theta_4 \\ \theta_5 \end{bmatrix} = \frac{M_o\ell}{24EI} \begin{bmatrix} 7 \\ 0 \\ 2 \\ -3 \end{bmatrix} . \qquad (54)$$

These rotations have the correct signs, and provided that no hinge has unloaded, these are the correct rotation values.

Displacements can now be found. If, for example, we require the horizontal displacement u_H at the point of application of the horizontal load, we use the moment system m* shown in Figure 5, which is in equilibrium with a unit horizontal load acting at the eaves. The principle of virtual work gives

$$1.u_H = \int m* \frac{M}{EI} \, dS + \ell\theta_1 \quad ,$$

or

$$u_H = \frac{5}{8} \frac{M_o\ell^2}{EI} \quad . \qquad (55)$$

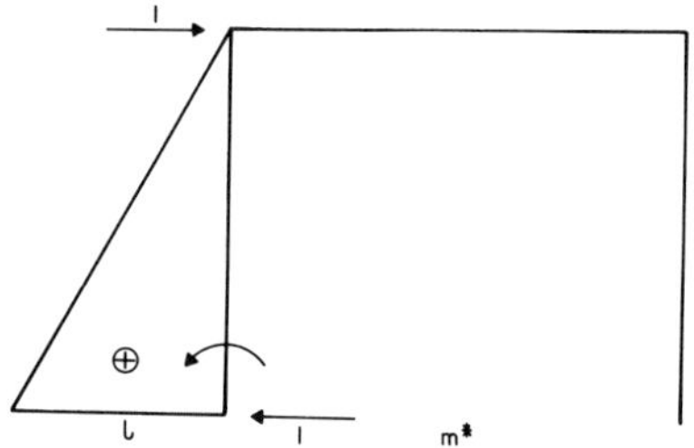

Figure 5. Dummy load system

20.4 Historical and Bibliographical Remarks

The work bounds developed in Section 20.1 were given by Koiter [1960]. The displacement bound described in Section 20.2 follows the work of Ponter [1972], although it is given here in a slightly restricted form.

The problem of the determination of the displacements at incipient flow in framed structures has received considerable attention. Representative papers are those due to Symonds and Neal [1951, 1952], Stevens [1960], and Heyman [1961, 1968].

References

J.Heyman	1961	"On the estimation of deflexions in elastic-plastic framed structures", Proc. I.C.E., $\underline{19}$, 39.
J.Heyman	1968	"Bending moment distribution in collapsing frames", in *Engineering Plasticity* (edited by J.Heyman and F.A.Leckie), Cambridge University Press, 219.
W.T.Koiter	1960	"General theorems for elastic-plastic solids", *Progress in Solid Mechanics*, Vol.2 (edited by I.N.Sneddon and R. Hill), North Holland Press, Chap.4.
A.R.S.Ponter		"An upper bound on the small displacements of elastic, perfectly plastic structures", J.Appl.Mech., $\underline{39}$, 959.
L.K.Stevens	1960	"Direct design by limiting deformations", Proc. I.C.E., $\underline{17}$, 431.
P.S.Symonds and B.G.Beal	1951	"Recent progress in the plastic methods of structural analysis", J. Franklin Inst., $\underline{252}$, 383 and 469.
P.S.Symonds and B.G.Neal	1952	"The interpretation of failure loads in the plastic theory of continuous beams and frames", J.Aero.Sci., $\underline{19}$, 15.

PART V

DEFORMATION THEORIES AND INCREMENTAL ANALYSIS

Chapter 21

THE BOUNDING THEOREM FOR COMPLEMENTARY WORK

21.1 Introductory Remarks

In this and the following two chapters we shall consider a further approach to elastic-plastic structural problems which will permit us to bound a particular property of the solution without the necessity of carrying out an incremental analysis. We shall not be restricted to elastic, perfectly plastic materials (as in Parts III and IV) in this discussion, but will treat constitutive relations for both hardening and perfectly plastic cases.

For simplicity, we consider a structure which is initially stress-free and in its virgin state. Over some time period $0 \leq \tau \leq t$ loads $\hat{\underline{p}}(s,\tau)$ are applied on S_p, body forces $\hat{\underline{F}}(s,\tau)$ are applied on V, and displacements $\hat{\underline{u}}(s,\tau)$ are imposed on S_u. In order to be consistent with the stress free initial condition we assume that

$$\hat{\underline{p}}(s,0) = \hat{\underline{F}}(s,0) = \hat{\underline{u}}(s,0) = 0 \quad . \tag{1}$$

The three functions $\hat{\underline{p}}$, $\hat{\underline{F}}$, $\hat{\underline{u}}$ constitute the *path of loading*, and it is essential that they should be known throughout the interval $0 \leq \tau \leq t$ if the exact solution to the problem at time t is to be obtained.

Let us suppose, however, that in addition to the initial values of the loads (equation 1) only the *final values* at time t are known. Thus at the end of the interval

$$\hat{\underline{p}}(s,t) = \hat{\underline{p}}^t \text{ on } S_p \quad ,$$

$$\hat{\underline{F}}(s,t) = \hat{\underline{F}}^t \text{ on } V \quad , \tag{2}$$

$$\hat{\underline{u}}(s,t) = \hat{\underline{u}}^t \text{ on } S_u \quad .$$

Thus we treat a problem in which the *terminal values of loading are given* (i.e. the initial and final values) but the *path of loading is unspecified.*

We shall suppose that if the exact solution at time t for *any particular loading path* were determined by an incremental analysis it could be characterized by generalized stresses $Q_j(s,t) = Q_j^t$, generalized strains $q_j(s,t) = q_j^t$, displacements $\underline{u}(s,t) = \underline{u}^t$ on S_p and V and reactions $\underline{p}(s,t) = \underline{p}^t$ on S_u.

Two major results will be presented. First, it will be shown that the *complementary work function* $\bar{U}_c$ can be bounded from above. $\bar{U}_c$ is defined by the expression

$$\bar{U}_c(Q_j^t) = \int_V \bar{\Omega}(Q_j^t)dV - \int_{S_u} \underline{p}^t \cdot \hat{\underline{u}}^t dS \ , \tag{3}$$

where

$$\bar{\Omega}(Q_j^*) = \int_o^{Q_j^*} q_j dQ_j \ . \tag{4}$$

If the material is *elastic* $\bar{\Omega}$ becomes the complementary energy density Ω and $\bar{U}_c$ becomes the complementary energy function U_c (see Chapter 1). The function $\Omega(Q_j^*)$ is path independent; $\bar{\Omega}(Q_j^*)$ on the other hand involves plastic strains and is in general a path dependent function. We shall demonstrate a bound on $\bar{U}_c$ which does not require that the stress path $Q_j(s,\tau)$ should be known.

Secondly, it will be shown that the potential work function $\bar{U}_p$ can be bounded from below. $\bar{U}_p$ is defined by the expression

$$\bar{U}_p(q_j^t) = \int_V \bar{W}(q_j^t)dV - \int_{S_p} \hat{\underline{p}}^t \cdot \underline{u}^t dS - \int_V \hat{\underline{F}}^t \cdot \underline{u}^t dV \ , \tag{5}$$

where

$$\bar{W}(q_j^*) = \int_o^{q_j^*} Q_j dq_j \ . \tag{6}$$

If the material is elastic $\bar{W}$ becomes the potential energy U_p
(see Chapter 1). The function $W(q_j^*)$ is path independent, but
$\bar{W}(q_j^*)$ depends on the strain path. We shall again show that $\bar{U}_p$
can be bounded by means of an expression which does not require
that the strain paths be known.

The bound on $\bar{U}_c$ will be given in terms of pertinent statically admissible fields $Q_j^s(s)$, $\underline{p}^s(s)$ on S_u which are in equili-
brium with the final loads $\hat{\underline{p}}^t$ on S_p and $\hat{\underline{F}}^t$ on V. In that $\bar{U}_c$ is
bounded from above, the result can be considered as a valid ge-
neralization of the minimum complementary energy theorem of
elasticity to elastic-plastic materials. The generalization
must be distinguished from a true minimum principle by the fact
that the best or optimal bound which can be obtained will still,
except for special cases, be larger than $\bar{U}_c$. This motivates the
choice of the term *bounding principle* for results of this type.

The bound on $\bar{U}_p$ will be given in terms of pertinent kinematically admissible fields $\underline{u}^c(s)$ on S_p and V, $q_j^c(s)$, which are
compatible with the displacement boundary conditions $\hat{\underline{u}}^t$ on S_u.
Inasmuch that $\bar{U}_p$ is bounded from *below*, our result *cannot* be
considered as a generalization of the minimum potential energy
theorem of elasticity. The new result is again a bounding theorem, and it gives a trivial result in the case of elasticity.

The boundary theorems are obtained by introducing *extremal*
paths in stress and strain space. The extremal paths depend on
the terminal value of stress and strain only, and are the device which permits us to compute bounds on quantities which involve unspecified stress paths. The extremal paths have some
interesting properties, and in particular it will be shown that
the final total strain (elastic and plastic strain) can be derived from a potential function of stress when the stress follows an extremal path in stress space, and that the final stress
can be derived from a potential function of total strain when

the strain follows an extremal path in strain space. This suggests treating the material, for the purpose of computing the bounds, as if it were *elastic*, with a non-linear but reversible relation between stress and total strain. Such relations are usually termed *deformation theories of plasticity*.

Partly because it is consistent with our previous development and partly because the complementary work bounding theorem appears at the present time to be the more powerful, we shall proceed with emphasis on the extremal path in stress space and the proof of the complementary work result. It will then be shown that the existence of an extremal path in stress space implies the existence of an extremal path in strain space and the potential work result follows.

21.2 The Extremal Path in Stress Space

The formulation and exploitation of the bounding theorems make use of the postulates introduced in Chapter 2, Section 2. In our present context the first postulate of Section 2.4 may be stated in the following terms. Let an element of material be subject to some state of stress Q_j^t. This implies some previous history of stress in the element, but this history is not specified. Let the state of stress in the element now be changed from Q_j^t to some other state of stress along a monotonic *straight line path* or proportional path in the stress space, as shown diagrammatically in Figure 1. The first postulate (Section 2.4, equation 2) requires that

$$(Q_j^s - Q_j^t)(q_j^s - q_j^t) \geq 0 \quad , \tag{7}$$

where q_j^t and q_j^s are respectively the total strains at the beginning and end of the straight line path. The strain q_j^t, and consequently q_j^s, depends on the prior history of stress, but inequality (7) is required to hold for all possible such prior histories.

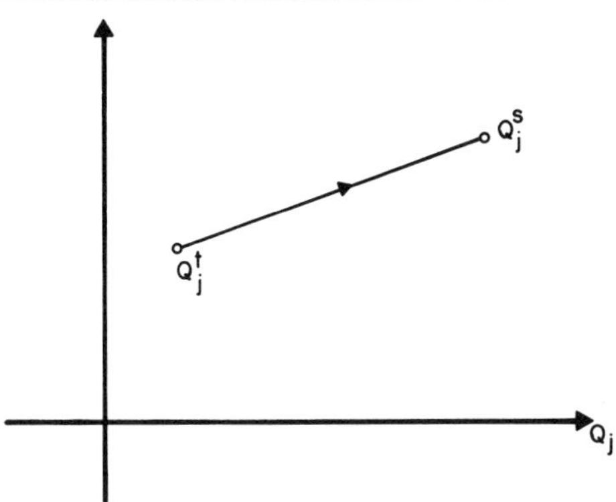

Figure 1. Straight line path

In addition, it was shown in Section 2.4 that inequality (7) implies that the net complementary work and the net work along the straight line path should both be non-negative. Thus

$$\int_{Q_j^t}^{Q_j^s} (q_j - q_j^t)\, dQ_j \geq 0 \quad , \qquad \text{and} \tag{8a}$$

$$\int_{q_j^t}^{q_j^s} (Q_j - Q_j^t)\, dq_j \geq 0 \quad . \tag{8b}$$

In the integrands of these expressions Q_j, q_j represent the stress and strain as the stress point moves monotonically along the path.

We now introduce the concept of an *extremal path in stress space*. Suppose that an element of material, initially stress free and in the virgin state, is subjected to some stress history which terminates with the stress given by Q_j^*. The complementary work along any such path can be computed as

$$\bar{\Omega}(Q_j^*) = \int_0^{Q_j^*} q_j\, dQ_j \quad . \tag{9}$$

$\bar{\Omega}(Q_j^*)$ is thus a scalar quantity which is a function of the stress path between the origin and the terminal state of stress Q_j^*. We define an *extremal path in the stress space as a particular stress path such that the complementary work along this path, $\Omega^0(Q_j^*)$, is not less than the complementary work between the origin and the terminal stress along any other path.* Thus

$$\Omega^0(Q_j^*) \geq \bar{\Omega}(Q_j^*) \; . \tag{10}$$

The alternative paths are shown diagrammatically in Figure 2.

The introduction of the extremal path does not imply any further restrictions on the material; we simply identify a particular path with special properties. The problem of the *determination* of the extremal path for any given material will be deferred for the present. We shall instead proceed to show how such a path can be used in the formulation of the bounding theorem.

21.3 The Complementary Work Inequality

Consider an element of the material which is subject to the following *alternative* stress histories. First, starting from the stress free virgin state, the stress history follows the *extremal path* to a terminal state of stress Q_j^s. Alternatively, the material is subjected to some *unspecified stress history* which leads to stress Q_j^t; thereafter the stress is changed proportionally and monotonically between Q_j^t and Q_j^s. Thus as shown diagrammatically in Figure 3, one stress history follows the

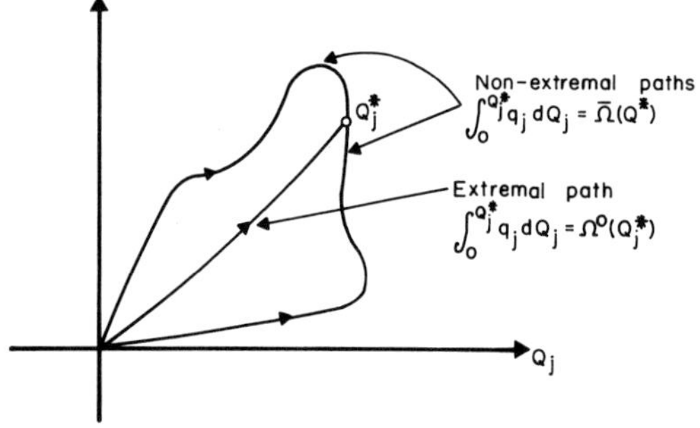

Figure 2. Extremal stress path

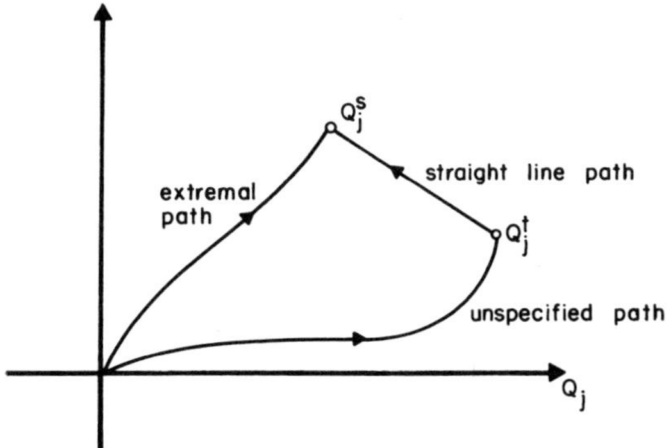

Figure 3. Alternate stress paths

extremal path from the origin to Q_j^s, while the other follows an unspecified path between the origin and Q_j^t and then a straight line path between Q_j^t and Q_j^s. Clearly, as a result of our definition of an extremal path, equation (10) may be applied to these two paths. Additional information can be gained by breaking the integral along the two paths into two parts, and carrying out the following manipulations;

$$\Omega^o(Q_j^s) \geq \bar{\Omega}(Q_j^s)$$

$$= \int_o^{Q_j^t} q_j dQ_j + \int_{Q_j^t}^{Q_j^s} q_j dQ_j \qquad (11)$$

$$= \int_o^{Q_j^t} q_j dQ_j + \int_{Q_j^t}^{Q_j^s} (q_j - q_j^t) dQ_j + q_j^t(Q_j^s - Q_j^t) \quad .$$

Now we observe that the second term on the right hand side of this expression,

$$\int_{Q_j^t}^{Q_j^s} (q_j - q_j^t) dQ_j \quad ,$$

is an integral along a monotonic straight line path. As a

result of equation (8a), therefore, it is non-negative and hence may be dropped from inequality (11). Rearranging, inequality (11) becomes

$$\Omega^{\circ}(Q_j^s) - Q_j^s q_j^t \geq \bar{\Omega}(Q_j^t) - Q_j^t q_j^t \quad . \tag{12}$$

This is the fundamental inequality which is required to establish the bounding theorem. It relates the terminal stresses, strains and complementary work achieved along an extremal path terminating at Q_j^s and an unspecified path terminating at Q_j^t. The introduction of the straight line path between Q_j^t and Q_j^s is merely a device used to establish the inequality; integrals of the response along the straight line path do not appear in inequality (12). We can always *imagine* that an unspecified path terminating at Q_j^t can be extended along the straight line path between Q_j^t and Q_j^s, since the prior mechanical behavior is not affected, but we do not actually have to carry out this stress program.

By making use of the identity

$$\int_0^{Q_j^t} q_j \, dQ_j + \int_0^{q_j^t} Q_j \, dq_j = Q_j^t q_j^t \quad , \tag{13a}$$

and putting

$$\bar{W}(q_j^t) = \int_0^{q_j^t} Q_j \, dq_j \quad , \tag{13b}$$

the fundamental inequality can be written in an alternative form. Substituting from (13a), inequality (12) becomes

$$\Omega^{\circ}(Q_j^s) + \bar{W}(q_j^t) \geq Q_j^s q_j^t \quad . \tag{14}$$

An inequality of this form can be established for a wide range of mechanical behavior, including time dependent behavior.

21.4 The Bounding Theorem for Complementary Work

The complementary energy bounding theorem follows immediately from inequality (12). As set out in Section 21.1, we consider a body subjected to loads $\hat{\underline{p}}^t$ on S_p, body forces $\hat{\underline{F}}^t$ on V and displacements $\hat{\underline{u}}^t$ on S_u. The path of loading is not given, although it is assumed that the body was originally in the stress-free virgin state. We suppose that the exact solution to the problem (obtained by an incremental analysis if necessary) is characterized by stresses Q_j^t, reactions $\underline{p}^t$ on S_u, strains q_j^t and displacements $\underline{u}^t$. The stress history at each point in the body will depend on the load history, and is not known.

Let stresses $Q_j^s(s)$ and reactions $\underline{p}^s(s)$ on S_u constitute a *pertinent statically admissible field*, i.e. these quantities are in internal and external equilibrium with loads $\hat{\underline{p}}^t(s)$ on S_p and body forces $\hat{\underline{F}}^t$ on V. No restriction is placed on the stress paths which terminate in Q_j^s by the equilibrium requirements; we may assume that they are extremal paths at each point in the body. Inequality (12) thus applies at each point in the body; it may be integrated over the volume to give

$$\int_V \Omega^o(Q_j^s)\,dV - \int_V Q_j^s q_j^t\,dV \ge \int_V \bar{\Omega}(Q_j^t)\,dV - \int_V Q_j^t q_j^t\,dV \quad . \tag{15}$$

Because they are taken from the exact solution, $\hat{\underline{p}}^t$ on S_p, $\underline{p}^t$ on S_u, $\hat{\underline{F}}^t$ on V and Q_j^t are statically admissible and $q_j^t, \underline{u}^t$ are kinematically admissible. Thus, by the principle of virtual work,

$$\int_V Q_j^s q_j^t\,dV = \int_{S_p} \hat{\underline{p}}^t \cdot \underline{u}^t\,dS + \int_{S_u} \underline{p}^s \cdot \hat{\underline{u}}^t\,dS + \int_V \hat{\underline{F}}^t \cdot \underline{u}^t\,dV \quad , \tag{16a}$$

$$\int_V Q_j^t q_j^t\,dV = \int_{S_p} \hat{\underline{p}}^t \cdot \underline{u}^t\,dS + \int_{S_u} \underline{p}^t \cdot \hat{\underline{u}}^t\,dS + \int_V \hat{\underline{F}}^t \cdot \underline{u}^t\,dV \quad . \tag{16b}$$

Substituting equations (16) into (15), and eliminating the terms

$$\int_{S_p} \hat{\underline{p}}^t \cdot \underline{u}^t \, dS + \int_V \hat{\underline{F}}^t \cdot \underline{u}^t \, dV$$

which appear on both sides of the inequality, the *complementary work bounding theorem* is obtained;

$$U_c^o(Q_j^s) = \int_V \Omega^o(Q_j^s) \, dV - \int_{S_u} \underline{p}^s \cdot \hat{\underline{u}}^t \, dS \geq \int_V \bar{\Omega}(Q_j^t) \, dV - \int_{S_u} \underline{p}^t \cdot \hat{\underline{u}}^t \, dS \tag{17}$$

$$= \bar{U}_c(Q_j^t) \quad .$$

The left hand side of this expression can be computed in terms of the assumed pertinent statically admissible field Q_j^s, $\underline{p}^s$ on S_u and the extremum function $\Omega^o(Q_j^*)$. The right hand side is the actual complementary work which depends on the path of loading. Thus, *even when the path of loading is not known* an upper bound on $\bar{U}_c$ can be obtained.

The relation between the result given in inequality (17) and the elastic minimum complementary energy theorem can be readily understood if we consider the meaning of the extremal path in the case of elastic behavior. The generalized strain q_j can be divided into the elastic and plastic parts

$$q_j = q_j^e + q_j^p \quad , \tag{18}$$

so that the complementary work done over a stress path which originates in the stress free virgin state and terminates at Q_j^* may be written as

$$\bar{\Omega}(Q_j^*) = \int_o^{Q_j^*} q_j \, dQ_j$$

$$= \int_{o}^{Q_j^*} q_j^e dQ_j \; + \; \int_{o}^{Q_j^*} q_j^p dQ_j \tag{19}$$

$$= \Omega(Q_j^*) + \Omega^p(Q_j^*) \quad .$$

The relation between q_j^e and Q_j is by definition elastic and path independent, and hence the elastic contribution $\Omega(Q_j^*)$ is simply the elastic complementary energy density. It is also path independent, and depends only on the terminal stress Q_j^*. This is a sufficient condition that $\Omega(Q_j^*)$ should be a potential function, and the elastic component of strain is given by

$$q_j^{e*} = \left. \frac{\partial \Omega}{\partial Q_j} \right|_{Q_j = Q_j^*} \quad . \tag{20}$$

It is now evident that if the material behaves entirely elastically so that q_j^p is zero along *all* possible paths, the complementary work is the same for all paths. Thus in the elastic case,

$$\Omega^o(Q_j^*) = \Omega(Q_j^*) \quad , \tag{21}$$

and the complementary work along the extremal path is simply the complementary energy. The extremum path is not unique (which may also be true in more complex materials) but $\Omega^o(Q_j^*)$ is of course uniquely determined.

In the elastic case, therefore, both Ω^o and $\bar{\Omega}$ in inequality (17) are replaced by the complementary energy Ω as a consequence of path independence. The bounding theorem reduces to the minimum complementary energy theorem of elasticity (Section 1.6, equation 30), and can thus be considered an as appropriate generalization of this theorem to path dependent materials.

One other feature in the comparison between the elastic and elastic-plastic cases should be emphasized. Suppose that we put $Q_j^s = Q_j^t$, $\underline{p}^s = \underline{p}^t$ on S_u in inequality (17). If the material is elastic, the left hand and right hand sides of inequality (17) will be equal. We can in fact determine Q_j^t, $\underline{p}^t$ on S_u by finding the absolute minimum value of the left hand side of (17). When the material may undergo plastic deformation, however, the left hand and right hand sides can be equal for $Q_j^s = Q_j^t$ only if

$$\Omega^o(Q_j^t) = \bar{\Omega}(Q_j^t) \quad , \tag{22}$$

i.e., only if the actual paths are such that the complementary work is a maximum. This may happen under some circumstances, but in the general case it cannot be expected.

As a result of this we can conclude that the best or optimum bound on $\bar{U}_c$ (not necessarily given by Q_j^t) cannot be expected to coincide with $\bar{U}_c$. Alternatively, Q_j^t, $\underline{p}^t$ on S_u *cannot* be found by minimizing the bound on $\bar{U}_c$ as can be done in the elastic case. This can be regarded as the cost of determining a bound which does not depend on the loading and stress paths when the material is in reality path dependent.

21.5 Extremal Paths for Elastic, Perfectly Plastic Materials

In the previous section it was seen that when the material is elastic the complementary energy and the maximum complementary work are indistinguishable, and that any path may in consequence be chosen as an extremal path. A stronger result regarding elastic paths (i.e. paths which lie entirely within the yield surface) can readily be established; this result holds for all elastic-plastic materials but it is of particular significance for elastic, perfectly plastic materials.

Suppose that for an elastic-plastic material we wish to

determine the extremal path between the virgin state and a final stress state Q_j^* such that $\bar{\phi}(Q_j^*) < 0$, when $\bar{\phi}(Q_j) = 0$ is the *initial* yield surface. The extremal path may be either an elastic path (i.e. one which lies entirely within the yield surface and for which $q_j^p = 0$) or it may be a path which passes through the initial yield surface so that plastic deformation occurs, as shown diagrammatically in Figure 4.

It is appropriate at this point to emphasize an implicit assumption in the concept of extremal paths; the extremal path is a member of the class of permissible paths, so that it must be possible for the extremal stress history to actually be applied to an element of material. Thus, if there exists a limit surface $\psi(Q_j) = 0$ in stress space, clearly the terminal state of stress and the stress paths under consideration must all lie within or on the limit surface.

Returning to Figure 4, no plastic deformation and hence no change in the recorded history occurs along the elastic path. We can thus imagine that the stress point follows the elastic path to Q_j^*, and then traverses a *cycle* in stress space by returning to the origin along the elastic path and then following the inelastic path to Q_j^*. For this cycle, from the second fundamental postulate (Section 2.4, equation 12) in its strong form,

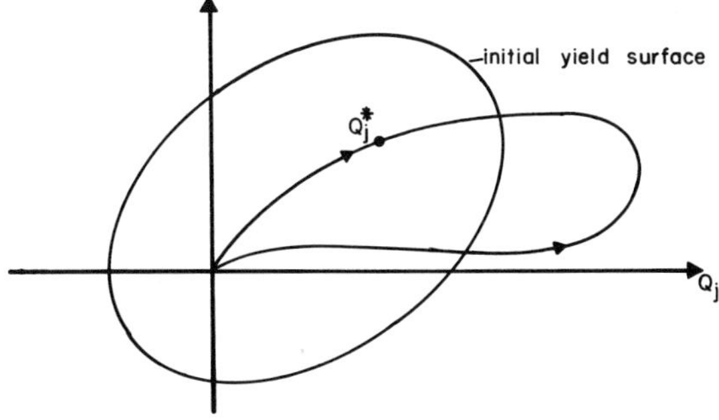

Figure 4. Elastic and inelastic paths

$$\oint q_j\, dQ_j \leq 0 \quad . \tag{23}$$

Breaking the path into two parts, this gives

$$\int_{Q_j^*}^{o} q_j\, dQ_j + \int_{o}^{Q_j^*} q_j\, dQ_j \leq 0 \quad . \tag{24}$$

The first integral is carried out along an elastic path for a material in its virgin state. Hence $q_j^p = 0$, $q_j = q_j^e$ and because of the elastic relation between q_j^e and Q_j we may write

$$\int_{Q_j^*}^{o} q_j\, dQ_j = \int_{Q_j^*}^{o} q_j^e\, dQ_j = -\int_{o}^{Q_j^*} q_j^e\, dQ_j = -\Omega(Q_j^*) \quad . \tag{25}$$

Because elastic behavior does not affect the recorded history the second term in inequality (24) is the complementary work along the inelastic path, $\bar{\Omega}(Q_j^*)$. Hence, from (24) and (25)

$$\Omega(Q_j^*) \geq \bar{\Omega}(Q_j^*) \quad . \tag{26}$$

Thus the elastic complementary energy is not less than the complementary work done along any inelastic path. This identifies the elastic path as an extremal path; the extremal path is again not unique, since any elastic path may be chosen, but the maximum complementary work is uniquely determined. The result can be summarized as follows; *if an elastic path between the origin and Q_j^* can be found it is an extremal path, and*

$$\Omega^o(Q_j^*) = \Omega(Q_j^*) \quad . \tag{27}$$

It may be noted that this result also applies when $\bar{\phi}(Q_j^*) = 0$, where $\bar{\phi}$ is again the *initial* yield function, since Q_j^* can always be reached by a path which approaches it from within the yield surface and does not cause plastic deformation.

This is also true for an *elastic, perfectly plastic material,*
where the yield function ϕ and the limit function ψ coincide;
in fact *any accessible stress state* in an elastic, perfectly
plastic material *can be reached by an elastic path* since no
stress states lie without the initial yield surface. This is
of particular significance, since it means that for an elastic,
perfectly plastic material $\Omega^o(Q_j^*)$ is given by the elastic complementary energy $\Omega(Q_j^*)$ for all stress states.

21.6 The Haar-Kármán Principle

The complementary work bounding principle (equation 17) *for an
elastic, perfectly plastic material* thus becomes

$$\int_V \Omega(Q_j^s)dV - \int_{S_u} \underline{p}^s \cdot \hat{\underline{u}}^t dS \geq \int_V \bar{\Omega}(Q_j^t)dV - \int_{S_u} \underline{p}^t \cdot \hat{\underline{u}}^t dS \quad . \qquad (28)$$

The quantities Q_j^s, $\underline{p}^s$ on S_u represent a pertinent statically admissible field, with the restriction that

$$\phi(Q_j^s) \leq 0 \qquad (29)$$

because $\phi = 0$ coincides with a limit surface.

This expression is of importance in understanding the his-
torically significant *Haar-Kármán principle*. Haar and von
Kármán proposed heuristically that in an elastic, perfectly
plastic structure subjected to loads $\hat{\underline{p}}^t$ on S_p, body forces $\hat{\underline{F}}^t$
on V and displacements $\hat{\underline{u}}^t$ on S_u, the actual stresses and reactions are given by those values of $Q_j^s, \underline{p}^s$ on S_u which minimize

$$\int_V \Omega(Q_j^s)dV - \int_{S_u} \underline{p}^s \cdot \hat{\underline{u}}^t dS \quad , \qquad (30)$$

where Q_j^s, $\underline{p}^s$ on S_u, constitute a pertinent statically admissible
field, and $\phi(Q_j^s) \leq 0$.

The function which Haar and von Kármán considered, expression

(30), and the left hand side of (28) are thus identical. We have already seen that the minimum value of the left hand side of (28) need not necessarily occur for $Q_j^s = Q_j^t$; the actual stresses Q_j^t cannot thus be found by minimizing expression (30) and the Haar-Kármán principle does not hold in general.

The cases where the Haar-Kármán principle will be certain to hold are those where the left hand and the right hand sides of inequality (28) are equal when $Q_j^s = Q_j^t$; we expect this to happen when

$$\Omega(Q_j^t) = \bar{\Omega}(Q_j^t) \quad . \tag{31}$$

Since (equation 19)

$$\bar{\Omega}(Q_j^t) = \Omega(Q_j^t) + \int_o^{Q_j^t} q_j^p dQ_j \quad , \tag{32}$$

equation (31) implies that

$$\int_o^{Q_j^t} q_j^p dQ_j = 0 \quad . \tag{33}$$

If $q_j^p = 0$, so that the stress path lies in the elastic region, equation (33) is of course satisfied. If $q_j^p \neq 0$ and the plastic deformation takes place, equation (33) can be satisfied in certain limited cases. First, it will be satisfied if the stress remains constant after the stress point first reaches the yield surface (Figure 5). Any plastic deformation that occurs does so with $dQ_j = 0$, and consequently equation (33) is satisfied.

It is clear that plastic deformation followed by unloading from the yield surface can never satisfy equation (33). In fact the only circumstance in which equation (33) can be satisfied when $dQ_j \neq 0$ after the stress point first makes contact with the yield surface occurs when the stress point moves along a flat

region of the yield or limit surface. In this case, shown diagrammatically in Figure 6, it is evident that the integrand of
equation (33) is identically zero because q_j^p and dQ_j are orthogonal.

If it can be shown for any particular structure that the
Haar-Kármán principle will hold (i.e. the exact stress field Q_j^t
can be found by minimizing expression 30), we do of course have
a powerful tool for analyzing the structure without the necessity of an incremental analysis. For this reason circumstances
under which the Haar-Kármán principle is applicable have continued to receive attention. Whether or not the principle is
valid depends partly on the loading, partly on the configuration of the structure and partly on the material. The dependence of the path of loading is in a sense the most fundamental.
Loading programs which cause continuous yielding in load space

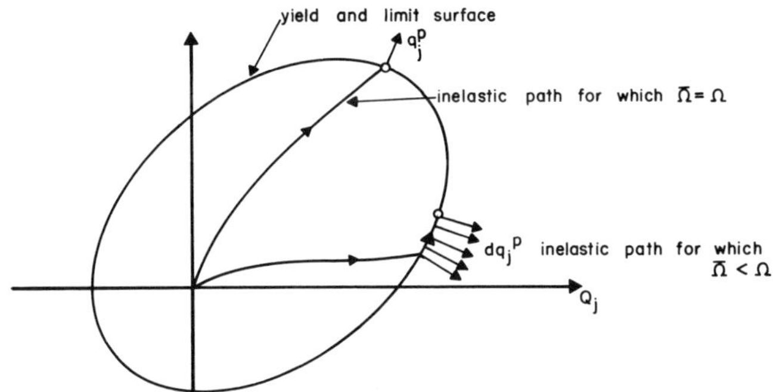

Figure 5. Paths for an elastic, perfectly plastic material

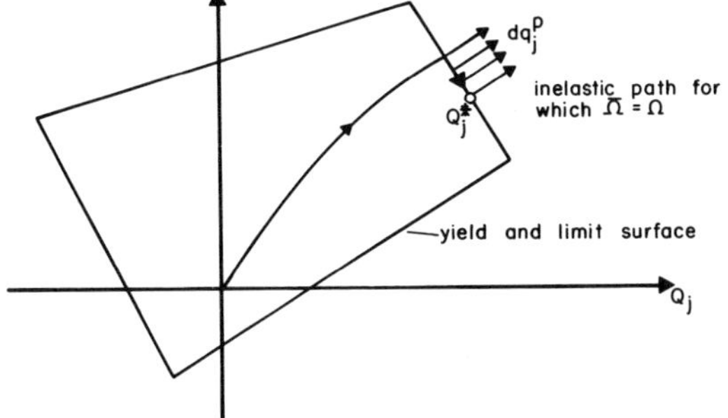

Figure 6. Paths along flats in the limit surface

are necessary for the Haar-Kármán principle to hold; loading programs which involve unloading from a yield surface in load space by definition involve unloading from the yield surface in stress space at some points in the body. The structural configuration, i.e. the type of structure and its shape, determines what the stress components are and influences the stress history. For example, pin-jointed trusses and beam problems in which only one generalized stress component appears are such that the Haar-Kármán principle is applicable if no unloading occurs at any point in the structure; the difficulty of determining whether a stress point can move around the yield surface is eliminated because there is only one stress component. However, unloading at a point in such a structure can occur even when the loads increase proportionally and monotonically; this again depends on the type of structure under consideration. Finally the choice of a yield condition determines whether or not the Haar-Kármán principle is valid if the stress point moves along the yield surface; in cases where a number of flats occur, as occur when the Tresca yield condition is used, or where a yield surface can be appropriately linearized by approximating it with a number of flats, the principle may be valid.

All these factors mean that determining whether or not the Haar-Kármán principle is valid in any one problem may be almost as difficult as carrying out the incremental analysis.

21.7 Bibliographical and Historical Remarks

The complementary work bounding theorem was first given for elastic, perfectly plastic materials by Hodge [1966]. The concept of extremal paths in stress space for elastic-plastic materials was introduced by Martin [1966a], [1966b], and the complementary work bounding theorem was given in the form presented in this Chapter by Martin [1970]. Contemporaneously, the complementary work bounding theorem was developed by Ponter [1968] and Soechting and Lance [1969].

Haar and von Kármán [1909] obtained their principle by heuristically extending the elastic minimum complementary energy principle. The conditions on its validity given in this Chapter were given, by somewhat different arguments, by Greenberg [1950] and have been restated by other writers. It can be shown in certain examples (e.g. Prager and Symonds [1950], Prager [1959]) that the principle can hold when the conditions stated are not met, leaving open the possibility that less stringent sufficient conditions for the validity of the principle could be stated. Such extended conditions have not been found. A further discussion of the principle has been given by Koiter [1960].

References

H.J.Greenberg — 1949 — "Complementary minimum principles for an elastic-plastic material", Quart. Appl.Math., $\underline{7}$, 85.

A.Haar and T. von Kármán — 1909 — "Zur Theorie der Spannungzustände in plastischen und sandartigen Medren", Nachr.der Wiss.zu Gottingen, Math.-phys. Klasse, 204.

P.G.Hodge,Jr. — 1966 — "A deformation bounding theorem for flow-law plasticity", Quart.Appl. Math., $\underline{24}$, 171.

W.T.Koiter — 1960 — "General theorem for elastic-plastic solids",*Progress in Solids Mechanics,* $\underline{1}$ (edited by I.N.Sneddon and R.Hill), Chapter 4, North Holland, 167.

J.B.Martin — 1966a — "Extended displacement bound theorems for work hardening continua subjected to dynamic loading", Int.J.Solids & Structures, $\underline{2}$, 9.

J.B.Martin — 1966b — "The determination of upper bounds on displacement resulting from static and dynamic loading by the application of energy methods", Proc.5th U.S.Nat. Congr.Appl.Mech., ASME (N.Y.), 221.

J.B.Martin   1970   "A complementary energy bounding
                              theorem for time-independent ma-
                              terials", *Developments in Theore-
                              tical and Applied Mechanics*, $\underline{8}$
                              (edited by D.Frederick and E.H.
                              Harris), Pergamon Press, 517.

A.R.S.Ponter   1968   "Convexity conditions and energy
                              theorems for time-independent ma-
                              terials", J.Mech.Phys.Sol.,$\underline{16}$,283.

W.Prager   1959   *An Introduction to Plasticity*, Ad-
                              dison-Wesley (Reading, Mass).

W.Prager and   1950   "Stress analysis in elastic-plastic
P.S.Symonds   structures", Proc.3rd Symp.Appl.
                              Math., N.Y., 187.

J.F.Soechting and   1969   "A bounding principle in the theory
R.H.Lance   of work hardening plasticity", J.
                              Appl.Mech., $\underline{36}$, 228.

Chapter 22

PROPERTIES OF EXTREMAL PATHS AND EXTENDED BOUNDING THEOREMS

22.1 Maximum Complementary Work Paths

In the preceding Chapter we discussed the particular case of elastic, perfectly plastic materials, giving the extremal path as an elastic path, the maximum complementary work as the elastic complementary energy and presenting the appropriate form of the complementary energy bounding theorem. In this section we shall discuss some general properties of elastic, plastic materials which fall within the framework of Chapter 2. This discussion may be given in general terms, and includes the special case of elastic, perfectly plastic materials.

If we suppose that an extremal path in stress space is known for all terminal states of stress, it is evident that we may associate with each stress state Q_j^* a state of *total strain* q_j^* which is the terminal strain when the stress path follows the extremal path from the origin to Q_j^*. It can be shown that there exists a relation between q_j^*, Q_j^* and the maximum complementary work function $\Omega^0(Q_j^*)$, which is also defined for each state of stress. To do so we demonstrate the following result

A sufficient condition that

$$q_j^* = \left.\frac{\partial \bar{\Omega}}{\partial Q_j}\right|_{Q_j = Q_j^*} \quad , \quad \bar{\Omega}(Q_j^*) = \int_o^{Q_j^*} q_j \, dQ_j \quad , \tag{1}$$

for a class of paths between $Q_j = 0$ and $Q_j = Q_j^*$ and defined in terms of Q_j^* is that the path makes $\bar{\Omega}$ an extremum.

Let us consider a class of extremal paths. As before, we denote the maximum complementary work along an extremal path by $\Omega^0(Q_j^*)$. If we permit small variations in the stress history (leaving the initial and final states unchanged) in an arbitrary path the first variation of $\bar{\Omega}(Q_j^*)$ may have any value. However, if the path is an extremum, so that $\bar{\Omega}(Q_j^*) = \Omega^o(Q_j^*)$, the first variation in $\bar{\Omega}(Q_j^*)$, $\delta\bar{\Omega}(Q_j^*)$, will be zero.

Consider two adjacent states of stress Q_j^* and $Q_j^* + a_j d\theta$, where a_j is a unit vector in stress space in some prescribed direction. We assume that we may find extremal paths to these two states of stress which become coincident as $d\theta \to 0$. The stress history defined by the extremal path to Q_j^* followed by the change from Q_j^* to $Q_j^* + a_j d\theta$ will give complementary work

$$\bar{\Omega}(Q_j^* + a_j d\theta) = \Omega^o(Q_j^*) + q_j^* a_j d\theta + 0(d\theta^2) \quad , \tag{2}$$

where q_j^* denotes the strain at Q_j^* after following the extremal path. Further, we define $\Delta\bar{\Omega}$ by the expression

$$\Omega^o(Q_j^* + a_j d\theta) = \bar{\Omega}(Q_j^* + a_j d\theta) + \Delta\bar{\Omega} \quad , \tag{3}$$

so that $\Delta\bar{\Omega}$ becomes the difference in the complementary work between the extremal and non-extremal paths to $(Q_j^* + a_j d\theta)$. Because the difference between these two paths is characterized by the parameter $d\theta$ in such a way that the two paths become identical for $d\theta \to 0$,

$$\lim_{d\theta \to 0} \{\frac{\Delta\bar{\Omega}}{d\theta}\} = \delta\bar{\Omega} = 0 \quad . \tag{4}$$

Combining (4) with equations (2) and (3) we obtain

$$\lim_{d\theta \to 0} \{\frac{\Omega^o(Q_j^* + a_j d\theta) - \Omega^o(Q_j^*)}{d\theta} - q_j^* a_j\} = 0 \quad . \tag{5}$$

By choosing a_j parallel to each of the stress axes in turn we have

$$q_j^* = \frac{\partial\Omega^o}{\partial Q_j}\bigg|_{Q_j = Q_j^*} \quad , \quad \Omega^o = \Omega^o(Q_j^*) \quad , \tag{6}$$

and the result is established.

It should be noted that we make use only of the requirement
that $\bar{\Omega}$ should be a *stationary* value (rather than an absolute
maximum) in order to obtain this result, so that the result is
in fact more generally applicable. It cannot be proven in general, however, that a stationary or extremal value of $\bar{\Omega}$ is
necessary for (1) to hold. By retracing the argument given
above we can only show that (1) implies that $\bar{\Omega}(Q_j^*)$ is stationary
for a restricted class of variations.

This result implies that $\Omega^0(Q_j^*)$ is a potential function from
which the total strain q_j^* resulting from an extremal stress history can be derived by means of equation (6). This type of potential function is the characteristic of a path independent (or
elastic) material; equation (6) thus tells us that we can *associate* with the path dependent elastic, plastic material a *new
material* whose constitutive equation is given by equation (6).
This new material is path independent, or elastic, since it is
evident from equation (6) that q_j^* depends only on Q_j^* *and* that

$$\int_{Q_j^{*(1)}}^{Q_j^{*(2)}} q_j^* \, dQ_j^* = \Omega^0(Q_j^{*(2)}) - \Omega^0(Q_j^{*(1)}) \quad , \tag{7}$$

which is independent of the path.

22.2 Minimum Work Paths

A dual result may be obtained in terms of the work done along
some path in strain space between the virgin state ($q_j = 0$) and
some given terminal value q_j^*,

$$\bar{W}(q_j^*) = \int_0^{q_j^*} Q_j \, dq_j \quad . \tag{8}$$

The function $\bar{W}(q_j^*)$ will in general be path dependent, i.e. it
depends not only on the final strain q_j^* but also on the strain

path. The work done along any path starting at the virgin state
is required to be positive definite in our general framework
(cf. Section 2.4, equation 71). It is then evident that $\bar{W}(q_j^*)$
can be expected to have a positive definite *minimum* value. A
maximum value will not in general exist, since it is always possible to increase the work along a given strain path to q_j^* by
adding a small cycle of strain at the end of the path. We thus
define an extremal path in strain space between $q_j = 0$ and $q_j = q_j^*$ as that path along which the work $W^o(q_j^*)$ is not greater than
the work $\bar{W}(q_j^*)$ along any other path between the same terminal
values,

$$W^o(q_j^*) \leq \bar{W}(q_j^*) \quad . \tag{9}$$

The following result may then be given.

A sufficient condition that

$$Q_j^* = \left. \frac{\partial \bar{W}}{\partial q_j} \right|_{q_j = q_j^*} \quad , \quad \bar{W}(q_j^*) = \int_o^{q_j^*} Q_j \, dq_j \quad , \tag{10}$$

for a class of paths between $q_j = 0$ *and* $q_j = q_j^*$ *defined in terms*
of q_j^*, *is that the path makes* $\bar{W}$ *an extremum.*

In this expression Q_j^* is the terminal stress.

The proof of this result can be obtained by arguments which
are identical to those given for extremum paths in stress space.
Note again that we make use only of the fact that $\bar{W}$ should have
a stationary value.

We see again that $W^o(q_j^*)$ is a potential function and that the
terminal stress Q_j^* at the end of an extremal path in strain space
is given by

$$Q_j^* = \left. \frac{\partial W^o}{\partial q_j} \right|_{q_j = q_j^*} \quad , \quad W^o = W^o(q_j^*) \quad . \tag{11}$$

This again implies that a path independent (or elastic) material can be associated with the elastic-plastic material. It is evident from the constitutive equation (11) for this new material that the stress Q_j^* is a function only of strain q_j^* for a given potential function $W^O(q_j^*)$ and that

$$\int_{q_j^{*(1)}}^{q_j^{*(2)}} Q_j^* dq_j^* = W^O(q_j^{*(2)}) - W^O(q_j^{*(1)}) \quad , \tag{12}$$

which is path independent.

<u>22.3 The Relation between Extremal Paths in Stress and Strain Space</u>

We have seen that the concepts of a maximum complementary work path and a minimum work path lead us to the constitutive equations (6) and (11) for two associated path independent materials. We can now proceed to derive, as a consequence of the first postulate of Section 2.4, two important further results. First, we can show that equations (6) and (11) are inverses with respect to each other, i.e. that paths which maximize $\bar{\Omega}$ also minimize $\bar{W}$, so that Ω^O and W^O are dual potential functions and equations (6) and (11) are *two forms of the constitutive equation for the same path independent material*. Second, we can show that Ω^O and W^O are *convex functions*.

The first result and the convexity of Ω^O follow from an earlier result given in Chapter 21. Consider two possible paths in stress space to a final state of stress $Q_j^{(2)}$ (Figure 1). Let one path be an extremal path to $Q_j^{(2)}$, and let the other be an arbitrary path to some other stress state $Q_j^{(1)}$ followed by a straight path from $Q_j^{(1)}$ to $Q_j^{(2)}$. Our earlier result (Chapter 21, equation 12) for this situation permits us to write

$$\Omega^O(Q_j^{(2)}) - Q_j^{(2)} q_j^{(1)} \geq \bar{\Omega}(Q_j^{(1)}) - Q_j^{(1)} q_j^{(1)} \quad , \tag{13}$$

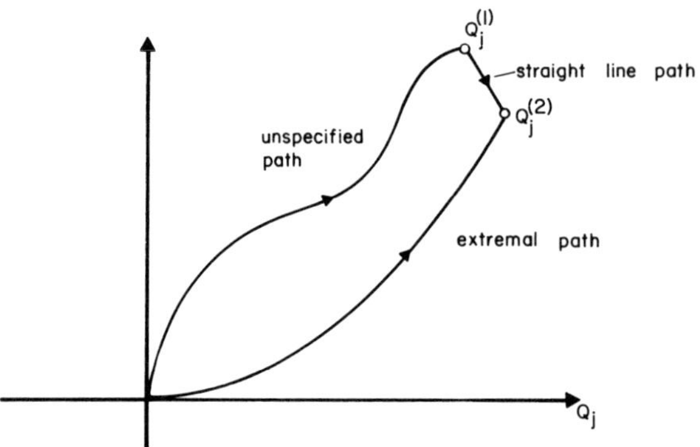

Figure 1.Alternative stress paths

where $q_j^{(1)}$ is the strain at the end of the unspecified path to $Q_j^{(1)}$. We now further specialize the paths shown in Figure 1.

First, let the path to $Q_j^{(1)}$ also be an extremal path. Inequality (13) then becomes

$$\Omega^o(Q_j^{(2)}) - Q_j^{(2)} q_j^{(1)} \geq \Omega^o(Q_j^{(1)}) - Q_j^{(1)} q_j^{(1)} \quad , \tag{14}$$

where $q_j^{(1)}$ is now the terminal strain after following an extremum path. Inequality (14) expresses the convexity of Ω^o. We may arrange (14) into a more conventional form by substituting from equation (6) for $q_j^{(1)}$;

$$\Omega^o(Q_j^{(2)}) - \Omega^o(Q_j^{(1)}) \geq (Q_j^{(2)} - Q_j^{(1)}) \left. \frac{\partial \Omega^o}{\partial Q_j}\right|_{Q_j = Q_j^{(1)}} \quad . \tag{15}$$

Secondly we note that the extremal path in stress space to $Q_j^{(2)}$ will map a path in strain space which terminates in strain $q_j^{(2)}$ (Figure 2). Let the unspecified path to $Q_j^{(1)}$ in stress space map into strain space a strain path which *also terminates at strain* $q_j^{(2)}$. We denote the work done along the strain path mapped by extremal path in stress space by $W^o(q_j^{(2)})$ and the work done along the unspecified strain path to $q_j^{(2)}$ mapped by the unspecified stress path to $Q_j^{(1)}$ by $\bar{W}(q_j^{(2)})$. Along these two paths we can write the identities

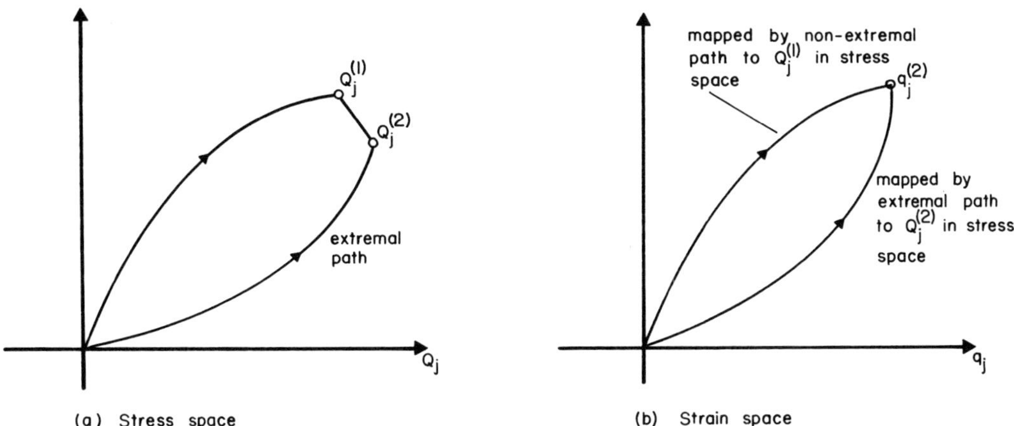

Figure 2. Stress and strain paths

$$\Omega^{o}(Q_j^{(2)}) + W^{o}(q_j^{(2)}) = Q_j^{(2)}q_j^{(2)} \quad , \tag{16a}$$

$$\bar{\Omega}\,(Q_j^{(1)}) + \bar{W}\,(q_j^{(2)}) = Q_j^{(1)}q_j^{(2)} \quad . \tag{16b}$$

Because $q_j^{(1)} = q_j^{(2)}$, inequality (13) becomes

$$\Omega^{o}(Q_j^{(2)}) - Q_j^{(2)}q_j^{(2)} \geq \bar{\Omega}(Q_j^{(1)}) - Q_j^{(1)}q_j^{(2)} \quad . \tag{17}$$

Substituting from equations (16a) and (16b), inequality (17) becomes

$$W^{o}(q_j^{(2)}) \leq \bar{W}(q_j^{(2)}) \quad . \tag{18}$$

The path mapped into strain space by the maximum complementary work path for $Q_j^{(2)}$ is thus the minimum work path for $q_j^{(2)}$, where $q_j^{(2)}$ and $Q_j^{(2)}$ are related through equation (6).

There remains the possibility that there exists a set of minimum work paths in strain space which map paths in stress space which are not maximum complementary work paths. However, provided that the states of strain which terminate the whole class of maximum complementary work paths cover the whole of the strain space, for each strain there exists at least one extremal path which maps an extremal path in stress space. Clearly we can expect these paths to be uniquely defined by the final states of stress and strain only in exceptional circumstances; any reversible behavior, for example, implies that there are elastic paths for at least some terminal states, and therefore guarantees non-uniqueness for we have seen (Chapter 21) that any elastic path is an extremal path.

It is evident that the extremal functions $\Omega^{\circ}(Q_j)$ and $W^{\circ}(Q_j)$ are uniquely determined for given states of stress and strain. However, the uniqueness of the terminal states of strain and stress derived from these functions by means of equations (6) and (11) depends on the derivatives of Ω° and W°. If Ω° and W° are everywhere continuously differentiable the terminal states of strain and stress associated with the terminal states of stress and strain and the respective extremal paths are certainly unique.

We can in fact give stronger results for the dependence of the terminal stress on the terminal strain and the minimum work path in strain space. Let us suppose that there exist extremal paths in stress space to distinct stress states $Q_j^{(1)}$ and $Q_j^{(2)}$ which both terminate in the same strain q_j^*. Using the more fundamental form of equation (13) (cf. Chapter 21, equations 11 and 12) we can write two inequalities;

$$\Omega^{\circ}(Q_j^{(2)}) - \Omega^{\circ}(Q_j^{(1)}) - (Q_j^{(2)} - Q_j^{(1)})q_j^* \geq \int_{Q_j^{(1)}}^{Q_j^{(2)}} (q_j - q_j^*)dQ_j \geq 0 \ , \qquad (19a)$$

$$\Omega^{\circ}(Q_j^{(1)}) - \Omega^{\circ}(Q_j^{(2)}) - (Q_j^{(1)} - Q_j^{(2)})q_j^* \geq \int_{Q_j^{(2)}}^{Q_j^{(1)}} (q_j - q_j^*)dQ_j \geq 0 \ , \qquad (19b)$$

where both integrals are carried out along straight line paths
in stress space. If these two inequalities are to be compatible
with each other

$$\int_{Q_j^{(1)}}^{Q_j^{(2)}} (q_j - q_j^*) dQ_j = \int_{Q_j^{(2)}}^{Q_j^{(1)}} (q_j - q_j^*) dQ_j = 0 \quad . \tag{20}$$

If a strict inequality holds in the first postulate of Section
2.4, so that

$$\int_{Q_j^{(1)}}^{Q_j^{(2)}} (q_j - q_j^*) dQ_j > 0 \quad , \quad \int_{Q_j^{(2)}}^{Q_j^{(1)}} (q_j - q_j^*) dQ_j > 0 \tag{21}$$

for $Q_j^{(1)} \neq Q_j^{(2)}$, equation (20) implies that $Q_j^{(1)} = Q_j^{(2)}$. This
in turn implies that the terminal state of stress is uniquely
determined by the terminal state of strain and the minimum work
path. No two extremal paths in stress space to distinct state
of stress give rise to the same state of strain.

If on the other hand the broader form of the postulate holds,
so that the integrals in (21) may be zero for $Q_j^{(1)} \neq Q_j^{(2)}$ (or
alternatively $dQ_j dq_j = 0$ for $dQ_j \neq 0$) equation (20) implies that
$q_j = q_j^*$ along the straight line path. Consequently all possible
terminal states of stress $Q_j^{(1)}$, $Q_j^{(2)}$ which lead to the same terminal strain when the respective stress histories are extremal
paths lie within a region in stress space in which the behavior
is *rigid*. Such a region must be bounded by a yield surface and
is consequently *convex*. In a rigid, perfectly plastic or rigid,
hardening material such regions will of course exist. If, however, the material is elastic-plastic, rigid behavior is excluded and the uniqueness result holds.

No similar result for the inverse uniqueness problem (that a
terminal state of stress and an extremal path in stress space

leads to a unique terminal state of strain) can be given. It is evident *a priori* that no such result can hold in the case of elastic, perfectly plastic behavior where as a result of plastic flow a given terminal state of stress on the limit surface can be associated with a range of terminal strains.

It remains to prove that the minimum work $W^O(q_j)$ is a convex function. This result follows immediately from the convexity of $\Omega^O(Q_j)$ for extremum paths in stress space if we substitute the identities

$$\Omega^O(Q_j^{(1)}) + W^O(q_j^{(1)}) = Q_j^{(1)} q_j^{(1)} \quad , \tag{22a}$$

$$\Omega^O(Q_j^{(2)}) + W^O(q_j^{(2)}) = Q_j^{(2)} q_j^{(2)} \quad , \tag{22b}$$

into inequality (14). This leads to

$$W^O(q_j^{(1)}) - Q_j^{(2)} q_j^{(1)} \geq W^O(q_j^{(2)}) - Q_j^{(2)} q_j^{(2)} \quad , \tag{23}$$

which establishes the convexity of W^O. In a more conventional form this inequality may be rewritten, after using equation (11) as,

$$W^O(q_j^{(1)}) - W^O(q_j^{(2)}) \geq (q_j^{(1)} - q_j^{(2)}) \left. \frac{\partial W^O}{\partial q_j} \right|_{q_j = q_j^{(2)}} \quad . \tag{24}$$

The results given in this section, involving the potential form of Ω^O and W^O, are of course consistent with elastic behavior. When either the material behaves entirely elastically or when we limit ourselves to stress states which lie within the initial yield surface (Chapter 21) any elastic path may be chosen as the extremum path and $\Omega^O = \Omega$, $W^O = W$. For linear elastic behavior equations (6) and (11) certainly hold, and the material satisfies the convexity requirements (15) and (24).

Another case where the determination of the extremum paths is

trivial is that where only one generalized stress component Q
associated with non-zero strain acts at a point in the structure. The inelastic constitutive equations provide a path dependent relation between Q and the generalized strain component
q. By a simple trial procedure it can readily be confirmed that
the extremum paths involve monotonic changes in the stress and
strain. Thus the maximum complementary work for a terminal
stress Q* is obtained when the stress is increased monotonically
from zero to Q*, as opposed to following paths which involve unloading. Similarly the minimum work path involves a monotonic
increase in strain from zero to the given terminal value q*. The
path independent material defined by equations (6) and (11) is
thus a non-linear elastic material which has a stress-strain
curve which coincides with the stress-strain curve obtained for
the elastic-plastic material by monotonically increasing and decreasing the generalized stress Q, starting in each case at the
virgin state. The convexity requirements (15) and (24) are met
provided that the slope of this stress-strain curve is everywhere positive.

It does not appear possible to give further results for the
extremal paths without considering more specific idealizations
of the material behavior. We shall consequently defer further
discussion on the determination of the extremum paths until
Chapter 23; in the remainder of this Chapter we shall consider
extensions of the bounding theorem of Chapter 21 which can be
given if we consider in addition solutions for the structural
problem derived using the constitutive relations (6) and (11).

22.4 Theorems for Extremal Paths

We return again to the structural problem set out in Chapter 21.
The structure is initially stress free and in its virgin state.
Over some time period $0 \leq \tau \leq t$ loads $\underline{p}(s,\tau)$ are applied on S_p,

body forces $\hat{\underline{F}}(s,\tau)$ are applied on V, and displacements $\hat{\underline{u}}(s,\tau)$ are applied on S_u. In order to conform with the stress free initial conditions

$$\hat{\underline{p}}(s,0) = \hat{\underline{F}}(s,0) = \hat{\underline{u}}(s,0) = 0 \quad . \tag{25}$$

At time t the current loading is described by

$$\hat{\underline{p}}^t(s) = \hat{\underline{p}}(s,t) \text{ on } S_p \quad ,$$

$$\hat{\underline{F}}^t(s) = \hat{\underline{F}}(s,t) \text{ on } V \quad , \tag{26}$$

$$\hat{\underline{u}}^t(s) = \hat{\underline{u}}(s,t) \text{ on } S_u \quad ,$$

The exact solution for the elastic-plastic structure, obtained incrementally and dependent on the loading history, is characterized at time t by generalized stresses Q_j^t, generalized strains q_j^t, displacements $\underline{u}^t$ on S_p and V and reactions $\underline{p}^t$ on S_u.

We confine ourselves in the first instance to the terminal values of the loading given in equation (26), and consider a structure identical in all respects to that which gave rise to the elastic-plastic solution *except* that the material is described by the time-independent constitutive relations (equations 6 and 11)

$$q_j = \frac{\partial \Omega^0}{\partial Q_j} \quad , \quad Q_j = \frac{\partial W^0}{\partial q_j} \tag{27}$$

where $\Omega^0(Q_j)$, $W^0(q_j)$ are obtained from the extremal paths for the elastic-plastic material. This new material is *path independent*, and consequently we can solve the boundary value problem set out in equations (26) for the material described by equations (27) *without reference to the previous history of loading*. Let the solution to this problem be characterized by stresses Q_j^*, strains q_j^*, displacements $\underline{u}^*$ on V and S_p and reactions $\underline{p}^*$ on S_u. This solution will be unique except in the case where the inelastic material from which Ω^0 and W^0 are

derived may undergo flow or exhibit some rigid behavior. In
this last case some uniqueness may be lost in the stress and
displacement fields. We shall not treat the uniqueness problem
directly; uniqueness may be inferred from the theorems which we
give below.

Because the potential functions $\Omega^O(Q_j)$, $W^O(q_j)$ are convex
the classical energy theorems of elasticity are applicable. Let
Q_j^S, $\underline{p}^S$ on S_u be a *pertinent statically admissible field* for the
structural problem at the current instant, and let q_j^c, $\underline{u}^c$ on V
and S_p be a *pertinent kinematically admissible field*. From the
convexity of Ω^O (equation 14, with $Q_j^{(1)} = Q_j^*$, $Q_j^{(2)} = Q_j^S$) we
have that

$$\Omega^O(Q_j^S) - Q_j^S q_j^* \geq \Omega^O(Q_j^*) - Q_j^* q_j^* \quad . \tag{28}$$

Integrating over the volume, applying the principle of virtual
work and eliminating the integral over S_p which appears on both
sides of the inequality, we find that

$$U_c^O(Q_j^S) \geq U_c^O(Q_j^*) \quad , \tag{29}$$

where

$$U_c^O(Q_j^S) = \int_V \Omega^O(Q_j^S)dV - \int_{S_u} \underline{p}^S \cdot \hat{u}dS \quad , \tag{30a}$$

$$U_c^O(Q_j^*) = \int_V \Omega^O(Q_j^*)dV - \int_{S_u} \underline{p}^* \hat{u}dS \quad . \tag{30b}$$

Similarly, from the convexity of W^O (equation 23, with $q_j^{(1)}$
$= q_j^c$, $q_j^{(2)} = q_j^*$) we have that

$$W^O(q_j^c) - Q_j^* q_j^c \geq W^O(q_j^*) - Q_j^* q_j^* \quad . \tag{31}$$

Integrating over the volume, using the principle of virtual work
and eliminating the integral over S_u which appears on both sides

of the inequality, we see that

$$U_p^O(q_j^c) \geq U_p^O(q_j^*) \quad , \tag{32}$$

where

$$U_p^O(q_j^c) = \int_V W^O(q_j^c)dV - \int_V \underline{\hat{F}}\cdot\underline{u}^c dV - \int_{S_p} \underline{\hat{p}}\cdot\underline{u}^c dS \quad , \tag{33a}$$

$$U_p^O(q_j^*) = \int_V W^O(q_j^*)dV - \int_V \underline{\hat{F}}\cdot\underline{u}^* dV - \int_{S_p} \underline{\hat{p}}\cdot\underline{u}^* dS \quad . \tag{33b}$$

Because the material defined by equations (27) is path independent it follows that the left and right hand sides of inequalities (29) and (32) become equal when $Q_j^S = Q_j^*$ and $q_j^c = q_j^*$ respectively. The solutions can thus be found by determining those fields Q_j^S, q_j^c which make $U_c^O(Q_j^S)$, $U_p^O(q_j^c)$ absolute minima. Uniqueness hinges upon whether there exists a unique stress field Q_j^S and a unique strain field q_j^c which give the absolute minima of $U_c^O(Q_j^S)$, $U_p^O(q_j^c)$. This in turn depends upon whether the functions Ω^O and W^O are strictly convex or not(i.e. whether an inequality occurs in inequalities (28) and (31) for all $Q_j^S \neq Q_j^*$, $q_j^c \neq q_j^*$). If the functions are not strictly convex, as may occur when either rigid behavior or flow may take place, the solutions may not be unique, but as before the lack of uniqueness under these conditions is readily understandable and is not necessarily disturbing.

<u>22.5 Bounding Theorems for Non-Extremal Paths</u>

Our next task is to relate the theorems for the path independent solution derived using equations (27) to the incremental solution for the material from which Ω^O and W^O are derived. A continued inequality involving the stress fields can be obtained by comparing the complementary work bounding theorem (Chapter 21, equation 17) with inequality (29). In terms of the complementary work bounding theorem of Chapter 21, Q_j^*, $\underline{p}^*$ on S_u is a *perti-nent statically admissible field for the incremental elastic, plastic problem.* Hence (from Chapter 21, equation 17)

$$U_c^o(Q_j^*) \geq \bar{U}_c(Q_j^t) \quad , \tag{34}$$

where

$$\bar{U}_c(Q_j^t) = \int_V \bar{\Omega}(Q_j^t)dV - \int_{S_u} \underline{p}^t \cdot \hat{\underline{u}}dV \quad . \tag{35}$$

The function $\bar{U}_c(Q_j^t)$ depends on the stress history at each point in the body and hence on the loading history. Combining inequalities (29) and (34) we find an extended result

$$U_c^o(Q_j^s) \geq U_c^o(Q_j^*) \geq \bar{U}_c(Q_j^t) \quad . \tag{36}$$

In terms of the complementary work bounding theorem, which is given in Chapter 21 (equation 17) as

$$U_c^o(Q_j^s) \geq \bar{U}_c(Q_j^t) \quad , \tag{37}$$

we have introduced a further expression $U_c^o(Q_j^*)$ which is the smallest attainable value of $U_c^o(Q_j^s)$. It will still be true that $U_c^o(Q_j^*) \neq \bar{U}_c(Q_j^t)$ unless $Q_j^* = Q_j^t$ and the stress history $Q_j(\tau)$, $0 \leq \tau \leq t$, follows an extremum path in stress space.

The dual result for the strain fields is a weaker result. In terms of the kinematic result (32), the strain field q_j^t, $\underline{u}^t$ on V and S_p (which is the solution to the incremental problem) is a pertinent kinematically admissible field for the path independent problem using equations (27). Hence from (32)

$$U_p^o(q_j^t) \geq U_p^o(q_j^*) \quad , \tag{38}$$

where

$$U_p^o(q_j^t) = \int_V W^o(q_j^t)dV - \int_V \hat{\underline{F}}\cdot\underline{u}^t dV - \int_{S_p} \hat{\underline{p}}\cdot\hat{\underline{u}}^t dS \quad . \tag{39}$$

We note now that by definition (equation 9) the work done along the strain path terminating in q_j^t in the path dependent elastic plastic body $\overline{W}(q_j^t)$, is not less than $W^o(q_j^t)$;

$$\overline{W}(q_j^t) \geq W^o(q_j^t) \quad . \tag{40}$$

It is then evident that

$$\overline{U}_p(q_j^t) \geq U_p^o(q_j^t) \quad , \tag{41}$$

where

$$\overline{U}_p(q_j^t) = \int_V \overline{W}(q_j^t)dV - \int_V \hat{\underline{F}}\cdot\underline{u}^t dV - \int_{S_p} \hat{\underline{p}}\cdot\underline{u}^t dS \quad . \tag{42}$$

The function $\overline{U}_p(q_j^t)$ may be referred to as the potential work done in the elastic -plastic body. Because $\overline{W}(q_j^t)$ is path dependent, $\overline{U}_p(q_j^t)$ depends on the history of loading. Finally, combining (38) and (41) we see that

$$\overline{U}_p(q_j^t) \geq U_p^o(q_j^t) \geq U_p^o(q_j^*) \quad . \tag{43}$$

This result is again a bounding theorem rather than a true extremum principle, since $\overline{U}_p(q_j^t)$ will be equal to $U_p^o(q_j^*)$ only if $q_j^t = q_j^*$ and the strain history $q_j(\tau)$, $0 \leq \tau \leq t$ is an extremum history in strain space.

The results given in this section are applicable to the special case of elastic, perfectly plastic materials because, as we have pointed out, the properties of W^o and Ω^o on which the theorems are based are themselves obtained from the general framework of Chapter 2. It should be appreciated, however, that in the theorems for statically admissible stress distributions (inequality 29) there is in the case of elastic, perfectly plastic behavior an *implicit* assumption that

$$\phi(Q_j^S) \leq 0 \quad , \tag{44}$$

because $\Omega^0(Q_j^S)$ is not defined for stress states which lie outside of the yield or limit surface. Furthermore, as shown in
Chapter 21

$$\Omega^0(Q_j) = \Omega(Q_j) \quad , \tag{45}$$

so that the extended inequality for complementary work (inequality 36) reduces to the result given earlier (Chapter 21, inequality 28).

The conclusions which can be drawn from the extended inequalities (36) and (43) can be set out from two slightly different
points of view. On one hand they permit us to find bounds on
$\bar{U}_c$ and $\bar{U}_p$ without carrying out an incremental analysis. Inequality (36) is a stronger result, since a bound can be found on
$\bar{U}_c$ in terms of *any* pertinent statically admissible stress field
Q_j^S. The *best* bound is given by the pertinent statically admissible stress field Q_j^* which is the solution of the path independent problem with a material defined by equation (27). Inequality (43) shows that a bound on $\bar{U}_p$ can be found *only* in
terms of the kinematically admissible strain field q_j^* which is
the solution of the path independent problem with a material
defined by equation (27).

On the other hand we can regard the determination of Q_j^*, q_j^*
as a simpler process than the determination by means of an incremental analysis of Q_j^t, q_j^t, and consider Q_j^*, q_j^* as an approximate solution of the incremental elastic, plastic problem. We
are then assured that the approximate solution bears a consistent relationship to the incremental solution in the sense given
in inequalities (36) and (43). This is the viewpoint of *deformation theories* of plasticity which we shall consider in greater

detail in Chapter 24. Considerable attention has also been
given to the problem of distinguishing those particular structural problems in which $Q_j^* = Q_j^t$, $q_j^* = q_j^t$ i.e. structural problems in which the loading history and constitutive relations
are such that the stress histories at each point in the body
are extremum histories. Clearly if this distinction can be made
those problems admit a simpler path independent solution and incremental analysis is unnecessary.

In the next Chapter we shall consider the problem of determining the extremum paths and the functions Ω^o and W^o for particular idealizations of material behavior.

22.6 Bibliographical and Historical Remarks

The bounding theorems in their present form were given by Ponter
and Martin [1972], being extensions of earlier work on complementary work bounding theorems (Ponter [1968], Soechting and
Lance [1969], Martin [1970]). Alternative arguments have been
used to establish restricted forms of these theorems. Hodge
[1966], [1968] gives results for elastic, perfectly plastic
materials, and Maier [1969a], [1969b], [1969c] gives results for
piecewise-linear elastic-plastic materials. Maier's results
appear in a rather different form, and will be derived from the
theorems given in this Chapter in Chapter 24.

References

P.G.Hodge,Jr.   1968   "Minimum principles in plasticity", *En-
                           gineering Plasticity* (edited by J.Hey-
                           man and F.A.Leckie), Cambridge, 237.

P.G.Hodge,Jr.   1966   "A deformation bounding theorem for flow
                           law plasticity", Q.Appl.Math., <u>24</u>, 171.

G.Maier   1969a   "Teorimi di minimo in termini finiti per
                           continui elastoplastici con leggi con-
                           stitutive linearizzate a tratti", Insti-
                           tuto Lombardo, Accademia di Scienzie
                           Lettere, Rendiconti <u>A104</u>, 1066.

G.Maier   1969b   "Some theorems for plastic strain rates
                         and plastic strains", J.de Mécanique, $\underline{8}$,
                         5.

G.Maier   1969c   "Complementary plastic work theorems in
                         piecewise-linear elastoplasticity", Int.
                         J.Solids & Structures, $\underline{5}$, 261.

J.B.Martin   1970   "A complementary energy bounding theorem
                         for time independent materials". *Develop-
                         ments in Theoretical and Applied Mecha-
                         nics* (ed. D.Frederick and E.H.Harris),
                         Pergamon Press, Oxford, $\underline{8}$, 517.

A.R.S.Ponter   1968   "Convexity conditions and energy theorems
                         for time-independent materials", J.Mech.
                         Phys.Sol., $\underline{16}$, 283.

A.R.Ponter and   1972   "Some extremal properties and energy
J.B.Martin   theorems for inelastic materials and
                         their relationship to the deformation
                         theory of plasticity", J.Mech.Phys.Sol.,
                         $\underline{20}$, 281.

J.F.Soechting   1969   "A bounding principle in the theory of
and R.H.Lance   work hardening plasticity", J.Appl.Mech.,
                         $\underline{36}$, 228.

THE DETERMINATION OF EXTREMAL PATHS

23.1 Plastic Work and Complementary Plastic Work

In this Chapter we shall discuss the determination of paths
which maximize the complementary work and minimize the work for
given terminal generalized stresses and strains for a variety
of specific forms of the elastic-plastic constitutive relations.
In our conventional representation the total strain q_j is divided into an elastic component q_j^e and a plastic component q_j^p;

$$q_j = q_j^e + q_j^p \; . \tag{1}$$

As in our previous descriptions, it will be assumed that elastic
strain and stress are related by the linear relation

$$q_j^e = C_{jk} Q_k \; , \tag{2}$$

where C_{jk} is symmetric and positive definite. The inverse of
this relationship is written as

$$Q_j = D_{jk} q_k^e \; , \tag{3}$$

and D_{jk} will also be symmetric and positive definite.

In Chapter 21 we established one important result for extremal paths in stress space: if an elastic path exists between the
origin and the specified terminal state of stress, that path is
an extremum path. For a material initially in its virgin state
the states of stress which can be reached by elastic paths are
those states of stress which satisfy the inequality

$$\bar{\phi}(Q_j) \leq 0 \; , \tag{4}$$

where $\bar{\phi}(Q_j) = 0$ is the initial yield surface. For these states
of stress,

$$\Omega^o(Q_j) = \Omega(Q_j) = \frac{1}{2} C_{jk} Q_j Q_k \text{ provided } \bar{\phi}(Q_j) \leq 0 \; . \tag{5}$$

It is evident that if the state of stress associated with a given state of strain q_j through the elastic relations satisfies inequality (4) the minimum work path mapped in strain space by the extremal path in stress space is one along which $q_j^p = 0$. Hence

$$W^o(q_j) = W(q_j) = \frac{1}{2} D_{jk} q_j q_k, \text{ provided } \bar{\phi}(Q_j) \leq 0, \ Q_j = D_{jk} q_k \quad (6)$$

Thus the cases of real interest are extremal paths along which $q_j^p \neq 0$. As a result of the separation of the strain into elastic and plastic components it is convenient to redefine the problem. Extremal paths in stress space are most easily dealt with; for a terminal state of stress Q_j^* we have

$$\bar{\Omega}(Q_j^*) = \int_0^{Q_j^*} q_j \, dQ_j = \int_0^{Q_j^*} q_j^e \, dQ_j + \int_0^{Q_j^*} q_j^p \, dQ_j$$

$$= \Omega(Q_j^*) + \bar{\Omega}^p(Q_j^*) \ . \quad (7)$$

By definition $\Omega(Q_j^*) = \frac{1}{2} C_{jk} Q_j^* Q_j^*$ is path independent. Consequently in order to find the maximum value of $\bar{\Omega}(Q_j^*)$ we must find the maximum value of the path dependent term $\bar{\Omega}^p(Q_j^*)$; the elastic part of the constitutive equations can be ignored in this process and we may treat only the plastic part of the constitutive equations. *This is equivalent to finding the maximum complementary work paths for a rigid-plastic material.*

The discussion given in Chapter 22 is applicable equally well to both elastic-plastic and rigid-plastic materials. We can thus draw the conclusion that paths in stress space which maximize $\bar{\Omega}^p(Q_j^*)$ map into the *plastic strain space* paths which minimize the plastic work $\bar{W}^p(q_j^{*p})$, where

$$\bar{W}^p(q_j^{*p}) = \int_0^{q_j^{*p}} Q_j \, dq_j^p \ . \quad (8)$$

Thus we may seek functions $\Omega^{op}(Q_j^*)$ and $W^{op}(q_j^{*p})$ which maximize and minimize respectively the complementary plastic work $\bar{\Omega}^p(Q_j^*)$ in stress space and the plastic work $\bar{W}^p(q_j^{*p})$ in plastic strain space. The maximum complementary work is obtained immediately when the maximum complementary plastic work is known, since from (7) and the path independence of $\Omega(Q_j^*)$.

$$\Omega^o(Q_j^*) = \Omega(Q_j^*) + \Omega^{op}(Q_j^*) \quad . \tag{9}$$

In order to minimize the work for a given terminal *total* strain q_j^* we must consider

$$\bar{W}(q_j^*) = \int_0^{q_j^*} Q_j dq_j = \int_0^{q_j^{*e}} Q_j dq_j^e + \int_0^{q_j^{*p}} Q_j dq_j^p = W(q_j^{*e}) + \bar{W}^p(q_j^{*p}) \quad . \tag{10}$$

For any given choice of *both* q_j^{*e} and q_j^{*p} it is evident that $W(q_j^{*e})$ is path independent and that the minimum value of $\bar{W}^p(q_j^{*p})$ is given by $W^{op}(q_j^{*p})$. It remains therefore to determine q_j^{*e} and q_j^{*p} for a given terminal total strain q_j^* so that the expression given in equation (10) is minimized. This question can be answered by considering the total differential of $\bar{W}$ considered as a function of the independent variables q_j^{*e} and q_j^{*p};

$$d\bar{W} = \left.\frac{\partial W}{\partial q_j^e}\right|_{q_j^e = q_j^{*e}} dq_j^{*e} + \left.\frac{\partial W^{op}}{\partial q_j^p}\right|_{q_j^p = q_j^{*p}} dq_j^{*p} \quad . \tag{11}$$

In addition, in order that q_j^* remains constant when q_j^{*e} and q_j^{*p} are varied,

$$dq_j^{*e} + dq_j^{*p} = 0 \quad . \tag{12}$$

Combining (11) and (12) we see that $d\bar{W} = 0$ if and only if

$$\left.\frac{\partial W}{\partial q_j^e}\right|_{q_j^e = q_j^{*e}} = \left.\frac{\partial W^{op}}{\partial q_j^p}\right|_{q_j^p = q_j^{*p}} \quad . \tag{13}$$

Together with the equation

$$q_j^* = q_j^{*e} + q_j^{*p} \quad , \tag{14}$$

equation (13) is sufficient to determine q_j^{*e} and q_j^{*p}. Equation (13) requires that the generalized stresses associated with the potential functions at the points q_j^{*e} in the elastic strain space and q_j^{*p} in the plastic strain space should be identical; any other way of choosing q_j^{*e} and q_j^{*p} would be physically un-acceptable.

Our immediate problem is thus to determine extremal paths in stress and strain space which respectively maximize $\bar{\Omega}^p(Q_j^*)$ and minimize $\bar{W}^p(q_j^{*p})$. We shall consider a number of particular examples, since it does not appear possible to derive a criterion for the extremal paths in general terms. We shall first treat some examples of hardening materials.

23.2 Extremal Paths for a Class of Hardening Materials

Let us consider in the first instance the formulation which includes both isotropic hardening and kinematic hardening as special cases (Chapter 7, equation 32 *et seq.*). We suppose that the current yield surface is given by

$$\phi(Q_j; q_j^p; \bar{W}^p) = 0 \quad , \tag{15}$$

where $\bar{W}^p$ is the plastic work computed over the entire stress history. ϕ is a convex function of Q_j, and reduces to the initial yield function $\bar{\phi}(Q_j)$ at the virgin state;

$$\phi(Q_j; 0; 0) = \bar{\phi}(Q_j) \tag{16}$$

If ϕ is a continuously differentiable function in stress space the plastic constitutive equations take the form

$$\dot{q}_j^p = cG \frac{\partial \phi}{\partial Q_j} \frac{\partial \phi}{\partial Q_k} \dot{Q}_k \quad , \tag{17a}$$

where

$$c = +1 \quad \text{if } \phi = 0 \text{ and } \frac{\partial \phi}{\partial Q_k} \dot{Q}_k \geq 0 \quad , \tag{17b}$$

$$c = 0 \quad \text{if } \phi < 0 \text{ or } \phi = 0 \text{ and } \frac{\partial \phi}{\partial Q_k} \dot{Q}_k \leq 0 \quad . \tag{17c}$$

G is a non-negative scalar hardening function. When loading occurs,

$$\dot{\phi} = \frac{\partial \phi}{\partial Q_j} \dot{Q}_j + \frac{\partial \phi}{\partial Q_j^p} \dot{q}_j^p + \frac{\partial \phi}{\partial \overline{W}^p} Q_j \dot{q}_j^p = 0 \quad . \tag{18}$$

Substituting (18) into (17) we find that

$$\frac{1}{G} = - \frac{\partial \phi}{\partial Q_j} \left(\frac{\partial \phi}{\partial q_j^p} + \frac{\partial \phi}{\partial \overline{W}^p} Q_j \right) \quad . \tag{19}$$

Consider first the problem of determining the extremum value of the total plastic work subject to fixed initial conditions $(q_j^p\{\tau=0\} = 0)$ and fixed final conditions $(q_j^p\{\tau=t\} = q_j^{*p})$ at times $\tau = 0$ and $\tau = t$ respectively. We write

$$\overline{W}^p(q_j^{*p}) - \int_0^t Q_j(\tau)\dot{q}_j^p(\tau)d\tau \quad . \tag{20}$$

We now permit variations in the strain path $\delta q_j^p(\tau)$ subject to $\delta q_j^p(0) = \delta q_j^p(t) = 0$. The corresponding variation in $\overline{W}^p$ is given by

$$\delta \overline{W}^p = \int_0^t (\dot{q}_j^p \delta Q_j + Q_j \delta \dot{q}_j^p)d\tau \quad . \tag{21}$$

Integrating by parts (21) becomes

$$\delta \overline{W}^p = \int_0^t (\dot{q}_j^p \delta Q_j - \dot{Q}_j \delta q_j^p)d\tau \quad . \tag{22}$$

In equations (21) and (22) δQ_j denotes the first variation of $Q_j(\tau)$ which must be related to $\delta q_j(\tau)$ through the constitutive relation (17).

Consider now the dual problem of finding the extremum value of the complementary plastic work

$$\bar{\Omega}^p(Q_j^*) = \int_0^t q_j^p(\tau)\dot{Q}_j(\tau)d\tau \tag{23}$$

subject to fixed initial and final stresses $Q_j(0) = 0$, $Q_j(t) = Q_j^*$. Permitting a variation in the stress path $Q_j(\tau)$, with $\delta Q_j(0) = \delta Q_j(t) = 0$, we obtain

$$\delta\bar{\Omega}^p = \int_0^t (q_j^p\delta\dot{Q}_j + \delta q_j^p\dot{Q}_j)d\tau \quad , \tag{24}$$

which, upon integrating by parts yields

$$\delta\bar{\Omega}^p = -\int_0^t (\dot{q}_j^p\delta Q_j - \dot{Q}_j\delta q_j^p)d\tau \quad . \tag{25}$$

It is now seen that (22) and (25) are formally identical. They differ, however, in the final conditions to which the variations are subjected ($\delta q_j^p(t) = 0$ in 22, $\delta Q_j(t) = 0$ in 25). We may now pose the *broader* variational problem of finding histories of stress and strain which give

$$\int_0^t (\dot{q}_j^p\delta Q_j - \dot{Q}_j\delta q_j^p)d\tau = 0 \quad , \tag{26}$$

where δQ_j, δq_j^p are related through the constitutive equations (17) but are not subjected to any conditions at time $\tau=t$. If this broader problem possesses a unique solution we then have the coincident unique solutions to the variational problems (22) and (25).

The variation in the yield function ϕ associated with a variation in the strain path is given by

$$\delta\phi(\tau) = \frac{\partial\phi}{\partial Q_j}\,\delta Q_j(\tau) + \frac{\partial\phi}{\partial q_j^p}\,\delta q_j^p(\tau) + \frac{\partial\phi}{\partial \bar{w}^p}\,\delta\bar{w}^p(\tau) \quad . \tag{27}$$

For the extremal plastic strain path $\delta\bar{w}^p = 0$ for variations satisfying the conditions $\delta q_j^p(0) = \delta q_j^p(t) = 0$. For the wider class of variations over the interval $[0,\tau]$ implicit in (27) $(\delta q_j^p(0) = 0,\ \delta q_j^p(\tau) \neq 0)$ we have

$$\delta\bar{w}^p(\tau) = Q_j(\tau)\delta q_j^p(\tau) \quad . \tag{28}$$

Without ambiguity the state of stress at all times along the extremal path may be assumed to lie on the yield surface, and therefore, from (27) with $\delta\phi = 0$,

$$\frac{\partial\phi}{\partial Q_j}\,\delta Q_j + \left(\frac{\partial\phi}{\partial q_j^p} + \frac{\partial\phi}{\partial\bar{w}^p}\,Q_j\right)\delta q_j^p = 0 \quad . \tag{29}$$

Substituting for $\dot{q}_j^p$ in equation (27) from (17), and then substituting for $(\delta\phi/\delta Q_j)\delta Q_j$ from (29), we obtain

$$\int_0^t \left\{ -G\left(\frac{\partial\phi}{\partial Q_k}\,\dot{Q}_k\right)\left(\frac{\partial\phi}{\partial q_j^p} + \frac{\partial\phi}{\partial\bar{w}_j^p}\,Q_j\right) - \dot{Q}_j \right\}\delta q_j^p\,d\tau = 0 \quad . \tag{30}$$

By the fundamental theorem of the calculus of variations the extremum path must satisfy the Euler equation,

$$\dot{Q}_j = -G\left(\frac{\partial\phi}{\partial Q_k}\,\dot{Q}_k\right)\left(\frac{\partial\phi}{\partial q_j^p} + \frac{\partial\phi}{\partial\bar{w}^p}\,Q_j\right) \quad . \tag{31}$$

Provided that the right hand side of equation (31) has a unique value, $\dot{Q}_j$ is determined for any prescribed values of $Q_j, q_j^p, \bar{w}^p$. Equation (31) thus gives a set of paths in stress space emanating from various stress states on the *initial* yield surface $\bar{\phi} = 0$.

In two particular cases (31) provides a particularly simple solution.

(a) Isotropic hardening. In this case ϕ depends only on Q_j and $\overline{W}^p$, the most familiar example being

$$\phi(Q_j,\overline{W}^p) = g(Q_j) - h(\overline{W}^p) , \tag{32}$$

where $g(Q_j)$ is homogeneous in the components of Q_j so that subsequent yield surfaces are geometrically similar to the initial yield surface. With $\partial\phi/\partial q_j^p = 0$ the Euler equation (31) becomes

$$\dot{Q}_j = -G\left(\frac{\partial\phi}{\partial Q_k}\,\dot{Q}_k\right)\frac{\partial\phi}{\partial\overline{W}^p}\,Q_j \quad . \tag{33}$$

It is evident that $\dot{Q}_j$ and Q_j are in the same direction in stress space and therefore that the solution to (31) yields radial or proportional loading paths in stress space (Figure 1).

(b) Linear kinematic hardening. A linear kinematic hardening material is described by equations (17) if the yield function is given by

$$\phi = \phi(Q_j - cq_j^p) , \tag{34}$$

where c is a constant. Changes in q_j^p cause the yield surface to translate in stress space; alternatively a subsequent yield

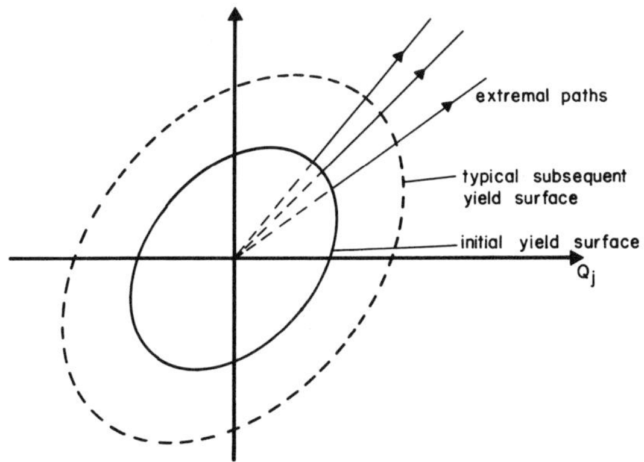

Figure 1. Extremal paths for isotropic hardening

surface in stress space with the coordinate system translated to cq_j^p is identical to the initial yield surface. On noting that

$$\frac{\partial \phi}{\partial q_j^p} = -c \, \frac{\partial \phi}{\partial Q_j} \quad , \tag{35}$$

equation (31) becomes

$$\dot{Q}_j = c \dot{q}_j^p \quad . \tag{36}$$

Since $\dot{q}_j^p$ is always normal to the yield surface the extremal path in stress space always intersects the subsequent yield surface in the direction of the outward normal. The extremal paths in stress space are thus straight line paths in the directions of the outward normals to the initial yield surface (Figure 2). In the case where

$$\phi(Q_j - cq_j^p) = (Q_j - cq_j^p)(Q_j - cq_j^p) - Q_o^2 \quad , \tag{37}$$

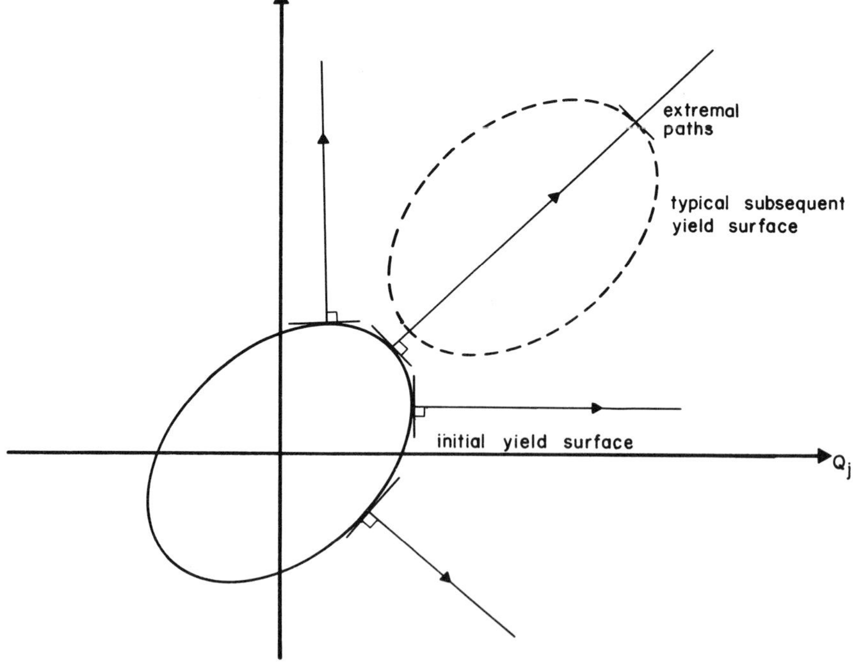

Figure 2. Extremal paths for kinematic hardening

the extremal paths in stress space are also proportional loading
paths. Equation (37) is applicable, for example, in the case of
a von Mises initial yield surface when Q_j is the stress deviator
s_{ij} and q_j^p is the plastic strain deviator in continuum plasticity.

In the case of isotropic hardening with the yield function
given by equation (32) (i.e. geometrically similar yield surfaces) and linear kinematic hardening the extremal paths in
stress space map radial straight line paths in the plastic strain
space.

The extremal paths for any other yield function which falls
into the class definied by equation (15) may be obtained from
equation (31). The extremal paths will not in general be as
simple as those found in the examples given above. One other
simple case of particular interest, however, is that where equations (15) and (17) are generalized to represent a number of
yield functions

$$\phi^r = \phi^r(Q_j ; q_j^{pr}) \quad , \quad r=1,\ldots,n \quad , \tag{38}$$

where

$$q_j^p = \sum_{r=1}^{n} q_j^{pr} \quad , \tag{39a}$$

$$q_j^{pr} = c_r G_r\left(\frac{\partial \phi^r}{\partial Q_k} \dot{Q}_k\right) \frac{\partial \phi^r}{\partial Q_j} \quad , \tag{39b}$$

$$c_r = +1 \quad \text{if} \quad \phi^r=0 \quad \text{and} \quad \frac{\partial \phi^r}{\partial Q_k} \dot{Q}_k \geq 0 \quad , \tag{39c}$$

$$c_r = 0 \quad \text{if} \quad \phi^r < 0 \quad \text{or} \quad \phi^r=0 \quad \text{and} \quad \frac{\partial \phi^r}{\partial Q_k} \dot{Q}_k \leq 0 \quad . \tag{39d}$$

The variational arguments above may be extended to this more extensive class of materials without difficulty. A case of special interest arises when

$$\phi^r = \pm\, g_r(Q_j - cq_j^p) \quad , \tag{40}$$

where g_r is linear and homogeneous in its argument. Each yield surface is then an independent pair of parallel hyperplanes in stress space. As n becomes very large such yield functions describe a particular case of slip theory (Chapter 7) in continuum plasticity. This material is path independent to a considerable degree. If we consider one yield function in isolation, loading into the yield surface at any point on one hyperplane (Figure 3) leads to identical plastic strain increments. It is clear then that any path to a given terminal state of stress or strain which involves continuous loading provides the same values of

$$\int q_j^{rp}\,dQ_j \quad \text{and} \quad \int Q_j\,dq_j^{rp}$$

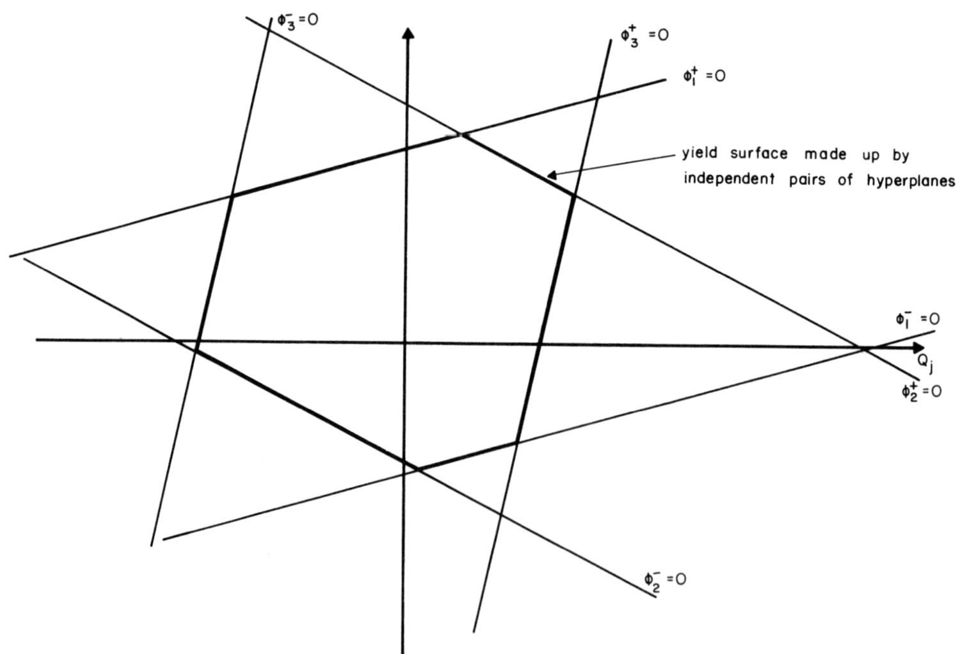

Figure 3.Yield surface made up of independent pairs of hyperplanes

and that these values are maximum and minimum values of these integrals. For a set of yield functions any path which does not involve unloading from *any* yield surface gives maximum and minimum values repectively of

$$\bar{\Omega}^p = \sum_{r=1}^{n} \int q_j^{rp} dQ_j \quad ,$$

$$\bar{W}^p = \sum_{r=1}^{n} \int Q_j dq_j^{rp} \quad .$$

$$(41)$$

Radial straight line paths in both stress and strain space will satisfy this condition and may thus be taken as the extremal paths.

To complete this study of hardening materials we shall return briefly to the case of linear kinematic hardening and the special case of linear isotropic hardening in order to demonstrate the potential nature of the functions Ω^{op} and W^{op}.

For the case of linear kinematic hardening let us put

$$\phi(Q_j - cq_j^p) = g(Q_j - cq_j^p) - g_o \tag{42}$$

where c and g_o are constants. From equation (36) $\dot{Q}_j = c\dot{q}_j^p$ on the extremal stress path and hence, on integrating over the path to some terminal state of stress Q_j,

$$q_j^p = \frac{1}{c}(Q_j - Q_j^o) \tag{43}$$

where Q_j^o lies on the *initial* yield surface and that outward normal to the initial yield surface which passes through Q_j (Figure 4). Further

$$\Omega^{op}(Q_j) = Q_j q_j^p - \int_o^t Q_j \dot{q}_j d\tau$$

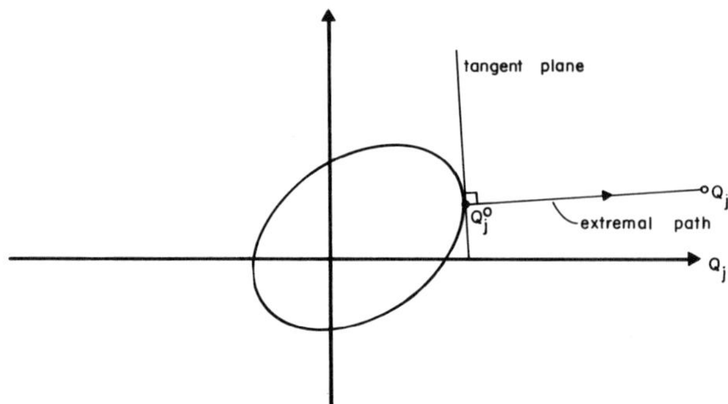

Figure 4. Extremal path for a kinematic hardening model

$$= \frac{1}{c} \, Q_j (Q_j - Q_j^O) - \frac{1}{2c} (Q_j Q_j - Q_j^O Q_j^O) \tag{44}$$

$$= \frac{1}{2c} \, (Q_j - Q_j^O)(Q_j - Q_j^O) \quad .$$

Now

$$\frac{\partial \Omega^{op}}{\partial Q_j} = \frac{1}{c} (Q_j - Q_j^O) + \frac{1}{c} (Q_k - Q_k^O) \frac{\partial Q_k^O}{\partial Q_j} \tag{45}$$

For a fixed value of the index j in this expression the last
term involves the scalar product of the vectors $(Q_k - Q_k^O)$ and
$\partial Q_k^O / \partial Q_j$. However $(Q_k - Q_k^O)$ is normal to the initial yield surface
and $\partial Q_k^O / \partial Q_j$ lies in the initial yield surface because a change
in Q_j can only cause Q_k^O to change to an adjacent point on the
initial yield surface. The last term in (45) is thus zero, and
from (43) and (45)

$$\frac{\partial \Omega^{op}}{\partial Q_j} = q_j^p \quad , \tag{46}$$

confirming that the plastic strains may be derived from Ω^{op}.
The plastic work function W^{op} can be found from the identity

$$W^{op} = Q_j q_j^p - \Omega^{op} \quad , \tag{47}$$

which from equations (43) and (44) gives

$$W^{op} = \frac{c}{2} q_j^p q_j^p + Q_j^o q_j^p \quad . \tag{48}$$

The stress Q_j^o is completely defined by the terminal strain q_j^p, since it is a point on the initial yield surface where the normal vector has the same direction as q_j^p. Q_j^o can only be given explicitly, however, when the function g is known. From equation (48),

$$\frac{\partial W^{op}}{\partial q_j^p} = cq_j^p + Q_j^o + q_k^p \frac{\partial Q_k^o}{\partial q_j^p} \quad . \tag{49}$$

As before, q_j^p and $\partial Q_k^o/\partial q_j^p$ are vectors which are perpendicular to each other for any choice of the subscript j. Thus, from (43) and (40)

$$\frac{\partial W^{op}}{\partial q_j^p} = Q_j \tag{50}$$

confirming that stress may be derived from the minimum plastic work function W^{op}.

Consider the particular case of an isotropic hardening material for which subsequent yield surfaces are given by

$$\phi(Q_j;\overline{W}^p) = g(Q_j) - (\frac{2}{h} \overline{W}^p + g_o)^{1/2} \quad , \tag{51}$$

where h, g_o are constant and $g(Q_j)$ is homogeneous and of degree one in the components of Q_j. The plastic strain rates are given by

$$\dot{q}_j^p = h \frac{\partial g}{\partial Q_j} (\frac{\partial g}{\partial Q_k} \dot{Q}_k) \tag{52}$$

for $\phi=0$ and $\dot{Q}_k \, \partial g/\partial Q_k > 0$. The extremum stress history is the radial path described by

$$Q_j(\tau) = \frac{\tau}{t} Q_j(t) \quad , \tag{53}$$

where $Q_j(t)$ denotes the terminal stress. Substituting (53) into (52) and integrating, and noting the property of the homogeneous function g,

$$\frac{\partial g}{\partial Q_j} Q_j = g \quad , \tag{54}$$

we obtain

$$q_j^p = h \{g(Q_j) - g_o\} \frac{\partial g}{\partial Q_j} \quad , \tag{55}$$

so that the terminal strain $q_j^p(t)$ is given in terms of the terminal stress $Q_j(t)$. Further, we find that

$$\Omega^{op}(Q_j) = \frac{1}{2} h \{g(Q_j) - g_o\}^2 \quad . \tag{56}$$

Differentiating this expression it is easily verified that

$$\frac{\partial \Omega^{op}}{\partial Q_j} = h \{g(Q_j) - g_o\} \frac{\partial g}{\partial Q_j} = q_j^p \quad . \tag{57}$$

Using the identity (47) and (56) we find that

$$W^{op} = \frac{h}{2} \{g^2(Q_j) - g_o^2\} \quad . \tag{58}$$

This expression is of course in terms of the terminal stress rather than the terminal strain. In equation (55) we may note that $\partial g/\partial Q_j$, which is homogeneous and of order zero in stress, is simply a direction in stress space which can be considered as a function of the terminal strain; it fixes the direction of the radial stress path which leads to the plastic strain increments which have the same direction as q_j^p (Figure 5). Hence by multiplying equation (55) by $\partial g/\partial Q_j$ we can find $g(Q_j)$ as a function of q_j^p and substitute this result into (58). Unfortunately we cannot proceed with this substitution if $g(Q_j)$ is not given

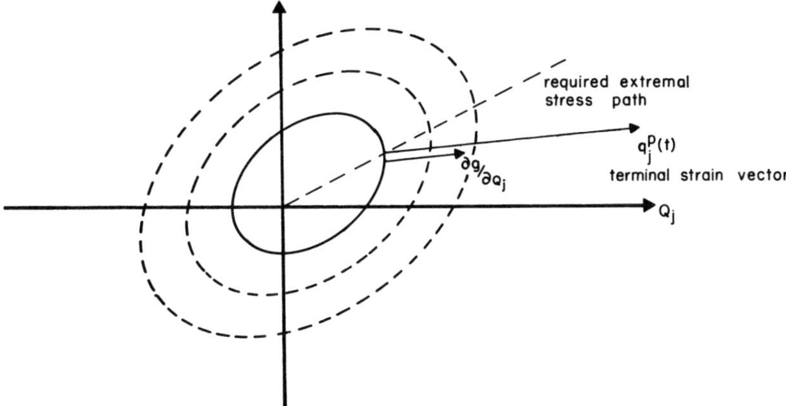

Figure 5.   Isotropic hardening model

explicitly without great complexity; the reader is invited to
take a particular example and confirm that

$$Q_j = \frac{\partial W^{op}}{\partial q_j^p} \quad . \tag{59}$$

23.3 Elastic, Perfectly Plastic Materials

Finally we may briefly reconsider elastic, perfectly plastic
materials in the light of the more general discussion of Chapters 22 and 23. It was established in Chapter 21 that the extremum path in stress space for an elastic, perfectly plastic
material is the elastic path, so that

$$\Omega^o(Q_j) = \Omega(Q_j) \tag{60}$$

for all stress states which lie on or within the limit or yield
surface. In the context of equation (9) this means that

$$\Omega^{op}(Q_j) = 0 \quad \text{for} \quad \phi(Q_j) \leq 0 \tag{61}$$

for an elastic, perfectly plastic material. On differentiating
(61) we see that

$$\frac{\partial \Omega^{op}}{\partial Q_j} = q_j^p = 0 \quad . \tag{62}$$

Clearly when the terminal stress state lies on the yield surface $\phi(Q_j) = 0$, $\dot{Q}_j = 0$, plastic strain of undetermined magnitude but restricted direction may take place, and the terminal state of strain is not unique. Thus (62) gives the correct terminal plastic strain for $\phi(Q_j) < 0$, but only one of the possible terminal strains when $\phi(Q_j) = 0$. This is consistent with our formulation, since the proof of the potential property of $\Omega^o(Q_j)$ requires that for each terminal stress Q_j there must exist an unrestricted neighborhood in stress space which can be reached by small increments of stress (Chapter 22, equation (1), *et seq.*). In a perfectly plastic material this condition is not met when $\phi(Q_j) = 0$ and the terminal stress state lies on the yield surface, since stress states such that $\phi(Q_j) > 0$ are not admitted. Alternatively, it can be considered that Ω^{op} is discontinuous on the surface $\phi(Q_j) = 0$, becoming infinitely large for $\phi(Q_j) > 0$. This is consistent with the notion that stresses such that $\phi(Q_j) > 0$ can be achieved only when q_j^p is infinitely large. The component of the gradient $\partial\Omega^{op}/\partial Q_j$ across the limit or yield surface is consequently undefined in magnitude.

The proof given in Chapter 22 that the maximum complementary work path in stress space maps a minimum work path in strain space (Chapter 22, equation 18) depends only on the postulate

$$dQ_j dq_j \geq 0 \; , \tag{63}$$

and is therefore valid for perfectly plastic materials. Further, if the yield surface is a *closed figure* in stress space the terminal strains which may be associated with extremal paths in stress space map into the whole strain space, and there is consequently a minimum work path for each terminal strain which is also a maximum complementary work path.

The minimum plastic work $W^{op}(q_j^p)$ corresponding to a terminal plastic strain q_j^p is easily computed from the identity (47) and equation (61);

$$W^{op} = Q_j q_j^p - \Omega^{op}$$

$$= Q_j q_j^p \quad , \tag{64}$$

where Q_j is the stress point on the yield surface at that point
where the outward normal vector $\partial\phi/\partial Q_j$ has the same direction
as q_j^p. This stress $Q_j(q_k^p)$ is uniquely defined if the yield surface is *strictly convex*. However, if the yield surface has
flat regions (i.e. it does not have a continuously turning tangent hyperplane), $Q_j(q_k^p)$ is not uniquely determined. However,
if Q_j and $Q_j + \Delta Q_j$ are two stress states on such a region,

$$(Q_j + \Delta Q_j) q_j^p = Q_j q_j^p \tag{65}$$

because ΔQ_j is normal to q_j^p, and W^{op} is consequently uniquely
determined.

On differentiating equation (64) we see that

$$\frac{\partial W^{op}}{\partial q_j^p} = Q_j(q_j^p) + \frac{\partial Q_k}{\partial q_j^p} q_k^p \quad . \tag{66}$$

For any given choice of the subscript j, the vectors $\partial Q_k / \partial q_j^p$
and q_k^p are normal to each other, since the terminal stress point
is constrained to lie in the yield surface and $\partial Q_k / q_j^p$ consequently lies in the tangent hyperplane. At a singular point or
corner in the yield surface where the tangent hyperplane is not
unique $\partial Q_k / \partial q_j^p$ is zero if q_j^p lies between the adjacent normals.
Thus the second term in equation (66) is zero, and the potential
nature of W^{op} is confirmed.

As we have just noted, $\partial W^{op}/\partial q_j^p$ is not unique when the yield
surface has flats. In fact a yield surface with flat regions
gives rise to level surfaces of W^{op} in plastic strain space

which have singular points or corners. This can be easily demonstrated with an example; consider a Tresca yield condition
for the principal stresses σ_x,σ_y in plane stress, shown in Figure 6(a). Let us compute, using equation (64), the level surface $W^{op} = a$ in plastic strain space. When the terminal strain
is ε_x^p, $\varepsilon_y^p = 0$ the terminal stress lies on AB, and is given by
$\sigma_x = \sigma_o$, $0 \leq \sigma_y \leq \sigma_o$. Hence

$$W^{op} = \sigma_x \varepsilon_x^p + \sigma_y \varepsilon_y^p = a \tag{67}$$

gives

$$\varepsilon_x^p = \frac{a}{\sigma_o} \tag{68}$$

which is the single point A' in plastic strain space (Figure 6b).
If $\varepsilon_x^p > 0$, $\varepsilon_y^p > 0$ then the terminal stress lies at B, with $\sigma_x = \sigma_y = \sigma_o$. Hence equation (67) gives

$$\varepsilon_x^p + \varepsilon_y^p = \frac{a}{\sigma_o} \tag{69}$$

which is a line A'B' in plastic strain space. We may continue
in a like manner for all other terminal strain states and plot
the level surface $W^{op} = a$ shown in Figure 6(b). It is seen
that flat regions on the yield surface lead to corners on the

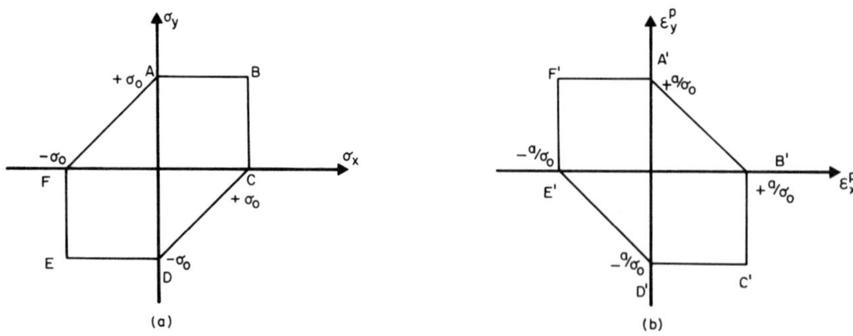

Figure 6. Tresca yield surface

level surface W^{op} = a, while corners on the yield surface lead to flat regions on the level surface of W^{op}.

When the elastic component of strain is included the elastic and plastic components of the total terminal strain are again determined from equation (13) which requires that the stresses associated with W^{op} and W are identical. The strict inequality $dQ_j dq_j > 0$ holds for all stress increments, and the proof of the uniqueness of the terminal stress for a given extremal path in strain space holds. The uniqueness of the terminal stress associated with q_j^p also implies the uniqueness of q_j^e and q_j^p, and we see that subdivision described by equation (13) will always provide a unique solution.

An alternative, and simpler, method of determining W^{op} for an elastic, perfectly plastic material may be given in terms of the dissipation function $D(\dot{q}_j^p)$ introduced in Chapter 9. It will be recalled that for given $\dot{q}_j^p$ we have

$$D(\dot{q}_j^p) = Q_j \dot{q}_j^p \ , \tag{70}$$

where Q_j is the stress associated with $\dot{q}_j^p$ through the flow rule. The dissipation function $D(\dot{q}_j^p)$ is convex, and is homogenous and of degree one in the components of $\dot{q}_j^p$. Further,

$$Q_j = \frac{\partial D}{\partial \dot{q}_j^p} \ . \tag{71}$$

We know that along the extremal path in stress space the plastic strain rate vanishes in the elastic region, and that for a perfectly plastic material the stress remains constant while the plastic strain changes. It may be imagined without loss in generality that the plastic strain rate is constant, so that the terminal plastic strain q_j^p can be written as

$$q_j^p = \beta \dot{q}_j^p \ , \tag{72}$$

where β is a constant. Hence

$$W^{op} = \beta D(\dot{q}_j^p) \quad , \tag{73}$$

and since D is homogeneous and of degree one,

$$W^{op} = D(\beta \dot{q}_j^p) = D(q_j^p) \quad . \tag{74}$$

The convexity of W^{op} follows immediately, and it is also evident
that

$$\frac{\partial W^{op}}{\partial q_j^p} = \frac{\partial D(q_j^p)}{\partial q_j^p} = \frac{\partial D(\dot{q}_j^p)}{\partial \dot{q}_j^p} = Q_j \quad . \tag{75}$$

23.4 Bibliographical and Historical Remarks

Extremal paths for elastic, perfectly plastic materials and for
elements in which only one component of stress acts were given
by Martin [1966], for isotropic hardening by Ponter [1968] and
Martin [1970], and for kinematic hardening by Soechting and
Lance [1969]. These results were extended by Ponter and Martin
[1972], whose development is presented in this Chapter.

References

J.B.Martin   1966   "Extended displacement bound theorems
for work hardening continua subjected
to impulsive loading", Int.J.Solids
and Structures, $\underline{2}$, 9.

J.B.Martin   1970   "A complementary energy boundary
theorem for time-independent materi-
als", *Developments in Theoretical and
Applied Mechanics* (ed. by D.Frederick
and E.H.Harris), $\underline{8}$, 517.

A.R.S.Ponter   1968   "Convexity conditions and energy theorem for time-independent materials",
J.Mech.Phys.Sol., $\underline{16}$, 283.

A.R.S.Ponter and   1972   "Some extremal properties and energy
J.B.Martin   theorems for inelastic materials and
                               their relationship to the deforma-
                               tion theory of plasticity", J.Mech.
                               Phys.Sol., $\underline{20}$, 281.

J.F.Soechting and   1969   "A bounding principle in the theory
R.H.Lance   of work hardening plasticity", J.
                               Appl.Mech., $\underline{36}$, 228.

Chapter 24

DEFORMATION THEORIES OF PLASTICITY

24.1 Remarks on Deformation Theories

The term *deformation theory* was given to theories of plasticity
in which a one to one relation between stress Q_j and total
strain q_j is assumed;

$$q_j = q_j(Q_k) \ . \tag{1}$$

Such theories have been used for a great many studies in plasticity. The material described in equation (1) is reversible
and path independent in the sense that strain is a function of
the current state of stress and not on stress history. Clearly
the phenomena associated with loading and unloading from the
yield surface cannot be adequately described by a deformation
theory; such phenomena, as we have seen in Chapter 2, lead directly to an incremental theory in which strain increments or
strain rates resulting from stress increments or stress rates
are studied. Nevertheless, deformation theories have been used
extensively for the solution of elastic-plastic problems largely
because the resulting boundary value problems are more tractable than in incremental theories.

An arbitrarily formulated deformation theory of the type
described by equation (1) is such that strain is a function
only of current stress; however, in one important respect it
may be *path dependent* in that the work or complementary work
between two given terminal states of strain or stress may depend on the strain or stress path between the terminal states.
It does not necessarily follow that constitutive relations of
the type

$$q_j = \frac{\partial \Omega}{\partial Q_j} \ , \quad \Omega = \int_o^{Q_j} q_j \, dQ_j \ , \tag{2a}$$

$$Q_j = \frac{\partial W}{\partial q_j} \quad , \quad W = \int_0^{q_j} Q_j \, dq_j \quad , \tag{2b}$$

can be written for the material described by equation (1).

A very simple example of this can be given if we consider a linear relation

$$q_j = B_{jk} Q_{jk} \tag{3}$$

where B_{jk} is *not* a symmetric matrix. For two stress components Q_1, Q_2 we might have

$$q_1 = B_{11} Q_1 + B_{12} Q_2 \quad ,$$
$$\tag{4}$$
$$q_2 = B_{21} Q_1 + B_{22} Q_2 \quad ,$$

with $B_{21} \neq B_{12}$. Consider, as an example, the complementary work done in changing the stresses from $Q_1 = Q_2 = 0$ to a final state $Q_1 = Q_1^*$, $Q_2 = Q_2^*$. If we carry out this change by changing first Q_1 and then Q_2 (path 1 in Figure 2) it is easily shown that

$$\int_0^{Q_j^*} q_j \, dQ_j = \frac{1}{2} B_{11} Q_1^{*2} + B_{21} Q_1^* Q_2^* + \frac{1}{2} B_{22} Q_2^{*2} \; . \tag{5}$$

If, on the other hand, we change first Q_2 and then Q_1 (path 2

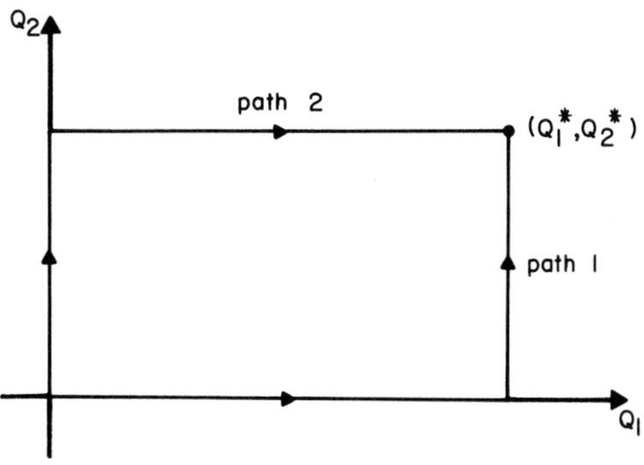

Figure 1. Alternate paths in stress space

in Figure 1),

$$\int_{0}^{Q_{j}^{*}} q_{j} dQ_{j} = \frac{1}{2} B_{11} Q_{1}^{*2} + B_{12} Q_{1}^{*} Q_{2}^{*} + \frac{1}{2} B_{22} Q_{2}^{*2} . \tag{6}$$

The complementary work is consequently not unique, but depends on the path. In the linear case (equation 3) the complementary work and work integrals will in fact be uniquely given in terms of the terminal states only if B_{jk} is symmetric.

A necessary and sufficient condition that equations (1) can be expressed in the form of equations (2) is that the work or complementary work around any cycle in strain or stress space respectively should be identically zero. Taking the case of work, for example, if the work around the cycle shown in Figure 2 passing through q_{j}^{*} is zero we can write

$$\oint Q_{j} dq_{j} = \int_{0}^{q_{j}^{*}} Q_{j} dq_{j} + \int_{q_{j}^{*}}^{0} Q_{j} dq_{j}$$

$$= \int_{0}^{q_{j}^{*}} Q_{j} dq_{j} - \int_{0}^{q_{j}^{*}} Q_{j} dq_{j} = 0 . \tag{7}$$

Since the two integrals in the penultimate expression in equation (7) are taken along two distinct but unspecified paths, the work integral must be path independent.

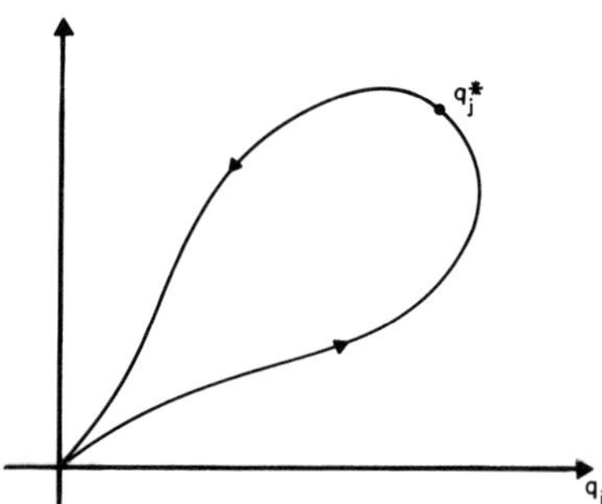

Figure 2. Cycle in strain space

Deformation theories for realistic plastic materials have frequently been formulated by adopting the stress, total strain relation along radial paths in stress space. We have seen, however, in Chapter 22 that radial paths may not be extremal paths for the material and consequently a deformation theory so formulated may not have associated with it work and complementary work integrals which are potential functions. The absence of potential functions in a material described by equation (1) is in general a severe limitation. We shall consider it desirable that a deformation theory should satisfy equation (7); *this leads us to the conclusion that a deformation theory can only be sensibly formulated if equation* (1) *gives the relation between stress and total strain along extremal paths.* In this case we are assured that

$$q_j = \frac{\partial \Omega^{\mathrm{o}}}{\partial Q_j} \quad , \quad Q_j = \frac{\partial W^{\mathrm{o}}}{\partial q_j} \quad . \tag{8}$$

The extremal functions $\Omega^{\mathrm{o}}(Q_j)$, $W^{\mathrm{o}}(q_j)$ are of course convex functions for materials satisfying the postulates of Chapter 2.This, together with the extremal nature of the functions, means that we have available minimum principles for the solution of boundary values problems using equations (8) as the constitutive relations (Chapter 22, inequalities 28 and 32), and we have a consistent relation between the solution obtained by deformation theory and the solution obtained by incremental theory (Chapter 22, inequalities 36 and 43). The determination of the solution of the deformation theory problem thus becomes in many cases a useful approximation to the incremental solution; it may in fact be exact under the special circumstances where the stress paths in the incremental solution are extremal paths.

24.2 Minimum Principles for the Deformation Theory Problem

As we have pointed out in the preceding section, a deformation

theory which has associated with it convex potential functions
(equations 8) has the advantage that we may employ the classical energy theorems for a reversible non-linear material. These
energy theorems were derived in Chapter 22; let us now restate
these theorems, *considering the deformation theory problem in
isolation*, and give alternative formulations which take into account the special nature of deformation theory.

We are given the boundary value problem characterized by
loads $\hat{\underline{p}}(s)$ on S_p, body forces $\hat{\underline{F}}(s)$ on V and displacements $\hat{\underline{u}}(s)$
on S_u. Since we are not treating the incremental solution to
this problem in the present context, let us with a slight change
in notation characterize the deformation theory solution to this
problem by generalized stresses $Q_j(s)$, generalized strains $q_j(s)$,
reactions $\underline{p}(s)$ on S_u, and displacements $\underline{u}(s)$ on S_p and V.   The
constitutive equations for this problem are of course those
given in equations (8).

If W^o and Ω^o are convex functions, we may give two complementary energy theorems for this problem.

First, let Q_j^S, $\underline{p}^S$ on S_u constitute a pertinent statically admissible field. Then (Chapter 22, equation 29) we have

$$U_c^o(Q_j^S) \geq U_c^o(Q_j) \quad , \tag{9}$$

where

$$U_c^o(Q_j^S) = \int_V \Omega^o(Q_j^S)dV - \int_{S_u} \underline{p}^S \cdot \hat{\underline{u}}dS \quad , \tag{10a}$$

$$U_c^o(Q_j) = \int_V \Omega^o(Q_j)dV - \int_{S_u} \underline{p} \cdot \hat{\underline{u}}dS \quad . \tag{10b}$$

Second, let q_j^c, $\underline{u}^c$ on S_p and V constitute a pertinent kinematically admissible field. Then (Chapter 22, equation 32) we have

$$U_p^o(q_j^c) \geq U_p^o(q_j) \quad , \tag{11}$$

where

$$U_p^o(q_j^c) = \int_V W^o(q_j^c)dV - \int_V \underline{\hat{F}}\cdot\underline{u}^c dV - \int_{S_p} \underline{\hat{p}}\cdot\underline{u}^c dS \quad , \tag{12a}$$

$$U_p^o(q_j) = \int_V W^o(q_j)dV - \int_V \underline{\hat{F}}\cdot\underline{u}dV - \int_{S_p} \underline{\hat{p}}\cdot\underline{u}dS \quad . \tag{12b}$$

In the case of an *elastic, perfectly plastic material* it must be recalled that we add the implied assumption that the stress states must be realizable. This affects only the static theorem, where we add the additional constraint that $\phi(Q_j^s) \leq 0$.

A modified form of the kinematic theorem can be derived if we assume that the plastic part of the total strain is given by q_j^{pc}, and that the elastic part q_j^{ec} is hence

$$q_j^{ec} = q_j^c - q_j^{pc} \quad . \tag{13}$$

We are assured that (Chapter 23, equation 10 *et seq.*)

$$W(q_j^c - q_j^{pc}) + W^{op}(q_j^{pc}) \geq W^o(q_j^c) \quad , \tag{14}$$

with equality occurring when

$$\left.\frac{\partial W}{\partial q_j^e}\right|_{q_j^{ec}} = \left.\frac{\partial W^{op}}{\partial q_j^p}\right|_{q_j^{pc}} \quad . \tag{15}$$

It follows then from equations (11) and (14) that

$$\bar{U}_p^o(q_j^c, q_j^{pc}) \geq U_p^o(q_j^c) \geq U_p^o(q_j) \quad , \tag{16}$$

where

$$\bar{U}_p^o(q_j^c, q_j^{pc}) = \int_V W(q_j^c - q_j^{pc})dV + \int_V W^{op}(q_j^{pc})dV$$

$$\tag{17}$$

$$- \int_V \underline{\hat{F}}\cdot\underline{u}^c dV - \int_{S_p} \underline{\hat{p}}\cdot\underline{u}^c dS \quad .$$

Further

$$\bar{U}_p^O(q_j^c, q_j^{pc}) = U_p^O(q_j^c) = U_p^O(q_j) \tag{18}$$

when $\quad q_j^c = q_j, \; q_j^{pc} = q_j^p.$

We thus seek a solution of the programming problem

$$\text{minimize} \quad \bar{U}_p^O(q_j^c, q_j^{pc}) \tag{19}$$

subject to the constraint that q_j^c, $\underline{u}^c$ on S_p and V, $\hat{\underline{u}}$ on S_u constitute a pertinent kinematically admissible field.

In the class of problems under consideration the generalized strains q_j may be written in the form

$$q_j = L_j(\underline{u}) \quad , \tag{20}$$

where $L_j(\;)$ is a vector valued linear operator. The constraint on the programming problem is removed if we treat $\bar{U}_p^O$ as a function of $\underline{u}^c$, q_j^{pc}.

Let us suppose that we first find the solution to the linear elastic problem characterized by loads $\hat{\underline{p}}(s)$ on S_p, body forces $\hat{\underline{F}}(s)$ on V, displacements $\hat{\underline{u}}(s)$ on S_u, with the constitutive equations

$$q_j = \frac{\partial \Omega}{\partial Q_j} = C_{jk} Q_k \quad , \tag{21a}$$

$$Q_j = \frac{\partial W}{\partial q_j} = D_{jk} q_k \quad , \tag{21b}$$

Let this solution be characterized by generalized stresses $Q_j^E(s)$, generalized strains $q_j^E(s)$, reactions $\underline{p}^E(s)$ on S_u, and displacements $\underline{u}^E(s)$ on S_p and V.

Using this solution we can generate a pertinent statically admissible field Q_j^S, $\underline{p}^S$ on S_u;

$$Q_j^S = Q_j^E + \rho_j^S \quad , \tag{22a}$$

$$\underline{p}^S = \underline{p}^E + \underline{p}^{rs} \quad \text{on } S_u \quad , \tag{22b}$$

where $\rho_j^S(s)$ is a statically admissible *residual stress field* in equilibrium with zero loads on S_p and reactions $\underline{p}^{rs}$ on S_u. Substituting equations (22) into equation (10a), we see that

$$U_c^O(Q_j^S) = \int_V \Omega^O(Q_j^E + \rho_j^S)dV - \int_{S_u} (\underline{p}^E + \underline{p}^{rs}) \cdot \underline{\hat{u}}dS \quad . \tag{23}$$

Recalling that $\Omega^O = \Omega + \Omega^{OP}$, we may write

$$\Omega^O(Q_j^E + \rho_j^S) = \Omega(Q_j^E + \rho_j^S) + \Omega^{OP}(Q_j^E + \rho_j^S)$$

$$= \Omega(Q_j^E) + C_{jk}Q_j^E\rho_k^S + \Omega(\rho_j^S) + \Omega^{OP}(Q_j^E + \rho_j^S) \tag{24}$$

$$= \Omega(Q_j^E) + \rho_j^S q_j^E + \Omega(\rho_j^S) + \Omega^{OP}(Q_j^E + \rho_j^S) \quad .$$

Substituting equation (24) into equation (23), and rearranging,

$$U_c^O(\rho_j^S) = \{\int_V \Omega^{OP}(Q_j^E + \rho_j^S)dV + \int_V \Omega(\rho_j^S)dV\}$$

$$+ \{\int_V \Omega(Q_j^E)dV - \int_{S_u} \underline{p}^E \cdot \underline{\hat{u}}dS\} \tag{25}$$

$$+ \{\int_V \rho_j^S q_j^E dV - \int_{S_u} \underline{p}^{rs} \cdot \underline{\hat{u}}dS \quad .$$

Of the three expressions on the right hand side of equation (25), the second is independent of ρ_j^S and the third is zero, since $\underline{p} = 0$ on S_p, $\underline{p}^{rs}$ on S_u, ρ_j^S are statically admissible, and $\underline{u}^E$ on S_p and V, $\underline{\hat{u}}$ on S_u, q_j^E are kinematically admissible. We may carry out a similar argument for the actual residual stress field ρ_j, $\underline{p}^r$ on S_u, defined by

$$Q_j = Q_j^E + \rho_j \quad , \tag{26a}$$

$$\underline{p} = \underline{p}^E + \underline{p}^r \text{ on } S_u \quad . \tag{26b}$$

It follows therefore that

$$U_c^{ro}(\rho_j^S) = \int_V \Omega^{op}(Q_j^E + \rho_j^S)dV + \int_V \Omega(\rho_j^S)dV$$

$$\geq \int_V \Omega^{op}(Q_j^E + \rho_j)dV + \int_V \Omega(\rho_j)dV = U_c^{ro}(\rho_j) \quad , \tag{27}$$

providing a minimum principle for the residual field. Note that in the case of an elastic, perfectly plastic material we impose the further constraint that $\phi(Q_j^S) = \phi(Q_j^E + \rho_j^S) \leq 0$.

In a similar manner we can generate a kinematically admissible field by using the elastic solution;

$$q_j^c = q_j^E + q_j^{rc} \quad , \tag{28a}$$

$$\underline{u}^c = \underline{u}^E + \underline{u}^{rc} \text{ on } S_p \text{ and } V \quad , \tag{28b}$$

where $q_{,j}^{rc}$, $\underline{u}^{rc}$ on S_p and V, $\underline{u} = 0$ on S_u are kinematically admissible. With q_j^{pc} representing as before the plastic component of q_j^c, equation (17) becomes

$$\bar{U}_p^O = \int_V W(q_j^E + q_j^{rc} - q_j^{pc})dV + \int_V W^{op}(q_j^{pc})dV$$

$$- \int_V \underline{\hat{F}}.(\underline{u}^E + \underline{u}^{rc})dV - \int_{S_p} \underline{\hat{p}}.(\underline{u}^E + \underline{u}^{rc})dV \quad . \tag{29}$$

We may write

$$W(q_j^E + q_j^{rc} - q_j^{pc}) = W(q_j^E) + D_{jk}q_j^E(q_k^{rc} - q_k^{pc}) + W(q_j^{rc} - q_j^{pc})$$

$$= W(q_j^E) + Q_j^E q_j^{rc} - Q_j^E q_j^{pc} + W(q_j^{rc} - q_j^{pc}) \quad . \tag{30}$$

Substituting equation (30) into equation (29), and rearranging, we see that

$$\bar{U}_p^{O}(q_j^{rc}, q_j^{pc}) = \{\int_V W^{OP}(q_j^{pc})dV + \int_V W(q_j^{rc} - q_j^{pc})dV - \int_V Q_j^E q_j^{pc}dV\}$$

$$+ \{\int_V W(q_j^{E})dV - \int_V \hat{\underline{F}}\cdot\underline{u}^E dV - \int_{S_p} \hat{\underline{p}}\cdot\underline{u}^E dS\} \tag{31}$$

$$+ \{\int_V Q_j^E q_j^{rc}dV - \int_V \hat{\underline{F}}\cdot\underline{u}^{rc}dV - \int_{S_p} \hat{\underline{p}}\cdot\underline{u}^{rc}dS\} \ .$$

Of the three terms on the right hand side of equation (31), the second is independent of q_j^{rc}, q_j^{pc}, and the third vanishes, since $\hat{\underline{p}}$ on S_p, $\underline{p}^E$ on S_u, $\hat{\underline{F}}$ on V, Q_j^E are kinematically admissible, and $\underline{u}^{rc}$ on S_p and V, $\underline{u} = 0$ on S_u, q_j^{rc} are kinematically admissible.

A similar argument may be carried out for the field q_j^r, u^r on S_p and V defined as the difference between q_j, $\underline{u}$ and q_j^E, $\underline{u}^E$;

$$q_j = q_j^E + q_j^r \ , \tag{32a}$$

$$\underline{u} = \underline{u}^E + \underline{u}^r \text{ on } S_p \text{ and V} . \tag{32b}$$

It follows therefore that

$$\bar{U}_p^{ro}(q_j^{rc}) = \int_V W^{OP}(q_j^{pc})dV + \int_V W(q_j^{rc} - q_j^{pc})dV - \int_V Q_j^E q_j^{pc}dV$$

$$\geq \int_V W^{OP}(q_j^{p})dV + \int_V W(q_j^{r} - q_j^{p})dV - \int_V Q_j^E q_j^{p}dV = \bar{U}_p^{ro}(q_j^{r}) \tag{33}$$

A further simplification of the form of the kinematic theorem can be given in the case of certain hardening materials, where we can express the plastic strains in terms of plastic strain multipliers. For brevity we shall consider only *linear kinematic hardening materials* in which the yield surface is composed of a number of independent hyperplanes in stress space.

Assume first that the yield surface consists of a *single hyperplane* in stress space; the yield function may then be written in the form

$$\phi = N_j(Q_j - hq_j^p) - Q_o \quad , \tag{34}$$

where N_j is a vector valued constant, and h is constant. First yield occurs for

$$N_j Q_j = Q_o \tag{35}$$

and

$$N_j = \frac{\partial \phi}{\partial Q_j} \tag{36}$$

is the outward normal to the yield surface and is independent of Q_j and q_j^p. The plastic strain rates are given by

$$\dot{q}_j^p = cG \frac{\partial \phi}{\partial Q_j} \frac{\partial \phi}{\partial Q_k} \dot{Q}_k \tag{37a}$$

$$= cG N_j N_k \dot{Q}_k$$

where

$$c = +1 \quad \text{if} \quad \phi = 0 \quad \text{and} \quad N_j \dot{Q}_j \geq 0 \tag{37b}$$

$$c = 0 \quad \text{if} \quad \phi = 0 \quad \text{and} \quad N_j \dot{Q}_j \leq 0 \tag{37c}$$

$$\text{or} \quad \phi < 0 \quad ,$$

and G is also constant.

The extremal path, as given in Chapter 23, involves continuous loading, so that in evaluating W^{op} we concern ourselves only with condition (37b). We also make use of the fact that $\dot{q}_j^p$ has a fixed direction in stress space; we can thus put

$$q_j^p = \Lambda N_j \quad , \quad \Lambda \geq 0 \tag{38}$$

for continuous loading. Further, for loading,

$$\dot{\phi} = \frac{\partial \phi}{\partial Q_j} \dot{Q}_j + \frac{\partial \phi}{\partial q_j^p} \dot{q}_j^p \tag{39}$$

$$= N_j \dot{Q}_j - \dot{\Lambda} h N_j N_j = 0 \quad .$$

In addition, from equation (37a) and (39),

$$\dot{q}_j^p = \dot{\Lambda} N_j = G\, N_j N_k \dot{Q}_k \tag{40}$$

$$= G\, N_j (\dot{\Lambda}\, h\, N_k N_k) \quad .$$

Equation (40) implies that

$$G = \frac{1}{h\, N_j N_j} \quad , \tag{41}$$

and consequently equation (39) gives

$$N_j \dot{Q}_j = \frac{\dot{\Lambda}}{G} \quad . \tag{42}$$

Thus we see that

$$W^{op} = \int_0^t Q_j \dot{q}_j^p d\tau = \int_0^t Q_j N_j \dot{\Lambda} d\tau \tag{43}$$

$$= \frac{\Lambda^2}{2G} + Q_o \Lambda$$

for continuous loading with $N_j Q_j = Q_o$ when yielding begins.

Expressing q_j^{pc} in the kinematic minimum principles as

$$q_j^{pc} = \Lambda^c N_j \quad , \tag{44}$$

the results given in inequalities (16) and (33) may be cast in
programming form and given respectively as

$$\text{minimize} \quad \int_V \frac{(\Lambda^c)^2}{2G}\, dV + \int_V Q_o \Lambda^c dV + \int_V W(q_j^c - \Lambda^c N_j)\, dV \tag{45a}$$

$$- \int_V \hat{\underline{F}}\cdot\underline{u}^c dV - \int_{S_p} \hat{\underline{p}}\cdot\underline{u}^c dS \quad,$$

and

$$\text{minimize} \quad \int_V \frac{(\Lambda^c)^2}{2G}\, dV + \int_V (Q_o - Q_j^E N_j)\Lambda^c dV + \int_V W(q_j^{rc} - \Lambda^c N_j)\, dV \tag{45b}$$

subject to $\Lambda^c \geq 0$.

The extension of these programming problems to the case where a number of independent hyperplanes make up the yield surface is straightforward. In this case the yield functions are

$$\phi^\alpha = N_j^\alpha(Q_j - h^\alpha q_j^p) - Q_o^\alpha, \quad \alpha = 1,2,\ldots,\nu \quad . \tag{46}$$

The plastic strain rates are written as the sum

$$\dot{q}_j^p = \sum_{\alpha=1}^\nu \dot{q}_j^{p\alpha} \quad, \tag{47}$$

where

$$\dot{q}_j^{p\alpha} = c^\alpha G^\alpha N_j^\alpha N_k^\alpha \dot{Q}_k \quad, \tag{48a}$$

and

$$c^\alpha = +1 \quad \text{if} \quad \phi^\alpha = 0 \quad \text{and} \quad N_j^\alpha \dot{Q}_j \geq 0 \quad, \tag{48b}$$

$$c^\alpha = 0 \quad \text{otherwise.}$$

Admissible plastic strains are now written in terms of a number of multipliers $\Lambda^{c\alpha}$, $\alpha = 1,\ldots,\nu$;

$$q_j^{pc} = \sum_{\alpha=1}^\nu \Lambda^{c\alpha} N_j^\alpha \quad . \tag{49}$$

The extremal paths (cf. Chapter 23) are paths which involve continuous loading on all planes. It is readily established by arguments similar to those which appear above that the programming problems (45a) and (45b) become respectively

$$\text{Minimize} \sum_{\alpha=1}^{\nu} \left\{ \int_V \frac{(\Lambda^{c\alpha})^2}{2G} \, dV + \int_V Q_o^{\alpha} \Lambda^{c\alpha} dV \right\}$$

$$+ \int_V W(q_j^c - \sum_{\alpha=1}^{\nu} \Lambda^{c\alpha} N_j^{\alpha}) dV - \int_V \hat{\underline{F}} \cdot \underline{u}^c dV - \int_{S_p} \hat{\underline{p}} \cdot \underline{u}^c dS \quad , \tag{50a}$$

and

$$\text{minimize} \sum_{\alpha=1}^{\nu} \left\{ \int_V \frac{(\Lambda^{c\alpha})^2}{2G} \, dV + \int_V (Q_o^{\alpha} - Q_j^E N_j^{\alpha}) \Lambda^{c\alpha} dV \right\}$$

$$\tag{50b}$$

$$+ \int_V W(q_j^{rc} - \sum_{\alpha=1}^{\nu} \Lambda^{c\alpha} N_j^{\alpha}) dV \quad ,$$

subject to $\Lambda^{c\alpha} \geq 0$, $\alpha = 1, \ldots, \nu$.

24.3 Bibliographical and Historical Remarks

Stress, total strain relations were introduced by Hencky [1924] for a von Mises material. A great deal of discussion followed on the question of whether a deformation theory could adequately represent an elastic, plastic material. While it is now evident that it cannot, the comparative simplicity of deformation theory as opposed to an incremental theory saw a number of applications of the theory, particularly in the Soviet Union (see, for example, Ilyushin [1943], [1945], [1948], Kliushnikov [1959], Sokolovsky [1969] and Kachanov [1969]). Attempts to assess the capability of deformation theory to approximate, rather than replace, incremental theory were made later (see, for example, Budiansky [1959], Ponter and Martin [1972]).

The theorems stated in Section 24.2 and derived in Chapter 22

are linked in this monograph with extremal paths and are taken
from the work of Ponter and Martin [1972]. However, when deformation theory is considered in isolation, the energy principles are the classical energy principles and are well established (see, for example, Greenberg [1949] and Hill [1956]).

The alternative formulations of the theorems given in Section
24.2 are simpler versions of results given by Maier [1968],
[1969a], [1969c] for discrete structures and by De Donato [1968]
for continua. These results were obtained from quadratic programming theory; it was shown that they can be derived from the
classical principles by Martin and Ponter [1972].

References

B. Budiansky   1959   "A reassessment of deformation theories
                            of plasticity", J.Appl.Mech.,$\underline{26}$, 259.

O. De Donato   1968   "Extension to continua of some minimum
                            theorems of elastoplastic theory",
                            Meccanica, $\underline{3}$, 1.

H.J.Greenberg   1949   "On the variational principles of plas-
                            ticity", Tech.Rept.All-54, Div.of Appl.
                            Math., Brown University.

H.Hencky   1924   "Zur theorie plastischer Deformationen
                            und der hierdurch im Material herror-
                            gerufenen Nachspannungen", Z.A.M.M.,
                            $\underline{4}$, 323.

R.Hill   1956   "New horizons in the mechanics of
                            solids", J.Mech.Phys.Sol., $\underline{5}$, 66.

A.A.Ilyushin   1943   "Nekotoryie voprosy teoriji plastiches-
                            kich deformacij", Prikl.Mech.Mat., $\underline{7}$,
                            245.

A.A.Ilyushin   1945   "Svjaz mezhdu teoriej Sen-Venana, Levi,
                            Mezesa i teoriej malych uprygo-plas-
                            ticheskich deformacij", Prikl.Mech.Mat.,
                            $\underline{9}$, 207.

A.A.Ilyushin   1948   *Plastichnost*, Moscow. (*Plasticity*
                            (Eyrolles) Paris, 1956).

L.M.Kachanov   1969   *Introduction to the Theory of Plasti-
                            city*, Nauka (Moscow). Republished as
                            *Foundations of the Theory of Plasti-
                            city*, North Holland (Amsterdam) 1971.

V.D.Kliushnikov   1959   "Noryje predstavlenija v plastichnosti
                            i deformacionnaja teorija", Prikl.Mech.
                            Mat., $\underline{23}$, 282.

G.Maier   1968   "Quadratic programming and the theory
                            of elastic, perfectly plastic struc-
                            tures", Meccanica, $\underline{3}$, 1.

G.Maier   1969a   "Teorimi di minimo in termini finiti
                            per continui elastoplastici con leggi
                            constitutive linearizzate a tratti",
                            Instituto Lombardo, Accademia di
                            Scienze e Lettere, Rendiconti Al03,
                            1066.

G.Maier   1969b   "Some theorems for plastic strain rates
                            and plastic strains", J. de Mécanique,
                            $\underline{8}$, 5.

G.Maier   1969c   "Complementary plastic work theorems
                            in piecewise-linear elasto-plasticity",
                            Int.J.Solids and Structures, $\underline{5}$, 261.

J.B.Martin and   1972   "On dual energy theorems for a class
A.R.S.Ponter   of elastic/plastic problems due to G.
                            Maier", J.Mech.Phys.Sol., $\underline{20}$, 301.

A.R.S.Ponter and   1972   "Some extremal properties and energy
J.B.Martin   theorems for inelastic materials and
                            their relationship to the deformation
                            theory of plasticity", J.Mech.Phys.
                            Sol., $\underline{20}$, 281.

V.V.Sokolovsky   1969   *The Theory of Plasticity*, Vyshaja Shkola
                            (Moscow).

Chapter 25

MINIMUM PRINCIPLES FOR THE RATE PROBLEM

25.1 Formulation of the Rate Problem

Since the point at which the limit theorems were introduced in
Chapter 9, we have dealt with methods by means of which it is
possible to obtain some information about the response of an
elastic-plastic structure when the loading history is not taken
into account or is only partially specified. Some results of
great practical importance can be obtained by these methods;
they nevertheless represent special cases and frequently do not
provide sufficient information for design purposes. When this is
the case, resort must be made to the most fundamental problem
in elastic-plastic analysis; that of determining the response
of the structure to successive increments of load and thus
taking into account the entire stress history.

For convenience in this discussion we again replace increments by rates in the sense introduced in Chapter 9. The incremental problem then becomes the *rate problem* and is defined as
follows: at an instant t in the loading program we are given
load rates $\dot{\hat{p}}(s)$ on S_p, body force rates $\dot{\hat{F}}(s)$ on V and displacement rates $\dot{\hat{u}}(s)$ on S_u. The solution can be characterized by
stress rates $\dot{Q}_j(s)$, strain rates $\dot{q}_j(s)$, displacement rates $\dot{u}(s)$
on S_p and V, and reaction rates $\dot{p}(s)$ on S_u.

The solution must satisfy the rate form of the equilibrium
equations and the kinematic relations. In this monograph we
are concerned with bodies for which the equilibrium and kinematic relations are linear in force, stress, displacement and
strain. It follows then that the incremental or rate form of
these relations will be linear in force rates, stress rates,
displacement rates and strain rates.

The constitutive relations will be given for a smooth yield
surface. There is, however, no difficulty in extending the

results which will be derived in this Chapter to the cases of two or more independent yield functions. Thus *for hardening behavior*

$$\dot{q}_j^e = C_{jk}\dot{Q}_k \quad , \tag{1a}$$

$$\dot{q}_j^p = cG \frac{\partial \phi}{\partial Q_j} \frac{\partial \phi}{\partial Q_k} \dot{Q}_k \quad , \tag{1b}$$

where $\phi = \phi(Q_j, H_\alpha)$ and is a convex function,

$\quad\quad G = G(Q_j, H_\alpha)$ and is positive definite,

$\quad\quad c = +1$ if $\phi = 0$ and $\frac{\partial \phi}{\partial Q_j} \dot{Q}_j \geq 0$

$\quad\quad c = 0$ otherwise.

For the case of flow

$$\dot{q}_j^e = C_{jk}\dot{Q}_k \quad , \tag{2a}$$

$$\dot{q}_j^p = \lambda \frac{\partial \psi}{\partial Q_j} \quad , \tag{2b}$$

where $\psi = \psi(Q_j)$ is a convex function with $\psi(0) < 0$,

$\quad\quad \lambda \geq 0$ for $\psi = 0$ and $\frac{\partial \psi}{\partial Q_j} \dot{Q}_j = 0$,

$\quad\quad \lambda = 0$ otherwise,

$\quad\quad \psi > 0$ is not permitted.

These relations are linear in the stress rates and strain rates, but because we do not know *a priori* whether loading or unloading will occur at any given point in the body, the rate problem is not linear.

In general, the rate problem at time t will have been preceded by the solution of a succession of rate problems, so that the response of the structure will be known for all times τ such that $0 \leq \tau \leq t$; this response may be external forces $\underline{p}(s,\tau)$, body forces $\underline{F}(s,\tau)$, stresses $Q_j(s,\tau)$, displacements $\underline{u}(s,\tau)$ and strains $q_j(s,\tau)$. In addition, at time $t = 0$ the stress field $Q_j(s,0)$ and the internal variables H_α are given. In most cases it is assumed that the body is initially stress free and in its virgin state. With this information given, H_α, ϕ and G are all known at time t, and, as shown in Chapter 3, the solution to the rate problem is unique if hardening occurs everywhere in the body. If flow takes place in parts of the body the strain rates and displacement rates may not be unique, but this is not of any real concern and does not affect any subsequent incremental or rate problems.

Thus, depending on the type of structure under discussion, the rate problem consists of a set of ordinary or partial differential equations with appropriate boundary conditions. We may subdivide the body into regions where $\phi < 0$, denoted by V_e and regions where $\phi = 0$, denoted by V_p. We are assured that $\dot{q}_j^p = 0$ in V_e. In V_p there may be regions of loading or flow where $\dot{q}_j^p \neq 0$, and regions of unloading (or neutral loading in hardening materials) where $\dot{q}_j^p = 0$.

Associated with the rate problem are two complementary minimum principles, one in terms of stress rates and one in terms of strain rates. Before presenting these principles we shall discuss briefly the problem of the *inversion* of the constitutive equations (1) and (2). This discussion is necessary if the minimum principles are to be set up in a general form.

25.2 Inversion of the Constitutive Equations

The constitutive equations given in (1) and (2) provide the strain rates in terms of the stress rates. We shall also require that the stress rate $\dot{Q}_j$ and other quantities be given in terms of the *total strain rate* $\dot{q}_j$. This requires the inversion of equations (1) and (2), and in this regard it is important to show that the general form of the constitutive equations admits a unique inversion.

In the elastic case this is clearly true, for when $\dot{q}_j^p = 0$ and

$$\dot{q}_j = C_{jk},\dot{Q}_k \quad , \tag{3a}$$

the equations can be inverted to give

$$\dot{Q}_j = D_{jk}\dot{q}_k \quad , \tag{3b}$$

where D_{jk} is in the inverse of C_{jk}.

In the more general case we write the constitutive equations for both hardening and flow in the form

$$\dot{q}_j = C_{jk}\dot{Q}_k + \dot{q}_j^p \tag{4}$$

Suppose that for a given total strain rate $\dot{q}_j$ we can find two stress rates $\dot{Q}_j^{(1)}$, $\dot{Q}_j^{(2)}$ which satisfy equation (4). This implies that there are two elastic strain rates

$$\dot{q}_j^{e(1)} = C_{jk}\dot{Q}_k^{(1)} \quad , \quad \dot{q}_j^{e(2)} = C_{jk}\dot{Q}_k^{(2)} \quad , \tag{5}$$

and two plastic strain rates $\dot{q}_j^{p(1)}$, $\dot{q}_j^{p(2)}$, *both in the direction of the outward normal to the yield surface, and such that*

$$\dot{q}_j = \dot{q}_j^{e(1)} + \dot{q}_j^{p(1)} = \dot{q}_j^{e(2)} + \dot{q}_j^{p(2)} \quad . \tag{6}$$

From equation (6) we can write

$$(\dot{q}_j^{e(1)} - \dot{q}_j^{e(2)}) + (\dot{q}_j^{p(1)} - \dot{q}_j^{p(2)}) = 0 \quad . \tag{7}$$

Multiplying by the stress rate difference $(\dot{Q}_j^{(1)} - \dot{Q}_j^{(2)})$, we see that

$$(\dot{Q}_j^{(1)} - \dot{Q}_j^{(2)})(\dot{q}_j^{e(1)} - \dot{q}_j^{e(2)}) + (\dot{Q}_j^{(1)} - \dot{Q}_j^{(2)})(\dot{q}_j^{p(1)} - \dot{q}_j^{p(2)}) = 0. \tag{8}$$

However,

$$(\dot{Q}_j^{(1)} - \dot{Q}_j^{(2)})(\dot{q}_j^{e(1)} - \dot{q}_j^{e(2)}) = C_{jk}(\dot{Q}_j^{(1)} - \dot{Q}_j^{(2)})(\dot{Q}_k^{(1)} - \dot{Q}_k^{(2)}) \geq 0 \ .$$

$$(9)$$

Further

$$(\dot{Q}_j^{(1)} - \dot{Q}_j^{(2)})(\dot{q}_j^{p(1)} - \dot{q}_j^{p(2)}) \geq 0 \tag{10}$$

for either hardening behavior (Chapter 2, equation 108) or flow (Chapter 2, equation 137). This applies for both zero and non-zero plastic strain rates, and plays a central part in establishing the uniqueness of the solution of the rate problem.

In view of (9) and (10), equation (8) can hold if and only if

$$\dot{Q}_j^{(1)} = \dot{Q}_j^{(2)} \ . \tag{11}$$

Thus there cannot exist two distinct stress rates associated with a given total strain rate, and the uniqueness of the inversion is guaranteed in the general case.

In the case of hardening behavior we are required to invert the equations

$$\dot{q}_j = C_{jk}\dot{Q}_k + G\,\frac{\partial\phi}{\partial Q_j}\,\frac{\partial\phi}{\partial Q_k}\,\dot{Q}_k$$

$$= (C_{jk} + G\,\frac{\partial\phi}{\partial Q_j}\,\frac{\partial\phi}{\partial Q_k})\,\dot{Q}_k \tag{12a}$$

for $\phi = 0$ and $\dfrac{\partial\phi}{\partial Q_j}\dot{Q}_j \geq 0$,

$$\dot{q}_j = C_{jk}\dot{Q}_k \ , \tag{12b}$$

for $\phi = 0$ and $\dfrac{\partial\phi}{\partial Q_j}\dot{Q}_j \leq 0$,

or $\phi < 0$.

Denoting the inverse of C_{jk} by D_{jk}, equation (12b) yields

$$\dot{Q}_j = D_{jk}\dot{q}_k \quad , \tag{13}$$

for $\phi = 0$ and $\dfrac{\partial\phi}{\partial Q_j}\dot{Q}_j = D_{jk}\dfrac{\partial\phi}{\partial Q_j}\dot{q}_k \leq 0 \quad ,$

or $\phi < 0.$

When loading occurs, we note from equation (12a) that

$$C_{jk}\dot{Q}_k = \dot{q}_j - G\,\frac{\partial\phi}{\partial Q_j}\frac{\partial\phi}{\partial Q_k}\dot{Q}_k \quad , \tag{14}$$

and hence

$$\dot{Q}_j = D_{jk}\left(\dot{q}_k - G\,\frac{\partial\phi}{\partial Q_k}\frac{\partial\phi}{\partial Q_\ell}\dot{Q}_\ell\right) \quad . \tag{15}$$

Multiplying by $\partial\phi/\partial Q_j$, it follows that

$$\frac{\partial\phi}{\partial Q_j}\dot{Q}_j = \frac{D_{jk}\,\dfrac{\partial\phi}{\partial Q_j}\dot{q}_k}{\left(1 + G\,D_{pr}\,\dfrac{\partial\phi}{\partial Q_p}\dfrac{\partial\phi}{\partial Q_r}\right)} \quad . \tag{16}$$

This expression confirms that the sign of

$$D_{jk}\,\frac{\partial\phi}{\partial Q_j}\dot{q}_k$$

determines whether or not loading occurs, since the denomina-
tion of the term on the right hand side of equation (16) is
positive definite. Substituting equation (16) into equation
(15), we see that

$$\dot{Q}_j = D_{jk}\dot{q}_k - D_{jk}\frac{\partial\phi}{\partial Q_k}\left[\frac{G\,D_{\ell m}\,\dfrac{\partial\phi}{\partial Q_\ell}\dot{q}_m}{1 + GD_{pr}\,\dfrac{\partial\phi}{\partial Q_p}\dfrac{\partial\phi}{\partial Q_r}}\right] \tag{17}$$

for $\phi = 0$ and $D_{jk}\dfrac{\partial\phi}{\partial Q_j}\dot{q}_k \geq 0.$

Equations (13) and (17) are thus the inverted form of the

constitutive equation for hardening behavior.

In the case of flow we are required to invert the relations

$$\dot{q}_j = C_{jk}\dot{Q}_k + \lambda \frac{\partial \psi}{\partial Q_j} \quad , \tag{18a}$$

for $\psi = 0$ and $\frac{\partial \psi}{\partial Q_j}\dot{Q}_j = 0$,

$$\dot{q}_j = C_{jk}\dot{Q}_k \quad , \tag{18b}$$

for $\psi = 0$ and $\frac{\partial \psi}{\partial Q_j}\dot{Q}_j < 0$,

or $\psi < 0$.

Equation (18b) may be inverted, as before. If the strain rate
is assumed to be elastic, and provides a stress rate which constitutes either unloading or neutral loading, the result is admissible and the strain rate will in fact be elastic. Hence
the inverted form of equation (18b) will apply for $\psi < 0$ and
for $\psi = 0$ and $D_{jk}(\partial \psi/\partial Q_j)\dot{q}_k \leq 0$.
When $\dot{q}_j^p \neq 0$, we must eliminate λ in the inversion. Equation
(18a) is written as

$$(\dot{q}_j - \lambda \frac{\partial \psi}{\partial Q_j}) = C_{jk}\dot{Q}_k \tag{19}$$

Multiplying by D_{jk}, we find

$$\dot{Q}_j = D_{jk}(\dot{q}_k - \lambda \frac{\partial \psi}{\partial Q_k}). \tag{20}$$

We next consider the scalar product $(\partial \psi/\partial Q_j)\dot{Q}_j$, which must be
zero if flow is to occur. Thus

$$0 = \frac{\partial \psi}{\partial Q_j}\dot{Q}_j = D_{jk}\frac{\partial \psi}{\partial Q_j}(\dot{q}_k - \lambda \frac{\partial \psi}{\partial Q_k}) \quad . \tag{21}$$

Solving for λ, we find

$$\lambda = (D_{jk} \frac{\partial \psi}{\partial Q_j} \dot{q}_k) / (D_{jk} \frac{\partial \psi}{\partial Q_j} \frac{\partial \psi}{\partial Q_k}) \quad . \tag{22}$$

The denominator of the right hand side of this expression is positive definite, and thus the sign of λ is controlled by the numerator. This confirms that $\dot{q}_j^p \neq 0$ if

$$D_{jk} \frac{\partial \psi}{\partial Q_j} \dot{q}_k > 0 \quad . \tag{23}$$

The stress rate is now found by substituting equation (22) into equation (20). The complete inverted equations are

$$\dot{Q}_j = D_{jk}\dot{q}_k - D_{jk} \frac{\partial \psi}{\partial Q_k} \{(D_{\ell m} \frac{\partial \psi}{\partial Q_\ell} \dot{q}_m) / (D_{pr} \frac{\partial \psi}{\partial Q_p} \frac{\partial \psi}{\partial Q_r})\} \quad , \tag{24a}$$

for $\psi = 0$ and $D_{jk} \frac{\partial \psi}{\partial Q_j} \dot{q}_k \geq 0$,

$$\dot{Q}_j = D_{jk}\dot{q}_k \quad , \tag{24b}$$

for $\psi = 0$ and $D_{jk} \frac{\partial \psi}{\partial Q_j} \dot{q}_k \leq 0$,

or $\psi < 0$.

In the statement of the minimum principles, which will be presented in the following section, it is convenient to express the product $\frac{1}{2} \dot{Q}_j \dot{q}_j$, for stress rates and total strain rates which satisfy the constitutive relations, in terms of either the components of stress rate or the components of strain rate. These functions will be denoted by

$$\dot{\Omega}^{\circ}(\dot{Q}_j) = \frac{1}{2} \dot{Q}_j \dot{q}_j \quad , \tag{25}$$

where $\dot{q}_j$ is the strain rate associated with the stress rate $\dot{Q}_j$, and

$$\dot{W}^{\circ}(\dot{q}_j) = \frac{1}{2} \dot{Q}_j \dot{q}_j \quad , \tag{26}$$

where $\dot{Q}_j$ is the stress rate associated with strain rate $\dot{q}_j$. It is presumed again that Q_j, ϕ and G are known. Note that

$$\dot{\Omega}^{o}(\dot{Q}_j) + \dot{W}^{o}(\dot{q}_j) = \dot{Q}_j \dot{q}_j \quad . \tag{27}$$

First consider $\dot{\Omega}^{o}(\dot{Q}_j)$. In the case of hardening behavior, from equations (12a) and (12b),

$$\dot{\Omega}^{o}(\dot{Q}_j) = \frac{1}{2} \left(C_{jk} + G \frac{\partial \phi}{\partial Q_j} \frac{\partial \phi}{\partial Q_k} \right) \dot{Q}_j \dot{Q}_k \quad , \tag{28a}$$

if $\phi = 0$ and $\dfrac{\partial \phi}{\partial Q_j} \dot{Q}_j \geq 0$,

$$\dot{\Omega}^{o}(\dot{Q}_j) = \frac{1}{2} C_{jk} \dot{Q}_j \dot{Q}_k \quad , \tag{28b}$$

if $\phi = 0$ and $\dfrac{\partial \phi}{\partial Q_j} \dot{Q}_j \leq 0$,

or $\phi = 0$.

In the case of flow, $\dot{q}_j^{p} \neq 0$ only when neutral loading occurs and

$$\frac{\partial \psi}{\partial Q_j} \dot{Q}_j = 0 \quad . \tag{29}$$

It follows then from equations (18a) and (18b) that

$$\dot{\Omega}^{o}(\dot{Q}_j) = \frac{1}{2} C_{jk} \dot{Q}_j \dot{Q}_k \tag{30}$$

under all conditions. Note that $(\partial \psi / \partial Q_j) \dot{Q}_j > 0$ is not permitted.

It is also convenient to introduce

$$\dot{\Omega}(\dot{Q}_j) = \frac{1}{2} \dot{Q}_j \dot{q}_j^{e} \quad , \tag{31}$$

where $\dot{q}_j^{e}$ is the elastic component of strain rate associated with $\dot{Q}_j$ and

$$\dot{\Omega}^{\text{op}}(\dot{Q}_j) = \frac{1}{2} \dot{Q}_j \dot{q}_j^p \quad , \tag{32}$$

where $\dot{q}_j^p$ is the plastic component of strain rate associated with $\dot{Q}_j$. It is evident then that

$$\dot{\Omega}^{\text{o}}(\dot{Q}_j) = \dot{\Omega}(\dot{Q}_j) + \dot{\Omega}^{\text{op}}(\dot{Q}_j) \quad , \tag{33a}$$

$$\dot{\Omega}(\dot{Q}_j) = \frac{1}{2} C_{jk} \dot{Q}_j \dot{Q}_k \quad , \tag{33b}$$

$$\dot{\Omega}^{\text{op}} = \frac{1}{2} G \frac{\partial \phi}{\partial Q_j} \frac{\partial \phi}{\partial Q_k} \dot{Q}_j \dot{Q}_k \tag{33c}$$

for loading in hardening materials, but zero otherwise.

Now consider $\dot{W}^{\text{o}}(\dot{q}_j)$. From equations (13) and (17) it can be seen that for hardening behavior

$$\dot{W}^{\text{o}}(\dot{q}_j) = \frac{1}{2} D_{jk} \dot{q}_j \dot{q}_k - \frac{G}{2} \frac{[D_{jk} \frac{\partial \phi}{\partial Q_j} \dot{q}_k]^2}{(1 + G D_{pr} \frac{\partial \phi}{\partial Q_p} \frac{\partial \phi}{\partial Q_r})} \quad , \tag{34a}$$

$$\text{for } \phi = 0 \text{ and } D_{jk} \frac{\partial \phi}{\partial Q_j} \dot{q}_k \geq 0, \tag{34b}$$

$$\dot{W}^{\text{o}}(\dot{q}_j) = \frac{1}{2} D_{jk} \dot{q}_j \dot{q}_k \quad ,$$

$$\text{for } \phi = 0 \text{ and } D_{jk} \frac{\partial \phi}{\partial Q_j} \dot{q}_k \leq 0,$$

or $\phi < 0$.

In the case of flow, from equations (24a) and (24b),

$$\dot{W}^{\text{o}}(\dot{q}_j) = \frac{1}{2} D_{jk} \dot{q}_j \dot{q}_k - \frac{1}{2} \frac{[D_{jk} \frac{\partial \psi}{\partial Q_j} \dot{q}_k]^2}{(D_{pr} \frac{\partial \psi}{\partial Q_p} \frac{\partial \psi}{\partial Q_r})} \quad , \tag{35a}$$

$$\text{for } \psi = 0 \text{ and } D_{jk} \frac{\partial \psi}{\partial Q_j} \dot{q}_k \geq 0,$$

$$\dot{W}^{o}(\dot{q}_j) = \frac{1}{2} D_{jk}\dot{q}_j\dot{q}_k \quad , \tag{35b}$$

for $\psi = 0$ and $D_{jk}\dfrac{\partial \psi}{\partial Q_j}\dot{q}_k \leq 0$,

or $\psi < 0$.

The function $\dot{W}^{o}(\dot{q}_j)$ is expressed in terms of the total strain rate for both hardening behavior and flow. Consider a situation in which plastic deformation is taking place; in this case the total strain rate $\dot{q}_j$ can be divided into an elastic component $\dot{q}_j^{e}$ and a plastic component $\dot{q}_j^{p}$ by first determining $\dot{Q}_j$ from the inverted constitutive relations, and then determining $\dot{q}_j^{e}$ and $\dot{q}_j^{p}$ from the original equations. Hence we can write

$$\dot{W}^{o}(\dot{q}_j) = \frac{1}{2}\dot{Q}_j\dot{q}_j = \frac{1}{2}\dot{Q}_j\dot{q}_j^{e} + \frac{1}{2}\dot{Q}_j\dot{q}_j^{p} \quad . \tag{36}$$

Let us put

$$\dot{W}(\dot{q}_j^{e}) = \frac{1}{2}\dot{Q}_j\dot{q}_j^{e} = \frac{1}{2}D_{jk}\dot{q}_j^{e}\dot{q}_k^{e} \quad , \tag{37}$$

and

$$\dot{W}^{op}(\dot{q}_j^{p}) = \frac{1}{2}\dot{Q}_j\dot{q}_j^{p} \quad . \tag{38}$$

Since $\dot{q}_j^{p}$ is constrained to have the direction of the outward normal to the yield or limit surface, let us further put

$$\dot{q}_j^{p} = \lambda \frac{\partial \phi}{\partial Q_j} \tag{39}$$

in the case of hardening behavior, and

$$\dot{q}_j^{p} = \lambda \frac{\partial \psi}{\partial Q_j} \tag{40}$$

in the case of flow.

When hardening is taking place, we see from equation (1) that

$$\dot{q}^p_j = G \, \frac{\partial \phi}{\partial Q_j} \, \frac{\partial \phi}{\partial Q_k} \, \dot{Q}_k \quad . \tag{41}$$

Denoting the inverse of $(\partial\phi/\partial Q_j)(\partial\phi/\partial Q_k)$ by B_{jk}, it follows that

$$\dot{Q}_j = \frac{1}{G} \, B_{jk} \dot{q}^p_k \quad . \tag{42}$$

Hence

$$\dot{W}^{op}(\dot{q}^p_j) = \frac{1}{2G} \, B_{jk} \dot{q}^p_j \dot{q}^p_k \quad . \tag{43}$$

Substituting now from equation (39), this can be written as

$$\dot{W}^{op}(\dot{q}^p_j) = \frac{1}{2G} \, B_{jk} \, (\lambda^2 \, \frac{\partial \phi}{\partial Q_j} \, \frac{\partial \phi}{\partial Q_k} \,)$$

$$= \frac{\lambda^2}{2G} \tag{44}$$

in view of the definition of B_{jk}.

When flow occurs, the stress rate $\dot{Q}_j$ constitutes neutral loading, while $\dot{q}^p_j$ is normal to the limit surface. Hence the product $\dot{Q}_j \dot{q}^p_j$ vanishes, and

$$\dot{W}^{op}(\dot{q}^p_j) = 0 \quad . \tag{45}$$

In the kinematic approach to the rate problem the total strain rate $\dot{q}_j$ is unknown, and hence we cannot *a priori* divide it into elastic and plastic parts. Suppose, however, that we make this division arbitrarily by setting

$$\dot{q}_j = \dot{q}^{eo}_j + \dot{q}^{po}_j \quad , \tag{46a}$$

with

$$\dot{q}^{po}_j = \lambda^o \, \frac{\partial \phi}{\partial Q_j} \tag{46b}$$

for hardening and

$$\dot{q}_j^{po} = \lambda^o \frac{\partial \psi}{\partial Q_j} \tag{46c}$$

for flow. We can then show that provided $\phi = 0$,

$$\dot{W}^o(\dot{q}_j) \leq \dot{W}(\dot{q}_j - \lambda^o \frac{\partial \phi}{\partial Q_j}) + \dot{W}^{op}(\lambda^o) \tag{47a}$$

subject to $\lambda^o \geq 0$ for the case of hardening, and

$$\dot{W}^o(\dot{q}_j) \leq \dot{W}(\dot{q}_j - \lambda^o \frac{\partial \psi}{\partial Q_j}) \tag{47b}$$

subject to $\lambda^o \geq 0$ for the case of flow. In each case equality occurs when $\lambda = \lambda^o$, i.e. when we achieve the correct division of $\dot{q}_j$ into its elastic and plastic components.

Consider first the case of hardening, and the expression

$$\frac{1}{2} D_{jk} (\dot{q}_j - \lambda^o \frac{\partial \phi}{\partial Q_j})(\dot{q}_k - \lambda^o \frac{\partial \phi}{\partial Q_k}) + \frac{1}{2} \frac{\lambda^{o^2}}{G} \; . \tag{48}$$

Differentiating with respect to λ^o, the stationary value of this expression occurs when

$$- D_{jk}(\dot{q}_j - \lambda^o \frac{\partial \phi}{\partial Q_j}) \frac{\partial \phi}{\partial Q_k} + \frac{\lambda^o}{G} = 0,$$

or

$$\lambda^o = \frac{G \, D_{jk} \frac{\partial \phi}{\partial Q_j} \dot{q}_k}{1 + G \, D_{jk} \frac{\partial \phi}{\partial Q_j} \frac{\partial \phi}{\partial Q_k}} \; . \tag{49}$$

Noting from equations (39) and (41) that

$$\lambda = G \frac{\partial \phi}{\partial Q_j} \dot{Q}_j \; , \tag{50}$$

we see from equation (16) that equation (49) implies that $\lambda^o = \lambda$ provided that

$$D_{jk} \frac{\partial \phi}{\partial Q_j} \dot{q}_k > 0 \; . \tag{51}$$

The second derivative of expression (48) with respect to λ^o is

$$D_{jk} \frac{\partial \phi}{\partial Q_j} \frac{\partial \phi}{\partial Q_k} + \frac{1}{G}$$

which is positive definite, indicating that the stationary value
is a minimum.
When

$$D_{jk} \frac{\partial \phi}{\partial Q_j} \dot{q}_k \leq 0 \tag{52}$$

the least value of expression (48) for which $\lambda^o \geq 0$ will occur
for $\lambda^o = 0$. This establishes the inequality given in (47a).

For the case of flow, consider the expression

$$\frac{1}{2} D_{jk} \left(\dot{q}_j - \lambda^o \frac{\partial \psi}{\partial Q_j} \right) \left(\dot{q}_k - \lambda^o \frac{\partial \psi}{\partial Q_k} \right) \ . \tag{53}$$

Differentiating with respect to λ^o, the stationary value of the
expression occurs when

$$- D_{jk} \left(\dot{q}_j - \lambda^o \frac{\partial \psi}{\partial Q_j} \right) \frac{\partial \psi}{\partial Q_k} = 0$$

or

$$\lambda^o = \frac{D_{jk} \dfrac{\partial \psi}{\partial Q_j} \dot{q}_k}{D_{jk} \dfrac{\partial \psi}{\partial Q_j} \dfrac{\partial \psi}{\partial Q_k}} \ . \tag{54}$$

Comparing equations (22) and (54), it is seen that equation
(54) implies that $\lambda = \lambda^o$ provided that

$$D_{jk} \frac{\partial \psi}{\partial Q_j} \dot{q}_k > 0 \ . \tag{55}$$

The second derivative of expression (53) with respect to λ^o is
$D_{jk} \frac{\partial \psi}{\partial Q_j} \frac{\partial \psi}{\partial Q_k}$, which is positive definite. If

$$D_{jk} \frac{\partial \psi}{\partial Q_j} \dot{q}_k \leq 0 \tag{56}$$

the least value of expression (53) occurs when $\lambda^o = 0$. This
establishes the inequality given in (47b).

The functions $\dot{\Omega}^o(Q_j)$ and $\dot{W}^o(q_j)$ have properties similar to
the convexity properties of the Ω^o and W^o functions associated
with maximum complementary work and minimum work paths. These
properties form the basis of the minimum principles for the
rate problem.

Consider two alternative stress rates $\dot{Q}_j^s$, $\dot{Q}_j^c$ imposed at a
point in a body where Q_j, ϕ and G are given, and their associated strain rates $\dot{q}_j^s$, $\dot{q}_j^c$. Consider the expression

$$\frac{1}{2} \dot{Q}_j^s \dot{q}_j^s + \frac{1}{2} \dot{Q}_j^c \dot{q}_j^c - \dot{Q}_j^s \dot{q}_j^c \quad . \tag{57}$$

We can show that this expression is non-negative for any choice
of $\dot{Q}_j^s$, $\dot{Q}_j^c$, and is zero if and only if $\dot{Q}_j^s = \dot{Q}_j^c$. In order to demonstrate this result we divide the strains into their elastic
and plastic parts.

Using the reciprocal theorem ($\dot{Q}_j^s \dot{q}_j^{ec} = \dot{Q}_j^c \dot{q}_j^{es}$), the elastic
part of (57) gives

$$\frac{1}{2} \dot{Q}_j^s \dot{q}_j^{es} + \frac{1}{2} \dot{Q}_j^c \dot{q}_j^{ec} - \dot{Q}_j^s \dot{q}_j^{ec}$$

$$= \frac{1}{2} \dot{Q}_j^s \dot{q}_j^s + \frac{1}{2} \dot{Q}_j^c \dot{q}_j^c - \frac{1}{2} \dot{Q}_j^s \dot{q}_j^{ec} - \frac{1}{2} \dot{Q}_j^c \dot{q}_j^{es}$$

$$\tag{58}$$

$$= \frac{1}{2} (\dot{Q}_j^s - \dot{Q}_j^c)(\dot{q}_j^{es} - \dot{q}_j^{ec})$$

$$= \frac{1}{2} C_{jk}(\dot{Q}_j^s - \dot{Q}_j^c)(\dot{Q}_k^s - \dot{Q}_k^c) \geq 0 \quad .$$

The plastic part of expression (57) is

$$\frac{1}{2} \dot{Q}_j^s \dot{q}_j^{ps} + \frac{1}{2} \dot{Q}_j^c \dot{q}_j^{pc} - \dot{Q}_j^s \dot{q}_j^{pc} \quad . \tag{59}$$

Consider first the case where $\psi = 0$, so that the plastic strain rates are the result of flow if they are non-zero. Whether unloading or neutral loading occurs, the product of stress rate and the associated plastic strain rate is zero. Hence

$$\dot{Q}_j^s \dot{q}_j^{ps} = \dot{Q}_j^c \dot{q}_j^{pc} = 0 \quad . \tag{60}$$

Further

$$\dot{Q}_j^s \dot{q}_j^{pc} \leq 0 \quad , \tag{61}$$

since $\dot{Q}_j^s$ constitutes either unloading or neutral loading, and $\dot{q}_j^{pc}$ is either zero or has the direction of the outward normal to the limit surface. Thus for $\psi = 0$

$$\frac{1}{2} \dot{Q}_j^s \dot{q}_j^{ps} + \frac{1}{2} \dot{Q}_j^c \dot{q}_j^{pc} - \dot{Q}_j^s \dot{q}_j^{pc} \geq 0 \tag{62}$$

for any choice of $\dot{Q}_j^s$, $\dot{Q}_j^c$.

Consider next the case where the current stress lies on a yield surface $\phi = 0$, and $\dot{q}_j^{ps}$, $\dot{q}_j^{pc}$ result from hardening if they are non-zero. Using the constitutive equation (1) we write expression (59) as

$$\frac{c^s}{2} G \left(\frac{\partial \phi}{\partial Q_j} \dot{Q}_j^s \right)^2 + \frac{c^c}{2} G \left(\frac{\partial \phi}{\partial Q_j} \dot{Q}_j^c \right)^2 - c^c G \left(\frac{\partial \phi}{\partial Q_j} \dot{Q}_j^s \right) \left(\frac{\partial \phi}{\partial Q_k} \dot{Q}_k^c \right) \quad , \tag{63}$$

where $c^s = +1$ if $\dfrac{\partial \phi}{\partial Q_j} \dot{Q}_j^s \geq 0$

$\qquad\quad c^c = +1$ if $\dfrac{\partial \phi}{\partial Q_j} \dot{Q}_j^c \geq 0$

$\qquad\quad c^s, c^c = 0$ otherwise.

Note that $\partial \phi / \partial Q_j$ and G have the same values in each term in this expression.

If both stress rates constitute loading, so that $c^S = c^C = +1$, expression (63) becomes

$$\frac{1}{2} G\left(\frac{\partial \phi}{\partial Q_j} \dot{Q}_j^S - \frac{\partial \phi}{\partial Q_j} \dot{Q}_j^C\right)^2 \geq 0 \quad . \tag{64}$$

If $\dot{Q}_j^S$ represents loading ($c^S = +1$) and $\dot{Q}_j^C$ unloading ($c^C = 0$), expression (63) becomes

$$\frac{1}{2} G\left(\frac{\partial \phi}{\partial Q_j} \dot{Q}_j^S\right)^2 \geq 0 \quad . \tag{65}$$

Finally, if $\dot{Q}_j^S$ constitutes unloading, and $\dot{Q}_j^C$ loading ($c^S = 0$, $c^C = +1$), expression (63) becomes

$$\frac{1}{2} G\left(\frac{\partial \phi}{\partial Q_j} \dot{Q}_j^C\right)^2 - G\left(\frac{\partial \phi}{\partial Q_j} \dot{Q}_j^S\right)\left(\frac{\partial \phi}{\partial Q_j} \dot{Q}_j^C\right) \geq 0 \quad , \tag{66}$$

since $(\partial \phi / \partial Q_j)\dot{Q}_j^S \leq 0$ and $(\partial \phi / \partial Q_j)\dot{Q}_j^C \geq 0$.

Thus expression (63) is non-negative for all conditions under which the plastic strain rates are non-zero, and it is of course zero if $\dot{q}_j^{ps} = \dot{q}_j^{pc} = 0$. Adding the results for the elastic and plastic parts, we then see that

$$\frac{1}{2} \dot{Q}_j^S \dot{q}_j^S + \frac{1}{2} \dot{Q}_j^C \dot{q}_j^C - \dot{Q}_j^S \dot{q}_j^C \geq 0 \tag{67}$$

for any choice of $\dot{Q}_j^S, \dot{Q}_j^C$ for both hardening and flow.

Inequality (67) may be arranged in two forms. First, put

$$\frac{1}{2} \dot{Q}_j^S \dot{q}_j^S - \dot{Q}_j^S \dot{q}_j^C \geq -\frac{1}{2} \dot{Q}_j^C \dot{q}_j^C = \frac{1}{2} \dot{Q}_j^C \dot{q}_j^C - \dot{Q}_j^C \dot{q}_j^C \quad . \tag{68}$$

Expressing $\frac{1}{2} \dot{Q}_j^S \dot{q}_j^S$, $\frac{1}{2} \dot{Q}_j^C \dot{q}_j^C$ in terms of stress rates, this becomes

$$\dot{\Omega}^O(\dot{Q}_j^S) - \dot{Q}_j^S \dot{q}_j^C \geq \dot{\Omega}^O(\dot{Q}_j^C) - \dot{Q}_j^C \dot{q}_j^C \quad . \tag{69}$$

Secondly, put

$$\frac{1}{2} \dot{Q}_j^c \dot{q}_j^c - \dot{Q}_j^s \dot{q}_j^c \geq -\frac{1}{2} \dot{Q}_j^s \dot{q}_j^s = \frac{1}{2} \dot{Q}_j^s \dot{q}_j^s - \dot{Q}_j^s \dot{q}_j^s \quad . \tag{70}$$

Expressing $\frac{1}{2} \dot{Q}_j^c \dot{q}_j^c$, $\frac{1}{2} \dot{Q}_j^s \dot{q}_j^s$ in terms of strain rates, this be-comes

$$\dot{W}^0(\dot{q}_j^c) - \dot{Q}_j^s \dot{q}_j^c \geq \dot{W}^0(\dot{q}_j^s) - \dot{Q}_j^s \dot{q}_j^s \quad . \tag{71}$$

We shall make direct use of inequalities (69) and (71) in es-tablishing the minimum principles for the rate problem.

25.3 Minimum Principles for the Rate Problem

We now reconsider the rate problem outlined in Section 25.1. At time t we are given load rates $\dot{\hat{p}}(s)$ on S_p, body force rates $\dot{\hat{F}}(s)$ on V, and displacement rates $\dot{\hat{u}}(s)$ on S_u. From the initial conditions at time t = 0 and the solutions of subsequent rate problems, at time t we know the stresses Q_j and the internal parameters at each point, so that the functions G, ϕ in the case of hardening and ψ in the case of flow can be determined. The solution to this problem is characterized by stress rates $\dot{Q}_j(s)$, strain rates $\dot{q}_j(s)$, reaction rates $\dot{p}(s)$ on S_u, and dis-placement rates $\dot{u}(s)$ on S_p and V.

A *safe, pertinent statically admissible stress rate field* $\dot{Q}_j^s(s)$ will be defined as a set of stress rates which are in in-ternal equilibrium and external equilibrium with load rates $\dot{\hat{p}}(s)$ on S_p, body force rates $\dot{\hat{F}}(s)$ on V and reaction rates $\dot{p}^s(s)$ on S_u. In addition, where a limit surface ψ is defined for the material, at each point in the body where $\psi(Q_j) = 0$ we require that $(\partial\psi/\partial Q_j)\dot{Q}_j^s \leq 0$. Thus $\dot{Q}_j^s$ cannot cause changes in the stress field which would lead to inadmissible states of stress. It is evident that $\dot{Q}_j$, $\dot{p}$ on S_u is a member of the set $\dot{Q}_j^s$, $\dot{p}^s$ on S_u.

In a similar way we define a *pertinent kinematically admis-sible strain rate field* $\dot{q}_j^c(s)$ as a strain rate field which is derived from an admissible displacement rate field $\dot{u}^c(s)$ such that $\dot{\underline{u}}^c(s) = \dot{\hat{\underline{u}}}(s)$ on S_u. Again, it is clear that $\dot{q}_j$, $\dot{\underline{u}}$ is a member of the set $\dot{q}_j^c$, $\dot{\underline{u}}^c$.

The minimum principles are now stated in terms of the pertinent statically and kinematically admissible fields. Each of these fields represents a partial solution to the problem. In one case, only the equilibrium requirements, the limit condition and the force boundary conditions are considered; in the other only the kinematic requirements and displacement boundary conditions are satisfied.

The minimum principle for stress rates states that the expression

$$\dot{U}_c^O(\dot{Q}_j^S) = \int_V \dot{\Omega}^O(\dot{Q}_j^S)dV - \int_{S_u} \dot{\underline{p}}^S \cdot \dot{\hat{\underline{u}}}dS \tag{72}$$

attains an absolute minimum when $\dot{Q}_j^S = \dot{Q}_j$, $\dot{\underline{p}}^S = \dot{\underline{p}}$ *on* S_u.

The minimum principle for strain rates states that the expression

$$\dot{U}_p^O(\dot{q}_j^c,\underline{u}^c) = \int_V \dot{W}^O(\dot{q}_j^c)dV - \int_{S_p} \dot{\hat{\underline{p}}} \cdot \dot{\underline{u}}^c dS - \int_V \dot{\hat{\underline{F}}} \cdot \dot{\underline{u}}^c dV \tag{73}$$

attains an absolute minimum when $\dot{q}_j^c = \dot{q}_j$, $\dot{\underline{u}}^c = \dot{\underline{u}}$.

Applying the principle of virtual work to the stress rate and strain rate fields which comprise the actual solution, we see that

$$\int_{S_p} \dot{\hat{\underline{p}}} \cdot \dot{\underline{u}}dS + \int_{S_u} \dot{\underline{p}} \cdot \dot{\hat{\underline{u}}}dS + \int_V \dot{\hat{\underline{F}}} \cdot \dot{\underline{u}}dV = \int_V \dot{Q}_j\dot{q}_j dV \quad . \tag{74}$$

This equation may be rewritten as

$$\{ \frac{1}{2} \int_V \dot{Q}_j\dot{q}_j dV - \int_{S_u} \dot{\underline{p}} \cdot \dot{\hat{\underline{u}}}dS \}$$

$$= -\{\frac{1}{2} \int_V \dot{Q}_j\dot{q}_j dV - \int_{S_p} \dot{\hat{\underline{p}}} \cdot \dot{\underline{u}}dS - \int_V \dot{\hat{\underline{F}}} \cdot \dot{\underline{u}}dV\} \quad , \tag{75a}$$

or

$$\dot{U}_c^O(\dot{Q}_j) = - \dot{U}_p^O(\dot{q}_j,\underline{\dot{u}}) \quad . \tag{75b}$$

Hence from the statements of the minimum principles we may write a continued inequality:

$$\dot{U}_c^O(\dot{Q}_j^S) \geq \dot{U}_c^O(\dot{Q}_j) = - \dot{U}_p^O(\dot{q}_j,\underline{\dot{u}}) \geq - \dot{U}_p^O(\dot{q}_j^C,\underline{\dot{u}}^C) \quad . \tag{76}$$

It is convenient to divide the body V into two regions; V_e where $\phi < 0$ and plastic deformation cannot occur, and V_p where $\phi = 0$ and plastic deformation may occur. The functions $\dot{\Omega}^O$ and $\dot{W}^O$ are unambiguously defined in V_e. However, the form of the functions in V_p will depend on whether loading or unloading occurs at each point.

The minimum principles may be established directly from inequalities (69) and (71). In the first case we associate $\dot{Q}_j^S$ in inequality (69) with the pertinent statically admissible field, and associate $\dot{Q}_j^C,\dot{q}_j^C$ with the solution $\dot{Q}_j,\dot{q}_j$. Integrating over the body, inequality (69) becomes

$$\int_V \dot{\Omega}^O(\dot{Q}_j^S)dV - \int_V \dot{Q}_j^S\dot{q}_j dV \geq \int_V \dot{\Omega}^O(\dot{Q}_j)dV - \int_V \dot{Q}_j\dot{q}_j dV \quad . \tag{77}$$

From the principle of virtual work, we have

$$\int_{S_p} \underline{\dot{\hat{p}}}\cdot\underline{\dot{u}}dS + \int_{S_u} \underline{\dot{p}}\cdot\underline{\hat{\dot{u}}}dS + \int_V \underline{\hat{\dot{F}}}\cdot\underline{\dot{u}}dV = \int_V \dot{Q}_j\dot{q}_j dV \quad , \tag{78a}$$

$$\int_{S_p} \underline{\dot{\hat{p}}}\cdot\underline{\dot{u}}dS + \int_{S_u} \underline{\dot{p}}^S\cdot\underline{\hat{\dot{u}}}dS + \int_V \underline{\hat{\dot{F}}}\cdot\underline{\dot{u}}dV = \int_V \dot{Q}_j^S\dot{q}_j dV \quad . \tag{78b}$$

Substituting equations (78a) and (78b) into inequality (77), we recover the static minimum principle (72) after eliminating terms which appear on both sides of the inequality.

Secondly, we associate $\dot{q}_j^C$, $\underline{\dot{u}}^C$ in inequality (71) with the pertinent kinematically admissible field, and associate $\dot{Q}_j^s, \dot{q}_j^s$ with the solution $\dot{Q}_j, \dot{q}_j$. Integrating over the volume, inequality (71) becomes

$$\int_V \dot{W}^O(\dot{q}_j^c)dV - \int_V \dot{Q}_j \dot{q}_j^c dV \geq \int_V \dot{W}^O(\dot{q}_j)dV - \int_V \dot{Q}_j \dot{q}_j dV \quad . \tag{79}$$

From the principle of virtual work.

$$\int_{S_p} \dot{\hat{p}} \cdot \underline{\dot{u}}^c dS + \int_{S_u} \dot{p} \cdot \underline{\hat{\dot{u}}} dS + \int_V \underline{\hat{F}} \cdot \underline{\dot{u}}^c dV = \int_V \dot{Q}_j \dot{q}_j^c dV \quad . \tag{80}$$

Substituting equations (78a) and (80) into inequality (79), and eliminating like terms from each side of the inequality, we recover the kinematic minimum principle (73).

An alternative form of the kinematic minimum principle can be given if we divide $\dot{q}_j^c$ into elastic and plastic parts, by putting

$$\dot{q}_j^{pc} = \lambda^c \frac{\partial \phi}{\partial Q_j} \quad , \tag{81a}$$

$$\dot{q}_j^{ec} = \dot{q}_j^c - \lambda^c \frac{\partial \phi}{\partial Q_j} \quad , \tag{81b}$$

in V_p. We then define

$$\dot{\bar{U}}_p^O(\dot{q}_j^c, \lambda^c) = \int_{V_p} \dot{W}^{OP}(\lambda^c)dV + \int_{V_p} \dot{W}(\dot{q}_j^c - \lambda^c \frac{\partial \phi}{\partial Q_j})dV + \int_{V_e} \dot{W}(\dot{q}_j^c)dV$$

$$-\int_V \underline{\hat{F}} \cdot \underline{\dot{u}}^c dV - \int_{S_p} \dot{\hat{p}} \cdot \underline{\dot{u}}^c dS \quad . \tag{82}$$

The function $\dot{W}^{OP}$ is given in equations (44) and (45).

From inequalities (47a) and (47b) we see that

$$\dot{\bar{U}}_p^O(\dot{q}_j^c, \lambda^c) \geq \dot{U}_p^O(\dot{q}_j^c) \geq \dot{U}_p^O(\dot{q}_j) \quad , \tag{83}$$

with equality signs occurring when q_j^c, λ^c take on the values of the correct solution.

A further modification of the minimum principles may be given if we first find the solution of the linear problem characterized by load rates $\hat{\dot{p}}(s)$ on S_p, body forces $\hat{\dot{F}}(s)$ on V and displacement rates $\hat{\dot{u}}(s)$ on S_u, with constitutive relations

$$\dot{q}_j = C_{jk}\dot{Q}_k \quad , \quad \dot{Q}_j = D_{jk}\dot{q}_k \quad . \tag{84}$$

Let this solution be characterized by generalized stress rates $\dot{Q}_j^E(s)$, generalized strain rates $\dot{q}_j^E(s)$, reaction rates $\dot{p}^E(s)$ on S_u and displacement rates $\dot{u}^E(s)$ on S_p and V.

Using this solution we can generate a pertinent statically admissible stress rate field $\dot{Q}_j^S$, $\dot{p}^S$ on S_u;

$$\dot{Q}_j^S = \dot{Q}_j^E + \dot{\rho}_j^S \quad , \tag{85a}$$

$$\dot{p}^S = \dot{p}^E + \dot{p}^{rs} \quad \text{on} \quad S_u \quad , \tag{85b}$$

where $\dot{\rho}_j^S$ is a statically admissible *residual stress rate field* in equilibrium with zero load rates on S_p and $\dot{p}^{rs}$ on S_u. In the case of an elastic, perfectly plastic material we impose the additional constraint

$$\frac{\partial \phi}{\partial Q_j} (\dot{Q}_j^E + \dot{\rho}_j^S) \leq 0 \quad \text{in} \quad V_p \quad . \tag{86}$$

Substituting equations (85) into expression (72), we find

$$\dot{U}_c^O(\dot{Q}_j^S) = \int_V \dot{\Omega}^O(\dot{Q}_j^E + \dot{\rho}_j^S)dV - \int_{S_u} (\dot{p}^E + \dot{p}^{rs}).\hat{\dot{u}}dS \quad . \tag{87}$$

From equations (33), we may put in V_p

$$\dot{\Omega}^O(\dot{Q}_j^E + \dot{\rho}_j^S) = \dot{\Omega}(\dot{Q}_j^E + \dot{\rho}_j^S) + \dot{\Omega}^{Op}(\dot{Q}_j^E + \dot{\rho}_j^S)$$

$$= \dot{\Omega}(\dot{Q}_j^E) + C_{jk}\dot{Q}_j^E\dot{\rho}_k^S + \dot{\Omega}(\dot{\rho}_j^S) + \dot{\Omega}^{OP}(\dot{Q}_j^E + \dot{\rho}_j^S) \tag{88}$$

$$= \dot{\Omega}(\dot{Q}_j^E) + \dot{\rho}_j^S\dot{q}_j^E + \dot{\Omega}(\dot{\rho}_j^S) + \dot{\Omega}^{OP}(\dot{Q}_j^E + \dot{\rho}_j^S) \ .$$

The same manipulation may be carried out in V_e except that $\dot{\Omega}^{OP}$ will always be zero. Substituting equation (88) into equation (87), and rearranging,

$$\dot{U}_c^O(\dot{\rho}_j^S) = \{\int\limits_{V_p} \dot{\Omega}^{OP}(\dot{Q}_j^E + \dot{\rho}_j^S)dV + \int\limits_{V} \dot{\Omega}(\dot{\rho}_j^S)dV\}$$

$$+ \{\int\limits_{V} \dot{\Omega}(\dot{Q}_j^E)dV - \int\limits_{S_u} \dot{\underline{p}}^E\cdot\hat{\underline{u}}dS\} \tag{89}$$

$$+ \{\int\limits_{V} \dot{\rho}_j^S\dot{q}_j^E dV - \int\limits_{S_u} \dot{\underline{p}}^{rs}\cdot\hat{\underline{u}}dS\}$$

Of the three expressions on the right hand side of equation (89), the second is independent of $\dot{\rho}_j^S$ and the third is zero, since $\dot{\underline{p}} = 0$ on S_p, $\dot{\underline{p}}^{rs}$ on S_u, $\dot{\rho}_j^S$ are statically admissible and $\dot{\underline{u}}^E$ on S_p and V, $\hat{\underline{u}}$ on S_u, $\dot{q}_j^E$ are kinematically admissible. We may carry out a similar argument for the actual residual stress rate field $\dot{\rho}_j$, $\dot{\underline{p}}^r$ on S_u, defined by

$$\dot{Q}_j = \dot{Q}_j^E + \dot{\rho}_j \ , \tag{90a}$$

$$\dot{\underline{p}} = \dot{\underline{p}}^E + \dot{\underline{p}}^r \text{ on } S_u. \tag{90b}$$

It follows therefore that

$$\dot{U}_c^{ro}(\dot{\rho}_j^S) = \int\limits_{V_p} \dot{\Omega}^{OP}(\dot{Q}_j^E + \dot{\rho}_j^S)dV + \int\limits_{V} \dot{\Omega}(\dot{\rho}_j^S)dV \geq \tag{91}$$

$$\int\limits_{V_p} \dot{\Omega}^{OP}(\dot{Q}_j^E + \dot{\rho}_j)dV + \int\limits_{V} \dot{\Omega}(\dot{\rho}_j) = \dot{U}_c^{ro}(\dot{\rho}_j) \ ,$$

providing a minimum principle for the residual stress rate field.

Using the elastic solution, we may also define a kinematically admissible strain rate field;

$$\dot{q}^c_j = \dot{q}^E_j + \dot{q}^{rc}_j \quad , \tag{92a}$$

$$\underline{\dot{u}}^c = \underline{\dot{u}}^E + \underline{\dot{u}}^{rc} \text{ on } S_p \text{ and } V \quad , \tag{92b}$$

where $\dot{q}^{rc}_j$, $\underline{\dot{u}}^{rc}$ on S_p and V, $\underline{\dot{u}} = 0$ on S_u are kinematically admissible. With $\dot{q}^{pc}_j$ defined as in equations (81), we see that on substituting equations (92) into equation (82),

$$\overset{+O}{\bar{U}}_p = \int_{V_p} \dot{w}^{Op}(\lambda^c) dV + \int_{V_p} \dot{W}(\dot{q}^E_j + \dot{q}^{rc}_j - \lambda^c \frac{\partial \phi}{\partial Q_j}) dV + \int_{V_e} \dot{W}(\dot{q}^E_j + \dot{q}^{rc}_j) dV$$

$$- \int_{V} \underline{\dot{\hat{F}}} \cdot (\underline{\dot{u}}^E + \underline{\dot{u}}^{rc}) dV - \int_{S_p} \underline{\dot{\hat{p}}} \cdot (\underline{\dot{u}}^E + \underline{\dot{u}}^{rc}) dS \quad . \tag{93}$$

In V_p we may put

$$\dot{W}(\dot{q}^E_j + \dot{q}^{rc}_j - \lambda^c \frac{\partial \phi}{\partial Q_j}) = \dot{W}(\dot{q}^E_j) + D_{jk} \dot{q}^E_j (\dot{q}^{rc}_k - \lambda^c \frac{\partial \phi}{\partial Q_j})$$

$$+ \dot{W}(\dot{q}^{rc}_j - \lambda^c \frac{\partial \phi}{\partial Q_j}) \tag{94}$$

$$= \dot{W}(\dot{q}^E_j) + Q^E_j \dot{q}^{rc}_j - \lambda^c Q^E_j \frac{\partial \phi}{\partial Q_j} + \dot{W}(\dot{q}^{rc}_j - \lambda^c \frac{\partial \phi}{\partial Q_j})$$

The same equation, with $\lambda^c = 0$, holds in V_e. Hence equation (93) becomes

$$\overset{\bullet O}{\bar{U}}_p = \{ \int_{V_p} [\dot{W}^{Op}(\lambda^c) - \lambda^c \dot{Q}^E_j \frac{\partial \phi}{\partial Q_j}] dV + \int_{V_p} \dot{W}(\dot{q}^{rc}_j - \lambda^c \frac{\partial \phi}{\partial Q_j}) dV$$

$$+ \int_{V_e} \dot{W}(\dot{q}^{rc}_j) dV \}$$

$$+ \left\{ \int_{V_p} \dot{W}(\dot{q}_j^E) \, dV - \int_V \hat{\dot{F}} \cdot \dot{\underline{u}}^E \, dV - \int_{S_p} \hat{\dot{\underline{p}}} \cdot \dot{\underline{u}}^E \, dS \right\}$$

$$+ \left\{ \int_V \dot{Q}_j^E \dot{q}_j^{rc} \, dV - \int_V \hat{\dot{F}} \cdot \dot{\underline{u}}^{rc} \, dV - \int_{S_p} \hat{\dot{\underline{p}}} \cdot \dot{\underline{u}}^{rc} \, dS \right\} \ . \tag{95}$$

Of the three expressions on the right hand of equation (95), the second is independent of $\dot{q}_j^{rc}$, λ^c, and the third is zero, since $\dot{Q}_j^E$, $\hat{\dot{p}}$ on S_p, $\dot{p}^E$ on S_u, $\hat{\dot{F}}$ on V is a statically admissible set and $\dot{\underline{u}}^{rc}$ on S_p and V, $\dot{\underline{u}} = 0$ on S_u, $\dot{q}_j^{rc}$ is a kinematically admissible set.

Hence we may define

$$\dot{\bar{U}}_p^{ro}(\dot{q}_j^{rc}, \lambda^c) = \int_{V_p} \left[\dot{W}^{op}(\lambda^c) - \lambda^c \dot{Q}_j^E \frac{\partial \phi}{\partial Q_j} \right] dV$$

$$+ \int_{V_p} \dot{W}\left(\dot{q}_j^{rc} - \lambda^c \frac{\partial \phi}{\partial Q_j}\right) dV + \int_{V_e} \dot{W}(\dot{q}_j^{rc}) \, dV \ , \tag{96}$$

and note that *the solution of the rate problem,*

$$\dot{q}_j = \dot{q}_j^E + \dot{q}_j^r \ , \tag{97a}$$

$$\dot{\underline{u}} = \dot{\underline{u}}^E + \dot{\underline{u}}^r \text{ on } S_p \text{ and } V \ , \tag{97b}$$

is given *by the least value of* $\dot{\bar{U}}_p^{ro}$, *subject to* $\lambda^c \geq 0$.

25.4 Bibliographical and Historical Remarks

The classical minimum principles were first established in a weak form by Prager [1942, 1946]. A generalization was achieved by Hodge and Prager [1948], and a proof of the principles in their present form was given by Greenberg [1949a, 1949b] for a smooth yield surface. Koiter [1953] further generalized the principles to cover singular yield surfaces.

Further discussions of the minimum principles have been given by Hill [1956], Drucker [1958], Hodge [1958], Koiter [1960] and Hodge [1968].

The alternative form of the kinematic theorem employing multipliers for the plastic strain rates was given by Martin [1975], although this follows the work of Ceradini [1966] and Maier [1969] who gave the static and kinematic theorems for the residual field respectively. Maier's kinematic principle was based on quadratic programming arguments, and did not make direct use of the inequality arising from the arbitrary division of the strain rate into elastic and plastic parts.

References

G.Ceradini	1966	"A maximum principle for the analysis of elastic-plastic systems", Meccanica, $\underline{1}$, 77.
D.C.Drucker	1958	"Variational principles in the mathematical theory of plasticity", Proc. Symp.Appl.Math., $\underline{8}$, 7.
H.J.Greenberg	1949a	"On the variational principles of plasticity", Tech.Rept. A11-54, Div. of Appl.Math., Brown University.
H.J.Greenberg	1949b	"Complementary minimum principles for an elastic-plastic material", Quart. Appl.Math., $\underline{7}$, 85.
R.Hill	1956	"New horizons in the mechanics of solids", J.Mech.Phys.Sol., $\underline{5}$, 66.
P.G.Hodge	1958	"The mathematical theory of plasticity", *Surveys in Applied Mathematics*, (Wiley, N.Y.), $\underline{1}$, 49.
P.G.Hodge	1968	"Numerical applications of minimum principles in plasticity", *Engineering Plasticity*, edited by J.Heyman and F. A.Leckie, Cambridge University Press, 237.
P.G.Hodge and W.Prager	1948	"A variational principle for plastic materials with strain hardening", J. Math. and Phys., $\underline{27}$, 1.

W.T.Koiter   1953   "Stress-strain relations, uniqueness and variational theorems for elastic-plastic materials with a singular yield surface", Quart.Appl.Maths., $\underline{11}$, 350.

.W.T.Koiter   1960   "General theorems for elastic-plastic solids", *Progress in Solid Mechanics*, edited by R.Hill and I.Sneddon, North-Holland Press, 167.

G.Maier   1969   "Some theorems for plastic strain rates and plastic strains", J.de Mécanique, $\underline{8}$, 5.

J.B.Martin   1975   "On the Kinematic Minimum Principle in Classical Plasticity", J.Mech.Phys. Sol., $\underline{23}$, to appear.

W.Prager   1942   "Fundamental theorems of a new mathematical theory of plasticity", Duke Math.J., $\underline{9}$, 228.

W.Prager   1946   "Variational principles in the theory of plasticity", Proc. 6th Int.Congr. Appl.Mech. (Paris).

NUMERICAL APPLICATIONS OF THE MINIMUM PRINCIPLES

26.1 Introduction: A Truss Example

It will be clear from the complexity of both deformation theory
and incremental theory that few analytical solutions of realistic boundary value problems can be expected. A great deal of
progress has been made, however, in the numerical analysis of
elastic-plastic problems. A detailed development and description of this work is beyond the scope of this monograph, but we
shall attempt to give a brief indication of the basis of the
numerical approach.

In order to achieve this we shall proceed from the particular to the general in three stages. First, we shall consider a
specific three-bar truss example in order to clarify the form
of the problem and the manner of solution. Secondly, we shall
treat a plane truss of arbitrary configuration in matrix terms,
showing how more complex examples can be efficiently handled.
Finally, we shall consider as an example of the discretization
of a continuous body the problem of plane stress, generalizing
the matrix approach to truss problems.

In each case we shall consider the elastic problem and both
the deformation theory problem and the incremental theory problem for perfectly plastic and hardening materials. The particular minimum principles we shall treat are listed for reference
purposes in Table 1.

In the case of a pin-jointed truss each bar will be assumed
to be of uniform cross-sectional area A along its length ℓ. It
is convenient to give directly the relation between the *extension* of the bar δ and the axial tension N.

For an *elastic* material

$$\delta = (\frac{\ell}{AE})N \quad . \tag{1}$$

Table 1. Kinematic Minimum Principles under Consideration

Elastic Problem Minimize $U_p^O(q_j^c, \underline{u}^c) = \int_V W(q_j^c)dV - \int_{S_p} \hat{\underline{p}} \cdot \underline{u}^c dS$	subject to q_j^c, $\underline{u}^c$ on S_p and V, $\hat{u}$ on S_u kinematically admissible
Deformation Theory Problem Minimize $U_p^O(q_j^c, \underline{u}^c) = \int_V W^O(q_j^c)dV - \int_{S_p} \hat{\underline{p}} \cdot \underline{u}^c dS$	subject to q_j^c, $\underline{u}^c$ on S_p and V, $\hat{\underline{u}}$ on S_u k.a.
Minimize $\bar{U}_p^O(q_j^c, q_j^{pc}, \underline{u}^c) = \int_V W^{op}(q_j^{pc})dV + \int_V W(q_j^c - q_j^{pc})dV - \int_{S_p} \hat{\underline{p}} \cdot \underline{u}^c dS$	subject to q_j^c, $\underline{u}^c$ on S_p and V, $\hat{u}$ on S_u k.a.
Incremental Theory Problem Minimize $\dot{U}_p^O(\dot{q}_j^c, \underline{\dot{u}}^c) = \int_V \dot{W}^O(\dot{q}_j^c)dV - \int_{S_p} \hat{\underline{\dot{p}}} \cdot \underline{\dot{u}}^c dS$	subject to $\dot{q}_j^c$, $\underline{\dot{u}}^c$ on S_p and V, $\hat{\underline{u}}$ on S_u k.a.
Minimize $\bar{\dot{U}}_p^O(\dot{q}_j^c, \lambda^c, \underline{\dot{u}}^c) = \int_{V_p} \dot{W}^{op}(\lambda^c)dV + \int_{V_p} \dot{W}(\dot{q}_j^c - \lambda\frac{\partial\phi}{\partial Q_j})dV + \int_{V_e} \dot{W}(\dot{q}_j^c)dV - \int_{S_p} \hat{\underline{\dot{p}}} \cdot \underline{\dot{u}}^c dS$	subject to $\dot{q}_j^c$, $\underline{\dot{u}}^c$ on S_p and V, $\hat{\underline{u}}$ on S_u k.a., $\lambda^c \geq 0$.

For a *perfectly plastic* material

$$\delta = \delta^e + \delta^p \ ,$$

$$\delta^e = \left(\frac{\ell}{AE}\right) N \ , \tag{2}$$

$$\dot{\delta}^p = \lambda \ \langle N \rangle \ ,$$

where $\langle N \rangle = +1$ if $N = +N_o$ and $\dot{N} = 0$

$$\langle N \rangle = \ 0 \ \ \text{if} \begin{cases} N = +N_o \ \text{ and } \ \dot{N} < 0 \\[4pt] -N_o < N < N_o \\[4pt] N = -N_o \ \text{ and } \ \dot{N} > 0 \end{cases}$$

$\langle N \rangle = -1$ if $N = -N_o$ and $\dot{N} = 0$,

$$\lambda \geq 0 \ .$$

As an example of an elastic-plastic material we will adopt a linear kinematic hardening material. In this case

$$\delta^e = \left(\frac{\ell}{AE}\right) N \ ,$$

$$\dot{\delta}^p = \left(\frac{\ell}{AE_p}\right) \dot{N} \ \text{ for } \ N = N_o + \frac{AE_p}{\ell} \delta^p \ \text{ and } \ \dot{N} \geq 0 \ ,$$

$$\text{or} \ \ N = -N_o + \frac{AE_p}{\ell} \delta^p \ \text{ and } \ \dot{N} \leq 0 \ ,$$

$$\dot{\delta}^p = 0 \ \ \ \ \ \text{ for } \ N = N_o + \frac{AE_p}{\ell} \delta^p \ \text{ and } \ \dot{N} < 0 \ , \tag{3}$$

$$\text{or} \ -N_o + \frac{AE_p}{\ell} \delta^p < N < N_o + \frac{AE_p}{\ell} \delta^p \ ,$$

$$\text{or} \ \ N = -N_o + \frac{AE_p}{\ell} \delta^p \ \text{ and } \ \dot{N} > 0 \ .$$

The relation between N and δ for monotonic increase in δ is shown in Fig.1. In the case of the hardening material it is convenient to define E_T, where, for loading

$$\dot{\delta} = \dot{\delta}^e + \dot{\delta}^p = \left(\frac{\ell}{AE}\right)\dot{N} + \left(\frac{\ell}{AE_p}\right)\dot{N}$$

$$= \frac{\ell}{A}\left(\frac{1}{E} + \frac{1}{E_p}\right)\dot{N} = \left(\frac{\ell}{AE_T}\right)\dot{N} \quad . \tag{4}$$

We may now define the functions W, W^o, W^{op}, $\dot{W}$, $\dot{W}^o$, $\dot{W}^{op}$ as applying to an entire bar, thus integrating the specific functions over the length of the bar. The integrals in Table 1 will then simply become sums over the bars of the truss. The functions are easily computed and are given in Tables 2 and 3. It will be recalled that the minimum work path involves continuous loading for a relation between a single generalized stress component and a single generalized strain component.

Consider the specific truss example (which is identical to that introduced in Chapter 6) shown in Figure 2. The three bars will be assumed to have identical properties. The vertical

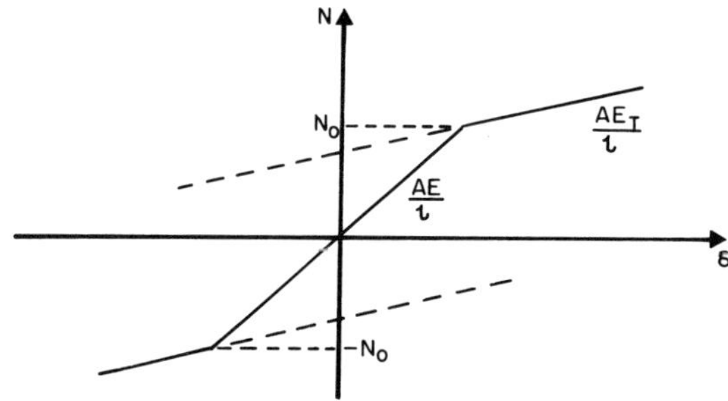

Figure 1. Relation between axial force and extension

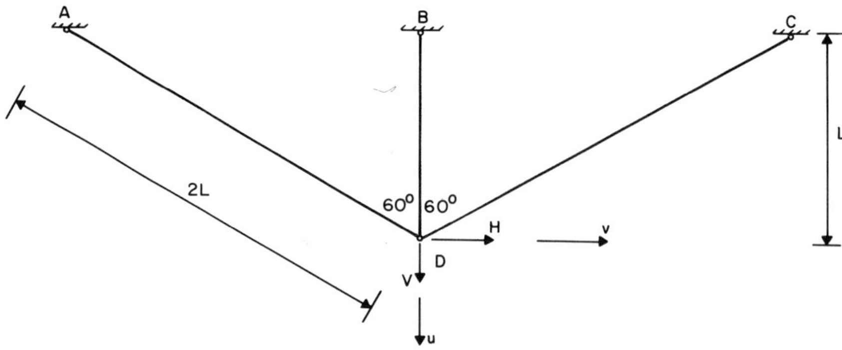

Figure 2. Three bar truss

Table 2. Functions for Elastic and Deformation Theory Analysis
of Trusses

Elastic

$$W = \frac{1}{2}\left(\frac{AE}{\ell}\right)\delta^2$$

Elastic, perfectly plastic

$$W^o = \begin{cases} N_o\delta - \frac{1}{2}\left(\frac{\ell}{AE}\right)N_o^2 & \text{for } \delta \geq \left(\frac{\ell}{AE}\right)N_o \\[2ex] \frac{1}{2}\left(\frac{AE}{\ell}\right)\delta^2 & \text{for } -\left(\frac{\ell}{AE}\right)N_o \leq \delta \leq \left(\frac{\ell}{AE}\right)N_o \\[2ex] -N_o\delta - \frac{1}{2}\left(\frac{AE}{\ell}\right)N_o^2 & \text{for } \delta \leq -\left(\frac{\ell}{AE}\right)N_o \end{cases}$$

$$W^{op} = \begin{cases} N_o\delta^p & \text{for } \delta^p > 0 \\[2ex] 0 & \text{for } \delta^p = 0 \\[2ex] -N_o\delta^p & \text{for } \delta^p < 0 \end{cases}$$

Linear kinematic hardening

$$W^o = \begin{cases} N_o\delta - \frac{1}{2}\left(\frac{\ell}{AE}\right)N_o^2 + \frac{1}{2}\left(\frac{AE_T}{\ell}\right)\left\{\delta - \frac{\ell}{AE}N_o\right\}^2 & \text{for } \delta \geq \left(\frac{\ell}{AE}\right)N_o \\[2ex] \frac{1}{2}\left(\frac{AE}{\ell}\right)\delta^2 & \text{for } -\left(\frac{\ell}{AE}\right)N_o \leq \delta \leq \left(\frac{\ell}{AE}\right)N_o \\[2ex] -N_o\delta - \frac{1}{2}\left(\frac{\ell}{AE}\right)N_o^2 + \frac{1}{2}\left(\frac{AE_T}{\ell}\right)\left\{\delta + \frac{\ell}{AE}N_o\right\}^2 & \text{for } \delta \leq -\left(\frac{\ell}{AE}\right)N_o \end{cases}$$

$$W^{op} = \begin{cases} N_o\delta^p + \frac{1}{2}\left(\frac{AE_p}{\ell}\right)(\delta^p)^2 & \text{for } \delta^p > 0 \\[2ex] 0 & \text{for } \delta^p = 0 \\[2ex] -N_o\delta^p + \frac{1}{2}\left(\frac{AE_p}{\ell}\right)(\delta^p)^2 & \text{for } \delta^p < 0 \end{cases}$$

Table 3. Functions for Incremental Analysis of Trusses

<u>Elastic</u>

$$\dot{W} = \frac{1}{2} \left(\frac{AE}{\ell}\right)\dot{\delta}^2$$

<u>Elastic, perfectly plastic</u>

$$\dot{W}^O = \begin{cases} 0 \text{ for } N = N_o, \ \dot{\delta} > 0 \\[2ex] \frac{1}{2}\left(\frac{AE}{\ell}\right)\dot{\delta}^2 \text{ for} \begin{cases} N = N_o, \ \dot{\delta} < 0 \\[1ex] -N_o < N < N_o \\[1ex] N = -N_o, \ \dot{\delta} > 0 \end{cases} \\[5ex] 0 \text{ for } N = -N_o, \ \dot{\delta} < 0 \end{cases}$$

$$\dot{W}^{OP} = 0 \ , \qquad \dot{\delta}^p = \lambda \ \langle N \rangle$$

<u>Linear kinematic hardening</u>

$$\dot{W}^O = \begin{cases} \frac{1}{2}\left(\frac{AE_T}{\ell}\right)\dot{\delta}^2 \text{ for } N = N_o + \frac{AE_p}{\ell}\delta^p, \ \dot{\delta} \geq 0 \\[3ex] \frac{1}{2}\left(\frac{AE}{\ell}\right)\dot{\delta}^2 \text{ for} \begin{cases} N = N_o + \frac{AE_p}{\ell}\delta^p, \ \dot{\delta} \leq 0 \\[2ex] -N_o + \frac{AE_p}{\ell}\delta^p \leq N \leq N_o + \frac{AE_p}{\ell}\delta^p \\[2ex] N = -N_o + \frac{AE_p}{\ell}\delta^p, \ \dot{\delta} \geq 0 \end{cases} \\[3ex] \frac{1}{2}\left(\frac{AE_T}{\ell}\right)\dot{\delta}^2 \text{ for } N = -N_o + \frac{AE_p}{\ell}\delta^p, \ \dot{\delta} \leq 0 \end{cases}$$

$$\dot{W}^{OP} = \frac{1}{2}\left(\frac{AE_p}{\ell}\right)\lambda^2 \ , \qquad \dot{\delta}^p = \lambda \ \langle N \rangle$$

load at node D is denoted by V, and the horizontal load by H.

The displacements of the node D, denoted by u, v in the x and y directions respectively, are the primary unknowns in the displacement method. The extensions of the bars are

$$\delta_A = \frac{1}{2} u + \frac{\sqrt{3}}{2} v \quad ,$$
$$\delta_B = u \quad , \tag{5}$$
$$\delta_C = \frac{1}{2} u - \frac{\sqrt{3}}{2} v \quad .$$

The equilibrium equations at node D are

$$N_A + 2N_B + N_C = 2V \quad ,$$
$$N_A \qquad - N_C = \frac{2}{\sqrt{3}} H \quad . \tag{6}$$

The elastic solution may be obtained formally from the principle of minimum potential energy by formulating

$$U_p = \frac{1}{2} \left(\frac{AE}{2L}\right)\delta_A^{\,2} + \frac{1}{2} \left(\frac{AE}{L}\right)\delta_B^{\,2} + \frac{1}{2} \left(\frac{AE}{2L}\right)\delta_C^{\,2} - Vu - Hv \tag{7}$$
$$= \frac{1}{2} \left(\frac{AE}{2L}\right)\left(\frac{u}{2} + \frac{\sqrt{3}}{2} v\right)^2 + \frac{1}{2} \left(\frac{AE}{L}\right)u^2 + \frac{1}{2} \left(\frac{AE}{2L}\right)\left(\frac{u}{2} - \frac{\sqrt{3}}{2} v\right)^2 - Vu - Hv \quad .$$

The minimum value of U_p is given by the conditions

$$\frac{\partial U_p}{\partial u} = 0 \quad , \qquad \frac{\partial U_p}{\partial v} = 0 \quad . \tag{8}$$

This provides two simultaneous equations, which are fortuitously uncoupled in this example:

$$\left(\frac{AE}{\ell}\right)\begin{bmatrix} \frac{5}{4} & 0 \\ 0 & \frac{3}{4} \end{bmatrix}\begin{bmatrix} u \\ v \end{bmatrix} = \begin{bmatrix} V \\ H \end{bmatrix} \tag{9}$$

These equations are of course identical to what we would obtain

if we had used the elastic constitutive relation and equations
(5) to substitute for the bar forces in equations (6). We can
now work back from the solution of equations (9) to obtain the
extensions and, using the constitutive relations, the bar
forces.

Consider now the deformation theory problem. We formulate
first

$$U_p^O = W_A^O(\frac{u}{2} + \frac{\sqrt{3}}{2} v) + W_B^O(u) + W_C^O(\frac{u}{2} - \frac{\sqrt{3}}{2} v) - Vu - Hv \quad . \qquad (10)$$

The least value of U_p^O provides the correct values of u, v for
given V, H. This is an unconstrained minimization problem, but
it must be noted that W^O is not continuously differentiable.
There are a variety of ways in which we may approach the solution, but the conceptually simplest method is to guess (possibly on the basis of the elastic solution) into which of the
ranges

$$\delta \geq \frac{\ell}{AE} N_o \quad ,$$

$$-\frac{\ell}{AE} N_o < \delta < \frac{\ell}{AE} N_o \quad , \qquad (11)$$

$$\delta \leq -\frac{\ell}{AE} N_o \quad ,$$

each particular bar falls. We may then treat the functions W^O
in each bar as continuously differentiable, and put

$$\frac{\partial U_p^O}{\partial u} = 0 \quad , \qquad \frac{\partial U_p^O}{\partial v} = 0 \quad . \qquad (12)$$

From these equations we determine u, v and hence δ_A, δ_B, δ_C. We
must then ascertain whether the values of δ_A, δ_B, δ_C in fact
fall within the ranges in which they were assumed to lie. If
so, the solution is correct. If not, we must adjust our

assumption, and repeat the procedure.

An iterative procedure based on placing an incorrectly assigned bar strain into the range in which it is found to be
after the previous iteration will generally lead to fairly rapid convergence. Alternatively, if we seek solutions for a path
of loading in which a sequence of values of H, V are taken, difficulties in the iterative procedure can be avoided if the difference between one set of load values and the next is small.
When it is necessary to iterate this approach has the disadvantage that the entire set of equations (12) must be reformulated.

Suppose we now introduce plastic extension components δ_A^p,
δ_B^p, δ_C^p. We may then formulate $\bar{U}_p^O$. For the case of a perfectly
plastic material, for example, this takes the form

$$\bar{U}_p^O = N_o |\delta_A^p| + N_o |\delta_B^p| + N_o |\delta_C^p|$$

$$+ \frac{1}{2} \left(\frac{AE}{2L}\right)\left(\frac{u+v\sqrt{3}}{2} - \delta_A^p\right)^2 + \frac{1}{2}\left(\frac{AE}{L}\right)(u-\delta_B^p)^2 + \frac{1}{2}\left(\frac{AE}{2L}\right)\left(\frac{u-v\sqrt{3}}{2} - \delta_C^p\right)^2$$

$$- Vu - Hv \quad . \tag{13}$$

The solution of the problem is given by the least value of $\bar{U}_p^O$.
The minimization problem is unconstrained, but the functions
$|\delta_A^p|$, $|\delta_B^p|$, $|\delta_C^p|$ have discontinuous derivatives at the origin.

We may proceed by differentiating $\bar{U}_p^O$ and formulating the
five equations

$$\frac{\partial \bar{U}_p^O}{\partial u} = \frac{\partial \bar{U}_p^O}{\partial v} = \frac{\partial \bar{U}_p^O}{\partial \delta_A^p} = \frac{\partial \bar{U}_p^O}{\partial \delta_B^p} = \frac{\partial \bar{U}_p^O}{\partial \delta_C^p} = 0 \quad . \tag{14}$$

Expressed in matrix form, these equations are

$$\frac{AE}{L}\begin{bmatrix} \frac{5}{4} & 0 & -\frac{1}{4} & -1 & -\frac{1}{4} \\ 0 & \frac{3}{4} & -\frac{\sqrt{3}}{4} & 0 & \frac{\sqrt{3}}{4} \\ -\frac{1}{4} & -\frac{\sqrt{3}}{4} & \frac{1}{2} & 0 & 0 \\ -1 & 0 & 0 & 1 & 0 \\ -\frac{1}{4} & \frac{\sqrt{3}}{4} & 0 & 0 & \frac{1}{2} \end{bmatrix}\begin{bmatrix} u \\ v \\ \delta^p_A \\ \delta^p_B \\ \delta^p_C \end{bmatrix} = \begin{bmatrix} V \\ H \\ -\langle N_o \rangle \\ -\langle N_o \rangle \\ -\langle N_o \rangle \end{bmatrix} . \tag{15}$$

In these equations

$$\begin{aligned} \langle N_o \rangle &= +N_o & \text{if } \delta^p &> 0 \;, \\ -N_o \leq \langle N_o \rangle &\leq N_o & \text{if } \delta^p &= 0 \;, \\ \langle N_o \rangle &= -N_o & \text{if } \delta^p &< 0 \;, \end{aligned} \tag{16}$$

arising from the derivative of $N_o|\delta^p|$ with respect to δ^p.

It will be appreciated that when, say, δ^p_B is zero, the corresponding $\langle N_o \rangle$ takes on a value such that the fourth equation is satisfied, the remaining variables being determined by the other four equations. The limits on $\langle N_o \rangle$ in this case simply express the limits of elastic behavior.

We may thus adopt the following procedure for the solution of the equations. We guess whether each of δ^p_A, δ^p_B, δ^p_C is positive, zero or negative. For each plastic extension which is assumed to be zero, we delete the appropriate equation, and we assign the value $+N_o$ or $-N_o$ to the remaining $\langle N_o \rangle$ terms in the vector on the right hand side of equation (15). The remaining equations are solved, and the extensions and bar forces are found. Provided that in each bar $\delta \geq N_o \ell/AE$ if δ^p was assumed positive, $-N_o \ell/AE \leq \delta < N_o \ell/AE$ if δ was assumed zero, or $\delta \leq -N_o \ell/AE$ if δ^p was assumed negative, the solution is correct. If not, the assumptions must be modified, and a further iteration carried out.

This is essentially the same procedure that was adopted in minimizing U_p^o, with two major differences. First, there is the disadvantage that the number of variables in the second procedure is larger. Secondly, there is the advantage that when an iteration is necessary the enlarged stiffness matrix is unaffected, whereas in the first procedure the entire set of equations must be reformulated.

The incremental theory problem is conceived as a series of rate problems. The loading path $V(t)$, $H(t)$ is divided into a series of increments dV, dH each of which occurs in time dt. After each increment the solution is updated, providing information for the next rate problem. We shall consider only the form of any one individual rate problem.

Consider in this case a truss composed of a linear kinematic hardening material. It is assumed that N, δ^p are known in each bar. We formulate first

$$\dot{U}_p^o = \dot{W}_A^o(\dot{\delta}_A) + \dot{W}_B^o(\dot{\delta}_B) + \dot{W}_C^o(\dot{\delta}_C) - \dot{V}\dot{u} - \dot{H}\dot{v} \quad . \tag{17}$$

If $\phi < 0$ in any bar, the choice of $\dot{W}^o$ is unambiguous. If $\phi = 0$, $\dot{W}^o$ will depend on whether loading or unloading occurs. Hence it is again necessary to assume whether loading or unloading is occurring in each plastic bar if we are to follow the procedure of setting

$$\frac{\partial \dot{U}_p^o}{\partial \dot{u}} = \frac{\partial \dot{U}_p^o}{\partial \dot{v}} = 0 \quad . \tag{18}$$

We again obtain a set of linear equations. The solution must be checked to determine whether or not the assumptions hold; if not, the assumptions must be adjusted.

Since the problem of guessing whether or not loading is occurring in any rate problem is far simpler than determining whether or not plastic deformation occurs in the deformation

theory problem, the number of occasions on which an iteration
is necessary is greatly reduced. However, whenever an additional bar becomes plastic in going from one rate problem to
another, the coefficients of the variables in equation (18)
must be changed.

Alternatively, consider the formulation of $\dot{\bar{U}}_p^o$. Here there is
an advantage in assuming that all bars are plastic. In this
case we introduce $\dot{\delta}_A^p$, $\dot{\delta}_B^p$, $\dot{\delta}_C^p$, and find

$$\dot{\bar{U}}_p^o = \frac{1}{2}\,(\frac{AE_p}{2L})(\dot{\delta}_A^p)^2 + \frac{1}{2}\,(\frac{AE_p}{L})(\dot{\delta}_B^p)^2 + \frac{1}{2}\,(\frac{AE_p}{2L})(\dot{\delta}_C^p)^2$$

$$+ \frac{1}{2}\frac{AE}{2L}\,(\frac{\dot{u}+\sqrt{3}\dot{v}}{2} - \dot{\delta}_A^p)^2 + \frac{1}{2}\frac{AE}{L}\,(\dot{u} - \dot{\delta}_B^p)^2 \qquad (19)$$

$$+ \frac{1}{2}\frac{AE}{2L}\,(\frac{\dot{u}-\sqrt{3}\dot{v}}{2} - \dot{\delta}_C^p)^2 - V\dot{u} - H\dot{v}\ .$$

This function must be minimized subject to the constraints that
$\dot{\delta}^p \geq 0$ if yield occurs in tension, $\dot{\delta}^p = 0$ if the bar is elastic, and $\dot{\delta}^p \leq 0$ if yield occurs in compression.

Ignoring the constraints, we put the partial derivatives of
$\dot{\bar{U}}_p^o$ equal to zero. In matrix form, this gives

$$\frac{AE}{L}
\begin{bmatrix}
\frac{5}{4} & 0 & -\frac{1}{4} & -1 & -\frac{1}{4} \\[6pt]
0 & \frac{3}{4} & -\frac{\sqrt{3}}{4} & 0 & \frac{\sqrt{3}}{4} \\[6pt]
-\frac{1}{4} & -\frac{\sqrt{3}}{4} & (\frac{1}{2} + \frac{1}{2}\frac{E_p}{E}) & 0 & 0 \\[6pt]
-1 & 0 & 0 & (1 + \frac{E_p}{E}) & 0 \\[6pt]
-\frac{1}{4} & \frac{\sqrt{3}}{4} & 0 & 0 & (\frac{1}{2} + \frac{1}{2}\frac{E_p}{E})
\end{bmatrix}
\begin{bmatrix}
\dot{u} \\[6pt] \dot{v} \\[6pt] \dot{\delta}_A^p \\[6pt] \dot{\delta}_B^p \\[6pt] \dot{\delta}_C^p
\end{bmatrix}
=
\begin{bmatrix}
V \\[6pt] H \\[6pt] 0 \\[6pt] 0 \\[6pt] 0
\end{bmatrix}\ .$$

$$(20)$$

We now proceed as follows. First, we delete those $\dot{\delta}^p$ equations for which the bar is elastic. Second, we guess whether loading or unloading occurs in the remaining bars. If we assume that unloading occurs, we delete the appropriate $\dot{\delta}^p$ equation. The remaining equations are solved, and the solution checked to ascertain whether the assumptions were correct.

By comparison with the minimization of $\dot{U}^o_p$, this problem has the disadvantage that the number of variables is increased, but the advantage that the matrix occurring in equation (20) applies for *all* rate problems, and need not be reformulated in passing from one increment to another.

The reader is invited to carry through some specific solutions involving the three bar truss problem to familiarize himself with the differences between the various approaches. In the next Section we shall generalize these results to an arbitrary truss.

26.2 Matrix Formulation of the General Plane Truss Problem

Consider the problem of a plane pin-jointed truss made up of m members of uniform cross-section. We introduce a global X-Y coordinate system and resolve the forces and the displacements of the joints in this system. For simplicity we shall assume that certain X or Y components of the joint displacements are specified zero to make up the displacement boundary conditions. The remaining n unknown displacement components are ordered to form the displacement vector $\{u\}$. The n given force components acting at the nodes are similarly ordered to form the load vector $\{P\}$.

The extension of a bar connecting the i-th and j-th nodes can be expressed in terms of the displacements $\underline{u}^i$, $\underline{u}^j$ of the nodes and the unit vector $\underline{n}^{ij}$ in the direction i-j,

$$\delta^{ij} = (\underline{u}^j - \underline{u}^i).\underline{n}^{ij} \quad . \tag{21}$$

We may then make up an extension vector $\{\delta\}$, ordering the extensions of the m members, and relate it to the displacement vector $\{u\}$;

$$\{\delta\} = [B]\{u\} \quad , \tag{22}$$

where $[B]$ is the *deformation matrix*.

We may also order the bar forces in the same manner and make up the axial force vector $\{N\}$. The node equilibrium equations will take the form

$$\{P\} = [A]\{N\} \quad , \tag{23}$$

where $[A]$ is the *statics matrix*.

The deformation and statics matrices are transposes of each other, since by the principle of virtual work

$$\{P\}^T\{u\} = \{N\}^T\{\delta\} \quad . \tag{24}$$

Recalling that we may write

$$\{P\}^T = ([A]\{N\})^T = \{N\}^T[A]^T \quad , \tag{25}$$

and using equation (22), equation (24) becomes

$$\{N\}^T[A]^T\{u\} = \{N\}^T[B]\{u\} \quad , \tag{26}$$

showing that

$$[A]^T = [B] \quad . \tag{27}$$

In the analysis of an elastic truss, the extension of any bar is related to the force in that bar by the relation

$$N = \frac{AE}{\ell} \delta \quad , \tag{28}$$

where A, E and ℓ are respectively the area, elastic modulus and length of the bar. We may form the *member stiffness matrix* $[S]$ by writing

$$\{N\} = [S]\{\delta\} \quad . \tag{29}$$

The matrix $[S]$ is diagonal and square.

If we substitute equation (29) and then equation (22) into equation (23), we see that

$$[A][S][B]\{u\} = \{P\} \quad . \tag{30}$$

This equation is generally written as

$$[K]\{u\} = \{P\} \quad , \tag{31}$$

where

$$[K] = [A][S][B] = [B]^T[S][B] \tag{32}$$

is the *system stiffness matrix*. $[K]$ is a square nxn matrix, and is symmetric.

The potential energy of the system may be written as

$$U_p = \frac{1}{2} \{N\}^T\{\delta\} - \{u\}^T\{P\}$$

$$= \frac{1}{2} \{\delta\}^T[S]\{\delta\} - \{u\}^T\{P\}$$

$$= \frac{1}{2} \{u\}^T[B]^T[S][B]\{u\} - \{u\}^T\{P\} \quad . \tag{33}$$

If we now put

$$\frac{\partial U_p}{\partial\{u\}} = 0 \tag{34}$$

we recover equation (31).

The solution of the elastic problem is accomplished by determining the inverse of $[K]$,

$$\{u\} = [K]^{-1}\{P\} \quad . \tag{35}$$

Consider now the deformation theory analysis of the truss. In minimizing $\overset{\mathrm{o}}{U}_p$ let us anticipate the approach adopted in the previous section. We guess whether δ in each bar lies in range $\delta \geq (\ell/AE)N_o$, $-(\ell/AE)N_o < \delta < (\ell/AE)N_o$ or $\delta \leq -(\ell/AE)N_o$, and determine the partial derivatives of U_p^o with respect to the elements of $\{u\}$. This implies that in the materials under consideration we select a constitutive relation for each bar from one of the three linear possibilities. In the case of a perfectly plastic material, for example, we choose either

$$N = \frac{\partial W^o}{\partial \delta} = N_o \quad ,$$

$$N = \frac{\partial W^o}{\partial \delta} = (\frac{AE}{\ell})\delta \quad , \tag{36}$$

$$N = \frac{\partial W^o}{\partial \delta} = -N_o \quad ,$$

while in the case of a linear kinematic hardening material we choose either

$$N = \frac{\partial W^o}{\partial \delta} = (\frac{AE_T}{\ell})\{\delta - \frac{\ell}{AE} N_o\} + N_o \quad ,$$

$$N = \frac{\partial W^o}{\partial \delta} = (\frac{AE}{\ell})\delta \quad , \tag{37}$$

$$N = \frac{\partial W^o}{\partial \delta} = (\frac{AE_T}{\ell})\{\delta + \frac{\ell}{AE} N_o\} - N_o \quad .$$

In view of the linearity of these equations, it is clear that on the basis of our choice we may write

$$\{N\} = [S^o]\{\delta\} + \{C^o\} \quad , \tag{38}$$

where $[S^o]$ is again a diagonal matrix and $\{C^o\}$ is a vector. From equations (22) and (23) we thus see that

$$[B]^T[S^o][B]\{u\} = \{P\} - [B]^T\{C^o\} \quad . \tag{39}$$

The symmetric matrix

$$[K^o] = [B]^T[S^o][B] \tag{40}$$

is the system stiffness matrix, and the solution is

$$\{u\} = [K^o]^{-1}(\{P\} - [B]^T\{c^o\}) \ . \tag{41}$$

In the event that each element of $\{\delta\} = [B]\{u\}$ does not fall within the range to which it was assigned, the $[S^o]$ and $\{c^o\}$ matrices must be reformulated on the basis of a revised assignment, and the process continued.

In the alternative formulation, involving the minimization of $\bar{U}^o_p$, consider again a perfectly plastic material. We introduce the plastic extension vector $\{\delta^p\}$, and we are required to guess whether each element of $\{\delta^p\}$ is positive, zero or negative. Let us introduce a column vector $\{<N_o>\}$ whose elements are either $+N_o$, 0, or $-N_o$ depending on whether the corresponding element of $\{\delta^p\}$ is guessed to be positive, zero or negative. Noting the form of equation (33), it may be seen that

$$\bar{U}^o_p = \{\delta^p\}^T\{<N_o>\} + \frac{1}{2}(\{\delta\}^T - \{\delta^p\}^T)[S](\{\delta\} - \{\delta^p\}) - \{u\}^T\{P\}$$

$$= \{\delta^p\}^T\{<N_o>\} + \frac{1}{2}\{\delta\}^T[S]\{\delta\} - \{\delta\}^T[S]\{\delta^p\}$$

$$+ \frac{1}{2}\{\delta^p\}^T[S]\{\delta^p\} - \{u\}^T\{P\} \tag{42}$$

$$= \{\delta^p\}^T\{<N_o>\} + \frac{1}{2}\{u\}^T[B]^T[S][B]\{u\} - \{u\}^T[B]^T[S]\{\delta^p\}$$

$$+ \frac{1}{2}\{\delta^p\}^T[S]\{\delta^p\} - \{u\}^T\{P\} \ .$$

We now equate the partial derivatives of this expression with respect to the elements of $\{u\}$ and $\{\delta^p\}$ to zero. This gives a set of linear equations of the form

$$[K^*]\{u \mid \delta^p\} = \{P \mid - <N_o>\} \ . \tag{43}$$

The modified $(m+n) \times (m+n)$ system matrix $[K^*]$ has the form

$$[K^*] = \begin{bmatrix} [B]^T[S][B] & \vdots & -[B]^T[S] \\ \hdashline -[S][B] & \vdots & [S] \end{bmatrix} \ . \tag{44}$$

We eliminate from equation (43) those equations corresponding to elements of $\{\delta^p\}$ which have been assigned to be zero,
and invert to determine $\{u\}$ and the remaining elements of $\{\delta^p\}$.
Where elements of $\{\delta^p\}$ were assumed to be non-zero their signs
must be checked, and where an element of $\{\delta^p\}$ was assumed zero
the magnitude of the corresponding element of $\{\delta\}$ must be within the range $-(\ell/AE)N_o \leq \delta \leq (\ell/AE)N_o$. If these checks reveal
an inconsistency, new assumptions must be made. It will be
necessary only to change elements of $\{<N_o>\}$ to implement the
revised assumptions.

In the rate problem we define the extension rate vector $\{\dot{\delta}\}$,
the displacement rate vector $\{\dot{u}\}$, the axial force rate vector
$\{\dot{N}\}$ and the load rate vector $\{\dot{P}\}$. Equations (22) and (23) hold
in the rate form, giving

$$\{\dot{\delta}\} = [B]\{\dot{u}\} \ , \tag{45a}$$

$$\{\dot{P}\} = [A]\{\dot{N}\} = [B]^T\{\dot{N}\} \ . \tag{45b}$$

If the behavior of each bar is elastic, from equations (28) and
(29) we see that

$$\dot{N} - \frac{AE}{\ell} \dot{\delta} \ , \tag{46a}$$

$$\{\dot{N}\} = [S]\{\dot{\delta}\} \ , \tag{46b}$$

where the member stiffness matrix $[S]$ is defined as before.

Let us consider a linear kinematically hardening material.
In minimizing $\dot{U}_p^o$ we must formulate $\dot{W}^o$ (Table 3) for each bar.
If the bar is elastic $(-N_o +AE_p \delta^p/\ell \leq N \leq N_o +AE_p \delta^p/\ell)$, $\dot{W}^o$ is unambiguously defined. If on the other hand the bar is plastic
$(N = \pm N_o)$ we must guess whether loading or unloading is taking
place. We put

$$N = \frac{\partial W^o}{\partial \dot{\delta}} = \frac{AE}{\ell} \dot{\delta} \tag{47a}$$

if we elect to assume that unloading occurs, and

$$\dot{N} = \frac{\partial \dot{W}^{O}}{\partial \dot{\delta}} = \frac{AE_{T}}{\ell}\,\dot{\delta} \tag{47b}$$

if loading is assumed.

In view of the linearity of these equations, on the basis of our choice we may put

$$\{\dot{N}\} = [\bar{S}^{O}]\{\dot{\delta}\} \quad, \tag{48}$$

where $[\bar{S}^{O}]$ is again a diagonal matrix, with the diagonal elements consisting of the appropriate values of (AE/ℓ) or (AE_{T}/ℓ). As before, we may define a system stiffness matrix

$$[\bar{K}^{O}] = [B]^{T}[\bar{S}^{O}][B] \quad, \tag{49}$$

$$\{\dot{u}\} = [\bar{K}^{O}]^{-1}\{\dot{P}\} \quad. \tag{50}$$

The extension rates in each plastic bar must be checked to ensure that the assumption of loading or unloading is justified; if not, the member stiffness matrix $[\bar{S}^{O}]$ must be modified.

The member stiffness matrix, and hence the system stiffness matrix, must be reformulated for each successive rate problem in the complete incremental solution. Care must also be taken to ensure that in a discrete load increment the stress in an elastic bar which is close to yielding does not at the end of the increment exceed the yield stress.

In the alternative approach to the rate problem, involving the minimization of $\dot{U}_{p}^{O}$, it is again convenient to define plastic extension rates for every bar in the truss. Let this vector be $\{\dot{\delta}^{p}\}$. We may then express $\dot{W}^{Op}$ as

$$\dot{W}^{Op} = \frac{1}{2}\,\{\dot{\delta}^{p}\}^{T}[\bar{S}^{Op}]\{\dot{\delta}^{p}\} \quad, \tag{51}$$

where $[\bar{S}^{Op}]$ is a diagonal matrix whose diagonal elements are the appropriate values of (AE_p/ℓ). It follows then that we may write

$$\dot{U}_p^o = \frac{1}{2}\{\dot{\delta}^p\}^T[\bar{S}^{op}]\{\dot{\delta}^p\} + \frac{1}{2}(\{\dot{\delta}\}^T - \{\dot{\delta}^p\}^T)[S](\{\dot{\delta}\} - \{\dot{\delta}^p\})$$

$$- \{\dot{u}\}^T\{\dot{P}\} \quad . \tag{52}$$

Proceeding as if all bars are plastic, we minimize this expression with respect to the elements of $\{\dot{u}\}$ and $\{\dot{\delta}^p\}$. Using the same arguments which led to equation (43) in the deformation theory problem, we find that the minimum of this unconstrained problem is given by

$$[\bar{K}^*]\{\dot{u} \mid \dot{\delta}^p\} = \{\dot{P} \mid 0\} \quad , \tag{53}$$

where

$$[\bar{K}^*] = \begin{bmatrix} [B]^T[S][B] & \vdots & - [B]^T[S] \\ \hline - [S][B] & \vdots & [S]+[\bar{S}^{op}] \end{bmatrix} \quad . \tag{54}$$

The system matrix $[\bar{K}^*]$ is a square $(m+n) \times (m+n)$ symmetric matrix, and is defined for *all* rate problems for the structure.

To determine the solution of any one rate problem we proceed as follows. We identify all the elastic bars in the truss, and those plastic bars in which we assume that unloading will occur. The rows and columns of $[\bar{K}^*]$ corresponding to the elements of $\{\dot{\delta}^p\}$ which are thus zero or assumed zero are eliminated. The system matrix is then inverted to provide $\{\dot{u}\}$ and the remaining elements of $\{\dot{\delta}^p\}$. The solution will be correct if the signs of the non-zero elements of $\{\dot{\delta}^p\}$ are consistent (i.e. positive if the bar is yielding in tension and negative if the bar is yielding in compression) and the signs of the elements of $\{\dot{N}\}$ corresponding to unloading plastic bars are also consistent (i.e. negative if $N = +N_o$ and positive if $N = -N_o$). If this is not so, the assumption regarding the identification of plastic bars which are unloading must be revised, and the process repeated.

26.3 The Plane Stress Problem

We may now proceed to formulate the finite element analysis of
continuum problems as generalizations of the truss problem. The
plane stress problem will be treated as a specific example.

The discretization of the continuum problem is accomplished
by dividing the body into regions, or finite elements, within
which we make simplifying assumptions about the displacement
field. In the kinematic approach we must also ensure that the
approximate displacement field is continuous over the whole
body, implying that there must be continuity of displacements
along the common boundaries of adjacent elements.

The simplest element employed in the plane stress problem is
the constant strain triangular element. Consider a generic element, as shown in Figure 3, with apices having coordinates $(x_1,
y_1)$, (x_2, y_2), (x_3, y_3). We require that the strains

$$\varepsilon_x = \frac{\partial u}{\partial x} \quad , \quad \varepsilon_y = \frac{\partial v}{\partial y} \quad ,$$

$$\varepsilon_{xy} = \frac{\partial u}{\partial y} + \frac{\partial v}{\partial x} \quad , \tag{55}$$

should be constant over the element. Thus the displacement
field u,v must have the form

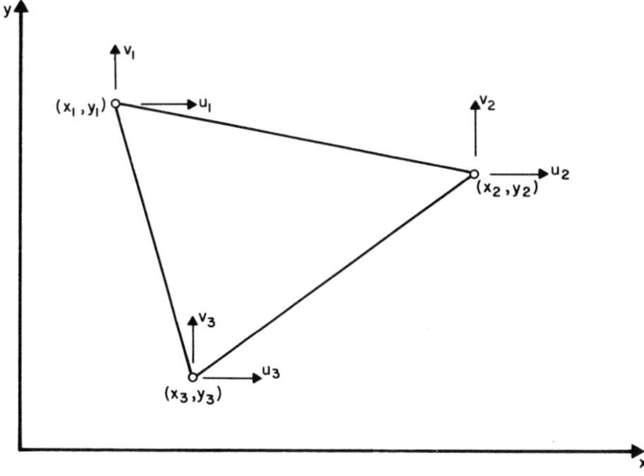

Figure 3. Triangular plane stress element

$$u = \varepsilon_x \, x + \frac{1}{2}(\varepsilon_{xy} + a)y + b \quad ,$$

$$v = \frac{1}{2}(\varepsilon_{xy} - a)x + \varepsilon_y \, y + c \quad ,$$

$$(56)$$

where a, b, c are constants.

Let the displacements of the apices of the element, or the nodes, be respectively (u_1, v_1), (u_2, v_2), u_3, v_3). From equations (56) we see that

$$
\begin{bmatrix} u_1 \\ v_1 \\ u_2 \\ v_2 \\ u_3 \\ v_3 \end{bmatrix}
=
\begin{bmatrix}
x_1 & 0 & \frac{1}{2}y_1 & \frac{1}{2}y_1 & 1 & 0 \\
0 & y_1 & \frac{1}{2}x_1 & -\frac{1}{2}x_1 & 0 & 1 \\
x_2 & 0 & \frac{1}{2}y_2 & \frac{1}{2}y_2 & 1 & 0 \\
0 & y_2 & \frac{1}{2}x_2 & -\frac{1}{2}x_2 & 0 & 1 \\
x_3 & 0 & \frac{1}{2}y_3 & \frac{1}{2}y_3 & 1 & 0 \\
0 & y_3 & \frac{1}{2}x_3 & -\frac{1}{2}x_3 & 0 & 1
\end{bmatrix}
\begin{bmatrix} \varepsilon_x \\ \varepsilon_y \\ \varepsilon_{xy} \\ a \\ b \\ c \end{bmatrix}
\qquad (57)
$$

These equations may be inverted to give

$$
\begin{bmatrix} \varepsilon_x \\ \varepsilon_y \\ \varepsilon_{xy} \\ a \\ b \\ c \end{bmatrix}
= [B^*]
\begin{bmatrix} u_1 \\ v_1 \\ u_2 \\ v_2 \\ u_3 \\ v_3 \end{bmatrix}
=
\begin{bmatrix} B_e \\ \hline B_r \end{bmatrix}
\begin{bmatrix} u_1 \\ v_1 \\ u_2 \\ v_2 \\ u_3 \\ v_3 \end{bmatrix} \quad ,
\qquad (58)
$$

and hence

$$
\begin{bmatrix} \varepsilon_x \\ \varepsilon_y \\ \varepsilon_{xy} \end{bmatrix} = [B_e] \{u_1 \ v_1 \ u_2 \ v_2 \ u_3 \ v_3\}^T \ . \tag{59}
$$

We have thus expressed the strains in the element in terms of the displacements at the nodes.

Since the displacement field in the element is linear, it follows that the displacements along any one side depend only on the displacements of the bounding nodes. Hence if we divide any sheet into triangular elements, with nodes occurring only at the apices of any one triangular element, the displacement field will be continuous across the boundaries of adjacent elements. A typical subdivision is shown in Figure 4.

It is convenient to introduce for each element a weighted strain defined by

$$
\begin{bmatrix} \delta_x \\ \delta_y \\ \delta_{xy} \end{bmatrix} = \Delta_e \begin{bmatrix} \varepsilon_x \\ \varepsilon_y \\ \varepsilon_{xy} \end{bmatrix} , \tag{60}
$$

where Δ_e is the area of the element. Thus

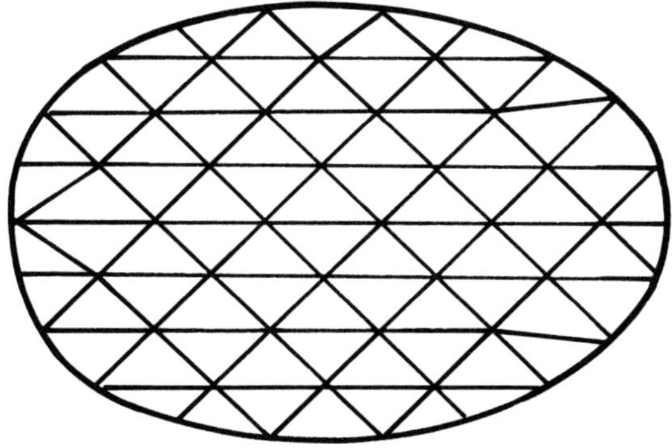

Figure 4. Finite element net

$$
\begin{bmatrix} \delta_x \\ \delta_y \\ \delta_{xy} \end{bmatrix} = \Delta_e \, [B_e] \, \{u_1 \ v_1 \ u_2 \ v_2 \ u_3 \ v_3\}^T \ .
\tag{61}
$$

Consider now the assembly of elements. Ordering the nodes and considering the x and y components in turn, we may define a displacement vector $\{u\}$. Similarly, ordering the elements and taking δ_x, δ_y, δ_{xy} in turn, we may define a weighted strain vector $\{\delta\}$. Using the relation between the weighted strains and the node displacements for each element, given by equation (61) we may assemble a deformation matrix for the assembly of elements

$$
\{\delta\} = [B]\{u\} \ .
\tag{62}
$$

Returning to an individual element, the stress components σ_x, σ_y, σ_{xy} will be constant over the element. The constitutive equations for an isotropic elastic material provide

$$
\begin{bmatrix} \sigma_x \\ \sigma_y \\ \sigma_{xy} \end{bmatrix} = \frac{E}{(1-\nu^2)} \begin{bmatrix} 1 & \nu & 0 \\ \nu & 1 & 0 \\ 0 & 0 & \frac{(1-\nu)}{2} \end{bmatrix} \begin{bmatrix} \varepsilon_x \\ \varepsilon_y \\ \varepsilon_{xy} \end{bmatrix}
$$

$$
= \frac{E}{\Delta_e(1-\nu^2)} \begin{bmatrix} 1 & \nu & 0 \\ \nu & 1 & 0 \\ 0 & 0 & \frac{(1-\nu)}{2} \end{bmatrix} \begin{bmatrix} \delta_x \\ \delta_y \\ \delta_{xy} \end{bmatrix} \ .
\tag{63}
$$

Ordering the stress components for individual elements in the same manner as the strain components, and defining a stress vector $\{\sigma\}$, we may use equation (63) for individual elements to assemble an element stiffness matrix $[S]$;

$$\{\sigma\} = [S]\{\delta\} \quad . \tag{64}$$

The matrix $[S]$ will be symmetric, and will be composed of 3×3 submatrices along a diagonal band.

The strain energy of any element can be now written as

$$W_e = \frac{\Delta_e}{2} (\varepsilon_x \ \varepsilon_y \ \varepsilon_{xy}) \begin{bmatrix} \sigma_x \\ \sigma_y \\ \sigma_{xy} \end{bmatrix} \tag{65}$$

$$= \frac{1}{2} (\delta_x \ \delta_y \ \delta_{xy}) \begin{bmatrix} \sigma_x \\ \sigma_y \\ \sigma_{xy} \end{bmatrix} \quad .$$

The strain energy of the entire assembly is then

$$W = \frac{1}{2} \{\delta\}^T \{\sigma\}$$

$$= \frac{1}{2} \{\delta\}^T [S] \{\delta\} \tag{66}$$

$$= \frac{1}{2} \{u\}^T [B]^T [S] [B] \{u\}$$

$$= \frac{1}{2} \{u\}^T [K] \{u\} \quad ,$$

where $[K] = [B]^T [S] [B]$ is the system stiffness matrix.

Displacements on the boundary of the plane sheet can be expressed as linear functions of the nodes on the boundary. The potential energy of the tractions acting on the boundary can thus be written in terms of the boundary node displacements. We shall not develop these expressions in detail, but simply note that this potential energy will take the form $\{u\}^T \{P\}$, where the elements of the load vector $[P]$ are calculated by equating

$$\{u\}^T\{P\} = \int_S \underline{p} \cdot \underline{u} \, dS \quad . \tag{67}$$

The potential energy of the body can now be written as

$$U_p = \frac{1}{2}\{u\}^T[K]\{u\} - \{u\}^T\{P\} \quad . \tag{68}$$

The least value of this expression is given by

$$[K]\{u\} = \{P\} \quad ,$$
$$\{u\} = [K]^{-1}\{P\} \quad . \tag{69}$$

Since it will only fortuitously occur that the exact displacement field is contained within the class of displacement fields defined by $\{u\}$, equations (69) will not yield the absolute minimum value of U_p, but merely the member of the class $\{u\}$ which most closely approaches this minimum. However, it is apparent that we can approach the absolute minimum value of U_p as closely as we wish by refining the finite element net. The approximation is thus consistent, and will converge on the correct solution as we reduce the element size.

In extending the formulation to the deformation theory, we shall consider for simplicity only the minimization of $\bar{U}_p^o$ for an elastic, perfectly plastic material. Let us adopt the von Mises yield condition,

$$\phi(\sigma_x, \sigma_y, \sigma_{xy}) = \sigma_x^2 - \sigma_x\sigma_y + \sigma_y^2 + 3\sigma_{xy}^2 - \sigma_o^2 \quad . \tag{70}$$

In the case of flow, for stress states satisfying the initial yield condition and for neutral loading,

$$\dot{\varepsilon}_x^p = \lambda(2\sigma_x - \sigma_y) \quad ,$$

$$\dot{\varepsilon}_y^p = \lambda(-\sigma_x + 2\sigma_y) \quad , \tag{71}$$

$$\dot{\varepsilon}_{xy}^p = \lambda(6\sigma_{xy}) \quad .$$

Inverting these equations, and substituting into equation (70) to eliminate λ,

$$\sigma_x = \frac{\sigma_o}{k\sqrt{3}}(2\dot{\varepsilon}_x^p + \dot{\varepsilon}_y^p) \quad ,$$

$$\sigma_y = \frac{\sigma_o}{k\sqrt{3}}(\dot{\varepsilon}_x^p + 2\dot{\varepsilon}_y^p) \quad , \tag{72}$$

$$\sigma_{xy} = \frac{\sigma_o}{2k\sqrt{3}}\dot{\varepsilon}_{xy}^p \quad ,$$

where

$$k = \sqrt{(\dot{\varepsilon}_x^{p^2} + \dot{\varepsilon}_x^p\dot{\varepsilon}_y^p + \dot{\varepsilon}_y^{p^2} + \tfrac{1}{4}\dot{\varepsilon}_{xy}^{p^2})} \quad .$$

The dissipation function $D(\dot{\varepsilon}_x^p, \dot{\varepsilon}_y^p, \dot{\varepsilon}_{xy}^p)$ is

$$D = \frac{2\sigma_o}{\sqrt{3}}\sqrt{(\dot{\varepsilon}_x^{p^2} + \dot{\varepsilon}_x^p\dot{\varepsilon}_y^p + \dot{\varepsilon}_y^{p^2} + \tfrac{1}{4}\dot{\varepsilon}_{xy}^{p^2})} \quad . \tag{73}$$

For a perfectly plastic material the minimum work path involves plastic deformation at constant stress, and

$$W^{op}(\varepsilon_x^p, \varepsilon_y^p, \varepsilon_{xy}^p) = D(\varepsilon_x^p, \varepsilon_y^p, \varepsilon_{xy}^p) \quad . \tag{74}$$

Following the development in the truss problem in Section 26.2, it is convenient to express W^{op} in terms of a product of the plastic strain and an assumed stress. Let us then assume that the stress in any one element is σ_x^o, σ_y^o, σ_{xy}^o, with $\phi(\sigma_x^o, \sigma_y^o, \sigma_{xy}^o) = 0$. In this element we then put

$$W^{op} = \sigma_x^o\varepsilon_x^p + \sigma_y^o\varepsilon_y^p + \sigma_{xy}^o\varepsilon_{xy}^p \quad . \tag{75}$$

In view of the fact that stress and strain are constant over the element

$$\int_{\text{element}} W^{op}dV = \Delta_e W^{op}$$

$$= \sigma_x^o \delta_x^p + \sigma_y^o \delta_y^p + \sigma_{xy}^p \delta_{xy}^p \quad , \tag{76}$$

where

$$\begin{bmatrix} \delta_x^p \\ \delta_y^p \\ \delta_{xy}^p \end{bmatrix} = \Delta_e \begin{bmatrix} \varepsilon_x^p \\ \varepsilon_y^p \\ \varepsilon_{xy}^p \end{bmatrix} . \tag{77}$$

If we then form the column vectors $\{\delta^p\}$ and $\{\sigma^o\}$ for the entire assembly, we see that

$$\bar{U}_p^o = \{\delta^p\}^T \{\sigma^o\} + \frac{1}{2} (\{\delta\} - \{\delta^p\})^T [S](\{\delta\} - \{\delta^p\}) - \{u\}^T \{P\} \quad . \tag{78}$$

Using precisely the same argument which followed equation (42), we may write

$$[K^*]\{u \mid \delta^p\} = \{P \mid \sigma^o\} \quad . \tag{79}$$

The system stiffness matrix is again

$$[K^*] = \begin{bmatrix} [B]^T[S][B] & \mid & -[B]^T[S] \\ -[S][B] & \mid & [S] \end{bmatrix} . \tag{80}$$

The solution may now proceed as follows. On the basis, say, of an elastic solution, certain elements are assumed to be plastic. The stresses in these elements are also assumed; it may for example be convenient to choose the stress states at the intersection of the elastic stress vector and the yield surface. These choices are used to make up $\{\sigma^o\}$. In the remaining elements δ_x^p, δ_y^p, δ_{xy}^p are assumed to be zero, and the corresponding equations are eliminated from equations (79). The modified stiffness matrix $[K^*]$ is inverted, and the values of $\{u\}$ and the non-zero elements of $\{\delta^p\}$ are found. From equations

(62) and (63), we may then find

$$\{\sigma\} = [S](\{\delta\} - \{\delta^p\}) \ . \tag{81}$$

Each element is then reexamined. If $\phi(\sigma_x, \sigma_y, \sigma_{xy}) < 0$ the element is taken to be elastic, and δ_x^p, δ_y^p, δ_{xy}^p are now assumed zero. If $\phi(\sigma_x, \sigma_y, \sigma_{xy}) \geq 0$, the weighted plastic strains are retained. New choices are made for the stresses $\{\sigma^o\}$, again based say on the intersection of the stress vector and the yield surface. A further iteration is carried out, and the process is continued until a satisfactory solution is obtained. A number of devices may be employed to speed convergence, but the basic prodecure can be seen as a generalization of that for the truss problem.

In treating the incremental theory problem, let us treat a linear kinematic hardening material. Let the yield function be

$$\phi = (\sigma_x - h\varepsilon_x^p)^2 - (\sigma_x - h\varepsilon_x^p)(\sigma_y - h\varepsilon_y^p) + (\sigma_y - h\varepsilon_y^p)^2$$
$$+ 3(\sigma_{xy} - h\varepsilon_{xy}^p)^2 - \sigma_o^2 \ , \tag{82}$$

where h is a constant.

We assume that in each element we know at the beginning of the increment the stresses, which we shall write as

$$\{\sigma\}_e = \{\sigma_x \ \sigma_y \ \sigma_{xy}\}^T \ , \tag{83}$$

and the plastic strains, which will be given as

$$\{\varepsilon^p\}_e = \{\varepsilon_x^p \ \varepsilon_y^p \ \varepsilon_{xy}^p\}^T = \frac{1}{\Delta_e} \{\delta_x^p \ \delta_y^p \ \delta_{xy}^p\}^T \ . \tag{84}$$

If $\phi < 0$ in the element, equation (63) holds. We shall put

$$\{\dot{\sigma}\}_e = [S]_e\{\dot{\delta}\}_e \ , \tag{85a}$$

where $\ \{\delta\}_e = \{\delta_x \ \delta_y \ \delta_{xy}\}^T \ ,$ \hfill (85b) and

$$[S]_e = \frac{E}{\Delta_e(1-\nu^2)} \begin{bmatrix} 1 & \nu & 0 \\ \nu & 1 & 0 \\ 0 & 0 & \frac{1-\nu}{2} \end{bmatrix} . \tag{85c}$$

If $\phi = 0$ in the element, we assume either that unloading occurs, in which case equation (85a) holds, or that loading occurs.

In the latter case, in the notation we have so far used,

$$\dot{q}_j^p = \frac{1}{h \frac{\partial\phi}{\partial Q_p}\frac{\partial\phi}{\partial Q_p}} \frac{\partial\phi}{\partial Q_j}\frac{\partial\phi}{\partial Q_k}\dot{Q}_k . \tag{86}$$

To write this equation in matrix form, it is convenient to introduce the column vector

$$\{\frac{\partial\phi}{\partial\sigma}\} = \begin{bmatrix} \frac{\partial\phi}{\partial\sigma_x} \\ \frac{\partial\phi}{\partial\sigma_y} \\ \frac{\partial\phi}{\partial\sigma_{xy}} \end{bmatrix} = \begin{bmatrix} 2(\sigma_x - h\varepsilon_x^p) - (\sigma_y - h\varepsilon_y^p) \\ -(\sigma_x - h\varepsilon_x^p) + 2(\sigma_y - h\varepsilon_y^p) \\ 6(\sigma_{xy} - h\varepsilon_{xy}^p) \end{bmatrix} . \tag{87}$$

We see then that

$$\{\dot{\delta}^p\}_e = \frac{\{\frac{\partial\phi}{\partial\sigma}\}\{\frac{\partial\phi}{\partial\sigma}\}^T\{\dot{\sigma}\}_e}{\frac{h}{\Delta_e}\{\frac{\partial\phi}{\partial\sigma}\}^T\{\frac{\partial\phi}{\partial\sigma}\}} . \tag{88}$$

The inverted constitutive relations have been formerly written as (see Chapter 25, equation 17)

$$\dot{Q}_j = D_{jk}\dot{q}_k - D_{jk}\frac{\partial\phi}{\partial Q_k}\;\frac{D_{\ell m}\frac{\partial\phi}{\partial Q_\ell}\dot{q}_m}{[h\,\frac{\partial\phi}{\partial Q_p}\frac{\partial\phi}{\partial Q_p} + D_{rs}\frac{\partial\phi}{\partial Q_r}\frac{\partial\phi}{\partial Q_s}\,]}\quad . \tag{89}$$

In matrix form, this becomes

$$\{\dot{\sigma}\}_e = [S]_e\{\dot{\delta}\}_e - \frac{[S]_e\{\frac{\partial\phi}{\partial\sigma}\}\{\frac{\partial\phi}{\partial\sigma}\}^T[S]_e\{\dot{\delta}\}_e}{\frac{h}{\Delta_e}\{\frac{\partial\phi}{\partial\sigma}\}^T\{\frac{\partial\phi}{\partial\sigma}\} + \{\frac{\partial\phi}{\partial\sigma}\}^T[S]_e\{\frac{\partial\phi}{\partial\sigma}\}}$$

$$= \left[\,[I] - \frac{[S]_e\{\frac{\partial\phi}{\partial\sigma}\}\{\frac{\partial\phi}{\partial\sigma}\}^T}{\frac{h}{\Delta_e}\{\frac{\partial\phi}{\partial\sigma}\}^T\{\frac{\partial\phi}{\partial\sigma}\} + \{\frac{\partial\phi}{\partial\sigma}\}^T[S]_e\{\frac{\partial\phi}{\partial\sigma}\}}\,\right][S]_e\{\dot{\delta}\}_e \quad . \tag{90}$$

In this equation $[I]$ is a 3x3 unit matrix.

Considering now the assembly of elements, on the basis of the value of ϕ and the assumption of loading or unloading in elements where $\phi = 0$, we may form

$$\{\dot{\delta}\} = [\bar{S}^O]\{\dot{\delta}\} \quad . \tag{91}$$

The matrix $[\bar{S}^O]$ will be symmetric, and is composed of 3x3 sub-matrices along a diagonal band. The system stiffness matrix now follows as

$$[\bar{K}^O] = [B]^T[\bar{S}^O][B] \quad , \tag{92}$$

and

$$\{\dot{u}\} = [\bar{K}^O]^{-1}\{\dot{P}\} \quad . \tag{93}$$

In those elements where $\phi = 0$, the sign of

$$\frac{\partial\phi}{\partial Q_j}\dot{Q}_j = \{\frac{\partial\phi}{\partial\sigma}\}^T\{\dot{\sigma}\}_e \tag{94}$$

must be checked in order to confirm the assumption made initially. When the assumptions are satisfied, the stresses $\{\sigma\}$
and the plastic strains $\{\varepsilon^p\}$ are updated and the next increment
may be treated. Care must be taken to ensure that this process
does not lead to stress states and plastic strains which do not
satisfy the yield condition, but the principle behind the process is again clearly seen as a generalization of the treatment
of the truss problem.

In the alternative approach to the incremental theory prob-
lem we must formulate an expression for $\dot{\overline{U}}{}^{o}_{p}$. This is somewhat
simpler than the expression for $\overline{U}{}^{o}_{p}$, since we may make use of
the restriction

$$\dot{q}^p_j = \lambda \, \frac{\partial \phi}{\partial Q_j} \quad , \tag{95}$$

and put

$$\dot{W}^{op} = \frac{1}{2} \frac{\lambda^2}{G} = \frac{1}{2} \, h \, \frac{\partial \phi}{\partial Q_j} \frac{\partial \phi}{\partial Q_j} \lambda^2 \tag{96}$$

To obtain a suitable matrix form let us put

$$\{\dot{\delta}^p\}_e = \Delta_e \lambda \{\tfrac{\partial \phi}{\partial \sigma}\} = \Lambda \{\tfrac{\partial \phi}{\partial \sigma}\} \quad , \tag{97}$$

where $\Lambda = \Delta_e \lambda$ is a weighted plastic strain rate multiplier in
an individual element. We may also form the column vector $\{\Lambda\}$,
being the ordered multipliers for the assembly of elements.
The relation between $\{\delta^p\}$ and $\{\Lambda\}$ may then be written as

$$\{\dot{\delta}^p\} = [T]\{\Lambda\} \quad . \tag{98}$$

The matrix $[T]$ has in its first column the submatrix $\{\partial \phi / \partial \sigma\}$
for the first element appearing in the first three rows, in
its second column the submatrix $\{\partial \phi / \partial \sigma\}$ for the second element
in the fourth to the sixth rows, and so on. All other elements
of $[T]$ are zero.

Let us also introduce a diagonal matrix $[\bar{S}^{op}]$ whose non-zero elements consist of the terms $h\{\frac{\partial\phi}{\partial\sigma}\}^T\{\frac{\partial\phi}{\partial\sigma}\}/\Delta_e$. Assuming that all elements undergo plastic deformation, it follows that

$$\overset{\bullet}{U}{}^o_p = \frac{1}{2}\{\Lambda\}^T[\bar{S}^{op}]\{\Lambda\} + \frac{1}{2}(\{\delta\}^T - \{\Lambda\}^T[T]^T)[S](\{\delta\} - [T]\{\Lambda\})$$
$$- \{\overset{\bullet}{u}\}^T\{\overset{\bullet}{P}\} \quad . \tag{99}$$

Setting the partial derivatives with respect to the elements of $\{\overset{\bullet}{u}\}$ and $\{\Lambda\}$ equal to zero, we find by similar arguments to those we have employed previously,

$$[\bar{K}^*]\{\overset{\bullet}{u} \mid \Lambda\} = \{\overset{\bullet}{P} \mid 0\} , \tag{100}$$

where

$$[\bar{K}^*] = \left[\begin{array}{c|c} [B]^T[S][B] & -\ [B]^T[S][T] \\ \hline -[T]^T[S][B] & [T]^T[S][T] + [\bar{S}^{op}] \end{array}\right] . \tag{101}$$

To determine the solution to any one rate problem we proceed as follows. We identify all elements in which $\phi < 0$ and those elements in which $\phi = 0$ and in which we suspect that unloading will occur. The rows and columns of $[\bar{K}^*]$ corresponding to the elements of $\{\Lambda\}$ which are thus zero or assumed zero are eliminated. The system matrix is then inverted to find $\{\overset{\bullet}{u}\}$ and the remaining elements of $\{\Lambda\}$. The solution will be correct if on one hand elements of $\{\Lambda\}$ are non-negative where they were assumed to be non-zero, and on the other the product $\{\partial\phi/\partial\sigma\}^T\{\overset{\bullet}{\sigma}\}_e$ is non-positive in elements which are plastic but assumed to be unloading. $\{\overset{\bullet}{\sigma}\}_e$ is found from equation (84a). If these conditions are not met, the assumptions regarding the identification of unloading elements must be revised and the process repeated.

In passing from one incremental problem to another the system matrix $[\bar{K}^*]$ must in part be reformulated, since the elements of $[T]$ and $[\bar{S}^{op}]$ depend on $\{\partial\phi/\partial\sigma\}$ and hence the current stress and plastic strain.

26.4 Bibliographical and Historical Remarks

It should be emphasized that in the development of the numerical methods of analysis in this Chapter attention has been devoted to the generalisation of principles from a simple example with the purpose of demonstrating concepts rather than of concern with efficient computational schemes. In particular, more efficient methods of assembling system stiffness matrices can be given. The reader is referred to the texts by Zienkiewicz [1971] and Desai and Abel [1972], for example, for a general outline of the finite element method.

Initial work in the finite element analysis of elastic, plastic problems made use of the incremental form of the stress, total strain relations. Representative papers are those of Argyris [1965], Marcal and King [1967], Zienkiewicz, Valliappan and King [1969], Argyris and Scharpf [1969] and Zienkiewicz and Valliappan [1971].

More recently alternative computational schemes have been investigated, particularly non-linear programming techniques. Representatives papers are those by Hodge, Belytschko and Herakovich [1969], Capurso and Maier [1970], Belytschko and Velebit [1972], Maier [1972], Sayegh and Rubinstein [1972] and Hodge [1973].

References

J.H.Argyris   1965   "Elasto-plastic matrix displacement analysis of three-dimensional continua", J.Roy.Aero.Soc., $\underline{69}$, 633.

J.H.Argyris and   1969   "Methods of elasto-plastic analysis", D.W.Scharpf   *Symposium on Finite Element Techniques*, Stuttgart.

T.Belytschko and   1972   "Finite element method for elastic-plastic plates", J.Eng.Mech.Div., ASCE, M.Velebit $\underline{98}$ (EM1), 227.

M.Capurso and   1970   "Incremental elasto-plastic analyses G.Maier   and quadratic programming", Meccanica,$\underline{5}$.

C.S.Desai and   1972   *Introduction to the Finite Element*
J.F.Abel   *Method*, Van Nostrand Reinhold.

P.G.Hodge   1973   "Complete solutions for elastic-plastic
                             trusses", SIAM J.Appl.Math., $\underline{25}$, 435.

P.G.Hodge,   1969   "Plasticity and quadratic programming,
T.Belytschko and   *"Proc. on Computational Approaches in*
C.Herakovich   *Applied Mechanics*, ASME, 73.

G.Maier   1972   "Mathematical programming methods in
                             structural mechanics", *Intern.Symposium*
                             on Variational Methods in Engineering,
                             Southampton.

P.V.Marcal and   1967   "Elastic-plastic analysis of two di-
I.P.King   mensional stress systems by the finite
                             element method", Int.J.Mech.Sci., $\underline{9}$,
                             143.

A.F.Sayegh and   1972   "Elastic-plastic analysis by quadratic
M.Rubinstein   programming", Proc.Eng.Mech.Div., ASCE,
                             $\underline{98}$ (EM6), 1547.

O.C.Zienkiewicz   1971   *The Finite Element Method in Engineering*
                             Science, McGraw-Hill.

O.C.Zienkiewicz,   1969   "Elasto-plastic solutions of engineer-
S.Valliappan and   ing problems. Initial stress, finite
I.P.King   element approach", Int.J.Num.Meth. in
                             Engineering, $\underline{1}$, 75.

O.C.Zienkiewicz   1971   "Analysis of real structures for creep,
and   plasticity and other complex constitu-
S.Valliappan   tive laws", *Structure, Solid Mechanics*
                             and Engineering Design (edited by M.
                             Te'eni), Wiley-Interscience, 27.

PART VI

THERMODYNAMIC FORMULATION OF THE CONSTITUTIVE EQUATIONS

BASIC CONCEPTS IN THERMODYNAMICS

27.1 Introductory Remarks

In principle the constitutive equations of inelastic solids are quantitative descriptions of deformation mechanisms which operate on the microstructural scale. In recent years the framework of plastic constitutive relations has been reformulated on the basis of thermodynamic potential functions which include internal variables. This development has provided a more rigorous physical basis to the classical theory, and promises further progress in deriving a quantitative relation between macroscopic deformation and the microstructural behavior of polycrystalline metals and other solids.

In Part VI we aim to explore some of the implications of the thermodynamic framework. Chapter 27 is devoted to a brief review of those basic concepts of thermodynamics which are required for the discussion. Thereafter we shall consider elastic behavior from the thermodynamic viewpoint, and expand the framework to include internal thermodynamic variables. In keeping with the scope of this volume we shall limit our discussion to infinitesimal strains and isothermal behavior, although these restrictions are by no means necessary.

27.2 Characterization of Equilibrium States

The starting point in classical thermodynamic analysis is the concept of a *system*. A system is a set of specified material particles contained within a *boundary*. The boundary may be real, in the sense of a containing vessel or an interface between one material and another, or it may be hypothetical in the sense that we arbitrarily isolate an element of material from a larger body in a discussion of mechanical behavior. The system is *closed* when matter does not cross the boundary. The fundamental principles of thermodynamics deal with closed

systems, and it will not be necessary for us to discuss systems
which are not closed.

We assume the existence of *equilibrium states* characterized
completely by the *internal energy* u of the system and by independent *thermodynamic variables* x_j (j = 1,...,n). The number
of independent thermodynamic variables depends on the nature of
the system. Because of the limited applications for which we
shall use the concepts of thermodynamics, the thermodynamic
variables x_j can be conceived of as displacement variables. The
internal energy of a system can be changed only by adding heat
or, in what will be sufficient for our purposes, doing mechanical work on the boundaries of the system. The conservation
law, which is usually written in incremental form, requires
that the increase in internal energy du should be equal to the
sum of the heat crossing the boundary dQ and the mechanical
work done on the system dW. Figure 1 shows diagrammatically
the sign convention for dQ and dW which we shall adopt; this
convention agrees with common practice in continuum mechanics
rather than in thermodynamics; thus

$$du = dQ + dW \ . \tag{1}$$

Equation (1) is generally referred to as the *first law of thermodynamics.*

A system is said to be *isolated* if it is such that no heat

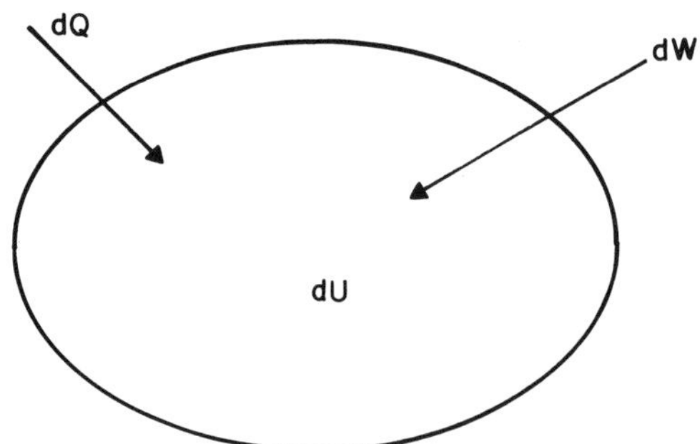

Figure 1. Sign convention for first law

can pass through the boundary of the system and no mechanical
work can be done on it. Alternatively, an isolated system is
enclosed by a *rigid, adiabatic boundary*. Clearly for an isolated system dQ = dW = 0 under all circumstances, and du = 0.
Consequently *the internal energy of an isolated system remains
constant*.

Two or more systems may interact with each other; the study
of such interactions is indeed the essential problem under consideration. It is convenient to separate a set of interacting
systems with which we are concerned from other systems with
which they do not have any mechanical or thermal interaction by
conceiving of the interacting systems as *subsystems of an isolated system*. Such subsystems may be distinct from each other
or they may occupy the same space; in the latter case they are
said to be *coexisting*.

Constraints may exist between subsystems of an isolated system. Such constraints may prevent thermal interactions (adiabatic walls) or mechanical interaction by fixing the deformation variables in various subsystems (rigid walls). Classical
thermodynamics is concerned with the equilibrium state associated with any given set of internal constraints in an isolated
system. It is assumed that any isolated system will reach an
equilibrium state after the passage of sufficient time; in the
equilibrium state the constituent subsystems will have ceased
to interact with one another.

When some or all of the internal constraints in an isolated
system are changed, the state of the system will change and the
system is said to undergo a *process*. Eventually the system
will reach a new equilibrium state; the determination of this
new equilibrium state is the fundamental problem in classical
thermodynamics. A change from one set of constraints to another
can often be conceived as occurring in infinitesimally small

steps, the isolated system achieving an equilibrium state after each step. Such an idealized process, in which *the system passes through a continuous sequence of equilibrium states*, is termed a *quasi-static process*, although it need not necessarily be slow.

The internal energy of a system is the sum of the internal energies of the constituent subsystems. It follows then that when an isolated system undergoes a process the sum of the internal energies of the constituent subsystems remains constant. The internal energy of any one subsystem may of course change.

When an isolated system undergoes a process initiated by the relaxation of some or all of the internal constraints the resulting equilibrium state can be characterized by a maximum principle. The quantity which achieves its maximum value in the equilibrium state is called the *entropy*. In an isolated system (in which the internal energy is constant) the entropy depends on the displacement variables of the system. Denoting the entropy of the isolated system by s,

$$s = s(x_j) \ . \tag{2}$$

The entropy maximum principle then states that in the equilibrium state which terminates a process the *unconstrained displacement parameters* take on values which render the entropy s a maximum.

Two implicit statements are contained in this maximum principle. Since at the initiation of the process, at the instant at which constraints are removed, the system is not in equilibrium and the entropy does not have the largest value compatible with the unconstrained displacement parameters, the entropy increases during a real process in an isolated system. A process in which the entropy increases is termed *irreversible*; all real processes are irreversible. Further, since we are dealing with an isolated system, it is implicit that the process takes place at constant internal energy.

27.3 The Fundamental Equation and Equations of State

In practice we normally compute the entropy of a system by considering its constitutive subsystems. Entropy is an extensive property; the entropy of a system is the *sum* of the entropies of its subsystems. During a process in an isolated system, as we have remarked above, the internal energy of any constituent subsystem need not remain constant. Consequently we must recognize that for any system or subsystem the entropy depends on the internal energy as well as the displacement parameters;

$$s = s(u, x_j) \ . \tag{3}$$

This equation is called the *fundamental equation* of the system. When it is known all thermodynamic information about equilibrium states of the system can be deduced.

The entropy of a system or subsystem is a single valued, continuous and continuously differentiable function of its arguments. Further, the entropy is a monotonically increasing function of internal energy for any fixed set of displacement parameters. Thus

$$\frac{\partial s}{\partial u} > 0 \tag{4}$$

for all values of u and x_j.

It is frequently more convenient to write the fundamental equation in the form

$$u = u(s, x_j) \ , \tag{5}$$

giving the internal energy as a function of entropy and the displacement variables. That s is a single valued continuous function of u and x_j is sufficient to permit us to solve equation (3) for u and write equation (5). It follows that u will be a single valued, continuous and continuously differentiable function of s and x_j.

Consider now an infinitesimal change in the entropy s and the displacement parameters x_j of a system, as might occur during a process. Our discussion is intended to apply to either an isolated system or a subsystem of an isolated system, so that we do not limit ourselves to systems in which the internal energy remains constant. The infinitesimal change in internal energy which takes place as a result of a change in entropy ds and a change in the displacement parameters dx_j may be written as

$$du = \frac{\partial u}{\partial s}\, ds + \frac{\partial u}{\partial x_j}\, dx_j \; , \tag{6}$$

with the summation convention implied.

The thermodynamic temperature T is given by

$$T = \frac{\partial u}{\partial s} = T(s,x_j) \; , \tag{7}$$

and the forces X_j conjugate to the displacement parameters x_j by

$$X_j = \frac{\partial u}{\partial x_j} = X_j(s,x_k) \; . \tag{8}$$

Equation (6) may then be written as

$$du = Tds + X_j\, dx_j \; . \tag{9}$$

Equations (7) and (8) are referred to as the *equations of state* for the system. They can of course be derived from the fundamental equation, as we have done, or they could be determined experimentally. Once all the equations of state are known, the fundamental equation can be inferred, at least to within an arbitrary constant.

It must be emphasized that equation (9) describes the change
in internal energy as a function of the changes in the arguments of the fundamental equation (5); it is quite distinct
from equation (1) which gives the change in internal energy
when an infinitesimal amount of heat dQ is added to the system
and an infinitesimal amount of work dW is done on the system.
In order to emphasize this distinction it is sometimes convenient to write equation (9) in the form

$$du = dQ^R + dW^R \quad , \tag{10}$$

where

$$dQ^R = Tds \quad , \quad \text{and} \tag{11a}$$

$$dW^R = X_j dx_j \quad . \tag{11b}$$

The terms *reversible heat* and *reversible work* are applied to
dQ^R and dW^R respectively. Since they are derived from the fundamental equation, $(dQ^R + dW^R)$ is integrable. A comparison of
equations (1) and (10) when they refer to the same small change
in energy during a process shows that

$$dQ + dW = dQ^R + dW^R \quad . \tag{12}$$

An alternative form of this equation, obtained by substituting
equation (11a) and rearranging, gives the change in entropy and
will be used in our discussion;

$$ds = \frac{dQ}{T} + \frac{dW - dW^R}{T} \quad . \tag{13}$$

Note that we are not referring at present to situations in
which the entropy maximum principle is applicable; it does not
follow, therefore, that ds in equation (13) is necessarily
positive.

Equation (11a) is referred to as the *first part of the second law of thermodynamics*, and equation (13) as the second part of the second law.

27.4 Heat and Work Sources

It is convenient to define certain idealized systems (which we invariably use as subsystems of a larger system) in which the fundamental equation (5) has a particularly simple form. These idealized systems are *reversible sources*. First, consider a system for which

$$u = T^{o}s + A \tag{14}$$

where T^{o} is a constant reference thermodynamic temperature and A is an unknown constant. This system is a *reversible heat source*. The internal energy depends only on a single thermo-dynamic parameter, the entropy of the system, and it is *linear* in this parameter. The equation of state (equation 7) clearly gives

$$T^{o} = \frac{\partial u}{\partial s} \; , \tag{15}$$

so that the thermodynamic temperature remains constant. Heat can be added to the system without changing the temperature. A reversible heat source is used to model the atmosphere or a large body of fluid which might surround a second system and maintain it at an ambient temperature.

It should be emphasized that the absolute value of the initial energy u is never regarded as defined by the fundamental equation. In all cases we accept the existence of an unknown constant. From the mathematical point of view, u is a potential function and without loss in generality we may drop the unknown constant in equation (14) and give the internal energy of a reversible heat source as

$$u = T^{o}s \; . \tag{16}$$

The second class of idealized systems which we define are *reversible work sources*. The internal energy of a set of reversible work sources has the form

$$u = X_j^O \, x_j \quad , \tag{17}$$

where again an unknown constant is neglected. The x_j again represent displacement parameters, and the X_j^O are *constant* forces. The internal energy of the work sources is linear in the displacement parameters x_j, and again the equations of state give

$$X_j^O = \frac{\partial u}{\partial x_j} \quad . \tag{18}$$

We shall use reversible work sources to model conservative loads applied to a deformable body: neither the magnitude nor the direction of a set of conservative loads is a function of the deformation of the body, and the potential energy of the loads is directly represented by the internal energy of the reversible work source.

27.5 Transformations of the Fundamental Equation

In many situations, and particularly for isothermal processes, it is convenient to transform the fundamental equation (5) into a form where the independent variables are temperature T and the displacement variables x_j. Since T is the partial derivative of internal energy u with respect to entropy s, this is rigorously accomplished by means of the theory of the Legendre transformation. However, the main features of the transformation can be appreciated with a simpler and less rigorous argument. Equation (9) was established by considering a small change in the internal energy u as a result of small changes in s and x_j. The argument may be reversed; if

$$T ds + X_j dx_j = du \tag{19}$$

we have a *sufficient* condition that u is a potential function, that

$$u = u(s, x_j) \quad , \tag{20}$$

and that equations (7) and (8) hold. This argument offers us the opportunity of determining other potential functions. Consider

$$- sdT + X_j dx_j = (- sdT - Tds) + (Tds + X_j dx_j)$$
$$= - d(Ts) + du \tag{21}$$
$$= d(u - Ts) \quad .$$

This identity is a sufficient condition to establish that

$$f = u - Ts \tag{22}$$

is a potential function of the form

$$f = f(T, x_j) \quad , \tag{23}$$

and that

$$-s = \frac{\partial f}{\partial T} \quad , \quad \text{and} \quad X_j = \frac{\partial f}{\partial x_j} \tag{24}$$

The function f is called the *Helmholtz potential* or the *free energy* of the system whose fundamental equation is given by equation (20). Equation (23) contains precisely the same information as the fundamental equation (20); it is simply a transformation of the fundamental equation with arguments T, x_j instead of s, x_j.

In the same manner we may establish the existence of two other potential functions. The first is

$$h = h(s, X_j) = u - X_j x_j \quad , \tag{25}$$

where

$$T = \frac{\partial h}{\partial s} \quad , \quad X_j = - \frac{\partial h}{\partial X_j} \quad . \tag{26}$$

The second is the Gibb's function

$$g = g(T, X_j) = u - Ts - X_j x_j \quad , \tag{27}$$

$$= f - X_j x_j \quad .$$

where

$$s = - \frac{\partial g}{\partial T} \quad , \quad x_j = - \frac{\partial g}{\partial X_j} \quad . \tag{28}$$

We shall have occasion to make extensive use of the Helmholtz free energy; in the description of strained solids the other transformations have less utility.

27.6 The Energy Minimum Principles

Let us now return to a consideration of the implications of the entropy maximum principle. In physical terms, the equilibrium state of an unconstrained isolated system whose fundamental equation is

$$s = s(u, x_j) \tag{29}$$

is characterized by an entropy which is larger than that of adjacent states. The entropy in the unconstrained equilibrium state may be either a global or a local maximum, but for the present we do not need to distinguish between these two possibilities. A mathematical statement of the maximum principle may be given in terms of the derivatives of equation (29). Since u is held constant, potential equilibrium states of the system are characterized by the requirement that

$$\frac{\partial s}{\partial x_j} = 0 \quad , \quad u \text{ constant} \quad . \tag{30}$$

This expresses the condition that the entropy s should be stationary for arbitrary variations in x_j about the unconstrained equilibrium state with u held constant. That s should be a maximum requires that for an unconstrained equilibrium state

$$\frac{\partial^2 s}{\partial x_j \partial x_k}$$

should be negative definite. This in turn ensures that the second variation of s,

$$\delta^2 s = \frac{\partial^2 s}{\partial x_j \partial x_k} \, \delta x_j \, \delta x_k \leq 0 \tag{31}$$

for arbitrary variations δx_j about the equilibrium state with u held constant.

These expressions, together with the assumption that $\partial s/\partial u$ is positive (equation 4), permit us to draw conclusions for variations in u and x_j about an equilibrium state when *the entropy s is held constant*. If s is fixed, equation (29) implies that

$$s(u, \, x_j) = \text{const.} \tag{32}$$

It follows then that

$$\frac{\partial s}{\partial u} \, du + \frac{\partial s}{\partial x_j} \, dx_j = 0 \quad , \tag{33}$$

or alternatively, that

$$\left(\frac{\partial u}{\partial x_j}\right)_{s \ \text{const.}} = - \frac{\partial s/\partial x_j}{\partial s/\partial u} \tag{34}$$

However, from equation (30), $\partial s/\partial x_j = 0$ for an equilibrium state, and thus

$$\left(\frac{\partial u}{\partial x_j}\right)_{s \ \text{const.}} = 0 \quad . \tag{35}$$

It follows then that u is stationary for arbitrary variations of x_j about an equilibrium state with s held constant. The second variation of u, again with s constant, is given by

$$\delta^2 u = \left(\frac{\partial^2 u}{\partial x_j \partial x_k}\right)_{s\ \mathrm{const.}} \delta x_j\, \delta x_k \quad . \tag{36}$$

In order to compute the sign of this term we write

$$\left(\frac{\partial^2 u}{\partial x_j \partial x_k}\right)_{s\ \mathrm{const.}} = \frac{\partial}{\partial x_j}\left(\frac{\partial u}{\partial x_k}\right)_{s\ \mathrm{const.}}$$

$$= \frac{\partial}{\partial x_j}\left(-\frac{\partial s/\partial x_k}{\partial s/\partial u}\right) + \frac{\partial}{\partial u}\left(-\frac{\partial s/\partial x_k}{\partial s/\partial u}\right)\left(\frac{\partial u}{\partial x_j}\right)_{s\ \mathrm{const.}} \tag{37}$$

$$= -\frac{\left[\dfrac{\partial s}{\partial u}\dfrac{\partial^2 s}{\partial x_j \partial x_k} - \dfrac{\partial s}{\partial x_k}\dfrac{\partial^2 s}{\partial u \partial x_j}\right]}{\left(\dfrac{\partial s}{\partial u}\right)^2} + \frac{\left[\dfrac{\partial s}{\partial u}\dfrac{\partial^2 s}{\partial u \partial x_k} - \dfrac{\partial s}{\partial x_k}\dfrac{\partial^2 s}{\partial u^2}\right]}{\left(\dfrac{\partial s}{\partial u}\right)^2}\frac{\left(\dfrac{\partial s}{\partial x_j}\right)}{\left(\dfrac{\partial s}{\partial u}\right)} \quad .$$

At an unconstrained equilibrium state, however, we know from equation (30) that $\partial s/\partial x_j = 0$. Substituting this into equation (37), we see that

$$\left(\frac{\partial^2 u}{\partial x_j \partial x_k}\right)_{s\ \mathrm{const.}} = -\frac{\left(\dfrac{\partial^2 s}{\partial x_j \partial x_k}\right)}{\left(\dfrac{\partial s}{\partial u}\right)} \quad . \tag{38}$$

Since $\partial s/\partial u$ is positive (equation 4), and we have shown that $\partial^2 s/\partial x_j \partial x_k$ is negative definite, it follows then that

$$\left(\frac{\partial^2 u}{\partial x_j \partial x_k}\right)_{s\ \mathrm{const.}}$$

is positive definite, and that $\delta^2 u$ in equation (36) is positive.

This establishes the *energy minimum principle*: unconstrained equilibrium states are characterized by the condition that for fixed s the displacement variables x_j take on values such that the internal energy u is a (local or global) minimum.

An alternative form of the energy minimum principle can be given in terms of a quasi-static process originating at an

unconstrained equilibrium state. Let u°, s°, x_j° be the internal energy, entropy and displacement variables characterizing
an unconstrained equilibrium state. Let the system pass
through a series of equilibrium states (i.e. states satisfying
the fundamental equation) with the entropy remaining constant,
and terminating at a state characterized by u^{*}, s°, x_j^{*}. It
follows from the minimum principle that

$$\Delta u = u^{*}(s^{\circ}, x_j^{*}) - u^{\circ}(s^{\circ}, x_j^{\circ}) > 0 \quad . \tag{39}$$

Using equation (8) to introduce the forces X_j conjugate to the
displacements x_j, we can write

$$u^{*}(s^{\circ}, x_j^{*}) - u^{\circ}(s^{\circ}, x_j^{\circ}) = \int_{x_j^{\circ}}^{x_j^{*}} X_j dx_j > 0 \quad . \tag{40}$$
$$\text{s const.}$$

Useful results can also be obtained by applying the energy
minimum principle to a variety of special systems. Consider a
system, characterized by internal energy U and entropy S, which
consists of an arbitrary subsystem in diathermal contact with a
heat source. The arbitrary subsystem is supposed to have internal energy u, entropy s and displacement variables x_j and a fundamental equation

$$u = u(s, x_j) \quad . \tag{41}$$

The heat source is at temperature T°, and has internal energy
u_H and entropy s_H. From equation (16), we see that we may put

$$u_H = T^{\circ} s_H \quad . \tag{42}$$

The restriction that the arbitrary subsystem and the heat
source are in diathermal contact implies that the temperature
of the arbitrary system is T°, and hence

$$\frac{\partial u}{\partial s} = T^{O} \quad . \tag{43}$$

Using the additive properties of energy and entropy, it is evident that for the combined system

$$S = s + s_{H} \quad , \tag{44a}$$

$$U = u + u_{H} = u(s, x_{j}) + T^{O}s_{H} \quad . \tag{44b}$$

Unconstrained equilibrium states of the combined system will be characterized by minimum U for fixed S. Consider small changes in the thermodynamic variables. For fixed total entropy

$$dS = ds + ds_{H} = 0 \quad , \tag{45a}$$

and hence

$$ds_{H} = - ds \quad . \tag{45b}$$

Further,

$$dU = (T^{O}ds + X_{j}dx_{j}) + T^{O}ds_{H}$$
$$= X_{j}dx_{j} \tag{46}$$

after using equation (45b).

Thus, if x_{j}^{O}, T^{O} characterize an unconstrained equilibrium state, and the system passes through a sequence of equilibrium states to a final state x_{j}^{*}, T^{O}, the energy minimum principle requires that

$$\Delta U = \int_{x_{j}^{O}}^{x_{j}^{*}} X_{j}dx_{j} > 0 \quad . \tag{47}$$
$$T \text{ const.}$$

It may be noted that, in view of equation (44a), we may write equation (44b) as

$$U = u(s, x_j) - T^o s + S$$

$$ = f(T, x_j) + S \quad, \tag{48}$$

where $f(T, x_j)$ is the free energy of the arbitrary subsystem
(cf. equations 22 and 23). Since S is fixed, equation (48)
shows that the internal energy of the combined system differs
from the Helmholtz free energy of the arbitrary subsystem by a
constant. It follows then that unconstrained equilibrium
states of the combined system are characterized by the condition that the free energy of the arbitrary subsystem should be
a minimum at constant T. This result is known as the *Helmholtz
free energy minimum principle.*

A second system which is of particular interest in the problems we shall discuss is a system which is comprised of an
arbitrary subsystem combined with a reversible heat source and
a number of reversible work sources. Let the arbitrary subsystem again be characterized by the fundamental equation

$$u = u(s, x_j) \quad . \tag{49}$$

The heat source is in diathermal contact with the arbitrary
subsystem. Its fundamental equation is

$$u_H = T^o s_H \quad, \tag{50}$$

and it imposes a constraint on the arbitrary subsystem by requiring that

$$\frac{\partial u}{\partial s} = T^o \quad . \tag{51}$$

The reversible work sources are coupled to the arbitrary subsystem in such a way that the displacements of the work sources
are $(- x_j)$. The internal energy u_w of the work sources is thus
given by

$$u_w = - X_j^o x_j \tag{52}$$

where the forces X_j^o are constant.

It follows that the total entropy of the combined system is

$$S = s + s_H \tag{52a}$$

and that its internal energy is

$$U = u(s, x_j) + T^o s_H - X_j^o x_j \quad . \tag{52b}$$

Since S is regarded as fixed, we can eliminate s_H and write

$$U = f(T^o, x_j) - X_j^o x_j + S \quad . \tag{53}$$

Unconstrained equilibrium states of the combined system are thus characterized by the condition that U should have a minimum value with T^o and X_j^o fixed. Alternatively,

$$U_p = f(T^o, x_j) - X_j^o x_j \tag{54}$$

should have a minimum value, since S is constant.

With $T = T^o$, changes in U_p caused by changes in x_j are given by

$$dU_p = \frac{\partial f}{\partial x_j} dx_j - X_j^o dx_j \tag{55}$$

$$= (X_j - X_j^o) dx_j \quad .$$

This indicates that unconstrained equilibrium states are such that $X_j = X_j^o$, since the first variation of U_p must vanish if U_p is a minimum.

If T^o, x_j^o characterize an unconstrained equilibrium state, and the combined system passes through a sequence of adjacent equilibrium states to a final state T^o, x_j^*, the minimum principle requires that

$$\Delta U_p = \{f(T^o, x_j^*) - X_j^o x_j^*\} - \{f(T^o, x_j^o) - X_j^o x_j^o\}$$

$$= \{f(T^o, x_j^*) - f(T^o, x_j^o)\} - X_j^o(x_j^* - x_j^o) \qquad (56)$$

$$= \int_{x_j^o}^{x_j^*} (X_j - X_j^o)dx_j > 0 \quad .$$

T const.

In this development we have assumed implicitly that all
displacement variables are unconstrained, and have termed the
equilibrium state characterized by the extremum principles the
unconstrained equilibrium state. If some of the displacement
variables are constrained the entropy maximum principle and
the energy minimum principle apply subject to the constraints.
The equilibrium state characterized by the extremum principles
are then *partially constrained.*

It should perhaps be reemphasized that our concern in this
aspect of the thermodynamic problem is to determine the values
of the displacement parameters in a closed system at equilibrium. Satisfaction of the fundamental equation is a necessary but not sufficient condition for equilibrium. States
which satisfy the fundamental equation thus represent the
class of states which contain the actual equilibrium state. If
constraints are present, the class of states from which the
actual equilibrium state is to be selected (representing states
which satisfy the fundamental equation and the constraints) is
reduced. The actual equilibrium state is always distinguished
from the states which satisfy the fundamental equation and the
constraints by the requirement that it maximizes s at constant
u or minimizes u at constant s. We see also that in a totally
constrained system, where all the displacements x_j are fixed,
there is only one possible equilibrium state and the fundamental equation simply provides a relation between s and u.

27.7 The Second Part of the Second Law of Thermodynamics

Consider now a closed system which is undergoing a quasi-static process. This closed system is regarded as part of a larger isolated system, and the process is initiated by relaxing constraints within the larger system. The closed subsystem is governed by the fundamental equation

$$u = u(s, x_j) \quad . \tag{57}$$

In some small interval of time during this process the internal energy, entropy and displacement variables of the closed subsystem change by du, ds, dx_j respectively, heat dQ enters the subsystem and work dW is done on the system, the second part of the second law permits us to write (cf. equation 13)

$$ds = \frac{dQ}{T} + \frac{dW - dW^R}{T} \quad , \tag{58}$$

where

$$dW^R = \frac{\partial u}{\partial x_j}\, dx_j = X_j\, dx_j \quad . \tag{59}$$

This expression provides the change in entropy in the closed subsystem. The change in entropy consists of two parts. The contribution from

$$\frac{dQ}{T}$$

is the change in entropy due to the heat flux dQ into the closed subsystem. The second part, proportional to the difference of work done on the system and change in internal energy due to the change in the displacement variables x_j, is the *entropy production* $d\zeta$ within the closed subsystem. The second part of the second law of thermodynamics requires that the entropy production over any small interval during a quasi-static process should be non-negative, i.e.

$$d\zeta = \frac{dW - dW^R}{T} \geq 0 \quad . \tag{60}$$

We may also speak of the *entropy production rate* $\dot{\zeta}$ in the closed subsystem, defined in terms of the rate of work done on the system $\dot{W}$ and the rate of reversible work $\dot{W}^R$ defined by

$$\dot{W}^R = X_j \dot{x}_j \quad . \tag{61}$$

Hence the second part of the second law can equivalently be written as the requirement that

$$\dot{\zeta} = \frac{\dot{W} - \dot{W}^R}{T} \geq 0 \tag{62}$$

for a closed system during a quasi-static process.

Inequalities (60) and (62) are used to distinguish between two kinds of processes. First, a process in which the entropy production rate is zero is called a *reversible process*. It follows that a necessary and sufficient condition for a reversible quasi-static process in a closed system is that

$$dW = dW^R \quad \text{or} \quad \dot{W} = \dot{W}^R \tag{63}$$

throughout the process. If

$$\dot{W} > \dot{W}^R \tag{64}$$

at any instant during the quasi-static process, the process is said to be *irreversible*.

We may describe a quasi-static process in a closed subsystem by plotting the trajectory of the equilibrium state on a hyper-surface defined by the fundamental equation $s = s(u, x_j)$ in an s, u, x_j space. Let the initial state be A, and the final state be B, so that the process proceeds along the trajectory from A to B. If the process is reversible, and the changes in entropy of the closed subsystem can be entirely ascribed to the heat flux dQ, it will be possible to retrace the process from B to A by suitable manipulations of the larger system in which our closed subsystem is embedded. If, however, entropy is produced within the subsystem it will not be possible to reverse the process.

A relationship between the entropy maximum principle and non-negative sign of the entropy production rate can be demonstrated with sufficient generality for the applications which follow. We note first that it is possible to construct a system which will always behave reversibly in certain classes of processes. The reversible heat and work sources defined in Section 27.4 are examples. Let us consider such a reversible system whose fundamental equation is

$$u^{(1)} = u^{(1)}(s^{(1)}, x_j^{(1)}) \quad . \tag{65}$$

Let us place this system in diathermal contact with the arbitrary closed subsystem whose fundamental equation is given by equation (51) and isolate this combination. Initially we have constraints imposed on the displacement variables x_j, $x_j^{(1)}$, but the two systems are at the same temperature. We now relax the constraints very slowly, permitting the entire isolated system to pass through a sequence of partially constrained equilibrium states until all the constraints are relaxed. The process thus initiated and controlled will be quasi-static.

Now consider two adjacent partially constrained states, with changes du, ds, dx_j and $du^{(1)}$, $ds^{(1)}$, $dx_j^{(1)}$ occurring as we pass from one to another. Let heat dQ be transferred from the reversible system to the arbitrary system, and let the reversible system do work dW on the second system. We now write the second part of the second law for the reversible system. Since dQ is the heat transferred out of the system,

$$ds^{(1)} = -\frac{dQ}{T} \quad . \tag{66}$$

For the arbitrary system,

$$ds = \frac{dQ}{T} + \frac{dW - dW^R}{T} \; .$$

(67)

Adding equations (66) and (67), the change in entropy for the complete system is

$$dS = ds^{(1)} + ds = \frac{dW - dW^R}{T} \; .$$

(68)

The entropy maximum principle requires that $dS \geq 0$, since dS is the change in entropy which occurs after a change in the constraint and the system again achieves equilibrium. This implies that inequalities (60) or (62) hold for the arbitrary system as it undergoes a quasi-static process.

27.8 Bibliographical Remarks

A number of texts on classical thermodynamics exist to which the reader may be referred for further background. The summary given in this Chapter is greatly influenced by Callen [1960] and Kestin [1966], [1968].

References

H.B.Callen   1960   *Thermodynamics*, Wiley.

J.Kestin   1966   *A Course in Thermodynamics.* Vol. I,
                          Blaisdell.

J.Kestin   1968   *A Course in Thermodynamics,* Vol. II,
                          Blaisdell.

Chapter 28

APPLICATION OF THERMODYNAMIC CONCEPTS TO THE CONSTITUTIVE EQUATIONS OF INELASTIC SOLIDS

28.1 Elastic Solids

Let us consider a system which comprises a representative macroscopic sample of the elastic solid in diathermal contact with a reversible heat source. Let the generalized volume of the solid be V. Further, let the sample be subject to boundary loading which causes macroscopically homogeneous deformation. The macroscopic deformation is characterized by generalized strains q_j and the boundary loading by generalized stresses Q_j.

As a result of a small change in the deformation dq_j the boundary loads Q_j will perform work

$$dW = VQ_j dq_j \quad . \tag{1}$$

Elastic behavior is *quasi-static* (i.e. at each instant during the process of deformation the fundamental equation is satisfied) and *reversible*. Hence (Chapter 27, equation 63)

$$dW^R = dW = VQ_j dq_j \quad . \tag{2}$$

The internal energy of the system under consideration is measured by the Helmholtz free energy f of the unit mass sample. Since

$$df\big|_{T \text{ constant}} = dW^R = VQ_j dq_j \quad , \tag{3}$$

it follows that

$$f = f(T, q_j) \quad , \tag{4a}$$

and

$$Q_j = \frac{1}{V} \frac{\partial f}{\partial q_j} \quad . \tag{4b}$$

Since we are dealing with infinitesimal deformations, and the

density of the sample may be assumed to remain constant, it is
convenient to introduce the Helmholtz free energy per unit
generalized volume, $\bar{f}$. Then

$$\bar{f} = \bar{f}(T, q_j) \tag{5a}$$

and

$$Q_j = \frac{\partial \bar{f}}{\partial q_j} \quad . \tag{5b}$$

At a given temperature $\bar{f}(T, q_j)$ can thus be seen to differ only
by a constant from the specific elastic strain energy $W(q_j)$.

   We may also introduce the Gibb's function per unit volume,
$\bar{g}$ (Chapter 27, equation 27),

$$\bar{g}(T, Q_j) = \bar{f}(T, q_j) - Q_j q_j \quad , \tag{6a}$$

where

$$q_j = - \frac{\partial \bar{g}}{\partial Q_j} \quad . \tag{6b}$$

At a given temperature, $-\bar{g}(T, Q_j)$ and the specific complementary
energy $\Omega(Q_j)$ can differ only by a constant.

   If equation (6b) describes a conventional linear elastic
solid, it has the form

$$q_j = C_{jk} Q_k + a_j (T - T^o) \quad , \tag{7}$$

where C_{jk} is an array of constant coefficients, a_j is a vector
of coefficients of linear expansion, and T^o is a reference temperature. It follows that

$$-\bar{g}(T, Q_j) = \frac{1}{2} C_{jk} Q_j Q_k + (T - T^o) a_j Q_j + A(T) \quad , \tag{8}$$

and that the array C_{jk} is symmetric. Note also that

$$C_{jk} = - \frac{\partial^2 \bar{g}}{\partial Q_j \partial Q_k} \quad . \tag{9}$$

If D_{jk} is the inverse of C_{jk}, equation (7) can be written as

$$Q_j = D_{jk}q_k - (T - T^o)D_{jk}a_k \quad , \tag{10}$$

and hence

$$\bar{f}(T,q_j) = \frac{1}{2}D_{jk}q_jq_k - (T - T^o)D_{jk}q_ja_k - A(T) \quad . \tag{11}$$

It is seen that

$$\bar{f} - \bar{g} = Q_jq_j \quad ,$$

corresponding to

$$W(q_j) + \Omega(Q_j) = Q_jq_j \quad . \tag{12}$$

We note also that

$$D_{jk} = \frac{\partial^2 \bar{f}}{\partial q_j \partial q_k} \tag{13}$$

and is symmetric.

The requirements of thermodynamic stability can be expressed in terms of a change from an unconstrained equilibrium state T^o,q_j^o to a second state T^o,q_j^* (Chapter 27, equation 56). We must have

$$\int_{q^o}^{q_j^*} (Q_j - Q_j^o)dq_j > 0 \quad , \tag{14a}$$
$$T \text{ constant}$$

where

$$Q_j^o = \left.\frac{\partial \bar{f}}{\partial q_j}\right|_{q_j^o} \quad . \tag{14b}$$

Inequality (14a) may be written as

$$\bar{f}(T^o,q_j^*) - \bar{f}(T^o,q_j^o) > (q_j^* - q_j^o)\left.\frac{\partial \bar{f}}{\partial q_j}\right|_{q_j^o} \quad , \tag{15}$$

which for an arbitrary choice of q_j^*, q_j^o implies that $\bar{f}(T, q_j)$ is strictly convex for T constant. If $\bar{f}(T, q_j)$ has the quadratic form given in equation (11) it will be strictly convex if and only if D_{jk} is positive definite.

28.2 Internal Variables

Consider now a system identical in all respects to that introduced in Section 28.1 except that the representative macroscopic sample may behave inelastically. Macroscopic deformation is still measured by the generalized strains q_j, and boundary loading by the generalized stresses Q_j. Hence we still have

$$dW = VQ_j dq_j \tag{16}$$

for the work performed by the boundary loads during a small change in deformation.

We introduce a set of *internal thermodynamic variables* χ_α $(\alpha = 1, \ldots, \nu)$ which describe microstructural rearrangements within the sample. The Helmholtz free energy per unit volume is then given by

$$\bar{f} = \bar{f}(T, q_j, \chi_\alpha) \quad . \tag{17}$$

The change in the Helmholtz free energy resulting from a small change in the thermodynamic variables is then

$$d\bar{f} = -\frac{1}{V} sdT + \frac{\partial \bar{f}}{\partial q_j} dq_j + \frac{\partial \bar{f}}{\partial \chi_\alpha} d\chi_\alpha \quad , \tag{18}$$

where s is the entropy per unit mass, and

$$dW^R = V\left\{ \frac{\partial \bar{f}}{\partial q_j} dq_j + \frac{\partial \bar{f}}{\partial \chi_\alpha} d\chi_\alpha \right\} \quad . \tag{19}$$

The forces conjugate to the generalized strains are VQ_j, and hence

$$Q_j = \frac{\partial \bar{f}}{\partial q_j} \quad . \tag{20}$$

The specific agency responsible for the microstructural rearrangement will apply a thermodynamic force $\partial \bar{f}/\partial \chi_\alpha$, conjugate to χ_α, on the sample. It is convenient to work with the equal and opposite force X_α applied by the sample on the agency by means of which the rearrangement occurs. Thus

$$X_\alpha = -\frac{\partial \bar{f}}{\partial \chi_\alpha} \quad . \tag{21}$$

Equation (19) then becomes

$$dW^R = V\{Q_j dq_j - X_\alpha d\chi_\alpha\} \quad . \tag{22}$$

Equations (20) and (21) are the equations of state of the sample of the inelastic solid. To complete the description we must introduce a kinetic relation specifying the rates of change of the internal variables χ_α. This equation may be assumed to take the form

$$\dot{\chi}_\alpha = \dot{\chi}_\alpha(T, X_\beta, \chi_\beta) \quad . \tag{23}$$

The rate of entropy production due to dissipation is

$$\dot{\zeta} = \frac{\dot{W} - \dot{W}^R}{T}$$
$$= V X_\alpha \dot{\chi}_\alpha \tag{24}$$

from equations (16) and (22). Since $\dot{\zeta}$ must be non-negative, the kinetic equation (23) is subject to the thermodynamic restriction

$$X_\alpha \dot{\chi}_\alpha \geq 0 \quad . \tag{25}$$

Together equations (20), (21) and (23) define a very wide class of inelastic solids. Elimination of X_α and χ_α from the three sets of equations will lead to what are generally referred to as the constitutive relations. This very wide class may be restricted by requiring that the response of the solid should be elastic if no microstructural rearrangements take place. This implies that for T, χ_α constant

$$dq_j = C_{jk} dQ_k \quad , \tag{26}$$

where the coefficients of C_{jk} are constant, and for Q_j, χ_α constant,

$$dq_j = a_j (T - T^\circ) \quad , \tag{27}$$

where the coefficients a_j are constant. This implies that

$$q_j = C_{jk} Q_k + a_j (T - T^\circ) + q_j^p (\chi_\alpha) \quad , \tag{28}$$

where q_j^p may be termed the *inelastic component* of strain. We may further introduce

$$q_j^e (T, Q_j) = C_{jk} Q_k + a_j (T - T^\circ) \quad , \tag{29}$$

the *elastic component* of strain.

It follows then that

$$dq_j = dq_j^e + \frac{\partial q_j^p}{\partial \chi_\alpha} d\chi_\alpha \quad , \tag{30}$$

and that equation (18) may be rewritten as

$$d\bar{f}(T, q_j^e, \chi_\alpha) = -\frac{1}{V} s dT + Q_j dq_j^e + \left(Q_j \frac{\partial q_j^p}{\partial \chi_\alpha} - X_\alpha \right) d\chi_\alpha \quad . \tag{31}$$

From equation (29) it is apparent that Q_j depends only on q_j^e and T. Hence there exists a potential function of the form

$$\bar{f}^e = \bar{f}^e (T, q_j^e) \quad , \tag{32a}$$

$$d\bar{f}^e = -\frac{1}{V} s^e dT + Q_j dq_j^e \quad . \tag{32b}$$

On comparing equation (31) and (32b), we may put

$$\bar{f} = \bar{f}^e(T, q_j^e) + \bar{f}^p(T, \chi_\alpha) \quad , \tag{33}$$

where

$$d\bar{f}^p = - \frac{1}{V} s^p dT + \frac{\partial \bar{f}^p}{\partial \chi_\alpha} d\chi_\alpha \quad . \tag{34}$$

The entropy s is then the sum of two parts

$$s = s^e + s^p \quad , \tag{35}$$

and

$$Q_j \frac{\partial q_j^p}{\partial \chi_\alpha} - X_\alpha = \frac{\partial \bar{f}^p}{\partial \chi_\alpha} \quad , \tag{36}$$

in order that the sum of equations (32b) and (34) lead to (31).
Equation (36) shows that

$$X_\alpha = Q_j \frac{\partial q_j^p}{\partial \chi_\alpha} - \frac{\partial \bar{f}^p}{\partial \chi_\alpha} \quad , \tag{37}$$

indicating that the thermodynamic forces acting on the agency
by means of which microstructural rearrangement occurs are
linear in the generalized stress Q_j. Further, on differentiating equation (37),

$$\frac{\partial X_\alpha}{\partial Q_j} = \frac{\partial q_j^p}{\partial \chi_\alpha} \quad . \tag{38}$$

Thus if the dependence of X_α on Q_j can be determined, the dependence of q_j^p on χ_α can be predicted, and *vice versa*.
The inelastic behavior of the material is governed by the
form of the kinetic equation (23). If it is possible to derive this relation from a potential function of X_α for fixed T,
χ_α an important general result follows. Suppose that (23) can
be written as

$$\dot{\chi}_\alpha = \frac{\partial P(T,X_\alpha,\chi_\alpha)}{\partial X_\alpha} \quad . \tag{39}$$

Making use of equation (38), it follows that

$$\dot{q}_j^p = \frac{\partial q_j^p}{\partial \chi_\alpha}\dot{\chi}_\alpha = \frac{\partial P}{\partial X_\alpha}\frac{\partial q_j^p}{\partial \chi_\alpha} = \frac{\partial P}{\partial X_\alpha}\frac{\partial X_\alpha}{\partial Q_j}$$

$$\tag{40}$$

$$= \frac{\partial P\left(T,Q_j\dfrac{\partial q_j^p}{\partial \chi_\alpha} - \dfrac{\partial \bar{f}^p}{\partial \chi_\alpha},\chi_\alpha\right)}{\partial Q_j} \quad .$$

The plastic strain rate can thus be expressed as the gradient of a potential function in stress space for given T,χ_α. This implies that $\dot{q}_j^p$ is normal to the level surfaces of the potential function P if $\partial P/\partial Q_j$ is continuous.

28.3 Some Specific Models of the Kinetic Equations

Both the macroscopic deformation measure, the strain q_j, and the internal variables χ_α may be interpreted in a variety of ways. At the most basic level, the sample may be a cube of unit mass, with the q_j representing the infinitesimal strain tensor ε_{ij}, and the generalized stresses Q_j representing the stress tensor σ_{ij}. Again on the most basic level, the internal variables in a crystalline solid may be discrete slips occurring on glide planes as a result of dislocation motion.

In this case the thermodynamic forces X_α represent the component of the resolved shear stress acting on the glide plane in the slip direction. It is made up of two parts (equation 37); the first, characterized by $Q_j\,\partial q_j^p/\partial \chi_\alpha$ being the shear on the slip plane due to the macroscopically applied stress, and the second, characterized by $-\partial \bar{f}^p/\partial \chi_\alpha$, being the restoring force applied to the slip system resulting from the local elastic

distortion of the crystal lattice due to slip. In the case of
a polycrystal, slip on a family of slip systems in any one
crystal will also result in elastic deformation of adjacent
crystals. The Helmholtz free energy is similarly made up of
two independent parts; $\bar{f}^e$ being the energy associated with the
elastic distortion of the entire sample due to the applied
stress and $\bar{f}^p$ being the energy associated with elastic distortion due to slip.

The kinetic relation is then the relation between dislocation velocity and resolved shear stress. This relation is
generally highly non-linear; in the absence of cross effects
which restrict dislocation motion the dislocation velocity and
the resolved shear stress on any one glide plane may be related
by a curve of the type shown in Figure 1.

A set of internal variables made up of discrete slips is not
feasible for determining constitutive relations for structural
problems, since the number of dislocation sites is very large
and in any case depends on the size of the sample. A restricted number of averaging variables must replace the discrete
slips; such averaging variables will be independent of the
size of the sample and lead to a continuum dislocation model.
Nevertheless the discrete slip model is useful in suggesting

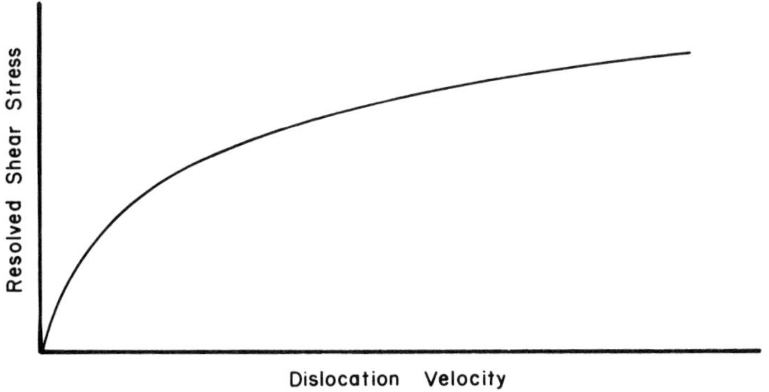

Figure 1. Relation between resolved shear stress and
dislocation velocity

the general framework of the governing equations and the likely
form of the kinetic equations.

The macroscopic deformation measure may also be generalized
strains, as in the case of a plate or bar element. We may extend this further and interpret the generalised displacements
of a structure as the macroscopic measure. Provided that the
structure is considered under isothermal conditions, the generalized loads Γ_α and the generalized displacements ξ_α may replace Q_j and q_j. In this latter the elastic Helmholtz free
energy provides the elastic solution of the problem, and the
plastic free energy represents the effects of inelastic deformation throughout the structure.

This broad interpretation of the thermodynamic potential
functions emphasizes again that the essential features of plastic behavior are independent of scale. This point has been
made throughout this volume, both in treating the formulation
in terms of generalized stresses and strains, and specifically
in Chapter 3 and subsequent applications of general theorems.
It also permits us to understand the essential features of constitutive relations by studying the behavior of simple structures. We shall make use of this in the ensuing discussion.

The kinetic equations (23) or (39) may be coupled (i.e. each
component of $\dot{X}_\alpha$ may depend on every component of X_α and χ_α) or
uncoupled (i.e. each component of $\dot{X}_\alpha$ may depend only on one
associated component of X_α and on one component of χ_α). As we
have seen in our previous studies of the general behavior of
structures, the essential features of the response will be the
same whether the equations are coupled or not. We shall ig-
nore the distinction between these cases in what follows, and
draw on both possibilities as examples.

In order to understand the place of time dependent plasticity within the general framework we have so far derived, let

us consider a specific model of the kinetic equations. Suppose
that $P = P(T, X_\alpha)$, and let us consider a fixed temperature T.
Further, let

$$P(X_\alpha) = \frac{1}{n+1} \phi^{n+1} \quad , \tag{41}$$

where $\phi = \phi(X_\alpha)$ is homogeneous, continuously differentiable and
of degree one in X_α, and n is an odd positive integer.

Thus

$$\dot{X}_\alpha = \frac{\partial P}{\partial X_\alpha} = \{\phi\}^n \frac{\partial \phi}{\partial X_\alpha} \quad , \tag{42}$$

and, using equation (40),

$$\dot{q}^p_j = \dot{X}_\alpha \frac{\partial q^p_j}{\partial X_\alpha} = \{\phi\}^n \frac{\partial \phi}{\partial X_\alpha} \frac{\partial X_\alpha}{\partial Q_j} = \{\phi(Q_j \frac{\partial q^p_j}{\partial X_\alpha} - \frac{\partial \bar{f}^p}{\partial X_\alpha})\}^n \frac{\partial \phi}{\partial Q_j} \quad . \tag{43}$$

If $n = 1$, equation (42) provides a linear relation between
X_α and X_α. Making some further simplifying assumptions, the
constitutive equation for a class of linear viscoelastic materials may be derived. Suppose that equation (42) can be
written as

$$\dot{X}_\alpha = A_{\alpha\beta} X_\beta \quad , \tag{44}$$

where the $A_{\alpha\beta}$ are constants. Suppose further that $\bar{f}^p(X_\alpha)$ is
quadratic (as we might expect if it represents the energy as-
sociated with local distortions in an elastic material), and
put

$$\frac{\partial \bar{f}^p}{\partial X_\alpha} = a_{\alpha\beta} X_\beta \quad . \tag{45}$$

Finally, let the components of the array $\partial q^p_j / \partial X_\alpha$ also be constants, and put

$$\frac{\partial q_j^p}{\partial \chi_\alpha} = b_{j\alpha} \quad .$$
(46)

Equation (42) will then take the form

$$\dot{\chi}_\alpha = A_{\alpha\beta}(b_{j\beta}Q_j - a_{\beta\gamma}\chi_\gamma)$$
(47a)

or

$$\dot{\chi}_\alpha + A_{\alpha\beta}a_{\beta\gamma}\chi_\gamma = A_{\alpha\beta}b_{j\beta}Q_j \quad .$$
(47b)

This is a set of linear simultaneous first order equations.
The general solution of these equations may be represented by
the convolution integral

$$\chi_\alpha = \int_{-\infty}^{t} h_{j\alpha}(t-\tau)\dot{Q}_j(\tau)d\tau \quad ,$$
(48)

where the $h_{j\alpha}(t)$, $j = 1,2,\ldots$, are solutions of equation (47b)
for a unit step function of $Q_j(t)$, $j = 1,2,\ldots$, with the
initial conditions $\chi_\alpha(-\infty) = 0$. Hence

$$q_j = q_j^e + \frac{\partial q_j^p}{\partial \chi_\alpha}\chi_\alpha = C_{jk}Q_k + \int_{-\infty}^{t} b_{j\alpha}h_{k\alpha}(t-\tau)\dot{Q}_j(\tau)d\tau \quad .$$
(49)

This is the conventional hereditary integral characterization
of a linear viscoelastic material, where the creep functions
$b_{j\alpha}h_{k\alpha}(t-\tau)$ can be measured macroscopically. While the restrictive assumptions of equations (44) - (46) are not essential, it
is clearly seen that general framework includes the description
of linear viscoelastic materials.

More generally, in studying the inelastic behavior of metals
we would be concerned with larger values of the index n. Indeed, with values of n $\simeq$ 5 we could recover, with appropriate
simplifications the constitutive equations usually employed for
creep (quasi-static behavior at high temperatures) and visco-

plasticity (behavior at high strain rates). In these cases the sharp knee which appears in Figure 1 is reproduced, and the level surfaces of the potential function $\{\phi\}^{n+1}/n+1$ plotted in X_α space are closely spaced in a region corresponding to the knee. The behavior on either side of this region of closely spaced level surfaces is sharply different. This distinction can be studied by considering the limiting case $n \to \infty$.

Consider, then,

$$\dot{q}_j^p = \{\phi(Q_j \frac{\partial q_j^p}{\partial X_\alpha} - \frac{\partial \bar{f}^p}{\partial X_\alpha})\}^{n \to \infty} \frac{\partial \phi}{\partial Q_j} \quad . \tag{50}$$

We see immediately that

$$\dot{q}_j^p = 0 \quad \text{if} \quad \phi(Q_j \frac{\partial q_j^p}{\partial X_\alpha} - \frac{\partial \bar{f}^p}{\partial X_\alpha}) < 1 \quad , \tag{51a}$$

$$\dot{q}_j^p = \lambda \frac{\partial \phi}{\partial Q_j} \quad \text{if} \quad \phi(Q_j \frac{\partial q_j^p}{\partial X_\alpha} - \frac{\partial \bar{f}^p}{\partial X_\alpha}) = 1 \quad , \tag{51b}$$

$$\dot{q}_j^p \to \infty \quad \text{if} \quad \phi(Q_j \frac{\partial q_j^p}{\partial X_\alpha} - \frac{\partial \bar{f}^p}{\partial X_\alpha}) > 1 \quad , \tag{51c}$$

where $\lambda \geq 0$.

The last case, equation (51c), is physically unacceptable, and may be reinterpreted as a restriction that

$$\phi(Q_j \frac{\partial q_j^p}{\partial X_\alpha} - \frac{\partial \bar{f}^p}{\partial X_\alpha}) \leq 1 \quad . \tag{52}$$

Considering the second case, equation (51b), we attempt to find the value of λ when $\phi = 1$, $\dot{\phi} = 0$. We note that

$$\dot{\phi} = \frac{\partial \phi}{\partial X_\alpha} \frac{\partial q_j^p}{\partial X_\alpha} \dot{Q}_j - \frac{\partial \phi}{\partial X_\alpha} \frac{\partial^2 \bar{f}^p}{\partial X_\alpha \partial X_\beta} \dot{X}_\beta \quad , \tag{53}$$

if we make the assumption that $\partial q_j^p/\partial X_\alpha$ is an array of constant coefficients. Using equation (38)

$$\frac{\partial \phi}{\partial X_\alpha} \frac{\partial q_j^p}{\partial X_\alpha} \dot{Q}_j = \frac{\partial \phi}{\partial X_\alpha} \frac{\partial X_\alpha}{\partial Q_j} \dot{Q}_j = \frac{\partial \phi}{\partial Q_j} \dot{Q}_j \quad . \tag{54}$$

Further, from the kinetic equation (42) and equation (51b),

$$\frac{\partial \phi}{\partial X_\alpha} \frac{\partial^2 \bar{f}^p}{\partial X_\alpha \partial X_\beta} \dot{\chi}_\beta = \lambda \frac{\partial^2 \bar{f}^p}{\partial X_\alpha \partial X_\beta} \frac{\partial \phi}{\partial X_\alpha} \frac{\partial \phi}{\partial X_\beta} \quad . \tag{55}$$

Hence the condition $\dot{\phi} = 0$ implies that

$$\frac{\partial \phi}{\partial Q_j} \dot{Q}_j = \lambda \frac{\partial^2 \bar{f}^p}{\partial X_\alpha \partial X_\beta} \frac{\partial \phi}{\partial X_\alpha} \frac{\partial \phi}{\partial X_\beta} \quad . \tag{56}$$

If $\bar{f}^p(\chi_\alpha)$ is quadratic and positive definite (again expected if
it results from distortion of the crystal lattice resulting
from local slips), equation (56) implies that

$$\frac{\partial \phi}{\partial Q_j} \dot{Q}_j \geq 0 \quad , \tag{57a}$$

and that

$$\lambda = \frac{1}{\dfrac{\partial^2 \bar{f}^p}{\partial X_\alpha \partial X_\beta} \dfrac{\partial \phi}{\partial X_\alpha} \dfrac{\partial \phi}{\partial X_\beta}} \frac{\partial \phi}{\partial Q_j} \dot{Q}_j \quad . \tag{57b}$$

Substituting back, it is evident that

$$\dot{q}_j^p = \frac{1}{\dfrac{\partial^2 \bar{f}^p}{\partial X_\alpha \partial X_\beta} \dfrac{\partial \phi}{\partial X_\alpha} \dfrac{\partial \phi}{\partial X_\beta}} \frac{\partial \phi}{\partial Q_j} \frac{\partial \phi}{\partial Q_k} \dot{Q}_k \quad \text{for} \quad \phi = 1, \ \frac{\partial \phi}{\partial Q_j} \dot{Q}_j \geq 0 \quad . \tag{58a}$$

The argument may be continued to show that $\lambda = 0$ and

$$\dot{q}_j^p = 0 \quad \text{for} \quad \phi = 1 \quad \text{and} \quad \frac{\partial \phi}{\partial Q_j} \dot{Q}_j \leq 0 \quad . \tag{58b}$$

These equations will be recognized as characterizing a kinematic

hardening material (cf. Chapter 2, equation 96). The level
surface $\phi = 1$ plotted in stress space for given χ_α is the current yield surface.

If $\bar{f}^p$ is identically zero, so that the local slips do not
change the internal energy of the sample, the equations give
simply

$$\dot{q}_j^p = 0 \quad \text{if} \quad \phi(Q_j \; \frac{\partial q_j^p}{\partial \chi_\alpha}) < 0 \tag{59a}$$

$$\text{or} \quad \phi = 1 \text{ and } \frac{\partial \phi}{\partial Q_j} \dot{Q}_j < 0$$

$$\dot{q}_j^p = \lambda \frac{\partial \phi}{\partial Q_j} \quad \text{if} \quad \phi(Q_j \; \frac{\partial q_j^p}{\partial \chi_\alpha}) = 1 \quad , \quad \frac{\partial \phi}{\partial Q_j} \dot{Q}_j = 0 \quad . \tag{59b}$$

These are the constitutive equations for an elastic, perfectly
plastic material.

A material which exhibits both hardening behavior and flow
is described if $\bar{f}^p$ depends only on a subset of the χ_α; we shall
not consider this case in detail, but it corresponds to the
case discussed earlier (Chapter 2, equations 132 - 135).

We see then that time independent plasticity can be recovered as a mathematical limit of the non-linear kinetic equation. It is also of interest to enquire whether this is a
physically reasonable limit to adopt under the conditions when
the equations of time independent plasticity are applicable. To
study this point, let us consider a simple structural model of
a relation between a single component of generalized strain q
and generalized stress Q.

The model is shown in Figure 2. It consists of two members,
one a spring of modulus E_1 and the other a spring of modulus E_2
and a viscous element in series. The two members are connected
in parallel by a rigid bar which displaces without rotating.

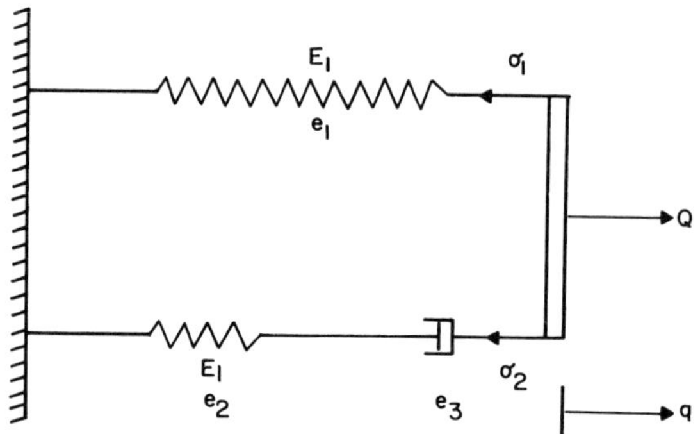

Figure 2. Structural model of the constitutive relation

Let the force applied externally to the bar be Q, and the displacement of the rigid bar be q. The forces in the members are σ_1 and σ_2, and the extensions of the elements are e_1, e_2 and e_3, as shown in Figure 2. Equilibrium requires that

$$Q = \sigma_1 + \sigma_2 \; , \tag{60}$$

and compatibility that

$$q = e_1 = e_2 + e_3 \; . \tag{61}$$

If the viscous element is rigid, $e_3 = 0$, and equation (61) gives

$$\frac{\sigma_1}{E_1} = \frac{\sigma_2}{E_2} = q \; . \tag{62}$$

Combining this with equation (60), we see that

$$Q = (E_1 + E_2)q = Eq \; . \tag{63}$$

More generally we put

$$q^e = \frac{Q}{E} \; , \tag{64}$$

and note that

$$\bar{f}^e = \frac{1}{2} E(q^e)^2 \; . \tag{65}$$

The strain e_3 is the single internal variable, and we shall adopt a kinetic equation of the form

$$\dot{e}_3 = \dot{e}_o \left(\frac{\sigma}{\sigma_o}\right)^5 \quad , \tag{66}$$

where $\dot{e}_o$ and σ_o are constants with the dimensions of $\dot{e}$ and σ respectively. Equation (61) now gives

$$q = \frac{\sigma_1}{E_1} = \frac{\sigma_2}{E_2} + e_3 \quad , \tag{67}$$

and on consideration of the deformation of the springs,

$$\bar{f}(q, e_3) = \frac{1}{2} E_1 q^2 + \frac{1}{2} E_2 (q - e_3)^2 \quad . \tag{68}$$

It may be confirmed that, on using equation (61),

$$\frac{d\bar{f}}{dq} = E_1 q + E_2 (q - e_3)$$

$$= E_1 e_1 + E_2 e_2$$

$$= \sigma_1 + \sigma_2 \tag{69}$$

$$= Q \quad .$$

Similarly,

$$\frac{d\bar{f}}{de_3} = E_2 (q - e_3) = \sigma_2 \quad , \tag{70}$$

which is the force acting on the viscous element. From equation (69) we also see that

$$q = \frac{Q}{(E_1 + E_2)} + \frac{E_2}{(E_1 + E_2)} e_3$$

$$= \frac{Q}{E} + \frac{E_2}{E} e_3 \quad . \tag{71}$$

From equation (64), it follows that

$$q^p = \frac{E_2}{E} e_3 \quad \text{and} \tag{72}$$

$$\frac{dq^p}{de_3} = \frac{E_2}{E} \quad . \tag{73}$$

Using equation (71), we may rewrite equation (68) as

$$\bar{f} = \frac{1}{2} \frac{Q^2}{E} + \frac{1}{2} E_2 (1 - \frac{E_2}{E}) e_3^2 \quad , \tag{74}$$

and hence

$$\bar{f}^p = \frac{1}{2} E_2 (1 - \frac{E_2}{E}) e_3^2 \quad . \tag{75}$$

It is seen that the force acting on the viscous element is

$$\sigma_2 = Q \frac{dq^p}{de_3} - \frac{d\bar{f}^p}{de_3}$$

$$= \frac{E_2}{E} Q - E_2 (1 - \frac{E_2}{E}) e_3 \quad . \tag{76}$$

This result may be confirmed by combining equations (70) and (71).

In order to give a specific result let us put

$$E_1 = \frac{1}{10} E \; , \quad E_2 = \frac{9}{10} E \quad . \tag{77}$$

The differential equation governing the behavior of the model is obtained from equations (71) and the kinetic equation (66). It is

$$\dot{q} = \frac{\dot{Q}}{10} + \frac{9}{10} \dot{e}_3$$

$$= \frac{\dot{Q}}{E} + \frac{9}{10} \dot{e}_o (\frac{\sigma_2}{\sigma_o})^5 \quad . \tag{78}$$

Substituting from equation (76), with equation (77), this

becomes

$$\dot{q} = \frac{\dot{Q}}{E} + \frac{9}{10}\,\dot{e}_o\,\{\frac{9}{10\sigma_o}\,(Q - \frac{1}{10}\,Ee_3)\}^5 \quad . \tag{79}$$

It is convenient to adopt the dimensionless variables

$$\bar{Q} = \frac{Q}{\sigma_o}\quad,\quad \bar{q} = \frac{Eq}{\sigma_o}\quad,\quad \tau = \frac{E\dot{e}_o}{\sigma_o}\,t \quad . \tag{80}$$

The governing equation then becomes, using equation (71),

$$\frac{d\bar{q}}{d\tau} = \frac{d\bar{Q}}{d\tau} + \frac{9}{10}\,\{\bar{Q} - \frac{1}{10}\,\bar{q}\}^5 \quad . \tag{81}$$

It may be noted that

$$\frac{d\bar{q}}{d\tau} = \frac{\dot{q}}{\dot{e}_o} \quad . \tag{82}$$

The constant $\dot{e}_o$ is a characteristic strain rate for the model.
Suppose that we carry out the following program on the model.
Beginning at an undeformed state ($Q = q = e_3 = 0$), we increase
$\bar{q}$ at a constant rate for a period of time τ, decrease $\bar{q}$ at the
same rate until $\bar{Q} = 0$ and again increase $\bar{q}$ at the same rate.
This corresponds to loading, unloading and reloading in a tension test. Equation (81) will give $\bar{Q}(\tau)$ for this program, and
we may eliminate τ and plot $\bar{Q}$ against $\bar{q}$.

The results are shown in Figure 3 for three values of $\dot{q}/\dot{e}_o$;
0.001, 0.1 and 10.0. These figures correspond to very slow
loading at one extreme and fairly rapid loading at the other.
The curves show similar characteristics, and each appears to be
similar to the quasi-static test we described in Chapter 2. At
slow rates we find the most sharply defined knee in the curve,
and unloading which is virtually elastic. It is also important
to note the 'yield' stress $\bar{Q}$, or Q/σ_o, where σ_o is a characteristic stress level for the model. At the highest rate yielding

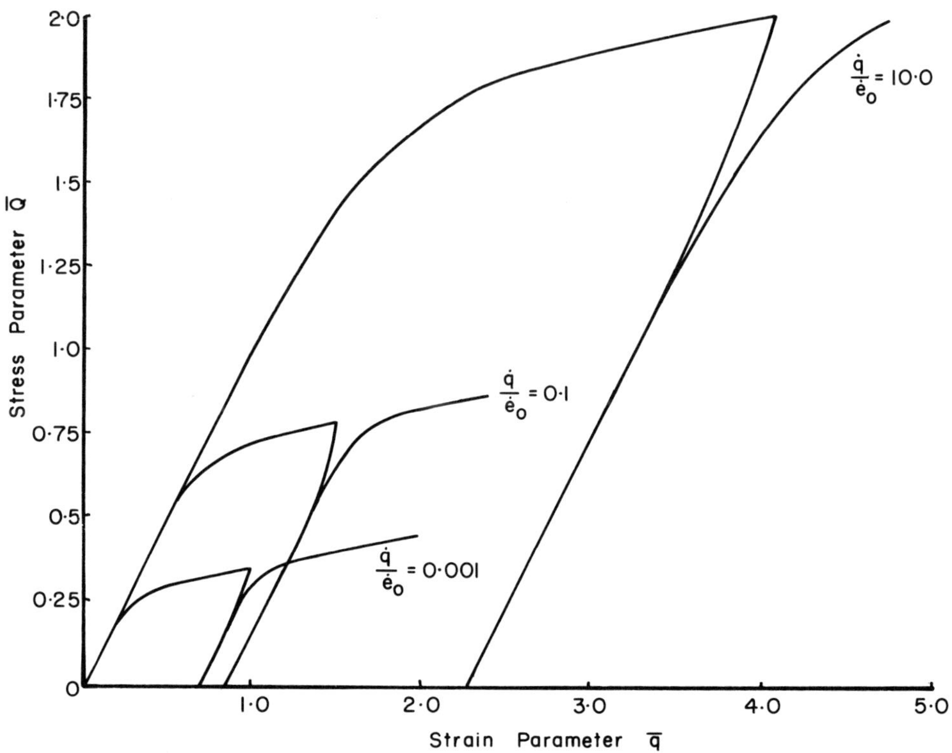

Figure 3. Loading-unloading tests at different rates

occurs at $\bar{Q} \simeq 1.3$, while at the slowest rate yielding occurs
for $\bar{Q} \simeq 0.3$.

The model is, of course, not time independent. However,
consider a creep test in which we suddenly apply a stress $\bar{Q}$ to
the undeformed model and hold the stress constant. The strain
response $\bar{q}(\tau)$ may be found from equation (81). In general the
initial response will be a jump in strain given by

$$\bar{q}(0^{+}) = \bar{Q} \ . \tag{83}$$

The viscous element will relax, and the final response will be

$$\bar{q}(\infty) = 10\bar{Q} \ . \tag{84}$$

This is a considerable change in strain.

The creep response is plotted in Figure 4 for $\bar{Q} = 0.5$, 1.0,
2.0 and 5.0. We note that if $\bar{Q} > 1$ the relaxation of the viscous element is rapid, while for $\bar{Q} = 0.5$ it is extremely slow.

$$X_\alpha - X_\alpha^* = (Q_j - Q_j^*) \frac{\partial q_j^p}{\partial X_\alpha} \; , \tag{106}$$

and hence

$$(X_\alpha - X_\alpha^*)\dot{X}_\alpha = (Q_j - Q_j^*) \frac{\partial q_j^p}{\partial X_\alpha} \dot{X}_\alpha \tag{107}$$

$$= (Q_j - Q_j^*) \, \dot{q}_j^p \; .$$

The principle of maximum plastic work follows from inequality (102).

It is seen then that the basic postulates of plasticity either in the form of inequality (103) for an arbitrary path or the principle of maximum plastic work, do not follow directly from thermodynamic stability and the consequences of the second law, but depend also on the form of the kinetic equation. However, if the kinetic relation satisfies inequality (104), which is the case in most models of plasticity, the second postulate of Section 2.4 (Chapter 2, equation 63) will hold in its strong form.

<u>28.5 Thermodynamic Basis of Structural Theorems</u>

It has been noted above that the thermodynamic framework under discussion may be applied to a structure as opposed to a continuum element. The macroscopic variables become the generalized displacements and the generalized loads, and the internal variables are related to the plastic strain distribution in the structure. In this interpretation of the thermodynamic formulation, the energy minimum principles provide the classical elastic energy theorems directly, and also the rate theorems for elastic-plastic structures.

In order to illustrate this point, we shall derive the minimum potential energy theorem for an *elastic* truss and a form of the kinematic rate theorem for an *elastic, perfectly plastic* pin-jointed truss, using the notation of Chapter 26, Section 26.2. Thus we treat a truss with m members and n unconstrained node displacement components. The extensions in the bars are represented by $\{\delta\}$, the bar forces by $\{N\}$, the node displacements by $\{u\}$, and the conjugate external force components at the nodes by $\{P\}$. We consider first the form of the fundamental equation for the truss and the equations of state, and secondly the equilibrium of the structure when it is coupled to a reversible work source applying conservative loads at the nodes. Isothermal conditions will be considered, and hence all reference to temperature will be omitted.

If the truss is composed of *linear elastic* bars, the Helmholtz free energy of the system is the sum of the strain energies of the individual bars. Denoting the Helmholtz free energy of the system of bars by F, we have

$$F = \frac{1}{2} \{\delta\}^T [S]\{\delta\}$$

$$= \frac{1}{2} \{u\}^T [B]^T [S][B]\{u\} \tag{108}$$

$$= \frac{1}{2} \{u\}^T [K]\{u\} \quad ,$$

where $[K] = [B]^T [S][B]$ is the system stiffness matrix. We also note that

$$dF = d\{u\}^T [K]\{u\} = d\{u\}^T \{P\} \quad , \tag{109a}$$

so that

$$\{P\} = [K]\{u\} \quad . \tag{109b}$$

The individual bar forces are not given directly by an equation of state, but may be obtained, using the inverse of equation (109a), from the displacements. Hence

$$\{N\} = [S]\{\delta\} = [S][B]\{u\} = [S][B][K]^{-1}\{P\} \quad . \tag{110}$$

Consider now an elastic, perfectly plastic truss. The plastic extensions of the bars are identified as the internal variables. Denoting the plastic extensions by $\{\delta^P\}$, the Helmholtz free energy of the system is again found by summing the strain energies of the bars. Hence

$$F = \frac{1}{2}\,\{\delta-\delta^P\}^T[S]\{\delta-\delta^P\}$$

$$= \frac{1}{2}\,([B]\{u\} - \{\delta^P\})^T[S]([B]\{u\} - \{\delta\}^P) \tag{111}$$

$$= \frac{1}{2}\,\{u\}^T[K]\{u\} - \{u\}^T[B]^T[S]\{\delta^P\} + \frac{1}{2}\,\{\delta^P\}^T[S]\{\delta^P\} \quad .$$

We see then that

$$dF = d\{u\}^T([K]\{u\} - [B]^T[S]\{\delta^P\}) \tag{112}$$

$$+ d\{\delta^P\}(- [S][B]\{u\} + [S]\{\delta^P\}) \quad .$$

The forces conjugate to the internal variables are equal and opposite to the forces themselves. Hence equation (112) gives

$$\{P\} = [K]\{u\} - [B]^T[S]\{\delta^P\} \quad , \tag{113a}$$

$$\{N\} = [S][B]\{u\} - [S]\{\delta^P\} \quad . \tag{113b}$$

We now divide the displacement $\{u\}$ into an elastic and a plastic part,

$$\{u\} = \{u^E\} + \{u^P\} \quad , \tag{114}$$

where $\{u^P\}$ depends only on $\{\delta^P\}$. We also expect that F may be divided into an elastic part F^E which depends on $\{u^E\}$ and a plastic part F^P which depends on $\{\delta^P\}$. Since $\{P\} = \partial F^E/\partial\{u^E\}$, from equation (113a) we note that

$$\{P\} = [K]\{u^E\} + [K]\{u^P\} - [B]^T[S]\{\delta^P\} \tag{115}$$

must be independent of $\{\delta^P\}$, and hence

$$\{u^P\} = [K]^{-1}[B]^T[S]\{\delta^P\} \quad . \tag{116}$$

Substituting equations (114) and (116) into equation (112), we find after some manipulation that

$$dF = d\{u^E\}^T[K]\{u^E\} + d\{\delta^P\}^T([S] - [S][B][K]^{-1}[B]^T[S])\{\delta^P\} \quad . \tag{117}$$

It follows then that

$$F^E = \frac{1}{2}\{u^E\}[K]\{u^E\} \quad , \tag{118a}$$

$$F^P = \frac{1}{2}\{\delta^P\}^T([S] - [S][B][K]^{-1}[B]^T[S])\{\delta^P\} \quad . \tag{118b}$$

From equations (116) and (118b) we may also write

$$\{N\} = (\frac{\partial\{u^P\}}{\partial\{\delta^P\}})^T \{P\} - \frac{\partial F^P}{\partial\{\delta^P\}} \tag{119}$$

$$= [S][B][K]^{-1}\{P\} - ([S] - [S][B][K]^{-1}[B]^T[S])\{\delta^P\} \quad .$$

This result may also be obtained from equations (113) - (116) directly.

Suppose now that the truss is coupled to a reversible work source which applies conservative loads at the nodes. The internal energy of the work source is

$$U_w = - \{u\}^T\{P^O\} \quad . \tag{120}$$

Taking first the case of an *elastic* truss, the Helmholtz free energy of the enlarged system is

$$U_p = \frac{1}{2}\{u\}^T[K]\{u\} - \{u\}^T\{P^O\} \quad . \tag{121}$$

For equilibrium, $\{u\}$ will be such that U_p has its least value. This occurs for

$$[K]\{u\} = \{P^O\} \quad , \quad \text{or} \quad \{u\} = \{u^O\} = [K]^{-1}\{P^O\} \quad . \tag{122}$$

This result we see to be the classical minimum potential energy theorem in a form appropriate for the truss problem.

It is also of interest to consider the incremental problem

in the elastic truss. Suppose the truss is in the equilibrium state defined above, and that the work source is replaced by one which applies conservative loads $\{P^O\} + \{\Delta P^O\}$. Let the truss undergo a process which leads to equilibrium in a state characterized by $\{u^O\} + \{\Delta u\}$. The Helmholtz free energy of the system may be written as

$$U_p = \frac{1}{2}(\{u^O\} + \{\Delta u\})^T[K](\{u^O\} + \{\Delta u\})$$

$$- (\{u^O\} + \{\Delta u\})^T(\{P^O\} + \{\Delta P^O\})$$

$$= (\frac{1}{2}\{u^O\}^T[K]\{u^O\} - \{u^O\}^T\{P^O\} - \{u^O\}^T\{\Delta P^O\}) \tag{123}$$

$$+ \{\Delta u\}^T([K]\{u^O\} - \{P^O\})$$

$$+ \frac{1}{2}\{\Delta u\}^T[K]\{\Delta u\} - \{\Delta u\}^T\{\Delta P^O\} \quad .$$

The first term in this expression is independent of $\{\Delta u\}$, and the second vanishes in view of equation (122). Furthermore, we may replace increments by rates, and consequently the displacement rates are found by minimizing

$$U_p(\{\dot{u}\}) = \frac{1}{2}\{\dot{u}\}^T[K]\{\dot{u}\} - \{\dot{u}\}^T\{\dot{P}^O\} \quad . \tag{124}$$

This is the result expected from static and kinematic arguments.

Suppose now that the expanded system consists of the work source and a truss composed of an elastic, perfectly plastic material. The Helmholtz free energy of the system is

$$U_p = F(\{u\}, \{\delta^P\}) - \{u\}^T\{P^O\}$$

$$= F^E(\{u^E\}) - \{u^E\}^T\{P\}^O + F^P(\{\delta^P\}) - \{u^P\}^T\{P^O\} \quad . \tag{125}$$

We expect to find a *constrained equilibrium state* for the system. The elastic displacements are unconstrained, and hence

$$U_p^E = F^E(\{u^E\}) - \{u^E\}^T\{P^O\} \tag{126}$$

will take its least value, characterized by

$$[K]\{u^E\} = \{P\} = \{P^o\} \quad . \tag{127}$$

The internal variables may be constrained in the sense that elements of $\{\delta^P\}$ cannot change if the associated element of $\{N\}$ lies within, say, the range $-N_o < N < N_o$. Furthermore, if $N = +N_o$, δ^P may not decrease, and if $N = -N_o$, δ^P may not increase. For this limited class of variations, however,

$$U^P_p = \frac{1}{2} \{\delta^P\}^T ([S] - [S][B][K]^{-1}[B]^T[S])\{\delta\}^P$$
$$- \{\delta^P\}^T [S][B][K]^{-1}\{P^o\} \tag{128}$$

will be stationary. These conditions are not sufficient to determine the constrained state uniquely; the solution is path dependent.

Consider now a small change in the work source such that conservative loads $\{P^o\} + \{\Delta P^o\}$ are applied to the truss. Let $\{u^E\} = \{u^{Eo}\} + \{\Delta u^E\}$, $\{\delta^P\} = \{\delta^{Po}\} + \{\Delta\delta^P\}$ characterize the new constrained equilibrium state. From our earlier argument regarding the elastic truss, it is evident that $\{\Delta u^E\}$ minimizes

$$U^E_p(\{\Delta u^E\}) = \frac{1}{2} \{\Delta u^E\}^T[K]\{\Delta u^E\} - \{\Delta u^E\}^T\{\Delta P^o\} \quad . \tag{129}$$

The plastic part of the total Helmholtz free energy, after putting for convenience

$$[\hat{K}] = [S] - [S][B][K]^{-1}[B]^T[S] \quad , \tag{130}$$

is given by

$$U^P_p = \frac{1}{2} (\{\delta^{Po}\} + \{\Delta\delta^P\})^T[\hat{K}](\{\delta^{Po}\} + \{\Delta\delta^P\})$$
$$- (\{\delta^{Po}\} + \{\Delta\delta^P\})^T[S][B][K]^{-1}(\{P^o\} + \{\Delta P^o\})$$
$$= (\frac{1}{2} \{\delta^{Po}\}^T[\hat{K}]\{\delta^{Po}\} - \{\delta^{Po}\}^T[S][B][K]^{-1}\{P^o + \Delta P^o\})$$

$$+ \{\Delta\delta^P\}^T([\hat{K}]\{\delta^{Po}\} - [S][B][K]^{-1}\{P^o\})$$

$$+ (\frac{1}{2}\{\Delta\delta^P\}^T[\hat{K}]\{\Delta\delta^P\} - \{\Delta\delta^P\}^T[S][B][K]^{-1}\{\Delta P^o\}) \quad . \tag{131}$$

We place constraints on $\{\Delta\delta^P\}$ in view of the kinetic relation for perfectly plastic bar;

$$\Delta\delta^P \geq 0 \quad \text{if} \quad N = +No \quad ,$$

$$\Delta\delta^P = 0 \quad \text{if} \quad -No < N < No \quad , \tag{132}$$

$$\Delta\delta^P \leq 0 \quad \text{if} \quad N = -No \quad .$$

The first term on the right hand side of equation (131) is independent of $\{\Delta\delta^P\}$. Given the constraints of equations (132), the second term can be interpreted as a first variation of equation (128) with respect to the *unconstrained* parameters; consequently it must vanish. Converting back to a rate form, the solution is given by the least value of

$$\dot{U}_p^P = \frac{1}{2}\{\dot{\delta}^P\}^T([S] - [S][B][K]^{-1}[B]^T[S])\{\dot{\delta}^P\}$$
$$- \{\dot{\delta}^P\}[S][B][K]^{-1}\{\dot{P}^o\} \quad , \tag{133}$$

subject to the rate form of equations (132).

This result is a less general form of the theorem given in Chapter 25, equation (96). It can be suitably generalized and combined with the elastic part of the total Helmholtz free energy to recover all the various forms of the kinematic rate principle we have discussed in Chapter 25 for elastic,perfectly plastic materials.

A more general approach must be taken for hardening materials, since the plastic extensions $\{\delta^P\}$ are not the internal variables. Nevertheless, the same arguments may be followed with suitably redefined internal variables. The static rate principles will also follow from transformations of the energy minimum principles.

28.6 Minimum Work Paths

The potential functions associated with minimum work paths may also be derived from the thermodynamic framework. We consider a macroscopic sample of the inelastic solid, and consider the quasi-static processes which lead from the unstrained state $(q_j = 0, \chi_\alpha = 0)$ to a fixed terminal strain q_j^*. The work done per unit volume by the boundary forces in such a process is

$$\bar{W} = \int_0^{q_j^*} Q_j dq_j \quad .$$

$$(134)$$

The least value of $\bar{W}$ is the minimum work $W^o(q_j^*)$.

From equation (134) we see that

$$d\bar{W} = Q_j dq_j \quad ,$$

$$(135)$$

and, if $\bar{W}^R$ is the reversible work per unit volume, from equation (19) we have

$$d\bar{W}^R = d\bar{f} = Q_j dq_j - X_\alpha d\chi_\alpha \quad .$$

$$(136)$$

Hence

$$d\bar{W} = Q_j dq_j = d\bar{f} + X_\alpha d\chi_\alpha \quad .$$

$$(137)$$

Integrating over the process which terminates at $q_j = q_j^*$,

$$\bar{W} = \int_0^{q_j^*} Q_j dq_j = \bar{f}(q_j^*, \chi_\alpha) + \bar{\zeta} \quad ,$$

$$(138)$$

where $\bar{\zeta} = \int X_\alpha d\chi_\alpha$ is the entropy per unit volume produced in the process. The internal variables χ_α are not known, and depend on the details of the process.

To determine $W^o(q_j^*)$, we must find the terminal internal variables χ_α^* which minimize $\bar{W}$ in equation (138). It is evident that for each choice of terminal χ_α we choose that process which minimizes $\bar{\zeta}(\chi_\alpha)$. Assuming that the kinetic relation is the perfectly plastic form given by $n \to \infty$ in equation (42), we invert the relation and put

$$X_\alpha = \frac{\partial D}{\partial \dot{\chi}_\alpha} \tag{139}$$

where $D = D(\dot{\chi}_\alpha)$ is homogeneous and of degree one in the components of $\dot{\chi}_\alpha$. Then

$$\bar{\zeta} = \int X_\alpha \dot{\chi}_\alpha dt = \int D(\dot{\chi}_\alpha) dt \quad . \tag{140}$$

This expression must be minimized subject to the constraint that the terminal value of χ_α is fixed. We put

$$\delta\bar{\zeta} = \delta\{\int X_\alpha \dot{\chi}_\alpha dt - b_\alpha \int \dot{\chi}_\alpha dt\} \tag{141}$$

$$= \int (X_\alpha - b_\alpha) \delta\dot{\chi}_\alpha dt = 0 \quad .$$

The Lagrange multipliers b_α are constant, and hence if $\bar{\zeta}$ is to be stationary the X_α must be constant. This implies that we choose a process in which $\dot{\chi}_\alpha$ is itself constant. Because D is homogeneous and of degree one, we may then put, from equation (140),

$$\bar{\zeta} = D(\chi_\alpha) \quad . \tag{142}$$

Equation (138) is now replaced by

$$\bar{W} = \bar{f}(q_j^*, \chi_\alpha) + D(\chi_\alpha) \quad . \tag{143}$$

We now seek stationary values of $\bar{W}$ with respect to variations in χ_α. The first variation of $\bar{W}$ is

$$\delta\bar{W} = \frac{\partial \bar{f}}{\partial \chi_\alpha} \delta\chi_\alpha + \frac{\partial D}{\partial \chi_\alpha} \delta\chi_\alpha \quad . \tag{144}$$

The condition that $\delta\bar{W} = 0$ requires simply that X_α calculated from the gradient of $\bar{f}$ and from the gradient of D should be the same. This permits us to associate a terminal $\chi_\alpha = \chi_\alpha^*$ with each strain q_j^*. We can readily show that the second variation of $\bar{W}$ is non-negative if $\bar{f}$ is convex, and hence $\bar{W}$ will have a minimum value. We thus put

$$W^O(q_j^*) = \bar{f}(q_j^*, \chi_\alpha^*) + D(\chi_\alpha^*) \quad . \tag{145}$$

For a small change in q_j^*, the change in W^O is given by

$$dW^O = \frac{\partial \bar{f}}{\partial q_j^*} dq_j^* + \frac{\partial \bar{f}}{\partial \chi_\alpha} d\chi_\alpha^* + \frac{\partial D}{\partial \chi_\alpha} d\chi_\alpha^* = Q_j^* dq_j^* \tag{146}$$

in view of the requirement that the variation in equation (144) vanishes. The stress Q_j^* associated with q_j^* after following a minimum work path is

$$Q_j^* = \frac{\partial W^O}{\partial q_j^*} \quad . \tag{147}$$

Furthermore, we see that we may put

$$W^O(q_j^*) = \min \{\bar{f}(q_j^*, \chi_\alpha) + D(\chi_\alpha)\}$$

$$\qquad = \min \{\bar{f}^e(q_j^* - q_j^p) + \bar{f}^p(\chi_\alpha) + D(\chi_\alpha)\} \quad , \tag{148}$$

where q_j^p is the plastic strain. This result shows that an arbitrary division of a terminal strain into an elastic part $(q_j^* - q_j^p)$ and a plastic part $q_j^p(\chi_\alpha)$ can be used to bound $W^O(q_j^*)$ from above. This result parallels the development given in Chapter 23. Indeed, if the internal variables are the plastic strains themselves, the result given in equation (148) reduces to the expression for W^O for a linear kinematically hardening material.

28.7 Bibliographical and Historical Remarks

The successful understanding of the behavior of inelastic strained solids in terms of classical thermodynamics is a

comparatively recent development, but is one which promises
further progress in the general theory of plasticity and in
the development of constitutive equations in particular. The
approach adopted in this Chapter is taken from the work of
Kestin [1968], Kestin and Rice [1970] and Rice [1970], [1971].
Other representative contributions are those of Valanis [1966],
[1968], Coleman and Gurtin [1967], Nemat-Nasser [1972], Lubli-
ner [1972], Kestin [1973] and Hill and Rice [1973]. The discussion on thermodynamic stability is taken from Martin [1974].

A topic of great importance in the development of constitutive equations in plasticity, and which has not been covered in
this volume, is the study of the relation between macroscopic
stress-strain relations and microstructural deformation mechanisms in crystals and polycrystals. While this problem can
be understood in general terms as a problem of structural
analysis on the microscale, attention must be given to suitable
averaging techniques in view of the large number of microstructural deformation sites in a macroscopic sample. Recent reviews of work in this area are those of Hutchinson [1970] and
Rice [1973].

References

B.D.Coleman and M.E.Gurtin — 1967 — "Thermodynamics with Internal State Variables", J.Chem.Phys., *47*, 597.

R.Hill and J.R.Rice — 1973 — "Elastic potentials and the structure of inelastic constitutive laws", SIAM J.Appl.Math., *25*, 448.

J.W.Hutchinson — 1970 — "Elastic-plastic behaviour of poly-crystalline metals and composites", Proc.Roy.Soc., *A319*, 247.

J.Kestin — 1968 — "On the Application of the Principles of Thermodynamics to Strained Solid Materials", *Irreversible Aspects of Continuum Mechanics and Transfer of Physical Characteristics in Moving Fluids* (edited by H.Parkus and L.I.Sedov), Springer, 177.

J.Kestin   1973   "Thermodynamics in Thermoplasticity",
                                 Lecture notes for Summer School of
                                 the Polish Academy of Sciences,
                                 Jablonna, June 25-27, 1973.

J.Kestin and   1970   "Paradoxes in the Application of Ther-
J.R.Rice   modynamics to Strained Solids", *A
                                 Critical Review of Thermodynamics*
                                 (edited by E.B.Stuart, B.Gal-Or and
                                 A.J.Brainard), Mono Book Corp., 275.

J.Lubliner   1972   "On the Thermodynamic Foundations of
                                 Non-Linear Solids", Int.J.Non-Linear
                                 Mechanics, $\underline{7}$, 237.

J.B.Martin   1974   "A note on the implications of thermo-
                                 dynamic stability in the internal
                                 variable theory of inelastic solids",
                                 Int.J.Solids and Structures, to appear.

S.Nemat-Nasser   1972   "On nonequilibrium thermodynamics of
                                 continua", Lecture delivered at the
                                 ASME Winter Annual Meeting, November
                                 26-30, 1972.

J.R.Rice   1970   "On the Structure of Stress-Strain Re-
                                 lations for Time-Dependent Plastic De-
                                 formation of Metals", J.Appl.Mech.,
                                 $\underline{37}$, 328.

J.R.Rice   1971   "Inelastic constitutive relations for
                                 solids: an internal variable theory
                                 and its application to metal plasti-
                                 city", J.Mech.Phys.Sol., 19, 433.

J.R.Rice   1973   "Continuum Plasticity in Relation to
                                 Microscale Deformation Mechanisms",
                                 *Metallurgical Effects at High Strain
                                 Rates* (edited by R.W.Rohde, B.M.
                                 Butcher, J.R.Holland and C.H.Karnes),
                                 Plenum Publishing Corporation, 93.

K.C.Valanis   1966   "Thermodynamics of large viscoelastic
                                 deformations", J.Math.Phys., $\underline{45}$, 197.

K.C.Valanis   1968   "Unified Theory of Thermomechanical
                                 Behavior of Viscoelastic Materials",
                                 *Mechanical Behavior of Materials under
                                 Dynamic Loads* (edited by U.S.Lindholm),
                                 Springer-Verlag, 343.

NAME INDEX

SUBJECT INDEX

Jann-Nan Yang   1977

Department of Civil, Mechanical
and Environmental Engineering
George Washington University
Washington, D. C. 20052

Jann-Nu Yang

The George Washington University
CMEE Department   1976

DR. JANN N. YANG